Fundamentals of Plasma Physics

FUNDAMENTALS OF PLASMA PHYSICS

S. R. Seshadri
The University of Wisconsin
Madison, Wisconsin

American Elsevier Publishing Company Inc.
New York London Amsterdam

AMERICAN ELSEVIER PUBLISHING COMPANY, INC.
52 Vanderbilt Avenue, New York, N.Y. 10017

ELSEVIER PUBLISHING COMPANY
335 Jan Van Galenstraat, P.O. Box 211
Amsterdam, The Netherlands

Library of Congress Cataloging in Publication Data

Seshadri, S. R. 1925-
Fundamentals of plasma physics.

Bibliography: p.
1. Plasma (Ionized gases) I. Title.
QC718.S44 530.4'4 72-77559
ISBN 0-444-00125-5

Manufactured in the United States of America

To Ronold Wyeth Percival King

CONTENTS

Chapter 3 INTERACTIONS OF CHARGED PARTICLES WITH ELECTROMAGNETIC FIELDS

Chapter 4 CLASSICAL DYNAMICS OF COLLISIONS

Chapter 5 SMALL AMPLITUDE WAVES IN A PLASMA

S. R. SESHADRI has been a professor of electrical engineering at the University of Wisconsin since 1967. During 1970-1971 he was a National Science Foundation Senior Postdoctoral Fellow at the California Institute of Technology. He has conducted research in electromagnetic scattering and diffraction, surface waves, wave propagation in the ionosphere, and radiation in plasmas. Among other activities, he has served as a Principal Scientific Officer at the Electronics Research and Development Establishment in Bangalore, Senior Engineering Specialist at the Sylvania Applied Research Laboratory in Massachusetts, Visiting Professor at the University of Toronto, and Postdoctoral and Honorary Research Fellow at the Gordon McKay Laboratory of Harvard University. He is a member of the American Physical Society and the U.S. Commission 6 of the International Union of Radio Science, and a Senior Member of the Institute of Electrical and Electronics Engineers, Inc. Dr. Seshadri received his B.Sc. and M.A. degrees at Madras University in India and his Ph.D. degree at Harvard University. He holds a diploma in Electrical Communication Engineering from the Indian Institute of Science in Bangalore. Dr. Seshadri is also the author of *Fundamentals of Transmission Lines and Electromagnetic Fields.*

PREFACE

Dreams, books, are each a world; and books, we know,
Are a substantial world, both pure and good.
Round these, with tendrils strong as flesh and blood,
Our pastime and our happiness will grow.

William Wordsworth

An interdisciplinary program in plasma physics is in existence at the University of Wisconsin. The principal part of this program is the controlled thermonuclear research supported by the United States Atomic Energy Commission and carried out under the leadership of Professor Donald W. Kerst. As a part of this program, a number of undergraduate and graduate courses are offered to train the students in the general area of plasma physics. This textbook on the fundamentals of plasma physics was developed for the introductory course in this interdisciplinary plasma physics program suitable for students majoring in electrical engineering, nuclear engineering and physics, and, is based largely on the course of lectures given by the author in the spring semester of 1970.

This book is specifically written for senior level students. Detailed explanations for the various derivations are provided and the physical interpretations are emphasized. These features inject into the book a certain amount of clarity and completeness which make it suitable for self-study by the students. Although written for senior level students, this textbook can be used with advantage by first-year graduate students. This book contains more material than can be covered in a semester thus permitting sufficient freedom in the selection of topics for a one-semester course in plasma physics depending on the level and the interests of the students.

In view of the interdisciplinary character of the subject, even in an introductory textbook on plasma physics, there is a wide choice of topics that can be included. The topics chosen for treatment in this book are, no doubt, to some extent governed by the interests of the author but, to a large extent, dictated by the needs and the interests of the students from electrical engineering, nuclear engineering and physics departments.

The book begins by introducing the concept of velocity-distribution function, the Boltzmann equation satisfied by the velocity-distribution function and the characteristics of the Maxwell-Boltzmann distribution function and its modified form in the presence of a conservative force. After a discussion of the particle current density and the kinetic pressure, the first three moments of the Boltzmann equation are deduced to yield the equation of continuity, the momentum transport equation and the transport equation for the kinetic pressure. The approximations involved in the cold and the warm plasma models are enumerated and lumped macroscopic parameters are then introduced for a plasma which contains more than one species of particles. Chapter 1 concludes with a derivation of the hydrodynamical equations satisfied by these lumped parameters. In chapter 2 the basic plasma phenomena such as the electron plasma oscillations, the Debye shielding, the plasma sheath, the plasma probe, the effect of thermal motions on the electron plasma oscillations, the ion plasma oscillations, the magnetohydrodynamic equations and their simple applications, the pinch effect and the configuration- and the velocity-space instabilities are treated. A detailed discussion of the interactions of charge particles with electromagnetic fields is presented in chapter 3 which includes the treatment of cyclotron resonance, magnetic mirror effect, Fermi acceleration and magnetic pumping.

The development of the classical dynamics of collisions with particular emphasis on the collision between two perfectly elastic, hard spheres applicable to a weakly ionized gas, and, the scattering by Coulomb potential is contained in chapter 4. A detailed treatment of the characteristics of plane waves in a magnetoionic medium, magnetoionic theory at hydromagnetic frequencies and magnetohydrodynamic waves is presented in chapter 5 which also includes a discussion of such applications as whistlers, helicons, Faraday rotation, the problem of radio communication black-out during re-entry of a space vehicle into the earth's atmosphere and Cerenkov radiation. Chapter 6 contains a treatment of several applications of the Boltzmann equation. The characteristics of plane waves in a hot, isotropic plasma as well as the characteristics of plane waves propagating along and across the magnetostatic field in a hot, anisotropic plasma are developed starting from the Boltzmann-Vlasov equation. In this development attention is paid to the examination of such phenomena as Landau damping and cyclotron damping. Using a relaxation model for the collision term, conductivity, free and ambipolar diffusion are discussed for the case of a collision frequency that is not dependent on the particle velocity. A study of conductivity and diffusion for a velocity-dependent collision frequency is then presented. The book concludes with a derivation of an integral expression for the Boltzmann collision term, and this integral expression is used for deducing the Maxwell-Boltzmann velocity distribution function as well as for providing a justification for the relaxation model for the collision term for the case of a weakly ionized gas. Rationalized MKS units are used throughout the book.

A useful feature of the book is the set of problems of which there are about 150. These problems are designed partly to provide clarifications of the topics, and partly to extend the treatment contained in the text. The answers to many problems are contained in their statements. The answers to the remaining problems as well as the method of solution to several problems are presented at the end of the book.

For the sake of completeness, numerical values of physical parameters of some typical plasmas, a summary of vector analysis, useful vector relations, a brief review of the relevant properties of dyads, Bessel functions and Legendre polynomials, and a collection of useful integral relations are included in appendices.

The prerequisites for the use of this book are courses on mechanics and electromagnetic theory beyond the sophomore level. A knowledge of complex variables is helpful but is not required; a development of the necessary amount of complex variables is incorporated at the appropriate place in chapter 6 of this book.

This book is concerned largely with only the fundamentals of plasma physics and every effort has been directed into making the book self-contained. Therefore, an exhaustive list of references is not provided. However, at the end of each chapter I have given some references which I have found very useful. During the time of preparation of the manuscript of this book, I have read with advantage the publications of several previous authors whose works are listed at the end of this book. Moreover, it is only appropriate to point out here that my interest in plasma physics was inspired by the following works of three well-known plasma physicists whose expositions are as superb as their treatments are concise: S. Chandrasekhar, *Plasma Physics*, The University of Chicago Press, Chicago, 1960; J.L. Delcroix, *Introduction to the Theory of Ionized Gases*, Interscience publishers, Inc., New York, 1960; L. Spitzer, *Physics of Fully Ionized Gases*, Interscience publishers, Inc., New York, 1956.

A note on the method of numbering of the equations is in order. When reference is made to an equation as (4.5.6), the first two numbers indicate respectively the chapter and the section to which the equation pertains and the last gives the number of the equation. When reference is made to an equation in the same chapter, the first number is omitted. If reference is made to a section, figure or problem as 1.2, the first number gives the chapter number and the second the section, figure, or problem number as the case may be.

I am grateful to Dale Vaslow for his valuable assistance with the preparation of the manuscript and for all his efforts in improving the book. His carefully prepared notes of my lectures given during the spring semester of 1970 formed the nucleus around which the manuscript of this book was developed. He read the entire manuscript, worked out all the problems and offered many useful suggestions for improvement. I thank Professor Hang-Sheng Tuan and Dr. Douglas Preis for their kind interest as well as for their critical reading of the manuscript.

During the time of preparation of the manuscript of this book, the administration of the University of Wisconsin extended valuable and much-needed support without which this book could not have been written. The interest and the enthusiasm of the students who took the course on which this book is based were crucial for the initiation of the writing of this book as well as for completing it without delay. Mrs. Karen Jerry and Mrs. Judith Mohr, assisted ably by Mrs. Lynda Vaslow and Mrs. Jeanne Leschinsky, typed the entire manuscript with patience, promptitude and perfection. The original illustrations were prepared with the help of Mr. Thomas Freeman and Mrs. Helga A. Fack. My wife Susheela gave valuable assistance throughout the preparation of the manuscript and the production of the book.

I reviewed the manuscript when I was a visitor at California Institute of Technology. I am grateful to my host Professor Charles H. Papas for his kind hospitality as well as for the invigorating intellectual atmosphere that he provided.

Ronold Wyeth Percival King, Gordon McKay Professor of Applied Physics, Harvard University, has won international reputation as a brilliant researcher, a superb teacher and an outstanding author. To this illustrious son of the University of Wisconsin, I respectfully dedicate this book.

S. R. Seshadri

Madison, Wisconsin

CHAPTER 1

Macroscopic Kinetic Theory

1.1. Introduction

A classical gas consists of neutral particles and is a poor conductor of electricity because of the absence of free charged particles. The particle motions are neither affected by nor do they give rise to electromagnetic fields. In contrast a plasma, which may be loosely defined as a collection of charged particles, is a good conductor of electricity. In a volume which is large enough to contain a great number of particles but small compared to the scale length of spatial variation of macroscopic quantities of interest such as pressure and temperature, the resultant net electric charge is zero; that is, a plasma is macroscopically neutral. Moreover, the large Coulomb forces that are set up due to any significant charge separation enable the plasma to maintain its electrical neutrality in the presence of both static and dynamic disturbances. A plasma exhibits several novel features that are not present in an ordinary gas. This is because a plasma is at once a dynamical fluid and a good electrical conductor; it is capable of both interacting with electromagnetic fields and of creating them. The novelty of plasma phenomena is accentuated by the presence of a magnetic field which is used for the confinement and heating of plasmas in controlled thermonuclear research. Therefore, with a view to exploring all interesting phenomena, wherever possible, the plasma behavior is studied in the presence of the Lorentz force due to both the electric and the magnetic fields.

In a plasma only macroscopic phenomena due to the cumulative action of a large number of particles are observed and for the purpose of explaining and predicting these phenomena, it is sufficient to introduce and understand certain macroscopic variables such as number density, average velocity, pressure, and temperature. In this chapter, the elements of macroscopic kinetic theory of plasmas are developed and this development is intertwined with a brief review of the essentials of the kinetic theory of matter. The concept of the velocity distribution function necessary for a statistical characterization of the dynamics of the plasma particles is introduced and a systematic method for obtaining the average values of the particle properties is presented. The Boltzmann equation satisfied by the velocity distribution function is then deduced. The velocities of particles of a gas in a state of thermal equilibrium and in the absence of external forces are described by a simple and a very useful distribution function known as the Maxwell-Boltzmann distribution function. This distribution function is stated without proof and several of its important properties are treated. The Maxwell-Boltzmann distribution function is multiplied by a simple factor called the Boltzmann factor if a conservative force such as that due to an electrostatic field is present. The Boltzmann factor is also introduced without proof and its validity is verified. A detailed discussion of the concepts of particle current density and kinetic pressure is given.

The macroscopic variables such as the number density, the average velocity, the kinetic pressure, and the thermal energy flux density are all *moments* of the velocity distribution function. The partial differential equations specifying these macroscopic variables are determined by evaluating the various moments of the Boltzmann equation. The first three moments of the Boltzmann equation are deduced to yield the first three hydrodynamical equations which are, respectively, the equation of continuity, the momentum transport equation and the equation of transport of the kinetic pressure. At no stage is the hierarchy of hydrodynamical equations complete in the sense that there is a sufficient number of these equations for the determination of all the macroscopic variables appearing in them. It is necessary to truncate the system of hydrodynamical equations at some stage, introduce some simplifying approximation for the highest-order moment of the velocity distribution function appearing in these equations, and thus obtain a closed and complete set of hydrodynamical equations sufficient for the determination of all the relevant macroscopic variables. Two such simple sets of closed hydrodynamical equations are widely used and these characterize the so-called cold and warm plasma models, respectively. The governing equations for these two plasma models are deduced and the approximations involved in these models are discussed.

In a plasma which consists of more than one species of particles, a separate set of hydrodynamical equations is obtained for each species of particles. For example, in a fully ionized plasma consisting of electrons and singly charged hydrogen ions, there is a set of hydrodynamical equations for the electrons and another set for the ions. These two sets of hydrodynamical equations are not independent but are coupled. It is useful to introduce lumped macroscopic parameters which are combinations of the corresponding parameters relating to each species of particles contained in a plasma. The corresponding hydrodynamical equations pertaining to each species of particles are combined to yield the governing equations for these lumped macroscopic parameters. In particular, the equations of conservation of mass and electric charge of the plasma as a whole, treated as a conducting fluid, are derived. Also, the evolution equation for the average velocity of mass flow is deduced. This chapter is concluded with a discussion of the generalized Ohm's law which is just the evolution equation for the lumped electric current density.

1.2. Distribution function and average values

A plasma is a gas of electrons, ions and neutral particles. The various individual motions of the particles govern the microscopic state of the plasma. In classical mechanics, for a given observation time t, the motion of a particle is specified by its position vector $\mathbf{r}$ and its velocity vector $\mathbf{u}$. If the position and the velocity vectors of all the particles constituting a plasma are known at a given initial time, their values for a later time can be evaluated in principle. However, since there are a large number of particles, it is difficult to specify the initial position and velocity vectors let alone follow individually the trajectories of all the particles in order to determine the properties of a plasma. In the laboratory only macroscopic phenomena which depend on the average motions of the particles are observed. For the purpose of explaining and predicting macroscopic phenomena, it is sufficient to resort to a less detailed, statistical procedure. The concept of a velocity distribution function, which is important in the kinetic theory of matter, is necessary for carrying out the statistical procedure. From a knowledge of the distribution function, the average values of functions of the particle velocities can be systematically deduced and applied in the study of the collective behavior of the plasma particles.

Phase space

Let each particle be located by a position vector $\mathbf{r}$ drawn from the origin of the space coordinates x, y, and z. A small elemental volume around the terminal point of $\mathbf{r}$ is $d\mathbf{r} = dx\,dy\,dz$. Let the linear velocity of the center of mass of the particle be $\mathbf{u} = \hat{\mathbf{x}}u_x + \hat{\mathbf{y}}u_y + \hat{\mathbf{z}}u_z$; the particle speed is $|\mathbf{u}| = u$. In analogy with *the configuration space* defined by the coordinates x, y, and z, it is convenient to introduce a *velocity space* defined by the coordinates u_x, u_y, and u_z. Then, the velocity vector $\mathbf{u}$ may be regarded as a position vector drawn from the origin of the velocity space coordinates u_x, u_y, and u_z. A small elemental volume around the terminal point of $\mathbf{u}$ in the velocity space is $d\mathbf{u} = du_x\,du_y\,du_z$. It is convenient to consider *the phase space* defined by the six coordinates x, y, z, u_x, u_y, and u_z since at a specified time the position $\mathbf{r}$ and the velocity $\mathbf{u}$ of a particle can be represented together by an appropriate point in the phase space.

It is assumed that the elemental volume $d\mathbf{r}$ can be chosen large enough to contain a great many particles but yet is small compared to the scale lengths of spatial variation of the physical parameters of interest such as pressure and temperature. The types of plasmas for which the elemental volume $d\mathbf{r}$ cannot be chosen as indicated require a different formulation. Consider a time interval dt, centered around the observation time t, that is long compared to the times taken to cross the elemental volume for the particles which are undeflected during transit and yet short compared to the scale length of temporal variation of the relevant macroscopic parameters. The number of particles in the elemental volume $d\mathbf{r}$ when averaged over the time interval dt is proportional to the size and not the shape of the elemental volume; it may be therefore denoted by $N(\mathbf{r}, t)\, d\mathbf{r}$ where $N(\mathbf{r}, t)$ is the number density of particles.

Velocity distribution function

The large number of particles $N(\mathbf{r}, t)\, d\mathbf{r}$ have speeds varying from zero to a large value which is considerably less than the speed of light in free space. The distribution of the particle velocities is conveniently represented by the terminal points of the position vectors $\mathbf{u}$ in the velocity space. Note that the total number of points in the velocity space is equal to $N(\mathbf{r}, t)\, d\mathbf{r}$. The distribution of velocity points varies with time since the particle velocities change due to the forces of mutual interaction and external fields and since the velocity points appear or disappear according as the particles enter or leave the elemental volume $d\mathbf{r}$. Consider an elemental volume $d\mathbf{u}$ around the terminal point of the position vector $\mathbf{u}$ in the velocity space. The number of velocity points in the elemental volume $d\mathbf{u}$ after an averaging over the time interval dt is proportional to the size and not the shape of the elemental volume $d\mathbf{u}$. The time average of the number of the velocity points in the elemental volume $d\mathbf{u}$ in the velocity space is also proportional to the elemental volume $d\mathbf{r}$ in the configuration space. Moreover, this number, in general, depends on the position $\mathbf{r}$ of the volume $d\mathbf{r}$ in the configuration space and the position $\mathbf{u}$ of the volume $d\mathbf{u}$ in the velocity space. Hence the number of velocity points may be denoted by $f(\mathbf{r}, \mathbf{u}, t)\, d\mathbf{r}\, d\mathbf{u}$ where $f(\mathbf{r}, \mathbf{u}, t)$ is known as the velocity distribution function.

The following alternative interpretation of $f(\mathbf{r}, \mathbf{u}, t)$ is useful. Consider an elemental volume $d\mathbf{r}\, d\mathbf{u}$ around the position vectors $\mathbf{r}$ and $\mathbf{u}$ in the six-dimensional phase space defined by the coordinates x, y, z, u_x, u_y, and u_z. The time average over the interval dt of the number of representative points in the elemental volume $d\mathbf{r}\, d\mathbf{u}$ is independent of the shape but is proportional to the size of the volume. Therefore, this number may be denoted by $f(\mathbf{r}, \mathbf{u}, t)\, d\mathbf{r}\, d\mathbf{u}$. Thus the velocity distribution function $f(\mathbf{r}, \mathbf{u}, t)$ is the probable density of the representative points in the six-dimensional phase space.

If $f(\mathbf{r}, \mathbf{u}, t)$ is a function of the position vector $\mathbf{r}$, the corresponding plasma is inhomogeneous. Suppose that the plasma contained in a vessel is initially inhomogeneous. If there are no external forces, a state of equilibrium is created by the mutual interactions amongst the particles and the plasma becomes homogeneous in course of time. The velocity distribution function can be either isotropic or anisotropic in the velocity space. It is isotropic if $f(\mathbf{r}, \mathbf{u}, t)$ is not a function of the orientation of the velocity vector $\mathbf{u}$ but is anisotropic if $f(\mathbf{r}, \mathbf{u}, t)$ is dependent on the orientation of the velocity vector $\mathbf{u}$. The description of different types of plasmas requires homogeneous as well as inhomogeneous and isotropic as well as anisotropic velocity distribution functions.

If the number of velocity points $f(\mathbf{r}, \mathbf{u}, t)\, d\mathbf{r}\, d\mathbf{u}$ contained in the elemental volume $d\mathbf{u}$ is summed up for all possible velocities, the result is the total number $N(\mathbf{r}, t)\, d\mathbf{r}$ of velocity points in the entire velocity space. Therefore, it follows that

$$N(\mathbf{r}, t) = \int_{-\infty}^{\infty} f(\mathbf{r}, \mathbf{u}, t)\, d\mathbf{u} \tag{2.1}$$

In Eq. (2.1) it is implied that the integration is carried out with respect to each of the variables u_x, u_y, and u_z from $-\infty$ to ∞. From its definition, it follows that the distribution function is finite, continuous and positive for all values of time t. Physical considerations require that the distribution function tends to zero as the velocity becomes infinitely large.

Average values

With the help of the velocity distribution function $f(\mathbf{r}, \mathbf{u}, t)$, it is possible to obtain the average values of functions of particle velocity. Consider a function $g(\mathbf{r}, \mathbf{u}, t)$ associated with a particle and dependent on its position $\mathbf{r}$, its velocity $\mathbf{u}$, and the observation time t. The probable number of velocity points in the elemental volume $d\mathbf{u}$ around the position vector $\mathbf{u}$ in the velocity space is $f(\mathbf{r}, \mathbf{u}, t)\, d\mathbf{r}\, d\mathbf{u}$. Therefore, the value of $g(\mathbf{r}, \mathbf{u}, t)$ when summed over all the velocity points in the elemental volume $d\mathbf{u}$ is $g(\mathbf{r}, \mathbf{u}, t)\, f(\mathbf{r}, \mathbf{u}, t)\, d\mathbf{r}\, d\mathbf{u}$. The total value of this function for all the velocity points in the velocity space is obtained by an integration over all possible velocities. The result is

$$d\mathbf{r} \int_{-\infty}^{\infty} g(\mathbf{r}, \mathbf{u}, t)\, f(\mathbf{r}, \mathbf{u}, t)\, d\mathbf{u} \tag{2.2}$$

When the expression (2.2) is divided by the total number $N(\mathbf{r}, t)\, d\mathbf{r}$ of velocity points, the average of $g(\mathbf{r}, \mathbf{u}, t)$ over the entire distribution of velocities is obtained:

$$\langle g(\mathbf{r}, \mathbf{u}, t) \rangle = \frac{1}{N(\mathbf{r}, t)} \int_{-\infty}^{\infty} g(\mathbf{r}, \mathbf{u}, t)\, f(\mathbf{r}, \mathbf{u}, t)\, d\mathbf{u} \tag{2.3}$$

Peculiar velocity

For example, if $g(\mathbf{r}, \mathbf{u}, t)$ is equal to the velocity $\mathbf{u}$ of the particles in the vicinity of $\mathbf{r}$ at the instant of time t, the average velocity $\mathbf{v}$ is found from Eq. (2.3) as

$$\langle \mathbf{u} \rangle = \mathbf{v} = \frac{1}{N(\mathbf{r}, t)} \int_{-\infty}^{\infty} \mathbf{u}\, f(\mathbf{r}, \mathbf{u}, t)\, d\mathbf{u} \tag{2.4}$$

The velocity $\mathbf{u}$ of a particle may be specified by the velocity $\mathbf{w}$ relative to the average velocity as

$$\mathbf{u} = \mathbf{v} + \mathbf{w} \tag{2.5}$$

The velocity $\mathbf{w}$ is called the *peculiar velocity* and its magnitude $w = |\mathbf{w}|$ is called the *peculiar speed.* When Eq. (2.5) is substituted in Eq. (2.1), the result is

$$N(\mathbf{r}, t) = \int_{-\infty}^{\infty} f(\mathbf{r}, \mathbf{w}, t)\, d\mathbf{w} \tag{2.6}$$

where in accordance with the usual practice $f(\mathbf{r}, \mathbf{v} + \mathbf{w}, t)$ is abbreviated as $f(\mathbf{r}, \mathbf{w}, t)$. The elemental volume $d\mathbf{u}$ situated around the terminal point of $\mathbf{u}$ in the velocity space is denoted by $d\mathbf{w}$ and is situated around the terminal point of $\mathbf{w}$ if the origin of the velocity space is shifted to coincide with the terminal point of the average velocity vector $\mathbf{v}$. The distribution of velocity points is unaffected by the shift of the origin and therefore it follows that the number density given by the distribution function $f(\mathbf{r}, \mathbf{w}, t)$ is equal to that given by $f(\mathbf{r}, \mathbf{u}, t)$ as obtained in Eq. (2.6).

By its definition, the average of the peculiar velocity of all the particles in the elemental volume $d\mathbf{r}$ is zero. This result can be verified from Eqs. (2.4) and (2.5).

1.3. The Boltzmann equation

The statistical behavior of an ensemble of each species of particles is described by a velocity distribution function which is the density of the representative points in the phase space. The velocity distribution function satisfies an equation known as the Boltzmann equation. The positions of the particles vary in time in view of their velocity. The forces acting on the particles change their velocity in the course of time. Thus the representative points in the phase space move as a function of time; the motion of each point represents the evolution of an individual particle. The Boltzmann equation is the law governing the variation of the distribution function with time. Therefore, in principle, the Boltzmann equation enables the calculation of the evolution of the distribution function in time.

One-dimensional Boltzmann equation

It is instructive to deduce the Boltzmann equation first for the one-dimensional velocity distribution function which depends on the position coordinate x, the velocity coordinate u_x and the observation time t. For this case, the phase space is two-dimensional as shown in Fig. 1.1. As indicated in Fig. 1.1, the elemental volume in the phase space for this case reduces to an elemental area. If the particles represent the centers of mass of perfectly elastic, impenetrable spheres, the interaction between the particles takes place instantaneously. But for the charged particles, the interaction takes place over a finite time interval which becomes smaller as the distance between the initial trajectories of the particles becomes smaller. Consider a time interval dt that is long compared to the average time of interaction between two particles so that most interactions which begin in the interval dt are also completed in dt. Also, let dt be short compared to the average

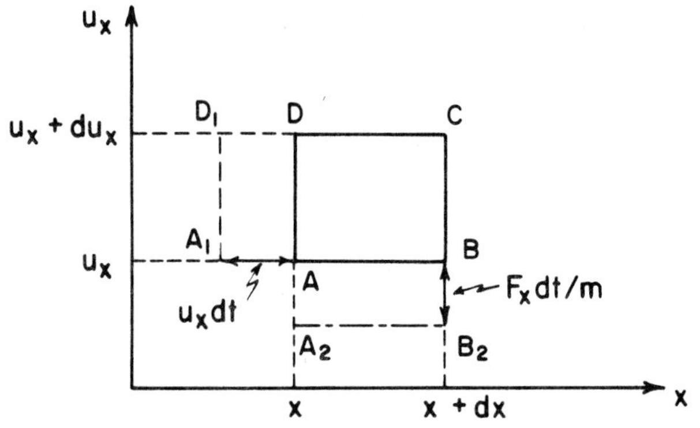

Fig. 1.1. Phase space for a one-dimensional velocity distribution function.

time between interactions so that each particle interacts at most once with another particle in the time interval dt. If these assumptions are valid, the trajectory of a particle may be considered to be composed of segments, where only the external forces act, joined by very short trajectories, during which there is interaction. These conditions must be fulfilled for the validity of the Boltzmann equation. In dense fluids and where the force of interaction varies slowly with the distance of separation between the particles, the above conditions are not satisfied and the applicability of the Boltzmann equation becomes questionable.

Let m be the mass of each particle. If the force acting on the particle is $\mathbf{F}$, its acceleration is given by $\mathbf{F}/m$. For the present case of one-dimensional distribution of velocities, the force has only an x-component F_x. From Fig. 1.1, it is seen that on the average in the time interval dt, the representative points in the area $A\,D\,D_1\,A_1$ of sides $u_x\,dt$ and du_x enter through the side AD the elemental area $A\,B\,C\,D$ of the phase space. Therefore, the number of representative points entering the elemental area $A\,B\,C\,D$ through the side AD in the time interval dt is

$$f(x, u_x, t)u_x\,dt\,du_x \tag{3.1}$$

The number of representative points leaving the elemental area $A\,B\,C\,D$ through the side BC in the same time interval dt is obtained from the expression (3.1) as

$$[f(x, u_x, t)u_x + (\partial/\partial x)\{f(x, u_x, t)u_x\}dx]\,dt\,du_x \tag{3.2}$$

Hence, the net number of representative points entering the elemental area $A\,B\,C\,D$ in the phase space in the time interval dt due to their inherent velocity is found by subtracting the expression (3.2) from the expression (3.1) with the following result:

$$-(\partial/\partial x)\{f(x, u_x, t)u_x\}\, dx\, du_x\, dt \tag{3.3}$$

Since x and u_x are independent variables, the expression (3.3) becomes

$$-u_x(\partial/\partial x)\{f(x, u_x, t)\}\, dx\, du_x\, dt \tag{3.4}$$

In a similar manner, it is seen from Fig. 1.1 that on the average in the time interval dt, the representative points in the area $A\,B\,B_2\,A_2$ of sides $F_x\,dt/m$ and dx enter through the side AB the elemental area $A\,B\,C\,D$ of the phase space. Therefore, the number of representative points entering the elemental area $A\,B\,C\,D$ through the side AB in the time interval dt is

$$f(x, u_x, t)(F_x/m)\, dt\, dx \tag{3.5}$$

The number of representative points leaving the elemental area $A\,B\,C\,D$ through the side CD in the same time interval dt is obtained from the expression (3.5) as

$$[f(x, u_x, t)(F_x/m) + (\partial/\partial u_x)\{f(x, u_x, t)F_x/m\}\, du_x]\, dt\, dx \tag{3.6}$$

Hence, the net number of representative points entering the elemental area $A\,B\,C\,D$ in the phase space in the time interval dt due to the changes in the velocity of the particles is determined by subtracting the expression (3.6) from the expression (3.5) with the following result:

$$-(\partial/\partial u_x)\{f(x, u_x, t)F_x/m\}\, dx\, du_x\, dt \tag{3.7}$$

It is assumed that the force F_x is independent of the velocity u_x and therefore the expression (3.7) may be rewritten as

$$-(F_x/m)(\partial/\partial u_x)\{f(x, u_x, t)\}\, dx\, du_x\, dt \tag{3.8}$$

In the time interval dt, the particles in the range dx at x interact or collide at most once with other particles and have their velocities changed. The velocities of some of the particles in the range dx at x change their original values between u_x and $u_x + du_x$ to leave the range du_x at u_x. Also, the velocities of some other particles in the range dx at x change their original values to enter the range du_x at u_x from outside this range. Therefore, due to the particle interactions, there is a net gain in the number of representative points in the elemental area $A\,B\,C\,D$ of the phase space. This gain is proportional to the time interval dt and the elemental area $dx\,du_x$ of the phase space under consideration; therefore, the net increase in the number of representative points in the elemental area $A\,B\,C\,D$ may be assumed to be of the form

$$(\partial f/\partial t)_{\text{coll}}\, dx\, du_x\, dt \tag{3.9}$$

If $(\partial/\partial t) f(x, u_x, t)$ is the rate of increase of the density of the representative points in the phase space, the increase in the number of the representative points in the elemental area $ABCD$ in the time interval dt is

$$(\partial/\partial t)\{f(x, u_x, t)\}\, dx\, du_x\, dt \tag{3.10}$$

From the expressions (3.4) and (3.8)–(3.10), it follows that

$$\frac{\partial}{\partial t} f(x, u_x, t) + u_x \frac{\partial}{\partial x} f(x, u_x, t) + \frac{F_x}{m} \frac{\partial}{\partial u_x} f(x, u_x, t) = \left(\frac{\partial f}{\partial t}\right)_{\text{coll}} \tag{3.11}$$

which is the one-dimensional Boltzmann equation. It is now clear that the Boltzmann equation is a statement of conservation of the number of representative points in the phase space. Since the acceleration has been set equal to F_x/m, it follows that Eq. (3.11) is nonrelativistic.

Three-dimensional Boltzmann equation

The three-dimensional Boltzmann equation can be deduced using a procedure similar to that employed for the one-dimensional equation. For this purpose consider an elemental volume $d\mathbf{r}\, d\mathbf{u} = dx\, dy\, dz\, du_x\, du_y\, du_z$ in the six-dimensional phase space. The cross-sectional "area" of the volume $d\mathbf{r}\, d\mathbf{u}$ by a plane perpendicular to the x-coordinate is $dy\, dz\, du_x\, du_y\, du_z$. Therefore, the number of representative points entering the elemental volume $d\mathbf{r}\, d\mathbf{u}$ through the side perpendicular to the x-coordinate at x in the time interval dt is obtained similarly to that in the expression (3.1) as

$$f(\mathbf{r}, \mathbf{u}, t) u_x\, dt\, dy\, dz\, du_x\, du_y\, du_z \tag{3.12}$$

The number of representative points leaving the elemental volume $d\mathbf{r}\, d\mathbf{u}$ through the side perpendicular to the x-coordinate at $x + dx$ in the same time interval dt is deduced from the expression (3.12) as

$$[f(\mathbf{r}, \mathbf{u}, t) u_x + (\partial/\partial x)\{f(\mathbf{r}, \mathbf{u}, t) u_x\}\, dx]\, dt\, dy\, dz\, du_x\, du_y\, du_z \tag{3.13}$$

Hence, the net number of representative points entering the elemental volume $d\mathbf{r}\, d\mathbf{u}$ along the direction of the x-coordinate in the time interval dt is found from the expressions (3.12) and (3.13) as

$$-(\partial/\partial x)\{f(\mathbf{r}, \mathbf{u}, t) u_x\}\, d\mathbf{r}\, d\mathbf{u}\, dt \tag{3.14}$$

Since u_x is independent of x, the expression (3.14) becomes

$$-u_x(\partial/\partial x)\{f(\mathbf{r}, \mathbf{u}, t)\}\, d\mathbf{r}\, d\mathbf{u}\, dt \tag{3.15}$$

Similarly, the net number of representative points entering $d\mathbf{r}\, d\mathbf{u}$ in the time interval dt due to the drifts parallel to the y- and the z-coordinates can be obtained to be given respectively by

$$-u_y(\partial/\partial y)\{f(\mathbf{r}, \mathbf{u}, t)\}\, d\mathbf{r}\, d\mathbf{u}\, dt \tag{3.16}$$

and

$$-u_z(\partial/\partial z)\{f(\mathbf{r}, \mathbf{u}, t)\}\, d\mathbf{r}\, d\mathbf{u}\, dt \tag{3.17}$$

The sum of the expressions (3.15), (3.16) and (3.17) yields the following result for the net number of representative points entering $d\mathbf{r}\, d\mathbf{u}$ in the time interval dt due to the drifts parallel to the space coordinates:

$$-(\mathbf{u} \cdot \nabla_r) f(\mathbf{r}, \mathbf{u}, t)\, d\mathbf{r}\, d\mathbf{u}\, dt \tag{3.18}$$

where

$$\nabla_r = \hat{\mathbf{x}}\frac{\partial}{\partial x} + \hat{\mathbf{y}}\frac{\partial}{\partial y} + \hat{\mathbf{z}}\frac{\partial}{\partial z}$$

is the gradient operator in the configuration space.

As before, the cross-sectional "area" of the volume $d\mathbf{r}\, d\mathbf{u}$ by a plane perpendicular to the u_x-coordinate is $dx\, dy\, dz\, du_y\, du_z$. Therefore, the number of representative points entering $d\mathbf{r}\, d\mathbf{u}$ through the side perpendicular to the u_x-coordinate at u_x in the time interval dt is deduced similarly to that in the expression (3.5) as

$$f(\mathbf{r}, \mathbf{u}, t)(F_x/m)\, dt\, dx\, dy\, dz\, du_y\, du_z \tag{3.19}$$

The number of representative points leaving $d\mathbf{r}\, d\mathbf{u}$ through the side perpendicular to the u_x-coordinate at $u_x + du_x$ in the same time interval dt is determined from the expression (3.19) as

$$[f(\mathbf{r}, \mathbf{u}, t)F_x/m + (\partial/\partial u_x)\{f(\mathbf{r}, \mathbf{u}, t)F_x/m\}\, du_x]\, dt\, dx\, dy\, dz\, du_y\, du_z \tag{3.20}$$

From the expressions (3.19) and (3.20), the net number of representative points entering $d\mathbf{r}\, d\mathbf{u}$ along the direction of the u_x-coordinate in the time interval dt is found to be given by

$$-(\partial/\partial u_x)\{f(\mathbf{r}, \mathbf{u}, t)F_x/m\}\, d\mathbf{r}\, d\mathbf{u}\, dt \tag{3.21}$$

If F_x is assumed to be independent of u_x, the expression (3.21) simplifies to

$$-(F_x/m)(\partial/\partial u_x)\{f(\mathbf{r}, \mathbf{u}, t)\}\, d\mathbf{r}\, d\mathbf{u}\, dt \tag{3.22}$$

Similarly, the net number of representative points entering $d\mathbf{r}\, d\mathbf{u}$ in the time interval dt due to the drifts parallel to the u_y- and the u_z-coordinates is evaluated to be given respectively by

$$-(F_y/m)(\partial/\partial u_y)\{f(\mathbf{r}, \mathbf{u}, t)\}\, d\mathbf{r}\, d\mathbf{u}\, dt \tag{3.23}$$

and

$$-(F_z/m)(\partial/\partial u_z)\{f(\mathbf{r},\mathbf{u},t)\}\,d\mathbf{r}\,d\mathbf{u}\,dt \tag{3.24}$$

In obtaining the expressions (3.23) and (3.24), F_y and F_z have been assumed to be independent of u_y and u_z, respectively. The sum of the expressions (3.22), (3.23) and (3.24) is conveniently written as

$$-\left(\frac{\mathbf{F}}{m}\cdot\nabla_u\right)f(\mathbf{r},\mathbf{u},t)\,d\mathbf{r}\,d\mathbf{u}\,dt \tag{3.25}$$

where

$$\nabla_u = \hat{\mathbf{x}}\frac{\partial}{\partial u_x} + \hat{\mathbf{y}}\frac{\partial}{\partial u_y} + \hat{\mathbf{z}}\frac{\partial}{\partial u_z}$$

is the gradient operator in the velocity space. Note that the expression (3.25) gives the net number of representative points entering $d\mathbf{r}\,d\mathbf{u}$ in the time interval dt due to the drifts parallel to the velocity coordinates. Moreover, it has been assumed that the force in a given direction is independent of the particle velocity in that direction, i.e.

$$\partial F_i/\partial u_i = 0 \qquad i = x, y, z \tag{3.26}$$

Due to the particle interactions, as before, there is a net gain in the number of representative points in the elemental volume $d\mathbf{r}\,d\mathbf{u}$ in the phase space and this number is proportional to the time interval dt and the elemental volume $d\mathbf{r}\,d\mathbf{u}$ of the phase space. Therefore, the net increase in the number of representative points in $d\mathbf{r}\,d\mathbf{u}$ in the time interval dt may be assumed to be of the form

$$\left(\frac{\partial f}{\partial t}\right)_{\text{coll}} d\mathbf{r}\,d\mathbf{u}\,dt \tag{3.27}$$

If $(\partial/\partial t)\,f(\mathbf{r},\mathbf{u},t)$ is the rate of increase of the density of the representative points in the phase space, the increase in the number of the representative points in $d\mathbf{r}\,d\mathbf{u}$ in the time interval dt is

$$(\partial/\partial t)\{f(\mathbf{r},\mathbf{u},t)\}\,d\mathbf{r}\,d\mathbf{u}\,dt \tag{3.28}$$

It follows from the expressions (3.18), (3.25), (3.27), and (3.28) that

$$\frac{\partial}{\partial t}f(\mathbf{r},\mathbf{u},t) + (\mathbf{u}\cdot\nabla_r)\,f(\mathbf{r},\mathbf{u},t) + \left(\frac{\mathbf{F}}{m}\cdot\nabla_u\right)f(\mathbf{r},\mathbf{u},t) = \left(\frac{\partial f}{\partial t}\right)_{\text{coll}} \tag{3.29}$$

which is the three-dimensional Boltzmann equation.

Several remarks in connection with the Boltzmann equation are appropriate at this stage. The collision term appearing on the right side of Eq. (3.29) is schematic and an explicit expression for it remains to be developed. It is possible to express

$(\partial f/\partial t)_{coll}$ as a multiple integral with $f(\mathbf{r}, \mathbf{u}, t)$ appearing in the integrand. The Boltzmann equation is therefore an integro-differential equation. The force acting on a charged particle in a plasma is the Lorentz force:

$$\mathbf{F} = q[\mathbf{E} + \mathbf{u} \times \mathbf{B}] \tag{3.30}$$

Here, the electric field $\mathbf{E}$ and the magnetic flux density $\mathbf{B}$ are the continuous macroscopic fields obtained after averaging over an elemental volume whose dimensions are large enough to contain a great many particles but small compared to the scale length of spatial variation of the physical quantities of interest. In Eq. (3.30) q is the charge carried by the particle. Only electromagnetic forces are to be considered with the result that there is no force acting on a neutral particle. An important feature of the force in Eq. (3.30) is that it is dependent on the particle velocity $\mathbf{u}$. Since $\mathbf{F}$ given by Eq. (3.30) satisfies the criterion (3.26), the validity of the expression (3.25) is unimpaired for the Lorentz force. In a plasma there are several species of particles such as electrons, ions, and neutral particles. Each species of particles is characterized by a separate distribution function. Hence, there is a Boltzmann equation similar to Eq. (3.29) specifying the velocity distribution function associated with each species of particles. The three Boltzmann equations for the electrons, the ions, and the neutral particles are coupled; the collision term in the Boltzmann equation pertaining to one species contains the velocity distribution functions associated with the other species as well because collisional interactions occur also between the different species of particles. The solutions of the Boltzmann equations for a plasma are therefore a matter of great difficulty.

1.4. Maxwell-Boltzmann distribution function

Let a gas of particles be injected through a hole in the walls of a container and let the hole be then closed. Initially, the particles are near the hole and there are no particles in the regions of the container remote from the hole with the result that the velocity distribution function is inhomogeneous. Also, initially the particle velocities are predominantly in the directions in which they were injected into the container and the other directions of particle velocities are less probable; therefore, the velocity distribution function is highly anisotropic. In course of time, this unstable situation is changed as a result of interactions of the particles with themselves and with the walls of the container. After sufficiently long time, the particles are uniformly distributed within the container resulting in a homogeneous velocity distribution function. Moreover, the speeds of the particles passing through any point are uniformly distributed in all directions, that is, the velocity distribution function becomes isotropic. Thus, an inhomogeneous and anisotropic velocity distribution function evolves towards a homogeneous and isotropic distribution function.

Maxwell-Boltzmann velocity distribution

Suppose that the container is isothermal, that is, it is maintained at a definite temperature. In course of time, the gas attains a thermal equilibrium and its temperature becomes equal to that of the container. The increase in the temperature of the gas is brought about by an increase in the speed of the particles and therefore a corresponding increase of kinetic energy due to their random motions. In the state of thermal equilibrium between the gas and the container, the velocity distribution function is well defined. This so-called Maxwell-Boltzmann distribution function is applicable to a gas of particles which (i) are in the form of impenetrable spheres with no internal structure, (ii) have only thermal energy due to translational motion, and (iii) are subject to no external forces. This distribution function is of the form

$$f(\mathbf{r}, \mathbf{u}, t) = f_M(u) = \alpha e^{-\beta u^2} \tag{4.1}$$

where

$$u^2 = u_x^2 + u_y^2 + u_z^2 \tag{4.2}$$

In Eq. (4.1) α and β are constants independent of $\mathbf{r}$ and $\mathbf{u}$ and can be determined in terms of the number density N_0, the particle mass m and the temperature T_0 of the gas. Note that N_0 and T_0 are the same in all parts of the gas in the container.

Number density

When the velocity distribution function is isotropic, it is convenient to define a speed distribution function. For this purpose, the spherical polar coordinates u, θ, and φ are introduced in the velocity space with the result that the elemental volume $d\mathbf{u}$ in the velocity space in the range from (u, θ, φ) to $(u + du, \theta + d\theta, \varphi + d\varphi)$ is obtained as

$$d\mathbf{u} = u^2 \sin\theta \, du \, d\theta \, d\varphi \tag{4.3}$$

The number density N_0 is seen from Eqs. (2.1), (4.1), and (4.3) to be given by

$$N_0 = \int_0^\infty du \int_0^\pi d\theta \int_0^{2\pi} d\varphi \, f_M(u) u^2 \sin\theta \tag{4.4}$$

Since $f_M(u)$ is independent of θ and φ, the integrations with respect to the angles are easily carried out. The result on using Eq. (4.1) is

$$N_0 = \int_0^\infty 4\pi u^2 \, f_M(u) \, du \tag{4.5a}$$

$$= \int_0^\infty \alpha 4\pi u^2 e^{-\beta u^2} \, du \tag{4.5b}$$

From Eq. (4.5a), it is clear that the number of particles in a unit volume having speeds between u and $u + du$ is

$$f_{sp}(u)\,du = 4\pi u^2 f_M(u)\,du \tag{4.6}$$

where $f_{sp}(u)$ is called the speed distribution function.

Integrals of the form (4.5b) arise in the evaluation of the averages for a Maxwell-Boltzmann distribution of velocities and the results of the commonly occurring integrals are given as follows:

$$\begin{aligned} &h(0) = \tfrac{1}{2}\sqrt{\pi/a} \qquad h(1) = \frac{1}{2a} \qquad h(2) = \tfrac{1}{4}\sqrt{\pi/a^3} \\ &h(3) = \frac{1}{2a^2} \qquad h(4) = \tfrac{3}{8}\sqrt{\pi}\,/a^{5/2} \qquad h(5) = a^{-3} \end{aligned} \tag{4.7a}$$

where

$$h(n) = \int_0^\infty x^n e^{-ax^2}\,dx \tag{4.7b}$$

With the help of Eqs.(4.7a), Eq. (4.5b) may be shown to yield

$$N_0 = \alpha(\pi/\beta)^{3/2} \quad \text{or} \quad \alpha = N_0(\beta/\pi)^{3/2} \tag{4.8}$$

Together with Eq. (4.8), Eq. (4.1) becomes

$$f_M(u) = N_0\left(\frac{\beta}{\pi}\right)^{3/2} e^{-\beta u^2} \tag{4.9}$$

where the constant β remains to be evaluated.

Temperature

The average kinetic energy due to the motion of the particles in the x-direction is deduced from Eqs. (2.3) and (4.9) as

$$\frac{1}{2}m\langle u_x^2\rangle = \frac{m}{2}(\beta/\pi)^{3/2}\int_{-\infty}^{\infty} du_x\, u_x^2 e^{-\beta u_x^2}\int_{-\infty}^{\infty} du_y\, e^{-\beta u_y^2}\int_{-\infty}^{\infty} du_z\, e^{-\beta u_z^2} \tag{4.10}$$

The three integrals in Eq. (4.10) are evaluated with the help of Eqs. (4.7a) and the result is

$$\tfrac{1}{2}m\langle u_x^2\rangle = m/4\beta \tag{4.11}$$

Since the velocity distribution function is isotropic, it follows that

$$\tfrac{1}{2}m\langle u_x^2\rangle = \tfrac{1}{2}m\langle u_y^2\rangle = \tfrac{1}{2}m\langle u_z^2\rangle = m/4\beta \tag{4.12}$$

and

$$\tfrac{1}{2}m\langle u^2\rangle = \tfrac{1}{2}m\langle u_x^2 + u_y^2 + u_z^2\rangle = 3m/4\beta \tag{4.13}$$

There are several ways of defining temperature. The average kinetic energy due to the random motion of the particles is proportional to the temperature. In kinetic theory, the temperature is defined by the relation

$$\frac{1}{2}m\langle u^2\rangle = \frac{3}{2}KT \tag{4.14}$$

where K is the Boltzmann constant and is equal to 1.3805×10^{-23} J/°K. Since the temperature of the gas and the container has been assumed to be T_0, β can be evaluated in terms of T_0 from Eqs. (4.13) and (4.14) as

$$\beta = m/2KT_0 \tag{4.15}$$

The substitution of Eq. (4.15) in Eq. (4.9) yields the Maxwell-Boltzmann distribution function in terms of N_0 and T_0 as

$$f_M(u) = N_0\left(\frac{m}{2\pi KT_0}\right)^{3/2} e^{-mu^2/2KT_0} \tag{4.16}$$

If the number density N_0 and the temperature T_0 of the gas are assigned, there is only one permanent mode of distribution of velocities, namely that given by Eq. (4.16). All velocity distributions which are different from Eq. (4.16) approach it in course of time if the gas is enclosed in a container and has no external forces acting upon it.

Properties of Maxwellian distribution of velocities

In view of their importance it is worthwhile to discuss some of the properties associated with the equilibrium distribution function given by Eq. (4.16). The number of particles in unit volume whose velocities are in the range from (u_x, u_y, u_z) to $(u_x + du_x, u_y + du_y, u_z + du_z)$ is determined from Eq. (4.16) as

$$f_M(u)\,d\mathbf{u} = N_0\left(\frac{m}{2\pi KT}\right)^{3/2} e^{-mu_x^2/2KT_0}\,du_x\,e^{-mu_y^2/2KT_0}\,du_y\,e^{-mu_z^2/2KT_0}\,du_z \tag{4.17}$$

The particles described by the Maxwell-Boltzmann distribution function are said to be in the Maxwellian state. It is seen from Eq. (4.17) that for the Maxwellian state, the distribution of u_x is independent of the values of u_y and u_z; the probability that a component of particle velocity in a given direction lies between specified limits is independent of the value of the component perpendicular to that direction. If

$$s_x^2 = mu_x^2/2KT_0 \tag{4.18}$$

the distribution of the x-component u_x of the velocity is seen to be proportional to

$\exp(-s_x^2)$. As shown in Fig. 1.2, this distribution is proportional to the exponential function, has its maximum value at $u_x = 0$ and reveals that positive and the corresponding negative values of u_x are equally probable. It is to be noted that the distributions of the velocity components u_y and u_z are also the same as that shown in Fig. 1.2.

Most probable speed

For the Maxwellian state, the isotropic speed distribution function is seen from Eqs. (4.6) and (4.16) to be given by

$$f_{sp}(u) = N_0 \sqrt{\frac{2}{\pi}} \left(\frac{m}{KT_0} \right)^{3/2} u^2 e^{-mu^2/2KT_0} \tag{4.19}$$

On setting $s^2 = mu^2/2KT_0$, the speed distribution function is seen to be proportional to $s^2 \exp(-s^2)$ which is also shown in Fig. 1.2. The speed distribution function has its maximum value at $s = 1$ and the corresponding particle speed is

$$u_{mp} = (2KT_0/m)^{1/2} \tag{4.20}$$

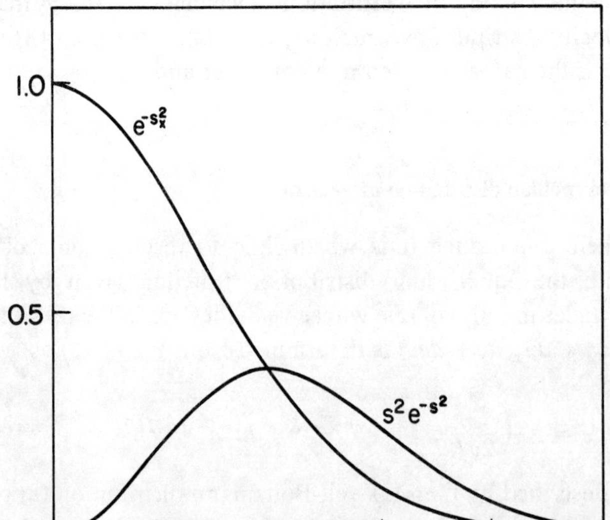

Fig. 1.2. Maxwell-Boltzmann distributions of velocity components and speed.

Since $f_{sp}(u)$ is the probability that the particles have a speed u, it is clear that u_{mp} given by Eq. (4.20) is the most probable speed.

Average speed and velocities

The average velocity of the particles in the Maxwellian state is found from Eq. (4.16) as

$$\langle \mathbf{u} \rangle = \left(\frac{m}{2\pi K T_0} \right)^{3/2} \int_{-\infty}^{\infty} du_x \int_{-\infty}^{\infty} du_y \int_{-\infty}^{\infty} du_z \times (\hat{\mathbf{x}} u_x + \hat{\mathbf{y}} u_y + \hat{\mathbf{z}} u_z) e^{-m(u_x^2+u_y^2+u_z^2)/2KT_0} = 0 \tag{4.21}$$

Since the integrand is of odd degree in all the velocity components u_x, u_y, and u_z, the value of the integral is zero. Hence $\langle \mathbf{u} \rangle = 0$ as indicated in Eq. (4.21). The physical implication is that the velocity vector has no preferred orientations. The average value of the velocity is zero for every isotropic velocity distribution function. The average *speed* of the particles is evaluated with the help of Eqs. (4.7) and (4.19) to be given by

$$\langle u \rangle = u_{av} = \sqrt{\frac{2}{\pi}} \left(\frac{m}{KT_0} \right)^{3/2} \int_0^{\infty} u^3 e^{-mu^2/2KT_0} \, du = \left(\frac{8KT_0}{\pi m} \right)^{1/2} \tag{4.22}$$

The root mean square velocity is found from Eq. (4.14) as

$$u_{rms} = \sqrt{\langle u^2 \rangle} = (3KT_0/m)^{1/2} \tag{4.23}$$

It is verified from Eqs. (4.20), (4.22), and (4.23) that

$$u_{av} = 1.128 u_{mp} \qquad u_{rms} = 1.225 u_{mp} \qquad u_{mp} < u_{av} < u_{rms} \tag{4.24}$$

The results given by Eqs. (4.12)–(4.14) are consistent with the principle of equipartition of energy. This principle states that when a large number of particles enclosed in a container of constant temperature attain a state of thermal equilibrium due to their interactions with themselves and with the walls of the container, each of the particles in this state of equilibrium has a random energy of thermal agitation whose average value is proportional to the absolute temperature T_0 and is equal to $nKT_0/2$ where n is the number of degrees of freedom in the system. Each of the particles possesses three independent translational degrees of freedom. The average value of the random energy of thermal agitation associated with each degree of freedom is seen from Eqs. (4.12) and (4.15) to be $KT_0/2$ in accordance with the equipartition principle. The average of the total random energy is equal to $3KT_0/2$ as indicated in Eq. (4.14). If the particles have additional degrees of freedom such as due to rotation and vibration, the random kinetic energy of the particles can exceed $3KT_0/2$.

Shifted Maxwell-Boltzmann velocity distribution

The gas can have a nonvanishing average velocity $\mathbf{v} = \langle \mathbf{u} \rangle$ if it is not contained in a vessel. In such a case, the Maxwellian distribution of velocities is given by

$$f_{MS}(\mathbf{u}) = N_0 \left(\frac{m}{2\pi K T_0} \right)^{3/2} \exp\left[-\frac{m}{2KT_0} \{ (u_x - v_x)^2 + (u_y - v_y)^2 + (u_z - v_z)^2 \} \right] \tag{4.25}$$

The average value of u_x is obtained from Eq. (4.25) as

$$\langle u_x \rangle = \left(\frac{m}{2\pi K T_0} \right)^{3/2} \int_{-\infty}^{\infty} du_x \, u_x \, e^{-m(u_x - v_x)^2/2KT_0} \int_{-\infty}^{\infty} du_y \, e^{-m(u_y - v_y)^2/2KT_0} \int_{-\infty}^{\infty} du_z \, e^{-m(u_z - v_z)^2/2KT_0} \tag{4.26}$$

On changing the variable of integration to $\mathbf{u} - \mathbf{v} = \mathbf{w}$ or $u_x - v_x = w_x$, $u_y - v_y = w_y$ and $u_z - v_z = w_z$, Eq. (4.26) becomes

$$\langle u_x \rangle = \left(\frac{m}{2\pi K T_0} \right)^{3/2} \int_{-\infty}^{\infty} dw_x (v_x + w_x) e^{-mw_x^2/2KT_0} \int_{-\infty}^{\infty} dw_y \, e^{-mw_y^2/2KT_0} \int_{-\infty}^{\infty} dw_z \, e^{-mw_z^2/2KT_0} \tag{4.27}$$

The integrals with respect to w_y and w_z are evaluated with the help of Eqs. (4.7a). The integral with respect to w_x has two terms; the term containing w_x yields an integrand which is an odd function of w_x and hence does not contribute to the integral. The other integral is evaluated again with the help of Eqs. (4.7a). The final result is

$$\langle u_x \rangle = v_x \tag{4.28}$$

Similarly, it can be shown that

$$\langle u_y \rangle = v_y \qquad \langle u_z \rangle = v_z \tag{4.29}$$

Therefore,

$$\langle \mathbf{u} \rangle = \langle \hat{\mathbf{x}} u_x + \hat{\mathbf{y}} u_y + \hat{\mathbf{z}} u_z \rangle = \hat{\mathbf{x}} v_x + \hat{\mathbf{y}} v_y + \hat{\mathbf{z}} v_z = \mathbf{v} \tag{4.30}$$

Thus it is verified that the distribution of velocities given by Eq. (4.25) gives a nonvanishing value $\mathbf{v}$ for the average velocity. If the number density N_0, the average velocity $\mathbf{v}$ and the temperature T_0 are assigned, there is only one permanent mode of distribution of velocities as given by the *shifted* Maxwell-Boltzmann distribution function $f_{MS}(\mathbf{u})$. Any distribution function which differs from $f_{MS}(\mathbf{u})$ approaches it eventually.

The average value of u_x^2 can be shown from Eq. (4.25) to be given by

$$\begin{aligned}\langle u_x^2 \rangle &= \left(\frac{m}{2\pi K T_0}\right)^{3/2} \int_{-\infty}^{\infty} dw_x (v_x + w_x)^2 e^{-mw_x^2/2KT_0} \\ &\quad \int_{-\infty}^{\infty} dw_y\, e^{-mw_y^2/2KT_0} \int_{-\infty}^{\infty} dw_z\, e^{-mw_z^2/2KT_0} \\ &= \left(\frac{m}{2\pi K T_0}\right)^{1/2} \int_{-\infty}^{\infty} dw_x (v_x^2 + 2v_x w_x + w_x^2) e^{-mw_x^2/2KT_0} \\ &= v_x^2 + KT_0/m = v_x^2 + \langle w_x^2 \rangle\end{aligned} \tag{4.31}$$

Similarly, it is obtained that

$$\langle u_y^2 \rangle = v_y^2 + \langle w_y^2 \rangle \qquad \langle u_z^2 \rangle = v_z^2 + \langle w_z^2 \rangle \tag{4.32}$$

Therefore,

$$\langle u^2 \rangle = v^2 + \langle w^2 \rangle \tag{4.33}$$

The amount of translatory kinetic energy possessed by the particles in unit volume is deduced from Eq. (4.33) as

$$N_0(m/2)\langle u^2 \rangle = N_0(m/2)v^2 + N_0(m/2)\langle w^2 \rangle \tag{4.34}$$

The first term in Eq. (4.34) represents the visible or mass motion of the gas and the second term gives the kinetic energy of the invisible peculiar motion. In kinetic theory, it is this hidden particle energy or rather the part which is communicable between the particles through their interactions that is identified with the heat energy of the gas. Thus it is the heat energy $(m/2)\langle w^2 \rangle$ of a particle acting as a point-center of force that is identified with $\frac{3}{2}KT$ for the purpose of defining temperature from the point of view of kinetic theory. Note that in Eq. (4.14) $(m/2)\langle u^2 \rangle$ has been set equal to $\frac{3}{2}KT$; this is because in that case the average velocity is zero and the actual velocity $\mathbf{u}$ of the particle becomes equal to the peculiar velocity $\mathbf{w}$.

Local Maxwellian distribution of velocities

The Maxwell-Boltzmann distribution function given by Eq. (4.16) and the shifted Maxwell-Boltzmann distribution function given by Eq. (4.25) are referred to as overall Maxwell-Boltzmann distributions since the number density N_0, the average velocity $\mathbf{v}$ and the temperature T_0 are constants throughout the gas. Sometimes, a locally Maxwellian state is defined by the distribution function

$$f_{lM}(\mathbf{u}) = N\left(\frac{m}{2\pi KT}\right)^{3/2} e^{-mw^2/2KT} \tag{4.35}$$

where the number density $N = N(\mathbf{r}, t)$, the average velocity $\mathbf{v} = \mathbf{v}(\mathbf{r}, t)$ and the temperature $T = T(\mathbf{r}, t)$ depend on the space coordinates and the time but not on the velocity. The local Maxwell-Boltzmann distribution function (4.35) is a solution of the Boltzmann equation with the collision term equal to zero.

In a plasma there are different species of particles such as electrons, ions, and neutral particles. The velocities of each species of particles are characterized by a separate distribution function such as those given by Eqs. (4.16), (4.25), and (4.35).

1.5. Steady state in the presence of an external force

Consider an *equilibrium* or steady-state distribution of velocities of particles in a *homogeneous* gas with *no external forces*. The requirements of (i) steady state, (ii) homogeneity, and (iii) absence of external forces yield respectively $(\partial/\partial t) f(\mathbf{r}, \mathbf{u}, t) = 0$, $(\mathbf{u} \cdot \nabla_r) f(\mathbf{r}, \mathbf{u}, t) = 0$, and $[(\mathbf{F}/m) \cdot \nabla_u] f(\mathbf{r}, \mathbf{u}, t) = 0$ in the Boltzmann equation (3.29) with the result that

$$(\partial f/\partial t)_{\text{coll}} = 0 \tag{5.1}$$

It follows from Eq. (5.1) that for an equilibrium state, in the time interval dt which is long compared to the average time of duration of interaction between two particles but short compared to the mean time interval between two interactions, as many representative points enter an elemental volume in the phase space due to collisional interactions as they leave. A detailed consideration of the mechanics of two-particle collisions in which the total energy is conserved together with Eq. (5.1) is shown in Sec. 6.13 to yield a Maxwell-Boltzmann distribution function (4.16) for the description of particle velocities. In other words, $f_M(u)$ given by Eq. (4.16) satisfies Eq. (5.1). In a similar manner, in the special case of a conservative force field acting on a gas in which the average velocity of the particles is zero, there is an equilibrium state characterized by a distribution function which differs from the Maxwellian distribution by a factor known as the *Boltzmann factor*.

Let the conservative force field be specified by

$$\mathbf{F} = -\nabla\Phi(\mathbf{r}) \tag{5.2}$$

where Φ is the potential energy. Consider the following distribution function

$$f(\mathbf{r}, \mathbf{u}, t) = f_M(u) e^{-\Phi(\mathbf{r})/KT} \tag{5.3}$$

where $f_M(u)$ is the Maxwellian distribution function as given by

$$f_M(u) = N_0 \left(\frac{m}{2\pi KT}\right)^{3/2} e^{-mu^2/2KT} \tag{5.4}$$

For the equilibrium state $(\partial/\partial t) f(\mathbf{r}, \mathbf{u}, t) = 0$. As was mentioned previously, the collision term $(\partial f/\partial t)_{\text{coll}}$ can be expressed as an integral over the distribution of velocities. Since it is independent of the velocity, the factor $\exp(-\Phi(\mathbf{r})/KT)$ in Eq. (5.3) can be taken outside the collision integral leaving only $f_M(u)$ inside. But for the Maxwellian distribution, the collision integral is zero. Therefore, even for the distribution function given by Eq. (5.3), the condition (5.1) holds. Hence, in the presence of a conservative force field, the steady state is reached as a result of the net number of representative points entering an elemental volume in the phase space in the time interval dt parallel to the coordinates of the configuration and the velocity spaces being equal to zero. The Boltzmann equation (3.29) then yields

$$(\mathbf{u} \cdot \nabla_r) f(\mathbf{r}, \mathbf{u}, t) + \left(\frac{\mathbf{F}}{m} \cdot \nabla_u\right) f(\mathbf{r}, \mathbf{u}, t) = 0 \tag{5.5}$$

It is an easy matter to verify that $f(\mathbf{r}, \mathbf{u}, t)$ in Eq. (5.3) satisfies Eq. (5.5) for a force $\mathbf{F}$ defined by Eq. (5.2). It may be therefore concluded that in the presence of a conservative force as specified by Eq. (5.2), there is an equilibrium state characterized by an inhomogeneous distribution function as given by Eqs. (5.3) and (5.4). The factor $\exp(-\Phi(\mathbf{r})/KT)$ by which this inhomogeneous distribution function differs from the Maxwellian distribution function $f_M(u)$ is called the Boltzmann factor.

The number density $N(\mathbf{r})$ of the particles of a gas described by the distribution function (5.3) is obtained as

$$N(\mathbf{r}) = \int_{-\infty}^{\infty} f(\mathbf{r}, \mathbf{u}, t)\, d\mathbf{u} = e^{-\Phi(\mathbf{r})/KT} \int_{-\infty}^{\infty} f_M(u)\, d\mathbf{u} = N_0 e^{-\Phi(\mathbf{r})/KT} \tag{5.6}$$

In Eq. (5.6) N_0 is clearly the number density in a region where the potential energy $\Phi(\mathbf{r})$ vanishes.

An example of a conservative force is that due to an electrostatic field $\mathbf{E}(\mathbf{r}) = -\nabla\phi(\mathbf{r})$ where $\phi(\mathbf{r})$ is the scalar potential. Since the force due to an electric field is given by $\mathbf{F} = q\mathbf{E}(\mathbf{r})$ where q is the electric charge on the particle, the potential energy $\Phi(\mathbf{r})$ is obtained from Eq. (5.2) as

$$\Phi(\mathbf{r}) = q\phi(\mathbf{r}) \tag{5.7}$$

In an equilibrium state of a gas acted on by an electrostatic field, there is a nonuniform distribution of particles and the number density is found from Eqs. (5.6) and (5.7) as

$$N(\mathbf{r}) = N_0 e^{-q\phi(\mathbf{r})/KT} \tag{5.8}$$

The expression (5.8) for $N(\mathbf{r})$ is useful in the study of electrostatic shielding in a plasma.

1.6. Particle current density

If a material body is immersed in a gas, the particles impinge on the sides of the body on account of their random motions. For certain applications, it is desirable to know the current density Γ of the particles due to their random motions. The particle current density is defined as the number of particles impinging on unit area of the body per unit time. This definition is also applicable to any unit area inside the gas provided only the particles crossing the unit area from one side only is taken into account. If the distribution of velocities is isotropic, the particle current density is independent of the orientation of the area but for an anisotropic distribution of velocities, Γ is a function of the orientation of the area. If the particles are electrically charged such as the electrons and the ions in a plasma, the passage of particles through an area constitutes a flow of charges or equivalently an electric current through the area. In the study of interaction of a material body with a plasma, a knowledge of the particle current density Γ and the accompanying electric current density is important.

Let $d\mathbf{S}$ be an infinitesimal area near the origin of the coordinates x, y, and z and oriented with its plane perpendicular to the z-axis as shown in Fig. 1.3. Consider the particles with their velocity in the range from $\mathbf{u}$ to $\mathbf{u} + d\mathbf{u}$ crossing the area $d\mathbf{S}$ from the side $z < 0$ in a time interval dt which is very short compared to the average time interval between two interactions of the particles. It is convenient to introduce the spherical coordinates u, θ, and φ in the velocity space so that the velocity $\mathbf{u}$ is specified completely by the speed u and its direction (θ, φ). Consider a cylinder with the infinitesimal area dS as the base and slanted in the direction of the velocity as illustrated in Fig. 1.3. Let the slant height of the cylinder be $u\,dt$ which is the distance a particle with speed u covers in the time interval dt; then the infinitesimal volume $d\mathbf{r}$ of the cylinder is

$$d\mathbf{r} = dS\,u\,dt\cos\theta \tag{6.1}$$

All the particles inside the infinitesimal volume $d\mathbf{r}$ of the cylinder with their velocities equal to $\mathbf{u} = (u, \theta, \varphi)$ pass through the infinitesimal area $d\mathbf{S}$ in the time interval dt. In view of the nature of the time interval dt, on the average no particles inside $d\mathbf{r}$ with velocities $\mathbf{u}$ are deflected due to the collisional interactions from their original direction (θ, φ). There are other particles inside the cylinder and these may or may not cross the surface area $d\mathbf{S}$ in the time dt. The particles which are inside $d\mathbf{r}$ and whose velocities are not in the direction (θ, φ) do not cross the area $d\mathbf{S}$. But the particles inside $d\mathbf{r}$ with their velocities in the direction (θ, φ) and their speeds different from u may or may not cross $d\mathbf{S}$ and those which cross $d\mathbf{S}$ in the time interval dt are taken into account subsequently by varying u to cover all possible speeds of the particles. It should be noted that the particles inside $d\mathbf{r}$ with their velocities in the direction (θ, φ) and their speeds given by u cross $d\mathbf{S}$ in the time interval dt.

If $f(\mathbf{r}, \mathbf{u}, t)$ is the velocity distribution function, the number of particles in the infinitesimal volume $d\mathbf{r}$ of the cylinder with velocities in the range from (u, θ, φ) to $(u + du, \theta + d\theta, \varphi + d\varphi)$ is obtained from Eq. (6.1) to be given by

$$f(\mathbf{r}, \mathbf{u}, t)\, d\mathbf{u}\, d\mathbf{r} = f(\mathbf{r}, \mathbf{u}, t) u^3 \cos\theta \sin\theta \, du\, d\theta\, d\varphi\, dS\, dt \tag{6.2}$$

since $d\mathbf{u} = u^2 \sin\theta \, du\, d\theta\, d\varphi$. The contribution $d\Gamma$ to the particle current density from those particles with velocities in the range from (u, θ, φ) to $(u + du, \theta + d\theta, \varphi + d\varphi)$ is determined by dividing the expression (6.2) by $dS\, dt$ with the following result

$$d\Gamma = f(\mathbf{r}, \mathbf{u}, t) u^3 \cos\theta \sin\theta \, du\, d\theta\, d\varphi \tag{6.3}$$

In order to find the particle current density Γ, the contributions due to the particles in the half space $z < 0$ traveling in all possible directions and with all possible speeds have to be summed up. Hence, Γ is obtained by integrating $d\Gamma$ given by Eq. (6.3) over the ranges: $0 < u < \infty$, $0 < \theta < \pi/2$ and $0 < \varphi < 2\pi$. The result is

$$\Gamma = \int_0^\infty du \int_0^{\pi/2} d\theta \int_0^{2\pi} d\varphi\, f(\mathbf{r}, \mathbf{u}, t) u^3 \cos\theta \sin\theta \tag{6.4}$$

An explicit and useful expression for the particle current density Γ can be found if the particles are assumed to be in the Maxwellian state. Then the distribution function $f(\mathbf{r}, \mathbf{u}, t)$ reduces to the Maxwellian distribution $f_M(u)$ as given by Eq. (5.4). Note that $f_M(u)$ is independent of the position $\mathbf{r}$, the direction (θ, φ) of the particle velocity and the time t. If $f(\mathbf{r}, \mathbf{u}, t)$ in Eq. (6.4) is replaced by $f_M(u)$, the integrations

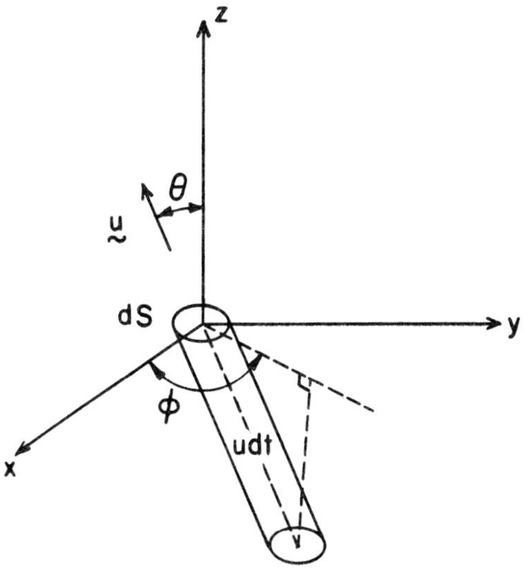

Fig. 1.3. Slanted cylinder containing the particles of velocity close to **u** which cross the infinitesimal area dS in the time dt.

with respect to θ and φ are easily carried out to yield

$$\Gamma = \pi \int_0^\infty f_M(u)u^3\,du \tag{6.5}$$

The substitution of the speed distribution function $f_{sp}(u)$ as defined by Eq. (4.6) in Eq. (6.5) is found to give

$$\Gamma = \tfrac{1}{4} \int_0^\infty u\, f_{sp}(u)\,du \tag{6.6}$$

Since the average speed $\langle u \rangle$ is defined by

$$\langle u \rangle = \int_0^\infty u\, f_{sp}(u)\,du \Big/ \int_0^\infty f_{sp}(u)\,du = (1/N) \int_0^\infty u\, f_{sp}(u)\,du \tag{6.7}$$

where N is the uniform number density, Γ given by Eq. (6.6) becomes

$$\Gamma = N\langle u \rangle/4 \tag{6.8}$$

Note that for the Maxwellian distribution of speeds, the average speed $\langle u \rangle$ as given by Eq. (4.22) is

$$\langle u \rangle = (8KT/\pi m)^{1/2} \tag{6.9}$$

where T is the uniform temperature and m is the particle mass.

In a plasma, the particle current density for each species of particles is given by an expression similar to Eq. (6.8). If the plasma consists of only electrons and singly charged hydrogen ions and if the temperature of the electron gas and the ion gas are equal, it follows from Eqs. (6.8) and (6.9) that the particle current density, being inversely proportional to the square root of the particle mass, is much greater for the electrons than for the ions since the electronic mass m_e is only 1/1836 times the ionic mass m_i. This difference in the particle current density of the electrons and the ions plays a role in the interaction of a material body with a plasma.

1.7. Kinetic pressure

The particles on account of their random motion impinge on the sides of a material body immersed in the gas and transfer a certain amount of momentum to the walls of the body; therefore, the body experiences a force. The force per unit area exerted by the particles due to their random motion on the walls of the body is defined as the kinetic pressure. This force is computed under the hypothetical assumption that the particles are elastically reflected from the wall. In an elastic reflection, the component of momentum of the particles parallel to the wall is conserved and only a part of the component perpendicular to the wall is transferred with the result that the force associated with the kinetic pressure is normal to the wall. Since the rate of transfer of momentum is expected to be proportional to the particle mass m and its number density N, it follows that the kinetic pressure may be anticipated to be proportional to the particle mass and its number density. Also, it is to be expected that the kinetic pressure is proportional to the average of some function of the particle speed.

Let the infinitesimal area $d\mathbf{S}$ of Fig. 1.3 be considered now to be a part of the wall. The wall is assumed to be of infinite mass and hence does not move when the particles impinge on it. Therefore, the kinetic energy of the particles remains unchanged after impact with the wall and hence the speed of the particles is unaltered on reflection. The wall is assumed to be perfectly smooth. Since by hypothesis the reflections are elastic, it follows that the particles are specularly reflected with the angle of incidence $\theta^i = \theta$ equal to the angle of reflection θ^r, as shown in Fig. 1.4. The component of the momentum of a particle parallel to the wall has the same value, $mu \sin \theta$, before and after the impact and hence no momentum is transferred parallel to the wall. The component of the momentum of the particle normally towards the wall is equal to $mu \cos \theta$ before reflection and $-mu \cos \theta$ after reflection. The change in the perpendicular component of the momentum of the particle on impact is $-2mu \cos \theta$ and therefore from the momentum conservation principle, the momentum imparted to the wall in the perpendicular direction by one particle is obtained as $2mu \cos \theta$.

As established in the previous section, all the particles inside the infinitesimal volume $d\mathbf{r}$ of the slanted cylinder of Fig. 1.3 having their velocities equal to $\mathbf{u} = (u, \theta, \varphi)$ impinge on the infinitesimal area $d\mathbf{S}$ in the time interval dt. As before, the time interval dt is chosen to be short enough that the probability of a particle being deflected away from its initial direction by collision is negligible. The number

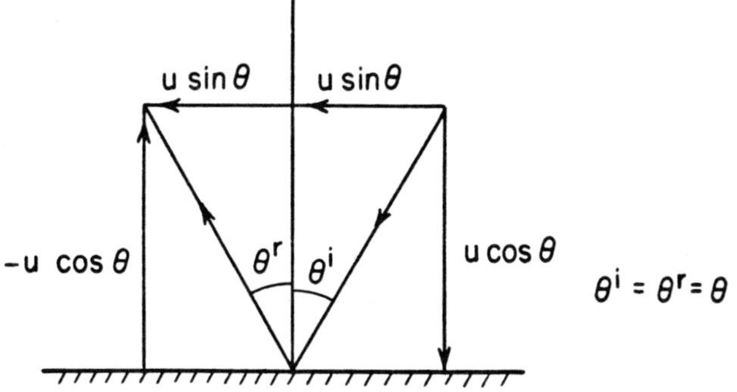

Fig. 1.4. Vector velocities before and after an elastic collision by a particle with a plane smooth wall of infinite mass.

of particles in the infinitesimal volume $d\mathbf{r}$ of the cylinder with velocities in the range from (u, θ, φ) to $(u + du, \theta + d\theta, \varphi + d\varphi)$ is given by Eq. (6.2). Each one of these particles transfers a momentum $2mu \cos\theta$ to the wall in the perpendicular direction. Hence the contribution dP to the momentum transferred to the wall in the perpendicular direction per unit area in unit time from those particles with velocities in the range from (u, θ, φ) to $(u + du, \theta + d\theta, \varphi + d\varphi)$ is determined by multiplying the expression (6.2) by $2mu \cos\theta$ and dividing it by $dS\,dt$ with the following result

$$dP = f(\mathbf{r}, \mathbf{u}, t) 2mu^4 \cos^2\theta \sin\theta \, du \, d\theta \, d\varphi \tag{7.1}$$

Since it is the rate of transfer of momentum per unit area, Eq. (7.1) gives the contribution to the kinetic pressure from the particles with the velocities in the range from (u, θ, φ) to $(u + du, \theta + d\theta, \varphi + d\varphi)$. As in the case of the particle current density Γ, the pressure P is obtained by integrating dP given by Eq. (7.1) over the ranges: $0 < u < \infty$, $0 < \theta < \pi/2$ and $0 < \varphi < 2\pi$. The result is

$$P = \int_0^\infty du \int_0^{\pi/2} d\theta \int_0^{2\pi} d\varphi \, f(\mathbf{r}, \mathbf{u}, t) 2mu^4 \cos^2\theta \sin\theta \tag{7.2}$$

A simple expression can be found for the pressure P if the particles are assumed to be in the Maxwellian state. If $f(\mathbf{r}, \mathbf{u}, t)$ in Eq. (7.2) is replaced by the Maxwellian distribution $f_M(u)$ as given by Eq. (5.4), the integrations with respect to θ and φ are carried out in a straightforward manner to yield

$$P = (4\pi m/3) \int_0^\infty u^4 f_M(u) \, du \tag{7.3}$$

The substitution of the speed distribution function $f_{sp}(u)$ as specified by Eq. (4.6) in Eq. (7.3) gives

$$P = (m/3) \int_0^\infty u^2 f_{sp}(u) \, du \tag{7.4}$$

Since the average of the square of the speed $\langle u^2 \rangle$ is defined by

$$\langle u^2 \rangle = (1/N) \int_0^\infty u^2 f_{sp}(u) \, du \tag{7.5}$$

where N is the number density, P given by Eq. (7.4) simplifies to

$$P = (mN/3) \langle u^2 \rangle \tag{7.6}$$

As anticipated, the pressure P is proportional to the particle mass m and its number density N; also, it is proportional to the mean squared speed. For the Maxwellian distribution of speeds, it is found from Eq. (4.23) that

$$\langle u^2 \rangle = 3KT/m \tag{7.7}$$

where T is the uniform temperature. Together with Eq. (7.7), Eq. (7.6) gives the "ideal gas" law

$$P = NKT \tag{7.8}$$

for a system of particles in the Maxwellian state.

The preceding calculation of pressure is not very satisfactory since experiments have indicated that the particles in a gas are, in general, not elastically reflected from a wall. Also, if the velocity distribution function is anisotropic, it is possible for the force to have a component tangential to the wall giving rise to a viscous drag on the wall by the gas. Moreover, the force exerted on the wall depends on the nature of the wall. Hence the definition of pressure in terms of forces exerted on a wall becomes complicated. An alternative and a more general definition of pressure is given subsequently and this general definition of pressure reproduces Eq. (7.8) for the special case of Maxwellian distribution of velocities.

1.8. Equation of continuity

In a gas containing a very large number of particles, only certain macroscopic phenomena are observed and these can be described in terms of the macroscopic variables such as the number density N, the average velocity $\mathbf{v}$, the temperature T and the pressure P. A knowledge of the velocity distribution function enables the determination of these macroscopic variables and indeed, these macroscopic parameters have been explicitly evaluated for the special case of Maxwell-Boltzmann distribution of velocities. In principle, the velocity distribution function can be obtained from the Boltzmann equation. But, as has been pointed out previously, the Boltzmann equation is difficult to solve. However, it is not necessary to solve the Boltzmann equation first for the velocity distribution function in order to be able to determine the macroscopic variables, since the partial differential equations satisfied by the macroscopic variables can be determined directly from the Boltzmann equation without obtaining its solution first. These partial differential equations are known as the transport equations and these can be solved, under certain approximations, to yield directly the macroscopic variables. The first transport equation is the equation of continuity that embodies the principle of conservation of electric charge and mass. The equation of continuity which is also known as the mass transport equation is deduced here.

Moment of the Boltzmann equation

Let $g(\mathbf{u})$ be any property of the particles. It is assumed that $g(\mathbf{u})$ is independent of the position $\mathbf{r}$ and the time t but is dependent on the particle velocity $\mathbf{u}$. The Boltzmann equation (3.29) is multiplied by $g(\mathbf{u})\,d\mathbf{u}$ and integrated throughout the velocity space. On assuming that all the integrals are convergent, it is found that

$$\int g(\mathbf{u}) \frac{\partial}{\partial t} f(\mathbf{r}, \mathbf{u}, t)\, d\mathbf{u} + \int g(\mathbf{u})(\mathbf{u} \cdot \nabla_r) f(\mathbf{r}, \mathbf{u}, t)\, d\mathbf{u} + \int g(\mathbf{u})\left(\frac{\mathbf{F}}{m} \cdot \nabla_u\right) f(\mathbf{r}, \mathbf{u}, t)\, d\mathbf{u} = \int g(\mathbf{u})\left(\frac{\partial f}{\partial t}\right)_{\text{coll}} d\mathbf{u} \tag{8.1}$$

Since $g(\mathbf{u})$ is independent of $\mathbf{r}$ and t, it is seen with the help of Eq. (2.3) that

$$\int g(\mathbf{u}) \frac{\partial}{\partial t} f(\mathbf{r}, \mathbf{u}, t)\, d\mathbf{u} = \frac{\partial}{\partial t} \int g(\mathbf{u}) f(\mathbf{r}, \mathbf{u}, t)\, d\mathbf{u} = \frac{\partial}{\partial t}\{N(\mathbf{r}, t)\langle g(\mathbf{u})\rangle\} \tag{8.2}$$

In view of the fact that $\mathbf{u}$ and $\mathbf{r}$ are independent variables, it is obtained from Eq. (2.3) that

$$\int g(\mathbf{u}) u_x \frac{\partial}{\partial x} f(\mathbf{r}, \mathbf{u}, t)\, d\mathbf{u} = \frac{\partial}{\partial x} \int g(\mathbf{u}) u_x f(\mathbf{r}, \mathbf{u}, t)\, d\mathbf{u} = \frac{\partial}{\partial x}\{N(\mathbf{r}, t)\langle g(\mathbf{u}) u_x\rangle\} \tag{8.3}$$

If the two other terms which depend on the derivatives with respect to y and z and which are similar to Eq. (8.3) are added to Eq. (8.3), the result is

$$\int g(\mathbf{u})(\mathbf{u} \cdot \nabla_r) f(\mathbf{r}, \mathbf{u}, t)\, d\mathbf{u} = \nabla_r \cdot \{N(\mathbf{r}, t)\langle g(\mathbf{u})\mathbf{u}\rangle\} \tag{8.4}$$

Assuming that F_x is independent of u_x, it can be shown by using an integration by parts that

$$\begin{aligned}\int g(\mathbf{u}) \frac{F_x}{m} \frac{\partial}{\partial u_x} f(\mathbf{r}, \mathbf{u}, t)\, d\mathbf{u} &= \int \frac{F_x}{m}\, du_y\, du_z \int g(\mathbf{u}) \frac{\partial}{\partial u_x} f(\mathbf{r}, \mathbf{u}, t)\, du_x \\ &= \int \frac{F_x}{m}\, du_y\, du_z \Big[g(\mathbf{u}) f(\mathbf{r}, \mathbf{u}, t)\Big|_{-\infty}^{\infty} - \int f(\mathbf{r}, \mathbf{u}, t) \frac{\partial}{\partial u_x} g(\mathbf{u})\, du_x\Big]\end{aligned} \tag{8.5}$$

Since $f(\mathbf{r}, \mathbf{u}, t)$ vanishes faster at $u_x = \pm\infty$ than any algebraic function $g(\mathbf{u})$ of the particle velocity, it follows from Eqs. (8.5) and (2.3) that

$$\begin{aligned}\int g(\mathbf{u}) \frac{F_x}{m} \frac{\partial}{\partial u_x} f(\mathbf{r}, \mathbf{u}, t)\, d\mathbf{u} &= -\int f(\mathbf{r}, \mathbf{u}, t) \frac{\partial}{\partial u_x}\left\{\frac{F_x}{m} g(\mathbf{u})\right\} d\mathbf{u} \\ &= -N(\mathbf{r}, t)\left\langle \frac{\partial}{\partial u_x}\left\{\frac{F_x}{m} g(\mathbf{u})\right\}\right\rangle\end{aligned} \tag{8.6}$$

If the two other terms which depend on the derivatives with respect to u_y and u_z and which are similar to Eq. (8.6) are added to Eq. (8.6), it is found that

$$\int g(\mathbf{u})\left(\frac{\mathbf{F}}{m} \cdot \nabla_u\right) f(\mathbf{r}, \mathbf{u}, t)\, d\mathbf{u} = -N(\mathbf{r}, t)\left\langle \nabla_u \cdot \left(\frac{\mathbf{F}}{m} g(\mathbf{u})\right)\right\rangle \tag{8.7}$$

In the derivation of Eq. (8.7), it has been assumed that the force in a given direction is independent of the particle velocity in that direction, that is, the validity of Eq. (3.26) is assumed to hold. The only force that is proposed to be treated is the

Lorentz force given by Eq. (3.30) and this force satisfies the condition (3.26). Hence, Eq. (8.7) is valid for the Lorentz force.

Together with Eqs. (8.2), (8.4), and (8.7), Eq. (8.1) simplifies to

$$\frac{\partial}{\partial t}\{N(\mathbf{r},t)\langle g(\mathbf{u})\rangle\} + \nabla_r \cdot \{N(\mathbf{r},t)\langle g(\mathbf{u})\mathbf{u}\rangle\} - N(\mathbf{r},t)\left\langle \nabla_u \cdot \left[\frac{\mathbf{F}}{m} g(\mathbf{u})\right]\right\rangle = \int g(\mathbf{u})\left(\frac{\partial f}{\partial t}\right)_{\text{coll}} d\mathbf{u} \tag{8.8}$$

It is to be noted that the term on the right side of Eq. (8.8) represents the rate of change of the average value of $g(\mathbf{u})$ due to collisional interactions.

Equation of continuity

The equation of continuity is obtained by setting $g(\mathbf{u}) = 1$, so that $\langle g(\mathbf{u})\mathbf{u}\rangle = \langle \mathbf{u}\rangle = \mathbf{v}$, the average velocity. Also $\nabla_u \cdot [(\mathbf{F}/m)g(\mathbf{u})] = \nabla_u \cdot (\mathbf{F}/m) = 0$ in view of Eq. (3.26). It is assumed that there is no creation or annihilation of particles and that the coordinates of the particles in the configuration space are unaltered by collisional interactions. The term on the right side of Eq. (8.8) becomes $\int (\partial f/\partial t)_{\text{coll}}\, d\mathbf{u}$ and it is equal to the sum of the representative points scattered by collisions per unit volume in unit time into all parts of the phase space with the same coordinates in the configuration space. Since the coordinates of the particles in the configuration space are unaltered by collisions or equivalently since the number density in an element of volume in the configuration space cannot be changed by collisions, it follows that $\int (\partial f/\partial t)_{\text{coll}}\, d\mathbf{u} = 0$. Hence, for $g(\mathbf{u}) = 1$, Eq. (8.8) reduces to

$$\frac{\partial}{\partial t} N(\mathbf{r},t) + \nabla_r \cdot \{N(\mathbf{r},t)\mathbf{v}(\mathbf{r},t)\} = 0 \tag{8.9}$$

which is the equation of continuity.

Fluid dynamical method

Since $N(\mathbf{r},t)$ and $\mathbf{v}(\mathbf{r},t)$ are average quantities, it follows that Eq. (8.9) is a fluid equation and hence Eq. (8.9) can also be derived from fluid dynamical methods. Consider a volume V of the gas bounded by a closed surface S. Let $d\mathbf{S}$ be an infinitesimal area of the surface; its magnitude is dS and its direction is the outwardly drawn unit normal $\hat{\mathbf{n}}$ to dS as shown in Fig. 1.5. From an examination of Fig. 1.3 and Eq. (6.1), it is seen that on the average the number of particles leaving the volume V through an elemental area $d\mathbf{S}$ in the time interval dt is

$$N(\mathbf{r},t)\mathbf{v}(\mathbf{r},t) \cdot d\mathbf{S}\, dt \tag{8.10}$$

The total number of particles leaving the volume V in the time interval dt is obtained by integrating the expression (8.10) throughout the surface area S bounding the volume V with the following result:

$$dt \int_S N(\mathbf{r}, t)\mathbf{v}(\mathbf{r}, t) \cdot d\mathbf{S} \tag{8.11}$$

The divergence theorem

$$\int_S \mathbf{A} \cdot d\mathbf{S} = \int_V \nabla \cdot \mathbf{A}\, dV \tag{8.12}$$

enables the expression (8.11) to be written as a volume integral:

$$dt \int_V \nabla_r \cdot \{N(\mathbf{r}, t)\mathbf{v}(\mathbf{r}, t)\}\, dV \tag{8.13}$$

If there is no creation or annihilation of particles, the particles leaving the volume V should be accompanied by a decrease in the number density of the particles inside V. Let this decrease be denoted by $-(\partial/\partial t)N(\mathbf{r}, t)$. Then, the total number of particles lost from V in the time dt is given by

$$-dt \int_V (\partial N(\mathbf{r}, t)/\partial t)\, dV \tag{8.14}$$

Since the expressions (8.13) and (8.14) are equal, it is found that

$$dt \int_V [\nabla_r \cdot \{N(\mathbf{r}, t)\mathbf{v}(\mathbf{r}, t)\} + (\partial/\partial t)N(\mathbf{r}, t)]\, dV = 0 \tag{8.15}$$

The result given by Eq. (8.15) is applicable for any arbitrarily small volume V and this is possible only if the integrand in Eq. (8.15) is zero. This immediately yields the equation of continuity (8.9). It should be remembered that as soon as the

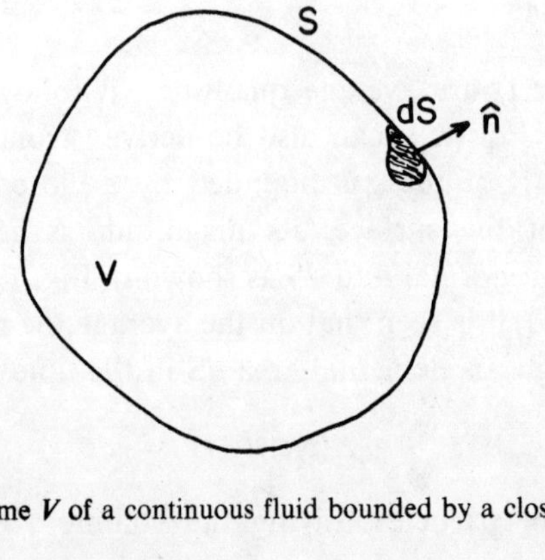

Fig. 1.5. Volume V of a continuous fluid bounded by a closed surface S.

average macroscopic parameters are defined, a gas of particles can be treated as a continuous fluid characterized by the macroscopic parameters such as $N(\mathbf{r}, t)$ and $\mathbf{v}(\mathbf{r}, t)$.

1.9. Momentum transport equation

The momentum transport equation is obtained by setting $g(\mathbf{u}) = m\mathbf{u}$. The arguments $\mathbf{r}$ and $\mathbf{t}$ are omitted here for convenience. It is then found that

$$\frac{\partial}{\partial t}\{N\langle m\mathbf{u}\rangle\} = \frac{\partial}{\partial t}(Nm\mathbf{v}) = Nm\frac{\partial \mathbf{v}}{\partial t} + \mathbf{v}\frac{\partial}{\partial t}(mN) \tag{9.1}$$

On setting

$$\mathbf{u} = \mathbf{v} + \mathbf{w} \tag{9.2}$$

where $\mathbf{w}$ is the peculiar velocity, it is obtained that

$$\nabla_r \cdot \{N\langle m\mathbf{uu}\rangle\} = \nabla_r \cdot [Nm\{\mathbf{vv} + \langle\mathbf{vw}\rangle + \langle\mathbf{wv}\rangle + \langle\mathbf{ww}\rangle] \tag{9.3}$$

because $\langle\mathbf{vv}\rangle = \mathbf{vv}$. Moreover, since $\langle\mathbf{vw}\rangle = \langle\mathbf{wv}\rangle = \mathbf{v}\langle\mathbf{w}\rangle = 0$, Eq. (9.3) may be rewritten as

$$\nabla \cdot \{Nm\langle\mathbf{uu}\rangle\} = \nabla \cdot (Nm\mathbf{vv}) + \nabla \cdot \mathbf{\Psi} \tag{9.4}$$

where for reasons to be explained subsequently

$$\mathbf{\Psi} = Nm\langle\mathbf{ww}\rangle \tag{9.5}$$

is known as the pressure dyad. When it appears without any subscript, the divergence operator is understood to refer to that of the configuration space. Therefore,

$$\nabla \cdot = \frac{\partial}{\partial x}\hat{\mathbf{x}} \cdot + \frac{\partial}{\partial y}\hat{\mathbf{y}} \cdot + \frac{\partial}{\partial z}\hat{\mathbf{z}} \cdot \tag{9.6}$$

It follows from Eq. (9.6) that

$$\begin{aligned}\nabla \cdot (Nm\mathbf{vv}) &= \frac{\partial}{\partial x}(Nmv_x\mathbf{v}) + \frac{\partial}{\partial y}(Nmv_y\mathbf{v}) + \frac{\partial}{\partial z}(Nmv_z\mathbf{v}) \\ &= Nmv_x\frac{\partial \mathbf{v}}{\partial x} + Nmv_y\frac{\partial \mathbf{v}}{\partial y} + Nmv_z\frac{\partial \mathbf{v}}{\partial z} + \mathbf{v}\frac{\partial}{\partial x}Nmv_x + \mathbf{v}\frac{\partial}{\partial y}Nmv_y \\ &\quad + \mathbf{v}\frac{\partial}{\partial z}Nmv_z = Nm(\mathbf{v} \cdot \nabla)\mathbf{v} + m\mathbf{v}(\nabla \cdot N\mathbf{v})\end{aligned} \tag{9.7}$$

From Eqs. (9.1), (9.4), and (9.7), it is seen that

$$\frac{\partial}{\partial t}\{N\langle m\mathbf{u}\rangle\} + \nabla \cdot \{Nm\langle \mathbf{u}\mathbf{u}\rangle\}$$

$$= Nm\{\frac{\partial}{\partial t} + (\mathbf{v} \cdot \nabla)\}\mathbf{v} + m\mathbf{v}\left\{\frac{\partial N}{\partial t} + \nabla \cdot (N\mathbf{v})\right\} + \nabla \cdot \boldsymbol{\Psi} \tag{9.8}$$

$$= Nm\frac{d\mathbf{v}}{dt} + \nabla \cdot \boldsymbol{\Psi}$$

In deducing Eq. (9.8), the equation of continuity (8.9) has been used. Note that the total or convective derivative as defined by

$$\frac{d}{dt} = \frac{\partial}{\partial t} + (\mathbf{v} \cdot \nabla) \tag{9.9}$$

is also used in Eq. (9.8).

The divergence operator in the velocity space is given by

$$\nabla_u \cdot = \frac{\partial}{\partial u_x}\hat{\mathbf{x}} \cdot + \frac{\partial}{\partial u_y}\hat{\mathbf{y}} \cdot + \frac{\partial}{\partial u_z}\hat{\mathbf{z}} \cdot \tag{9.10}$$

and therefore it follows that

$$\nabla_u \cdot \left(\frac{\mathbf{F}}{m} m\mathbf{u}\right) = \frac{\partial}{\partial u_x}(F_x \mathbf{u}) + \frac{\partial}{\partial u_y}(F_y \mathbf{u}) + \frac{\partial}{\partial u_z}(F_z \mathbf{u}) = F_x\hat{\mathbf{x}} + F_y\hat{\mathbf{y}} + F_z\hat{\mathbf{z}} = \mathbf{F} \tag{9.11}$$

The term on the right side of Eq. (8.8) becomes

$$\mathbf{P}_{\text{coll}} = \int m\mathbf{u}(\partial f/\partial t)_{\text{coll}}\, d\mathbf{u} \tag{9.12}$$

and it is equal to the rate of change of average momentum per unit volume due to collisional interactions. Since the total momentum is conserved during a collision, the momentum lost by one of the interacting particles is equal to that gained by the other particle. Thus for collisions involving particles of the same species, there is no change in the net momentum per unit volume and hence $\mathbf{P}_{\text{coll}} = 0$. If the gas consists of particles of more than one species such as in a plasma, there is a momentum transport equation for each species of particles. If particles of one species interact with those of another, there is a net nonvanishing momentum change in unit time per unit volume of particles of each species and this is denoted by $\mathbf{P}_{\text{coll}}$ as given by Eq. (9.12).

Let **F** be the Lorentz force as given by Eq. (3.30) where, as mentioned before, **E** and **B** are the continuous macroscopic fields obtained after averaging over an elemental volume. It is seen from Eq. (3.30) that

$$\langle \mathbf{F} \rangle = q[\mathbf{E} + \mathbf{v} \times \mathbf{B}] \tag{9.13}$$

where **v** is the average velocity. Only the forces due to the electric and the magnetic fields are proposed to be included in the present treatment. From Eqs. (9.8) and (9.11)–(9.13), it is found that Eq. (8.8) for $g(\mathbf{u}) = m\mathbf{u}$ simplifies to

$$Nm\, d\mathbf{v}/dt = -\nabla \cdot \boldsymbol{\Psi} + Nq[\mathbf{E} + \mathbf{v} \times \mathbf{B}] + \mathbf{P}_{\text{coll}} \tag{9.14}$$

which is the momentum transport equation.

Kinetic pressure dyad

The kinetic pressure dyad $\boldsymbol{\Psi}$ as defined by Eq. (9.5) can be written in the following component form:

$$\begin{aligned}\boldsymbol{\Psi} = {} & \hat{\mathbf{x}}\hat{\mathbf{x}}P_{xx} + \hat{\mathbf{x}}\hat{\mathbf{y}}P_{xy} + \hat{\mathbf{x}}\hat{\mathbf{z}}P_{xz} \\ & + \hat{\mathbf{y}}\hat{\mathbf{x}}P_{yx} + \hat{\mathbf{y}}\hat{\mathbf{y}}P_{yy} + \hat{\mathbf{y}}\hat{\mathbf{z}}P_{yz} \\ & + \hat{\mathbf{z}}\hat{\mathbf{x}}P_{zx} + \hat{\mathbf{z}}\hat{\mathbf{y}}P_{zy} + \hat{\mathbf{z}}\hat{\mathbf{z}}P_{zz}\end{aligned} \tag{9.15}$$

where

$$P_{ij} = Nm\langle w_i w_j \rangle \qquad i, j = x, y, z \tag{9.16}$$

It is common practice to omit the dyadic signs such as $\hat{\mathbf{x}}\hat{\mathbf{x}}$ as well as the addition signs and write the components of $\boldsymbol{\Psi}$ in a matrix form as

$$\boldsymbol{\Psi} = \begin{bmatrix} P_{xx} & P_{xy} & P_{xz} \\ P_{yx} & P_{yy} & P_{yz} \\ P_{zx} & P_{zy} & P_{zz} \end{bmatrix}$$

Since from Eq. (9.16) $P_{ij} = P_{ji}$, it follows that the matrix is symmetrical and there are six independent components of $\boldsymbol{\Psi}$. If all the particle velocities are equal to the average velocity, the peculiar velocities $\mathbf{w}$ of all the particles vanish and therefore the kinetic pressure dyad is zero. Consequently, it is clear that $\boldsymbol{\Psi}$ is a measure of the random deviation of the particle velocities from their average value.

It follows from Eq. (9.14) that the negative divergence of the kinetic pressure dyad is equal to the force exerted on a unit volume of the gas due to the random variations of the peculiar velocities of the particles. If $-\nabla \cdot \boldsymbol{\Psi}$ is integrated throughout a volume V of a gas bounded by a closed surface S and the divergence theorem similar to Eq. (8.12) is used, it is obtained that $-\hat{\mathbf{n}} \cdot \boldsymbol{\Psi}$ is the force acting on unit area of a surface normal to the unit vector $\hat{\mathbf{n}}$ due to the random variations of the peculiar velocities of the particles. Suppose that the unit area moves with a uniform velocity equal to the average velocity of the particles; then the particle velocities with reference to the unit area are identical to their peculiar velocities and therefore $-\hat{\mathbf{n}} \cdot \boldsymbol{\Psi}$ becomes the force acting on the unit area due to the actual particle motions. For example,

$$-\hat{\mathbf{x}} \cdot \boldsymbol{\Psi} = -\hat{\mathbf{x}}P_{xx} - \hat{\mathbf{y}}P_{xy} - \hat{\mathbf{z}}P_{xz}$$

is the force acting on a unit area normal to the x-coordinate as shown in Fig. 1.6. Note that $-P_{xx}$ is perpendicular to the surface area just like an hydrostatic pressure, whereas $-P_{xy}$ and $-P_{xz}$ are the shear forces which are tangential to the surface. All

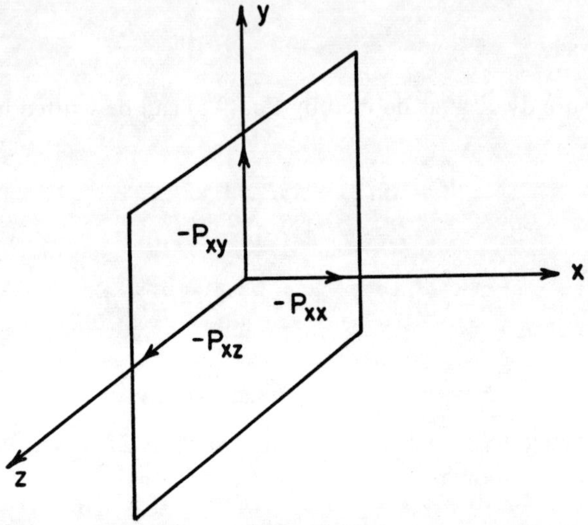

Fig. 1.6. Components of the kinetic pressure dyad corresponding to the forces acting on a unit area perpendicular to the x-axis; P_{xx}: normal stress; P_{xy}, P_{xz}: shear stresses.

the other components of $\mathbf{\Psi}$ are interpreted in a similar manner. The force P_{ij} is along the negative direction of the axis denoted by the second subscript. This force acts on a unit area whose outward normal is parallel to the axis indicated by the first subscript. If the outward normal to the unit area is in the negative direction of the axis indicated by the first subscript, the force P_{ij} is in the same direction as the axis denoted by the second subscript.

Scalar pressure

The momentum transport equation (9.14) which is exact for a nonrelativistic gas, is useful only when the distribution of the peculiar velocities is such as to yield a simple form for the kinetic pressure dyad. The simplest case corresponds to the isotropic Maxwell-Boltzmann distribution of peculiar velocities $\mathbf{w}$ as given by Eq. (5.4). Then, it can be verified that $\langle w_i w_j \rangle = 0$ for $i \neq j$ with the result that all the nondiagonal elements of $\mathbf{\Psi}$ are equal to zero. Also, it can be shown with the help of Eq. (4.14) that

$$\langle w_x^2 \rangle = \langle w_y^2 \rangle = \langle w_z^2 \rangle = \tfrac{1}{3}\langle w^2 \rangle = KT/m \tag{9.17}$$

where K is the Boltzmann constant and T is the uniform temperature. Therefore, it is obtained from Eq. (9.5) that

$$P_{xx} = P_{yy} = P_{zz} = P = NKT \tag{9.18}$$

Hence the forces acting on a unit area perpendicular to the x-, y-, or z-axis are always normal, directed towards the surfaces and have magnitudes equal to $P = NKT$. Since the forces acting along the principal coordinates x, y, and z have the same magnitude, it can be established that the force acting on unit area is always normal, directed towards the surface and has a magnitude equal to $P = NKT$ irrespective of the orientation of the area. Note that the principal coordinates are defined to be such that on the surfaces normal to them, there are no tangential forces. Thus for an isotropic distribution of peculiar velocities, the kinetic pressure dyad degenerates to the scalar kinetic pressure introduced previously. The scalar kinetic pressure was defined in terms of a force acting on a wall. Here a general, kinetic definition (9.5) of the pressure is given.

For an isotropic distribution of peculiar velocities, in view of Eq. (9.18), the kinetic pressure dyad becomes

$$\mathbf{\Psi} = (\hat{\mathbf{x}}\hat{\mathbf{x}} + \hat{\mathbf{y}}\hat{\mathbf{y}} + \hat{\mathbf{z}}\hat{\mathbf{z}})P = \mathbf{1}P \tag{9.19}$$

where $\mathbf{1}$ is known as the unit dyad. Therefore, it follows from Eqs. (9.6) and (9.19) that

$$\nabla \cdot \mathbf{\Psi} = \frac{\partial}{\partial x}\hat{\mathbf{x}}P + \frac{\partial}{\partial y}\hat{\mathbf{y}}P + \frac{\partial}{\partial z}\hat{\mathbf{z}}P = \nabla P \tag{9.20}$$

In the subsequent analysis, whenever the effect of kinetic pressure is to be included, only the simplified form (9.20) in terms of the scalar pressure P is to be adopted.

Another simplification of the kinetic pressure dyad is possible. Again the nondiagonal terms of $\mathbf{\Psi}$ vanish and the diagonal terms P_{xx}, P_{yy} and P_{zz} are all different. The pressure has different values in different directions; and this situation can arise in the absence of collisions, if the compression of the gas in one direction increases the average of the square of the peculiar velocity in that direction without affecting the corresponding average of the square of the peculiar velocities in the other directions. It is possible to have one- or two-dimensional compression resulting in the kinetic pressure in the direction of compression exceeding the pressure in another direction unaffected by compression. When the nondiagonal terms of $\mathbf{\Psi}$ are present, the shear stresses and the viscous drag do not vanish. The effect of viscosity is relatively unimportant in plasmas and therefore in the present treatment the nondiagonal terms of $\mathbf{\Psi}$ are ignored.

1.10. Energy transport equation

For the determination of the energy transport equation, it is necessary to begin by setting $g(\mathbf{r}, \mathbf{u}, t) = mw_x^2 = m(u_x - v_x)^2$. The average velocity $\mathbf{v}$ which is obtained by integrating $\mathbf{u}f(\mathbf{r}, \mathbf{u}, t)\, d\mathbf{u}/N(\mathbf{r}, t)$ throughout the velocity space is a function of $\mathbf{r}$ and t. Therefore, as indicated, $mw_x^2 = m(u_x - v_x)^2$ is a function of not only $\mathbf{u}$ but also of $\mathbf{r}$ and t. Previously Eq. (8.8) was deduced for the case in which g was a function of $\mathbf{u}$ only. Now, it is required to derive an equation similar to Eq. (8.8) for the general case in which g is a function of not only $\mathbf{u}$ but also of $\mathbf{r}$ and t. Such a generalization can be obtained by a procedure similar to that used in deducing Eq. (8.8) with the following result:

$$\frac{\partial}{\partial t}\{N\langle g\rangle\} - N\left\langle \frac{\partial g}{\partial t}\right\rangle + \nabla_r \cdot \{N\langle \mathbf{u}g\rangle\} - N\langle(\mathbf{u}\cdot\nabla_r)g\rangle - N\left\langle \nabla_u \cdot \left\{\frac{\mathbf{F}}{m}g\right\}\right\rangle = \int g\left(\frac{\partial f}{\partial t}\right)_{\text{coll}} d\mathbf{u} \tag{10.1}$$

For simplicity, in Eq. (10.1) the arguments $\mathbf{r}$, $\mathbf{u}$ and t are omitted.

Transport equation for the kinetic pressure dyad

If $g = mw_x^2$, it is found from Eqs. (9.5) and (9.15) that

$$(\partial/\partial t)\{N\langle mw_x^2\rangle\} = (\partial/\partial t)P_{xx} \tag{10.2}$$

Also,

$$N\left\langle \frac{\partial g}{\partial t}\right\rangle = N\left\langle \frac{\partial}{\partial t} mw_x^2\right\rangle = N\left\langle 2mw_x \frac{\partial}{\partial t}(u_x - v_x)\right\rangle = -2mN\left(\frac{\partial}{\partial t}v_x\right)\langle w_x\rangle = 0 \tag{10.3}$$

For $g = mw_x^2$, the third term on the left side of Eq. (10.1) can be expanded into the form:

$$\nabla_r \cdot \{N\langle \mathbf{u}mw_x^2\rangle\} = \frac{\partial}{\partial x}\{Nm\langle(v_x + w_x)w_x^2\rangle\} + \frac{\partial}{\partial y}\{Nm\langle(v_y + w_y)w_x^2\rangle\} + \frac{\partial}{\partial z}\{Nm\langle(v_z + w_z)w_x^2\rangle\} \tag{10.4}$$

It is convenient to introduce at this stage the thermal energy flux density triad $\mathbf{Q}$ as defined by

$$\mathbf{Q} = m\int \mathbf{www}f(\mathbf{r}, \mathbf{u}, t)\, d\mathbf{u} \tag{10.5}$$

The physical meaning of $\mathbf{Q}$ is explained subsequently; the components of $\mathbf{Q}$ are given explicitly by

$$Q_{ijk} = m\int w_i w_j w_k f(\mathbf{r}, \mathbf{u}, t)\, d\mathbf{u} = mN\langle w_i w_j w_k\rangle \qquad i, j, k = x, y, z \tag{10.6}$$

With the help of Eqs. (9.5) and (10.6), it follows that

$$(\partial/\partial x)\{Nm\langle(v_x + w_x)w_x^2\rangle\} = (\partial/\partial x)\{v_x P_{xx} + Q_{xxx}\}$$
$$= (\partial v_x/\partial x)P_{xx} + v_x(\partial/\partial x)P_{xx} + (\partial/\partial x)Q_{xxx} \tag{10.7}$$

$$(\partial/\partial y)\{Nm\langle(v_y + w_y)w_x^2\rangle\} = (\partial v_y/\partial y)P_{xx} + v_y(\partial/\partial y)P_{xx} + (\partial/\partial y)Q_{yxx} \tag{10.8}$$

and

$$(\partial/\partial z)\{Nm\langle(v_z + w_z)w_x^2\rangle\} = (\partial v_z/\partial z)P_{xx} + v_z(\partial/\partial z)P_{xx} + (\partial/\partial z)Q_{zxx} \tag{10.9}$$

The addition of Eqs. (10.7)–(10.9) yields

$$\nabla_r \cdot \{N\langle \mathbf{u} m w_x^2\rangle\} = (\nabla \cdot \mathbf{v})P_{xx} + (\mathbf{v} \cdot \nabla)P_{xx} + (\nabla \cdot \mathbf{Q})_{xx} \tag{10.10}$$

where

$$(\nabla \cdot \mathbf{Q})_{ij} = \hat{\mathbf{i}} \cdot \{\hat{\mathbf{j}} \cdot (\nabla \cdot \mathbf{Q})\} \tag{10.11}$$

From Eq. (10.11), it is verified that

$$(\nabla \cdot \mathbf{Q})_{xx} = (\partial/\partial x)Q_{xxx} + (\partial/\partial y)Q_{yxx} + (\partial/\partial z)Q_{zxx} \tag{10.12}$$

The sum of Eqs. (10.2), (10.3), and (10.10) can be written in a simpler form by using the equation of continuity (8.9) which can be recast with the help of Eq. (9.9) as

$$(\partial/\partial t)N + (\mathbf{v} \cdot \nabla)N + (\nabla \cdot \mathbf{v})N = (d/dt)N + N(\nabla \cdot \mathbf{v}) = 0 \tag{10.13}$$

Hence,

$$(\nabla \cdot \mathbf{v}) = -(1/N)(dN/dt) \tag{10.14}$$

On combining Eqs. (10.2), (10.3), and (10.10), it follows from Eqs. (9.9) and (10.14) that

$$\begin{aligned} &\frac{\partial}{\partial t}\{N\langle g\rangle\} - N\left\langle \frac{\partial g}{\partial t}\right\rangle + \nabla_r \cdot \{N\langle \mathbf{u} g\rangle\} \\ &= \frac{\partial}{\partial t}P_{xx} + (\mathbf{v} \cdot \nabla)P_{xx} + (\nabla \cdot \mathbf{v})P_{xx} + (\nabla \cdot \mathbf{Q})_{xx} \\ &= \frac{d}{dt}P_{xx} - \frac{P_{xx}}{N}\frac{dN}{dt} + (\nabla \cdot \mathbf{Q})_{xx} = N\frac{d}{dt}\left(\frac{P_{xx}}{N}\right) + (\nabla \cdot \mathbf{Q})_{xx} \end{aligned} \tag{10.15}$$

The fourth term on the left side of Eq. (10.1) can be expanded as follows:

$$N\langle(\mathbf{u} \cdot \nabla_r)g\rangle = N\left\langle\left(u_x\frac{\partial}{\partial x}mw_x^2 + u_y\frac{\partial}{\partial y}mw_x^2 + u_z\frac{\partial}{\partial z}mw_x^2\right)\right\rangle \tag{10.16}$$

The first term on the right side of Eq. (10.16) can be simplified as

$$
\begin{aligned}
N\left\langle u_x \frac{\partial}{\partial x} m w_x^2 \right\rangle &= N\left\langle (v_x + w_x) 2 m w_x \frac{\partial}{\partial x} w_x \right\rangle \\
&= -N\left\langle (v_x + w_x) 2 m w_x \frac{\partial}{\partial x} v_x \right\rangle \\
&= -2mNv_x \frac{\partial v_x}{\partial x} \langle w_x \rangle - 2mN\left(\frac{\partial}{\partial x} v_x\right) \langle w_x^2 \rangle = -2 \frac{\partial v_x}{\partial x} P_{xx}
\end{aligned}
\tag{10.17}
$$

In a similar manner, it can be shown that

$$
N\left\langle u_y \frac{\partial}{\partial y} m w_x^2 \right\rangle = -2 \frac{\partial v_x}{\partial y} P_{xy} \tag{10.18}
$$

and

$$
N\left\langle u_z \frac{\partial}{\partial z} m w_x^2 \right\rangle = -2 \frac{\partial v_x}{\partial z} P_{xz} \tag{10.19}
$$

The addition of Eqs. (10.17), (10.18), and (10.19) yields

$$
N\langle (\mathbf{u} \cdot \nabla_r) g \rangle = -2\left\{ \frac{\partial v_x}{\partial x} P_{xx} + \frac{\partial v_x}{\partial y} P_{xy} + \frac{\partial v_x}{\partial z} P_{xz} \right\} \tag{10.20}
$$

It can be verified that

$$
(\boldsymbol{\Psi} \cdot \nabla \mathbf{v})_{xx} = \hat{\mathbf{x}} \cdot \{\hat{\mathbf{x}} \cdot (\boldsymbol{\Psi} \cdot \nabla \mathbf{v})\} = P_{xx} \frac{\partial v_x}{\partial x} + P_{xy} \frac{\partial v_x}{\partial y} + P_{xz} \frac{\partial v_x}{\partial z} \tag{10.21}
$$

The dyad $(\boldsymbol{\Psi} \cdot \nabla \mathbf{v})^t$ is the transpose of the dyad $(\boldsymbol{\Psi} \cdot \nabla \mathbf{v})$. Therefore, $(\boldsymbol{\Psi} \cdot \nabla \mathbf{v})_{xx} = (\boldsymbol{\Psi} \cdot \nabla \mathbf{v})^t_{xx}$. Hence, Eq. (10.20) can be conveniently rewritten as

$$
N\langle (\mathbf{u} \cdot \nabla_r) g \rangle = -\{(\boldsymbol{\Psi} \cdot \nabla \mathbf{v})_{xx} + (\boldsymbol{\Psi} \cdot \nabla \mathbf{v})^t_{xx}\} \tag{10.22}
$$

The expansion of the fifth term on the left side of Eq. (10.1) gives

$$
\begin{aligned}
N\left\langle \nabla_u \cdot \left\{ \frac{\mathbf{F}}{m} g \right\} \right\rangle &= N\left\langle F_x \frac{\partial}{\partial u_x} (u_x - v_x)^2 + F_y \frac{\partial}{\partial u_y} (u_x - v_x)^2 \right. \\
&\quad \left. + F_z \frac{\partial}{\partial u_z} (u_x - v_x)^2 \right\rangle = N\langle 2F_x w_x \rangle
\end{aligned}
\tag{10.23}
$$

If the force $\mathbf{F}$ is independent of the velocity $\mathbf{u}$, it is seen that Eq. (10.23) is zero. Thus, it can be ascertained that the forces independent of the velocity do not contribute to the fifth term on the left side of Eq. (10.1). From Eq. (3.30), it is found that

$$
F_x = q[E_x + u_y B_z - u_z B_y] \tag{10.24}
$$

where B_x, B_y, and B_z are, respectively, the Cartesian components of the magnetic flux density $\mathbf{B}$. The result of the substitution of Eq. (10.24) in Eq. (10.23) is

$$N\left\langle \nabla_u \cdot \left\{\frac{\mathbf{F}}{m} g\right\}\right\rangle = 2qNE_x\langle w_x\rangle + 2qNB_z\langle(v_y + w_y)w_x\rangle$$
$$- 2qNB_y\langle(v_z + w_z)w_x\rangle = \frac{2q}{m} B_z P_{yx} - \frac{2q}{m} B_y P_{zx} \tag{10.25}$$

Let $\hat{\mathbf{e}}_B$ be the direction of the magnetic flux density $\mathbf{B}$. Then, by expansion, it can be verified that

$$(B\hat{\mathbf{e}}_B \times \Psi)_{xx} = \hat{\mathbf{x}} \cdot \{\hat{\mathbf{x}} \cdot (B\hat{\mathbf{e}}_B \times \Psi)\} = B_y P_{zx} - B_z P_{yx} \tag{10.26}$$

Also, since $(B\hat{\mathbf{e}}_B \times \Psi)_{xx} = (B\hat{\mathbf{e}}_B \times \Psi)^t_{xx}$, it follows from Eqs. (10.25) and (10.26) that

$$N\left\langle \nabla_u \cdot \left\{\frac{\mathbf{F}}{m} g\right\}\right\rangle = -\frac{q}{|q|}\omega_c\{(\hat{\mathbf{e}}_B \times \Psi)_{xx} + (\hat{\mathbf{e}}_B \times \Psi)^t_{xx}\} \tag{10.27}$$

where

$$\omega_c = |q|B/m \tag{10.28}$$

is known as the gyromagnetic angular frequency; its physical significance will be treated later.

Let

$$R_{ij} = \int m w_i w_j (\partial f/\partial t)_{\text{coll}}\, d\mathbf{u} \qquad \text{for } i, j = x, y, z \tag{10.29}$$

Together with Eqs. (10.15), (10.22), and (10.27), Eq. (10.1) yields

$$\frac{d}{dt}\left(\frac{P_{xx}}{N}\right) + \frac{1}{N}\{(\Psi \cdot \nabla\mathbf{v})_{xx} + (\Psi \cdot \nabla\mathbf{v})^t_{xx}\}$$
$$+ \frac{q}{|q|}\frac{\omega_c}{N}\{(\hat{\mathbf{e}}_B \times \Psi)_{xx} + (\hat{\mathbf{e}}_B \times \Psi)^t_{xx}\} + \frac{1}{N}(\nabla \cdot \mathbf{Q})_{xx} = \frac{R_{xx}}{N} \tag{10.30}$$

If the expanded forms of the second and the third terms on the left side of Eq. (10.30) are used, Eq. (10.30) becomes

$$\frac{d}{dt}\left(\frac{P_{xx}}{N}\right) + \frac{2}{N}\left\{P_{xx}\frac{\partial v_x}{\partial x} + P_{xy}\frac{\partial v_x}{\partial y} + P_{xz}\frac{\partial v_x}{\partial z}\right\} + \frac{1}{N}(\nabla \cdot \mathbf{Q})_{xx} = \frac{F_{xx}}{N} + \frac{R_{xx}}{N} \tag{10.31}$$

where

$$F_{xx} = \frac{2q}{m}\{B_z P_{yx} - B_y P_{zx}\} \tag{10.32}$$

In a similar manner, by successively setting g equal to mw_y^2, mw_z^2, $mw_x w_y$, $mw_y w_z$, and $w_z w_x$, the following results are deduced:

$$\frac{d}{dt}\left(\frac{P_{yy}}{N}\right) + \frac{2}{N}\left\{P_{yx}\frac{\partial v_y}{\partial x} + P_{yy}\frac{\partial v_y}{\partial y} + P_{yz}\frac{\partial v_y}{\partial z}\right\} + \frac{1}{N}(\nabla \cdot \mathbf{Q})_{yy} = \frac{F_{yy}}{N} + \frac{R_{yy}}{N} \tag{10.33}$$

$$\frac{d}{dt}\left(\frac{P_{zz}}{N}\right) + \frac{2}{N}\left\{P_{zx}\frac{\partial v_z}{\partial x} + P_{zy}\frac{\partial v_z}{\partial y} + P_{zz}\frac{\partial v_z}{\partial z}\right\} + \frac{1}{N}(\nabla \cdot \mathbf{Q})_{zz} = \frac{F_{zz}}{N} + \frac{R_{zz}}{N} \tag{10.34}$$

$$\frac{d}{dt}\left(\frac{P_{xy}}{N}\right) + \frac{P_{xy}}{N}\left(\frac{\partial v_x}{\partial x} + \frac{\partial v_y}{\partial y}\right) + \frac{P_{xx}}{N}\frac{\partial v_y}{\partial x} + \frac{P_{yy}}{N}\frac{\partial v_x}{\partial y} + \frac{P_{xz}}{N}\frac{\partial v_y}{\partial z} + \frac{P_{yz}}{N}\frac{\partial v_x}{\partial z} + \frac{1}{N}(\nabla \cdot \mathbf{Q})_{xy} = \frac{F_{xy}}{N} + \frac{R_{xy}}{N} \tag{10.35}$$

$$\frac{d}{dt}\left(\frac{P_{yz}}{N}\right) + \frac{P_{yz}}{N}\left(\frac{\partial v_y}{\partial y} + \frac{\partial v_z}{\partial z}\right) + \frac{P_{yy}}{N}\frac{\partial v_z}{\partial y} + \frac{P_{zz}}{N}\frac{\partial v_y}{\partial z} + \frac{P_{yx}}{N}\frac{\partial v_z}{\partial x} + \frac{P_{zx}}{N}\frac{\partial v_y}{\partial x} + \frac{1}{N}(\nabla \cdot \mathbf{Q})_{yz} = \frac{F_{yz}}{N} + \frac{R_{yz}}{N} \tag{10.36}$$

$$\frac{d}{dt}\left(\frac{P_{zx}}{N}\right) + \frac{P_{zx}}{N}\left(\frac{\partial v_z}{\partial z} + \frac{\partial v_x}{\partial x}\right) + \frac{P_{zz}}{N}\frac{\partial v_x}{\partial z} + \frac{P_{xx}}{N}\frac{\partial v_z}{\partial x} + \frac{P_{zy}}{N}\frac{\partial v_x}{\partial y} + \frac{P_{xy}}{N}\frac{\partial v_z}{\partial y} + \frac{1}{N}(\nabla \cdot \mathbf{Q})_{zx} = \frac{F_{zx}}{N} + \frac{R_{zx}}{N} \tag{10.37}$$

where

$$F_{yy} = (2q/m)\{B_x P_{zy} - B_z P_{xy}\} \tag{10.38}$$

$$F_{zz} = (2q/m)\{B_y P_{xz} - B_x P_{yz}\} \tag{10.39}$$

$$F_{xy} = (q/m)\{B_z(P_{yy} - P_{xx}) + B_x P_{zx} - B_y P_{yz}\} \tag{10.40}$$

$$F_{yz} = (q/m)\{B_x(P_{zz} - P_{yy}) + B_y P_{xy} - B_z P_{zx}\} \tag{10.41}$$

$$F_{zx} = (q/m)\{B_y(P_{xx} - P_{zz}) + B_z P_{yz} - B_x P_{xy}\} \tag{10.42}$$

Energy transport equation

The energy equation is obtained by taking one half of the sum of Eqs. (10.31), (10.33), and (10.34). For expressing the sum in a simple form, note that

$$P_{xx}\frac{\partial v_x}{\partial x} + P_{yx}\frac{\partial v_y}{\partial x} + P_{zx}\frac{\partial v_z}{\partial x} = (P_{xx}\hat{\mathbf{x}}\hat{\mathbf{x}} + P_{yx}\hat{\mathbf{y}}\hat{\mathbf{x}} + P_{zx}\hat{\mathbf{z}}\hat{\mathbf{x}}) \cdot \hat{\mathbf{x}}\frac{\partial}{\partial x} \cdot \mathbf{v} \tag{10.43}$$

$$P_{xy}\frac{\partial v_x}{\partial y} + P_{yy}\frac{\partial v_y}{\partial y} + P_{zy}\frac{\partial v_z}{\partial y} = (P_{xy}\hat{\mathbf{x}}\hat{\mathbf{y}} + P_{yy}\hat{\mathbf{y}}\hat{\mathbf{y}} + P_{zy}\hat{\mathbf{z}}\hat{\mathbf{y}}) \cdot \hat{\mathbf{y}}\frac{\partial}{\partial y} \cdot \mathbf{v} \tag{10.44}$$

$$P_{xz}\frac{\partial v_x}{\partial z} + P_{yz}\frac{\partial v_y}{\partial z} + P_{zz}\frac{\partial v_z}{\partial z} = (P_{xz}\hat{\mathbf{x}}\hat{\mathbf{z}} + P_{yz}\hat{\mathbf{y}}\hat{\mathbf{z}} + P_{zz}\hat{\mathbf{z}}\hat{\mathbf{z}}) \cdot \hat{\mathbf{z}}\frac{\partial}{\partial z} \cdot \mathbf{v} \tag{10.45}$$

The sum of Eqs. (10.43), (10.44), and (10.45) can be written compactly as

$$(\boldsymbol{\Psi} \cdot \boldsymbol{\nabla}) \cdot \mathbf{v} \tag{10.46}$$

From Eqs. (10.32), (10.38), and (10.39), it is seen that

$$F_{xx} + F_{yy} + F_{zz} = 0 \tag{10.47}$$

It follows from Eq. (10.29) that

$$\tfrac{1}{2}(R_{xx} + R_{yy} + R_{zz}) = (m/2) \int w^2 (\partial f/\partial t)_{\text{coll}}\, d\mathbf{u} = r \tag{10.48}$$

It is obtained from Eq. (10.6) that

$$q_i = \tfrac{1}{2}[Q_{ixx} + Q_{iyy} + Q_{izz}] = \tfrac{1}{2}m \int w_i w^2 f(\mathbf{r}, \mathbf{u}, t)\, d\mathbf{u} \tag{10.49}$$

for $i = x$, y, and z. Hence, it follows that

$$\mathbf{q} = \hat{\mathbf{x}} q_x + \hat{\mathbf{y}} q_y + \hat{\mathbf{z}} q_z = \tfrac{1}{2}m \int \mathbf{w} w^2 f(\mathbf{r}, \mathbf{u}, t)\, d\mathbf{u} = (N/2) m \langle \mathbf{w} w^2 \rangle \tag{10.50}$$

From Eqs. (10.11), (10.12), and (10.49), it is found that

$$\begin{aligned} \tfrac{1}{2}\{(\boldsymbol{\nabla} \cdot \mathbf{Q})_{xx} + (\boldsymbol{\nabla} \cdot \mathbf{Q})_{yy} + (\boldsymbol{\nabla} \cdot \mathbf{Q})_{zz}\} &= \frac{1}{2}\frac{\partial}{\partial x}\{Q_{xxx} + Q_{xyy} + Q_{xzz}\} \\ &+ \frac{1}{2}\frac{\partial}{\partial y}\{Q_{yxx} + Q_{yyy} + Q_{yzz}\} + \frac{1}{2}\frac{\partial}{\partial z}\{Q_{zxx} + Q_{zyy} + Q_{zzz}\} \\ &= \frac{\partial}{\partial x} q_x + \frac{\partial}{\partial y} q_y + \frac{\partial}{\partial z} q_z = \boldsymbol{\nabla} \cdot \mathbf{q} \end{aligned} \tag{10.51}$$

It can be shown with the help of Eq. (9.16) that

$$(1/2N)(P_{xx} + P_{yy} + P_{zz}) = \tfrac{1}{2}m\langle w^2 \rangle = U - \tfrac{1}{2}mv^2 \tag{10.52}$$

Since $\langle u^2 \rangle = \langle (\mathbf{v} + \mathbf{w}) \cdot (\mathbf{v} + \mathbf{w}) \rangle = \langle v^2 + 2\mathbf{v} \cdot \mathbf{w} + w^2 \rangle = v^2 + \langle w^2 \rangle$, the result given on the right side of Eq. (10.52) follows. It is to be noted that U is the average kinetic energy of a particle and $\tfrac{1}{2}mv^2$ is the kinetic energy of a particle associated with the mass motion of the gas.

The sum of Eqs. (10.31), (10.33), and (10.34) together with Eqs. (10.43)–(10.48), (10.51), and (10.52) may be shown to yield the following energy equation

$$N\frac{d}{dt}\left(U - \tfrac{1}{2}mv^2\right) + (\boldsymbol{\Psi} \cdot \boldsymbol{\nabla}) \cdot \mathbf{v} + \boldsymbol{\nabla} \cdot \mathbf{q} = r \tag{10.53}$$

It is possible to manipulate Eq. (10.53) into a form which is amenable to simple physical interpretation. It can be shown with the help of Eqs. (9.9) and (10.14) that

$$N\frac{d}{dt}\left(U-\frac{1}{2}mv^2\right)=N\frac{d}{dt}\left(\frac{1}{2}m\langle w^2\rangle\right)=\frac{d}{dt}\left(\frac{N}{2}m\langle w^2\rangle\right)-\frac{1}{2}m\langle w^2\rangle\frac{dN}{dt}$$
$$=\left[\frac{\partial}{\partial t}+(\mathbf{v}\cdot\nabla)\right]\frac{N}{2}m\langle w^2\rangle+\frac{N}{2}m\langle w^2\rangle\nabla\cdot\mathbf{v} \qquad (10.54)$$
$$=\frac{\partial}{\partial t}\left(\frac{N}{2}m\langle w^2\rangle\right)+\nabla\cdot\left(\frac{N}{2}m\langle w^2\rangle\mathbf{v}\right)$$

since

$$\nabla\cdot(a\mathbf{A})=(\mathbf{A}\cdot\nabla)a+a(\nabla\cdot\mathbf{A}) \qquad (10.55)$$

In view of the symmetry of the kinetic pressure dyad, it can be verified by actual expansion that

$$(\boldsymbol{\Psi}\cdot\nabla)\cdot\mathbf{v}=\nabla\cdot(\boldsymbol{\Psi}\cdot\mathbf{v})-(\nabla\cdot\boldsymbol{\Psi})\cdot\mathbf{v} \qquad (10.56)$$

If the momentum transport equation (9.14) is scalarly multiplied by $\mathbf{v}$, it is seen from Eqs. (9.9), (10.14), and (10.55) that

$$-(\nabla\cdot\boldsymbol{\Psi})\cdot\mathbf{v}=Nm\mathbf{v}\cdot\frac{d\mathbf{v}}{dt}-Nq\mathbf{v}\cdot(\mathbf{E}+\mathbf{v}\times\mathbf{B})-\mathbf{v}\cdot\mathbf{P}_{\text{coll}}$$
$$=\frac{d}{dt}\left(\frac{N}{2}mv^2\right)-\frac{1}{2}mv^2\frac{dN}{dt}-Nq\mathbf{v}\cdot(\mathbf{E}+\mathbf{v}\times\mathbf{B})-\mathbf{v}\cdot\mathbf{P}_{\text{coll}} \qquad (10.57)$$
$$=\frac{\partial}{\partial t}\left(\frac{N}{2}mv^2\right)+\nabla\cdot\left(\frac{N}{2}mv^2\mathbf{v}\right)-Nq\mathbf{v}\cdot(\mathbf{E}+\mathbf{v}\times\mathbf{B})-\mathbf{v}\cdot\mathbf{P}_{\text{coll}}$$

The rate of change of average energy per unit volume due to collisional interactions is given by

$$E_{\text{coll}}=\int\tfrac{1}{2}mu^2(\partial f/\partial t)_{\text{coll}}\,d\mathbf{u} \qquad (10.58)$$

Since $u^2=-v^2+2\mathbf{v}\cdot\mathbf{u}+w^2$, it follows from Eqs. (9.12), (10.48), and (10.58) that

$$E_{\text{coll}}=\mathbf{v}\cdot\mathbf{P}_{\text{coll}}+r \qquad (10.59)$$

since the rate of change of number density due to collisional interactions as given by $\int(\partial f/\partial t)_{\text{coll}}\,d\mathbf{u}$ is zero in view of the assumption that there is neither creation or annihilation of particles. The substitution of Eqs. (10.54), (10.56), (10.57), and (10.59) into Eq. (10.53) yields after some rearrangement

$$\frac{\partial}{\partial t}\left(\frac{N}{2}m\langle w^2\rangle\right)+\frac{\partial}{\partial t}\left(\frac{N}{2}mv^2\right)=-\nabla\cdot\left(\frac{N}{2}m\langle w^2\rangle\mathbf{v}\right)-\nabla\cdot\left(\frac{N}{2}mv^2\mathbf{v}\right)$$
$$-\nabla\cdot\left(\frac{N}{2}m\langle w^2\mathbf{w}\rangle\right)-\nabla\cdot(\boldsymbol{\Psi}\cdot\mathbf{v})+Nq\mathbf{v}\cdot\mathbf{E}+E_{\text{coll}} \qquad (10.60)$$

since $\mathbf{v}\cdot(\mathbf{v}\times\mathbf{B})=0$. In Eq. (10.60) the equivalent expression for $\mathbf{q}$ from Eq. (10.50) has also been inserted. The first term on the right side of Eq. (10.60) can be interpreted as the average rate of increase of thermal energy density due to the particles entering the volume with the average velocity **v**. The second term on the right side of Eq. (10.60) is the average rate of increase of the density of the visible energy associated with the mass motion of the gas due to the particles entering the volume with the average velocity **v**. In a similar manner, the third term on the right side of Eq. (10.60) is the average rate of increase of thermal energy density due to the particles entering the volume with the random thermal velocity **w**. It is this average rate of thermal energy transported across a unit area by the particles moving with random thermal velocity **w** that is identified with *the vector flux density of heat* **q** whose physical meaning is explained further subsequently. It is to be noted that the increase of the visible energy density due to the particles entering the volume with random thermal velocity averages to zero. Thus together the first three terms on the right side of Eq. (10.60) represent the average rate of increase of the total energy density convectively transported by the particles entering the volume.

Consider a unit surface area normal to $\hat{\mathbf{n}}$ moving uniformly with the average particle velocity **v**. Then, as has been pointed out previously, $-\hat{\mathbf{n}} \cdot \mathbf{\Psi}$ is the force acting on the unit area due to the actual particle motions. Since the unit area moves with the average velocity **v**, the dyadic kinetic pressure $\mathbf{\Psi}$ does work on the particles and hence increases the energy density of the particles. It can be established that the average rate of increase of the energy density due to the work done by the kinetic pressure on the particles is given by $-\nabla \cdot (\mathbf{\Psi} \cdot \mathbf{v})$, which is the fourth term on the right side of Eq. (10.60).

If the particles are charged, the force acting on a particle due to the electric field is $q\mathbf{E}$. Therefore, the average rate of work done on the charged particles in unit volume by the electric field is $Nq\mathbf{v} \cdot \mathbf{E}$ and this contributes to the average rate of increase of the energy density of the particles. The force acting on a charged particle due to the magnetic flux density **B** is $q\mathbf{v} \times \mathbf{B}$. Since this force is always perpendicular to the direction of motion of the particles, the magnetic flux density does no work on the particles and hence does not contribute to the increase of the kinetic energy of the particles. Thus the fifth term on the right side of Eq. (10.60) is the average rate of increase of the energy density of the particles due to the work done by the electromagnetic fields. Finally the sixth term E_{coll} on the right side of Eq. (10.60) is the average rate of increase of the energy density due to the collisional interactions with other species of particles. The sum of the six terms on the right side of Eq. (10.60) should give the total rate of increase of the energy density of the particles. This result on energy conservation is portrayed by Eq. (10.60) whose left side has two terms which are the rates of increase of the average densities of thermal and visible energies, respectively.

Physical meaning of q and Q

Both **q** and **Q** are seen from Eqs. (10.50) and (10.5) to have the dimensions of a rate of energy transport across unit area or energy flux density. Hence, **q** and **Q** are called vector and triadic flux densities of heat, respectively. For the purpose of obtaining a physical meaning of $\mathbf{q}$, it is desirable to consider the quantity $(N/2)m\langle u^2\mathbf{u}\rangle$. With the substitution of $\mathbf{u} = \mathbf{v} + \mathbf{w}$, this quantity can be expanded as

$$\begin{aligned}\frac{N}{2}m\langle u^2\mathbf{u}\rangle &= \frac{N}{2}m\langle v^2\mathbf{v} + 2(\mathbf{v}\cdot\mathbf{w})\mathbf{v} + w^2\mathbf{v} + v^2\mathbf{w} + 2\mathbf{v}\cdot\mathbf{ww} + w^2\mathbf{w}\rangle \\ &= \frac{N}{2}m(v^2 + \langle w^2\rangle)\mathbf{v} + \mathbf{v}\cdot Nm\langle\mathbf{ww}\rangle + \frac{Nm}{2}\langle w^2\mathbf{w}\rangle\end{aligned} \tag{10.61}$$

since $\langle(\mathbf{v}\cdot\mathbf{w})\rangle = \langle\mathbf{w}\rangle = 0$. Together with Eqs. (10.52), (10.50), and (9.5), Eq. (10.61) can be written as

$$(N/2)m\langle u^2\mathbf{u}\rangle = NU\mathbf{v} + \mathbf{v}\cdot\mathbf{\Psi} + \mathbf{q} \tag{10.62}$$

where U is the average kinetic energy of a particle. The left side of Eq. (10.62) gives the rate of average energy transported across a unit area. A part of this total energy flux density is due to the rate $\mathbf{v}\cdot\mathbf{\Psi}$ of work done by the dyadic kinetic pressure and another part is the energy flux density $NU\mathbf{v}$ convectively transported by the particles. The remainder of the energy flux density is the vector flux density of heat $\mathbf{q}$. It is to be noted that $\mathbf{q}$ is the average rate of *thermal* energy transported across unit area at *thermal* velocities. Just as the kinetic pressure dyad $\mathbf{\Psi}$, so the vector flux density of heat $\mathbf{q}$ is zero if all the particles have the same velocities with the result that there is no random distribution of velocities. To bring out the fact that $\mathbf{q}$ is the average rate of transport of thermal energy across unit area at thermal speeds, consider a unit area moving uniformly at the average particle velocity $\mathbf{v}$. Then with reference to the unit area, the particle velocities become identical to their thermal or peculiar velocities. Hence by setting $\mathbf{v} = 0$ on the right side of Eq. (10.62), the total energy flux density is seen to become identical to $\mathbf{q}$, as indicated.

If the thermal velocities are uniformly distributed in all directions, that is, if the distribution function for the peculiar velocities is isotropic, then it is seen from Eq. (10.50) that $\mathbf{q} = 0$ since the integrand is an odd function of $\mathbf{w}$. Thus $\mathbf{q}$ is at least a partial measure of the anisotropies in the distribution of the peculiar velocities.

It can be anticipated that the triadic flux density of heat $\mathbf{Q}$ is a complete measure of the anisotropies in the distribution of the thermal velocities. The relationship between $\mathbf{Q}$ and $\mathbf{q}$ is expressed in Eq. (10.49). The physical meaning of $\mathbf{Q}$ is somewhat similar to that of $\mathbf{q}$ and can be obtained in the following manner. It can be verified that

$$\begin{aligned}u_x u_y u_z &= v_x v_y u_z + v_y v_z u_x + v_z v_x u_y - 2v_x v_y v_z \\ &\quad + v_x w_y w_z + v_y w_z w_x + v_z w_x w_y + w_x w_y w_z\end{aligned} \tag{10.63}$$

Therefore,

$$Nm\langle u_x u_y u_z\rangle = Nmv_x v_y v_z + (\mathbf{v}, \mathbf{\Psi})_{xyz} + Q_{xyz} \tag{10.64}$$

where

$$(\mathbf{v}, \boldsymbol{\Psi})_{xyz} = v_x P_{yz} + v_y P_{zx} + v_z P_{xy} \tag{10.65}$$

From Eq. (10.64), the following triadic relation is obtained:

$$Nm\langle \mathbf{uuu} \rangle = Nm\mathbf{vvv} + (\mathbf{v}, \boldsymbol{\Psi}) + \mathbf{Q} \tag{10.66}$$

where the components of $(\mathbf{v}, \boldsymbol{\Psi})$ are determined from Eq. (10.65). As before with $\mathbf{q}$, the triadic flux density of heat $\mathbf{Q}$ can be considered as the difference between the total energy flux density $Nm\langle \mathbf{uuu} \rangle$ and the energy flux density of convective motion as given by the first two terms on the right side of Eq. (10.66). The triadic heat flux density $\mathbf{Q}$ considerably extends the concept of vector heat flux density $\mathbf{q}$ in the sense that it is a complete measure of the anisotropies of thermal motions of particles in a gas.

1.11. System of hydrodynamic equations

It is difficult to solve the Boltzmann equation (3.29) for the velocity distribution function. In principle, the velocity distribution function can be determined if all its moments are known. The moment of the velocity distribution function can be obtained by multiplying the velocity distribution function by an appropriate function of the particle velocity and integrating the result over the entire velocity space. The first four moments of the velocity distribution function are the number density N, the average velocity $\mathbf{v}$, the kinetic pressure dyad $\boldsymbol{\Psi}$, and the triad of heat flux density $\mathbf{Q}$. In order to determine the macroscopic parameters N, $\mathbf{v}$, $\boldsymbol{\Psi}$, and $\mathbf{Q}$, it is necessary to deduce the equations satisfied by these quantities. In the preceding three sections, these hydrodynamic equations were derived by taking the moments of the Boltzmann equation. The first moment is *the equation of continuity* (8.9) that relates the evolution of the number density N, which is a scalar, to the average particle velocity $\mathbf{v}$, which is a vector. For the purpose of determining the average particle velocity $\mathbf{v}$, the second moment of the Boltzmann equation is evaluated. The second moment is *the momentum transport equation* (9.14) that relates the evolution of the average particle velocity $\mathbf{v}$, which is a vector, to the kinetic pressure $\boldsymbol{\Psi}$, which is a dyad. In order to determine the kinetic pressure dyad $\boldsymbol{\Psi}$, the third moment of the Boltzmann equation is obtained. The third moment is *the transport equation for the kinetic pressure*; this equation relates the evolution of the kinetic pressure, which is a dyad, to the heat flux density $\mathbf{Q}$, which is a triad. It is seen that the set of hydrodynamic equations specifying the macroscopic parameters is never complete in the sense that the number of equations is never sufficient for the determination of all the macroscopic parameters appearing in them. Every time a higher moment of the Boltzmann equation is taken to obtain a closed system of hydrodynamic equations, another new macroscopic variable appears. Consequently, it is common to truncate the system of hydrodynamic equations arbitrarily at a particular moment in the hierarchy of moments of the Boltzmann equation and to make a simplifying assumption on the highest moment of the velocity distribution function appearing in the last of the series of hydrodynamic equations. Several different sets of such closed hydrodynamic equations are possible and we now discuss the two most commonly used sets of equations.

Cold plasma model

In the first closed set of hydrodynamic equations, the truncation is introduced at the momentum transfer equation (9.14). The highest moment of the velocity distribution function appearing in Eq. (9.14) is the kinetic pressure $\mathbf{\Psi}$ which is set equal to zero. In this system the thermal motions of the particles are neglected and therefore this system of closed hydrodynamic equations is said to describe a *cold plasma model*. The number density N and the average velocity $\mathbf{v}$ are the only two hydrodynamic variables. The theory based on the cold plasma model is also known as the *magnetoionic theory* and this theory has been successfully applied in the investigation of the properties of small-amplitude wave propagation in a plasma at phase velocities much larger than the thermal speeds of the plasma particles. For convenience, the two hydrodynamical equations pertaining to the cold plasma model are collected together as follows:

$$\partial N/\partial t + \nabla \cdot (N\mathbf{v}) = 0 \tag{11.1}$$

$$Nm\, d\mathbf{v}/dt = Nq[\mathbf{E} + \mathbf{v} \times \mathbf{B}] + \mathbf{P}_{\text{coll}} \tag{11.2}$$

It is necessary to be able to evaluate the collision term $\mathbf{P}_{\text{coll}}$ in order that Eqs. (11.1) and (11.2) be useful in the study of plasma properties.

Warm plasma model

In the second closed set of hydrodynamic equations, the truncation is introduced at the transport equations for the components of the dyadic kinetic pressure as given by Eqs. (10.31) and (10.33)–(10.37). The highest moment of the velocity distribution function appearing in these equations is the flux density of heat $\mathbf{Q}$ which enters in the form $\nabla \cdot \mathbf{Q}$. The simplifying approximation is to let

$$\nabla \cdot \mathbf{Q} = 0 \tag{11.3}$$

which is equivalent to assuming that the processes taking place in a plasma are such that there is no flow of heat. This is the so-called adiabatic approximation in which the heat conductivity is equal to zero. Further, the nondiagonal terms of the kinetic pressure dyad are assumed to vanish and the diagonal terms are assumed to be equal. In other words, instead of the dyadic kinetic pressure $\mathbf{\Psi}$, only the simple form of scalar pressure P is taken into account. This approximation of using only the scalar pressure is equivalent to neglecting the viscous forces. Then, as stated in Eq. (9.20), $\nabla \cdot \boldsymbol{\Psi}$ in Eq. (9.14) degenerates to ∇P.

The number density N, the average velocity $\mathbf{v}$, and the scalar pressure P are the three hydrodynamic variables. The first two hydrodynamical equations in this set are

$$\partial N/\partial t + \nabla \cdot N\mathbf{v} = 0 \tag{11.4}$$

$$Nm\, d\mathbf{v}/dt = -\nabla P + Nq[\mathbf{E} + \mathbf{v} \times \mathbf{B}] + \mathbf{P}_{\text{coll}} \tag{11.5}$$

Under an additional assumption that the energy interchange due to collisional interactions is negligible, the third hydrodynamical equation, which is the transport equation for the kinetic pressure, is shown later in this section to lead to the following adiabatic gas law:

$$PN^{-\gamma} = \text{constant} \qquad \gamma = 1 + 2/\delta \tag{11.6}$$

where δ is the number of degrees of freedom. The system of hydrodynamical equations (11.4)–(11.6) are the governing equations for the so-called *warm plasma model*. It is emphasized that Eq. (11.6) is valid only when heat conduction, viscosity, and energy interchange due to collisional interactions are all neglected. As before, it is required to evaluate $\mathbf{P}_{\text{coll}}$ in Eq. (11.5) in order that Eqs. (11.4)–(11.6) be useful in the investigation of the plasma behavior. In many cases, the warm plasma model gives a better approximation to the plasma behavior than the cold plasma model.

Adiabatic gas law

The adiabatic gas law (11.6) can be deduced from the transport equations (10.31) and (10.33)–(10.37) for the components of the kinetic pressure dyad. The adiabatic condition (11.3) is used. The energy interchange due to collisional interactions with other species of particles is neglected. Therefore,

$$R_{ij} = 0 \qquad \text{for } i, j = x, y, z \tag{11.7}$$

where R_{ij} is defined by Eq. (10.29). The magnetic flux density $\mathbf{B}$ is assumed to be uniform and directed along the z-axis with the result that

$$B_x = B_y = 0 \tag{11.8}$$

The distribution of velocities of the particles in every small elemental volume is assumed to have rotational symmetry about the direction of the magnetic flux density $\mathbf{B}$. The existence of rotational symmetry can be shown to lead to the following conditions among the components of the kinetic pressure dyad:

$$P_{xx} = P_{yy} = P_{\perp} \qquad P_{zz} = P_{\parallel} \qquad P_{xy} = P_{yx} = P_{yz} = P_{zy} = P_{zx} = P_{xz} = 0 \tag{11.9}$$

For convenience, the labels $\perp$ and $\|$ are introduced to denote respectively the directions perpendicular and parallel to the magnetic flux density $\hat{\mathbf{z}}B_z$. In view of Eqs. (11.8) and (11.9), it follows from Eqs. (10.32) and (10.38)–(10.42) that

$$F_{xx} = F_{yy} = F_{zz} = F_{xy} = F_{yz} = F_{zx} = 0 \tag{11.10}$$

With the help of Eqs. (11.3), (11.7), (11.9), and (11.10), the evolution equations for the components of the kinetic pressure dyad as given by Eqs. (10.31) and (10.33)–(10.37) can be simplified to yield the following results:

$$\frac{d}{dt}\left(\frac{P_\perp}{N}\right) + \frac{2}{N}P_\perp \frac{\partial v_x}{\partial x} = 0 \tag{11.11}$$

$$\frac{d}{dt}\left(\frac{P_\perp}{N}\right) + \frac{2}{N}P_\perp \frac{\partial v_y}{\partial y} = 0 \tag{11.12}$$

$$\frac{d}{dt}\left(\frac{P_\|}{N}\right) + \frac{2}{N}P_\| \frac{\partial v_z}{\partial z} = 0 \tag{11.13}$$

$$\frac{\partial v_y}{\partial x} + \frac{\partial v_x}{\partial y} = 0 \tag{11.14}$$

$$P_\perp \frac{\partial v_z}{\partial y} + P_\| \frac{\partial v_y}{\partial z} = 0 \tag{11.15}$$

$$P_\| \frac{\partial v_x}{\partial z} + P_\perp \frac{\partial v_z}{\partial x} = 0 \tag{11.16}$$

From Eqs. (11.11) and (11.12), it is found that

$$\partial v_x/\partial x = \partial v_y/\partial y \tag{11.17}$$

It is convenient to consider three cases separately.

Linear compression parallel to the B field

First case is the one-dimensional compression parallel to the magnetic flux density. There are no variations in the *x*- and *y*-directions, i.e.,

$$\frac{\partial}{\partial x} = \frac{\partial}{\partial y} = 0 \tag{11.18}$$

In view of Eq. (11.18), Eqs. (11.14), and (11.17) become trivial and Eqs. (11.15) and (11.16) reduce to

$$\partial v_x/\partial z = \partial v_y/\partial z = 0 \tag{11.19}$$

Also, Eqs. (11.11) and (11.12) give

$$P_\perp N^{-1} = \text{constant} \tag{11.20}$$

If Eqs. (11.18) and (10.14) are substituted into Eq. (11.13), it follows that

$$\frac{d}{dt}\left(\frac{P_{\|}}{N}\right) - \frac{2}{N^2}P_{\|}\frac{dN}{dt} = N^2\frac{d}{dt}\left(\frac{P_{\|}}{N^3}\right) = 0 \tag{11.21}$$

Therefore,

$$P_{\|}N^{-3} = \text{constant} \tag{11.22}$$

It is seen from Eqs. (11.20) and (11.22) that for a one-dimensional compression parallel to the magnetic field, the adiabatic ratio γ as given by Eq. (11.6) is equal to 1 for the perpendicular pressure and 3 for the parallel pressure. In accordance with the ideal gas law (7.8), it is useful to introduce a parallel and a perpendicular temperature by the two relations

$$P_{\perp} = NKT_{\perp} \qquad P_{\|} = NKT_{\|} \tag{11.23}$$

For the case of one-dimensional compression parallel to the magnetic field,

$$T_{\perp} = \text{constant} \qquad T_{\|} \propto N^2 \tag{11.24}$$

with the result that this type of compression is isothermal with regard to the transverse temperature. The variations in $P_{\perp}$ are due entirely to the changes in the number density N whereas those of $P_{\|}$ are contributed by the changes in N as well as in $T_{\|}$. It is ascertained from Eqs. (11.22) and (11.23) that $T_{\|}$ is proportional to the square of the number density N, as stated in the relation (11.24).

Cylindrical compression perpendicular to the B field

For this case, there are no variations in the z-direction, i.e.,

$$\frac{\partial}{\partial z} = 0 \tag{11.25}$$

It is obtained from Eqs. (11.15) and (11.16) that

$$\partial v_z/\partial y = \partial v_z/\partial x = 0 \tag{11.26}$$

Also, Eq. (11.13) gives

$$P_{\|}N^{-1} = \text{constant} \tag{11.27}$$

If Eqs. (11.11) and (11.12) are added and Eqs. (11.25) and (10.14) are substituted into the resulting equation, it is found that

$$\frac{d}{dt}\left(\frac{P_\perp}{N}\right) + \frac{1}{N}P_\perp(\nabla \cdot \mathbf{v}) = \frac{d}{dt}\left(\frac{P_\perp}{N}\right) - \frac{P_\perp}{N^2}\frac{dN}{dt} = N\frac{d}{dt}\left(\frac{P_\perp}{N^2}\right) = 0 \qquad (11.28)$$

It follows from Eq. (11.28) that

$$P_\perp N^{-2} = \text{constant} \qquad (11.29)$$

It is seen from Eqs. (11.27) and (11.29) that for a two-dimensional (cylindrically symmetrical) compression perpendicular to the magnetic field, the adiabatic ratio γ is equal to 1 for the parallel pressure and 2 for the perpendicular pressure. For this case, it is deduced from Eqs. (11.23), (11.27), and (11.29) that

$$T_\parallel = \text{constant} \qquad T_\perp \propto N \qquad (11.30)$$

This type of compression is therefore isothermal with regard to the parallel temperature. Also, for the two-dimensional compression perpendicular to the magnetic field, the variations in $P_\parallel$ are due entirely to the changes in N whereas those of $P_\perp$ are caused by the changes in both N and $T_\perp$.

Three-dimensional compression

If the compression is spherically symmetric,

$$P_\perp = P_\parallel = P \qquad (11.31)$$

and therefore Eqs. (11.15) and (11.16) yield

$$\partial v_z/\partial y + \partial v_y/\partial z = \partial v_x/\partial z + \partial v_z/\partial x = 0 \qquad (11.32)$$

Also, it is seen from Eqs. (11.11)–(11.13) and (11.31) that

$$\partial v_x/\partial x = \partial v_y/\partial y = \partial v_z/\partial z \qquad (11.33)$$

The addition of Eqs. (11.11), (11.12), and (11.13) and the use of Eq. (11.31) yield

$$(d/dt)(P/N) + \tfrac{2}{3}(P/N)(\nabla \cdot \mathbf{v}) = 0 \qquad (11.34)$$

Together with Eq. (10.14), Eq. (11.34) can be manipulated to give

$$\frac{d}{dt}\left(\frac{P}{N}\right) - \frac{2}{3}\frac{P}{N^2}\frac{dN}{dt} = N^{2/3}\frac{d}{dt}(PN^{-5/3}) = 0 \qquad (11.35)$$

It follows from Eq. (11.35) that

$$PN^{-5/3} = \text{constant} \tag{11.36}$$

From Eq. (11.6), it is verified that $\gamma = 3$, $\gamma = 2$, $\gamma = 5/3$ for $\delta = 1$, $\delta = 2$, and $\delta = 3$, respectively. Therefore, Eqs. (11.22), (11.29), and (11.36) are conveniently combined to yield the adiabatic gas law (11.6) where δ is the number of degrees of freedom affected by the compression. It is emphasized that the adiabatic gas law (11.6) is valid only if the heat conduction, viscosity, and energy interchange due to collisional interactions are all not taken into account.

The fluid has to be subjected to a system of forces in order to be able to achieve the desired type of adiabatic compression. For example, if a spherically symmetric, three-dimensional adiabatic compression is desired, the auxiliary conditions (11.14), (11.32), and (11.33) have to be used in conjunction with the momentum transport equation (9.14) to determine the required system of forces which are not explicitly investigated in this treatment.

1.12. Lumped macroscopic parameters and their governing equations

A plasma is a mixture of electrons, ions, and neutral particles. Let N_e, N_i, and N_n be the number densities of electrons, ions and neutral particles respectively. The existence of macroscopic electrical neutrality is expressed by the relation $N_e = N_i = N$. Although a detailed discussion is to be given subsequently, it is appropriate to point out here that the large electrostatic forces which are set up by any significant charge separation are responsible for maintaining this electrical neutrality. The degree of ionization α is defined by the relation $\alpha = N/(N + N_n)$ with the result that $\alpha = 0$ for an ordinary gas with no charged particles and $\alpha = 1$ for a fully ionized plasma which has no neutral particles. For a reason to be stated presently, the treatment in this section is specialized to apply only to fully ionized plasmas for which $\alpha = 1$.

There is a set of hydrodynamical equations corresponding to each species of particles in a plasma. Thus in a fully ionized plasma, one set of hydrodynamical equations applies to the electrons and another set to the ions. The macroscopic variables for the electrons are the number density N_e, the average velocity $\mathbf{v}_e$ and the kinetic pressure dyad $\boldsymbol{\Psi}_e$. For the ions, there is a corresponding set of macroscopic variables distinguished by the subscript i. For certain applications, it is advantageous to lump the macroscopic variables of the two constituents of the plasma in order to obtain certain macroscopic parameters applicable to the ionized gas as a whole instead of to the two constituents separately and to deduce the governing equations for these lumped macroscopic variables. The mass and the electric charge densities, the mass current and the electric current densities, and, the total kinetic and electrokinetic pressure dyads are the six lumped macroscopic variables to be considered. The mass (electric charge) density is the mass (electric charge) per unit volume. The mass (electric) current density is the rate of flow of mass (electric charge) across unit area. With these definitions, the following expressions for the lumped macroscopic variables are obtained:

$$\rho_m = N_e m_e + N_i m_i \qquad \text{mass density} \tag{12.1}$$

$$\rho = N_e q_e + N_i q_i \qquad \text{electric charge density} \tag{12.2}$$

$$\mathbf{J}_m = N_e m_e \mathbf{v}_e + N_i m_i \mathbf{v}_i \qquad \text{mass current density} \tag{12.3}$$

$$\mathbf{J} = N_e q_e \mathbf{v}_e + N_i q_i \mathbf{v}_i \qquad \text{electric current density} \tag{12.4}$$

$$\mathbf{\Psi} = \mathbf{\Psi}_e + \mathbf{\Psi}_i \qquad \text{total kinetic pressure dyad} \tag{12.5}$$

$$\mathbf{\Psi}_E = \frac{q_e}{m_e}\mathbf{\Psi}_e + \frac{q_i}{m_i}\mathbf{\Psi}_i \qquad \text{total electrokinetic pressure dyad} \tag{12.6}$$

Also, it is convenient to introduce the mass velocity $\mathbf{V}$ as defined by

$$\mathbf{V} = \frac{\mathbf{J}_m}{\rho_m} = \frac{N_e m_e \mathbf{v}_e + N_i m_i \mathbf{v}_i}{N_e m_e + N_i m_i} \tag{12.7}$$

The subscript e (i) in Eqs. (12.1)–(12.6) denotes that the corresponding quantity pertains to the electrons (ions).

If the plasma is partially ionized, Eqs. (12.1), (12.3), and (12.5) should, in addition, contain the terms corresponding to the neutral particles. In general, the summation in Eqs. (12.1)–(12.6) should contain the contributions from all the different species of particles in the gas mixture. In a fully ionized plasma, there are altogether six macroscopic variables ($N_e, N_i, \mathbf{v}_e, \mathbf{v}_i, \mathbf{\Psi}_e$ and $\mathbf{\Psi}_i$) associated with the constituent particles and the total number of lumped macroscopic variables is seen from Eqs. (12.1)–(12.6) to be also six. It can be ascertained that in a gas mixture containing only two different types of particles, the number of lumped macroscopic variables is equal to the number of macroscopic variables associated with the constituent particles. If more than two different species of particles are present, the number of lumped variables is less than the total number of partial variables with the result that some information is lost in using the lumped macroscopic variables. Since the same information is contained in the partial and the lumped macroscopic variables for a gas containing only two different kinds of particles, the lumped macroscopic variables have proved to be particularly useful in the study of a fully ionized gas.

The hydrodynamic equations governing the lumped macroscopic variables are deduced from the corresponding equations satisfied by the macroscopic variables pertaining to the constituent particles. These equations are complicated in view of their nonlinearity. For most purposes, it is sufficient to consider only the simplified version of these equations obtained by a suitable linearization. In this treatment, only the so-called linearized magnetohydrodynamic (MHD) equations specifying these lumped macroscopic variables are determined.

Equations of mass and charge conservation

The equation of continuity for the electrons is obtained from Eqs. (8.9) and (10.55) as

$$\partial N_e/\partial t + (\mathbf{v}_e \cdot \nabla)N_e + N_e(\nabla \cdot \mathbf{v}_e) = 0 \tag{12.8}$$

It is assumed that the average velocity, the electric field, and the kinetic pressure are first-order perturbations and that the second-order terms resulting from the product of two perturbation terms can be neglected. Also, let

$$N \to N_0 + N \qquad \mathbf{B} \to \mathbf{B}_0 + \mathbf{B} \tag{12.9}$$

where N_0 and $\mathbf{B}_0$ are ambient terms, and N and $\mathbf{B}$ are first-order perturbations which are very small in comparison with their respective ambient terms. Note that the relations (12.9) apply both to the electrons and the ions. If the relations (12.9) are substituted into Eq. (12.8) and the second-order terms are neglected, it is found that

$$\partial N_e/\partial t + N_{e0}(\nabla \cdot \mathbf{v}_e) = 0 \tag{12.10}$$

which is the linearized form of the equation of continuity for the electrons. The corresponding equation of continuity for the ions is derived as

$$\partial N_i/\partial t + N_{i0}(\nabla \cdot \mathbf{v}_i) = 0 \tag{12.11}$$

If Eqs. (12.10) and (12.11) are multiplied by m_e and m_i, respectively, and the results are added, it is seen on noting Eqs. (12.1), (12.3), and (12.7) that

$$\frac{\partial}{\partial t}\rho_m + \nabla \cdot (\rho_{m0}\mathbf{V}) = \frac{\partial}{\partial t}\rho_m + \rho_{m0}(\nabla \cdot \mathbf{V}) = 0 \tag{12.12}$$

where $\mathbf{V}$ is the linearized form of the mass velocity as given by

$$\mathbf{V} = \frac{\mathbf{J}_m}{\rho_{m0}} = \frac{N_{e0}m_e\mathbf{v}_e + N_{i0}m_i\mathbf{v}_i}{N_{e0}m_e + N_{i0}m_i} \tag{12.13}$$

In Eq . (12.12), ρ_{m0} is the ambient density and ρ_m is the perturbation. In a similar manner, if Eqs. (12.10) and (12.11) are multiplied by q_e and q_i, respectively, and the results are added, it is obtained that

$$(\partial/\partial t)\rho + \nabla \cdot \mathbf{J} = 0 \tag{12.14}$$

where $\mathbf{J}$ is the linearized form of the electric current density as given by

$$\mathbf{J} = N_{e0}q_e\mathbf{v}_e + N_{i0}q_i\mathbf{v}_i \tag{12.15}$$

The two equations (12.10) and (12.11) of continuity satisfied by the partial variables are now replaced by the equation (12.12) of mass conservation and the equation (12.15) of conservation of electric charge.

Equation of motion

The linearized form of the momentum transport equation for the electrons is found from Eq. (9.14) as

$$N_{e0} m_e \frac{\partial}{\partial t} \mathbf{v}_e = -\nabla \cdot \mathbf{\Psi}_e + N_{e0} q_e [\mathbf{E} + \mathbf{v}_e \times \mathbf{B}_0] + \mathbf{P}_{ei} \tag{12.16}$$

where $\mathbf{P}_{ei}(\mathbf{P}_{ie})$ is the total momentum transferred to the electrons (ions) per unit volume per unit time by collisional interactions with the ions (electrons). Similarly, the linearized equation of momentum transport for the ions is

$$N_{i0} m_i \frac{\partial}{\partial t} \mathbf{v}_i = -\nabla \cdot \mathbf{\Psi}_i + N_{i0} q_i [\mathbf{E} + \mathbf{v}_i \times \mathbf{B}_0] + \mathbf{P}_{ie} \tag{12.17}$$

The sum of Eqs. (12.16) and (12.17) together with Eqs. (12.13), (12.5), and (12.15) yields

$$\rho_{m0}\, \partial \mathbf{V}/\partial t = -\nabla \cdot \mathbf{\Psi} + \rho_0 \mathbf{E} + \mathbf{J} \times \mathbf{B}_0 \tag{12.18}$$

Since the total momentum is conserved in collisional interactions between the electrons and the ions, it follows that

$$\mathbf{P}_{ei} + \mathbf{P}_{ie} = 0 \tag{12.19}$$

This result has been used in obtaining Eq. (12.18). Also,

$$\rho_0 = N_{e0} q_e + N_{i0} q_i \tag{12.20}$$

is the ambient electric charge density in a plasma. Since a plasma is electrically neutral on a macroscopic scale, $\rho_0 = 0$ with the result that Eq. (12.18) becomes

$$\rho_{m0} (\partial/\partial t) \mathbf{V} = -\nabla \cdot \mathbf{\Psi} + \mathbf{J} \times \mathbf{B}_0 \tag{12.21}$$

which is the well-known equation of motion in conducting fluids.

Generalized Ohm's law

If Eqs. (12.16) and (12.17) are multiplied by q_e/m_e and q_i/m_i, respectively, and the results are added, it is seen from Eqs. (12.15) and (12.6) that

$$\begin{aligned} \frac{\partial}{\partial t} \mathbf{J} = -\nabla \cdot \mathbf{\Psi}_E &+ \left(\frac{N_{e0} q_e^2}{m_e} + \frac{N_{i0} q_i^2}{m_i} \right) \mathbf{E} \\ &+ \left(\frac{N_{e0} q_e^2}{m_e} \mathbf{v}_e + \frac{N_{i0} q_i^2}{m_i} \mathbf{v}_i \right) \times \mathbf{B}_0 + \frac{q_e}{m_e} \mathbf{P}_{ei} + \frac{q_i}{m_i} \mathbf{P}_{ie} \end{aligned} \tag{12.22}$$

From Eq. (12.13), it is found that

$$\rho_{m0} = N_{e0} m_e + N_{i0} m_i \tag{12.23}$$

If Eqs. (12.20) and (12.23) are solved simultaneously for N_{e0} and N_{i0}, it is obtained that

$$N_{e0} = (m_i \rho_0 - q_i \rho_{m0})/(q_e m_i - q_i m_e) \tag{12.24a}$$

and

$$N_{i0} = (-m_e \rho_0 + q_e \rho_{m0})/(q_e m_i - q_i m_e) \tag{12.24b}$$

It follows from Eqs. (12.24a) and (12.24b) that

$$\frac{N_{e0} q_e^2}{m_e} + \frac{N_{i0} q_i^2}{m_i} = \left(\frac{q_e}{m_e} + \frac{q_i}{m_i}\right)\rho_0 - \frac{q_e q_i}{m_e m_i}\rho_{m0} \tag{12.25}$$

For a macroscopically neutral plasma, $\rho_0 = 0$ and therefore Eq. (12.25) becomes

$$N_{e0} q_e^2/m_e + N_{i0} q_i^2/m_i = -(q_e q_i/m_e m_i)\rho_{m0} \tag{12.26}$$

In a similar manner, the simultaneous solution of Eqs. (12.13) and (12.15) for $\mathbf{v}_e$ and $\mathbf{v}_i$ gives

$$\mathbf{v}_e = \frac{\rho_{m0}}{m_e m_i N_{e0}(q_e/m_e - q_i/m_i)}\left[\frac{m_i}{\rho_{m0}}\mathbf{J} - q_i \mathbf{V}\right] \tag{12.27a}$$

and

$$\mathbf{v}_i = \frac{\rho_{m0}}{m_e m_i N_{i0}(q_e/m_e - q_i/m_i)}\left[-\frac{m_e}{\rho_{m0}}\mathbf{J} + q_e \mathbf{V}\right] \tag{12.27b}$$

With the help of Eqs. (12.27a) and (12.27b), it can be shown that

$$\frac{N_{e0} q_e^2}{m_e}\mathbf{v}_e + \frac{N_{i0} q_i^2}{m_i}\mathbf{v}_i = \left(\frac{q_e}{m_e} + \frac{q_i}{m_i}\right)\mathbf{J} - \frac{q_e q_i}{m_e m_i}\rho_{m0}\mathbf{V} \tag{12.28}$$

The substitution of Eqs. (12.19), (12.26), and (12.28) in Eq. (12.22) yields

$$\begin{aligned}\frac{\partial}{\partial t}\mathbf{J} = -\nabla \cdot \mathbf{\Psi}_E &+ \left(\frac{N_{e0} q_e^2}{m_e} + \frac{N_{i0} q_i^2}{m_i}\right)[\mathbf{E} + \mathbf{V} \times \mathbf{B}_0] \\ &+ \left(\frac{q_e}{m_e} + \frac{q_i}{m_i}\right)\mathbf{J} \times \mathbf{B}_0 + \left(\frac{q_e}{m_e} - \frac{q_i}{m_i}\right)\mathbf{P}_{ei}\end{aligned} \tag{12.29}$$

It is convenient to simplify Eq. (12.29) by making some additional approximations. The ion mass m_i is very much larger than the electron mass m_e with the result that the following approximations are in order:

$$q_e/m_e \pm q_i/m_i \approx q_e/m_e \tag{12.30}$$

and

$$N_{e0} q_e^2/m_e + N_{i0} q_i^2/m_i \approx N_{e0} q_e^2/m_e \tag{12.31}$$

It is assumed that the plasma is not far from thermodynamic equilibrium. Under this condition, the partial kinetic pressures of the electrons and the ions are expected to be approximately of the same order of magnitude. Then, it is seen from Eq. (12.6) that the following approximation is valid:

$$\mathbf{\Psi}_E = (q_e/m_e)\mathbf{\Psi}_e + (q_i/m_i)\mathbf{\Psi}_i \approx (q_e/m_e)\mathbf{\Psi}_e \tag{12.32}$$

With the use of Eqs. (12.30)–(12.32), Eq. (12.29) can be reduced to the following simplified form:

$$\frac{\partial}{\partial t}\mathbf{J} = -\frac{q_e}{m_e}\nabla\cdot\mathbf{\Psi}_e + \frac{N_{e0}q_e^2}{m_e}[\mathbf{E} + \mathbf{V}\times\mathbf{B}_0] + \frac{q_e}{m_e}\mathbf{J}\times\mathbf{B}_0 + \frac{q_e}{m_e}\mathbf{P}_{ei} \tag{12.33}$$

The collision term $\mathbf{P}_{ei}$ remains to be expressed in terms of the lumped macroscopic variables. Since $\rho_0 = 0$, it follows from Eqs. (12.15) and (12.20) that

$$\mathbf{J} = N_{e0}q_e(\mathbf{v}_e - \mathbf{v}_i) \tag{12.34}$$

It is reasonable to expect that the total momentum $\mathbf{P}_{ei}$ transferred to the electrons per unit volume in unit time is proportional to the relative average velocity $(\mathbf{v}_e - \mathbf{v}_i)$ and to the mass of the electrons per unit volume. Hence $\mathbf{P}_{ei}$ may be assumed to be of the form:

$$\mathbf{P}_{ei} = -N_{e0}m_e\nu_{ei}(\mathbf{v}_e - \mathbf{v}_i) \tag{12.35}$$

It is ascertained from Eq. (12.35) that ν_{ei} has the dimensions of a frequency; it is called the collision frequency for the momentum transfer to the electrons from the ions. In view of Eq. (12.34), Eq. (12.35) may be written as

$$\mathbf{P}_{ei} = -(m_e\nu_{ei}/q_e)\mathbf{J} \tag{12.36}$$

It is convenient to introduce a parameter η_0 which has the dimensions of *resistivity*. Then, in view of Eq. (12.36), $\mathbf{P}_{ei}$ may be expressed as

$$\mathbf{P}_{ei} = -N_{e0}q_e\eta_0\mathbf{J} \tag{12.37}$$

with the result that

$$\eta_0 = m_e\nu_{ei}/N_{e0}q_e^2 = 1/\sigma_0 \tag{12.38}$$

It is verified from Eq. (12.38) that η_0 and σ_0 have the dimensions of resistivity and conductivity, respectively.

The main assumption, therefore, is that $\mathbf{P}_{ei}$ is proportional to the electric current density $\mathbf{J}$. It is to be noted that both $\mathbf{P}_{ei}$ and $\mathbf{J}$ are first-order perturbations and that therefore they are linearly related. The most general linear relationship between them may be assumed in the form

$$\mathbf{P}_{ei} = -N_{e0}q_e\boldsymbol{\eta}\cdot\mathbf{J} \tag{12.39}$$

where $\boldsymbol{\eta}$ is known as the dyadic resistivity. Clearly, Eq. (12.37) is a special case of Eq. (12.39). The substitution of Eq. (12.39) in Eq. (12.33) gives

$$\frac{\partial}{\partial t}\mathbf{J} = -\frac{q_e}{m_e}\nabla \cdot \boldsymbol{\Psi}_e + \frac{N_{e0}q_e^2}{m_e}[\mathbf{E} + \mathbf{V} \times \mathbf{B}_0] + \frac{q_e}{m_e}\mathbf{J} \times \mathbf{B}_0 - \frac{N_{e0}q_e^2}{m_e}\boldsymbol{\eta} \cdot \mathbf{J} \quad (12.40)$$

For a steady current in a uniform plasma with no static magnetic field, $\partial\mathbf{J}/\partial t = 0$, $\nabla \cdot \boldsymbol{\Psi}_e = 0$, and $\mathbf{B}_0 = 0$, and the resistivity dyad $\boldsymbol{\eta}$ reduces to the scalar resistivity η_0. For this case, Eq. (12.40) yields

$$\mathbf{E} = \eta_0 \mathbf{J} \quad \text{or} \quad \mathbf{J} = \sigma_0 \mathbf{E} \quad (12.41)$$

which is the well-known Ohm's law. Consider a steady current in a uniform plasma with the static magnetic field present. Suppose that the fluid is perfectly conducting, that is, $\boldsymbol{\eta} = 0$. Then, Eq. (12.40) yields

$$\mathbf{E} + \mathbf{V} \times \mathbf{B}_0 + (1/N_{e0}q_e)\mathbf{J} \times \mathbf{B}_0 = 0 \quad (12.42)$$

The third term on the left side of Eq. (12.42) gives rise to a phenomenon called the Hall effect in magnetohydrodynamic flow problems and is therefore called the Hall effect term. If the Hall effect is negligible Eq. (12.42) becomes

$$\mathbf{E} + \mathbf{V} \times \mathbf{B}_0 = 0 \quad (12.43)$$

Thus Eqs. (12.42) and (12.43) are the Ohm's law for the steady current in a uniform, perfectly conducting plasma in a static magnetic field, respectively, with and without the Hall effect terms. Since it yields the Ohm's law for the special cases, Eq. (12.40) is called the *generalized Ohm's law*.

It is to be noted that the two equations (12.16) and (12.17) of momentum transport satisfied by the partial variables are now replaced by the equation of motion (12.21) and the generalized Ohm's law (12.40) in the treatment of the whole plasma as a conducting fluid.

References

1.1. J. L. Delcroix, *Plasma Physics*, Chapters 6, 8-10, John Wiley, New York, 1965.
1.2. R. Jancel and Th. Kahan, *Electrodynamics of Plasmas*, Chapters 5 and 6, John Wiley, New York, 1966.
1.3. S. Chapman and T. G. Cowling, *The Mathematical Theory of Nonuniform Gases*, Chapter 2, Cambridge University Press, London, 1960.
1.4. L. Spitzer, *Physics of Fully Ionized Gases*, 2nd ed., Chapter 2, Interscience Publishers, New York, 1962.

Problems

1.1. Verify that the average of the peculiar velocity of all the particles in an elemental volume $d\mathbf{r}$ is zero.

1.2. Let $\langle w_{z+} \rangle$ be the average value of the z-component of the peculiar velocity, the average being taken over all the particles for which this component is positive at a given point. Show that $\langle w_{z+} \rangle = \langle w \rangle/2$.

1.3. Show that in the Maxwellian state, the number of particles per unit volume with kinetic energies ranging between E and $E + dE$ is given by

$$f(E)\,dE = N_0 \sqrt{\frac{4}{\pi}}\,(KT)^{-3/2} \sqrt{E}\, e^{-E/KT}\, dE$$

Prove that the speed of the particles having the most probable energy is equal to $\sqrt{KT/m}$ where m is the mass of the particle.

1.4. The velocities of the particles lying in a plane ($z = 0$) are described by the distribution function

$$f(u) = N_0 \frac{m}{2\pi KT_0} e^{-m(u_x^2+u_y^2)/2KT_0}$$

where N_0 is the number of particles per unit area. Find the most probable speed of the particles.

1.5. What fraction of the number of particles in unit volume of a gas in the Maxwellian state has speeds greater than the most probable speed?

1.6. Consider the vertical distribution of neutral particles of mass m in the earth's atmosphere acted on by the conservative gravitational force. Let the origin of the space coordinates be at the earth's surface with the z-axis vertically upwards. Assume a uniform value T for the atmospheric temperature and a constant value g for the acceleration due to gravity. Deduce an expression for the number density $N(z)$ as a function of height z in terms of its value at the earth's surface.

1.7. Consider a two-dimensional gas whose particles are constrained to move in a plane. The particle velocities are characterized by a homogeneous, isotropic, two-dimensional Maxwell-Boltzmann distribution function. Show that the number of particles crossing unit length per unit time from one side only is given by $N\langle u \rangle/\pi$ where N is the number of particles per unit area.

1.8. Find an expression for the kinetic pressure for the two-dimensional gas of Problem 1.7.

1.9. A collimated beam of neutral particles is called a molecular beam and is produced by allowing the gas particles to escape through a small orifice in the walls of a container into a vacuum chamber in which the pressure is kept low by continuous pumping. The container is usually an oven maintained at high temperature in order to produce a high vapor pressure if a molecular beam of a material which is a solid at room temperature is desired. The gas particles in the oven are assumed to be in the Maxwellian state which is unperturbed as the gas escapes through the orifice.

Verify that the number of particles with speed u escaping through unit area of the orifice in unit time is equal to $uN/4$ where N is the number of particles in a unit volume (inside the oven) with speed u. Show that the root mean squared speed of the particles leaving the oven is greater than that of the particles inside the oven by the factor $\sqrt{4/3}$.

1.10. Show that the average rate of increase of momentum in an infinitesimal volume dV due to the particles entering the volume is given by $-\nabla \cdot \{Nm\mathbf{v}\mathbf{v}\}dV$ where N is the number density, m the particle mass, and $\mathbf{v}$ the average velocity. Employ the fluid dynamical method similar to that used in deducing the equation of continuity.

1.11. Consider an infinitesimal volume $dV = dx\,dy\,dz$ of a gas in the range from (x, y, z) to $(x + dx, y + dy, z + dz)$. If the components of the kinetic pressure dyad $\boldsymbol{\Psi}$ are interpreted as forces acting on unit area in accordance with a scheme indicated by the subscripts, as explained in Sec. 1.9, show that a spatial variation of $\boldsymbol{\Psi}$ results in a force $-(\nabla \cdot \boldsymbol{\Psi})\,dV$ acting on the infinitesimal volume dV.

1.12. The momentum of the particles inside an infinitesimal volume dV of a gas of charged particles changes due to (i) the momentum carried by the particles entering dV, (ii) the spatial variation of the kinetic pressure dyad, (iii) the Lorentz force acting on the particles in dV, and (iv) the momentum interchange on account of collisional interactions with other species of particles. The change in momentum of the particles in dV in unit time due to all these four factors results in an instantaneous rate of change of momentum of the particles in dV. Hence, with the help of the results of Problems 1.10 and 1.11, deduce the momentum transport equation as given by (9.14).

1.13. The distribution of the peculiar velocities of the particles are given by the following modified Maxwell-Boltzmann distribution function:

$$f(\mathbf{u}) = N\left(\frac{m}{2\pi K T_\perp}\right)\left(\frac{m}{2\pi K T_\parallel}\right)^{1/2} \exp\left[\frac{-m}{2K}\left\{\frac{u_x^2 + u_y^2}{T_\perp} + \frac{u_z^2}{T_\parallel}\right\}\right]$$

Verify that N represents correctly the number density of particles. Deduce the kinetic pressure dyad and show that it has only diagonal components two of which are equal. This type of pressure is known as uniaxially anisotropic pressure and indicates the presence of an anisotropic axis in the z-direction.

1.14. Let $g(\mathbf{r}, \mathbf{u}, t)$ be any property of the particles and as indicated, let it be a function of the position $\mathbf{r}$, the velocity $\mathbf{u}$ and the time t. Deduce the following result by multiplying the Boltzmann equation by $g(\mathbf{r}, \mathbf{u}, t)\,d\mathbf{u}$ and integrating throughout the velocity space:

$$\begin{aligned}
&\frac{\partial}{\partial t}\{N(\mathbf{r}, t)\langle g(\mathbf{r}, \mathbf{u}, t)\rangle\} - N(\mathbf{r}, t)\left\langle \frac{\partial}{\partial t} g(\mathbf{r}, \mathbf{u}, t)\right\rangle \\
&\quad + \nabla_r \cdot \{N(\mathbf{r}, t)\langle \mathbf{u} g(\mathbf{r}, \mathbf{u}, t)\rangle\} \\
&\quad - N(\mathbf{r}, t)\langle(\mathbf{u} \cdot \nabla_r) g(\mathbf{r}, \mathbf{u}, t)\rangle - N(\mathbf{r}, t)\left\langle \nabla_u \cdot \left\{\frac{\mathbf{F}}{m} g(\mathbf{r}, \mathbf{u}, t)\right\}\right\rangle \\
&\quad = \int g(\mathbf{r}, \mathbf{u}, t)(\partial f/\partial t)_{\text{coll}}\, d\mathbf{u}
\end{aligned}$$

Assume that **F** is the Lorentz force due to the electric and the magnetic fields. Verify that the above result reduces to Eq. (8.8) if $g(\mathbf{r}, \mathbf{u}, t)$ is independent of **r** and t.

1.15. Establish the following relation by setting $g = m w_x w_y$ in Eq. (10.1):

$$\frac{d}{dt}\left(\frac{P_{xy}}{N}\right) + \frac{P_{xy}}{N}\left(\frac{\partial v_x}{\partial x} + \frac{\partial v_y}{\partial y}\right) + \frac{P_{xx}}{N}\frac{\partial v_y}{\partial x} + \frac{P_{yy}}{N}\frac{\partial v_x}{\partial y}$$

$$+ \frac{P_{xz}}{N}\frac{\partial v_y}{\partial z} + \frac{P_{yz}}{N}\frac{\partial v_x}{\partial z} + \frac{1}{N}(\nabla \cdot \mathbf{Q})_{xy} = \frac{F_{xy}}{N} + \frac{R_{xy}}{N}$$

where **Q**, R_{ij}, and F_{xy} are given in Eqs. (10.5), (10.29), and (10.40) respectively. Show that the above relation can also be expressed as follows:

$$\frac{d}{dt}\left(\frac{P_{xy}}{N}\right) + \frac{1}{N}\{(\boldsymbol{\Psi} \cdot \nabla \mathbf{v})_{xy} + (\boldsymbol{\Psi} \cdot \nabla \mathbf{v})^t_{xy}\}$$

$$+ \frac{q}{|q|}\frac{\omega_c}{N}\{(\hat{\mathbf{e}}_B \times \boldsymbol{\Psi})_{xy} + (\hat{\mathbf{e}}_B \times \boldsymbol{\Psi})^t_{xy}\} + \frac{1}{N}(\nabla \cdot \mathbf{Q})_{xy} = \frac{R_{xy}}{N}$$

Hence, prove that Eqs. (10.31) and (10.33)–(10.37) can be succinctly written in the following dyadic notation

$$\frac{d}{dt}\left(\frac{\boldsymbol{\Psi}}{N}\right) + \frac{1}{N}\{(\boldsymbol{\Psi} \cdot \nabla \mathbf{v}) + (\boldsymbol{\Psi} \cdot \nabla \mathbf{v})^t\}$$

$$+ \frac{q}{|q|}\frac{\omega_c}{N}\{(\hat{\mathbf{e}}_B \times \boldsymbol{\Psi}) + (\hat{\mathbf{e}}_B \times \boldsymbol{\Psi})^t\} + \frac{1}{N}(\nabla \cdot \mathbf{Q}) = \frac{\mathbf{R}}{N}$$

1.16. Verify the following relation by actual expansion:

$$\nabla \cdot (\boldsymbol{\Psi} \cdot \mathbf{v}) = (\boldsymbol{\Psi} \cdot \nabla) \cdot \mathbf{v} + (\nabla \cdot \boldsymbol{\Psi}) \cdot \mathbf{v}$$

1.17. Consider an infinitesimal volume $dV = dx\,dy\,dz$ of a gas in the range from (x, y, z) to $(x + dx, y + dy, z + dz)$ and let this elemental volume dV be assumed to move uniformly with the average particle velocity **v**. If $\hat{\mathbf{n}}$ is a unit outwardly drawn normal to the surface bounding the elemental volume dV, then $-\hat{\mathbf{n}} \cdot \boldsymbol{\Psi}$ is the force acting on unit area due to the actual particle motions. Show that as a result of the work done by the dyadic kinetic pressure $\boldsymbol{\Psi}$ on the particles, the average rate of increase of the energy of the particles inside dV is equal to $-\nabla \cdot (\boldsymbol{\Psi} \cdot \mathbf{v})\, dV$.

1.18. Deduce the value of the thermal energy flux density triad **Q** for a gas of particles in the Maxwellian state.

1.19. Find the number of independent elements in the thermal energy flux density triad **Q** by noting that Q_{ijk} is symmetric under exchange of any two of its three indices.

1.20. Let $\mathbf{F}_x$ be the force acting on a unit area normal to the x-coordinate due to the random thermal motions of the particles and let $\mathbf{F}_y$ and $\mathbf{F}_z$ be similarly defined. Then, $\mathbf{F}_x$, $\mathbf{F}_y$ and $\mathbf{F}_z$ written as a column matrix can be expressed as follows:

$$[\mathbf{F}] = \begin{bmatrix} \mathbf{F}_x \\ \mathbf{F}_y \\ \mathbf{F}_z \end{bmatrix} = -\begin{bmatrix} P_{xx} & P_{xy} & P_{xz} \\ P_{yx} & P_{yy} & P_{yz} \\ P_{zx} & P_{zy} & P_{zz} \end{bmatrix} [\hat{\mathbf{n}}] \qquad [\hat{\mathbf{n}}] = \begin{bmatrix} \hat{\mathbf{x}} \\ \hat{\mathbf{y}} \\ \hat{\mathbf{z}} \end{bmatrix}$$

Let the original coordinates x, y and z be rotated through an arbitrary angle θ about the z-axis to obtain the new coordinates x', y' and z. A rotational operator $[R]$ is defined as follows:

$$[R] = \begin{bmatrix} \cos\theta & -\sin\theta & 0 \\ \sin\theta & \cos\theta & 0 \\ 0 & 0 & 1 \end{bmatrix}$$

Then, show that

$$[\mathbf{F}'] = \begin{bmatrix} \mathbf{F}_{x'} \\ \mathbf{F}_{y'} \\ \mathbf{F}_z \end{bmatrix} = [R]^{-1}[\mathbf{F}] \qquad [\hat{\mathbf{n}}'] = \begin{bmatrix} \hat{\mathbf{x}}' \\ \hat{\mathbf{y}}' \\ \hat{\mathbf{z}} \end{bmatrix} = [R]^{-1}[\hat{\mathbf{n}}]$$

Find the matrix relation connecting $[\mathbf{F}']$ and $[\hat{\mathbf{n}}']$. Assume that there is rotational symmetry about the z-axis with the result that $\mathbf{F}_{x'} = \mathbf{F}_x$ and $\mathbf{F}_{y'} = \mathbf{F}_y$. Hence, prove that the existence of rotational symmetry about the z-axis leads to the following conditions:

$$P_{xx} = P_{yy} \qquad P_{xy} = P_{yx} = P_{yz} = P_{zy} = P_{zx} = P_{xz} = 0$$

by making use of the fact that $P_{ij} = P_{ji}$.

1.21. Consider a fully ionized plasma consisting of electrons and singly charged hydrogen ions. Let m_e and m_i be the mass of an electron and an ion, respectively. The electron and the ion temperatures are each equal to T. The electron gas and the ion gas are each assumed to have no drift velocity. Both the gases have a Maxwellian distribution of velocities. Show that the square of the relative speed g between the two species of particles when averaged over both their velocity distributions is equal to $3KT/m_r$ where $m_r = m_e m_i/(m_e + m_i)$ and K is the Boltzmann constant.

1.22. The velocity distribution function of a system of particles which are uniformly distributed in space is a constant for $|u_x| \leqslant a$, $|u_y| \leqslant a$, $|u_z| \leqslant a$ and is zero for the components of particle velocity outside of the indicated ranges. Find the value of the distribution function in terms of N the number density.

Find the kinetic temperature T in terms of the particle mass m and the Boltzmann constant K.

What is the value of the vector flux density of heat?

1.23. Consider a gas whose particles are constrained to move in the x-direction. The particle velocities are characterized by a homogeneous, velocity distribution function which is proportional to $1/(u_x^2 + a_r^2)$. Find the constant of proportionality in terms of a_r and the number density N.

Find the average kinetic energy density. Is the given velocity distribution function physically realistic?

1.24. Show that the total number of particles crossing unit area per second lying within an element $d\Omega$ of solid angle is $(N/\pi)(KT/2\pi m)^{1/2}\cos\theta\, d\Omega$, where θ is the angle between the normal to the area and the direction of the solid angle. Also, m is the particle mass, N the number density of the particles, T the kinetic temperature and K the Boltzmann constant. Assume that the particle velocities are described by the Maxwell-Boltzmann distribution function.

1.25. Consider a unit area perpendicular to the x-axis in a gas whose particle velocities are characterized by the Maxwell-Boltzmann distribution function. Show that the fraction of the particles that cross the unit area in unit time from one side only, having the velocity components in the ranges du_x, du_y, and du_z, is given by

$$(1/2\pi)(m/KT)^2 u_x \exp[-m(u_x^2 + u_y^2 + u_z^2)/2KT]\, du_x\, du_y\, du_z$$

All the parameters have the usual meaning.

1.26. Consider a gas of electrons whose kinetic pressure P is a scalar quantity. Assume that there is neither annihilation nor creation of electrons. Multiply the Boltzmann equation by $mw^2/2$, where m is the electron mass and w its peculiar velocity, and integrate throughout the velocity space to obtain directly the following energy transport equation:

$$N(d/dt)(m\langle w^2\rangle/2) + P\nabla \cdot \mathbf{v} + \nabla \cdot \mathbf{q} = r \tag{A}$$

where N is the number density and $\mathbf{v}$ is the average velocity of the electrons. Also, r and $\mathbf{q}$ are the same as defined by Eqs. (10.48) and (10.50), respectively.

Recast (A) into the form

$$(d/dt)(PN^{-5/3}) = \tfrac{2}{3}N^{-5/3}[r - \nabla \cdot \mathbf{q}] \tag{B}$$

and, stating the assumptions, deduce the adiabatic gas law for the spherically symmetric three-dimensional compression.

1.27. In Problem 1.26, assume that the kinetic pressure is a constant, the average velocity $\mathbf{v}$ is equal to zero and there is no energy interchange due to the collisional interactions with the other species of particles. In Problem 6.22, it is shown that

$$\mathbf{q} = -\mathcal{K}\nabla T \qquad \mathcal{K} = 5KP/2m\nu$$

where $\mathcal{K}$ is the thermal conductivity, K is the Boltzmann constant and ν is the constant collision frequency. Note that it has been assumed that the temperature T is a function of position in the electron gas. Use (A) of Problem 1.26 and deduce the following *heat-conduction equation*:

$$NK(\partial/\partial t)T = \tfrac{2}{3}\mathcal{K}\nabla^2 T$$

CHAPTER 2

Basic Plasma Phenomena

2.1. Introduction

A number of important plasma phenomena can be studied by treating the plasma as a whole or any of its constituents as a fluid characterized by a certain number of macroscopic variables such as the mass density, the average velocity, the kinetic pressure, and the temperature. It was pointed out that a set of moments of the Boltzmann equation appropriately truncated and suitably completed by an assumption on the highest velocity moment of the distribution function enables the determination of the equations governing these macroscopic variables. The electrons and the ions form conducting fluids with the result that there are also macroscopic electromagnetic variables such as the charge density, the current density, the electric field, and the magnetic flux density. These electromagnetic variables are governed by the well-known Maxwell's equations. The behavior of the plasma can be analyzed in terms of the hydrodynamical and Maxwell's equations. However, a treatment of some of the plasma phenomena requires the use of the velocity distribution function. In this chapter, a variety of basic plasma phenomena is investigated largely with the help of the hydrodynamical and Maxwell's equations.

A fundamental property of a plasma is its tendency to preserve its macroscopic electrical neutrality. When this neutrality is disturbed, the electrons break into collective, high-frequency oscillations and the heavy positive ions remain immobile and provide the necessary neutralizing background. On an average over the time period of these high-frequency oscillations, the macroscopic electrical neutrality is maintained. A treatment of these electron plasma oscillations based on the cold plasma model is first given. If an electric field due to a test charge is imposed on a plasma, the plasma particles redistribute themselves around the test charge in such a manner as to shield its influence from a major portion of the plasma. This so-called Debye shielding is discussed and an expression is obtained for the Debye length which is the scale length over which significant departure from macroscopic electrical neutrality can occur in a plasma. When an external body is immersed in a plasma, it develops a negative potential which falls off very rapidly to the zero plasma potential in a distance of the order of a Debye length. This phenomenon is called the plasma sheath. A simple analysis of the structure of the plasma sheath as well as an estimate of the negative floating potential on the surface of the external body are given. The theory of the plasma sheath is then adapted to the treatment of the Langmuir probe which is very commonly used for the determination of the temperature and the number density of the electrons in a plasma.

The generalization of the Poynting vector for a warm plasma and the concept of the group velocity are developed as these background material is required for the further treatment of the theory of the plasma oscillations. The effect of thermal motions on the electron plasma oscillations is then analyzed from the viewpoint of the hydrodynamical equations. The theory of the plasma oscillations which includes the motion of both the constituents of a plasma is then presented and in this presentation emphasis is placed on the two low frequency phenomena, namely, the ion acoustic wave and the ion plasma oscillations.

The linearized magnetohydrodynamic equations are summarized and the various approximations implicit in the MHD equations are enumerated. The concepts of magnetic pressure and the freezing of magnetic flux lines in a perfectly conducting fluid are then developed as simple applications of the MHD equations. The confinement of high temperature plasma is an important branch of plasma physics in view of its application to controlled thermonuclear reactors. A sufficiently detailed theory of both the equilibrium and the dynamic pinch effects is given. The pinched plasma column, like the other plasma configurations, suffers from two kinds of instabilities. One is the configuration-space instability which depends on the geometrical structure of the plasma device and the other is the velocity-space instability which is governed by the forms of the velocity distribution functions of the plasma particles. This chapter is concluded with a qualitative discussion of the configuration-space instability with particular emphasis on the pinched plasma column and an introductory study of the velocity-space instability directed towards the elucidation of the underlying physical mechanism.

2.2. Electron plasma oscillations

Maxwell's equations

The equations satisfied by the fluid-dynamical variables such as the number density, the average velocity, and the kinetic pressure of each species of particles in a plasma were deduced in the previous chapter. Since the particles are charged, there are also associated with the dynamical motions of the plasma particles macroscopic variables such as the electric charge density ρ and the electric current density $\mathbf{J}$ which are electromagnetic in character. The definitions of ρ and $\mathbf{J}$ are reproduced from Eqs. (1.12.2) and (1.12.15) as

$$\rho = N_e q_e + N_i q_i \tag{2.1}$$

and

$$\mathbf{J} = N_{e0} q_e \mathbf{v}_e + N_{i0} q_i \mathbf{v}_i \tag{2.2}$$

where $\mathbf{J}$ is the linearized form of the electric current density. For convenience, the equation (1.12.14) embodying the principle of conservation of electric charge is also reproduced:

$$(\partial/\partial t)\rho + \nabla \cdot \mathbf{J} = 0 \tag{2.3}$$

The time-varying charge and current densities give rise to electromagnetic fields which satisfy the following Maxwell's equations:

$$\nabla \times \mathbf{E} = -\mu_0 \, \partial \mathbf{H}/\partial t \tag{2.4}$$

$$\nabla \times \mathbf{H} = \varepsilon_0 \, \partial \mathbf{E}/\partial t + \mathbf{J} \tag{2.5}$$

$$\nabla \cdot \mathbf{E} = \rho/\varepsilon_0 \tag{2.6}$$

$$\nabla \cdot \mathbf{H} = 0 \tag{2.7}$$

where μ_0 and ε_0 are the vacuum permeability and permittivity, respectively. The electromagnetic fields obtained from Eqs. (2.4)–(2.7) should be the same as those used for the Lorentz force in the Boltzmann equation. In other words, the electromagnetic fields are to be derived in a self-consistent manner. The plasma dynamics is governed by Eqs. (2.1)–(2.7) and the equations satisfied by the fluid-dynamical variables in Eqs. (2.1) and (2.2). As a simple example of the application of the equations of plasma dynamics, the theory of the electron plasma oscillations is treated in this section. Before proceeding with the theoretical treatment, it is instructive to consider a descriptive account of these oscillations.

Description of the plasma oscillations

The fundamental property of a plasma is its tendency to remain electrically neutral on a macroscopic scale. In the equilibrium state, let the plasma be uniform and at rest. Let every elemental volume in the plasma contain as many electrons as there are ions, that is, the plasma is macroscopically neutral in the equilibrium state. If there is a significant imbalance of charge, large electrostatic (Coulomb) forces are brought into play and these forces which cannot be ordinarily sustained in a plasma are responsible for the electron plasma oscillations. These oscillations enable the plasma to retain its electrical neutrality on an average over a short period of time. Suppose that in a small spherical region as shown in Fig. 2.1, a perturbation in the form of an excess of negative charge is introduced. It can be shown from Gauss's law (2.6) that the macroscopic electric field due to the resulting electric charge is

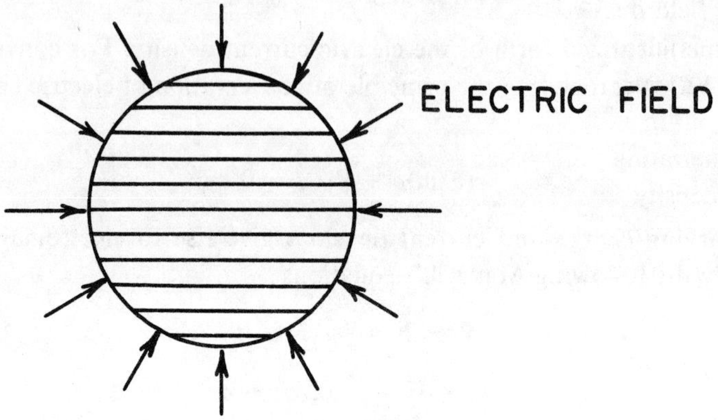

Fig. 2.1. Electric field produced by a spherically symmetric distribution of excess of negative charge represented by the shaded region.

radial because of the spherical symmetry and points towards the center of the spherical region. The electric field forces the electrons to move radially outwards. Since the electrons gain kinetic energy in the course of their motion, on account of their momentum, after a certain time, more electrons leave the spherical region than is necessary to resume a state of electrical neutrality. Now the spherical region has an excess of positive charge and therefore the direction of the electric field is reversed. The electrons are caused to move radially inward. This movement of electrons back and forth radially continues resulting in electron plasma oscillations. These are usually very rapid oscillations. The total charge inside the spherical region averaged over a period of these oscillations is zero. Thus the plasma preserves its electrical neutrality in the mean. In this discussion, the ions have been assumed to be immobile. This assumption is justifiable since the frequency of electron oscillations is very high and the ions in view of their heavy mass are unable to follow the rapidity of these oscillations.

Characteristics of the plasma oscillations

The properties of the electron plasma oscillations can be deduced with the help of the equations of plasma dynamics. As before, it is assumed that the ions are immobile. The ion density N_0 is constant in time and uniform in space. The equilibrium electron density is also equal to N_0 and let the perturbation of the electron density be given by $N(\mathbf{r}, t)$. It is assumed that the perturbation is very small in comparison with the ambient value. In a similar manner, it is assumed that the electric field $\mathbf{E}(\mathbf{r}, t)$ and the average velocity $\mathbf{v}(\mathbf{r}, t)$ are first-order perturbations with the result that it is reasonable to use the linearized form of the hydrodynamical equations. The cold plasma model is used; the thermal motions of the electrons and the kinetic pressure dyad $\boldsymbol{\Psi}$ are not taken into account. It is assumed that there is no magnetic field due to external sources. The magnetic field due to the motion of the electrons is also a first-order perturbation and it is multiplied by the average velocity in the Lorentz force term. Consequently, this term disappears in the process of linearization. In addition, it is assumed that the electron gas and the ion gas form two interpenetrating but noninteracting fluids. Therefore, the rate of loss of momentum density from the electron fluid is zero.

Under these various approximations, the linearized forms of the equation of continuity (1.12.10) and the momentum transport equation (1.12.16) become

$$(\partial/\partial t)N(\mathbf{r}, t) + N_0\{\nabla \cdot \mathbf{v}(\mathbf{r}, t)\} = 0 \tag{2.8}$$

$$(\partial/\partial t)\mathbf{v}(\mathbf{r}, t) = -(e/m_e)\mathbf{E}(\mathbf{r}, t) \tag{2.9}$$

where the electronic charge is set equal to $-e$. Since the ionic charge is equal to e, it follows from Eqs. (2.1) and (2.6) that

$$\rho(\mathbf{r}, t) = -[N_0 + N(\mathbf{r}, t)]e + N_0 e = -eN(\mathbf{r}, t) \tag{2.10}$$

and

$$\nabla \cdot \mathbf{E}(\mathbf{r}, t) = -(e/\varepsilon_0)N(\mathbf{r}, t) \tag{2.11}$$

If Eq. (2.8) is differentiated with respect to time and Eq. (2.9) is used, it can be shown that

$$(\partial^2/\partial t^2)N(\mathbf{r}, t) - (N_0 e/m_e)\nabla \cdot \mathbf{E}(\mathbf{r}, t) = 0 \tag{2.12}$$

Together with Eq. (2.11), Eq. (2.12) can be reduced to the form

$$(\partial^2/\partial t^2)N(\mathbf{r}, t) + \omega_{pe}^2 N(\mathbf{r}, t) = 0 \tag{2.13}$$

where

$$\omega_{pe} = \sqrt{N_0 e^2/m_e \varepsilon_0} \tag{2.14}$$

It is seen from Eq. (2.13) that the perturbation number density of the electrons has a simple harmonic variation with time. In the phasor notation, the solution of Eq. (2.13) is given by

$$N(\mathbf{r}, t) = N(\mathbf{r})\exp(-i\omega_{pe} t) \tag{2.15}$$

where ω_{pe} is called the electron plasma angular frequency.

It can be verified that all the first-order perturbations vary with time at the electron plasma angular frequency ω_{pe}. For this purpose, it is convenient to begin with the assumption that all the field quantities have a harmonic time variation as expressed by

$$f(\mathbf{r}, t) = f(\mathbf{r})\exp(-i\omega t) \tag{2.16}$$

where $f(\mathbf{r}, t)$ is any field quantity. Then, Eqs. (2.8), (2.9), and (2.11) can be reduced to yield

$$N(\mathbf{r}) = (-i/\omega)N_0 \nabla \cdot \mathbf{v}(\mathbf{r}) \tag{2.17}$$

$$\mathbf{v}(\mathbf{r}) = (-ie/\omega m_e)\mathbf{E}(\mathbf{r}) \tag{2.18}$$

and

$$\nabla \cdot \mathbf{E}(\mathbf{r}) = -(e/\varepsilon_0)N(\mathbf{r}) \tag{2.19}$$

With the help of Eqs. (2.17) and (2.18), Eq. (2.19) can be expressed in the form

$$(1 - \omega_{pe}^2/\omega^2)\{\nabla \cdot \mathbf{E}(\mathbf{r})\} = 0 \tag{2.20}$$

The nontrivial solution of Eq. (2.20) gives $\omega = \omega_{pe}$ which is equivalent to the statement that all the perturbations have a harmonic time variation at the electron plasma angular frequency ω_{pe}.

It is found from Eq. (2.18) that the electron velocity and hence electron motion are parallel to the electric field. Therefore, the electron plasma oscillations are longitudinal in character. From Eqs. (2.2) and (2.18), the current density is obtained as

$$\mathbf{J}(\mathbf{r}) = -N_0 e\mathbf{v}(\mathbf{r}) = (iN_0 e^2/\omega m_e)\mathbf{E}(\mathbf{r}) \tag{2.21}$$

The right side of Eq. (2.5) is $-i\omega\varepsilon_0 \mathbf{E}(\mathbf{r}) + \mathbf{J}(\mathbf{r})$ which together with Eqs. (2.14) and (2.21) can be written as

$$-i\omega\varepsilon_0 \mathbf{E}(\mathbf{r}) + \mathbf{J}(\mathbf{r}) = -i\omega\varepsilon_0[1 - \omega_{pe}^2/\omega^2]\mathbf{E}(\mathbf{r}) \tag{2.22}$$

For harmonic time variation as expressed in Eq. (2.16), Eqs. (2.4) and (2.5) can be simplified with the help of Eq. (2.22) to yield

$$\nabla \times \mathbf{E}(\mathbf{r}) = i\omega\mu_0 \mathbf{H}(\mathbf{r}) \tag{2.23}$$

$$\nabla \times \mathbf{H}(\mathbf{r}) = -i\omega\varepsilon_0 \varepsilon_r \mathbf{E}(\mathbf{r}) \tag{2.24}$$

where

$$\varepsilon_r = 1 - \omega_{pe}^2/\omega^2 \tag{2.25}$$

Under the various approximations stated previously, the plasma is seen from Eqs.

(2.23) and (2.24) to behave like an ordinary dielectric with a relative permittivity as given by Eq. (2.25). For the electron plasma oscillations, $\omega = \omega_{pe}$, $\varepsilon_r = 0$, and therefore Eq. (2.24) reduces to

$$\nabla \times \mathbf{H}(\mathbf{r}) = 0 \tag{2.26}$$

Since $\nabla \times (\nabla a) = 0$ for any scalar function a, it follows that the magnetic field can be sought in the form

$$\mathbf{H}(\mathbf{r}) = -\nabla\psi(\mathbf{r}) \tag{2.27}$$

where $\psi(\mathbf{r})$ is a scalar potential. Also, since $\nabla \cdot (\nabla \times \mathbf{A}) = 0$ for any vector function $\mathbf{A}$, it can be deduced with the help of Eqs. (2.23) and (2.27) that

$$\nabla \cdot \{\nabla\psi(\mathbf{r})\} = \nabla^2\psi(\mathbf{r}) = 0 \tag{2.28}$$

The only solution of Laplace's equation which is not singular at infinity is given by $\psi(\mathbf{r})$ = constant. Then Eq. (2.27) gives that

$$\mathbf{H}(\mathbf{r}) = 0 \tag{2.29}$$

The magnetic field associated with the electron plasma oscillations is seen from Eq. (2.29) to be zero with the result that these oscillations are electrostatic in character. Moreover, the perturbation number density and all the associated field variables at every point in space change with time in phase and there is no relative phase variation from point to point in space. This implies the absence of any wave propagation. Hence the electron plasma oscillations are also stationary oscillations. Summing up, the electron plasma oscillations are *longitudinal, electrostatic*, and *stationary*.

In the study of plasma phenomena whose time scales are much larger than the period of electron plasma oscillations, the net electric charge density ρ in Eq. (2.6) can be set equal to zero in the first approximation. As an example for the numerical value of the plasma frequency, consider a dense laboratory plasma for which the number density is approximately $N_0 = 10^{22}$ per cubic meter. By substituting the numerical values of the electron charge and mass as well as the vacuum permittivity, the plasma frequency is obtained as $f_{pe} = \omega_{pe}/2\pi \simeq 10^6$ MHz which is in the far infrared region.

2.3. The Debye length

If the equilibrium state in which the plasma is electrically neutral on a macroscopic scale is perturbed, the electron plasma oscillations were shown to be excited as a result of the effort of the plasma to assert its neutrality. It was established that on an average over the short time period of these high frequency oscillations, the macroscopic electrical neutrality of the plasma is preserved. Since there are no charge separations, in the equilibrium state, the plasma cannot sustain macroscopic electric fields. Suppose that the equilibrium state of the plasma is disturbed by the imposition of an electric field due to an external charged particle. This electric field may also be due to one of the plasma particles isolated for observation. The source of the electric field is called a test particle for convenience. For the sake of definiteness, the test particle is assumed to be a positive ion of charge $+q$. It is desired now to examine the mechanism by which the plasma strives to re-establish its macroscopic electrical neutrality in the presence of this disturbing electric field.

Correlations

The plasma state has so far been described by the velocity distribution functions associated individually with the different kinds of plasma particles. This description does not take into account the correlations that exist between the positions of the electrons and the ions. The interaction between the charged particles enters only through the collision terms that appear on the right side of the Boltzmann equation. This description is inadequate for the following reasons: Since the test particle attracts the electrons and repels the ions, it follows that the test particle influences the nature of the distribution of the particles in its close neighborhood. In other words, there is correlation between the positions of the charged particles. The existence of this correlation can be taken into account by the introduction of the joint velocity distribution function which gives the probability of finding simultaneously a particle of one kind in the range $d\mathbf{r}_1$ at $\mathbf{r}_1$ and $d\mathbf{u}_1$ at $\mathbf{u}_1$, and a particle of another kind in the range $d\mathbf{r}_2$ at $\mathbf{r}_2$ and $d\mathbf{u}_2$ at $\mathbf{u}_2$ at a given instant of time. The joint velocity distribution functions are conceptually difficult and consequently it is proposed to be content with a somewhat crude treatment of the correlations between the positions of the electrons and the ions.

The Debye potential

The test particle is assumed to coincide with the origin of a spherical coordinate system. The electrostatic potential $\phi(\mathbf{r})$ due to the combined effects of the test particle and the equilibrium distribution of the particles surrounding it is evaluated. At a large distance from the test particle, the electrostatic potential vanishes and therefore the average number densities of the electrons and the ions are each equal to N_0. Near the origin where there is an ion of charge $+q$, the number densities $N_e(\mathbf{r})$ of the electrons and $N_i(\mathbf{r})$ of the ions vary slightly. Suppose that the electrons and the ions are in thermodynamic equilibrium at the same temperature T under the action of the conservative electric field associated with the electrostatic potential $\phi(\mathbf{r})$. Since the electronic and the ionic charges are equal to $-e$ and e, respectively, it follows from Eq. (1.5.8) that

$$N_e(\mathbf{r}) = N_0 e^{e\phi(\mathbf{r})/KT} \tag{3.1}$$

and

$$N_i(\mathbf{r}) = N_0 e^{-e\phi(\mathbf{r})/KT} \tag{3.2}$$

It is assumed that the perturbing electrostatic potential is weak so that

$$e\phi(\mathbf{r})/KT \ll 1 \tag{3.3}$$

The approximation (3.3) enables Eqs. (3.1) and (3.2) to be simplified as

$$N_e(\mathbf{r}) = N_0[1 + e\phi(\mathbf{r})/KT] \tag{3.4}$$

and

$$N_i(\mathbf{r}) = N_0[1 - e\phi(\mathbf{r})/KT] \tag{3.5}$$

The macroscopic electric charge density ρ is deduced with the help of Eqs. (2.1), (3.4), and (3.5) as

$$\rho(\mathbf{r}) = -N_0 e[1 + e\phi(\mathbf{r})/KT] + N_0 e[1 - e\phi(\mathbf{r})/KT] = -2N_0 e^2\phi(\mathbf{r})/KT \tag{3.6}$$

In view of the following relation between the electric field and the scalar potential,

$$\mathbf{E}(\mathbf{r}) = -\nabla\phi(\mathbf{r}) \tag{3.7}$$

Eq. (2.6) gives Poisson's equation

$$\nabla^2\phi(\mathbf{r}) = -\rho(\mathbf{r})/\varepsilon_0 \tag{3.8}$$

which enables the determination of the scalar potential. There is spherical symmetry and the scalar potential depends only on the radial distance r from the position of the test particle. Therefore, it is found from Eqs. (3.6) and (3.8) that

$$(1/r^2)(d/dr)[r^2(d/dr)\phi(r)] = 2\phi(r)/\lambda_D^2 \tag{3.9}$$

where

$$\lambda_D = \sqrt{\varepsilon_0 KT/N_0 e^2} = (1/\omega_{pe})\sqrt{KT/m_e} \tag{3.10}$$

Note that Eq. (2.14) has been used in obtaining the right side of Eq. (3.10).

In the very close neighborhood of the test particle, the scalar potential should be the same as that for an isolated particle in free space. For this case, the scalar potential can be determined as follows. From the symmetry considerations, the electric field is seen to be directed radially outward from the test particle. If Eq. (2.6) is integrated throughout a very small spherical volume of radius r and centered on the location of the test particle and if the divergence theorem is used to simplify the result, it follows that the total outward electric flux $4\pi r^2 \varepsilon_0 E_r(r)$ through the bounding surface of the spherical region is numerically equal to the total electric charge q enclosed by the surface. Therefore, together with Eq. (3.7), it follows that

$$E_r(r) = -d\phi(r)/dr = q/4\pi\varepsilon_0 r^2 \tag{3.11}$$

which yields the Coulomb potential $\phi_c(r)$ due to the isolated particle as

$$\phi_c(r) = q/4\pi\varepsilon_0 r \tag{3.12}$$

Consequently the solution of Eq. (3.9) is sought in the form

$$\phi(r) = \phi_c(r)f(r) = (q/4\pi\varepsilon_0 r)f(r) \tag{3.13}$$

where, from what has been said previously, it is necessary that

$$f(r) \to 1 \quad r \to 0 \tag{3.14}$$

Also, the potential $\phi(r)$ is required to vanish at infinity, that is,

$$\phi(r) \to 0 \qquad \text{as } r \to \infty \tag{3.15}$$

The substitution of Eq. (3.13) in Eq. (3.9) may be shown to yield the following differential equation specifying $f(r)$:

$$d^2 f(r)/dr^2 = (\sqrt{2}/\lambda_D)^2 f(r) \tag{3.16}$$

Of the following two solutions which fulfill the condition (3.14), only the former is seen from Eq. (3.13) to satisfy the requirement (3.15):

$$f(r) = \exp(-\sqrt{2}\, r/\lambda_D) \qquad \text{or} \qquad \exp(\sqrt{2}\, r/\lambda_D) \tag{3.17}$$

Hence together with Eqs. (3.13) and (3.17), the solution of Eq. (3.9) is found to be

$$\phi(r) = (q/4\pi\varepsilon_0 r)\exp(-\sqrt{2}\, r/\lambda_D) \tag{3.18}$$

This nonrigorous derivation was first given by Debye and Hückel in their theory of electrolytes and Eq. (3.18) is commonly known as the *Debye potential.*

The presence of the $1/r$ term in Eq. (3.18) makes the Debye potential very large near the location of the test particle and hence the condition (3.3) is unlikely to be fulfilled for r tending to zero. This is not a serious limitation. The main purpose of the present development is to find the order of magnitude of the distance from the test particle where the potential becomes vanishingly small. At the distances where $\phi(r)$ is insignificantly small, the validity of the condition (3.3) is unimpaired and the conclusions drawn from Eq. (3.18) are legitimate.

In deducing the expression for the Debye potential, it is usual to ignore the motions of the ions and assume for the number density of ions a constant value N_0 equal to the unperturbed number density of the electrons. Under this assumption Eqs. (3.2) and (3.5) both reduce to $N_i(\mathbf{r}) = N_0$, and, the factor 2 disappears on the right sides of Eqs. (3.6) and (3.9). The Debye potential is then verified to be given by

$$\phi(r) = (q/4\pi\varepsilon_0 r)\exp(-r/\lambda_D) \tag{3.19}$$

which is the form in which it appears extensively in the literature.

The Debye sphere

Together with Eqs. (3.10) and (3.18), Eq. (3.6) becomes

$$\rho(r) = -(q/2\pi r\lambda_D^2)\exp(-\sqrt{2}\, r/\lambda_D) \tag{3.20}$$

The total charge q_t is obtained by integrating Eq. (3.20) throughout all space:

$$q_t = \int_0^\infty \rho(r)4\pi r^2\, dr = -2q\lambda_D^{-2} \int_0^\infty r\exp(-\sqrt{2}\, r/\lambda_D)\, dr \tag{3.21}$$

There is a spherical symmetry in the distribution of the charge as shown by Eq. (3.20). If the variable of integration is changed to $y = \sqrt{2}\, r/\lambda_D$, Eq. (3.21) simplifies to

$$q_t = -q \int_0^\infty ye^{-y}\, dy = -q \tag{3.22}$$

where the right side is obtained by an integration by parts. The charge $+q$ of the test particle is seen from Eq. (3.22) to be annulled by the resultant charge due to the plasma particles. Since for large r, $\rho(r)$ falls off exponentially with increasing distance, it follows from Eq. (3.20) that the principal contribution to q_t arises from the charges of the plasma particles in the very close neighborhood of the test particle. Even though the neutralization of the charge of the test particle is effected by all the plasma particles, the significant part of the neutralization takes place within a very short distance from the test particle. From Eq. (3.10), λ_D is seen to have the dimensions of a length and it is called *the Debye length.* It is convenient to define a *Debye sphere* with its center coinciding with the location of the test particle and a radius equal to the order of magnitude of the Debye length. The neutralization of the charge of the test particle, in effect, takes place on account of all the plasma particles within the Debye sphere.

It is seen from Eqs. (3.4) and (3.5) that the charge density of the electrons is larger than that of the ions in the neighborhood of the test particle. This is to be expected from simple physical grounds: The test particle attracts the electrons and repels the ions. Thus, in the neighborhood of the test particle, there is an imbalance of charge and as a consequence there exists a macroscopic electric field and a scalar potential.

The previous discussion has placed in evidence the fact that the potential is effective only within the Debye sphere. The electric field due to the charge of the test particle or for that matter, the microscopic electric field due to one of the plasma particles isolated for observation is excluded from the major portion of the plasma by a proper distribution of the plasma particles surrounding the test particle. This screening or shielding is almost completed within the Debye sphere. Thus any departures from macroscopic electrical neutrality can take place only over distances of the order of the Debye length.

The Debye length is an important physical parameter for a plasma. It is seen from Eq. (3.10) to be directly proportional to the square root of the temperature and inversely proportional to the square root of the number density. The magnitude of the Debye length is determined by the distance within which a balance is obtained between two counteracting forces. The two forces are those due to the thermal agitation which tends to disturb the macroscopic neutrality and the electrostatic forces resulting from any charge separation that strive to restore the neutrality. The Debye length is usually very small. As an example, consider the ionosphere where the particle number density is approximately equal to $N_0 = 10^{12}$ per cubic meter and the temperature is equal to $T = 1000°\text{K}$. The Debye length for this case is obtained from Eq. (3.10) to be approximately equal to 2×10^{-3} meter.

Quantitative criteria for a plasma

The concepts of the Debye shielding and the Debye length are important in emphasizing the restrictions under which the phenomenon of electron plasma oscillations are physically meaningful. It is to be noted that the electron plasma oscillations result from a collective behavior of the plasma particles. Therefore, it is necessary that a great number of particles interact with the potential arising from any charge separation. It has been established here that such potentials are significant only over distances of the order of a Debye length. Hence the electron plasma oscillations are physically meaningful only if the number of particles in a Debye sphere is much larger than unity, that is, $N_0\lambda_D^3 \gg 1$. As is pointed out presently, this is also one of the conditions necessary for a quantitative definition of a plasma. Under this stipulation, the statistical average of the microscopic variations of the potential arising from the discrete nature of the particles surrounding the test charge becomes negligible. The present treatment of plasma is based on the Boltzmann equation which rests on the assumption that the discrete distribution of the representative points in the phase space can be replaced by a continuous distribution. Such a replacement is justifiable only if the number of particles in every elemental volume is very large. Therefore for a plasma, it is required that $N_0\lambda_D^3 \gg 1$.

Another restriction applicable to plasmas can be obtained in the following manner. A plasma has been defined to be a macroscopically neutral mixture of charged particles of different kinds. It has been proved here that deviations from macroscopic electrical neutrality occur over distances of the order of a Debye length. In order that a plasma may be considered as a macroscopically neutral gas of charged particles, it is necessary that a typical dimension L of a plasma should be much larger than a Debye length. The two restrictions under consideration are conveniently expressed as follows:

$$L \gg \lambda_D \gg N_0^{-1/3} \tag{3.23}$$

which is a quantitative criterion in the definition of a plasma.

Summary

The Coulomb potential due to an isolated charge falls off inversely as the distance whereas the Debye potential due to a test particle in a plasma, for sufficiently large distances, falls off exponentially with increasing distance as illustrated in Fig. 2.2.

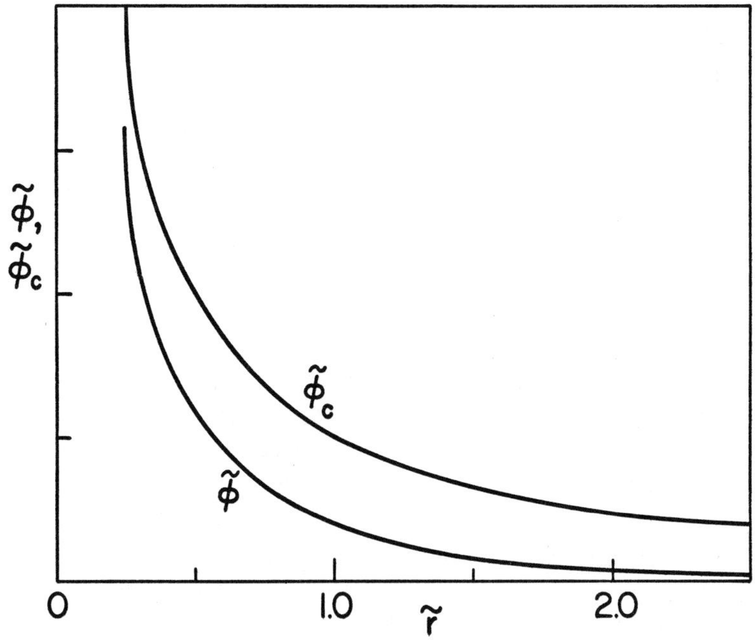

Fig. 2.2. Normalized Coulomb ($\tilde{\phi}_c$) and Debye ($\tilde{\phi}$) potentials as a function of the normalized distance $\tilde{r}$. $\tilde{r} = \sqrt{2}\, r/\lambda_D$; $\phi_0 = q\sqrt{2}/4\pi\varepsilon_0\lambda_D$; $\tilde{\phi}_c = \phi_c/\phi_0$; $\tilde{\phi} = \phi/\phi_0$.

It is this very rapid attenuation of the potential that screens every plasma particle from having any significant influence on a major portion of the plasma and restricts, in essence, its influence to only the particles lying within its Debye sphere. Thus every plasma particle may be considered (i) to have negligible influence on the particles lying outside its Debye sphere and (ii) to interact cumulatively with all the particles lying within its Debye sphere.

2.4. Plasma sheath

A negative potential is acquired by a material body immersed in a plasma. Near the wall of the body, there is a boundary layer in which the potential increases monotonically from a negative value on the wall to a zero value which corresponds to an unperturbed plasma. This boundary layer between the wall and the plasma is called the plasma sheath. The number densities of electrons and ions are not equal inside the plasma sheath. In the previous section, it was established that the departure from macroscopic electrical neutrality caused by the presence of a test charge exists over a distance of the order of a Debye length. The thickness of the plasma sheath where macroscopic electrical neutrality does not exist is also found to be of the order of a Debye length. In this section, a simple theoretical treatment of the formation and the structure of the plasma sheath is given.

Formation of the plasma sheath

It is instructive to begin with a descriptive account of the physical mechanism leading to the formation of the plasma sheath. For this purpose, it is necessary to argue that the average thermal speed of the electrons is usually greater than that of the ions. If the average energies of the electrons and the ions are equal, that is, if

$$\tfrac{1}{2}m_e\langle u_e^2\rangle = \tfrac{1}{2}m_i\langle u_i^2\rangle \tag{4.1}$$

it follows that the average thermal speed of the electrons is much greater than that of the ions since the ion mass m_i is much larger than the electron mass m_e. Even for the least heavy ion, hydrogen, m_i is greater than m_e by a factor of 1836. However, in general, the average thermal energy of the electrons is much higher than that for the positive ions. This result can be deduced in the following manner. Suppose that an electric field is present in a plasma. The electrons accelerated by the field gain energy, the energy gained being independent of the initial average energy of the electrons. When they interact with other particles, the electrons lose energy by an amount which is proportional to their initial energy. If the interaction is approximated as an elastic collision in which the total mass, momentum, and energy are conserved, it can be shown as in Eq. (4.6.10) that the proportionality factor averaged over all deflection angles is equal to $2m_e/m_i$. This result is true for electron-ion collisions. For collision with neutral particles, the result is $2m_e/m_n$ where m_n is the mass of the neutral particle. In either case, the proportionality factor is very small. If the plasma is in a state of equilibrium, the average energy of the electrons is time-invariant and this situation can occur only if the energy imparted to the electrons by the field is transferred to the ions and the neutral particles by collisional interactions. Since the energy gained is independent of the initial energy of the electrons but the energy lost is a very small fraction of the initial energy, in order for the entire energy gained from the field by the electrons to be lost by collisional interactions to the other particles, it is necessary that the average energy of the electrons be much larger than that of the ions. If the electric field is removed, the electrons and the ions relax through collisional interactions to the same temperature, that is to a state in which the average thermal energies of the electrons and the ions are equal. But the required relaxation time is very long compared to the time of existence of the ions. Therefore, it is concluded that for all practical purposes, the average thermal energy of the electrons is much greater than that of the positive ions. Consequently, the average thermal speed of the electrons is greater than that of the positive ions by a factor which is larger than would be the case if the average energies of the electrons and the ions are equal.

The particle current densities Γ_e and Γ_i for the electrons and the ions caused by their random motions are obtained from Eq. (1.6.8) as

$$\Gamma_e = N_e \langle u_e \rangle / 4 \qquad \Gamma_i = N_i \langle u_i \rangle / 4 \tag{4.2}$$

Suppose that initially the number densities of the electrons and the ions, N_e and N_i, are equal. Since $\langle u_e \rangle \gg \langle u_i \rangle$, it follows from Eqs. (4.2) that when an external body is immersed in a plasma, initially the particle current directed towards the wall contributed by the electrons is much greater than that due to the ions. The particles striking the wall are essentially lost from the plasma. In view of the fact that the particles reaching the wall are predominantly electrons, the wall acquires a negative potential with respect to the plasma. The negative potential on the wall counteracts the effect due to the initial preponderance of the electron current; it repels the electrons and attracts the ions as a consequence of which the electron particle current tends to decrease whereas the ion particle current increases. The negative potential on the wall increases up to a value which is sufficiently strong to equalize the particle currents due to the electrons and the ions so that the net current is zero and at this floating negative potential, the external body reaches a dynamical equilibrium with the plasma. Far away from the wall, the plasma potential is zero. Therefore, across the plasma sheath, as illustrated in Fig. 2.3, the potential increases from a negative value on the wall to a zero value in the plasma. As is to be deduced, this increase is monotonic and is completed in a distance of the order of a Debye length.

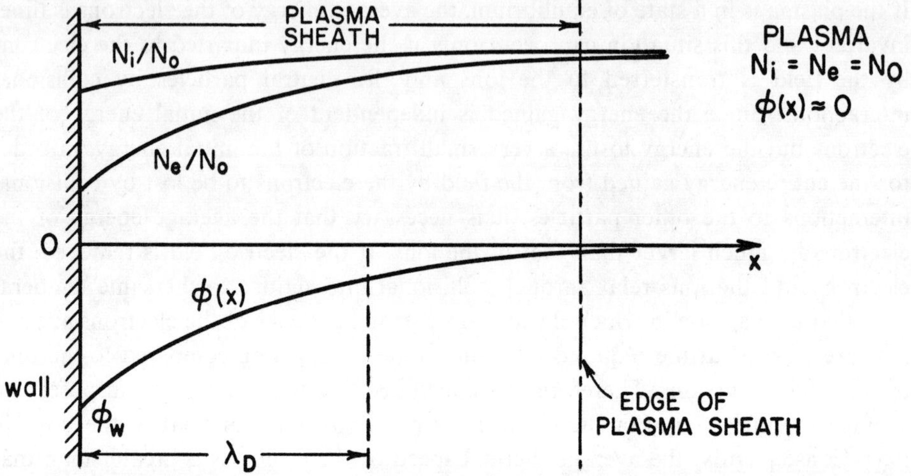

Fig. 2.3. Schematic diagram of the plasma sheath for an infinite plane wall showing the variation of the potential and the number densities inside the sheath.

Potential on the wall

A rigorous treatment of the plasma sheath problem is difficult. For the present purpose, the following simple treatment is sufficient to obtain an understanding of the salient features of the plasma sheath. To start with, a simple geometry is used for the wall; it is assumed to be an infinite plane bounding the plasma ($x > 0$) at $x = 0$. Only the one-dimensional problem in which there are no variations with respect to the y- and the z-coordinates is considered. Finally attention is restricted to the investigation of the equilibrium situation in which the plasma sheath is already formed.

Let $\phi(x)$ be the potential distribution. The values of $\phi(x)$ on the wall at $x = 0$ and at $x = \infty$ are assumed to be given by

$$\phi(0) = \phi_w \qquad \phi(\infty) = 0 \tag{4.3}$$

It is desired to find an expression for ϕ_w in terms of the various physical parameters. At $x = \infty$, the plasma is not perturbed and the number densities of the electrons and the ions are equal. Therefore, at $x = \infty$, the potential is a constant which is set equal to zero. In other words, the potential at infinity is taken as the reference potential. The electrons and the ions are assumed to be in thermodynamic equilibrium under the action of the conservative electric field specified by the potential. In order to simplify the details, the electrons and the ions are assumed to have the same temperature T. If the electronic and the ionic charges are denoted by $-e$ and e, respectively, the number densities of the electrons and the ions are obtained from Eq. (1.5.8) as

$$N_e(\mathbf{r}) = N_0 e^{e\phi(x)/KT} \qquad N_i(\mathbf{r}) = N_0 e^{-e\phi(x)/KT} \tag{4.4a, b}$$

For $x = \infty$, $\phi(x) = 0$ and Eqs. (4.4a,b) give $N_e(\mathbf{r}) = N_i(\mathbf{r}) = N_0$, as is to be expected.

The assumptions implied by Eqs. (4.4a,b) are best set forth at this stage. Since Eqs. (4.4a,b) are deduced from the Boltzmann equation, the same drawback that was discussed in connection with the derivation of the Debye potential is present. The wall with a negative potential attracts the ions and repels the electrons with the result that the velocity distribution functions of the charged particles have some correlation with the position of the wall. This correlation is ignored in the formulation of the Boltzmann equation and therefore also in Eqs. (4.4a,b). The electrons and the ions impinging on the wall are lost from the plasma. There is a steady stream of both species of particles towards the wall for providing the replenishment for this loss of the charged particles to the wall. The average velocity associated with the Maxwell-Boltzmann distribution function on which Eqs. (4.4a,b) are based is zero. Thus Eqs. (4.4a,b) do not take into account the drift velocity of the particles towards the wall. As is to be established, the thermal energy is much greater than the visible energy associated with the mass motion for the electrons and vice versa for the ions. Hence, Eq. (4.4a) still continues to hold but Eq. (4.4b) requires to be modified to include the effect of the drift motion of the ions. If the shifted Maxwell-Boltzmann distribution function (1.4.25) with $v_y = v_z = 0$ is used to incorporate the effect of the drift motion of the ions, the corresponding expression for $N_i(\mathbf{r})$ becomes less simple and the sheath equation which is to be derived presently becomes so complicated that no other than numerical techniques can be used for its solution. Therefore, in the following treatment the drift motion of the ions is taken into account in an approximate manner through the use of the hydrodynamical equations.

The potential on the wall can now be evaluated. Under the equilibrium conditions, the particle current densities due to the electrons and the ions are equal at the wall, i.e.,

$$\Gamma_e = \Gamma_i \qquad \text{at } x = 0 \tag{4.5}$$

With the use of Eqs. (4.2), (4.3), (4.4a,b), and (1.6.9), Eq. (4.5) is shown to give

$$m_e^{-1/2} e^{e\phi_w/KT} = m_i^{-1/2} e^{-e\phi_w/KT} \tag{4.6}$$

Therefore

$$e^{-2e\phi_w/KT} = (m_i/m_e)^{1/2} \tag{4.7}$$

It is an easy matter to obtain from Eq. (4.7) that

$$\phi_w = -(KT/4e)\ln(m_i/m_e) \tag{4.8}$$

The other sophisticated methods of evaluating the potential on the wall yield results which for $T_e = T_i$ are in qualitative agreement with that predicted by Eq. (4.8). It is noted from Eq. (4.8) that near the wall, the ratio of the magnitude of the potential energy $|e\phi_w|$ to the thermal energy KT of a particle is given by

$$|e\phi_w|/KT = \tfrac{1}{4}\ln(m_i/m_e) \tag{4.9}$$

For an hydrogen ion, $|e\phi_w|/KT$ is seen to be approximately equal to 2. Thus, the wall potential is of the same order of magnitude as the average thermal energy of the plasma particles in electron volts.

A modification of the result given by Eq. (4.8) is commonly suggested. It is based on the argument that the ions reaching the edge of the plasma sheath fall freely to the wall. Near the edge of the plasma sheath, the potential is practically zero and Eqs. (4.2), (4.4b), and (1.6.9) yield for the ion current at the wall the expression

$$\Gamma_i = N_0(KT/2\pi m_i)^{1/2} \tag{4.10}$$

Noting that the potential $\phi(x)$ is negative, it follows from Eqs. (4.2), (4.4b), and (4.10), that Γ_i given by Eq. (4.10) is less than that used previously. The electrons reaching the edge of the plasma sheath with insufficient energy are repelled back by the negative potential but those with sufficiently large energy are able to penetrate the potential barrier to reach the wall. The electron particle current therefore, is the same as that employed previously. The net result is that the potential on the wall is twice the value given by Eq. (4.8). The plasma at the edge of the sheath has a net drift velocity towards the wall. If an appropriately shifted Maxwell-Boltzmann distribution function is used for the evaluation of Γ_i, the result is found to be larger than that given by Eq. (4.10) which, therefore, is probably the lower bound to the values of Γ_i. Thus, ϕ_w given by Eq. (4.8) is a good approximation to the potential on the wall.

Structure of the plasma sheath

For the reasons mentioned previously, the hydrodynamical equations are used to obtain an improved expression for the number density of ions to replace Eq. (4.4b) which does not reflect correctly the actual situation, wherein for the ions the visible energy of mass motion is much greater than the thermal energy. The equation of continuity (1.8.9) for the present case reduces to

$$\frac{d}{dx}(N_i v_x) = N_i \frac{d}{dx}v_x + v_x \frac{d}{dx}N_i = 0 \tag{4.11}$$

The viscosity effects are neglected with the result that the kinetic pressure dyad can be approximated by the scalar pressure. The magnetic field effects are also ignored. The collisional interactions are also not taken into account; this is justifiable because the thickness of the plasma sheath is much less than the mean free path between collisional interactions. Moreover, the ideal gas law (1.7.8) can be used. With these assumptions, for the present case, the momentum transport equation (1.9.14) for the ions together with Eqs. (1.9.9) and (3.7) reduces to the form:

$$m_i v_x \frac{d}{dx} v_x = -\frac{KT}{N_i}\frac{d}{dx}N_i - e\frac{d}{dx}\phi(x) \tag{4.12}$$

where T is the ion temperature assumed to be a constant throughout the sheath. Note that the nonlinear form of the hydrodynamical equations is used. It follows from Eq. (4.11) that

$$\frac{|(KT/N_i)(d/dx)N_i|}{|m_i v_x (d/dx) v_x|} = KT/m_i v_x^2 \ll 1 \tag{4.13}$$

where the assumption

$$m_i v_x^2 \gg KT \tag{4.14}$$

is introduced but remains to be justified subsequently. In view of the relation (4.13), the first term on the right side of Eq. (4.12) can be neglected. The integrations of Eqs. (4.11) and (4.12) give

$$\tfrac{1}{2}m_i v_x^2 + e\phi(x) = C_1 \qquad N_i v_x = C_2 \tag{4.15a, b}$$

where C_1 and C_2 are constants. At $x = \infty$, $\phi(x) = 0$ and $N_i = N_0$; also, let $v_x = v_{x\infty}$ for $x = \infty$. These boundary conditions and Eqs. (4.15) enable the determination of the constants C_1 and C_2 with the result that

$$C_1 = \tfrac{1}{2}m_i v_{x\infty}^2 \qquad C_2 = N_0 v_{x\infty} \tag{4.16a, b}$$

The use of Eqs. (4.16) and the elimination of v_x from Eqs. (4.15a) and (4.15b) give the following expression for N_i:

$$N_i = N_0(1 - 2e\phi(x)/m_i v_{x\infty}^2)^{-1/2} \tag{4.17}$$

which is substantially different from that given by Eq. (4.4b). The ion number density N_i as given by Eq. (4.4b) increases monotonically towards the wall but that given by Eq. (4.17) decreases. The negative potential on the wall causes v_x to increase as the ions move towards the wall but in view of Eq. (4.11) the ion particle current density $N_i v_x$ due to the drift remains a constant with the result that N_i decreases towards the wall as predicted by Eq. (4.17).

With the help of Eqs. (2.1), (3.8), (4.4a), and (4.17), Poisson's equation satisfied by the potential distribution is found to be

$$\frac{d^2}{dx^2}\phi(x) = \frac{N_0 e}{\varepsilon_0}\left[\exp\left\{\frac{e\phi(x)}{KT}\right\} - \left\{1 - \frac{2e\phi(x)}{m_i v_{x\infty}^2}\right\}^{-1/2}\right] \tag{4.18}$$

where the ion drift velocity $v_{x\infty}$ far away from the sheath is as yet undetermined. The plasma sheath equation (4.18) is nonlinear and requires further approximation to facilitate its solution without resorting to numerical techniques. It is known that $|e\phi(x)|$ ranges from zero in the plasma to a value of the order of KT on the wall. In view of this and the condition (4.14), near the edge of the sheath adjacent to the plasma, $|e\phi(x)|$ is small compared to both KT and $m_i v_{x\infty}^2$. Suppose that the analysis is restricted to the edge of the plasma sheath. Then, the right side of Eq. (4.18) can be expanded for $|e\phi(x)/KT| \ll 1$ and $|e\phi(x)/m_i v_{x\infty}^2| \ll 1$ with the result that the plasma sheath equation reduces to

$$\frac{d^2\phi(x)}{dx^2} = \frac{\phi(x)}{h^2} \qquad h = \lambda_D\left(1 - \frac{KT}{m_i v_{x\infty}^2}\right)^{-1/2} \tag{4.19a, b}$$

where λ_D is the Debye length.

In view of the condition (4.14), h is seen to be real. The solution of Eq. (4.19a) subject to the boundary conditions (4.3) is evaluated as

$$\phi(x) = \phi_w e^{-x/h} \tag{4.20}$$

where the solution of $\phi(x)$ valid near the edge of the plasma sheath has been continued to apply throughout the sheath in order to be able to impose the boundary condition on the wall. From the condition (4.14) and Eq. (4.19b), it is noted that h is of the order of the Debye length. Since ϕ_w is negative, it is seen from Eq. (4.20) that the potential increases from the value ϕ_w on the wall to zero in the interior of the plasma and since it takes place exponentially, this increase is completed, in effect, within a distance of the order of h or the Debye length. Thus the departure from macroscopic electrical neutrality that exists near the boundary of an external body immersed in a plasma extends only to a distance of the order of a Debye length.

If $KT > m_i v_{x\infty}^2$, h turns out to be imaginary and the potential becomes an oscillatory function of distance near the wall. Hence it is required that

$$KT < m_i v_{x\infty}^2 \tag{4.21}$$

for the formation of the plasma sheath. The condition (4.21) is known as *the Bohm criterion.*

Validity of the approximations

It is difficult to evaluate the potential ϕ_w on the wall and the ion drift velocity $v_{x\infty}$ from the theory based on the hydrodynamical equations. However the approximation to the potential ϕ_w on the wall as given by Eq. (4.8) holds good reasonably well at least for the case in which the electron and the ion temperatures are equal. An approximate value for $v_{x\infty}$ can also be deduced as follows. Since the particle current density is a constant, its value $N_0 v_{x\infty}$ at $x = \infty$ can be equated to its value Γ_i at the wall obtained from the Boltzmann equation. With the help of Eqs. (4.2), (4.4b), and (1.6.9) it is found that

$$v_{x\infty} = (KT/2\pi m_i)^{1/2} \exp(-e\phi_w/KT) \tag{4.22}$$

It can be shown by using Eq. (4.22) that the Bohm criterion is satisfied and the conclusion regarding the formation of the plasma sheath is validated.

It is now possible to examine the validity of the approximation (4.14). From Eqs. (4.15b) and (4.16b), it is found that $v_x = N_0 v_{x\infty}/N_i$, and together with Eq. (4.22) and the fact that N_0 is the maximum value of N_i, it can be proved further that

$$KT/m_i v_x^2 < 2\pi \exp(2e\phi_w/KT) \approx 0.1 \tag{4.23}$$

For the hydrogen plasma, $e\phi_w = -2KT$ approximately and this result has been used in evaluating the right side of the relation (4.23). Clearly, the relation (4.23) reveals the validity of the assumption given by the condition (4.14). From the equation of continuity for the electrons similar to that given by Eq. (4.11) for the ions, the particle current density of the electrons is seen to be a constant; its value $N_0 \tilde{v}_{x\infty}$ at $x = \infty$ can be equated to its value Γ_e at the wall obtained from the Boltzmann equation. With the help of Eqs. (4.2), (4.4a), and (1.6.9), it is found that

$$\tilde{v}_{x\infty} = (KT/2\pi m_e)^{1/2} \exp(e\phi_w/KT) = v_{x\infty} \tag{4.24}$$

where the result on the right side follows from Eqs. (4.5) or (4.6) and (4.22). Note that $\tilde{v}_x = N_0 \tilde{v}_{x\infty}/N_e$, where $\tilde{v}_x$ is the drift velocity of the electrons and $\tilde{v}_{x\infty}$ is the value of $\tilde{v}_x$ at $x = \infty$. Since the minimum value of N_e is seen from Eq. (4.4a) to be $N_0 \exp(e\phi_w/KT)$, it follows that $\tilde{v}_x \leqslant \tilde{v}_{x\infty} \exp(-e\phi_w/KT)$. This result together with Eq. (4.24) yields

$$KT/m_e v_x^2 > 2\pi \tag{4.25}$$

In view of the condition (4.25), the thermal energy is seen to be much greater than the visible energy of mass motion for the electrons and therefore the use of Eq. (4.4a) which ignores the drift motion of the electrons is justified.

Summary

The wall of a material body immersed in a plasma acquires a negative potential. The potential increases monotonically from the negative value on the wall to zero inside the plasma within a distance of the order of a Debye length. This increase takes place exponentially near the edge of the plasma sheath and probably not too differently near the wall. The number densities of the electrons and the ions decrease towards the wall, the decrease being much larger for the electrons as illustrated in Fig. 2.3. Therefore there is charge imbalance and deviation from macroscopic electrical neutrality inside the plasma sheath. There is a steady drift of the plasma particles towards the wall to replenish the loss caused by the continuous absorption of the particles at the wall. The drift velocity is of the order of the thermal velocity of the ions.

2.5. Plasma probe

It is by now clear that a plasma is a complex state of matter and any theory can give only a partial understanding of its properties. It is best to resort to experimental techniques for probing the plasma properties as a supplementation to the theory. Langmuir and Mott-Smith have developed an electrostatic probe capable of measuring the temperature and the density of the particles in a plasma. The theory of the plasma sheath developed in the preceding section can be adapted to provide an approximate explanation of the physical mechanism of operation of an electrostatic probe. An electrode is immersed in a plasma, and the current collected by it is measured for various potentials applied to the electrode. From the current-potential characteristic of the electrode, it is possible to assess the temperature and the number density of the electrons.

The current-voltage characteristic of a metallic probe inserted in a plasma is illustrated in Fig. 2.4. When an electrode is immersed in a plasma, it is enveloped by a plasma sheath which shields the major portion of the plasma from the disturbing field caused by the electrode. If the probe is maintained negative, zero, or slightly positive with respect to the plasma potential, the shielding effect of the plasma sheath continues to be effective. The sheath is very thin—of the order of a Debye length—and the major portion of the plasma is undisturbed. It is this property which makes the plasma probe particularly useful.

Current-voltage characteristic

When the electrode does not draw any electric current, the electrode is at the floating potential ϕ_w discussed in the previous section. At this potential, the particle current density due to the electrons is the same as that due to the ions. The electric current associated with the flow of electrons is directed away from the electrode and this direction is assumed to be positive. Therefore, the electric current due to the flow of ions is negative. Since these electric current densities are of equal magnitudes but are in opposite directions, the resulting electric current drawn by the electrode is zero. If the potential ϕ is made more negative than ϕ_w, due to the additional repulsive force on the electrons, the electric current caused by the flow of electrons is reduced and this reduction increases as the potential is made more

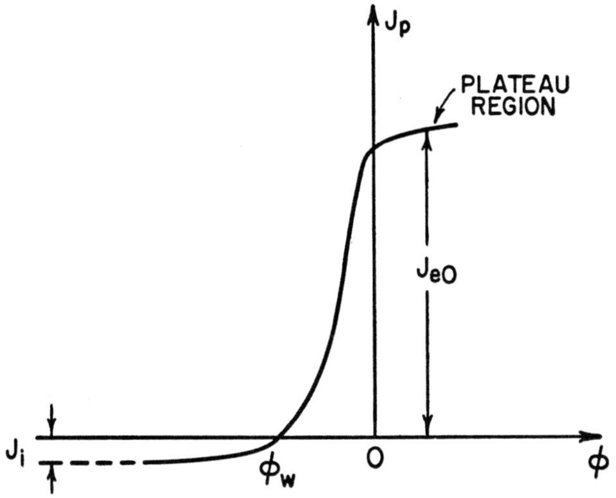

Fig. 2.4. The current-voltage characteristics of an electrostatic probe immersed in a plasma. ϕ_w: the floating potential of the probe with reference to the plasma potential.

negative until finally the contribution to the electric current arising from the electrons becomes negligible. The ions reaching the edge of the plasma sheath fall into the potential well and are practically unaffected by the electrode potential being made more negative than ϕ_w. Consequently, as the potential is made more negative, the electric current becomes negative and asymptotically approaches a constant negative value corresponding to the electric current density J_i due to the flow of ions only.

As the electrode potential ϕ is increased from the negative value ϕ_w, due to the decrease in the repulsive force on the electrons, the electric current caused by the flow of electrons increases. An expression for the magnitude of the electric current density due to the flow of electrons is obtained from Eq. (4.4a) as

$$J_e = J_{e0} \exp(e\phi/KT_e) \tag{5.1}$$

where e is the magnitude of the charge on an electron, T_e is the electron temperature and K is the Boltzmann constant. In Eq. (5.1), J_{e0} is the electric current density due to the electrons corresponding to the zero potential. If N_0 is the number density of the electrons in the unperturbed region of the plasma, it is found from Eqs. (1.6.9) and (4.2) that

$$J_{e0} = N_0 e(KT_e/2\pi m_e)^{1/2} \tag{5.2}$$

If ϕ is negative, the ions reaching the edge of the plasma sheath continue to fall into the potential well and hence the ion current density is a constant J_i in the negative potential region. The probe current density is obtained from Eq. (5.1) as

$$J_p = J_{e0}\exp(e\phi/KT_e) - J_i \qquad \text{for } \phi < 0 \tag{5.3}$$

For $\phi = 0$, the electrode is at the same potential as the plasma and there is no electric field near the electrode. Since the average thermal velocity of the electrons is much greater than that of the ions, the electron current density J_{e0} is much greater than the ion current density.

If the potential ϕ is made positive, due to the repulsive force now acting on the ions, the very small ion current density J_i present for $\phi = 0$ vanishes. Due to the attractive force now acting on the electrons, the electron current density increases and reaches a fairly constant value. This plateau region in the current-voltage characteristic of the probe is called the region of saturation electron current. If the electrode potential is made sufficiently positive, there are complications in the current-voltage characteristic of the probe due to the occurrence of another phenomenon.

Measurement of electron temperature and number density

From Eq. (5.3), it can be deduced that

$$T_e = (e/K)[(d/d\phi)\ln(J_p + J_i)]^{-1} \tag{5.4}$$

The relation (5.4) can be used to evaluate the electron temperature. The electrode potential is made sufficiently negative with respect to the plasma potential with the result that the probe current is due only to the ions. The measurement of the probe current yields immediately the ion current density J_i. Then, the current-voltage (J_p - ϕ) characteristic of the probe is measured. The plot of $\ln(J_p + J_i)$ as a function of ϕ has a straightline section corresponding to the probe potential less than the plasma potential. The slope of this straight line gives $(d/d\phi)[\ln(J_p + J_i)]$ which together with Eq. (5.4) enables the electron temperature to be determined. The probe current corresponding to the plateau region of the J_p - ϕ characteristic gives the electron current density J_{e0}. Since T_e is known Eq. (5.2) enables the number density N_0 of the electrons in the unperturbed region of the plasma to be evaluated.

It is to be noted that the J_p - ϕ characteristic and therefore Eq. (5.4) are independent of the absolute value of the probe potential. Hence the probe potential can be measured with respect to any fixed potential in the plasma.

2.6. Generalized Poynting vector and group velocity

The electron plasma oscillations were studied in Sec. 2.2, among others, under the assumption that the thermal motions of the electrons are negligible. As a result, it was found that the electron plasma oscillations are longitudinal, electrostatic, and stationary. It is now desired to investigate the effect of the thermal motions on the electron plasma oscillations. It is assumed that the thermal motions of the electrons are adequately described by a scalar pressure. This corresponds to the omission of viscosity effects. The other assumptions are the same as in Sec. 2.2. Let N_0 be the equilibrium number density of both the electrons and the ions, and, let $N(\mathbf{r}, t)$ be the perturbation on the electron number density. In a similar manner, let P_0 and $P(\mathbf{r}, t)$ be, respectively, the equilibrium and the perturbation pressures of the electrons. The equilibrium values are assumed to be constant and uniform and the perturbations small compared to their ambient values. The linearized form of the equation of continuity is the same as Eq. (2.8). The linearized form of the momentum transport equation (2.9) has to be modified to include the pressure term as indicated in Eq. (1.11.5). The result is

$$(\partial/\partial t)\mathbf{v}(\mathbf{r}, t) = -(e/m_e)\mathbf{E}(\mathbf{r}, t) - (1/N_0 m_e)\nabla P(\mathbf{r}, t) \tag{6.1}$$

Moreover, if heat conduction and energy interchange due to collisional interactions are negligible, the energy transport equation yields the following equation of state, as in Eq. (1.11.6):

$$P_t N_t^{-\gamma} = \text{constant} \tag{6.2}$$

where the subscript t has been added to indicate the total values. The logarithmic derivative of Eq. (6.2) may be shown to give

$$P(\mathbf{r}, t)/P_0 = \gamma N(\mathbf{r}, t)/N_0 \tag{6.3}$$

or equivalently

$$P(\mathbf{r}, t)/m_e N(\mathbf{r}, t) = \gamma P_0/m_e N_0 = a_e^2 \tag{6.4}$$

Since $P(\mathbf{r}, t)$ is the force per unit area, its dimensions are $ML^{-1}T^{-2}$. The dimensions of $m_e N(\mathbf{r}, t)$ are ML^{-3}. Hence $P(\mathbf{r}, t)/m_e N(\mathbf{r}, t)$ has the dimensions $L^2 T^{-2}$ and a_e, the dimensions of velocity. It is usual to designate a_e as *the sound velocity* in the electron gas. The reason for this is explained later. When Eq. (6.4) is used to eliminate $N(\mathbf{r}, t)$, Eq. (2.8) becomes

$$(\partial/\partial t)P(\mathbf{r}, t) + m_e a_e^2 N_0\{\nabla \cdot \mathbf{v}(\mathbf{r}, t)\} = 0 \tag{6.5}$$

Consider a gas of particles having the same mass as electrons but carrying no electric charge. Suppose that the gas has only this species of particles. Since the gas contains only one species of particles, the collision terms in the momentum and the energy transport equations are identically zero. Note that this does not imply the absence of collisional interactions amongst the particles. For this case, Eq. (6.5) is unaffected and Eq. (6.1) reduces to

$$(\partial/\partial t)\mathbf{v}(\mathbf{r}, t) = -(1/N_0 m_e)\nabla P(\mathbf{r}, t) \tag{6.6}$$

If Eq. (6.5) is differentiated with respect to t and $\mathbf{v}(\mathbf{r}, t)$ is eliminated by using Eq. (6.6), it is found that

$$\nabla^2 P(\mathbf{r}, t) = (1/a_e^2)(\partial^2/\partial t^2)P(\mathbf{r}, t) \tag{6.7}$$

which is the wave equation specifying the propagation of pressure in an *ideal*, uniform gas of neutral particles characterized by the electronic mass. It is for this reason that a_e is called the sound velocity in the electron gas. It is important to note that the mechanism of transfer of momentum and energy in this gas is the collisional interactions whereas in the electron gas of the plasma, the electromagnetic fields take over that role.

All the field quantities are assumed to have a harmonic time dependence as expressed in Eq. (2.16). Then, Eqs. (6.5) and (6.1) reduce respectively to

$$-i\omega P(\mathbf{r}) + m_e a_e^2 N_0\{\nabla \cdot \mathbf{v}(\mathbf{r})\} = 0 \tag{6.8}$$

and

$$\mathbf{v}(\mathbf{r}) = (-ie/\omega m_e)\mathbf{E}(\mathbf{r}) - (i/\omega N_0 m_e)\nabla P(\mathbf{r}) \tag{6.9}$$

Also, Maxwell's equations (2.4)–(2.7) together with Eqs. (2.1) and (2.2) simplify to the following form on linearization:

$$\nabla \times \mathbf{E}(\mathbf{r}) = i\omega\mu_0 \mathbf{H}(\mathbf{r}) \tag{6.10}$$

$$\nabla \times \mathbf{H}(\mathbf{r}) = -i\omega\varepsilon_0 \mathbf{E}(\mathbf{r}) - N_0 e\mathbf{v}(\mathbf{r}) \tag{6.11}$$

$$\nabla \cdot \mathbf{E}(\mathbf{r}) = -N(\mathbf{r})e/\varepsilon_0 \tag{6.12}$$

$$\nabla \cdot \mathbf{H}(\mathbf{r}) = 0 \tag{6.13}$$

All the field quantities in Eqs. (6.8)–(6.13) are in the complex phasor form.

Consider the following complex vector:

$$\mathbf{S}(\mathbf{r}) = \tfrac{1}{2}\,\text{Re}\,[\mathbf{E}(\mathbf{r}) \times \mathbf{H}^*(\mathbf{r}) + P(\mathbf{r})\mathbf{v}^*(\mathbf{r})] \tag{6.14}$$

where Re denotes the real part and * indicates the complex conjugate. Since the divergence is a real operator, it follows from Eq. (6.14) that

$$\begin{aligned}\nabla \cdot \mathbf{S}(\mathbf{r}) = \tfrac{1}{2}\,\text{Re}\,[&\mathbf{H}^*(\mathbf{r}) \cdot \{\nabla \times \mathbf{E}(\mathbf{r})\} - \mathbf{E}(\mathbf{r}) \cdot \{\nabla \times \mathbf{H}^*(\mathbf{r})\} \\ &+ \mathbf{v}^*(\mathbf{r}) \cdot \nabla P(\mathbf{r}) + P(\mathbf{r})\{\nabla \cdot \mathbf{v}^*(\mathbf{r})\}]\end{aligned} \tag{6.15}$$

In obtaining Eq. (6.15) the vector formulas (1.10.55) and

$$\nabla \cdot (\mathbf{A} \times \mathbf{B}) = \mathbf{B} \cdot (\nabla \times \mathbf{A}) - \mathbf{A} \cdot (\nabla \times \mathbf{B}) \tag{6.16}$$

are used. If $\nabla \times \mathbf{E}(\mathbf{r})$, $\nabla \times \mathbf{H}^*(\mathbf{r})$, $\nabla P(\mathbf{r})$, and $\nabla \cdot \mathbf{v}^*(\mathbf{r})$ are replaced by their equivalents from Eqs. (6.10), (6.11), (6.9), and (6.8), respectively, Eq. (6.15) becomes

$$\nabla \cdot \mathbf{S}(\mathbf{r}) = \tfrac{1}{2}\,\mathrm{Re}\left[i\omega\mu_0|\mathbf{H}(\mathbf{r})|^2 - i\omega\varepsilon_0|\mathbf{E}(\mathbf{r})|^2 + i\omega m_e N_0|\mathbf{v}(\mathbf{r})|^2 - \frac{i\omega|P(\mathbf{r})|^2}{m_e N_0 a_e^2}\right] \tag{6.17}$$
$$= 0$$

If Eq. (6.17) is integrated throughout a volume V bound by the closed surface S and if the divergence theorem is used, it is obtained that

$$\int_V \nabla \cdot \mathbf{S}(\mathbf{r})\,dV = \int_S \mathbf{S}(\mathbf{r}) \cdot d\mathbf{S} = 0 \tag{6.18}$$

In the warm plasma model under consideration there is no dissipative mechanism and there are no sources. Hence, on an average over the time period $2\pi/\omega$, the power flowing in through a closed surface should be equal to that flowing out or in other words the net time-averaged flow of power through a closed surface is zero. Consequently, Eq. (6.18) shows that $\mathbf{S}(\mathbf{r})$ can be interpreted as the time-averaged flow of power normally across unit area. When there are no charged particles, the hydrodynamical variables $P(\mathbf{r})$ and $\mathbf{v}(\mathbf{r})$ vanish and $\mathbf{S}(\mathbf{r})$ reduces to $\frac{1}{2}\,\mathrm{Re}\,\mathbf{E}(\mathbf{r}) \times \mathbf{H}^*(\mathbf{r})$ which is the well-known Poynting vector in electromagnetic theory. For this reason $\mathbf{S}(\mathbf{r})$ given by Eq. (6.14) is called the generalized Poynting vector.

Group velocity

Any field quantity $f(\mathbf{r}, t)$ which is an arbitrary function of time can be synthesized in terms of time-periodic quantities of all possible periods by using the relation

$$f(\mathbf{r}, t) = \int_{-\infty}^{\infty} f(\mathbf{r}, \omega)e^{-i\omega t}\,d\omega \tag{6.19}$$

For the sake of convenience, the field quantities are assumed to depend only on one spatial coordinate (x). The field quantity $f(x, \omega)$ which has an arbitrary spatial variation can also be synthesized in terms of space-periodic quantities of all possible periods with the help of the relation

$$f(x, \omega) = \frac{1}{2\pi}\int_{-\infty}^{\infty} f(k_x, \omega)e^{ik_x x}\,dk_x \tag{6.20}$$

The combination of Eqs. (6.19) and (6.20) yields

$$f(x, t) = \frac{1}{2\pi}\int_{-\infty}^{\infty}\int_{-\infty}^{\infty} dk_x\,d\omega\,f(k_x, \omega)e^{i(k_x x - \omega t)} \tag{6.21}$$

It is seen from Eq. (6.21) that any field quantity $f(x, t)$ having an arbitrary space and time dependence, x and t, can be synthesized from the field quantities which are periodic functions of space and time and are of the form

$$f_{k_x, \omega}(x, t) = e^{i(k_x x - \omega t)} \tag{6.22}$$

The quantity appearing in the exponent in Eq. (6.22) is called the *phase*, which at any time t is constant over parallel plane surfaces (x = constant) in space and therefore Eq. (6.22) is said to represent a plane wave. The amplitude of the plane wave is $f(k_x, \omega)$. It is seen from Eq. (6.21) that any field quantity can be built up in terms of the constituent plane waves.

If t changes by an integral multiple of $2\pi/\omega$, Eq. (6.22) is unchanged and therefore $2\pi/\omega$ is the time period. Similarly, if x changes by an integal multiple of $2\pi/k_x$, Eq. (6.22) is unaltered and therefore $2\pi/k_x$ is the space period or the wavelength. Thus ω and k_x are the parameters describing the time and the space behavior of the wave respectively.

The wavetrain of a plane wave such as in Eq. (6.22) is of infinite duration and the only identifiable quantity in Eq. (6.22) is its phase. The surfaces of constant state or phase of the wave motion are given by $k_x x - \omega t$ = constant. If this state obtains at two space-time coordinates (x_1, t_1) and (x_2, t_2), then the phase velocity v_{ph} which is the velocity of arrangement of phases is given by

$$v_{ph} = (x_2 - x_1)/(t_2 - t_1) = \omega/k_x \tag{6.23}$$

Note that Eq. (6.23) is not the velocity of physical propagation of any significant quantity.

Another velocity known as the group velocity is commonly used. It is defined as the velocity of propagation of the amplitude maximum of a quasimonochromatic wave which is also called a wave packet or a group. A spatially localized field can be constructed on the basis of the Fourier integral as in Eq. (6.21) by a suitable choice of the amplitude function, $f(k_x, \omega)$. It is usual to call $f(k_x, \omega)$ the Fourier transform of $f(x, t)$. All the field quantities can be represented as in Eq. (6.21). When these representations are substituted in the source-free equations of plasma dynamics, the various Fourier transforms satisfy a set of homogeneous algebraic equations. The determinant of this system of equations should vanish to yield a nontrivial solution. This condition gives rise to a functional relation between ω and k_x, the various physical quantities characterizing the medium appearing as parameters. This functional relation is known as the dispersion relation whose explicit form is derived subsequently under various auxiliary conditions. It is convenient to carry out the integration with respect to ω in Eq. (6.21). In doing so, it is found that for a given k_x, the contributions normally come from certain discrete values of $\omega(k_x)$ which are the solutions of the dispersion equation. Therefore, Eq. (6.21), after the integration with respect to ω is carried out, can be formally expressed as

$$f(x, t) = \int_{-\infty}^{\infty} dk_x\, f(k_x) e^{i\{k_x x - \omega(k_x)t\}} \tag{6.24}$$

It is now possible to describe a wave packet or group. Suppose that $f(k_x)$ is of significant magnitude only in the neighborhood of the wave number k_{x0} with the result that outside the range

$$k_{x0} - \delta k_x \leqslant k_x \leqslant k_{x0} + \delta k_x \tag{6.25}$$

$f(k_x)$ is negligible. If $\omega(k_x)$ is a slowly varying function of k_x, the condition (6.25) implies that $\omega(k_x)$ deviates only very slightly from its value $\omega(k_{x0})$ at k_{x0} with the result that a quasimonochromatic field is obtained. In view of the condition (6.25), Eq. (6.24) simplifies to

$$f(x, t) = \int_{k_{x0}-\delta k_x}^{k_{x0}+\delta k_x} dk_x\, f(k_x) e^{i\{k_x x - \omega(k_x) t\}} \tag{6.26}$$

Since the frequency spectrum is confined to a narrow band, Eq. (6.26) is called a "wave packet" or "group". Let $\omega(k_x)$ be expanded into the first two terms of a Taylor series around $k_x = k_{x0}$ as

$$\omega(k_x) = \omega(k_{x0}) + (k_x - k_{x0})\left(\frac{\partial\omega}{\partial k_x}\right)_{k_x=k_{x0}} \tag{6.27}$$

The approximation (6.27) enables the phase factor in Eq. (6.26) to be rearranged as in

$$k_x x - \omega(k_x)t = k_{x0} x - \omega(k_{x0})t + (k_x - k_{x0})\left\{x - t\left(\frac{\partial\omega}{\partial k_x}\right)_{k_x=k_{x0}}\right\} \tag{6.28}$$

which can be used to recast Eq. (6.26) into the form

$$f(x, t) = \tilde{f}(k_{x0}) e^{i\{k_{x0}x - \omega(k_{x0})t\}} \tag{6.29}$$

The amplitude $\tilde{f}(k_{x0})$ is seen to be given by

$$\tilde{f}(k_{x0}) = \int_{k_{x0}-\delta k_x}^{k_{x0}+\delta k_x} dk_x\, f(k_x) \exp\left[i(k_x - k_{x0})\left\{x - t\left(\frac{\partial\omega}{\partial k_x}\right)_{k_x=k_{x0}}\right\}\right] \tag{6.30}$$

which is found to be a constant at the space-time coordinates related by

$$x - t(\partial\omega/\partial k_x)_{k_x=k_{x0}} = 0 \tag{6.31}$$

The velocity of propagation of the amplitude of the wave packet is obtained from Eq. (6.31) as

$$v_g = x/t = \partial\omega/\partial k_x \tag{6.32}$$

which is the expression for the group velocity. It follows from Eq. (6.30) that when the expansion of $\omega(k_x)$ is carried out only up to a term which is linear in $(k_x - k_{x0})$ as in Eq. (6.27), at any time t, the spatial variation of $\tilde{f}(k_{x0})$ or the shape of the disturbance is the same. But if the higher order terms in $(k_x - k_{x0})$ are included, the shape of the disturbance changes with time; it spreads out and gets diffuse. The expansion (6.27) is valid only if δk_x is small and if $\omega(k_x)$ is a very smoothly varying function of k_x. Therefore, the concept of group velocity is strictly applicable only to a wavepacket with an extremely narrow frequency spectrum.

For the electron plasma oscillations treated in Sec. 2.2, there is no functional relation between ω and k; therefore, the group velocity is zero and there is no propagation of waves. It is shown in the following section that the inclusion of the thermal motions of the electrons results in a functional relationship between ω and k. As a consequence, the electron plasma oscillations propagate through the uniform and unbounded plasma with a finite group velocity.

2.7. Effect of thermal motions on electron plasma oscillations

It is known that electromagnetic waves can exist in an unbounded medium without any charged particles. It was pointed out that the stationary electron plasma oscillations that can exist in a cold plasma become propagating disturbances on account of the thermal motions of the electrons. The electromagnetic waves are known to be transverse and the electron plasma waves are expected to be longitudinal since they were shown to have that behavior in the absence of the thermal motions of the electrons. Therefore, it appears probable that both the transverse electromagnetic and the longitudinal plasma waves can exist in a warm plasma. The validity of this supposition is now established systematically.

Separation into electromagnetic and plasma waves

It is obtained from Eqs. (6.9), (2.14), and (2.25) that

$$-i\omega\varepsilon_0 \mathbf{E}(\mathbf{r}) - N_0 e\mathbf{v}(\mathbf{r}) = -i\omega\varepsilon_0\varepsilon_r \mathbf{E}(\mathbf{r}) + \frac{ie}{\omega m_e}\nabla P(\mathbf{r}) \tag{7.1}$$

which suggests the separation of the electric field into two parts as follows:

$$\mathbf{E}(\mathbf{r}) = \mathbf{E}_{em}(\mathbf{r}) + \mathbf{E}_p(\mathbf{r}) \qquad \mathbf{E}_p(\mathbf{r}) = \frac{e}{\omega^2 m_e \varepsilon_0 \varepsilon_r}\nabla P(\mathbf{r}) \tag{7.2a, b}$$

where the subscripts *em* and *p* stand for electromagnetic and plasma, respectively. The substitution of Eqs. (7.2) into Eq. (7.1) yields

$$-i\omega\varepsilon_0 \mathbf{E}(\mathbf{r}) - N_0 e\mathbf{v}(\mathbf{r}) = -i\omega\varepsilon_0\varepsilon_r \mathbf{E}_{em}(\mathbf{r}) \tag{7.3}$$

The right side of Eqs. (6.11) and (7.3) suggest the separation of the magnetic field also into two parts as follows:

$$\mathbf{H}(\mathbf{r}) = \mathbf{H}_{em}(\mathbf{r}) + \mathbf{H}_p(\mathbf{r}) \qquad \mathbf{H}_p(\mathbf{r}) = 0 \tag{7.4}$$

Since $\nabla \times [\nabla P(\mathbf{r})] = 0$, the substitution of Eqs. (7.2)–(7.4) into Eqs. (6.10) and (6.11) may be shown to give

$$\nabla \times \mathbf{E}_{em}(\mathbf{r}) = i\omega\mu_0 \mathbf{H}_{em}(\mathbf{r}) \tag{7.5}$$

$$\nabla \times \mathbf{H}_{em}(\mathbf{r}) = -i\omega\varepsilon_0\varepsilon_r \mathbf{E}_{em}(\mathbf{r}) \tag{7.6}$$

The result of taking the divergence of both sides of Eq. (7.6) is

$$\nabla \cdot \mathbf{E}_{em}(\mathbf{r}) = 0 \tag{7.7}$$

since $\nabla \cdot \{\nabla \times \mathbf{H}_{em}(\mathbf{r})\} = 0$. Together with Eqs. (7.4), Eq. (6.13) becomes

$$\nabla \cdot \mathbf{H}_{em}(\mathbf{r}) = 0 \tag{7.8}$$

If Eqs. (7.2) are substituted into Eq. (6.9) and the resulting equation simplified with the help of Eqs. (2.14) and (2.25), it is found that

$$\mathbf{v}(\mathbf{r}) = \mathbf{v}_{em}(\mathbf{r}) + \mathbf{v}_p(\mathbf{r}) \tag{7.9}$$

$$\mathbf{v}_{em}(\mathbf{r}) = -\frac{ie}{\omega m_e}\mathbf{E}_{em}(\mathbf{r}) \qquad \mathbf{v}_p(\mathbf{r}) = -\frac{i}{\omega N_0 m_e \varepsilon_r}\nabla P(\mathbf{r}) \tag{7.10a, b}$$

From Eqs. (7.7) and (7.10a) it is obtained that

$$\nabla \cdot \mathbf{v}_{em}(\mathbf{r}) = 0 \tag{7.11}$$

An examination of Eqs. (6.8) and (7.11) suggests the separation of $P(\mathbf{r})$ into the following two parts.

$$P(\mathbf{r}) = P_{em}(\mathbf{r}) + P_p(\mathbf{r}) \qquad P_{em}(\mathbf{r}) = 0 \tag{7.12}$$

The substitution of Eqs. (7.9)–(7.12) into Eq. (6.8) may be shown to yield the following result:

$$\nabla^2 P_p(\mathbf{r}) + \frac{\omega^2}{a_e^2}\left(1 - \frac{\omega_{pe}^2}{\omega^2}\right)P_p(\mathbf{r}) = 0 \tag{7.13}$$

With the help of Eqs. (7.2b), (7.12), and (7.13), it can be deduced that

$$\nabla \cdot \mathbf{E}_p(\mathbf{r}) = \frac{-e}{m_e \varepsilon_0 a_e^2} P_p(\mathbf{r}) \tag{7.14}$$

The preceding discussion shows that all the field variables—both the hydrodynamic and the electromagnetic quantities—are amenable to separation into two parts which are associated with (i) the electromagnetic and (ii) the plasma modes, respectively. The subscripts *em* and *p* denote the electromagnetic and the plasma modes, respectively. The electromagnetic mode consists of $\mathbf{H}_{em}(\mathbf{r})$, $\mathbf{E}_{em}(\mathbf{r})$, $\mathbf{v}_{em}(\mathbf{r})$, and with $P_{em}(\mathbf{r}) = 0$. These are governed by Eqs. (7.5)–(7.8), (7.10), and (7.11). The plasma mode consists of $\mathbf{E}_p(\mathbf{r})$, $\mathbf{v}_p(\mathbf{r})$, $P_p(\mathbf{r})$, and with $\mathbf{H}_p(\mathbf{r}) = 0$. These are governed by Eqs. (7.13), (7.2b), and (7.10b). The electromagnetic mode contains the entire magnetic field and in view of Eq. (7.7) has no charge accumulation. It is seen that Eqs. (7.5) and (7.6) are identical to Maxwell's equations in an ordinary dielectric medium with a relative permittivity ε_r. Therefore, it is clear that the electromagnetic mode represents waves which are transverse to the direction of propagation. The plasma mode contains all the charge accumulation and no magnetic field. If the plasma wave propagates only in a particular direction, then the electron pressure has variation only in that direction. As a consequence, Eqs. (7.2b) and (7.10b) show that $\mathbf{E}_p(\mathbf{r})$ and $\mathbf{v}_p(\mathbf{r})$ have only components in the direction of propagation. In other words, the plasma mode represents longitudinal waves, as anticipated. Moreover, since there is no magnetic field associated with it, the plasma mode is electrostatic in character. Thus the longitudinal, electrostatic, and stationary electron plasma oscillations that exist in a cold plasma retain their longitudinal and electrostatic character but become propagating disturbances if the effect of the thermal motions of the electrons is included.

So far the field variables in a warm plasma have been separated into two sets and the characteristics of these two sets of fields have been enumerated. Before these two sets of fields can be identified as two different modes, it is necessary to establish that there is no interchange of energy associated with these two sets of fields. The *modes* have a definite connotation. A set of fields forms a mode only if there is no exchange of energy associated with this set with that of any other set of fields. The two sets of fields in a warm plasma can be shown to form two modes in the strict sense of the term. The truth of this statement is demonstrated later for the case of plane waves.

Plane plasma wave

The plane wave characteristics of longitudinal plasma waves are first analyzed. Consider the following plane wave representation for the pressure

$$P_p(\mathbf{r}) = P_{p0}\, e^{ik_p(l_x x + l_y y + l_z z)} \tag{7.15}$$

where the direction cosines l_x, l_y, and l_z are such that

$$l_x^2 + l_y^2 + l_z^2 = 1 \tag{7.16}$$

The wave number of the longitudinal plasma wave is denoted by k_p whose properties are now desired. The substitution of Eq. (7.15) into Eq. (7.13) together with Eq. (7.16) yields

$$k_p = (\omega/a_e)(1 - \omega_{pe}^2/\omega^2)^{1/2} \tag{7.17}$$

Note that Eq. (7.17) which gives the functional relationship between k_p and ω is called a dispersion relation. For $\omega < \omega_{pe}$, k_p is seen to be imaginary from Eq. (7.17) and for this case, Eq. (7.15) shows that $P_p(\mathbf{r})$ is exponentially damped with distance, that is, the wave is evanescent. For $\omega > \omega_{pe}$, k_p is real and the longitudinal plasma wave exists. From Eq. (7.17), the phase velocity v_p is deduced as

$$\omega/k_p = v_p = a_e(1 - \omega_{pe}^2/\omega^2)^{-1/2} \tag{7.18}$$

The phase velocity v_p is infinite at $\omega = \omega_{pe}$, decreases monotonically as the frequency is increased and approaches asymptotically the sound velocity in the electron gas. Since a_e is of the order of the thermal velocities of the electrons, for very high frequencies the phase velocity of the wave becomes of the order of the thermal velocities of the electrons and resonant interactions between the wave and the particles are to be expected. But the hydrodynamical theory of the longitudinal plasma waves does not reveal the existence of these resonant interactions. The same problem is treated later on as an application of the Boltzmann equation which reveals the existence of these resonant interactions.

It is convenient to recast Eq. (7.17) into the following form

$$\omega^2 = \omega_{pe}^2[1 + a_e^2 k_p^2/\omega_{pe}^2] \tag{7.19}$$

In the limit of very low temperature, the thermal motions of the electrons can be neglected and Eq. (7.19) shows that the electron plasma oscillations can have only one frequency $\omega = \omega_{pe}$, regardless of the wave number—a result which is in complete accord with that deduced in Sec. 2.2. From Eq. (6.4) and the ideal gas law (1.7.8), it is found that $a_e^2 = \gamma KT/m_e$. For a plane wave, the compression is one-dimensional and hence it is obtained from Eq. (1.11.6) that $\gamma = 3$. Therefore, Eq. (7.19) becomes

$$\omega^2 = \omega_{pe}^2[1 + (3KT/m_e)k_p^2/\omega_{pe}^2] \tag{7.20}$$

The treatment of the electron plasma oscillations from the Boltzmann equation reproduces Eq. (7.20) as well as clarifies the condition under which Eq. (7.20) is valid.

Consider a plasma wave propagating in the x-direction. Then, since $l_y = l_z = 0$ and $l_x = 1$, Eq. (7.15) becomes

$$P_p(\mathbf{r}) = P_{p0}\, e^{ik_p x} \tag{7.21}$$

From Eqs. (7.2b), (7.10b), and (7.21), the other field components of the longitudinal plasma wave are evaluated as

$$\mathbf{E}_p(\mathbf{r}) = \frac{\hat{\mathbf{x}} iek_p}{\omega^2 m_e \varepsilon_0 \varepsilon_r} P_{p0}\, e^{ik_p x} \qquad \mathbf{v}_p(\mathbf{r}) = \frac{\hat{\mathbf{x}} k_p}{\omega N_0 m_e \varepsilon_r} P_{p0}\, e^{ik_p x} \tag{7.22}$$

Plane electromagnetic wave

Consider a plane electromagnetic wave propagating in the x-direction. The electric and the magnetic fields lie in a plane perpendicular to the direction of propagation. The electromagnetic fields are completely determined only if the polarization of the electric vector is specified. The electric vector is assumed to be in the y-direction. Then it is obtained from Eqs. (7.5) and (7.6) that

$$i\omega\mu_0 H_{em,z} = (d/dx)E_{em,y} \tag{7.23}$$

and

$$-i\omega\varepsilon_0\varepsilon_r E_{em,y} = -(d/dx)H_{em,z} \tag{7.24}$$

The argument $\mathbf{r}$ has been omitted in Eqs. (7.23) and (7.24). It is deduced from Eqs. (7.23), (7.24), and (2.25) that

$$(d^2/dx^2)E_{em,y} + (\omega^2/c^2)(1 - \omega_{pe}^2/\omega^2)E_{em,y} = 0 \tag{7.25}$$

where $c = 1/\sqrt{\mu_0\varepsilon_0}$ is the velocity of electromagnetic waves in free space. For a plane wave propagating in the x-direction, the electric field can be represented as

$$E_{em,y} = E_{y0}\,e^{ik_{em}x} \tag{7.26}$$

where k_{em} is the wave number of the electromagnetic wave. The substitution of Eq. (7.26) in Eq. (7.25) gives

$$k_{em} = (\omega/c)(1 - \omega_{pe}^2/\omega^2)^{1/2} \tag{7.27}$$

For $\omega < \omega_{pe}$, k_{em} is imaginary and just as the plasma wave, so also the electromagnetic wave is evanescent for $\omega < \omega_{pe}$. For $\omega > \omega_{pe}$, k_{em} is real and the transverse electromagnetic wave propagates in a warm electron plasma. The phase velocity v_{em} of the electromagnetic wave is obtained from Eq. (7.27) as

$$v_{em} = \omega/k_{em} = c(1 - \omega_{pe}^2/\omega^2)^{-1/2} \tag{7.28}$$

The phase velocity of the electromagnetic wave is infinite at $\omega = \omega_{pe}$, decreases monotonically as ω is increased and asymptotically approaches the velocity of electromagnetic waves in free space. Since the phase velocity of the transverse electromagnetic wave never approaches the thermal velocity of the electrons, it is to be anticipated that the resonant interactions between the wave and the particles are negligible for this case.

From Eqs. (7.23) and (7.10a), the other field components of the transverse electromagnetic wave are evaluated to be given by

$$H_{em,z} = \frac{k_{em}}{\omega\mu_0}E_{y0}\,e^{ik_{em}x} \qquad v_{em,y} = \frac{-ie}{m_e\omega}E_{y0}\,e^{ik_{em}x} \tag{7.29}$$

The two modes

As indicated earlier, it is now possible to show that the plane plasma wave and the plane electromagnetic wave form two independent modes. Since the waves are propagating in the x-direction, it is to be anticipated that the net power flow is in the x-direction. Therefore, it is sufficient to consider only the x-component of the generalized Poynting vector which is obtained from Eq. (6.14) as

$$S_x = \tfrac{1}{2}\,\mathrm{Re}\,[E_y H_z^* - E_z H_y^* + P v_x^*] \tag{7.30}$$

Only the electromagnetic wave contains the magnetic field which from Eqs. (7.29) is seen to be in the z-direction; therefore the second term on the right side of Eq. (7.30) is zero. Both the electromagnetic and the plasma waves contain the electric field but the y-component of the electric field which is required in Eq. (7.30) is seen from Eqs. (7.22) and (7.26) to be contributed only by the electromagnetic wave. Similarly the pressure P is contributed only by the plasma wave. The particle velocity $\mathbf{v}$ is contributed both by the electromagnetic and the plasma waves but the x-component of the particle velocity which is required in Eq. (7.30) is seen from Eqs. (7.22) and (7.29) to be contributed only by the plasma wave. Thus of the two nonvanishing terms in Eq. (7.30), $E_y H_z^*$ is associated *only* with the electromagnetic wave and $P v_x^*$ *only* with the plasma wave. This shows that the electromagnetic and the plasma waves do not exchange energy with each other. This result can be made even more explicit in the following manner. The substitution of P, v_x, E_y, and H_z from Eqs. (7.21), (7.22), (7.26), and (7.29) into Eq. (7.30) gives for $\omega > \omega_{pe}$:

$$S_x = (k_{em}/2\omega\mu_0)|E_{y0}|^2 + (k_p/2\omega N_0 m_e \varepsilon_r)|P_{p0}|^2 \tag{7.31}$$

It is known that S_x represents the time-averaged power flow through unit area perpendicular to the x-direction. Since S_x is independent of y and z, it follows that the time-averaged power flow through unit area is independent of the location of the unit area in a plane perpendicular to the x-direction. Consequently, S_x can be taken as a *measure* of the *total* time-averaged power carried by the plane wave. Note that P_{p0} and E_{y0} are the amplitudes of the plasma and the electromagnetic waves, respectively. When $E_{y0} = 0$, the time-averaged power is seen from Eq. (7.31) to be carried only by the plasma wave. Similarly, when $P_{p0} = 0$, Eq. (7.31) shows that the time-averaged power is carried only by the electromagnetic wave. But when both P_{p0} and E_{y0} are nonvanishing, the resultant time-averaged power is equal to the sum of the time-averaged powers transported by the two waves separately. There is no contribution to the resultant time-averaged power that depends on the amplitudes of both the plasma and the electromagnetic waves. Thus there is no interchange of energy between these two sets of waves. Hence the longitudinal plasma wave and the transverse electromagnetic wave strictly form the two modes in a uniform warm plasma of infinite extent.

2.8. Ion plasma oscillations

Separation of electromagnetic and plasma modes

The electron plasma oscillations treated in Sec. 2.2 take place at very high frequencies. For these high frequencies, the ions, in view of their heavy mass, remain stationary and provide the necessary neutralizing background for the electron plasma oscillations. However, for very low frequencies, even the ions are expected to play a role in the plasma oscillations. With a view to exploring the part played by the heavy ions, the treatment of the previous section is extended to include the motion of the heavy ions. In order to begin the analysis, the hydrodynamical equations pertaining to the electron gas and the ion gas are first written down. For the electrons, the equation of continuity combined with the equation of state as given by Eq. (6.8) and the momentum transport equation as given by Eq. (6.9) are reproduced as follows:

$$- i\omega P_e + m_e a_e^2 N_0(\nabla \cdot \mathbf{v}_e) = 0 \tag{8.1}$$

and

$$\mathbf{v}_e = -(ie/\omega m_e)\mathbf{E} - (i/\omega N_0 m_e)\nabla P_e \tag{8.2}$$

where the sound speed a_e in the electron gas is obtained from Eqs. (6.4) and (1.7.8) as

$$a_e^2 = \gamma_e KT_e/m_e \tag{8.3}$$

The subscript e denotes the quantities pertaining to the electrons. In a similar manner, the corresponding equations for the ions are:

$$- i\omega P_i + m_i a_i^2 N_0(\nabla \cdot \mathbf{v}_i) = 0 \tag{8.4}$$

and

$$\mathbf{v}_i = (ie/\omega m_i)\mathbf{E} - (i/\omega N_0 m_i)\nabla P_i \tag{8.5}$$

where the sound speed a_i in the ion gas is given by

$$a_i^2 = \gamma_i KT_i/m_i \tag{8.6}$$

The equilibrium number density N_0 is the same for the electrons and the ions. For the present case, Maxwell's equations (6.10) and (6.11) become

$$\nabla \times \mathbf{E} = i\omega\mu_0 \mathbf{H} \tag{8.7}$$

and

$$\nabla \times \mathbf{H} = -i\omega\varepsilon_0 \mathbf{E} + N_0 e(\mathbf{v}_i - \mathbf{v}_e) \tag{8.8}$$

It is obtained from Eqs. (8.2) and (8.5) that

$$-i\omega\varepsilon_0\mathbf{E} + N_0 e(\mathbf{v}_i - \mathbf{v}_e) = -i\omega\varepsilon_0\varepsilon_r\mathbf{E} + \frac{ie}{\omega m_e}\nabla P_e - \frac{ie}{\omega m_i}\nabla P_i \tag{8.9}$$

which suggests the separation of the electric field into two parts as follows:

$$\mathbf{E} = \mathbf{E}_{em} + \mathbf{E}_p \qquad \mathbf{E}_p = \frac{1}{\omega^2\varepsilon_r N_0 e}\{\omega_{pe}^2\nabla P_e - \omega_{pi}^2\nabla P_i\} \tag{8.10a, b}$$

where

$$\varepsilon_r = 1 - \omega_{pe}^2/\omega^2 - \omega_{pi}^2/\omega^2 \qquad \omega_{pi} = \sqrt{N_0 e^2/m_i\varepsilon_0} \tag{8.11}$$

and ω_{pe} is the same as defined by Eq. (2.14). The substitution of Eqs. (8.10) into Eq. (8.9) yields

$$-i\omega\varepsilon_0\mathbf{E} + N_0 e(\mathbf{v}_i - \mathbf{v}_e) = -i\omega\varepsilon_0\varepsilon_r\mathbf{E}_{em} \tag{8.12}$$

As before, the magnetic field is also separated as follows:

$$\mathbf{H} = \mathbf{H}_{em} + \mathbf{H}_p \qquad \mathbf{H}_p = 0 \tag{8.13}$$

The result of substituting Eqs. (8.10), (8.12), and (8.13) into Eqs. (8.7) and (8.8) is

$$\nabla \times \mathbf{E}_{em} = i\omega\mu_0\mathbf{H}_{em} \tag{8.14}$$

and

$$\nabla \times \mathbf{H}_{em} = -i\omega\varepsilon_0\varepsilon_r\mathbf{E}_{em} \tag{8.15}$$

If the divergences of both sides of Eqs. (8.14) and (8.15) are taken, it is found that

$$\nabla \cdot \mathbf{E}_{em} = 0 \tag{8.16}$$

and

$$\nabla \cdot \mathbf{H}_{em} = 0 \tag{8.17}$$

The substitution of Eqs. (8.10) into Eqs. (8.2) and (8.5) enables the separation of $\mathbf{v}_e$ and $\mathbf{v}_i$ into two parts as follows:

$$\mathbf{v}_e = \mathbf{v}_{e,em} + \mathbf{v}_{e,p} \qquad \mathbf{v}_i = \mathbf{v}_{i,em} + \mathbf{v}_{i,p} \tag{8.18}$$

where

$$\mathbf{v}_{e,em} = \frac{-ie}{\omega m_e}\mathbf{E}_{em} \qquad \mathbf{v}_{i,em} = \frac{ie}{\omega m_i}\mathbf{E}_{em} \tag{8.19}$$

$$\mathbf{v}_{e,p} = (-i/m_e N_0\varepsilon_r\omega^3)[(\omega^2 - \omega_{pi}^2)\nabla P_e - \omega_{pi}^2\nabla P_i] \tag{8.20}$$

$$\mathbf{v}_{i,p} = (-i/m_i N_0\varepsilon_r\omega^3)[-\omega_{pe}^2\nabla P_e + (\omega^2 - \omega_{pe}^2)\nabla P_i] \tag{8.21}$$

From Eqs. (8.16) and (8.19), it is obtained that

$$\nabla \cdot \mathbf{v}_{e,em} = 0 \qquad \nabla \cdot \mathbf{v}_{i,em} = 0 \tag{8.22}$$

The partial pressures are also separated into two parts as follows:

$$P_e = P_{e,em} + P_{e,p} \qquad P_{e,em} = 0 \tag{8.23}$$

$$P_i = P_{i,em} + P_{i,p} \qquad P_{i,em} = 0 \tag{8.24}$$

The substitution of Eqs. (8.18) and (8.20)–(8.24) into Eqs. (8.1) and (8.4) may be shown to give the following two coupled differential equations satisfied by the partial pressures $P_{e,p}$ and $P_{i,p}$:

$$(\omega^2 - \omega_{pi}^2)\nabla^2 P_{e,p} - \omega_{pi}^2 \nabla^2 P_{i,p} + \frac{\omega^4 \varepsilon_r}{a_e^2} P_{e,p} = 0 \tag{8.25}$$

$$-\omega_{pe}^2 \nabla^2 P_{e,p} + (\omega^2 - \omega_{pe}^2)\nabla^2 P_{i,p} + \frac{\omega^4 \varepsilon_r}{a_i^2} P_{i,p} = 0 \tag{8.26}$$

From Eqs. (8.8), (8.10a), (8.16), (8.23), (8.24), (8.1), and (8.4), it can be deduced that

$$\nabla \cdot \mathbf{E}_p = (e/\varepsilon_0)[P_{i,p}/m_i a_i^2 - P_{e,p}/m_e a_e^2] \tag{8.27}$$

It is ascertained from the foregoing development that all the field variables are separable into two parts which are associated with (i) the electromagnetic (*em*) mode and (ii) the plasma (*p*) mode. The electromagnetic mode contains the entire magnetic field and has no charge accumulation. It represents, as usual, waves which are transverse to the direction of propagation. The plasma mode contains all the charge accumulation and no magnetic field. As in the case of the electron plasma wave, the plasma mode even for the present case represents longitudinal waves.

Electromagnetic mode

A comparison of Eqs. (8.14) and (8.15), respectively, with Eqs. (7.5) and (7.6) shows that the electromagnetic mode for the present case is the same as that treated in the previous section except for the fact that ε_r which was previously $1 - \omega_{pe}^2/\omega^2$ now has become $1 - (\omega_{pe}^2 + \omega_{pi}^2)/\omega^2$. Since $\omega_{pi}^2/\omega_{pe}^2 = m_e/m_i \ll 1$, it follows that the change is extremely small. For this reason, the electromagnetic mode is not given further attention here. However, the inclusion of ion motion has introduced complexity into the plasma mode whose properties therefore require further examination.

Dispersion relation for the plasma mode

Consider the following plane wave representation for the partial pressures:

$$P_{e,p} = P_{e,p0}\, e^{ik_p(l_x x + l_y y + l_z z)} \tag{8.28}$$

$$P_{i,p} = P_{i,p0}\, e^{ik_p(l_x x + l_y y + l_z z)} \tag{8.29}$$

where the direction cosines satisfy the requirement stated in Eq. (7.16). Noting that $\nabla^2 P_{e,p} = -k_p^2 P_{e,p}$, $\nabla^2 P_{i,p} = -k_p^2 P_{i,p}$, the substitution of Eqs. (8.28) and (8.29) into Eqs. (8.25) and (8.26) gives the following simultaneous equations for the plasma mode amplitudes $P_{e,p0}$ and $P_{i,p0}$:

$$\left[\frac{\omega^4 \varepsilon_r}{a_e^2} - (\omega^2 - \omega_{pi}^2)k_p^2\right]P_{e,p0} + k_p^2 \omega_{pi}^2 P_{i,p0} = 0 \tag{8.30}$$

$$k_p^2 \omega_{pe}^2 P_{e,p0} + \left[\frac{\omega^4 \varepsilon_r}{a_i^2} - (\omega^2 - \omega_{pe}^2)k_p^2\right]P_{i,p0} = 0 \tag{8.31}$$

The set of simultaneous equations (8.30) and (8.31) can have a nontrivial solution only if the determinant of their coefficients vanishes. This requirement leads to the following dispersion relation:

$$\left[\frac{\omega^4 \varepsilon_r}{a_e^2} - (\omega^2 - \omega_{pi}^2)k_p^2\right]\left[\frac{\omega^4 \varepsilon_r}{a_i^2} - (\omega^2 - \omega_{pe}^2)k_p^2\right] - k_p^4 \omega_{pe}^2 \omega_{pi}^2 = 0 \tag{8.32}$$

The expansion of the left side of Eq. (8.32) together with Eq. (8.11) gives

$$\frac{\omega^8 \varepsilon_r^2}{a_e^2 a_i^2} - \omega^4 \varepsilon_r k_p^2\left[\frac{(\omega^2 - \omega_{pe}^2)}{a_e^2} + \frac{(\omega^2 - \omega_{pi}^2)}{a_i^2}\right] + k_p^4 \omega^4 \varepsilon_r = 0 \tag{8.33}$$

The cancellation of the common factor $\omega^4 \varepsilon_r$ enables the dispersion relation to be written as follows:

$$k_p^4 a_e^2 a_i^2 - k_p^2[a_i^2(\omega^2 - \omega_{pe}^2) + a_e^2(\omega^2 - \omega_{pi}^2)] + \omega^2(\omega^2 - \omega_{pe}^2 - \omega_{pi}^2) = 0 \tag{8.34}$$

It is possible to manipulate and recast Eq. (8.34) into the form

$$1 = \omega_{pe}^2/(\omega^2 - k_p^2 a_e^2) + \omega_{pi}^2/(\omega^2 - k_p^2 a_i^2) \tag{8.35}$$

in which it was obtained by Fried and Gould.

The dispersion equation (8.34) has two roots for k_p^2. The electron plasma mode discussed in the previous section is associated with one of the roots and the new mode associated with the other root is called the ion plasma mode. Before proceeding to analyze the dispersion relation (8.34) in detail, it is worthwhile to examine a special case which emphasizes the role of the ions in plasma phenomena.

Special case of $T_i = 0$

The ion temperature T_i is usually much less than the electron temperature T_e and therefore it is not unrealistic to neglect T_i in comparison with T_e. If this approximation is made, $a_i^2 = 0$ and then Eq. (8.34) can be rearranged as follows

$$\omega^2[\omega_{pe}^2 + \omega_{pi}^2 - \omega^2 + k_p^2 a_e^2] = k_p^2 a_e^2 \omega_{pi}^2 \tag{8.36}$$

Since $\omega_{pi}^2/\omega_{pe}^2 = m_e/m_i \ll 1$, ω_{pi}^2 can be neglected in comparison with ω_{pe}^2. If attention is restricted to only the very low frequency phenomena such that

$$\omega^2 \ll \omega_{pi}^2 \tag{8.37}$$

then ω^2 appearing inside the square brackets on the left side of Eq. (8.36) can also be neglected. As a consequence Eq. (8.36) can be simplified to yield

$$\omega^2 = \omega_{pi}^2/(1 + \omega_{pe}^2/k_p^2 a_e^2) \tag{8.38}$$

From Eqs. (2.14), (3.10), and (8.3), it follows that $\omega_{pe}^2/a_e^2 = 1/\gamma_e \lambda_{De}^2$ and therefore Eq. (8.38) can be written as

$$\omega^2 = \omega_{pi}^2/(1 + \lambda^2/4\pi^2 \gamma_e \lambda_{De}^2) \tag{8.39}$$

where λ_{De} is the Debye length for the electrons and $\lambda = 2\pi/k_p$ is the wavelength. It is convenient to rewrite Eq. (8.38) also as follows

$$v_p^2 = \omega^2/k_p^2 = (\gamma_e K T_e/m_i)/(1 + 4\pi^2 \gamma_e \lambda_{De}^2/\lambda^2) \tag{8.40}$$

where v_p is the phase velocity of the propagating low frequency disturbances in a plasma.

Ion acoustic wave

Suppose that $\lambda^2 \gg \lambda_{De}^2$; then Eq. (8.40) yields for the phase velocity of the wave $v_p = \sqrt{(\gamma_e K T_e/m_i)}$. From Eq. (6.4), it is seen that the phase velocity is expected to be proportional to the square root of the ratio of the pressure to the mass density. Interestingly enough, the expression for v_p obtained here depends on the *electron* pressure and the *ion* mass density. It is instructive to examine the physical basis of this result. Since the wavelength is much larger than the Debye length of the electrons, the ions are completely shielded by the electronic clouds with the result that macroscopic electrical neutrality exists to a high degree of approximation. The number density N_e of the electrons varies with the number density N_i of the ions and the macroscopic electric current vanishes. The generalized Ohm's law (1.12.40) shows that in the absence of external magnetic field and collisional interactions ($\mathbf{B}_0 = \eta = 0$), the electric current density $\mathbf{J}$ can vanish only if the electron pressure is counteracted by an electric field. In other words, an electric field $\mathbf{E}$ determined by the gradient of the electron pressure should be present and this electric field acts on the ions. The force of restitution for the ion motions is this electric field which is proportional to the electron pressure gradient. Consequently, the phase velocity of the low-frequency propagating ion oscillations is proportional to the square root of the ratio of the electron pressure to the ion mass density. These low-frequency waves are called *the ion acoustic waves.* There is a fundamental distinction between an acoustic wave in a neutral gas and an ion acoustic wave in a collisionless plasma in that the former propagates on account of the short-range intermolecular collisions whereas the potential energy required to drive an ion acoustic wave is electrostatic in origin and is due to the difference in the amplitudes of oscillations of the electrons and the ions.

Ion plasma oscillations

On the other hand, suppose that $\lambda^2 \ll \lambda_{De}^2$; Then Eq. (8.39) gives for the frequency of oscillation $\omega = \omega_{pi}$ where ω_{pi} is the ion plasma angular frequency. When the electron termperature is sufficiently high, the wavelength becomes much less than the Debye length with the result that the electrons do not exhibit their collective behavior but are dominated by their random thermal motions. For the electron plasma oscillations, the frequency is very high and the ions provide an immobile positive neutralizing background for the collective oscillations of the negative electrons. A similar role is played by the electrons in the low-frequency ion plasma oscillations. However, there is an essential difference. Since the electrons have a high temperature, they vibrate with high random speeds and thus provide a dynamic neutralizing background for the cooperative oscillations of the plasma ions.

Analysis of the dispersion relation for the plasma mode

It is convenient to carry out the detailed analysis of the dispersion relation (8.34) in terms of the phase velocity. Let the following normalized parameters be introduced: $\Omega = \omega/\omega_{pe}$, normalized frequency; $\tilde{v}_p = \omega/k_p a_e = v_p/a_e$, the normalized phase velocity. If Eq. (8.34) is multiplied throughout by $\omega^2 \omega_{pe}^{-2} k_p^{-4} a_e^{-4}$, the following result is obtained after some rearrangement:

$$\tilde{v}_p^4(\Omega^2 - 1 - m) - \tilde{v}_p^2[\Omega^2\{\tau m + 1\} - m(1 + \tau)] + \Omega^2 \tau m = 0 \tag{8.41}$$

where

$$m = m_e/m_i \qquad \tau = \gamma_i T_i/\gamma_e T_e \tag{8.42a}$$

In deducing Eq. (8.41), the following substitutions have been made:

$$\omega_{pi}^2/\omega_{pe}^2 = m_e/m_i = m \qquad a_i^2/a_e^2 = (\gamma_i T_i/\gamma_e T_e)m = \tau m \tag{8.42b}$$

Note that usually τ is less than unity. Since $m \ll 1$, m and τm can be neglected in comparison with unity. Therefore Eq. (8.41) simplifies to

$$A\tilde{v}_p^4 - 2B\tilde{v}_p^2 + C = 0 \tag{8.43}$$

where

$$A = (\Omega^2 - 1) \tag{8.44a}$$

$$B = [\Omega^2 - m(1 + \tau)]/2 \tag{8.44b}$$

and

$$C = \Omega^2 \tau m \tag{8.44c}$$

The solutions of Eq. (8.43) are given by

$$v_{pe}/a_e = \tilde{v}_{pe} = [(B + \sqrt{B^2 - AC})/A]^{1/2} \tag{8.45a}$$

$$v_{pi}/a_e = \tilde{v}_{pi} = [(B - \sqrt{B^2 - AC})/A]^{1/2} \tag{8.45b}$$

The additional subscripts *e* and *i* indicate electron plasma and ion plasma modes, respectively, for reasons which are clarified presently.

For $\Omega < 1$, $A < 0$, $C > 0$ and $|\sqrt{B^2 - AC}| > |B|$. The term within the parentheses in Eq. (8.45a) is positive and that within the square brackets is negative with the result that v_{pe} is imaginary. But the term within the parentheses in Eq. (8.45b) is negative and therefore that within the square brackets is positive with the result that v_{pi} is real. For $\Omega > 1$, $A > 0$, $C > 0$, and $|\sqrt{B^2 - AC}| < |B|$. Since for $\Omega > 1$, $B > 0$, it can be argued that both v_{pe} and v_{pi} are real. When v_p is real, the corresponding mode propagates and if v_p is imaginary, the corresponding mode is cut off or does not exist.

In view of the fact that $\tau m \ll 1$, very good approximations to v_{pe} and v_{pi} can be determined in the following manner. Since $\tau m \ll 1$, $B^2 \gg AC$ and hence the square root terms in Eqs. (8.45a) and (8.45b) can be expanded into binomial series, and if only the first two terms are retained, the following results are obtained:

$$v_{pe}/a_e = \tilde{v}_{pe} = [(B + |B| - AC/2|B|)/A]^{1/2} \tag{8.46a}$$

$$v_{pi}/a_e = \tilde{v}_{pi} = [(B - |B| + AC/2|B|)/A]^{1/2} \tag{8.46b}$$

Since $B^2 \gg AC$, Eq. (8.46a) can be simplified with the help of Eqs. (8.44a) and (8.44b) to yield for $\Omega > 1$:

$$v_{pe} = a_e(1 - \omega_{pe}^2/\omega^2)^{-1/2} \tag{8.47}$$

Note that Eqs. (8.47) and (7.18) are identical. Hence it is clear that v_{pe} is associated with the electron plasma mode.

Ion plasma mode

The new mode that comes into existence on account of the inclusion of the motion of the heavy ions has a phase velocity given by v_{pi} and this mode is designated as *the ion plasma mode* . Since $B \gtrless 0$ according as $\Omega \gtrless \sqrt{m(1+\tau)}$, it follows from Eqs. (8.44a)–(8.44c), and (8.46b) that

$$v_{pi} = a_e[m(1+\tau) - \Omega^2]^{1/2} \qquad \Omega < \sqrt{m(1+\tau)} \tag{8.48a}$$

and

$$v_{pi} = a_i \Omega[\Omega^2 - m(1+\tau)]^{-1/2} \qquad \Omega > \sqrt{m(1+\tau)} \tag{8.48c}$$

For $\Omega = \sqrt{m(1+\tau)}$, $B = 0$; for this case it is obtained directly from Eqs. (8.45b), (8.44a), and (8.44c) that

$$v_{pi} = \sqrt{a_e a_i}\,\{m(1+\tau)\}^{1/4} \qquad \Omega = \sqrt{m(1+\tau)} \tag{8.48b}$$

In the derivation of Eqs. (8.48b,c), Eqs. (8.42b) have been used. For $\Omega = 0$, it is evaluated from Eqs. (8.48a), (8.42b), (8.3), and (8.6) that

$$v_{pi} = [\{\gamma_e KT_e + \gamma_i KT_i\}/m_i]^{1/2} \tag{8.49}$$

which is the phase velocity of the *ion acoustic wave*. The ion acoustic wave has been previously considered for the special case of $T_i = 0$. It is worthwhile to remark at this stage that the analysis of the ion plasma wave treated from the standpoint of the Boltzmann equation reveals that the ion acoustic wave is severely damped unless the electron temperature is much larger than the ion temperature. For very high frequencies, Eq. (8.48c) shows that the phase velocity of the ion acoustic mode approaches the sound velocity a_i in a neutral particle gas characterized by the ionic mass.

The phase velocity of the ion plasma mode as evaluated from Eq. (8.45b) is depicted in Fig. 2.5 for the special case of $\tau = 1$ and $m = 1/1836$. It is seen from Fig. 2.5 that the phase velocity of the ion plasma mode is approximately equal to the value given by Eq. (8.49) or a_i depending on whether $\Omega < \sqrt{m(1+\tau)}$ or $\Omega > \sqrt{m(1+\tau)}$ and that the transition takes place rapidly near $\Omega = \sqrt{m(1+\tau)}$. In Fig. 2.5, the phase velocity of the electron plasma mode as given by Eq. (8.45a) and that of the electromagnetic mode as implied by Eqs. (8.14) and (8.15) are also illustrated as a function of frequency in the propagating region.

2.9. MHD equations and their simple applications

So far consideration has been given to only such plasma phenomena as the electron oscillations and the ion oscillations which, as their names imply, depend primarily on one of the constituents of the plasma. An exception was the ion acoustic waves occurring at very low frequencies where the partial pressures of the electrons and the ions combined in the determination of the nature of the phenomenon. It is now proposed to present systematically a theory applicable for the treatment of very low

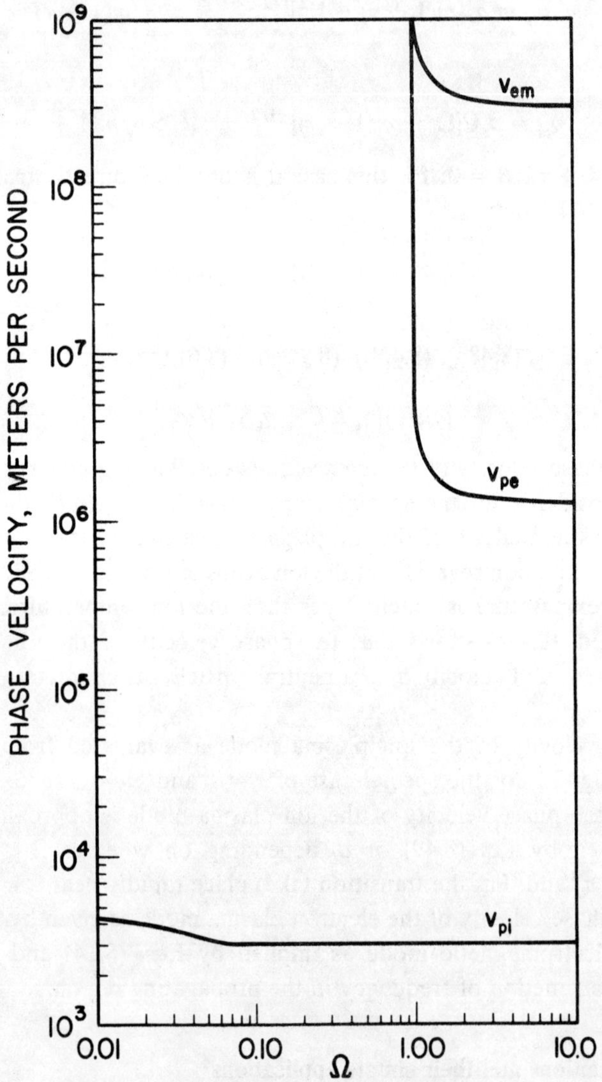

Fig. 2.5. Phase velocities of the electromagnetic, the electron plasma, and the ion plasma modes as a function of frequency in an unbounded collisionless plasma; $a_i = 3 \times 10^3$ m/sec, $c = (\mu_0 \varepsilon_0)^{-1/2} = 3 \times 10^8$ m/sec, $m = 1/1836$, $a_e = a_i/\sqrt{m}$, $\tau = 1$.

frequency phenomena in a plasma. For very low frequencies, it is convenient to treat a plasma as a classical fluid which is governed by the conventional hydrodynamical equations. However, the fluid is an electric conductor and therefore the electromagnetic forces must be taken into account explicitly. As was pointed out previously, for very low frequencies, the macroscopic electrical neutrality is maintained to a very high degree of approximation; therefore the electric charge and the force due to the electric field are negligible. Since the electromagnetic force is entirely due to the magnetic field, the subject of low frequency phenomena in a plasma is called magnetohydrodynamics or MHD for short.

MHD equations

In order to be able to treat a plasma containing different species of particles as a conducting fluid, it is necessary to introduce lumped macroscopic variables and deduce their governing equations. Some aspects of this subject were treated in the previous chapter. Here the magnetohydrodynamic equations are collected together and presented in the form in which they are commonly used. In addition, the approximations involved are enumerated and where possible, justifications are provided. The two equations of continuity pertaining separately to the electrons and the ions were combined to yield the equation (1.12.12) of mass conservation and the equation (1.12.14) of conservation of electric charge. Since the latter is implicit in the electromagnetic equations, only the linearized form of the equation of mass conservation is reproduced here:

$$(\partial/\partial t)\rho_m + \rho_{m0}(\nabla \cdot \mathbf{V}) = 0 \tag{9.1}$$

where $\mathbf{V}$ is the mass velocity of the plasma considered as one conducting fluid.

The two linearized forms of the momentum transport equations pertaining separately to the electrons and the ions were combined to yield the equation of motion (1.12.21) and the generalized Ohm's law (1.12.40) applicable to the treatment of the plasma as one conducting fluid. The viscosity effects are neglected with the result that the total kinetic pressure dyad simplifies to a scalar kinetic pressure P. The macroscopic electrical neutrality is assumed to be maintained to a very high degree of accuracy. Therefore, the electric charge density and the force due to the electric field both vanish. Under these approximations, the linearized equation of motion (1.12.21) becomes

$$\rho_{m0}(\partial/\partial t)\mathbf{V} = -\nabla P + \mathbf{J} \times \mathbf{B}_0 \tag{9.2}$$

Note that in Eq. (9.2) the gravitational force is also neglected.

Several approximations are introduced for the simplification of the generalized Ohm's law and it is difficult to justify some of these approximations. The advantage of these approximations is that they reduce substantially the complexity of the magnetohydrodynamic equations and thus facilitate the understanding of the physical processes that take place in a plasma at very low frequencies. Since for low frequencies the variation with respect to time is extremely small, the term $(\partial/\partial t)\mathbf{J}$ is ignored. It should be noted that the time-varying terms are retained in the other equations of magnetohydrodynamics. Insofar as the generalized Ohm's law is concerned, the plasma electrons are assumed to be cold and the kinetic pressure dyad of the electrons arising from their random thermal motions is set equal to zero. The resistivity dyad $\boldsymbol{\eta}$ is assumed to reduce to a scalar resistivity $\eta = 1/\sigma$ where σ is the scalar conductivity. If these approximations are incorporated, the generalized Ohm's law (1.12.40) reduces to

$$\mathbf{J} = \sigma[\mathbf{E} + \mathbf{V} \times \mathbf{B}_0 + (1/N_{e0}q_e)\mathbf{J} \times \mathbf{B}_0] \tag{9.3}$$

Except under special circumstances, it is usual to neglect the Hall effect term $(1/N_{e0}q_e)\mathbf{J} \times \mathbf{B}_0$ with the result that Eq. (9.3) becomes

$$\mathbf{J} = \sigma[\mathbf{E} + \mathbf{V} \times \mathbf{B}_0] \tag{9.4}$$

which is the form in which Ohm's law is most commonly used in magnetohydrodynamics. In the present treatment, it is desired to investigate only perfectly conducting ($\sigma = \infty$) plasmas and hence Eq. (9.4) is further simplified to

$$\mathbf{E} + \mathbf{V} \times \mathbf{B}_0 = 0 \tag{9.5}$$

In the following study, only the simplest form (9.5) of Ohm's law is to be adopted.

It appears that two equations of state corresponding to the kinetic pressures of the electrons and the ions are necessary. In approximating the generalized Ohm's law, the electron kinetic pressure which appears separately is omitted and only the total kinetic pressure of the electrons and the ions enters the hydrodynamical equations. Consequently, one equation of state relating the total kinetic pressure to the total density is sufficient. This equation of state can be deduced in an approximate manner from the two equations of state relating the electron and the ion pressures to their respective densities. Let P_{e0} and N_{e0} be the pressure and the number density of the electrons in the equilibrium state and let their perturbations be denoted by P_e and N_e, respectively. Let P_{i0}, N_{i0}, P_i, and N_i be the corresponding quantities for the ions. The equilibrium number densities of the electrons and the ions, N_{e0} and N_{i0} are equal. The adiabatic ratio is denoted by γ_e for the electrons and γ_i for the ions. It follows from Eq. (6.4) that

$$P = P_e + P_i = \gamma_e P_{e0}(N_e/N_{e0}) + \gamma_i P_{i0}(N_i/N_{i0}) \tag{9.6}$$

It is assumed that the adiabatic ratios γ_e and γ_i are equal, that is, $\gamma_e = \gamma_i = \gamma$. As for a perfect mixture of gases, it is assumed that in a plasma the fractional change in the number densities of each species of particles is nearly equal during a compression or a rarefaction, that is, $N_e/N_{e0} = N_i/N_{i0} = N/N_0$. Hence, Eq. (9.6) becomes

$$P = \gamma \frac{N}{N_0}(P_{e0} + P_{i0}) = \gamma P_0 \frac{N}{N_0} = a^2 N(m_e + m_i) \approx a^2 N m_i \tag{9.7}$$

Here a is the sound speed in a perfect gas consisting of a mixture of two kinds of neutral particles characterized by the electronic and the ionic masses respectively. It should be noted that Eq. (9.7), in the same manner as Eq. (9.6), implies the omission of heat conduction and the energy interchange due to collisional interactions.

In Maxwell's equations two approximations are made to facilitate their use in magnetohydrodynamics. First, as has been repeatedly pointed out, the electric charge density is set equal to zero. Second, in the generalized Ampère's law (2.5), the displacement current density is neglected in comparison with the current density contributed by the motion of the charged particles. The magnitude of the displacement current density is given by $\omega\varepsilon_0\mathbf{E}$. For the purpose of estimating the magnitudes, the current density contributed by the motion of the charged particles may be assumed from Eq. (9.4) to be $\sigma\mathbf{E}$. The ratio $\omega\varepsilon_0/\sigma$ of the displacement to the convection current densities is extremely small for the very low frequencies under consideration and for the very highly conducting fluids that are investigated in this treatment. This justifies the omission of the displacement current density. Together with these two approximations, Maxwell's equations become

$$\nabla \times \mathbf{E} = -(\partial/\partial t)\mathbf{B} \tag{9.8}$$

$$\nabla \times \mathbf{B} = \mu_0 \mathbf{J} \tag{9.9}$$

$$\nabla \cdot \mathbf{B} = 0 \tag{9.10}$$

$$\nabla \cdot \mathbf{E} = 0 \tag{9.11}$$

Note that Eq. (9.9) implies that $\nabla \cdot \mathbf{J} = 0$ which is the equation of conservation of electric charge in the absence of net macroscopic electric charge. It is for this reason that the equation (1.12.14) of conservation of electric charge has not been included explicitly. Thus Eqs. (9.1), (9.2), (9.5), (9.7), and (9.8)–(9.11) are the governing equations for the magnetohydrodynamics of perfectly conducting fluids.

Frozen magnetic flux lines

The first application of MHD equations is to prove that the lines of magnetic flux are frozen into the perfectly conducting fluid and therefore are transported with the fluid. If Eq. (9.10) is integrated throughout a volume $\tilde{V}$ bounded by the closed surface $\tilde{S}$ and if the divergence theorem is used, it is found that

$$\int_{\tilde{V}} \nabla \cdot \mathbf{B}(\mathbf{r}, t)\, d\tilde{V} = \int_{\tilde{S}} \mathbf{B}(\mathbf{r}, t) \cdot d\tilde{\mathbf{S}} = 0 \tag{9.12}$$

where $d\tilde{\mathbf{S}} = \hat{\mathbf{n}}\, d\tilde{S}$ and $\hat{\mathbf{n}}$ is an outwardly drawn unit normal to the surface. Consider a closed curve C bounding an open surface S in a time-varying, nonuniform magnetic flux density $\mathbf{B}(\mathbf{r}, t)$. Let the closed curve C move with a uniform velocity $\mathbf{V}(\mathbf{r})$. The velocity $\mathbf{V}(\mathbf{r})$ of the different portions of the curve C need not be the same with the result that the curve C may change in shape as well as undergo translational and rotational motion. Let C_1 and S_1 represent the curve C and the surface S at time t, and, C_2 and S_2 represent them at time $t + \Delta t$ where Δt is an infinitesimal time interval. Let the volume $\tilde{V}$ in Eq. (9.12) be the infinitesimal volume bounded by the surfaces S_1 and S_2 together with the cylindrical surface S_c traced by the curve C in the time interval Δt, as depicted in Fig. 2.6. With reference to Fig. 2.6, the vector element $d\mathbf{l}$ of the curve C_1 covers an area $d\mathbf{S}_c = \hat{\mathbf{n}}\, dS_c = \mathbf{V}(\mathbf{r}) \times d\mathbf{l}\,\Delta t$ in the time interval Δt where $\hat{\mathbf{n}}$ is the outwardly drawn unit normal to the surface dS_c. The resulting outwardly directed magnetic flux through the cylindrical surface S_c is $\int_{C_1} \mathbf{B}(\mathbf{r}, t) \cdot \{\mathbf{V}(\mathbf{r}) \times d\mathbf{l}\}\,\Delta t$ which by a vector formula is also equal to $-\int_{C_1}\{\mathbf{V}(\mathbf{r}) \times \mathbf{B}(\mathbf{r}, t)\} \cdot d\mathbf{l}\,\Delta t$. Together with Eq. (9.12), the total outwardly directed magnetic flux through the surface $\tilde{S} = S_1 + S_2 + S_c$ bounding the volume $\tilde{V}$ is evaluated as

$$\int_{S_1} \mathbf{B}(\mathbf{r}, t) \cdot d\mathbf{S} - \int_{S_2} \mathbf{B}(\mathbf{r}, t) \cdot d\mathbf{S} - \int_{C_1} [\mathbf{V}(\mathbf{r}) \times \mathbf{B}(\mathbf{r}, t)] \cdot d\mathbf{l}\,\Delta t = 0 \qquad (9.13)$$

The minus sign in the second term on the left side of Eq. (9.13) is due to the fact that the outwardly drawn unit normal to the surface S_2 is in a direction opposite to that of the surface S_1. Note that the outwardly drawn unit normal to the surface S_1 is related to the direction of integration around the curve C_1 by the right-hand rule.

The rate of change of magnetic flux through an open surface S is defined as

$$\frac{d}{dt}\left[\int_S \mathbf{B}(\mathbf{r}, t) \cdot d\mathbf{S}\right] = \lim_{\Delta t \to 0} \frac{1}{\Delta t}\left[\int_{S_2} \mathbf{B}(\mathbf{r}, t + \Delta t) \cdot d\mathbf{S} - \int_{S_1} \mathbf{B}(\mathbf{r}, t) \cdot d\mathbf{S}\right] \qquad (9.14)$$

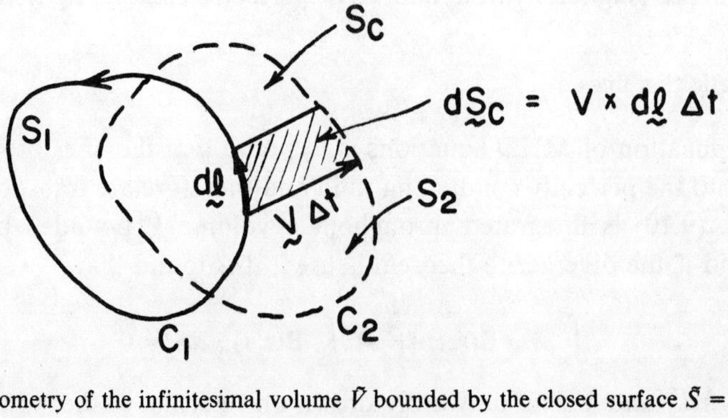

Fig. 2.6. Geometry of the infinitesimal volume $\tilde{V}$ bounded by the closed surface $\tilde{S} = S_1 + S_2 + S_c$.

since $S(t) = S_1$ and $S(t + \Delta t) = S_2$. If $\mathbf{B}(\mathbf{r}, t + \Delta t)$ is expanded into a Taylor series around $\mathbf{B}(\mathbf{r}, t)$, Eq. (9.14) reduces to

$$\frac{d}{dt}\left[\int_S \mathbf{B}(\mathbf{r}, t) \cdot d\mathbf{S}\right] = \lim_{\Delta t \to 0}\left[\int_{S_2} \frac{\partial}{\partial t}\mathbf{B}(\mathbf{r}, t) \cdot d\mathbf{S} + \frac{1}{\Delta t}\left\{\int_{S_2} \mathbf{B}(\mathbf{r}, t) \cdot d\mathbf{S} - \int_{S_1} \mathbf{B}(\mathbf{r}, t) \cdot d\mathbf{S}\right\}\right] \tag{9.15}$$

If Eq. (9.13) is substituted into Eq. (9.15) and the limit $\Delta t \to 0$ is evaluated, the result, on noting that $S_2 = S_1 = S(t)$ in the limit of $\Delta t \to 0$, is

$$\frac{d}{dt}\left[\int_S \mathbf{B}(\mathbf{r}, t) \cdot d\mathbf{S}\right] = \int_S \frac{\partial}{\partial t}\mathbf{B}(\mathbf{r}, t) \cdot d\mathbf{S} - \int_C [\mathbf{V}(\mathbf{r}) \times \mathbf{B}(\mathbf{r}, t)] \cdot d\mathbf{l} \tag{9.16}$$

The line integral in Eq. (9.16) is converted into a surface integral using Stokes' theorem to yield

$$\frac{d}{dt}\left[\int_S \mathbf{B}(\mathbf{r}, t) \cdot d\mathbf{S}\right] = \int_S \left[\frac{\partial}{\partial t}\mathbf{B}(\mathbf{r}, t) - \nabla \times \{\mathbf{V}(\mathbf{r}) \times \mathbf{B}(\mathbf{r}, t)\}\right] \cdot d\mathbf{S} \tag{9.17}$$

The use of Ohm's law (9.5) in its nonlinear form in Eq. (9.8) gives

$$(\partial/\partial t)\mathbf{B} - \nabla \times \{\mathbf{V} \times \mathbf{B}\} = 0 \tag{9.18}$$

Hence, it follows from Eqs. (9.17) and (9.18) that

$$\frac{d}{dt}\left(\int_S \mathbf{B}(\mathbf{r}, t) \cdot d\mathbf{S}\right) = 0 \tag{9.19}$$

which is a mathematical statement of the fact that the magnetic flux threading an open surface S is unchanged as the surface moves with the fluid velocity $\mathbf{V}$. This is equivalent to saying that the magnetic flux also moves with the fluid velocity or the lines of magnetic flux are frozen into the perfectly conducting fluid.

Magnetic pressure

Another simple but very useful application of the MHD equations is the development of the concept of *magnetic pressure* which is very useful in a qualitative discussion of the confinement of high temperature plasmas. The MHD equations deduced in this section are applicable to a fully ionized plasma. Since the MHD equations have been linearized, it follows that the mass velocity $\mathbf{V}$ is a first-order small quantity. The steady-state situation in which there is no time variation of either the lumped hydrodynamic variables or the electromagnetic fields corresponds to the so-called *magnetohydrostatics* for which the equation of motion (9.2) reduces to

$$\nabla P = \mathbf{J} \times \mathbf{B}_0 \tag{9.20}$$

The subscript 0 for **B** is used to indicate that the magnetic flux density does not vary with time. For the sake of convenience, the two magnetostatic equations (9.9) and (9.10) are reproduced here:

$$\nabla \times \mathbf{B}_0 = \mu_0 \mathbf{J} \tag{9.21}$$

and

$$\nabla \cdot \mathbf{B}_0 = 0 \tag{9.22}$$

The governing equations of magnetohydrostatics are given by Eqs. (9.20)–(9.22).

From Eqs. (9.20) and (9.21), it is obtained that

$$(1/\mu_0)(\nabla \times \mathbf{B}_0) \times \mathbf{B}_0 = \nabla P = \nabla \cdot (\mathbf{1}P) \tag{9.23}$$

where $\mathbf{1} = \hat{\mathbf{x}}\hat{\mathbf{x}} + \hat{\mathbf{y}}\hat{\mathbf{y}} + \hat{\mathbf{z}}\hat{\mathbf{z}}$ is a unit dyad. With the help of Eq. (9.22), it is possible to show that

$$(1/\mu_0)(\nabla \times \mathbf{B}_0) \times \mathbf{B}_0 = \nabla \cdot \mathbf{T}^{(m)} \tag{9.24}$$

where $\mathbf{T}^{(m)}$ is the magnetic part of the electromagnetic stress dyad whose expression is given by

$$T_{ij}^{(m)} = (1/\mu_0)[B_i B_j - \delta_{ij} B^2/2] \tag{9.25}$$

Note that $\delta_{ij} = 0$ for $i \neq j$ and $\delta_{ij} = 1$ for $i = j$. It follows from Eqs. (9.23) and (9.24) that

$$\nabla \cdot [\mathbf{1}P - \mathbf{T}^{(m)}] = 0 \tag{9.26}$$

The stress is assumed to be positive if it is tensile and negative if it is compressive. Therefore Eq. (9.26) shows that $-\mathbf{T}^{(m)}$ may be defined as the magnetic pressure dyad. It is instructive to consider a special case in which the static magnetic field is unidirectional and is in the z-direction as given by $\mathbf{B}_0 = \hat{\mathbf{z}}B_0$. The components of the total pressure dyad which is the sum of the kinetic pressure and the magnetic pressure dyads, when written in a matrix form, may be found to be given by

$$\begin{bmatrix} P + B_0^2/2\mu_0 & 0 & 0 \\ 0 & P + B_0^2/2\mu_0 & 0 \\ 0 & 0 & P - B_0^2/2\mu_0 \end{bmatrix} \tag{9.27}$$

It is clear from the matrix (9.27) that the stress caused by the magnetic flux is equivalent to a pressure $B_0^2/2\mu_0$ perpendicular to the lines of magnetic flux and a tension $B_0^2/2\mu_0$ along the magnetic flux lines. Alternatively, the stress caused by the magnetic flux can be thought of as an isotropic magnetic pressure $B_0^2/2\mu_0$ and a *tension* B_0^2/μ_0 along the magnetic flux lines, as illustrated in Fig. 2.7.

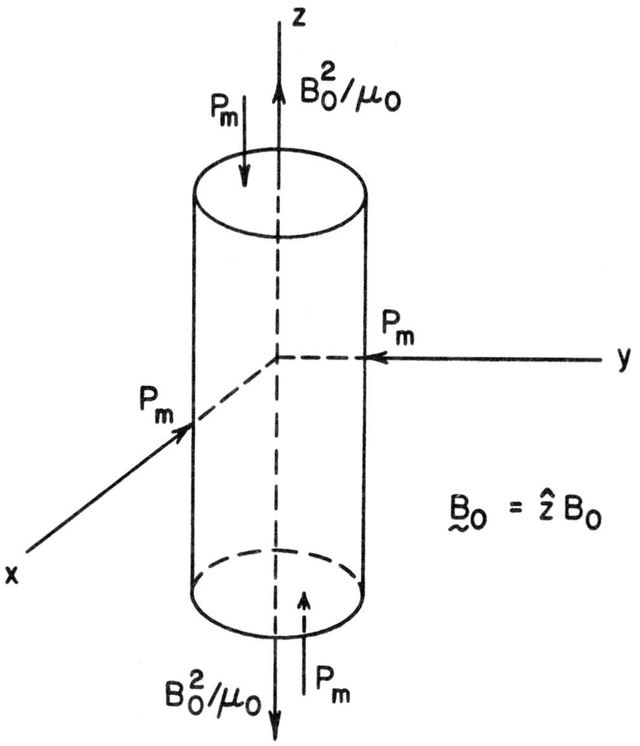

Fig. 2.7. Decomposition of the stress caused by the magnetic flux into an isotropic magnetic pressure $P_m = B_0^2/2\mu_0$ and a tension B_0^2/μ_0 along the magnetic flux lines.

It is usual to introduce in a plasma region hypothetical surfaces over which the kinetic pressure is a constant. These isobaric surfaces are therefore defined by $P = C$ where C is a constant. An isobaric surface is obtained corresponding to each constant value for C. At all points the vector ∇P is along the normal to the isobaric surface passing through the point under consideration. From Eq. (9.20), it is found that

$$\mathbf{B}_0 \cdot \nabla P = 0 \tag{9.28}$$

and

$$\mathbf{J} \cdot \nabla P = 0 \tag{9.29}$$

which show that both $\mathbf{B}_0$ and $\mathbf{J}$ lie on surfaces of constant pressure. Suppose that the isobaric surfaces are closed, concentric cylindrical surfaces as depicted in Fig. 2.8. Since according to Eqs. (9.28) and (9.29) neither the lines of magnetic flux nor those of electric current pass through the isobaric surfaces, it follows that these isobaric surfaces are formed by a network of lines of magnetic flux and electric current. It is assumed that the kinetic pressure increases in the direction towards the axis of the concentric cylindrical surfaces. Therefore ∇P is along a radial line directed towards the axis. In view of Eq. (9.20) it is obtained that $\mathbf{J} \times \mathbf{B}_0$ is also along a radial line directed towards the axis. Consequently, it follows that the lines of magnetic flux and electric current lying on the isobaric surfaces cross each other in such a manner that $\mathbf{J} \times \mathbf{B}_0$ is in the same direction as ∇P. The kinetic pressure is a maximum along the axis which also coincides with a line of magnetic flux. Therefore the axis of the concentric cylindrical surfaces is also called the *magnetic axis* of the plasma configuration.

If the kinetic pressure vanishes on one of the outer cylindrical surfaces, there is no plasma outside this surface. In other words, the plasma is said to be confined within this cylindrical surface. The subject of confinement of a plasma by a magnetic field is important in the theory of controlled fusion reactors and therefore it is desirable to study this subject further. For simplicity, consideration is given here only to the special case of unidirectional magnetic flux, that is, $\mathbf{B}_0 = \hat{\mathbf{z}}B_0$. It is obtained from Eq. (9.26) and the matrix (9.27) that

$$\frac{\partial}{\partial x}[P + B_0^2/2\mu_0] = 0 \quad \frac{\partial}{\partial y}[P + B_0^2/2\mu_0] = 0 \quad \frac{\partial}{\partial z}[P - B_0^2/2\mu_0] = 0 \qquad (9.30\text{a, b, c})$$

Also, Eq. (9.22) gives that

$$(\partial/\partial z)B_0 = 0 \qquad (9.31)$$

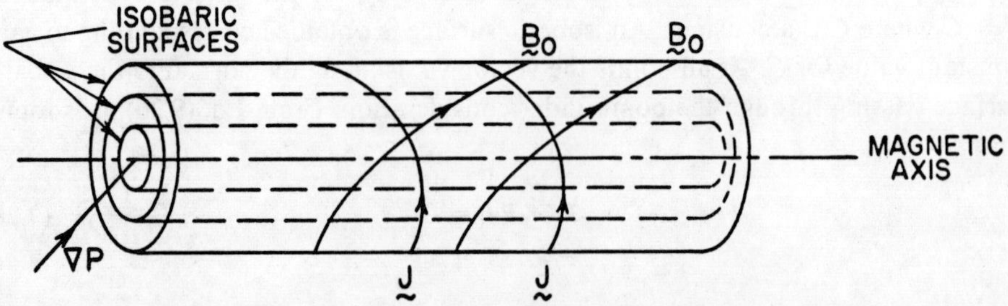

Fig. 2.8. A set of isobaric concentric cylindrical surfaces with the pressure gradient directed towards the axis and a network of lines of magnetic flux and electric current situated on the isobaric surfaces.

which together with Eq. (9.30c) shows that both P and B_0 do not vary in the direction of the magnetic flux density $\mathbf{B}_0$. This result when combined with the solutions of Eqs. (9.30a) and (9.30b) yields

$$P + B_0^2/2\mu_0 = \text{constant} \tag{9.32}$$

Suppose that the plasma is bounded. Then, the kinetic pressure P at the plasma boundary is zero. Let $\mathbf{B}_{0b}$ be the value of the magnetic flux density at the boundary of the plasma. The constant in Eq. (9.32) is determined by evaluating at the boundary of the plasma the sum of the kinetic and the magnetic pressures. Since this sum is equal to $B_{0b}^2/2\mu_0$, Eq. (9.32) becomes

$$P + B_0^2/2\mu_0 = B_{0b}^2/2\mu_0 \tag{9.33}$$

which enables a number of important conclusions to be reached.

Diamagnetism

The first conclusion to be drawn from Eq. (9.33) is that the magnetic flux density inside the plasma is less than its value at the plasma boundary, i.e., the plasma is *diamagnetic*. At a given position inside the plasma, if the kinetic pressure increases, the magnetic flux density decreases, thus enhancing the diamagnetic effect. The reasonableness of this result can be explained in the following manner. The particle motions give rise to electric current which induces a magnetic flux in a direction opposite to the applied magnetic flux and therefore the resultant magnetic flux density inside the plasma is less than its value on the plasma boundary. The electric current depends on the number density of the charged particles and their velocities. The kinetic pressure is also dependent on the same parameters. Therefore, as the kinetic pressure increases, the electric current, the induced magnetic flux density, and therefore also the diamagnetic effect all increase. Subsequently the diamagnetism of a plasma is re-examined in terms of the orbits of the individual particle in a magnetic field and the results deduced there are in accordance with the preceding discussion.

Plasma confinement in a magnetic field

For a unidirectional magnetic flux, Eq. (9.33) reveals the possibility of confinement of the plasma in a magnetic field. If an external magnetic flux is applied, the kinetic pressure of the plasma decreases from the axis radially outwards and the magnetic pressure increases in the same direction such that their sum is a constant in accordance with Eq. (9.33). If the applied magnetic flux is sufficiently large, the kinetic pressure can be forced to vanish on an outer surface and thus the plasma is confined within this surface. For the confinement of the plasma in a unidirectional magnetic flux, it is necessary that the maximum value of the kinetic pressure of the plasma should be less than the magnetic pressure at the plasma boundary.

In Fig. 2.9 is illustrated a device that can be used to confine a plasma by producing a unidirectional magnetic flux. The plasma is initially confined in a hollow cylindrical metal tube whose side is split in the axial direction to form a capacitor. A high voltage is discharged through the capacitor to produce a large azimuthal (θ) current in the metal tube. This current produces a magnetic flux in the axial direction and the electric current induced in the plasma is in the azimuthal direction but opposite to that on the metal tube. The $\mathbf{J} \times \mathbf{B}$ force pushes the plasma towards the axis until a balance is reached between the kinetic and the magnetic pressures. The kinetic pressure which is due to the random thermal motions of the plasma particles encourages the plasma to spread out whereas the magnetic pressure counteracts this effect and acts to constrict or pinch the plasma. Since the pinch effect is due to the electric currents in the θ-direction, this device is called a *theta-pinch.*

It is usual to introduce a parameter β which is a measure of the relative magnitudes of the kinetic and the magnetic pressures; it is defined as the ratio of the kinetic pressure at a location inside the plasma to the magnetic pressure at the plasma boundary:

$$\beta = \frac{P}{B_{0b}^2/2\mu_0} \tag{9.34}$$

Together with Eq. (9.33), β defined by Eq. (9.34) can also be expressed as

$$\beta = 1 - B_0^2/B_{0b}^2 \tag{9.35}$$

The value of β ranges between 0 and 1. A plasma confinement scheme is called a

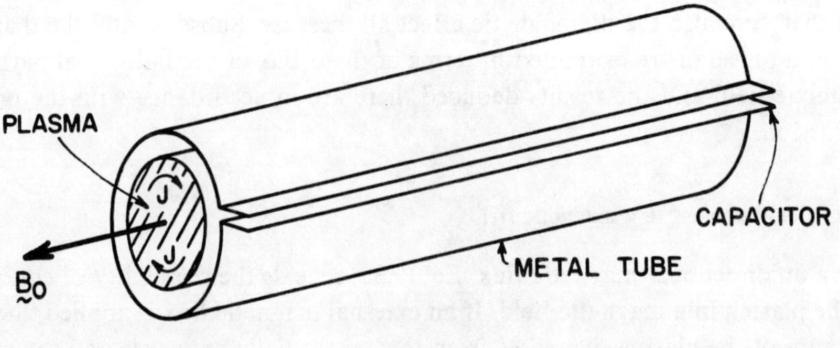

Fig. 2.9. Schematic diagram of a theta pinch device.

low β or a *high* β device according as to whether the kinetic pressure is small in comparison with or of an equal order of magnitude as the magnetic pressure at the plasma boundary.

In the equations of motion (9.2) for magnetohydrodynamics and (9.20) for magnetohydrostatics, the gravitational force has been neglected. This is a justifiable approximation if the plasma dimensions are small as is the case for the plasmas produced in laboratories. Suppose that D is the characteristic dimension of the plasma. The kinetic pressure varies significantly in a distance of the order of D. But, since D is usually very much smaller than the radius of the earth, the variation of the gravitational potential in a distance of the order of D is negligibly small. Consequently, the gravitational force which is proportional to the gradient of the gravitational potential is negligible and hence is not taken into account.

2.10. The pinch effect

A qualitative discussion of plasma confinement by a unidirectional magnetic field was given in the previous section. In view of its importance in controlled thermonuclear reactors, it is proposed to give here a detailed treatment of plasma confinement for the special case in which the confinement is effected by an azimuthal magnetic field produced by the plasma particles drifting in one direction due to an appropriately applied electric field. It is best to begin the treatment by examining the basic mechanism which causes the lateral constriction of the plasma column. Consider an infinitely long column of fully ionized plasma. Let an electric field be applied parallel to the axis of the column in the z-direction. The electrons and the ions drift in the negative and the positive z-directions, respectively, and the electric current due to both species of particles is in the z-direction. Thus the plasma column can be visualized as consisting of a bundle of parallel current filaments. Let two such filaments one of which lying on the axis and the other off the axis be isolated for examination. The magnetic field $B_{\theta 1}$ due to the first filament on the axis is in the θ-direction and the $\mathbf{J} \times \mathbf{B}$ force acting on the second filament carrying the current I_{z2} in the z-direction is $-I_{z2} B_{\theta 1} \hat{\mathbf{r}}$. The second filament is therefore forced to move towards the first filament. In a similar manner, it can be shown that the magnetic field due to the second filament creates a $\mathbf{J} \times \mathbf{B}$ force on the first filament in such a direction as to force it to move towards the second filament. Thus two parallel current carrying filaments tend to move towards each other. This is the underlying mechanism which forces the plasma column to contract laterally. This lateral constriction of the plasma column is known as the pinch effect.

The number density and the temperature of the plasma increase as the column is compressed laterally. The random thermal motions of the particles endow them with the property to stray away from high density to low density regions and from high temperature to low temperature regions. Thus the kinetic pressure of the plasma acts to deter the confinement of the plasma whereas the electromagnetic forces aid the confinement process. When these counteracting forces are balanced, an equilibrium state is reached in which the radius of the plasma column remains constant in time. This state of affairs is referred to as the *equilibrium pinch*. When the electromagnetic forces exceed the kinetic pressure, the radius of the plasma column changes with time resulting in what is known as *the dynamic pinch*. The equilibrium pinch is first investigated and it is then followed by a treatment of the dynamic pinch. In some cases, the dynamic pinch can be treated as a succession of equilibrium pinch states in each one of which the radius of the plasma column remains a constant whereas in the succeeding pinch states the radius changes by a small but a finite amount.

The equilibrium pinch

In an equilibrium pinch, none of the system parameters varies with time. As mentioned previously, the electrons and the ions drift in the negative and in the positive z-directions, respectively, under the action of an electric field in an axial direction. The pinch discharge is assumed to be infinitely long and circularly symmetrical with the result that none of the parameters varies with θ or z. The gravitational and the viscous forces are neglected and therefore the kinetic pressure reduces to a scalar. If $\mathbf{v}$ is the drift velocity, it follows that in the equilibrium state

$$(d/dt)\mathbf{v} = (\partial/\partial t)\mathbf{v} + (\mathbf{v} \cdot \nabla)\mathbf{v} = 0 \tag{10.1}$$

The first term vanishes because the parameters are independent of time. The second term vanishes because $\mathbf{v}$ is in the z-direction and is independent of z. Therefore, the momentum transport equations for the electrons and the ions are obtained from Eq. (1.9.14) as

$$0 = -\nabla P_e + N_e q_e(\mathbf{E} + \mathbf{v}_e \times \mathbf{B}) + \mathbf{P}_{ei} \tag{10.2}$$

and

$$0 = -\nabla P_i + N_i q_i(\mathbf{E} + \mathbf{v}_i \times \mathbf{B}) + \mathbf{P}_{ie} \tag{10.3}$$

where $\mathbf{P}_{ei}$ and $\mathbf{P}_{ie}$ are the collision terms. The addition of Eqs. (10.2) and (10.3) yields

$$\nabla P = \mathbf{J} \times \mathbf{B} \tag{10.4}$$

where

$$P = P_e + P_i \tag{10.5}$$

The resultant electric charge density $N_e q_e + N_i q_i$ is set equal to zero since the plasma is macroscopically neutral. Also Eqs. (1.12.4) and (1.12.19) have been used in deducing Eq. (10.4). If $q_e = -e$ and $q_i = e$ are respectively the charge on an electron and an ion, it follows that $N_e = N_i = N$. It is to be noted that the steady state, linearized equation of motion which is one of the MHD equations is of the same form as Eq. (10.4). Also, the kinetic pressure $P(r)$, the electric current density $\mathbf{J}(r)$ and the magnetic field $\mathbf{B}(r)$ are functions of only the radial distance r.

The equilibrium pinch is illustrated in Fig. 2.10. The equilibrium radius of the plasma column is denoted by R. The current density $\mathbf{J}(r)$ has only a z-component $J_z(r)$. Since all the parameters depend only on the radial distance r, the magnetic flux density has only a θ-component $B_\theta(r)$. Only the radial component of Eq. (10.4) is relevant and it is given by

$$dP(r)/dr = -J_z(r)B_\theta(r) \tag{10.6}$$

The total current $I_z(r)$ inside the cylinder of radius r is found as

$$I_z(r) = \int_0^r J_z(r')2\pi r'\,dr' \tag{10.7}$$

and hence

$$(d/dr)I_z(r) = 2\pi r J_z(r) \tag{10.8}$$

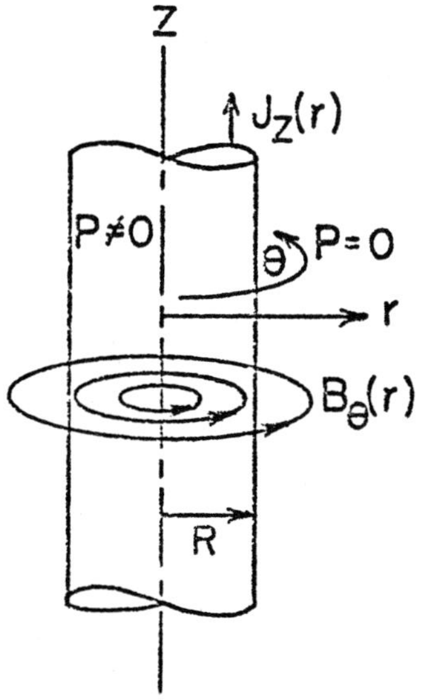

Fig. 2.10. Schematic diagram of the equilibrium pinch showing the various parameters.

From Ampère's law, the magnetic flux density B_θ is deduced as

$$B_\theta(r) = \mu_0 I_z(r)/2\pi r \tag{10.9}$$

The substitution of $B_\theta(r)$ and $J_z(r)$ from Eqs. (10.9) and (10.8), respectively, into Eq. (10.6) yields

$$4\pi^2 r^2 \frac{dP(r)}{dr} = -\mu_0 I_z(r) \frac{d}{dr} I_z(r) = -\frac{d}{dr}\left\{ \frac{\mu_0 I_z^2(r)}{2} \right\} \tag{10.10}$$

Both sides of Eq. (10.10) are integrated with respect to r from $r = 0$ to $r = R$. If the left side of Eq. (10.10) is simplified by an integration by parts, the result is

$$4\pi^2 r^2 P(r)\Big|_{r=0}^{r=R} - 4\pi \int_0^R 2\pi r P(r)\, dr = -\frac{\mu_0 I_0^2}{2} \tag{10.11}$$

where $I_0 = I_z(R)$ is the axial current through the entire cross section of the plasma column and obviously $I_z(0) = 0$. Since the plasma column is confined to the range $0 \leqslant r < R$, it follows that $P(r)$ is nonzero and finite for $0 \leqslant r < R$ and zero for $R \leqslant r \leqslant \infty$. Therefore, the first term on the left side of Eq. (10.11) vanishes. Let the partial pressures of the electrons and the ions be governed by the ideal gas law. Therefore,

$$P_e(r) = N(r)KT_e \qquad P_i(r) = N(r)KT_i \tag{10.12a}$$

$$P(r) = P_e(r) + P_i(r) = N(r)K(T_e + T_i) \tag{10.12b}$$

It is assumed that the electron and the ion temperatures, T_e and T_i are constants throughout the discharge. The number of particles per unit length of the plasma column is evaluated as

$$N_l = \int_0^R N(r) 2\pi r\, dr \tag{10.13}$$

which has the dimensions of L^{-1}. If Eqs. (10.12b) and (10.13) are substituted into Eq. (10.11), it is found that

$$I_0^2 = (8\pi/\mu_0)K(T_e + T_i)N_l \tag{10.14}$$

which is known as *the Bennett relation.*

The Bennett relation gives the current that must be discharged through the column in order to be able to contain a plasma at a specified temperature and a given number of particles N_l per unit length of the column. Note that N_l is unchanged as the plasma column is radially compressed. Suppose that $N_l = 4 \times 10^{19}$ particles per meter and that the plasma temperature $T_e + T_i = 10^8\,°\mathrm{K}$. Since $\mu_0 = 4\pi \times 10^{-7}$ H/m and $K = 1.38 \times 10^{-23}$ J/°K, it is found that I_0 is about a million amperes. Thus a very large current is required for the containment of the plasma. It appears from Eq. (10.14) that a plasma can be contained at any temperature by discharging a sufficiently large current through the column. The radiation losses in the plasma column increase with increasing current and hence impose a limit on the current that can be discharged. Therefore there is a limit on the plasma temperature that can be attained in a cylindrical pinch device. Numerous instabilities occur and force a limit on the time duration for which a high temperature plasma can be contained. A descriptive treatment of some of these instabilities is given in the following section.

Sheath current model

So far the radial distribution of $P(r)$, $J_z(r)$, and $B_\theta(r)$ have not been discussed. For this purpose, it is convenient to proceed from Eq. (10.6) in a different manner. From the law of magnetostatics, $\nabla \times \mathbf{B(r)} = \mu_0 \mathbf{J(r)}$, it is found that

$$J_z(r) = \frac{1}{\mu_0 r}\frac{d}{dr}\{rB_\theta(r)\} = \frac{1}{\mu_0}\frac{d}{dr}B_\theta(r) + \frac{B_\theta(r)}{\mu_0 r} \tag{10.15}$$

The substitution of $J_z(r)$ from Eq. (10.15) into Eq. (10.6) gives

$$\begin{aligned}(d/dr)P(r) &= -\frac{B_\theta(r)}{\mu_0}\frac{d}{dr}B_\theta(r) - \frac{B_\theta^2(r)}{\mu_0 r} \\ &= -\frac{1}{2\mu_0 r^2}\frac{d}{dr}\{r^2 B_\theta^2(r)\}\end{aligned} \tag{10.16}$$

The integration of Eq. (10.16) with respect to r from 0 to r yields

$$P(r) = P(0) - \frac{1}{2\mu_0}\int_0^r \frac{1}{r^2}\frac{d}{dr}\{r^2 B_\theta^2(r)\}\,dr \tag{10.17}$$

Since $P(R) = 0$, it is obtained from Eq. (10.17) that

$$P(0) = \frac{1}{2\mu_0}\int_0^R \frac{1}{r^2}\frac{d}{dr}\{r^2 B_\theta^2(r)\}\,dr \tag{10.18}$$

The result of substitution of Eq. (10.18) into Eq. (10.17) is

$$P(r) = \frac{1}{2\mu_0}\int_r^R \frac{1}{r^2}\frac{d}{dr}\{r^2 B_\theta^2(r)\}\,dr \tag{10.19}$$

From Eqs. (10.7), (10.9), and (10.19), the radial distribution of $B_\theta(r)$ and $P(r)$ can be deduced if the radial distribution of the current density $J_z(r)$ is known. The expressions for $B_\theta(r)$ and $P(r)$ can be evaluated if, for example, the current density $J_z(r)$ is a constant. It is useful to consider another radial distribution of $J_z(r)$ which is employed in the investigation of the dynamic pinch. If the plasma is perfectly conducting the current cannot penetrate the plasma and therefore exists only on the surface $r = R$ of the column. Then Eqs. (10.7) and (10.9) show that $B_\theta(r)$ exists only for $r > R$ and is given by

$$B_\theta(r) = \mu_0 I_0 / 2\pi r \qquad \text{for } R < r < \infty \tag{10.20}$$

where I_0 is the total axial current. It is then obtained from Eq. (10.17) that

$$P(r) = P(0) \qquad \text{for } 0 < r < R \tag{10.21}$$

Therefore, for a perfectly conducting plasma the current vanishes inside the column and exists only on the surface of the column. The magnetic flux density vanishes inside the column and falls off as r^{-1} exterior to the column; hence $B_\theta(r)$ has the maximum value on the surface of the column. The kinetic pressure is a constant inside the column and vanishes outside the column. These radial distributions are illustrated in Fig. 2.11. For this special case, the pinch effect can be thought of as due to the abrupt building up of the magnetic pressure in the region external to the column.

Another interesting result can be derived from Eq. (10.16). The average value of the kinetic pressure inside the column is determined as

$$\langle P(r)\rangle = \frac{1}{\pi R^2}\int_0^R 2\pi r P(r)\,dr \tag{10.22}$$

which can be simplified by an integration by parts to yield

$$\langle P(r)\rangle = \frac{1}{R^2}\left[r^2 P(r)\Big|_0^R - \int_0^R r^2 \frac{d}{dr}P(r)\,dr\right] \tag{10.23}$$

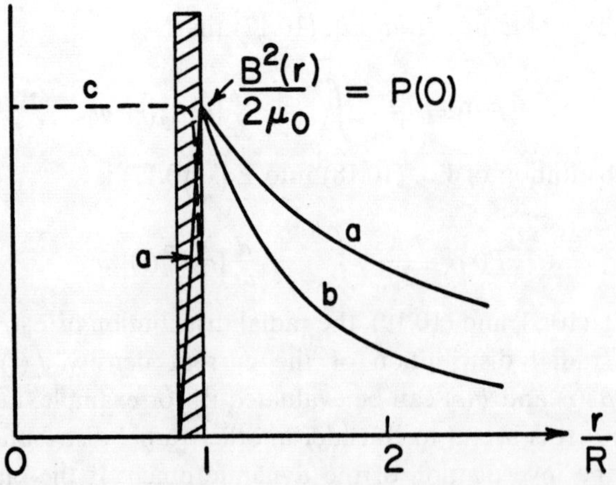

Fig. 2.11. Radial distribution of $B_\theta(r)$, $B_\theta^2(r)/2\mu_0$, and $P(r)$ for a plasma column with a current restricted to a small sheath region (indicated by shading with slanted lines) on the surface of the column: (a) $B_\theta(r)$, (b) $B_\theta^2(r)/2\mu_0$, (c) $P(r)$.

The integrated term is zero and together with Eq. (10.16), the integral in Eq. (10.23) can be evaluated with the following result:

$$\langle P(r) \rangle = B_\theta^2(R)/2\mu_0 \tag{10.24}$$

It is seen from Eq. (10.24) that the average kinetic pressure inside an equilibrium pinch is balanced by the magnetic pressure at the surface of the column. This result is to be anticipated since equilibrium is achieved as a result of the balance between these very two forces.

The Bennett pinch

In addition to the sheath model for the equilibrium pinch, any number of other models is possible depending on the radial distribution of $J_z(r)$, $B_\theta(r)$, and $P(r)$. Another model is that corresponding to a constant drift velocity of the plasma particles throughout the cross section of the column. It is instructive to investigate this model particularly because it was used by Bennett, the discoverer of the pinch effect. According to Eq. (1.12.4), the current density is given by

$$J_z(r) = N(r)e[v_{iz} + v_{ez}] \tag{10.25}$$

where $\mathbf{v}_i = \hat{\mathbf{z}}v_{iz}$ and $\mathbf{v}_e = -\hat{\mathbf{z}}v_{ez}$, and, $\mathbf{v}_i$ and $\mathbf{v}_e$ are the drift velocities of the ions and the electrons respectively. Since the electric field is in the z-direction, it follows that v_{ez} is positive. As indicated before, the drift velocities are constants independent of r. In view of the fact that the ion mass is much larger than the electron mass, and since they are produced by the same electric field, the drift velocity of the ions is much smaller than that of the electrons and hence can be neglected with the result that Eq. (10.25) becomes

$$J_z(r) = N(r)ev_{ez} \tag{10.26}$$

The use of Eqs. (10.12b) and (10.26) in Eq. (10.6) gives

$$(d/dr)N(r) + [ev_{ez}/K(T_e + T_i)]N(r)B_\theta(r) = 0 \tag{10.27}$$

If Eq. (10.27) is multiplied by $r/N(r)$, and then differentiated with respect to r, and if $(d/dr)\{rB_\theta(r)\}$ appearing in the resulting equation is replaced by its equivalent expression in terms of $N(r)$ as obtained from Eqs. (10.15) and (10.26), the result is

$$(d/dr)\{(r/N(r))(d/dr)N(r)\} + 8brN(r) = 0 \tag{10.28}$$

where

$$b = \mu_0 e^2 v_{ez}^2/8K(T_e + T_i) \tag{10.29}$$

The solution of the nonlinear differential equation (10.28) gives the radial distribution of the number density $N(r)$.

Since the number density has been assumed to be symmetrical about the z-axis, where $r = 0$, it follows that, if $N(r)$ is a smoothly varying function of r, $dN(r)/dr = 0$ for $r = 0$. Bennett has obtained the solution of Eq. (10.28) subject to this boundary condition to be given by

$$N(r) = N_0[1 + N_0 br^2]^{-2} \tag{10.30}$$

where N_0 is the number density on the axis corresponding to $r = 0$. The radial distribution of the number density of the particles as given by Eq. (10.30) is known as the *Bennett distribution* which is illustrated in Fig. 2.12. Although Eq. (10.30) shows that the particles are present up to infinity, since the number density $N(r)$ falls off very rapidly from the z-axis radially outwards as seen from Fig. 2.12, for all essential purposes, the plasma is concentrated symmetrically in a small neighborhood around the z-axis. Let $r = R$ be introduced arbitrarily as the cylindrical surface within which the plasma is confined. The precise meaning of confinement as used here is clarified presently.

The number of particles $N_l(R)$ per unit length contained in a cylinder of radius $r = R$ is determined from Eq. (10.30) as

$$N_l(R) = \int_0^R 2\pi r N(r)\, dr = \int_0^R 2\pi r N_0[1 + N_0 br^2]^{-2}\, dr \tag{10.31}$$

The integral in Eq. (10.31) is evaluated to yield

$$N_l(R) = N_0 \pi R^2[1 + N_0 bR^2]^{-1} \tag{10.32}$$

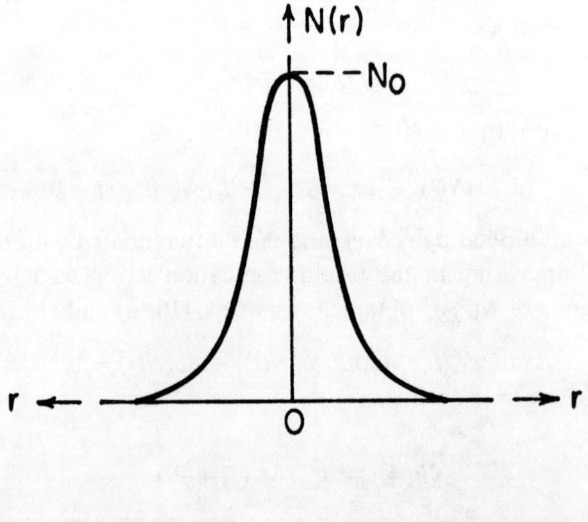

Fig. 2.12. The Bennett distribution of the number density $N(r)$ of particles in an equilibrium pinch.

Since the plasma extends up to $r = \infty$, the actual number of particles per unit length is deduced from Eq. (10.32) as

$$N_l(\infty) = \pi/b \tag{10.33}$$

where b is seen to have the dimensions of length. Let α be the fraction of the number of particles $N_l(\infty)$ per unit length that is contained in a cylinder of radius R:

$$N_l(R) = \alpha N_l(\infty) = \alpha\pi/b \tag{10.34}$$

The substitution of Eq. (10.34) into Eq. (10.32) yields after some rearrangement

$$\sqrt{N_0 b}\, R = [\alpha/(1-\alpha)]^{1/2} \tag{10.35}$$

Since $\sqrt{N_0 b}$ has the dimensions of an inverse length, $\sqrt{N_0 b}\, R$ can be thought of as the normalized radius of the plasma column in the equilibrium state. If 90% of the particles are contained within this column, that is, if $\alpha = 0.9$, $\sqrt{N_0 b}\, R = 3$. Similarly, if 99% of the particles are contained within the aforementioned column, it is found that $\sqrt{N_0 b}\, R = 9.95$. Thus even though they extend up to infinity, a major portion of these particles lie within a short distance from the z-axis. For the sake of definiteness, it is convenient to assume that a plasma is confined within a column of radius R if 90% of the particles are within this column. Therefore, the radius R introduced arbitrarily as the cylindrical surface within which the plasma is confined is now given a reasonable meaning and is defined by the relation

$$\sqrt{N_0 b}\, R = 3 \tag{10.36}$$

The Bennett relation (10.14) is generally valid. It is interesting to verify its validity for the special case of the Bennett pinch. The total current I_0 in the discharge is given by the relation

$$I_0^2 = [N_l(\infty) e v_{ez}]^2 \tag{10.37}$$

The substitution for $e^2 v_{ez}^2$ from Eq. (10.29) and b from Eq. (10.33) into Eq. (10.37) yields

$$I_0^2 = (8\pi/\mu_0) K (T_e + T_i) N_l(\infty) \tag{10.38}$$

which is identical to that in Eq. (10.14), as anticipated.

The dynamic pinch

The theory of the equilibrium pinch can be applied either when the radius of the plasma column is constant in time or when it is varying very slowly in comparison with the time required for the plasma particles to attain constant temperature. If the variation of the pinch current is sufficiently rapid that thermal equilibrium is not reached, the theory of the equilibrium pinch cannot be applied. In order to illustrate the general features of the time-varying pinch, the following simple theory of the dynamic pinch is presented.

Consider an infinitely long, hollow dielectric cylinder of inner radius r_0. A fully ionized plasma fills the interior region $0 < r < r_0$. For the sake of simplicity, the plasma is assumed to be perfectly conducting; also, the kinetic pressure of the plasma is ignored. An electric field is applied at time $t = 0$ parallel to the axis of the cylinder. Since the plasma is perfectly conducting, the current exists only on its external surface $r = r_0$. There is neither electric current nor magnetic flux inside the plasma, as illustrated in Fig. 2.13. It is to be noted that the physical parameters of the dynamic pinch are independent of θ and z. As before, the magnetic flux density is in the azimuthal direction and its magnitude just exterior to the current sheet at the radius r is given by

$$B_\theta(r) = \mu_0 I(t)/2\pi r \tag{10.39}$$

where $I(t)$ is the total axial current at time t. The initial value of $B_\theta(r)$ can be obtained from Eq. (10.39) by setting $t = 0$ and $r = r_0$. The magnetic pressure $P_m(r)$ acting on the current sheet radially inwards is obtained as

$$P_m(r) = B_\theta^2(r)/2\mu_0 \tag{10.40}$$

In view of Eq. (10.40), the total force $\mathbf{F}(r) = \hat{\mathbf{r}}F(r)$ acting radially inwards per unit length of the current sheet is deduced from Eqs. (10.39) and (10.40) as

$$F(r) = -2\pi r P_m(r) = -\mu_0 I^2(t)/4\pi r \tag{10.41}$$

As mentioned previously, the initial values of $P_m(r)$ and $F(r)$ are obtained by setting $r = r_0$ and $t = 0$. On account of the force given by Eq. (10.41), the current sheet

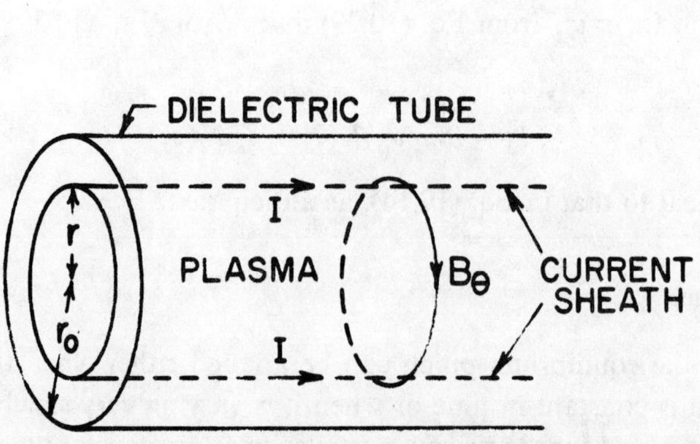

Fig. 2.13. Snowplow model for the dynamic pinch.

moves radially inwards with the result that the radius r of the plasma column is a function of time. As the current sheet moves inwards, it sweeps up all the plasma particles in front of it and these particles are assumed to adhere to the current sheet. For this reason, the present model of the dynamic pinch is called the *snowplow model.*

The instantaneous radius r of the pinch discharge can be determined by the solution of the equation of motion which can be set up in the following manner. If ρ_m is the original mass density of the plasma, the mass per unit length of the current sheet at time t when the radius of the current sheet is r, is given by $\pi\rho_m(r_0^2 - r^2)$. Since the kinetic pressure is zero, the balancing of the magnetic force as expressed by Eq. (10.41) with the rate of change of momentum of the current sheet of radius r gives the equation of motion as:

$$(d/dt)[\pi\rho_m(r_0^2 - r^2)dr/dt] = -\mu_0 I^2(t)/4\pi r \tag{10.42}$$

The radius of the pinch discharge can be evaluated from Eq. (10.42) as a function of time if the functional dependence of the pinch current is known.

The pinch current is determined by the resistance, the inductance and the capacitance of the external circuit which applies the electric field to the plasma column as well as the circuit parameters of the discharge itself. The plasma column which has been assumed to be infinite for theoretical analysis, is in practice finite in length and its terminals are connected to a bank of capacitors which are discharged to produce the pinch current. Thus the external circuit has associated with it a large capacitance. Suppose that the inductance of the external circuit is greater than that of the pinch discharge. Then the pinch current is determined primarily by the external circuit. Under these circumstances, the form of the pinch current is that of the initial portion of a damped sinusoid. The damping is usually negligible and the frequency depends on the external circuit parameters. Therefore, the pinch current may be assumed to be of the form

$$I(t) = I_0 \sin \omega t \approx I_0 \omega t \tag{10.43}$$

If the pinch discharge has a greater inductance than that of the external circuit, it is necessary to take into account the circuit parameters of the pinch discharge itself in the determination of the pinch current. This situation is not analyzed in the following treatment.

It is possible to recast Eq. (10.42) together with Eq. (10.43) in the following normalized form:

$$(d/dt_n)[(1 - r_n^2)dr_n/dt_n] = -t_n^2/r_n \tag{10.44}$$

where

$$r_n = r/r_0 \qquad t_n = t(\mu_0 I_0^2 \omega^2/4\pi^2 r_0^4 \rho_m)^{1/4} \tag{10.45}$$

It is necessary to solve Eq. (10.44) numerically. In Fig. 2.14, the normalized radius of the dynamic pinch is depicted as a function of the normalized time t_n.

A phenomenon that usually occurs in the dynamic pinch is not taken into account in this analysis. A qualitative discussion of this phenomenon is appropriate at this place. As the current sheet moves radially inwards, a wave motion is set up and this wave travels faster than the current sheet. These waves move radially inwards, get reflected off the axis, travel radially outwards and strike the current sheet whose motion is thus reversed. After a short interval of time, the magnetic forces predominate and the current sheet moves radially inwards again. This sequence of events takes place periodically with the result that the radius of the current sheet bounces about an equilibrium value as indicated in Fig. 2.15. The amplitude of each succeeding bounce becomes smaller and presumably the radius settles down to an equilibrium value. This phenomenon is known as *bouncing*.

2.11. Configuration-space instability

It is difficult to sustain a pinched plasma column for any reasonable time interval. The dynamic behavior of the plasma column follows the predictions of the previous section until approximately the first bounce period. Thereafter any small departure from the cylindrical geometry grows rapidly in time and as a result of the growth of these perturbations, the plasma column disintegrates. A satisfactory mathematical analysis of these instabilities is difficult. Indeed, the subject of plasma instabilities

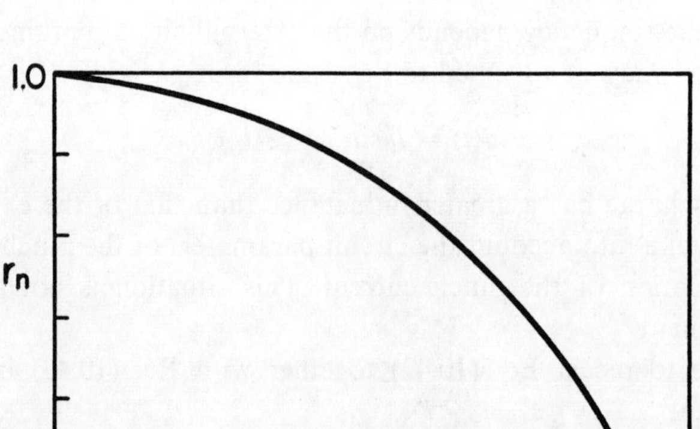

Fig. 2.14. Normalized radius r_n of the plasma column of the dynamic pinch as a function of the normalized time t_n.

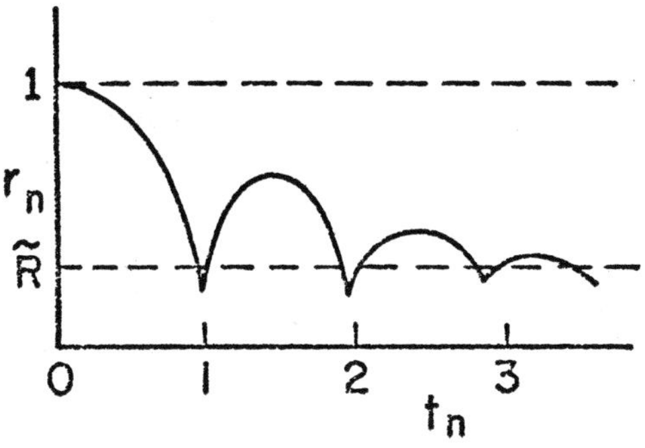

Fig. 2.15. Illustration of the bouncing of the normalized radius of the plasma column of the dynamic pinch as a function of the normalized time t_n; $\tilde{R} = R/r_0$.

is so intriguing that it has emerged as an important field of research in plasma physics. The instabilities in a plasma are conveniently divided into two broad categories. The first category is the configuration-space instability which depends on the geometrical structure of the plasma and the second is the velocity-space instability which is governed by the forms of the velocity distribution functions of the electrons and the ions. A qualitative discussion of the configuration-space instability is given in this section with particular reference to the pinched plasma column.

Before proceeding with a treatment of these instabilities, it is only proper to begin with a definition of instability. An equilibrium is said to be stable if a small perturbation of the equilibrium state is followed by either a return to the equilibrium state or a nongrowing oscillation about the equilibrium state. On the other hand, if a perturbation of the equilibrium state grows without limit, the corresponding equilibrium is said to be unstable. Any departure from the equilibrium state is accompanied by additional forces, and if these forces act to increase rather than diminish these deviations from equilibrium, instability is said to occur. The following discussion of instabilities is simplified by assuming a perfectly conducting plasma with the result that the magnetic pressure P_m exists only exterior to the plasma. The kinetic pressure P is uniform inside and has a zero value outside the plasma. If P_{mb} is the value of the magnetic pressure at the plasma boundary, the equilibrium state is defined by

$$P = P_{mb} = B_{0b}^2/2\mu_0 \tag{11.1}$$

where B_{0b} is the value of the magnetic flux density at the boundary of the plasma.

The sausage instability

Consider the equilibrium state of a pinched plasma column indicated by the dashed lines in Fig. 2.16. Let the equilibrium state be disturbed as shown in Fig. 2.16. It is assumed that the volume of the plasma and therefore the kinetic pressure are unchanged. The azimuthal symmetry is also assumed to be unimpaired. The perturbation is such as to constrict the plasma column at some locations and expand it at others. In other words, the perturbation is in the form of a radially symmetric wave-like disturbance with both crests and troughs on the surface of the plasma column. The magnetic flux lines which exist only external to the plasma column remain parallel to the plasma boundary.

The deviation from the equilibrium state alters the values of the magnetic flux density and the magnetic pressure at the boundary of the plasma. Since the axial current through the pinch discharge is unchanged and is the same throughout the length of the column, the resulting azimuthal magnetic flux density is inversely proportional to the radial distance r from the axis of the plasma column. At the locations where the radius has decreased from its equilibrium value, the magnetic flux density and the magnetic pressure are greater than their equilibrium values. Since the kinetic pressure is unchanged, it follows that the magnetic pressure which now exceeds the kinetic pressure forces the plasma column radially inwards, thus accentuating the constriction. At the locations where the radius has increased from its equilibrium value, the magnetic flux density and the magnetic pressure are less than their equilibrium values. The kinetic pressure which is unchanged is now greater than the magnetic pressure and therefore forces the plasma column radially outwards thus increasing the expansion of the plasma column. Therefore, the initial perturbations give rise to new forces which tend to increase the perturbations with the result that the plasma column is unstable for this kind of departure from the equilibrium state. After a time the constrictions reach the axis and the plasma column appears like a string of sausages. For this reason, this instability has been

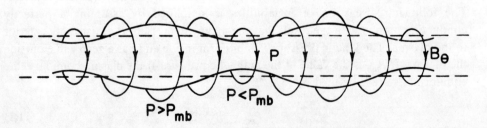

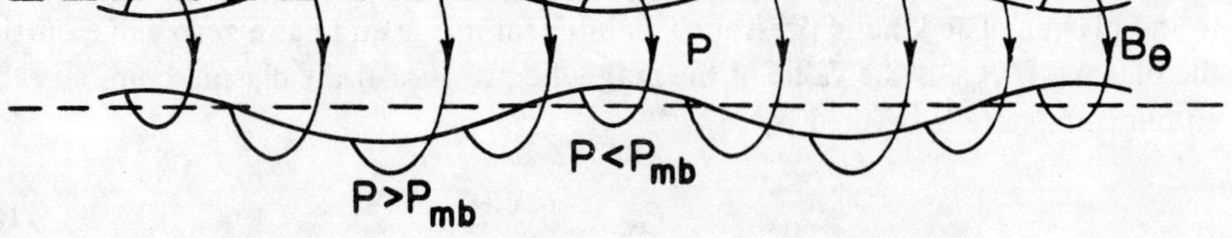

Fig. 2.16. The sausage instability of the pinched plasma column.

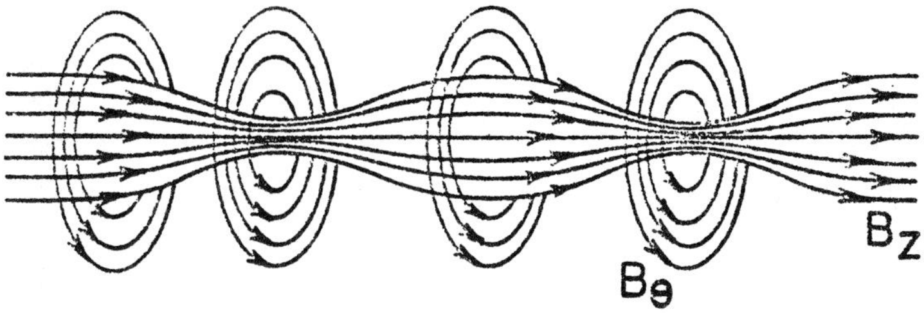

Fig. 2.17. Schematic diagram of the inhibition of the sausage instability by the trapped axial magnetic flux lines.

designated as *the sausage instability*.

The sausage instability can be inhibited by an axial magnetic flux which can be applied in the following manner. Before activating the pinch discharge, a current is passed through a solenoidal coil wound around the discharge tube and this produces an axial magnetic flux which is largely inside the plasma column. When the discharge current is passed, there is a lateral contraction of the plasma column as a result of the pinch effect and the axial lines of magnetic flux contract with the column since they are frozen inside the plasma. For this case, in the equilibrium state, the external magnetic pressure due to the azimuthal magnetic flux is balanced by the kinetic pressure and the internal magnetic pressure due to the axial magnetic flux. When the perturbations leading to the sausage instability grow, as shown in Fig. 2.17, the axial magnetic flux density increases near the constrictions and associated with it the internal magnetic pressure increases and forces the constriction to expand. At the locations where the plasma column bulges out, the internal magnetic pressure decreases and this creates a net force which tends to constrict the plasma column. Thus the axial magnetic flux has a stabilizing influence on the pinched plasma column against the setting in of the sausage instability.

The kink instability

Another main type of instability that leads to the destruction of the pinched plasma column is known as *the kink instability*. Consider a perturbation in the form of a bend or kink in the plasma column which is assumed to retain its uniform, circular cross section. It is usual for this type of perturbation to occur periodically along the length of the plasma column which is therefore bent successively upwards and downwards along its length. The kinetic pressure inside the plasma is assumed to be undisturbed. Since the same amount of magnetic flux is now spread over a shorter length, the magnetic flux density near the concave side of the bend is greater than its equilibrium value and therefore the magnetic pressure external to the column is greater than the kinetic pressure inside, near the concave side of the bend. Near the convex side of the bend, the same amount of magnetic flux is now spread over a longer distance with the result that the magnetic flux density is less than its equilibrium value. Since the kinetic pressure is unchanged, the internal kinetic pressure is greater than the external magnetic pressure. These results are illustrated in Fig. 2.18. The net effect is that the bend gets more pronounced and the kink instability sets in.

As in the case of the sausage instability, the kink instability can also be prevented by the application of an axial magnetic flux, as shown in Fig. 2.19. It was established that there is a tension acting along the axial magnetic flux lines which are frozen inside the plasma column before activating the pinch discharge. When

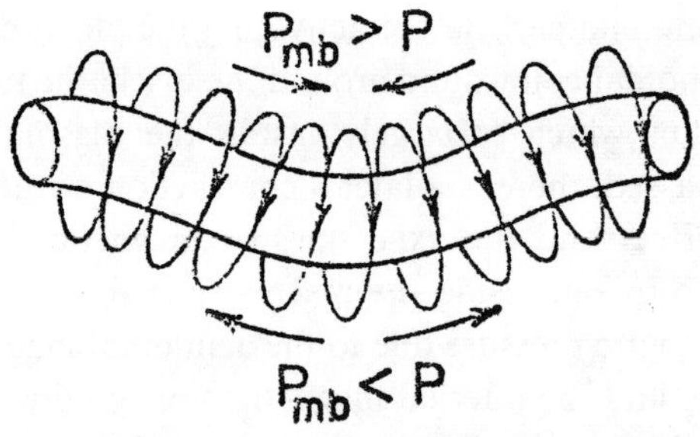

Fig. 2.18. The kink instability of the pinched plasma column.

the perturbations which grow into the kink instability set in, the axial magnetic flux lines are stretched and the resulting increase in the tension along the flux lines inside the plasma column opposes the external forces and tends to straighten out the plasma column. Thus the axial magnetic flux exerts a stabilizing influence on the pinched plasma column against the onset of the kink instability.

The cusp field

The common features of the sausage and the kink instabilities are that the external magnetic flux lines are azimuthal and are concave towards the plasma. In such cases, any perturbation of the equilibrium state grows for the same reasons which lead to the sausage and the kink instabilities. On the other hand, suppose that the magnetic flux lines are convex towards the plasma. It is to be argued presently that

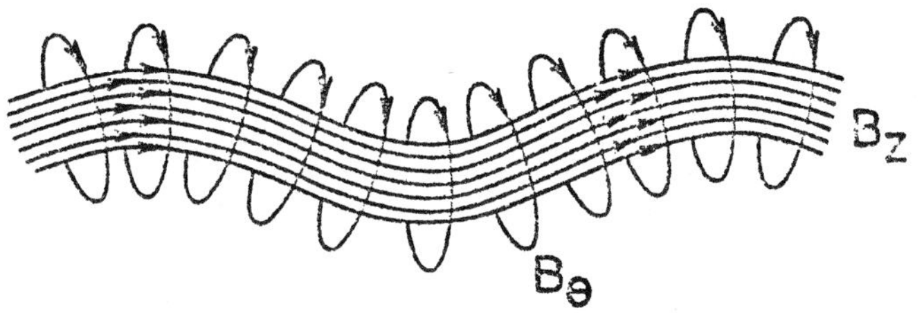

Fig. 2.19. Schematic diagram of the inhibition of the kink instability caused by the increased tension along the trapped axial magnetic flux lines.

for this case, any perturbation of the equilibrium state is diminished and is followed by a return to the equilibrium state. Consequently, for the confinement of plasma, it is desirable to use this stable configuration in which the magnetic flux lines are convex towards the plasma that is contained. An example of this type of configuration is shown in Fig. 2.20 which is an illustration of *the cusp field* produced by an array of four current-carrying, parallel wires at the corners of a square. The currents in any two adjacent wires are in opposite directions resulting in a distribution of magnetic flux lines which are as depicted in Fig. 2.20. The distinguishing feature of this geometry over that of the pinched plasma column is that in addition to the magnetic flux lines being convex towards the plasma, the magnetic flux density

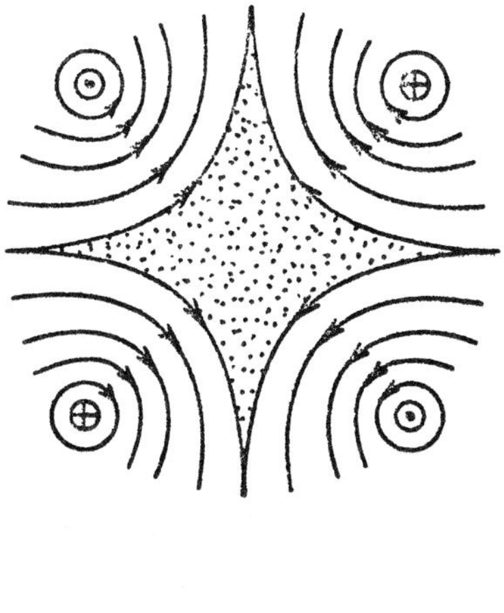

Fig. 2.20. The confinement of a plasma in a cusped magnetic field. ⊕ current into the plane of the paper; ⊙ current out of the plane of the paper.

increases from the plasma boundary in the direction away from the plasma. If there is a protuberance on the plasma boundary, the magnetic pressure on the protuberance is greater than its equilibrium value and therefore pushes the protuberance back to restore the equilibrium state. A higher order cusp field is obtained by two parallel rows of equidistant current-carrying wires. These wires are also parallel and the currents in the corresponding wires in the two rows as well as in any two adjacent wires in each row are in opposite directions. This so-called *picket fence* configuration is illustrated in Fig. 2.21.

If the magnetic flux lines are convex to the plasma, at the plasma boundary, there are sharp edges and cusps which form the avenues of escape for the plasma particles. Despite this difficulty, modifications of the cusp field are commonly used for the confinement of high temperature plasmas.

2.12. Velocity-space instability

In the previous section, a qualitative discussion of the instability arising from the nature of the geometrical configuration of the plasma was given. It is proposed to present here a semiquantitative treatment of the velocity-space instability which is governed by the characteristics of the velocity distribution function. These instabilities are also called the *microinstabilities* since their analysis is based on the microscopic description of the plasma. For the purpose of obtaining an understanding of the physical mechanism causing this instability, it is sufficient to consider only the dynamics of the electron gas in a plasma. The heavy ions are assumed to be stationary but they provide the necessary neutralizing background. The collisional interactions between the electrons and the ions are neglected.

It is known that two dynamical systems can interact if their velocities are approximately equal. One of the dynamical systems to be considered is the particles in the electron gas. The particle velocities may be assumed to be of the order of the

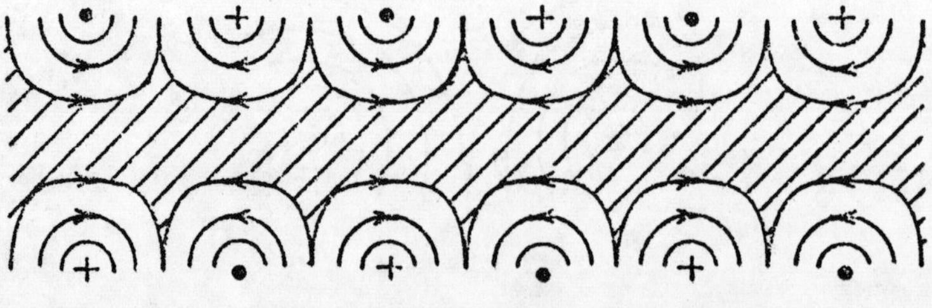

Fig. 2.21. The confinement of a plasma in a magnetic field having the configuration of a *picket fence*. ⊕ current into the plane of the paper; ⊙ current out of the plane of paper.

root mean square thermal velocity since the velocities of a large number of particles are in the neighborhood of this particular velocity. The other dynamical system that is to be explicitly examined here is the electron plasma wave. As was pointed out previously, the phase velocity of the electron plasma wave can become of the order of the root mean square thermal velocity. Thus there can be interaction between the plasma wave and the particles. The wave gets damped if the particles gain energy at the expense of the wave. On the other hand, if the wave gains energy at the expense of the particles, the wave amplitude grows in time signifying the occurrence of instability. The main objective of the following analysis is a semiquantitative examination of the physical processes that lead to the interchange of energy between the wave and the particles.

Trapping process

For this purpose, one-dimensional plane wave with the dependence on space and time as given by

$$f(X,t) = \exp\{i(k_x X - \omega t)\} = \exp\{ik_x(X - v_{ph} t)\} \tag{12.1}$$

is alone considered. In Eq. (12.1), k_x is the wavenumber, ω is the angular frequency and $v_{ph} = \omega/k_x$ is the phase velocity. The plane wave represented by Eq. (12.1) progresses in the X-direction. Note that Eq. (12.1) is in the phasor form. It is assumed that k_x is real and positive, that is, there is no spatial variation in the amplitude of the wave. If ω is real, the amplitude of the wave given by Eq. (12.1) is constant in time with the result that, on an average, there is no net interchange of energy between the wave and the particles. If ω is complex as expressed by

$$\omega = \omega_r + i\omega_i \tag{12.2}$$

where ω_r and ω_i are real, the amplitude of the wave as given by Eq. (12.1) is equal to $\exp(\omega_i t)$. The result is that the amplitude of the wave grows or decays in time according as $\omega_i > 0$ or $\omega_i < 0$. If $\omega_i > 0$, there is instability because of the temporal growth of the amplitude of the wave. As is to be shown presently, the nature of the velocity distribution function determines the temporal behavior of the wave. In the course of the following analysis an estimate of ω_i is obtained in terms of the relevant physical parameters.

It was established that for the electron plasma wave the electric field is in the direction of propagation. In order to be able to obtain the value of the net energy interchanged between the wave and the particles, and, to evaluate the time period over which this interaction is completed, it is necessary to treat the wave-particle interaction as an initial value problem. Consequently, it is assumed that the electric field of the longitudinal plasma wave as given by

$$E_x(X,t) = E_{x0}\sin(k_x X - \omega t) \tag{12.3}$$

is established at time $t = 0$. By analyzing the subsequent temporal behavior, the underlying physical mechanism for the energy interchange is determined.

It is convenient to analyze the problem in a coordinate system which is defined by $x = X - v_{ph}t$ and which moves uniformly at the phase velocity v_{ph} of the wave with reference to the plasma. The original and the moving coordinate systems are designated as the plasma and the wave frames respectively. In the wave frame, Eq. (12.3) simplifies to

$$E_x(x) = E_{x0} \sin k_x x \qquad (12.4)$$

where, without introducing any limitations, E_{x0} is assumed to be positive. Since the electric field as given by Eq. (12.4) is static in the wave frame, it can be obtained from a scalar potential $\phi(x)$ through the relation $E_x(x) = -d\phi(x)/dx$. The scalar potential is found by integrating Eq. (12.4) with respect to x. The result is

$$\phi(x) = -(E_{x0}/k_x)[1 - \cos k_x x] \qquad (12.5)$$

If the electronic charge is $-e$, the potential energy is obtained as

$$\Phi(x) = -e\phi(x) = (eE_{x0}/k_x)[1 - \cos k_x x] \qquad (12.6)$$

The constant of integration in Eq. (12.5) is chosen such that the minimum potential energy as given by Eq. (12.6) is zero. The potential energy $\Phi(x)$, as illustrated in Fig. 2.22, is a periodic distribution of potential wells. The bottom of the wells occur at $k_x x = \pm 2n\pi$ where n is a positive integer and the top of the wells occur at $k_x x = \pm(2n - 1)\pi$. The plasma is assumed to be uniform in the x-direction and the

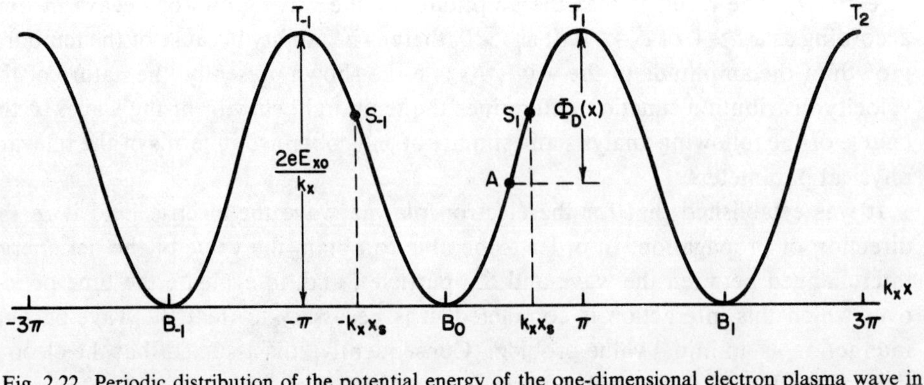

Fig. 2.22. Periodic distribution of the potential energy of the one-dimensional electron plasma wave in a reference frame moving uniformly with respect to the plasma at the phase velocity of the wave.

initial velocity distribution function $f(u_x)$ is independent of x. In view of the periodicity of the potential energy distribution and the homogeneity of the velocity distribution function $f(u_x)$, it is sufficient to focus attention on one period $(-\pi \leqslant k_x x \leqslant \pi)$ of the potential energy distribution.

The behavior of the electrons in the wave frame depends on their *initial* velocities. The electrons with velocities equal to the phase velocity of the wave are stationary in the wave frame and therefore they can be said to move with the wave. Let the electrons with their *initial* velocities *greater* than the phase velocity of the wave be considered next. In the wave frame, these electrons have their initial velocities in the positive x-direction. Let $W(t)$ be the kinetic energy of the electrons in the wave frame. Consider an electron situated at A at $t = 0$ in the region $0 < k_x x < \pi$. The potential energy at A is $\Phi(x)$ and the depth of the potential energy wells is equal to $2eE_{x0}/k_x$. Therefore the difference between the potential energies at the top of the well and that of A is given by $\Phi_D(x) = 2eE_{x0}/k_x - \Phi(x) = (eE_{x0}/k_x)(1 + \cos k_x x)$. Suppose that $W(0) > \Phi_D(x)$. In view of its initial kinetic energy, the electron moves up the potential barrier and reaches the top T_1 of the well at $k_x x = \pi$. In doing so it loses kinetic energy equal to the potential energy difference $\Phi_D(x)$. The electron has a residual kinetic energy $W(0) - \Phi_D(x)$ at T_1. Therefore, aided by the drop in the potential energy, the electron falls to the bottom $B_1(k_x x = 2\pi)$ of the potential well and then rises against the potential barrier to reach the top T_2 of the well at $k_x x = 3\pi$. The kinetic energy gained by the electron in the region $T_1 B_1$ is lost in the region $B_1 T_2$ with the result that the kinetic energy of the electron at T_2 is the same as at T_1. This process is repeated and the electron slides up and down the successive potential wells. When averaged over a wavelength $2\pi/k_x$, there is no net loss or gain of energy by the electrons. Thus the electrons with initial kinetic energy $W(0) > \Phi_D(x)$ do not interchange energy with the wave.

Suppose that $W(0) < \Phi_D(x)$. In this case, the electron moving up the potential barrier does not reach the top T_1 of the well since its kinetic energy is reduced to zero at some location S_1 specified by $k_x x_s < \pi$. Then the electron is forced by the electric field to fall to the bottom $B_0(k_x x = 0)$ of the potential well. In doing so, it gains kinetic energy whose value at B_0 is equal to $\Phi(x_s)$. The electron now rises against the potential barrier in the negative x-direction and reaches up to $S_{-1}(x = -x_s)$ where the kinetic energy is again equal to zero. The electron is then forced to travel in the reverse direction. It is easy to ascertain that the electron oscillates between S_{-1} and S_1 about the bottom B_0 of the potential well. In other words, the electron is *trapped* at the bottom of the potential well. The time taken for the electron to execute one oscillation, that is, the time taken for the electron to move from S_1 to S_{-1} and back to S_1 is called the trapping period. The above results are true, independently of the initial position of the electron.

Consider now an electron with its *initial* velocity *less* than the phase velocity of the wave. In the wave frame, this electron has its initial velocity in the negative x-

direction. If the initial kinetic energy $W(0)$ of this electron is such that $W(0) > \Phi_D(x)$, then from the arguments as before, this electron is seen to ride the successive crests and troughs of the periodic potential energy distribution in the negative x-direction. On the other hand, if $W(0) < \Phi_D(x)$, this electron is forced to oscillate about the bottom of the potential energy well.

It is clear that irrespective of whether the initial velocity of the electron is greater or less than the phase velocity of the wave, it is trapped if its initial kinetic energy $W(0)$ in the wave frame is such that $W(0) < \Phi_D(x)$. Since the electrons are distributed uniformly, $\Phi_D(x)$ can be averaged over a wavelength $2\pi/k_x$ to yield $\langle\Phi_D(x)\rangle = eE_{x0}/k_x$. Let $\tilde{u}_x = u_x - v_{ph}$ represent the electron velocity in the wave frame. Then, the condition for the trapping of the electrons can be assumed to be that for which the initial kinetic energy $\frac{1}{2}m_e\tilde{u}_x^2$ of the electron in the wave frame is less than or equal to the average potential energy difference $\langle\Phi_D(x)\rangle$, i.e.,

$$\tfrac{1}{2}m_e\tilde{u}_x^2 \leqslant eE_{x0}/k_x \quad \text{or} \quad |\tilde{u}_x| \leqslant \sqrt{2eE_{x0}/m_e k_x} = \tilde{u}_{x0} \tag{12.7}$$

Equivalently, on the average, electrons with velocities in the range

$$v_{ph} - \tilde{u}_{x0} < u_x < v_{ph} + \tilde{u}_{x0} \tag{12.8}$$

are trapped. It is interesting to observe that the electrons with initial velocities close to the phase velocity of the wave are trapped but those with velocities appreciably different from the phase velocity of the wave are not trapped and do not interchange energy with the wave.

Estimate of the energy interchanged

An estimate of the energy interchanged by the trapped electrons can be deduced in the following manner. The kinetic energy of the electron before it is trapped is given by $\frac{1}{2}m_e u_x^2$. The kinetic energy of the electron after it is trapped is equal to $\frac{1}{2}m_e(v_{ph} + \tilde{u}_x)^2 = \frac{1}{2}m_e(v_{ph}^2 + 2v_{ph}\tilde{u}_x + \tilde{u}_x^2)$. It is to be remembered that $\tilde{u}_x$ is a time-varying periodic function. In each oscillation, the electron passes the same location twice in opposite directions. For these two passages, the electron velocity has the same magnitude but opposite direction. Thus when the kinetic energy is averaged over a period of oscillation, the term linear in $\tilde{u}_x$ vanishes. Hence the kinetic energy of the electron after it is trapped is equal to $\frac{1}{2}m_e[v_{ph}^2 + \tilde{u}_x^2]$ which is equal to the sum of the electron kinetic energy $\frac{1}{2}m_e v_{ph}^2$ in the absence of oscillation and the kinetic energy $\frac{1}{2}m_e\tilde{u}_x^2$ of oscillation. The energy gained by the electron in the process of getting trapped is obtained as

$$\Delta E = \tfrac{1}{2}m_e[v_{ph}^2 + (u_x - v_{ph})^2 - u_x^2] = -m_e v_{ph}(u_x - v_{ph}) \tag{12.9}$$

It is to be noted that $\tilde{u}_x$ has to be interpreted as the root mean square value.

Since the number of electrons per unit distance in the x-direction having the velocities in the range du_x at u_x is equal to $f(u_x)du_x$, the energy gained by the electrons per unit distance in the x-direction or equivalently the gain in the energy density of the electrons in the trapping process is evaluated as

$$\Delta W = \int_{v_{ph}-\tilde{u}_{x0}}^{v_{ph}+\tilde{u}_{x0}} f(u_x)(\Delta E)\, du_x \tag{12.10}$$

The contribution to ΔW arises from the values of u_x in the neighborhood of v_{ph}. Therefore, it is legitimate to expand $f(u_x)$ in a Taylor series around $u_x = v_{ph}$ as

$$f(u_x) = f(v_{ph}) + (u_x - v_{ph})[\partial f(u_x)/\partial u_x]_{u_x = v_{ph}} \tag{12.11}$$

The result of substitution of Eqs. (12.9) and (12.11) into Eq. (12.10) is

$$\begin{aligned}\Delta W = & -f(v_{ph})v_{ph} m_e \int_{v_{ph}-\tilde{u}_{x0}}^{v_{ph}+\tilde{u}_{x0}} (u_x - v_{ph})\, du_x \\ & -f'(v_{ph})v_{ph} m_e \int_{v_{ph}-\tilde{u}_{x0}}^{v_{ph}+\tilde{u}_{x0}} (u_x - v_{ph})^2\, du_x\end{aligned} \tag{12.12}$$

where $f'(v_{ph}) = [\partial f(u_x)/\partial u_x]_{u_x = v_{ph}}$. The first integral in Eq. (12.12) does not give any contribution and the second integral can be evaluated with the following result:

$$\Delta W = -\tfrac{2}{3} m_e v_{ph} \tilde{u}_{x0}^3 f'(v_{ph}) \tag{12.13}$$

which shows that the electrons gain or lose energy according as $f'(v_{ph})$ is negative or positive.

The result implied by Eq. (12.13) is amenable to simple physical interpretation. After the electrons are trapped their average velocity in the wave frame is equal to zero and this is equivalent to the statement that in the trapping process the electrons having velocities in the neighborhood of the phase velocity of the wave, on the average, attain the value of the phase velocity of the wave. If the number of electrons with $u_x = v_{ph}+$ is less than the number of electrons with $u_x = v_{ph}-$, then $f'(v_{ph}) < 0$ and more electrons are speeded up than are slowed down with the result that the kinetic energy density of the electrons increases. This result is in accordance with Eq. (12.13) for $f'(v_{ph}) < 0$. On the other hand, if the number of electrons with $u_x = v_{ph}+$ is greater than the number of electrons with $u_x = v_{ph}-$, then $f'(v_{ph}) > 0$. In this case more electrons are slowed down than are speeded up. Consequently the kinetic energy density of the electrons decreases in conformity with Eq. (12.13) for $f'(v_{ph}) > 0$. It is therefore clear that the slope of the velocity distribution function at $u_x = v_{ph}$ determines whether the particles lose or gain energy in the trapping process.

The wave energy is decreased or increased according as the kinetic energy of the particles is increased or decreased on account of the principle of conservation of total energy. For a Maxwellian distribution of particles, $f'(v_{ph}) < 0$, the kinetic energy of the particles is increased and the wave energy is therefore decreased. Consequently, the amplitude of the wave is damped; this damping which occurs in the absence of collisions is known as *Landau damping*. A rigorous treatment of Landau damping based on the Boltzmann equation is given in Sec. 6.2. For a non-Maxwellian distribution of particles, it is possible to have $f'(v_{ph}) > 0$ and in such cases microinstability occurs. A velocity distribution function with a positive slope can be obtained, for example, in an electron beam in which the majority of the particles have velocities equal to the beam velocity. The number of particles with velocities different from the beam velocity decreases as the velocity difference increases with the result that the velocity distribution function has a hump at the beam velocity. Consequently, the distribution function characterizing an electron beam has a positive slope at a velocity less than the beam velocity. In such a case, the amplitude of a wave interacting with the particles can grow temporally leading to the microinstability.

Growth or damping rate

In order to be able to deduce the growth or the damping constant, it is necessary to evaluate the dynamics of the electron in the wave frame, at least approximately. The equation of motion of the electron in the wave frame is obtained from Eq. (12.4) as

$$m_e d^2x/dt^2 = -eE_{x0} \sin k_x x \tag{12.14}$$

whose linearized form

$$m_e d^2x/dt^2 = -eE_{x0} k_x x \tag{12.15}$$

is used in the following analysis. Clearly, Eq. (12.15) is valid only if $|k_x x| \ll 1$, that is, if the positions of the electrons do not deviate appreciably from the bottoms of the potential wells. In such cases, the particles execute simple harmonic motion about the troughs of the potential distribution with an angular frequency ω_t as given by

$$\omega_t = (eE_{x0} k_x / m_e)^{1/2} \tag{12.16}$$

The increase in the energy density of the wave is equal to $-\Delta W$ where ΔW is given by Eq. (12.13). This increase is assumed to take place in the time period $1/\omega_t$. Presumably, the electrons have to execute at least one oscillation before it can be concluded that the trapping process has begun. It is for this reason that the increase $-\Delta W$ in the energy density of the wave is assumed to take place in the time interval

$1/\omega_t$. The energy density U of the wave is obtained from Eq. (12.4) as

$$U = \tfrac{1}{2}\varepsilon_0 E_{x0}^2 \tag{12.17}$$

From the foregoing discussion, it follows that

$$dU/dt = -\Delta W/(1/\omega_t) = -\omega_t \Delta W \tag{12.18}$$

With the help of Eqs. (12.7), (12.13), (12,16), and (12,17), Eq. (12.18) can be simplified to yield

$$dU/dt = (8\sqrt{2}\,/3)(\omega_{pe}^2 \omega_r / N_0 k_x^2) f'(v_{ph}) U \tag{12.19}$$

where N_0 is the number density of the electrons and $\omega_{pe} = (N_0 e^2/m_e \varepsilon_0)^{1/2}$ is the electron plasma angular frequency.

The amplitude of the electric field has a temporal behavior of the form $\exp(\omega_i t)$ for $t > 0$ on account of the wave-particle interaction. In view of Eq. (12.17), the energy density of the wave has the time behavior of the form

$$U = U_0 e^{2\omega_i t} \tag{12.20}$$

where U_0 is the energy density at time $t = 0$. It follows from Eq. (12.20) that

$$dU/dt = 2\omega_i U \tag{12.21}$$

A comparison of Eq. (12.19) with Eq. (12.21) yields the following expression for the growth or decay factor:

$$\omega_i = (4\sqrt{2}\,/3)(\omega_{pe}^2 \omega_r / N_0 k_x^2) f'(v_{ph}) \tag{12.22}$$

In accordance with the previous discussion, Eq. (12.22) specifies a growing wave for $f'(v_{ph}) > 0$. It is to be noted that after the trapping is completed, the wave and the particles do not interchange any net energy and therefore the wave propagates without any further change in amplitude.

It is important to examine the conditions under which the preceding analysis is valid. It is assumed that the initial velocity distribution function of the particles is unchanged as a result of the trapping process. This is possible only if the net energy given to or taken away from the particles by the wave is small in comparison with the initial energy of the particles. Consequently it is required that the energy in the disorganized thermal motions of the electrons be much larger than that in the organized motion of the wave. In other words, only small amplitude waves are considered. This stipulation together with Eq. (12.16) shows that $\omega_i/\omega_t \gg 1$ or $1/\omega_t \gg 1/\omega_i$ where $1/\omega_i$ is the decay or the growth time. Hence, it is seen that the trapping time of the particles should be much larger than the growth or the decay time of the wave.

Summary

The particles with velocities sufficiently different from the phase velocity of the wave alternately give energy to and extract energy from the wave as they pass through the crests and the troughs of the wave resulting in no net interchange of energy between the particles and the wave. These crests and troughs, though stationary in the wave frame, progress at the phase velocity of the wave in the plasma frame. But the particles with initial velocities quite close to the phase velocity of the wave are trapped inside the moving potential wells of the wave and this trapping process results in a net interchange of energy between the particles and the wave. If the velocity distribution function has a positive slope at the particle velocity equal to the phase velocity of the wave, the particles lose energy and the amplitude of the wave grows in time, resulting in the velocity-space instability.

References

2.1. J. L. Delcroix, *Plasma Physics*, Chapters 8 and 9, John Wiley, New York, 1965.
2.2. L. Spitzer, *Physics of Fully Ionized Gases*, 2nd ed., Chapters 2 and 3, Interscience Publishers, New York, 1962.
2.3. D. J. Rose and M. Clark, *Plasmas and Controlled Fusion*, Chapters 7, 8, and 14, John Wiley, New York, 1961.
2.4. S. Glasstone and R. H. Lovberg, *Controlled Thermonuclear Reactions*, Chapters 3, 6, and 7, Van Nostrand Reinhold Company, New York, 1960.
2.5. J. D. Jackson, *Classical Electrodynamics*, Chapter 10, John Wiley, New York, 1962.
2.6. T. J. M. Boyd and J. J. Sanderson, *Plasma Dynamics*, Chapters 4 and 8, Thomas Nelson, London, 1969.
2.7. P. Debye and E. Hückel, Zur Theorie der Elektrolyte, *Phys. Z.* **24**, No. 9, 185–208 (1923).
2.8. H. M. Mott-Smith and I. Langmuir, The theory of collectors in gaseous discharges, *Phys. Rev.* **28**, 727–763 (1926).
2.9. L. Tonks and I. Langmuir, Oscillations in ionized gases, *Phys. Rev.* **33**, 195–210 (1929).
2.10. B. D. Fried and R. W. Gould, Longitudinal ion oscillations in a hot plasma, *Phys. Fluids* **4**, 139–147 (1961).
2.11. C. Walén, On the theory of sun-spots, *Arkiv. Mat.* **30A**, No.15 (1944);
C. Walén, On the distribution of the solar general magnetic field, *Arkiv. Mat.* **33A**, No.18 (1946).
2.12. W. H. Bennet, Magnetically self-focussing streams, *Phys. Rev.* **45**, 890–897 (1934).
2.13. J. D. Jackson, Longitudinal plasma oscillations, *J. Nucl. Energy Part C: Plasma Phys.* **1**, 171–189 (1960).

Problems

2.1. Consider the electron plasma oscillations as treated in Sec. 2.2 but without omitting the effect of the motion of the ions. Show that the natural frequency of oscillation is given by

$$\omega = \sqrt{\omega_{pe}^2 + \omega_{pi}^2}$$

where ω_{pe} and ω_{pi} are the electron and the ion plasma angular frequencies, respectively. Note that $\omega_{pi}^2 = N_0 e^2/m_i \varepsilon_0$ where e is the charge and m_i the mass of an ion.

2.2. Consider a plasma consisting of electrons of charge $-e$ and ions of charge Ze. Suppose that T_e and T_i are the temperatures of the electrons and the ions, respectively. Deduce the Debye potential for a test particle of charge q and verify that it is consistent with that given by Eq. (3.18).

2.3. Show that for a gas in thermodynamic equilibrium, the average energy per particle is equal to $1.292 \times 10^{-4} eV/°K$.

Consider a plasma with the number density of electrons equal to $10^{20}/m^3$. Suppose that in a uniform spherical volume of radius $r = 10^{-2}$ m, the instantaneous number density of ions is greater than that of electrons by 1%. Find the potential at the surface of the spherical volume and verify that the equivalent plasma temperature is of the order of millions of degrees Kelvin. Note that as a consequence the fluctuations in the thermal energies of the particles correspond to extremely small variations in the potential.

2.4. Consider a plasma consisting of electrons of charge $-e$ and ions of charge Ze. Let T_e and T_i denote the temperatures of the electrons and the ions, respectively. Evaluate the potential ϕ_w on an infinite plane immersed in a plasma under equilibrium conditions.

2.5. Consider the plasma sheath formed between an infinite plane and a semi-infinite plasma. With the help of the expressions for the number densities as given in Eqs. (4.4a,b), deduce the differential equation satisfied by the potential distribution $\phi(x)$ across the sheath. If ϕ_w is the potential on the wall and if the potential at infinity is zero, evaluate the expression for the potential distribution for the case in which $[e\phi(x)/KT] \ll 1$. Note that $-e$ and e are the charges on an electron and an ion, respectively.

2.6. Let the ions at the edge of the plasma sheath be described by a shifted Maxwell-Boltzmann distribution function with the drift velocity $\mathbf{v} = \hat{\mathbf{x}} v_x$. Show that the ion particle current density at the edge of the sheath is given as follows:

$$\Gamma_i = N_0(KT/2\pi m_i)^{1/2}[\exp(-y^2) + y\sqrt{\pi}\,\{1 + \mathrm{erf}(y)\}]$$

where $y = v_x\sqrt{m_i/2KT}$ is the normalized drift velocity and the error function is defined by

$$\mathrm{erf}(y) = (2/\sqrt{\pi}) \int_0^y e^{-s^2}\, ds$$

The error function is zero for $y = 0$, increases monotonically as y increases and asymptotically attains the value of unity for $y = \infty$. Also $(d/dy)\,\mathrm{erf}\,(y) = (2/\sqrt{\pi})e^{-y^2}$.

Verify that Γ_i increases monotonically with increasing normalized drift velocity y.

Note that N_0 is the number density, m_i the ion mass, T the ion temperature, and K is the Boltzmann constant.

2.7. The potential ϕ_w on an infinite plane immersed in a plasma is given by the relation $\phi_w = -2KT/e$ where K is the Boltzmann constant, T is the temperature of the electrons, and the ions, and $e(-e)$ is the charge on an ion (electron). Estimate the number densities of the electrons and the ions near the wall in terms of their values far away from the sheath region. Find the drift velocity of the plasma particles near the wall in terms of the thermal velocity $\sqrt{KT/m_i}$ of the ions.

2.8. Show that the plasma sheath equation deduced in Problem 2.5 can be recast into the form

$$d^2\tilde{\phi}/d\tilde{x}^2 = \sinh \tilde{\phi}$$

where $\tilde{\phi} = e\phi(x)/KT$ and $\tilde{x} = \sqrt{2}\, x/\lambda_D$. If at $x = \infty$, $N_i = N_e = N_0$, $\tilde{\phi} = 0$, and $d\tilde{\phi}/d\tilde{x} = 0$, show that

$$\tilde{\phi} = 4 \tanh^{-1}[\exp\{-(\tilde{x} - \tilde{x}_0)\}]$$

where $\tilde{x}_0$ is a constant. If $\tilde{\phi} = e\phi(x)/KT \ll 1$, prove that the expression for $\tilde{\phi}$ reduces to the approximate result obtained in Problem 2.5.

2.9. The current-voltage characteristic is measured for an electrostatic probe which is inserted into the plasma contained by a magnetic mirror machine. The collecting area of the probe is 0.644 cm^2. All voltages are measured with respect to a fixed reference potential. The data obtained are given in the following table:

ϕ(V)	I(mA)	ϕ(V)	I(mA)
− 30.0	− 0.9	1.0	0.1
− 18.5	− 0.8	1.8	0.5
− 8.0	− 0.6	3.0	0.8
− 3.0	− 0.4	5.5	1.2
− 0.5	− 0.2	32.0	2.3
0.5	0.0	72.0	2.5

Determine the electron temperature in the plasma, the electron number density, and the floating potential of the probe.

2.10. Consider the following Maxwell's equations satisfied by the electromagnetic fields in a plasma:

$$(a)\ \nabla \times \mathbf{E} = -\mu_0(\partial/\partial t)\mathbf{H} \qquad (c)\ \nabla \cdot \mathbf{H} = 0$$

$$(b)\ \nabla \times \mathbf{H} = \varepsilon_0(\partial/\partial t)\mathbf{E} + Nq\mathbf{v} \qquad (d)\ \nabla \cdot \mathbf{E} = Nq/\varepsilon_0$$

With the help of these equations prove that

$$\mu_0\varepsilon_0(\partial/\partial t)(\mathbf{E} \times \mathbf{H}) = \nabla \cdot \mathbf{T} - Nq[\mathbf{E} + \mathbf{v} \times \mathbf{B}] \qquad \text{(A)}$$

where the components of the electromagnetic stress dyad $\mathbf{T}$ are defined by

$$T_{ij} = \varepsilon_0 E_i E_j + \mu_0 H_i H_j - (\delta_{ij}/2)(\varepsilon_0 E^2 + \mu_0 H^2)$$

and $\delta_{ij} = 0$ for $i \neq j$ and $\delta_{ij} = 1$ for $i = j$. Note that $\mu_0 \varepsilon_0 \mathbf{E} \times \mathbf{H}$ is the electromagnetic momentum density.

Use (A) and the equation of continuity (1.8.9) to recast the momentum transport equation (1.9.14) with $\mathbf{P}_{\text{coll}} = 0$ into the following form:

$$(\partial/\partial t)\mathbf{G} = -\nabla \cdot \boldsymbol{\pi}$$

where the total momentum density $\mathbf{G}$ and the rate of flow of total momentum through unit area, $\boldsymbol{\pi}$, are given by

$$\mathbf{G} = Nm\mathbf{v} + \varepsilon_0 \mu_0 \mathbf{E} \times \mathbf{H}$$

and

$$\boldsymbol{\pi} = Nm\mathbf{v}\mathbf{v} + \boldsymbol{\Psi} - \mathbf{T}$$

2.11. Use Maxwell's equations (a) and (b) of Problem 2.10 and show that the energy transport equation (1.10.60) with $E_{\text{coll}} = 0$ can be rewritten as

$$(\partial/\partial t)W = -\nabla \cdot \mathbf{S}$$

where the total energy density W and the power flow per unit area, $\mathbf{S}$, are given by

$$W = \frac{\varepsilon_0}{2}E^2 + \frac{\mu_0}{2}H^2 + \frac{N}{2}mv^2 + \frac{N}{2}m\langle w^2 \rangle$$

and

$$\mathbf{S} = \mathbf{E} \times \mathbf{H} + \boldsymbol{\Psi} \cdot \mathbf{v} + \frac{N}{2}m\mathbf{v}(\langle w^2 \rangle + v^2) + \mathbf{q} \tag{A}$$

Assume that the vector flux density of heat $\mathbf{q}$ is zero and that the kinetic pressure dyad simplifies to a scalar pressure. Examine how the time average of $\mathbf{S}$ as given by (A) reduces to the generalized Poynting vector in Eq. (6.14).

2.12. The constituent plane wave of a field quantity $f(\mathbf{r}, t)$ with general space and time variation, depends on both the spatial coordinates x and y as given by

$$f_{k_x, k_y, \omega}(x, y, t) = e^{i(k_x x + k_y y - \omega t)}$$

Show that the group velocity is

$$\mathbf{v}_g = \hat{\mathbf{x}}\partial\omega/\partial k_x + \hat{\mathbf{y}}\partial\omega/\partial k_y$$

Note that Taylor series expansion of a function of two variables is given by

$$g(x,y) = g(x_0, y_0) + (x - x_0)\left(\frac{\partial g}{\partial x}\right)_{x_0, y_0} + (y - y_0)\left(\frac{\partial g}{\partial y}\right)_{x_0, y_0}$$

2.13. For the electron plasma mode with the phase velocity v_p defined by Eq. (7.18), find the group velocity v_g and show that $v_p v_g$ is independent of frequency.

2.14. With the help of Eqs. (8.20) and (8.21), prove that in the limit of very low frequencies ($\omega^2 \ll \omega_{pi}^2$), the electron and the ion velocities associated with the plasma mode are equal and, in effect, the two partial pressures add in the determination of the particle velocities. Obtain the equation satisfied by the total pressure. Deduce another equation satisfied by the total pressure by using Eqs. (8.1) and (8.4). Derive the wave equation satisfied by the total pressure by eliminating the particle velocity from these two sets of equations and hence reproduce the phase velocity of the ion acoustic wave as deduced in Eq. (8.49).

2.15. The low frequency ion oscillations were treated differently by Tonks and Langmuir whose theory is developed in this problem. Let N_0 be the equilibrium number density of both the electrons and the ions. In the perturbed state, let all the field quantities including the electrostatic potential ϕ vary as $\exp\{i(k_p x - \omega t)\}$. Therefore, the electric field $\mathbf{E} = -\nabla\phi$ has only an x-component. If the plasma is collisionless and the ion temperature is neglected, show that the ion velocity is given by

$$v_{ix} = (ek_p/\omega m_i)\phi$$

From the equation of continuity prove that the perturbation charge density of the ions is obtained as

$$(N_0 e^2 k_p^2/\omega^2 m_i)\phi$$

Tonks and Langmuir assumed that the ion oscillations are so slow that the electrons remain in a Maxwell-Boltzmann distribution. If $e\phi/KT_e \ll 1$, show that the perturbation charge density of the electrons is found as

$$-(N_0 e^2/KT_e)\phi$$

Finally use Poisson's equation to deduce the following dispersion relation:

$$k_p^2 = (N_0 e^2/\omega^2 m_i \varepsilon_0)k_p^2 - N_0 e^2/KT_e \varepsilon_0$$

Recast this dispersion relation into forms similar to Eqs. (8.39) and (8.40). Verify that the theory of Tonks and Langmuir assumes, essentially, an isothermal expansion of the plasma.

2.16. For a fully ionized plasma, show that the nonlinear form of the equation of continuity is given by

$$(\partial/\partial t)\rho_m + \nabla \cdot (\rho_m \mathbf{V}) = 0$$

where ρ_m is the mass density and $\mathbf{V}$ is the mass velocity and hence prove that for an incompressible fluid $\nabla \cdot \mathbf{V} = 0$.

2.17. For a perfectly conducting fluid, it is known that

$$(\partial/\partial t)\mathbf{B} = \nabla \times (\mathbf{V} \times \mathbf{B})$$

With the help of this relation and the nonlinear form of the equation of continuity as indicated in Problem 2.16, deduce that

$$\frac{d}{dt}\left(\frac{\mathbf{B}}{\rho_m}\right) = \left(\frac{\mathbf{B}}{\rho_m} \cdot \nabla\right)\mathbf{V}$$

This relation which was first derived by Walén can be used to establish that fluid elements which lie initially on a magnetic flux line continue to lie on a flux line in a perfectly conducting fluid. [See S. Goldstein, *Lectures on Fluid Mechanics*, Wiley, London, 1960.]

2.18. A plasma is confined by a unidirectional magnetic flux density of 5 Wb/m^2. Find the value of the magnetic pressure. If $\beta = 0.4$ and the particle temperature is 10 keV, what is the number density of particles? If the particle temperature is increased to 40 keV and if β remains the same, what is the new value of the magnetic flux density required for the confinement of the plasma?

2.19. Consider the generalized Ohm's law as stated in Eq. (9.3). Prove that

$$\mathbf{J}_\perp\left(1 + \frac{\sigma^2 B_0^2}{N_{e0}^2 q_e^2}\right) = \sigma\left[\mathbf{E}_\perp^* + \frac{\sigma}{N_{e0} q_e}\mathbf{E}^* \times \mathbf{B}_0\right]$$

where $\mathbf{E}^* = \mathbf{E} + \mathbf{V} \times \mathbf{B}_0$ and the subscript $\perp$ indicates the component perpendicular to $\mathbf{B}_0$. Examine the result for the limiting case of a perfectly conducting fluid.

2.20. Consider the steady state of a perfectly conducting plasma characterized by scalar pressures for the electrons and the ions. With the help of the equation of motion (1.12.21) and the generalized Ohm's law (1.12.40), deduce that

$$\mathbf{V}_\perp = -\frac{\mathbf{B}_0}{B_0^2} \times \left(\mathbf{E} + \frac{\nabla P_i}{N_e q_e}\right)$$

where, as before, the subscript $\perp$ indicates the component perpendicular to $\mathbf{B}_0$. For the theta pinch device, indicate the directions of $\mathbf{V}$, $\mathbf{B}_0$, $\mathbf{E}$, $\mathbf{J}$, and ∇P_i.

2.21. For the equilibrium pinch, assume that the current density $J_z(r)$ is a constant throughout the cross section of the plasma column. If I_0 is the total axial current, deduce the expressions for the radial distributions of the magnetic flux density $B_\theta(r)$ and the kinetic pressure $P(r)$. Draw a diagram illustrating these radial distributions.

2.22. For the Bennett pinch, obtain the expression for the magnetic flux density $B_\theta(r)$.

2.23. With the help of the expression indicated in Problem 2.19 find the radial electric field for the equilibrium pinch of Problem 2.21 and for the Bennett pinch.

2.24. Deduce the expression for the magnetic pressure used in the formulation of the dynamic pinch.

2.25. Consider the theta pinch and by using the snowplow model, deduce the differential equation specifying the time-dependent radius of the plasma column.

2.26. Consider a pinched plasma column with trapped axial magnetic flux lines. Let $P_{m\theta}$ and P_{mz} be the magnetic pressures due to the azimuthal and the axial magnetic flux densities, respectively. If x is a small inward displacement of the boundary of a plasma column of unperturbed radius R, show that the fractional changes in the two magnetic pressures are given by

$$\Delta P_{m\theta}/P_{m\theta} = 2x/R \qquad \Delta P_{mz}/P_{mz} = 4x/R$$

Hence prove that the plasma column is stable against the perturbations leading to the sausage instability if $B_z^2 > 0.5B_\theta^2$.

2.27. Show that a perfectly conducting hollow tube placed coaxially around a pinched plasma column exerts a stabilizing influence against the formation of the kink instability.

2.28. Demonstrate qualitatively that a pair of parallel and coaxial circular loops carrying oppositely directed currents produce a cusped magnetic field between them. Find the nature of the magnetic field in the neighborhood of the loops if the currents are in the same direction.

2.29. Examine the mechanism by which the plasma particles escape the confining regions of the cusped magnetic field illustrated in Fig. 2.20.

2.30. Find the requirement on the amplitude E_{x0} of the electric field of the longitudinal electron plasma wave in order that the expression deduced for the growth constant ω_i may remain valid. How does the limitation imposed on the amplitude of the electric field depend on the wavelength? What is the associated limitation on the depth of the potential energy wells?

2.31. Find the growth constant ω_i for an unstable one-dimensional electron plasma wave if the initial velocity distribution function of the particles is given by

$$f(u_x) = N_0\sqrt{\frac{m_e}{2\pi KT}}\exp\left[-\frac{m_e}{2KT}(u_x - v_x)^2\right]$$

where K is the Boltzmann constant and T is the electron temperature. Assume that v_x is equal to 1.5 times the phase velocity v_{ph} of the wave.

What is the growth constant if $v_x = v_{ph}$? Give a physical explanation for the result obtained in this case.

2.32. A gas of electrons is characterized by the following one-dimensional velocity distribution function:

$$f(u_x) = \frac{N_0}{\sqrt{2\pi}}\left(\frac{m_e}{KT}\right)^{3/2} u_x^2 \exp\left[-\frac{m_e}{2KT}u_x^2\right]$$

where m_e is the mass of an electron and K is the Boltzmann constant. Verify that N_0 represents correctly the number density of electrons. Find how T is related to the kinetic temperature.

A longitudinal plasma wave is set up initially in the plasma such that its real propagation coefficient k is in the x-direction. Find the frequency range for which the plasma wave is unstable and evaluate the growth constant.

CHAPTER 3

Interactions of Charged Particles with Electromagnetic Fields

3.1. Introduction

Plasma is a gas of particles consisting of electrons, ions and neutral particles. The ions may be singly or multiply charged; they may be positively or negatively charged. However, in this treatment, except when stated otherwise, positively charged hydrogen ions are assumed. The dynamics of the charged particles in a plasma is determined by the Lorentz force contributed by the electric and the magnetic fields arising from external sources and from their own motion as well as those due to the other charged particles. The Coulomb and the magnetic interaction between the charged particles is taken into account by the Lorentz force contributed by the motion of the other charged particles. If **E** and **B** are the electric field and the magnetic flux density, respectively, the motion of a typical particle of charge q and rest mass m is governed by the following Newton's law of motion:

$$\frac{d}{dt}\mathbf{r} = \mathbf{u} \tag{1.1}$$

and

$$m\frac{d}{dt}\mathbf{u} = q\left(1 - \frac{u^2}{c^2}\right)^{1/2}\left[\mathbf{E} + \mathbf{u} \times \mathbf{B} - \frac{\mathbf{u}}{c^2}(\mathbf{u} \cdot \mathbf{E})\right] \tag{1.2}$$

where **r** and **u** are the position and the velocity vectors of the particle. Also, u and c are the speeds of the particle and of an electromagnetic wave in vacuum, respectively. If u^2/c^2 is negligible in comparison with unity, Eq. (1.2) reduces to the following familiar nonrelativistic equation of motion:

$$m\frac{d}{dt}\mathbf{u} = q[\mathbf{E} + \mathbf{u} \times \mathbf{B}] \tag{1.3}$$

If the velocity obtained from Eq. (1.3) does not satisfy the criterion $u^2/c^2 \ll 1$, the corresponding result is not valid. However, in most practical situations, the restriction implicit in Eq. (1.3) is not violated and therefore, the use of the nonrelativistic form (1.3) of the equation of motion is justified.

When the position and the velocity vectors of all the particles are determined, the charge and the current distributions are evaluated by summing up the contributions from all the charged particles. If there is neither annihilation nor creation of particles, the charge density ρ and the current density $\mathbf{J}$ satisfy the following equation of continuity:

$$\nabla \cdot \mathbf{J} + (\partial/\partial t)\rho = 0 \tag{1.4}$$

The charge and the current densities, in turn, give rise to electric and magnetic fields which are governed by Maxwell's equations:

$$\nabla \times \mathbf{E} = -(\partial/\partial t)\mathbf{B} \tag{1.5}$$

$$\nabla \times \mathbf{B} = \mu_0 \varepsilon_0 (\partial/\partial t)\mathbf{E} + \mu_0 \mathbf{J} \tag{1.6}$$

The permeability and the permittivity of free space are denoted by μ_0 and ε_0, respectively. If divergence of both sides of Eqs. (1.5) and (1.6) are taken, the result on using Eq. (1.4) may be shown to be:

$$(\partial/\partial t)[\nabla \cdot \mathbf{B}] = 0 \tag{1.7}$$

$$(\partial/\partial t)\left[\nabla \cdot \mathbf{E} - \frac{\rho}{\varepsilon_0}\right] = 0 \tag{1.8}$$

Let

$$\nabla \cdot \mathbf{B} = 0 \tag{1.9}$$

and

$$\nabla \cdot \mathbf{E} = \rho/\varepsilon_0 \tag{1.10}$$

If Eqs. (1.9) and (1.10) are valid at time $t = 0$, it follows from Eqs. (1.7) and (1.8) that they are valid for all times. Thus Eqs. (1.9) and (1.10) serve as initial conditions which together with Eqs. (1.5) and (1.6) enable the determination of the electric and the magnetic fields associated with the charge and the current distributions. It is emphasized that additional requirements are necessary for obtaining uniquely the electric and the magnetic fields from Eqs. (1.5) and (1.6). The fields so obtained should be consistent with those originally used for the determination of the particle trajectories. Thus, in principle, a self-consistent formulation is available for the evaluation of the plasma dynamics.

The feasibility of the foregoing scheme is beset with formidable difficulties. Unique determination of the particle position and velocity vectors from Eqs. (1.1) and (1.3) requires a knowledge of the initial position and velocity vectors of all the particles. The initial position and velocity vectors of the particles can be specified only to within a certain accuracy. Moreover, the equations of which there are usually an unimaginably large number of them, are nonlinear. The fields associated with the charge and the current distributions determined by the particle trajectories, in turn, govern the trajectories. Thus, the particle motions and the fields are inextricably coupled. Also, if point charges are assumed, the fields become singular at the positions of the particles.

These difficulties indicate that in order to carry out the aforementioned scheme for the plasma dynamics, it is necessary to start with a simple approximation. Such an approximation is bound to shed some light on the plasma behavior; also, it can provide insight for obtaining a better approximation to the actual behavior of the plasma. A very useful approximation is one in which the motion of the individual charged particle is studied for *specified* electric and magnetic fields. In this approximation, the macroscopic fields due to the motion of the charged particles are not taken into account. The implication is that all interactions between the particles are neglected with the result that the plasma reduces to a free gas of electrons, ions, and neutral particles. The plasma of the ionosphere and the solar corona is highly rarefied and therefore, the interactions between the charged particles are of minor importance. Hence, the theory of a free gas of electrons, ions, and neutral particles is useful in predicting the behavior of a highly rarefied plasma, since the behavior of such a plasma is determined primarily by the interactions of the charged particles with the external electromagnetic fields rather than with themselves.

The *microscopic* motions of the individual particles are not by themselves very important. The main interest is in certain *macroscopic* effects due to the motion of a large number of charged particles. All the detailed microscopic motions of the individual particles do not contribute to the cumulative effect. In the motion of the individual particle, it is possible to isolate the parts which contribute to the collective behavior. Since information on certain collective effects in a plasma can be obtained, the study of the motion of the individual particle becomes important.

The trajectories of a single particle in various specified electric and magnetic fields are treated in this chapter. The simplest case in which the fields are constant in time and uniform in space is considered first. The motions of a particle in an electric field, a magnetic field, and in a combination of electric and magnetic fields are studied. The polarization drift arising in a combined electric and magnetic field due to a slow variation of the electric field with respect to time is then investigated. This investigation is extended to the case of an electric field which has a harmonic variation in time. In that case, the collective behavior of large number of similar particles can be isolated and described in terms of macroscopic mobility and conductivity. Explicit expressions are obtained for these macroscopic parameters which are valid if the frequency of the electric field is sufficiently removed from the gyromagnetic frequency of the charged particle. The phenomenon of cyclotron resonance that occurs when the frequency of the electric field is quite close to the gyromagnetic frequency of the charged particle is examined in detail.

A simple treatment of the influence of the variation of the magnetic field in space on the motion of the charged particle is given. The adiabatic invariance of the orbital magnetic moment is established and is used in the investigation of the magnetic mirror effect arising in an inhomogeneous magnetic field with converging tubes of magnetic flux. When two magnetic mirrors approach each other, a charged particle trapped between them acquires an acceleration. This so-called Fermi acceleration is then examined. The gradient drift caused by the variation of the magnetic field in a direction perpendicular to the field lines and the curvature drift resulting from the curved magnetic field lines are briefly studied.

The mechanism of heating a plasma based on the variation of the magnetic field in time is called magnetic pumping. The effect of the time variation of the magnetic field on the motion of a charged particle is considered and the results are applied for the development of a simplified theory of magnetic pumping.

3.2. Constant and uniform electric field

Consider the motion of a particle of charge q and mass m in an electric field which is constant in time and uniform in space. The direction of the electric field is assumed to be the same everywhere, namely parallel to the y-axis of the Cartesian coordinates x, y, and z. For this case, the equations of motion are given by Eqs. (1.1) and (1.3) with $\mathbf{B}$ set equal to zero. The equations of motion can be solved to yield the following result:

$$\mathbf{u} = (q/m)\mathbf{E}t + \mathbf{u}_0 \tag{2.1}$$

$$\mathbf{r} = (q/m)\mathbf{E}(t^2/2) + \mathbf{u}_0 t + \mathbf{r}_0 \tag{2.2}$$

where $\mathbf{r}_0$ and $\mathbf{u}_0$ are the initial position and velocity vectors of the particle. From Eq. (2.1), the velocity of the particle perpendicular to the direction of the electric field is seen to be unaffected whereas in the direction of the electric field, the particle is either accelerated or decelerated depending on whether it is positively or negatively charged.

If both sides of Eq. (1.3) are scalarly multiplied by $\mathbf{u}$, the result is

$$\mathbf{u} \cdot m(d/dt)\mathbf{u} = (d/dt)(\tfrac{1}{2}mu^2) = q\mathbf{u} \cdot \mathbf{E} \tag{2.3}$$

since $\mathbf{u} \cdot (\mathbf{u} \times \mathbf{B}) = 0$. In Eq. (2.3), $u = |\mathbf{u}|$ is the speed of the particle. For static fields, it follows from Eq. (2.3) that the kinetic energy of the particle is unaltered by the application of a magnetic field but is changed by the application of an electric field. Without loss of generality, the electric field may be taken to be in the y-direction. Let a scalar potential ϕ be defined such that $E = -\partial\phi/\partial y$. Then,

$$q\mathbf{u} \cdot \mathbf{E} = -q(\partial\phi/\partial y)(dy/dt) = -q(d\phi/dt) \tag{2.4}$$

If Eq. (2.4) is substituted into Eq. (2.3) and the resulting equation is integrated with respect to time, it is obtained that

$$\tfrac{1}{2}mu^2 + q\phi = \text{constant} \tag{2.5}$$

which is a statement of conservation of energy. The first term gives the kinetic energy of the particle and the second its potential energy.

3.3. Constant and uniform magnetic field

Helical orbit

Let the charge be acted on by a magnetic field which is constant in time and uniform in space. The direction of the magnetic field is assumed to be everywhere parallel to the z-axis. Since $\mathbf{E} = 0$, it follows from Eq. (2.3) that the speed of the particle is unchanged. This is due to the fact that for $\mathbf{E} = 0$, the Lorentz force has no component in the direction of motion of the particle and hence does no work on the particle. For the present case, the Cartesian components of Eq. (1.3) are

$$\frac{d}{dt}u_x = \frac{qB}{m}u_y \qquad \frac{d}{dt}u_y = -\frac{qB}{m}u_x \qquad \frac{d}{dt}u_z = 0 \tag{3.1a, b, c}$$

From Eq. (3.1c), it is clear that the component u_z of the velocity parallel to the static magnetic field is a constant equal to its initial value u_{z0}. If Eq. (3.1b) is multiplied by i and is added on to Eq. (3.1a) it is found that

$$\frac{d}{dt}U = -i\frac{qB}{m}U \tag{3.2}$$

where

$$U = u_x + iu_y \tag{3.3}$$

is a complex velocity. Let the gyromagnetic or the cyclotron angular frequency be defined as follows:

$$\omega_c = |q|B/m \tag{3.4}$$

The analysis is carried out only for a negatively charged particle or an electron and the corresponding results for a positively charged particle can be obtained by comparison. Note that ω_c defined by Eq. (3.4) is always a positive quantity. For an electron, $q = -e$ where e is the magnitude of the electronic charge and Eq. (3.2) together with Eq. (3.4) becomes

$$\frac{d}{dt}U = i\omega_c U \tag{3.5}$$

The solution of Eq. (3.5) is

$$U = u_{\perp 0} e^{i(\omega_c t + \theta_0)} \tag{3.6}$$

where $u_{\perp 0}$ and θ_0, which depend on the initial velocity of the particle, are constants. From Eqs. (3.3), (3.6), and (3.1c), it follows that

$$u_x = dx/dt = u_{\perp 0}\cos(\omega_c t + \theta_0) \tag{3.7}$$

$$u_y = dy/dt = u_{\perp 0}\sin(\omega_c t + \theta_0) \tag{3.8}$$

$$u_z = dz/dt = u_{z0} \tag{3.9}$$

It is verified from Eqs. (3.7) and (3.8) that $\sqrt{u_x^2 + u_y^2} = u_{\perp 0}$, which is the component of the velocity perpendicular to the magnetic field. Thus,

$$u^2 = u_x^2 + u_y^2 + u_z^2 = u_{\perp 0}^2 + u_{z0}^2 = \text{constant} \tag{3.10}$$

which is in accordance with the original deduction. The integration of Eqs. (3.7)–(3.9) yields

$$x = x_0 + (u_{\perp 0}/\omega_c)[\sin(\omega_c t + \theta_0) - \sin\theta_0] \tag{3.11}$$

$$y = y_0 - (u_{\perp 0}/\omega_c)[\cos(\omega_c t + \theta_0) - \cos\theta_0] \tag{3.12}$$

$$z = z_0 + u_{z0} t \tag{3.13}$$

where $\mathbf{r} = (x_0, y_0, z_0)$ denotes the initial position of the particle. It is seen from Eqs. (3.11) and (3.12) that

$$\sqrt{(x - \tilde{x}_0)^2 + (y - \tilde{y}_0)^2} = u_{\perp 0}/\omega_c = r_L \tag{3.14}$$

where

$$\tilde{x}_0 = x_0 - \frac{u_{\perp 0}}{\omega_c}\sin\theta_0 \qquad \tilde{y}_0 = y_0 + \frac{u_{\perp 0}}{\omega_c}\cos\theta_0 \tag{3.15}$$

It follows from Eq. (3.14) that the projection of the trajectory of the particle onto a plane perpendicular to the magnetic field is a circle centered at $(\tilde{x}_0, \tilde{y}_0)$ and radius equal to $r_L = u_{\perp 0}/\omega_c$ which is called the *Larmor radius*. Moreover, Eqs. (3.11) and (3.12) show that this circle is traced in the counterclockwise direction with a uniform angular velocity ω_c, as shown in Fig. 3.1. In view of Eq. (3.13), the coordinate of the particle parallel to the magnetic field is seen to change uniformly with time. Thus it is found that the trajectory of the particle is a helix with (a) its axis parallel to the magnetic field and passing through $(\tilde{x}_0, \tilde{y}_0)$, (b) its radius equal to $u_{\perp 0}/\omega_c$, and (c) its pitch equal to $2\pi u_{z0}/u_{\perp 0}$. The parameters of the helix depend on the initial position and velocity of the particle. In a plasma, the initial positions and velocities of the particles vary in a random manner and so are the parameters of their helical trajectories in a constant and uniform magnetic field. If $u_{\perp 0} = 0$, the trajectory is a straight line parallel to the direction of the magnetic field. If $u_{z0} = 0$, the trajectory is a circle of radius $u_{\perp 0}/\omega_c$ and center $(\tilde{x}_0, \tilde{y}_0)$ which is called *the guiding center*. In general the guiding center moves in the direction of the magnetic field with a constant velocity u_{z0}. The helical motion may be separated into a uniform motion along the magnetic field and a circular motion around the field lines.

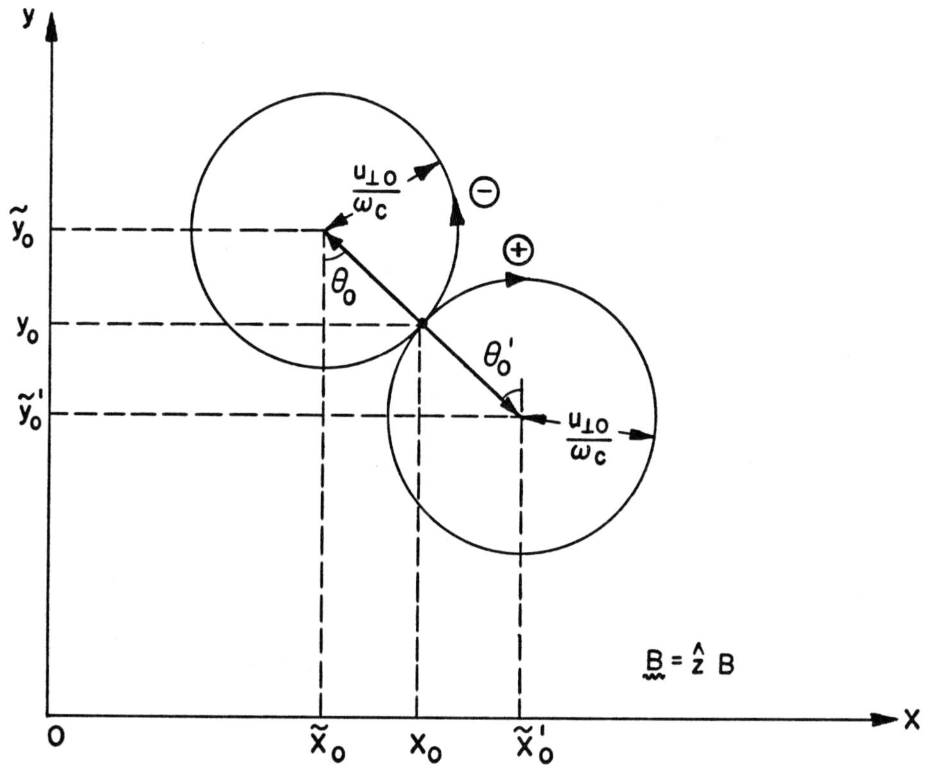

Fig. 3.1. Trajectory of a particle in a constant and uniform magnetic field.

For a positively charged particle, $q = e$ and since ω_c, by definition, is always positive, it is seen from Eq. (3.2) that ω_c in Eq. (3.5) should be changed to $-\omega_c$ provided it is assumed that the mass of the particle is unchanged. The trajectory of a particle of mass m and charge e is given by Eqs. (3.11)–(3.13) with ω_c replaced by $-\omega_c$. The projection of this trajectory of the particle onto a plane perpendicular to the magnetic field is also depicted in Fig. 3.1. It is seen that the circular motion perpendicular to the magnetic field is traced in the clockwise direction for a positively charged particle. It is found from Fig. 3.1 that on the application of a constant and uniform magnetic field, the trajectory of a particle in a plane perpendicular to the magnetic field becomes curved in one direction for a negative charge and in the opposite direction for a positive charge. The mass of an ion is much larger than that of an electron. The radius of curvature is seen from Eqs. (3.4) and (3.14) to be proportional to the particle mass. Consequently, the curvature of the trajectory of the particle in a plane perpendicular to the magnetic field is much smaller for an ion than that for an electron. Thus, in a plane perpendicular to the magnetic field, the trajectory of an electron becomes curved in one direction but that of an ion becomes curved in the opposite direction to a much smaller degree.

For the case in which the guiding center is stationary, the particle trajectories and the directions of the resulting circulating current are depicted in Fig. 3.2. The direction of the circulating current is independent of the sign of the charge and is given by the left-hand rule, that is, with the left thumb in the direction of the specified B-field, the left fingers point in the direction of the current.

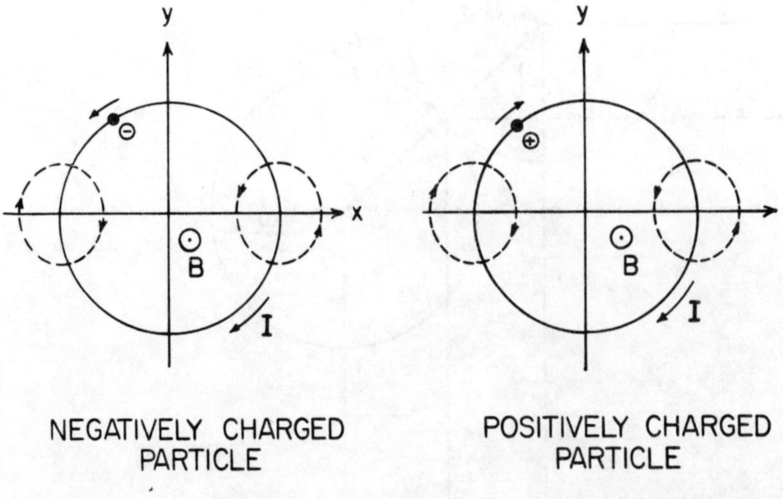

Fig. 3.2. Particle trajectories for $u_{z0} = 0$ indicating the directions of the current and the resulting magnetic field (shown by dashed lines); B: applied magnetic field, I: current.

Magnetic moment and magnetization

Suppose that the guiding center is stationary. The period of the resulting circular orbit is called the *Larmor period* and it is given by $T_L = 2\pi/\omega_c$. The magnitude of the circulating current is equal to the rate of flow of charge and hence $I = q/T_L = q\omega_c/2\pi$. This circulating current gives rise to a magnetic moment whose magnitude $\mathfrak{m}$ is obtained as

$$\mathfrak{m} = \text{current} \times \text{area} = (q\omega_c/2\pi)\pi r_L^2 \tag{3.16}$$

where the Larmor radius r_L is given by Eq. (3.14). The substitution of Eqs. (3.4) and (3.14) in Eq. (3.16) yields

$$\mathfrak{m} = w_\perp/B \tag{3.17}$$

where $w_\perp = \frac{1}{2}mu_{\perp 0}^2$ is the transverse kinetic energy of the particle. According to Ampère's law, the direction of the magnetic field due to the circulating current is given by the right-hand rule, that is, with the right thumb pointed in the direction of the current, the right fingers curl in the direction of the resulting magnetic field. The magnetic field of the particle is thus seen to oppose the applied field inside the orbit and augment it outside with the result that a plasma, a collection of charged particles, is diamagnetic. The direction of the magnetic moment is perpendicular to the area described by the circulating current and is given by the right-hand rule applied to the current. The direction of the magnetic moment is, therefore, antiparallel to that of the applied B-field. Thus, as a vector, the magnetic moment $\boldsymbol{\mathfrak{m}}$ is given by

$$\boldsymbol{\mathfrak{m}} = -(w_\perp/B^2)\mathbf{B} \tag{3.18}$$

The magnetic moment of a collection of particles in an applied magnetic field has fluctuations due to the distribution of the transverse kinetic energies as well as the phases of the orbital motions of the particles. If there are a large number of particles in a volume having the dimensions of the order of the Larmor radius and the distribution of the phases of the orbital motions is random, the fluctuations are small. Hence, the average magnetic field due to the particles can be computed from the average current due to the orbital motions of the particles.

The magnetic moments due to the orbital motion of a collection of charged particles act together to alter the applied magnetic field appreciably and a measure of the change can be obtained in the following manner. Consider a certain volume containing a large number of particles and consider its element of area S bounded by a closed curve C (Fig. 3.3). It is desired to find the net current crossing S due to the orbits which penetrate it. Orbits such as (1) which penetrate the area twice do not contribute to the net current whereas the orbits such as (2) which encircle the bounding curve contribute to the net current. Let $d\mathbf{l}$ be an element of length along the curve C. If there are N current loops per unit volume, each carrying a current I, and having an area $\mathbf{A}$, the number of current loops penetrated by $d\mathbf{l}$ is $N\mathbf{A} \cdot d\mathbf{l}$. Therefore, the net current I_n penetrating S is given by

$$I_n = \oint_C NI\mathbf{A} \cdot d\mathbf{l} = \oint_C \mathbf{M} \cdot d\mathbf{l} = \int_S \nabla \times \mathbf{M} \cdot d\mathbf{S} \tag{3.19}$$

where $d\mathbf{S}$ denotes an element of area and $\mathbf{M}$ the magnetic moment per unit volume, is known as the magnetization. The net current can be associated with an average current density $\mathbf{J}_M$, in terms of which the net current is given by

$$I_n = \int_S \mathbf{J}_M \cdot d\mathbf{S} \tag{3.20}$$

The magnetization current density $\mathbf{J}_M$ is obtained from Eqs. (3.19) and (3.20) to be

$$\mathbf{J}_M = \nabla \times \mathbf{M} \tag{3.21}$$

where

$$\mathbf{M} = -(Nw_\perp / B^2)\mathbf{B} = -(W_\perp / B^2)\mathbf{B} \tag{3.22}$$

and $W_\perp$ is the transverse kinetic energy per unit volume. The magnetization current density given by Eq. (3.21) corresponds to the case of a guiding center at rest. It is shown later that the motion of the guiding centers gives rise to other currents.

If the current density $\mathbf{J}$ in Eq. (1.6) is separated into $\mathbf{J}_M + \mathbf{J}_e$ where $\mathbf{J}_M$ and $\mathbf{J}_e$ are the magnetization and the external current densities, respectively, it follows that

$$\nabla \times \mathbf{B} = \mu_0 \nabla \times \mathbf{M} + \mu_0 \mathbf{J}_e + \mu_0 \varepsilon_0 (\partial/\partial t)\mathbf{E} \tag{3.23}$$

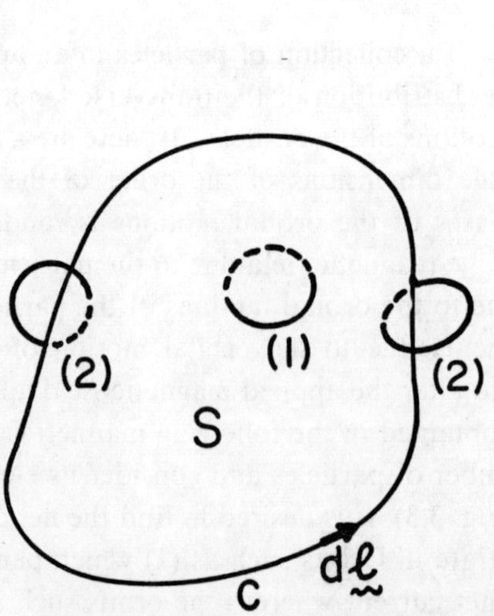

Fig. 3.3. Current loops and the resulting magnetization current.

On defining an effective magnetic field **H** as

$$\mathbf{H} = (1/\mu_0)\mathbf{B} - \mathbf{M} \tag{3.24}$$

Eq. (3.23) becomes

$$\nabla \times \mathbf{H} = \varepsilon_0(\partial/\partial t)\mathbf{E} + \mathbf{J}_e \tag{3.25}$$

The other Maxwell's equations (1.5), (1.9), and (1.10) are unchanged. In view of Eq. (1.4), the charge density ρ_M associated with the magnetization current density $\mathbf{J}_M$ is obtained from

$$(\partial/\partial t)\rho_M + \nabla \cdot \mathbf{J}_M = 0 \tag{3.26}$$

From Eqs. (3.21) and (3.26), it follows that ρ_M is a constant since $\nabla \cdot (\nabla \times \mathbf{M}) = 0$.

The treatment of magnetic materials is based on Eq. (3.24) with whose help an effective permeability is defined if the magnetization **M** is proportional to the applied B-field. In the present case, since **M** is inversely proportional to B, the introduction of an effective magnetic field **H** is not convenient.

3.4. Constant and uniform electric and magnetic fields

Consider a charged particle in a combination of electric and magnetic fields which are both constant and uniform. As before, the B-field is assumed to be everywhere parallel to the z-direction. The direction of the E-field does not change in space. It is convenient to separate the electric field and the particle velocity into two parts as follows

$$\mathbf{E} = \hat{\mathbf{z}}E_z + \mathbf{E}_\perp \qquad \mathbf{u} = \hat{\mathbf{z}}u_z + \mathbf{u}_\perp \tag{4.1}$$

The components perpendicular to the direction of the applied magnetic field are referred to as the perpendicular components. The parallel and the perpendicular components of Eq. (1.3) are

$$m(d/dt)u_z = qE_z \tag{4.2}$$

$$m(d/dt)\mathbf{u}_\perp = q[\mathbf{E}_\perp + \mathbf{u}_\perp \times \mathbf{B}] \tag{4.3}$$

The solution of Eq. (4.2) gives

$$u_z = (q/m)E_z t + u_{z0} \qquad z = (q/2m)E_z t^2 + u_{z0}t + z_0 \tag{4.4}$$

where u_{z0} and z_0 are, respectively, the initial values of u_z and z. The interpretation of Eqs. (4.4) is the same as that for a particle acted upon by a constant and uniform electric field alone.

Let a constant $\mathbf{u}_E$ be a particular solution of Eq. (4.3). The result of substitution of this solution in Eq. (4.3) is

$$0 = \mathbf{E}_\perp + \mathbf{u}_E \times \mathbf{B} \tag{4.5}$$

Note that without loss of generality, $\mathbf{u}_E$ can be assumed to be perpendicular to $\mathbf{B}$. On cross-multiplying Eq. (4.5) by $\mathbf{B}$, it is found that

$$0 = \mathbf{E}_\perp \times \mathbf{B} + (\mathbf{u}_E \times \mathbf{B}) \times \mathbf{B} = \mathbf{E}_\perp \times \mathbf{B} - \mathbf{u}_E B^2$$

with the result that

$$\mathbf{u}_E = \mathbf{E}_\perp \times \mathbf{B}/B^2 \tag{4.6}$$

is verified to be a particular solution of Eq. (4.3) and is perpendicular to both $\mathbf{E}_\perp$ and $\mathbf{B}$. Let the general solution of Eq. (4.3) be sought in the form

$$\mathbf{u}_\perp = \mathbf{u}_E + \mathbf{u}_l \tag{4.7}$$

The substitution of Eq. (4.7) in Eq. (4.3) and the use of Eq. (4.5) yields

$$m d\mathbf{u}_l/dt = q[\mathbf{E}_\perp + \mathbf{u}_E \times \mathbf{B} + \mathbf{u}_l \times \mathbf{B}] = q\mathbf{u}_l \times \mathbf{B} \tag{4.8}$$

where $\mathbf{u}_l$ corresponds to the orbital motion of the charged particle due to the application of the B-field alone. Thus, in a combination of electric and magnetic fields, the velocity perpendicular to the magnetic field, consists of a constant drift velocity which is also perpendicular to the electric field superimposed on an orbital motion which is the same as that due to the application of a constant and uniform magnetic field alone.

The projection of the trajectory of the particle onto a plane perpendicular to the magnetic field is the same as its motion with $E_z = 0$ and $u_{z0} = 0$. To facilitate the discussion of the particle trajectory, it is assumed, for the present, that $E_z = 0$, $u_{z0} = 0$, and $z_0 = 0$. Then, the motion of the particle is in the xy-plane. The electric field is assumed to be in the y-direction, that is $\mathbf{E} = \hat{\mathbf{y}}E_y$. Then, from Eqs. (3.7), (3.8), (4.6), and (4.7), it is obtained that

$$u_x = E_y/B + u_{\perp 0}\cos(\omega_c t + \theta_0) \tag{4.9}$$

$$u_y = u_{\perp 0}\sin(\omega_c t + \theta_0) \tag{4.10}$$

The expressions for $u_{\perp 0}$ and θ_0 can be evaluated in terms of the initial velocities u_{x0} and u_{y0} as

$$u_{\perp 0} = \left[\left(u_{x0} - \frac{E_y}{B}\right)^2 + u_{y0}^2\right]^{1/2} \qquad \tan\theta_0 = \frac{u_{y0}}{u_{x0} - (E_y/B)} \tag{4.11}$$

In particular, for a particle starting from rest ($u_{x0} = u_{y0} = 0$), it is found that

$$u_{\perp 0} = -E_y/B \qquad \theta_0 = 0 \tag{4.12}$$

It is emphasized that the magnitude of the applied electric field should be small enough so that the restriction $u^2/c^2 \ll 1$ is not violated; otherwise, the present analysis is not valid. This is because the nonrelativistic equation of motion, on which this analysis is based, is valid only if $u^2/c^2 \ll 1$.

The integration of Eqs. (4.9) and (4.10) yields

$$x = \frac{E_y}{B}t + \frac{u_{\perp 0}}{\omega_c}[\sin(\omega_c t + \theta_0) - \sin\theta_0] \tag{4.13}$$

$$y = -\frac{u_{\perp 0}}{\omega_c}[\cos(\omega_c t + \theta_0) - \cos\theta_0] \tag{4.14}$$

where the origin of the coordinates is chosen to coincide with the initial position of the particle. The trajectory of the particle represented by Eqs. (4.13) and (4.14) is called a cycloid. If a circular disc rolls without slipping along a straight line with a uniform angular velocity, the curve traced by any point on the disc is a cycloid. Different cycloidal trajectories are possible depending on different initial conditions. Several such trajectories are shown in Fig. 3.4 for an electron.

A. $u_{x0} = u_{y0} = 0$.

Consider first an electron starting from rest. From Eqs. (4.9), (4.10), (4.12)–(4.14), it can be shown that

$$\tilde{u}_x = 1 - \cos\omega_c t \qquad \tilde{u}_y = -\sin\omega_c t \tag{4.15}$$

$$\tilde{x} = \omega_c t - \sin\omega_c t \qquad \tilde{y} = \cos\omega_c t - 1 \tag{4.16}$$

where

$$\tilde{u}_x = u_x B/E_y \qquad \tilde{u}_y = u_y B/E_y \qquad \tilde{x} = xB\omega_c/E_y \qquad \tilde{y} = yB\omega_c/E_y \tag{4.17}$$

Since E_y/B has the dimensions of a velocity, it follows that the normalized velocities $\tilde{u}_x$ and $\tilde{u}_y$ are dimensionless. In a similar manner, $E_y/\omega_c B$ has the dimensions of a length and therefore the normalized displacements $\tilde{x}$ and $\tilde{y}$ are also dimensionless.

A physical understanding of the motion of the electron can be obtained in the following manner. Since the electron starts from rest, initially it is not subject to any $\mathbf{u} \times \mathbf{B}$ force; the electric field repels the electron in the negative y-direction. As $\mathbf{u}$ increases, the $\mathbf{u} \times \mathbf{B}$ force tends to curve the trajectory in the positive (counterclockwise) direction. The electron moves in the positive x- and negative y-directions. The y-component of the velocity becomes more and more negative, reaches a minimum, then becomes less and less negative and finally becomes zero at time

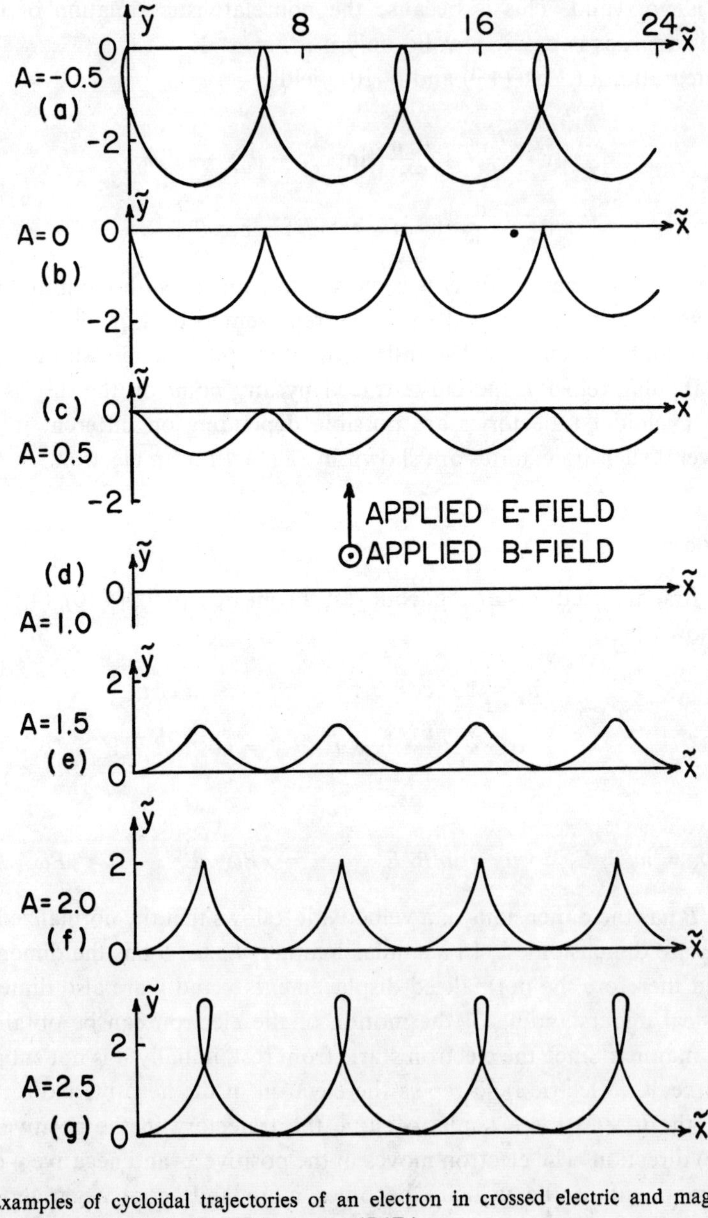

Fig. 3.4. Examples of cycloidal trajectories of an electron in crossed electric and magnetic fields: $\tilde{x} = x(B\omega_c/E_y)$; $\tilde{y} = y(B\omega_c/E_y)$; $A = u_{x0}(B/E_y)$.

$t = \pi/\omega_c$. In this time $(0 < t \leqslant \pi/\omega_c)$, the x-component of the velocity has continually increased to reach a maximum value. For $t > \pi/\omega_c$, the $\mathbf{u} \times \mathbf{B}$ force continues to curve the electron in the counterclockwise direction and since the y-component of the velocity is positive, the electron starts moving in the positive y-direction. The y-component of the velocity becomes more and more positive, reaches a maximum, then becomes less and less positive and finally reaches zero at time $t = 2\pi/\omega_c$. During this time $[(\pi/\omega_c) < t < (2\pi/\omega_c)]$, the x-component of the velocity continually decreases and becomes zero at time $t = 2\pi/\omega_c$. At $t = 2\pi/\omega_c$, the electron is at rest. In this time, the electron has effectively moved along the x-axis a distance of $(E_y/B)(2\pi/\omega_c)$. This cycloidal arc repeats itself every $2\pi/\omega_c$ seconds. The trajectory represented by Eqs. (4.16) is shown in Fig. 3.4b.

B. $u_{x0} = E_y/B;\ u_{y0} = 0.$

Consider an electron whose initial velocity is equal to the drift velocity, that is $u_{x0} = E_y/B$ and $u_{y0} = 0$. From Eqs. (4.9) and (4.10), it is seen that these initial conditions lead to the requirement that $u_{\perp 0} = 0$. Therefore, it follows from Eqs. (4.9), (4.10), (4.13), and (4.14) that

$$\tilde{u}_x = 1 \qquad \tilde{u}_y = 0 \tag{4.18}$$

$$\tilde{x} = \omega_c t \qquad \tilde{y} = 0 \tag{4.19}$$

Hence, the electron continues to move uniformly in the x-direction with the initial velocity which is equal to the drift velocity. For this case, the force due to the electric field is exactly annuled by the $\mathbf{u} \times \mathbf{B}$ force with the result that no net force acts on the electron. Consequently the initial velocity is maintained. This trajectory is depicted in Fig. 3.4d.

C. $0 < u_{x0} < E_y/B;\ u_{y0} = 0.$

Consider an electron whose initial velocity is in the x-direction but whose magnitude lies between zero and the drift velocity, i.e., $0 < u_{x0} < E_y/B$ and $u_{y0} = 0$. For these initial conditions, Eqs. (4.9), (4.10), (4.13), and (4.14) yield

$$\tilde{u}_x = 1 + (A - 1)\cos \omega_c t, \qquad \tilde{u}_y = (A - 1)\sin \omega_c t \tag{4.20}$$

$$\tilde{x} = \omega_c t + (A - 1)\sin \omega_c t \qquad \tilde{y} = (A - 1)(1 - \cos \omega_c t) \tag{4.21}$$

where

$$A = u_{x0} B/E_y \tag{4.22}$$

Note that $u_{x0} - (E_y/B) < 0$ and the minimum value attained by u_x is equal to u_{x0} whereas the variation of u_y is very similar to the case A. Initially the $\mathbf{u} \times \mathbf{B}$ force does not completely annul that due to the electric field with the result that the electron moves in the negative y-direction with the $\mathbf{u} \times \mathbf{B}$ force turning the orbit in the counterclockwise direction. The cycloidal motions are similar to the case A except that the minimum value of u_x is a nonzero positive quantity with the result that the electron never comes to rest but always keeps moving in the positive x-direction. The trajectory corresponding to Eqs. (4.21) is shown in Fig. 3.4c.

D. $u_{x0} > E_y/B$; $u_{y0} = 0$.

Consider now the case for which the initial velocity of the electron is in the x-direction but whose magnitude is greater than the drift velocity, i.e., $u_{x0} > E_y/B$ and $u_{y0} = 0$. The velocities and the trajectories are the same as those given in Eqs. (4.20) and (4.21), respectively. In the three previous cases, after the start the y-component of the velocity is either negative or zero with the result that the particle started moving in the negative y-direction or remained along the x-axis. In the present case, after the start the y-component of the velocity is positive with the result that the particle starts moving in the positive y-direction. Also in the neighborhood of $\cos \omega_c t$ equal to -1, u_x can be negative if $u_{x0} \geqslant 2(E_y/B)$. As a consequence, if these conditions are obtained, the electron in the vicinity of the top portions of the orbits has a velocity which has a negative x-component and hence loops back in the direction opposite to that of the drift. Since the x-component of the velocity in the bottom portion of the orbit is greater than twice the drift velocity, it is intuitively to be expected that at the top portions of the orbits the x-component of the velocity has to be negative in order that the overall drift may have the proper value.

The trajectory for the case D is depicted in Fig. 3.4e–g for various values of $u_{x0} B/E_y$. The following additional physical explanation for the trajectory shown in Fig. 3.4g is helpful. Since the initial velocity in the x-direction is greater than the drift velocity, the $\mathbf{u} \times \mathbf{B}$ force dominates over that due to the electric field and hence turns the electron in the counterclockwise direction; therefore, the particle starts moving in the positive y-direction. Since the electric field is in the y-direction, the negatively charged particle is decelerated in that direction. However, when the electron is moving in the negative y-direction it is accelerated. Consequently, the electron velocities are larger at the bottom portions of the orbits and smaller at the top. From Eq. (3.14), the Larmor radius is seen to be smaller at the top and larger at the bottom. Therefore, the radius of curvature of the trajectory is smaller at the top and larger at the bottom resulting in an overall drift in the positive x-direction.

If an electron had an initial velocity in the opposite direction to the drift velocity, it can be argued that the trajectory is the same as in the previous case with the following differences: (i) the whole trajectory is shifted in the negative y-direction such that the top portions of the trajectories are along the x-axis and (ii) the trajectory initially starts at the origin and begins moving in the negative y-direction. This type of trajectory is illustrated in Fig. 3.4a.

In Fig. 3.4, the trajectories of the negatively charged particle alone are given under various initial conditions. It can be shown that the reflection of these trajectories about the x-axis gives the trajectories for the corresponding positively charged particle. This is, however, true only if the masses are the same for the two kinds of particle. The positively charged ions are much heavier and therefore the Larmor radius and the Larmor period are correspondingly large. Consequently, the cycloidal arcs are larger but there are fewer of them per second such that the drift velocities of the electrons and the ions are equal.

It should be remembered that Fig. 3.4 depicts the projection of the actual trajectories onto the xy-plane. In addition, the particle is displaced in the z-direction in accordance with Eqs. (4.4).

Since it is independent of the initial parameters, $\mathbf{u}_E$ is the same for all the particles and therefore represents the collective motion of all the particles. But $\mathbf{u}_1$ which depends on the initial velocities of the particles is randomly distributed. The average of $\mathbf{u}$ yields $\mathbf{u}_E$. For the velocity $\mathbf{u}_E$ the force due to the electric field is exactly annuled by that due to the magnetic field. Thus the electric drift velocity $\mathbf{u}_E$ is such that the force due to the electric field is annuled by the average force due to the magnetic field. Note that the electric drift velocity $\mathbf{u}_E$ given by Eq. (4.6) is independent of the sign of the charge or the mass of the particle. Therefore in a neutral plasma, the electrons and the ions have the same drift velocity and hence, there is no net current. If the plasma is not neutral, the electric drift produces a net current in the direction perpendicular to both $\mathbf{E}$ and $\mathbf{B}$.

External force

If instead of an electric field, there is another field of force $\mathbf{F}_\perp$ such as that due to the gravitational field. Then $q\mathbf{E}_\perp$ has to be replaced by $\mathbf{F}_\perp$ to obtain the corresponding drift velocity. The result is

$$\mathbf{u}_F = \mathbf{F}_\perp \times \mathbf{B}/qB^2 \tag{4.23}$$

In this case, the direction of the drift velocity is changed on changing the sign of the charge. The force due to an electric field depends on the sign of the charge whereas $\mathbf{F}_\perp$ is independent of the sign of the charge. Since the positive and the negative charges move in opposite directions, the drift velocity given by Eq. (4.23) can produce a net current in the direction perpendicular to both $\mathbf{F}_\perp$ and $\mathbf{B}$.

3.5. Uniform and slowly time-varying electric field

The applied magnetic field is uniform, constant, and oriented everywhere parallel to the z-axis. The electric field is spatially uniform but varies slowly in time. As before, the direction of the electric field is the same everywhere in space. The parallel and the perpendicular components of the velocity satisfy Eqs. (4.2) and (4.3), respectively. The parallel component of the velocity is obtained from Eq. (4.2) but does not lead to any interesting result.

Since the electric field varies only very slowly in time, it is reasonable to expect the perpendicular component of the velocity to be not very different from that for a constant electric field. Consequently, the solution of $\mathbf{u}_\perp$ for Eq. (4.3) can be sought as a perturbation on Eq. (4.7). Thus

$$\mathbf{u}_\perp = \mathbf{u}_E + \mathbf{u}_P + \mathbf{u}_1 \tag{5.1}$$

where $\mathbf{u}_E$ is given by Eq. (4.6). Since $\mathbf{E}_\perp$ varies with time, it follows that $\mathbf{u}_E$ also varies slowly with time. The substitution of Eq. (5.1) into Eq. (4.3) together with Eqs. (4.5) and (4.6) yields

$$m\dot{\mathbf{E}}_\perp \times \mathbf{B}/B^2 + m\dot{\mathbf{u}}_P + m\dot{\mathbf{u}}_1 = q\mathbf{u}_P \times \mathbf{B} + q\mathbf{u}_1 \times \mathbf{B} \tag{5.2}$$

where the dot denotes differentiation with respect to time. On setting

$$\mathbf{u}_P = (m/q)(\dot{\mathbf{E}}_\perp/B^2) \tag{5.3}$$

Eq. (5.2) reduces to

$$m\dot{\mathbf{u}}_P + m\dot{\mathbf{u}}_1 = q\mathbf{u}_1 \times \mathbf{B} \tag{5.4}$$

Let the magnitudes of the first term on the left and that on the right be compared. With the help of Eq. (5.3), it is found that

$$\left|\frac{m\dot{\mathbf{u}}_P}{q\mathbf{u}_1 \times \mathbf{B}}\right| = \left|\frac{m^2\ddot{\mathbf{E}}_\perp}{q^2B^3\mathbf{u}_1}\right| = \left|\frac{\mathbf{E}_\perp}{\mathbf{u}_1 B}\frac{\omega^2}{\omega_c^2}\right| = \left|\frac{\mathbf{u}_E}{\mathbf{u}_1}\right|\left(\frac{\omega}{\omega_c}\right)^2 \tag{5.5}$$

In obtaining Eq. (5.5), $\mathbf{E}_\perp$ is assumed to have a harmonic time dependence with an angular frequency ω. Also Eqs. (3.4) and (4.6) have been used. If the characteristic frequency of the electric field is very much smaller than the gyromagnetic frequency, that is $\omega^2 \ll \omega_c^2$ and if $|\mathbf{u}_E/\mathbf{u}_1|$ is also small, then $m\dot{\mathbf{u}}_P$ in Eq. (5.4) can be neglected in comparison with the other terms with the result that $\mathbf{u}_1$ corresponds to the orbital motion of the charged particle due to the B-field alone. Thus Eq. (5.1) together with Eqs. (4.6) and (5.3) yields $\mathbf{u}_\perp$ for a slowly time-varying electric field.

The result of a slow temporal variation of the electric field is the addition of the velocity $\mathbf{u}_P$ perpendicular to the magnetic field. Unlike the electric drift velocity $\mathbf{u}_E$, the polarization drift velocity $\mathbf{u}_P$ is parallel to the electric field, depends on the sign of the charge and is proportional to the time derivative of the electric field. Since $\mathbf{u}_P$ is in opposite directions for charges of opposite signs, the polarization drift gives rise to a net current density $\mathbf{J}_P$ in a neutral plasma. The current density is the rate of flow of charges across unit area. All the charges in a rectangular volume of length

$\mathbf{u}_P$ pass through the unit area of the end face of the volume in unit time. The direction of the current is the same as that of $\mathbf{u}_P$. Hence,

$$\mathbf{J}_P = \sum q\mathbf{u}_P = \sum q\frac{m}{q}\frac{\dot{\mathbf{E}}_\perp}{B^2} = \frac{\rho_m}{B^2}\frac{\partial}{\partial t}\mathbf{E}_\perp \tag{5.6}$$

where the summation extends over both negatively and positively charged particles in unit volume. Therefore, ρ_m is the mass density of the plasma.

The polarization current density $\mathbf{J}_P$ can be regarded as a part of the total current density $\mathbf{J}$ given in Eq. (1.6) or taken into account through the introduction of an effective permittivity by combining it with the first term on the right side of Eq. (1.6). For this purpose, the total current density $\mathbf{J}$ is separated into the sum of the polarization current density $\mathbf{J}_P$ and the current density $\mathbf{J}_0$ due to the other sources. On combining $\mathbf{J}_P$ with the first term on the right side of Eq. (1.6), it is seen that

$$\mu_0\varepsilon_0\frac{\partial}{\partial t}\mathbf{E}_\perp + \mu_0\frac{\rho_m}{B^2}\frac{\partial}{\partial t}\mathbf{E}_\perp = \mu_0\varepsilon_0\left(1 + \frac{\rho_m}{\varepsilon_0 B^2}\right)\frac{\partial}{\partial t}\mathbf{E}_\perp \tag{5.7}$$

where

$$\varepsilon = \varepsilon_0\varepsilon_r = \varepsilon_0(1 + \rho_m/\varepsilon_0 B^2) \tag{5.8}$$

is the effective permittivity perpendicular to the magnetic field.

The charge density ρ_P which accumulates due to the polarization current is obtained from Eqs. (1.4) and (5.6), to be given by

$$\rho_P = -(\rho_m/B^2)\nabla\cdot\mathbf{E} \tag{5.9}$$

assuming that there is no parallel component of the electric field. Let the total charge density ρ be separated out as

$$\rho = \rho_P + \rho_0 \tag{5.10}$$

where ρ_0 corresponds to the current density $\mathbf{J}_0$. From Eqs. (1.10) and (5.8)–(5.10), it can be shown that

$$\nabla\cdot\mathbf{E} = \rho_0/\varepsilon \tag{5.11}$$

The remaining Maxwell's equations (1.5) and (1.9) are unaffected. Thus the polarization current density and the resulting charge density can be taken into account by the introduction of an effective permittivity ε.

It is instructive to examine how the particles acquire kinetic energy. For a change $\delta\mathbf{E}$ in the electric field, the displacement $\delta\mathbf{r}$ of the guiding center is obtained from Eq. (5.3) as

$$\delta\mathbf{r} = \mathbf{u}_P\,\delta t = (m/qB^2)\delta\mathbf{E}$$

The corresponding work done by the electric field is given by

$$\delta w = q\mathbf{E} \cdot \delta\mathbf{r} = \frac{m}{B^2}\mathbf{E} \cdot \delta\mathbf{E} = \delta\left[\frac{m}{2}\frac{E^2}{B^2}\right] = \delta\left[\frac{m}{2}u_E^2\right] \tag{5.12}$$

The implication is that the work done by the electric field on the particles in the act of polarization is equal to the increase in the kinetic energy of the particles due to their collective motion with the drift velocity $\mathbf{u}_E$. It is to be noted that the electric drift velocity $\mathbf{u}_E$ being perpendicular to the electric field does not lead to any change in energy due to variations in the electric field.

The correctness of the introduction of effective permittivity can be verified now by evaluating the total energy density associated with the electric field. The energy density in the electric field is given by

$$w_E = (\varepsilon_0/2)E^2 \tag{5.13}$$

The kinetic energy density due to the particles moving collectively with the drift velocity $\mathbf{u}_E$ is

$$w_u = (\rho_m/2)u_E^2 = (\rho_m/2)(E^2/B^2) \tag{5.14}$$

Since the kinetic energy density associated with the orbital motion of the particles is unaffected by the changes in the electric field, the total energy density associated with the electric field is

$$w_E + w_u = \frac{1}{2}\varepsilon_0\left(1 + \frac{\rho_m}{\varepsilon_0 B^2}\right)E^2 = \frac{\varepsilon}{2}E^2 \tag{5.15}$$

which is the same as the energy density in the electric field for a dielectric medium of effective permittivity ε. Thus, on account of the polarization current density, the plasma medium behaves like an ordinary dielectric. If the number density is $N = 10^{21}$ per cubic meter and $B = 1$ weber per square meter, ε_r is of the order of 10^5. Thus, at low frequencies, the relative permittivity of a plasma is very high.

3.6. Uniform electric field with arbitrary time variation and the conductivity dyad

The applied electric and magnetic fields are the same as in the preceding section where the electric field is assumed to be varying slowly with time. In this section, this restriction is relaxed and the time variation of the electric field is assumed to be arbitrary. The Cartesian components of the electric field

$$\mathbf{E} = \hat{\mathbf{x}}E_x + \hat{\mathbf{y}}E_y + \hat{\mathbf{z}}E_z \tag{6.1}$$

are conveniently rewritten in the following form

$$\mathbf{E} = E_+\frac{(\hat{\mathbf{x}} + i\hat{\mathbf{y}})}{\sqrt{2}} + E_-\frac{(\hat{\mathbf{x}} - i\hat{\mathbf{y}})}{\sqrt{2}} + \hat{\mathbf{z}}E_z \tag{6.2}$$

where

$$E_{\pm} = \frac{1}{\sqrt{2}}(E_x \mp iE_y) \tag{6.3}$$

and $(\hat{\mathbf{x}} + i\hat{\mathbf{y}})/\sqrt{2}$ and $(\hat{\mathbf{x}} - i\hat{\mathbf{y}})/\sqrt{2}$ are unit complex vectors. Without loss of generality, the time variation of the electric field is assumed to be periodic of angular frequency ω. For example,

$$E_+ = E_{+0}e^{-i\omega t} \tag{6.4}$$

where E_{+0}, the complex amplitude, is independent of time. Since Eq. (1.3) is linear, arbitrary time variation of the electric field can be synthesized by the superposition of terms similar to Eq. (6.4) corresponding to all possible values of ω. In accordance with the usual convention, only the real part of the right side of Eq. (6.4) and therefore, of Eq. (6.2) also, has to be taken. If $E_{+0} = |E_{+0}|e^{-i\theta_E}$, the first term on the right side of Eq. (6.2) together with Eq. (6.4) yields

$$2^{-1/2}|E_{+0}|[\hat{\mathbf{x}}\cos(\omega t + \theta_E) + \hat{\mathbf{y}}\sin(\omega t + \theta_E)]$$

which represents a circularly polarized field with the electric vector rotating in the counterclockwise direction. Similarly, the second term of Eq. (6.2) also represents a circularly polarized field with its electric vector rotating in the clockwise direction. Thus, the two linearly polarized perpendicular components of the electric field are recast in the form of two circularly polarized components with opposite directions of rotation. The advantage of using the circularly polarized components is that the perpendicular component of Eq. (1.3) decouples into two separate equations pertaining to the two circular polarizations.

The particle velocity is conveniently sought in the following form:

$$\mathbf{u} = a_+E_+\frac{(\hat{\mathbf{x}} + i\hat{\mathbf{y}})}{\sqrt{2}} + a_-E_-\frac{(\hat{\mathbf{x}} - i\hat{\mathbf{y}})}{\sqrt{2}} + \hat{\mathbf{z}}a_zE_z \tag{6.5}$$

where a_+, a_-, and a_z are functions of time. Also, E_+, E_-, and E_z have harmonic time dependence, as indicated in Eq. (6.4). It can be verified that

$$\frac{(\hat{\mathbf{x}} \pm i\hat{\mathbf{y}})}{\sqrt{2}} \times \hat{\mathbf{z}} = \pm i\frac{(\hat{\mathbf{x}} \pm i\hat{\mathbf{y}})}{\sqrt{2}} \qquad \frac{(\hat{\mathbf{x}} \pm i\hat{\mathbf{y}})}{\sqrt{2}} \cdot \hat{\mathbf{z}} = 0 \tag{6.6}$$

$$\frac{(\hat{\mathbf{x}} \pm i\hat{\mathbf{y}})}{\sqrt{2}} \cdot \frac{(\hat{\mathbf{x}} \pm i\hat{\mathbf{y}})}{\sqrt{2}} = 0 \qquad \frac{(\hat{\mathbf{x}} \pm i\hat{\mathbf{y}})}{\sqrt{2}} \cdot \frac{(\hat{\mathbf{x}} \mp i\hat{\mathbf{y}})}{\sqrt{2}} = 1 \tag{6.7}$$

If Eq. (6.5) is substituted in Eq. (1.3) with $q = -e$ and the scalar products of the resulting equation are taken successively with respect to $(\hat{\mathbf{x}} - i\hat{\mathbf{y}})/\sqrt{2}$, $(\hat{\mathbf{x}} + i\hat{\mathbf{y}})/\sqrt{2}$, and $\hat{\mathbf{z}}$ the following three separate equations are obtained:

$$\dot{a}_+ - i(\omega - \omega_c)a_+ = -e/m \tag{6.8}$$

$$\dot{a}_- - i(\omega + \omega_c)a_- = -e/m \tag{6.9}$$

$$\dot{a}_z - i\omega a_z = -e/m \tag{6.10}$$

It is assumed that the second terms on the left sides of Eqs. (6.8)–(6.10) do not vanish. In view of Eq. (6.5), the general solutions of Eqs. (6.8)–(6.10) are seen to depend on the initial velocities of the particles. However, it is possible to choose the following particular solutions of Eqs. (6.8)–(6.10):

$$a_+ = -ie/m(\omega - \omega_c) \tag{6.11}$$

$$a_- = -ie/m(\omega + \omega_c) \tag{6.12}$$

$$a_z = -ie/m\omega \tag{6.13}$$

These solutions are constants independent of time and can be shown to satisfy Eqs. (6.8)–(6.10), respectively. Also, these solutions, being independent of the initial velocities of the electrons, are therefore the same for all the electrons and hence represent the collective motion of the particles. Moreover, the associated particle velocity is seen from Eqs. (6.4) and (6.5) to vary in time with the same angular frequency as that of the applied electric field. Since Eq. (6.4) implies the existence of the impressed electric field for an indefinite period of time, it is natural to expect the forced collective oscillations of the particles to have the same frequency as that of the forcing electric field.

From Eqs. (6.5) and (6.11)–(6.13), it is found that

$$\mathbf{u} = \frac{(\hat{\mathbf{x}} + i\hat{\mathbf{y}})}{\sqrt{2}} u_+ + \frac{(\hat{\mathbf{x}} - i\hat{\mathbf{y}})}{\sqrt{2}} u_- + \hat{\mathbf{z}} u_z \tag{6.14}$$

where

$$\begin{aligned} u_+ &= \mu_+ E_+ = \frac{-ie}{m(\omega - \omega_c)} E_+ \\ u_- &= \mu_- E_- = \frac{-ie}{m(\omega + \omega_c)} E_- \\ u_z &= \mu_z E_z = \frac{-ie}{m\omega} E_z \end{aligned} \tag{6.15}$$

and u_+, u_-, and u_z are the right (counterclockwise) circularly polarized, the left (clockwise) circularly polarized, and the parallel components of the collective velocity of the electrons. When the electron velocity is expressed in terms of the circularly polarized components, it is seen from Eqs. (6.14) and (6.15) that each component of the electron velocity is proportional to the corresponding component of the electric field which causes it; the proportionality constant μ is called the mobility. Note that the mobility is different for the different components of the electric field emphasizing the anisotropic behavior of the plasma in a magnetostatic field. Since the foregoing analysis is specifically carried out for a negatively charged particle or electron, it is convenient to add a subscript e to $\mathbf{u}$, μ, m, and ω_c so that the corresponding quantities for a positively charged particle or ion can be distinguished by the subscript i.

If N_0 is the number density of electrons, the electron current density is obtained as

$$\mathbf{J}_e = -N_0 e\mathbf{u}_e = \frac{(\hat{\mathbf{x}} + i\hat{\mathbf{y}})}{\sqrt{2}} J_{+e} + \frac{(\hat{\mathbf{x}} - i\hat{\mathbf{y}})}{\sqrt{2}} J_{-e} + \hat{\mathbf{z}} J_{ze} \tag{6.16}$$

where

$$J_{+e} = -N_0 e\mu_+ E_+ = \frac{iN_0 e^2}{m_e(\omega - \omega_{ce})} E_+ = \sigma_{+e} E_+ \tag{6.17a}$$

$$J_{-e} = -N_0 e\mu_- E_- = \frac{iN_0 e^2}{m_e(\omega + \omega_{ce})} E_- = \sigma_{-e} E_- \tag{6.17b}$$

$$J_{ze} = -N_0 e\mu_z E_z = \frac{iN_0 e^2}{m_e \omega} E_z = \sigma_z E_z \tag{6.17c}$$

For the current expressed in terms of the circularly polarized components, the components of the current are proportional to the corresponding components of the electric field. The proportionality constant is the complex conductivity σ.

In view of the fact that the components of the current and the velocity are directly proportional to the corresponding components of the electric field in the coordinate system $(\hat{\mathbf{x}} + i\hat{\mathbf{y}})/\sqrt{2}$, $(\hat{\mathbf{x}} - i\hat{\mathbf{y}})/\sqrt{2}$, and $\hat{\mathbf{z}}$, these coordinates are called the natural or the canonical coordinates of the system. However, for the purpose of its use in subsequent analysis, it is necessary to express the relationship of the current to the electric field in the Cartesian coordinate system. If the expressions for J_{+e}, J_{-e}, and J_{ze} in terms of E_+, E_-, and E_z, respectively, are substituted from Eqs. (6.17) into Eq. (6.16) and if E_+ and E_- are replaced by their equivalents in terms of E_x and E_y from Eq. (6.3), the current density can be evaluated in terms of the Cartesian components of the electric field. The result is conveniently expressed in the matrix form as follows:

$$\begin{bmatrix} J_{xe} \\ J_{ye} \\ J_{ze} \end{bmatrix} = \frac{iN_0 e^2}{m_e \omega} \begin{bmatrix} \dfrac{\omega^2}{\omega^2 - \omega_{ce}^2} & \dfrac{-i\omega\omega_{ce}}{\omega^2 - \omega_{ce}^2} & 0 \\ \dfrac{i\omega\omega_{ce}}{\omega^2 - \omega_{ce}^2} & \dfrac{\omega^2}{\omega^2 - \omega_{ce}^2} & 0 \\ 0 & 0 & 1 \end{bmatrix} \begin{bmatrix} E_x \\ E_y \\ E_z \end{bmatrix} \tag{6.18}$$

In an isotropic medium, the current is in the direction of the electric field. From Eq. (6.18), it is found that for a plasma in a magnetic field, an electric field, in general, gives rise to a current not only in its own direction but also in a perpendicular direction. This result is the manifestation of the anisotropic behavior of the plasma.

In vector notation, Eq. (6.18) becomes

$$\mathbf{J}_e = \boldsymbol{\sigma}_e \cdot \mathbf{E} \tag{6.19}$$

where $\boldsymbol{\sigma}_e$ is a dyad. The components of the conductivity dyad $\boldsymbol{\sigma}_e$ are expressible in the following matrix form:

$$\boldsymbol{\sigma}_e = \frac{iN_0 e^2}{m_e \omega} \begin{bmatrix} \dfrac{\omega^2}{\omega^2 - \omega_{ce}^2} & \dfrac{-i\omega\omega_{ce}}{\omega^2 - \omega_{ce}^2} & 0 \\ \dfrac{i\omega\omega_{ce}}{\omega^2 - \omega_{ce}^2} & \dfrac{\omega^2}{\omega^2 - \omega_{ce}^2} & 0 \\ 0 & 0 & 1 \end{bmatrix} \tag{6.20}$$

It is instructive to examine the features of the conductivity dyad in certain limiting cases. In the absence of the magnetic field, the conductivities are seen from Eq. (6.20) to be the same in all directions. This results in an isotropic medium with a scalar conductivity which is imaginary. The electron conductivity given by Eq. (6.20) becomes

$$\boldsymbol{\sigma}_e = \frac{iN_0 e^2}{m_e \omega} \begin{bmatrix} 1 & 0 & 0 \\ 0 & 1 & 0 \\ 0 & 0 & 1 \end{bmatrix} \quad \text{for } B = 0 \tag{6.21}$$

If the magnetic field is sufficiently weak or if the frequency is quite high such that $\omega_{ce}^2/\omega^2 \ll 1$, Eq. (6.20) can be simplified to yield

$$\boldsymbol{\sigma}_e = \frac{iN_0 e^2}{m_e \omega} \begin{bmatrix} 1 + \dfrac{\omega_{ce}^2}{\omega^2} & -i\dfrac{\omega_{ce}}{\omega} & 0 \\ i\dfrac{\omega_{ce}}{\omega} & 1 + \dfrac{\omega_{ce}^2}{\omega^2} & 0 \\ 0 & 0 & 1 \end{bmatrix} \tag{6.22}$$

The conductivity in the direction of the magnetic field remains unchanged. The conductivities in the direction perpendicular to the magnetic field but parallel to the electric field, as given by the first two diagonal terms in Eq. (6.22), are changed from their values for no magnetic field by a second-order term in ω_{ce}/ω, giving rise to a phenomenon called *magnetoresistance*. In addition, a comparison of Eqs. (6.21) and (6.22) shows that the application of a magnetic field results in a current flow in a direction perpendicular to both the magnetic and the electric fields. These currents result in Eq. (6.22) having offdiagonal terms which are of first order in ω_{ce}/ω. These currents are known as *Hall-effect currents.*

The complex electrical conductivity $\boldsymbol{\sigma}_i$ due to the mobility of the ions can be deduced in the same manner as for electrons. Indeed, $\boldsymbol{\sigma}_i$ can be obtained from Eq. (6.20) by changing $-e$ and m_e to e and m_i respectively. Note that the number density N_0 is the same for electrons and ions. Also, since the gyromagnetic frequency is defined by Eq. (3.4) to be always positive and since the charges of an electron and an ion are of opposite signs, it is necessary that ω_{ce} should be changed to $-\omega_{ci}$. Hence, from Eq. (6.20), it follows that

$$\boldsymbol{\sigma}_i = \frac{iN_0 e^2}{m_i \omega} \begin{bmatrix} \dfrac{\omega^2}{\omega^2 - \omega_{ci}^2} & \dfrac{i\omega\omega_{ci}}{\omega^2 - \omega_{ci}^2} & 0 \\ \dfrac{-i\omega\omega_{ci}}{\omega^2 - \omega_{ci}^2} & \dfrac{\omega^2}{\omega^2 - \omega_{ci}^2} & 0 \\ 0 & 0 & 1 \end{bmatrix} \tag{6.23}$$

In obtaining the conductivities $\boldsymbol{\sigma}_e$ and $\boldsymbol{\sigma}_i$, only the currents due to the collective motions of the particles have been included. The random motion of the particles associated with those solutions of Eqs. (6.8)–(6.10) which depend on the initial velocities of the particles is assumed to contribute no net average current density.

Consider a neutral plasma in which the number density of the electrons and the ions are equal. The resultant conductivity is equal to the sum of the electron and the ion conductivities. Hence,

$$\boldsymbol{\sigma} = \boldsymbol{\sigma}_e + \boldsymbol{\sigma}_i \tag{6.24}$$

The conductivity in the absence of the magnetic field or that in the direction of the magnetic field is mainly contributed by the electrons since, as is evident from Eqs. (6.20) and (6.23) the contribution to $\boldsymbol{\sigma}$ is inversely proportional to the mass of the particle.

The current density due to the mobility of the charged particles can be regarded either as a part of the total current density $\mathbf{J}$ given in Eq. (1.6) or taken into account through the introduction of an effective permittivity by combining it with the first term on the right side of Eq. (1.6). On combining the current density due to the mobility of the charged particles with the first term on the right side of Eq. (1.6), it is found that

$$\begin{aligned} \mu_0 \varepsilon_0 \frac{\partial}{\partial t}\mathbf{E} + \mu_0 \boldsymbol{\sigma} \cdot \mathbf{E} &= -i\omega\mu_0 \varepsilon_0 \mathbf{E} + \mu_0 \boldsymbol{\sigma} \cdot \mathbf{E} \\ &= -i\omega\mu_0 \varepsilon_0 \left[\mathbf{1} + \frac{i}{\omega\varepsilon_0}\boldsymbol{\sigma}\right] \cdot \mathbf{E} \end{aligned} \tag{6.25}$$

where the unit dyad $\mathbf{1}$ written as a matrix has unity for its diagonal elements and zero for its nondiagonal elements. It follows from Eq. (6.25) that the plasma behaves like a dielectric characterized by the permittivity dyad $\varepsilon_0 \boldsymbol{\varepsilon}_r$. The relative permittivity dyad $\boldsymbol{\varepsilon}_r$ is obtained from Eq. (6.25) as

$$\boldsymbol{\varepsilon}_r = \mathbf{1} + i\boldsymbol{\sigma}/\omega\varepsilon_0 \tag{6.26}$$

. It is convenient to write $\boldsymbol{\varepsilon}_r$ in the following matrix form:

$$\varepsilon_r = \begin{bmatrix} \varepsilon_1 & i\varepsilon_2 & 0 \\ -i\varepsilon_2 & \varepsilon_1 & 0 \\ 0 & 0 & \varepsilon_3 \end{bmatrix} \tag{6.27}$$

If the mobility of the electrons alone are taken into account, it can be deduced from Eqs. (6.20), (6.26), and (6.27) that

$$\varepsilon_1 = 1 - \omega_{pe}^2/(\omega^2 - \omega_{ce}^2) \tag{6.28a}$$

$$\varepsilon_2 = \omega_{pe}^2\omega_{ce}/(\omega^2 - \omega_{ce}^2)\omega \tag{6.28b}$$

$$\varepsilon_3 = 1 - \omega_{pe}^2/\omega^2 \tag{6.28c}$$

where

$$\omega_{pe} = \sqrt{N_0 e^2/m_e \varepsilon_0} \tag{6.29}$$

is the electron plasma angular frequency. If the mobility of the ions are also included, it can be proved with the help of Eqs. (6.20), (6.23), (6.24), (6.26), and (6.27) that

$$\varepsilon_1 = 1 - \frac{\omega_{pe}^2}{\omega^2 - \omega_{ce}^2} - \frac{\omega_{pi}^2}{\omega^2 - \omega_{ci}^2} \tag{6.30a}$$

$$\varepsilon_2 = \frac{\omega_{pe}^2\omega_{ce}}{(\omega^2 - \omega_{ce}^2)\omega} - \frac{\omega_{pi}^2\omega_{ci}}{(\omega^2 - \omega_{ci}^2)\omega} \tag{6.30b}$$

$$\varepsilon_3 = 1 - \frac{\omega_{pe}^2}{\omega^2} - \frac{\omega_{pi}^2}{\omega^2} \tag{6.30c}$$

In Eqs. (6.30),

$$\omega_{pi} = \sqrt{N_0 e^2/m_i \varepsilon_0} \tag{6.31}$$

is the ion plasma angular frequency. Note that $\omega_{ce} = eB/m_e$ and $\omega_{ci} = eB/m_i$ are the gyromagnetic angular frequencies of the electrons and the ions, respectively.

For an electron, if $\omega = \omega_{ce}$ it is seen that Eq. (6.11) is not a valid solution of Eq. (6.8). For an ion, since ω_{ce} has to be replaced by $-\omega_{ci}$, it follows that Eq. (6.12) is not a valid solution of Eq. (6.9) for $\omega = \omega_{ci}$. In a similar manner, for $\omega = 0$, Eq. (6.13) is not a correct solution of Eq. (6.10). The particle velocity given by Eqs. (6.14) and (6.15) does not represent correctly the collective motion of the particles if $\omega = \omega_{ce}$, $\omega = \omega_{ci}$, or $\omega = 0$. If the time-varying electric field does not contain these singular frequencies, insofar as the collective behavior is concerned, the plasma can be adequately represented as a dielectric characterized by the relative permittivity dyad given by Eqs. (6.27), (6.28), and (6.30).

According to Maxwell's equations, a time-varying electric field is also accompanied by a time-varying magnetic field; also, any time variation in the fields results in the spatial variation of the fields as well. If the static magnetic field is very much larger than the time-varying part of the magnetic field, the latter can be neglected and the assumption of a B-field which is constant in time and uniform in space is a valid approximation. If the Larmor radius is very much small in comparison with a wavelength, i.e., the scale length of the variation of the electric field in space, then, in each orbit the charged particle sees essentially a constant electric field. Hence, the assumption of an electric field which is uniform in space and varying harmonically in time is a reasonable approximation. The local relation between the electric field and the current density obtained in Eq. (6.19) is valid only under the stated restrictions. Once the permittivity dyad is introduced for the adequate description of the macroscopic behavior of the plasma, then the self-consistent electromagnetic fields can be evaluated from the solution of Maxwell's equations. The electromagnetic theory of weakly ionized media based on Maxwell's equations together with the permittivity dyad is known as *magnetoionic theory.* A variety of problems of wave propagation in the ionosphere has been successfully solved by the application of magnetoionic theory.

3.7. Cyclotron resonance

In a constant and uniform magnetic field, the charged particles have a natural frequency of oscillation called the gyromagnetic frequency. The electrons orbit around the direction of the magnetic field in the counterclockwise direction with an angular frequency ω_c. In obtaining the conductivity dyad, the frequency of the forcing electric field is assumed to be sufficiently removed from the gyromagnetic frequency of the charged particles. Note that E_+ and E_- refer to the circularly polarized components of the electric field which are rotating in the counterclockwise and clockwise directions respectively. If $\omega = \omega_c$, it is seen from Eqs. (6.17) that the conductivity in the direction of the magnetic field and that associated with that circularly polarized component of the electric field which is rotating in a sense opposite to that of an electron are relatively unaffected whereas the conductivity associated with the other circularly polarized component of the electric field which is rotating in the same sense as that of an electron becomes infinite. When an electron and the forcing electric field rotate in the counterclockwise direction in synchronism, that is, with the same angular frequency, the particle is able to abstract energy from the field and thus has its speed increased continually and indefinitely with time. This phenomenon is called the cyclotron resonance. It is of interest to examine this phenomenon in detail to ascertain the mechanism of building up of this resonance.

For this purpose, it is necessary to have the complete solutions of Eqs. (6.8)–(6.10) and these can be derived to be

$$a_+ = a_{+0}e^{i(\omega-\omega_c)t} - ie/m(\omega - \omega_c) \tag{7.1}$$

$$a_- = a_{-0}e^{i(\omega+\omega_c)t} - ie/m(\omega + \omega_c) \tag{7.2}$$

$$a_z = a_{z0}e^{i\omega t} - ie/m\omega \tag{7.3}$$

where a_{+0}, a_{-0}, and a_{z0} are determined by the initial velocity of the particle. If the initial velocity of the particle is given by

$$\begin{aligned}(\mathbf{u})_{t=0} &= \hat{\mathbf{x}}u_{x0} + \hat{\mathbf{y}}u_{y0} + \hat{\mathbf{z}}u_{z0} \\ &= \frac{(\hat{\mathbf{x}} + i\hat{\mathbf{y}})}{\sqrt{2}}u_{+0} + \frac{(\hat{\mathbf{x}} - i\hat{\mathbf{y}})}{\sqrt{2}}u_{-0} + \hat{\mathbf{z}}u_{z0}\end{aligned} \tag{7.4}$$

where

$$u_{+0} = \frac{1}{\sqrt{2}}(u_{x0} - iu_{y0}) \qquad u_{-0} = \frac{1}{\sqrt{2}}(u_{x0} + iu_{y0}) \tag{7.5}$$

then it is deduced from Eqs. (6.5) and (7.1)–(7.4) that

$$\begin{aligned}a_{+0} &= \frac{u_{+0}}{E_{+0}} + \frac{ie}{m(\omega - \omega_c)} \\ a_{-0} &= \frac{u_{-0}}{E_{-0}} + \frac{ie}{m(\omega + \omega_c)} \\ a_{z0} &= \frac{u_{z0}}{E_{z0}} + \frac{ie}{m\omega}\end{aligned} \tag{7.6}$$

The use of Eqs. (6.4), (7.1)–(7.3), and (7.6) in Eq. (6.5) yields the following expression for the particle velocity:

$$\begin{aligned}\mathbf{u}(t) = {}& \frac{(\hat{\mathbf{x}} + i\hat{\mathbf{y}})}{\sqrt{2}}\left[\left\{u_{+0} + \frac{ieE_{+0}}{m(\omega - \omega_c)}\right\}e^{-i\omega_c t} - \frac{ieE_{+0}e^{-i\omega t}}{m(\omega - \omega_c)}\right] \\ &+ \frac{(\hat{\mathbf{x}} - i\hat{\mathbf{y}})}{\sqrt{2}}\left[\left\{u_{-0} + \frac{ieE_{-0}}{m(\omega + \omega_c)}\right\}e^{+i\omega_c t} - \frac{ieE_{-0}e^{-i\omega t}}{m(\omega + \omega_c)}\right] \\ &+ \hat{\mathbf{z}}\left[\left\{u_{z0} + \frac{ieE_{z0}}{m\omega}\right\} - \frac{ieE_{z0}}{m\omega}e^{-i\omega t}\right]\end{aligned} \tag{7.7}$$

When ω approaches ω_c, the terms containing $(\omega - \omega_c)$ in the denominator become very large with the result that in this limit the particle velocity $\mathbf{u}(t)$ may be approximated as follows:

$$\mathbf{u}(t) = -\frac{(\hat{\mathbf{x}} + i\hat{\mathbf{y}})}{\sqrt{2}}\frac{ieE_{+0}}{m(\omega - \omega_c)}e^{-i\omega t}[1 - e^{i(\omega-\omega_c)t}] \tag{7.8}$$

Except when $(\omega - \omega_c)t$ equals zero or integer multiples of 2π resulting in Eq. (7.8) becoming equal to zero, for $\omega \approx \omega_c$, Eq. (7.8) is a reasonably good approximation to Eq. (7.7). If $\omega = \omega_c$, Eq. (7.8) becomes

$$\mathbf{u}(t) = -\frac{(\hat{\mathbf{x}} + i\hat{\mathbf{y}})}{\sqrt{2}}\frac{e}{m}E_{+0}te^{-i\omega_c t} \tag{7.9}$$

It was assumed that Eq. (7.8) is the dominating term of the right side of Eq. (7.7) since $(\omega - \omega_c)$ appeared in the denominator. If $\omega = \omega_c$, it is seen from Eq. (7.9) that not only such is not the case but that for very small t, Eq. (7.9) may even be negligible in comparison with the terms of Eq. (7.7) which have not been included. Depending on the nature of the terms in Eq. (7.7) which have been omitted, $\mathbf{u}(t)$ may either decrease or increase initially as t is increased from zero. But when t is increased sufficiently, Eq. (7.9) soon becomes the dominating term in view of the fact that its magnitude increases linearly with time whereas those of the remaining terms in Eq. (7.7) merely oscillate as a function of time. Therefore, for $\omega = \omega_c$, a valid approximation for $\mathbf{u}(t)$ is given by Eq. (7.9), especially after a sufficient lapse of time. It is important to note that $\mathbf{u}(t)$ given in Eq. (7.9) is a circularly polarized component rotating in the same sense as that of an electron. The other circularly polarized component and the parallel component of the particle velocity do not undergo any significant changes as ω approaches and becomes equal to ω_c, as has been indicated earlier.

For large t, the magnitude of Eq. (7.9) is seen to be given by

$$u_+ = \frac{e}{m}|E_{\perp 0}|t \qquad |E_{\perp 0}| = \left[\frac{E_{x0}^2 + E_{y0}^2}{2}\right]^{1/2} \tag{7.10}$$

The corresponding Larmor radius r_L is

$$r_L = (e/m)(|E_{\perp 0}|t/\omega_c) \tag{7.11}$$

which is seen to be increasing linearly with time. Hence, the orbital motion is a counterclockwise spiral. From Eqs. (7.10), the kinetic energy of the particle is seen to increase indefinitely as a quadratic function of time. The energy in the time-varying electric field is continually transformed into the kinetic energy of the electrons. The factors which inhibit the continually growing spiral are (i) the dimensions of the container of the plasma in the direction perpendicular to the magnetic field and (ii) collisions with other particles.

When $\omega \neq \omega_c$, there is no cyclotron resonance. But if $\omega = \omega_c$, there is an outwardly spiralling motion of the electron and the consequent build-up of large kinetic energy of the electron resulting in cyclotron resonance. It is of interest to examine the behavior as the cyclotron resonance is approached. For ω tending to ω_c, Eq. (7.8) applies and its magnitude is given by

$$u_+ = \frac{2e|E_{\perp 0}|}{m|\omega - \omega_c|}\left|\sin\left[\frac{(\omega - \omega_c)}{2}t\right]\right| \tag{7.12}$$

and the corresponding Larmor radius is

$$r_L = \frac{2e|E_{\perp 0}|}{m\omega_c|\omega - \omega_c|}\left|\sin\left[\frac{(\omega-\omega_c)}{2}t\right]\right| \tag{7.13}$$

It is to be noted that Eqs. (7.12) and (7.13) are oscillatory quantities with the following maximum values:

$$(u_+)_{\max} = \frac{2e|E_{\perp 0}|}{m|\omega - \omega_c|} \qquad (r_L)_{\max} = \frac{2e|E_{\perp 0}|}{m\omega_c|\omega - \omega_c|} \tag{7.14}$$

When $|\omega - \omega_c|t$ equals zero or integer multiples of 2π, Eqs. (7.12) and (7.13) vanish and for this case Eq. (7.8) is not a good approximation to Eq. (7.7) which, in general, yields some minimum values of u_+ and r_L, say, $(u_+)_{\min}$ and $(r_L)_{\min}$. As t is increased from zero, the Larmor radius increases from $(r_L)_{\min}$ and reaches the maximum value $(r_L)_{\max}$ at $t = \pi/|\omega - \omega_c|$. For further increase in time, the Larmor radius decreases from $(r_L)_{\max}$ and attains the minimum value at $t = 2\pi/|\omega - \omega_c|$. This process repeats every $2\pi/|\omega - \omega_c|$ seconds. Since the period of an orbital motion of an electron as given by $2\pi/\omega_c$ is much smaller than the time $t = \pi/|\omega - \omega_c|$ required to reach the maximum Larmor radius, it follows that the electron spirals out and orbits several times before the Larmor radius equals $(r_L)_{\max}$. Then, it starts spiralling inwards and, as before, orbits several times before the Larmor radius becomes equal to $(r_L)_{\min}$. When the electron spirals out, its kinetic energy increases at the expense of the energy of the electric field and vice versa when it spirals inwards. Thus, the energy is interchanged between the particle and the wave at the beat frequency $|\omega - \omega_c|/2\pi$. The maximum radius of the spiral and therefore the maximum kinetic energy of the particle are increased and the beat frequency of energy exchange between the particle and the wave is reduced as the frequency of the electric field approaches more and more closely the gyromagnetic frequency of the electrons. Finally, when $\omega = \omega_c$, the energy is unilaterally and continually transferred from the time-varying electric field to the kinetic energy of the particle subject only to certain inhibiting factors already indicated.

The phenomenon of cyclotron resonance can be used to increase the particle speed and hence its kinetic temperature. This method of increasing the temperature of the plasma is known as radio frequency heating of the plasma by cyclotron resonance.

3.8. Magnetic mirror effect

It is proposed to study the effect of small spatial inhomogeneities of the magnetic field on the motion of a charged particle. The magnetic field is assumed to be in the z-direction at the origin of the coordinate system, that is, $B_x = 0$, $B_y = 0$, and $B_z = B$ at $x = y = z = 0$. Due to the spatial variation of the magnetic field, in addition to B_z, there are other components of the magnetic field in the neighborhood of the origin. Since each of the three components of the magnetic field can change with respect to the variation in each of the three coordinates x, y, and z, it follows that altogether nine parameters are needed for the complete specification of the spatial variation of the magnetic field. These nine parameters are conveniently grouped into the following four categories:

$$\text{(a) Divergence terms:} \quad \frac{\partial B_x}{\partial x} \quad \frac{\partial B_y}{\partial y} \quad \frac{\partial B_z}{\partial z} \tag{8.1a}$$

$$\text{(b) Gradient terms:} \quad \frac{\partial B_z}{\partial x} \quad \frac{\partial B_z}{\partial y} \tag{8.1b}$$

$$\text{(c) Curvature terms:} \quad \frac{\partial B_x}{\partial z} \quad \frac{\partial B_y}{\partial z} \tag{8.1c}$$

$$\text{(d) Shear terms:} \quad \frac{\partial B_x}{\partial y} \quad \frac{\partial B_y}{\partial x} \tag{8.1d}$$

It is proposed to examine the effect of the spatial variation of the magnetic field on the motion of a charged particle for each of the four groups of terms separately. Let L be the length in which the changes in the components of the magnetic field become of the order of B. It is assumed that L is very much larger than the Larmor radius. In other words, it is assumed that the changes in the components of the magnetic field in a distance of the order of the Larmor radius are very small in comparison with B. In general, more than one group of terms is present simultaneously. Since, in the region of interest, these groups of terms are small perturbations on the z-component of the magnetic field, it follows that the resultant effect is the sum of the separate effects due to the constituent group of terms. In view of this, it is justifiable to treat the effect of each group of terms separately. Since

$$\nabla \cdot \mathbf{B} = \partial B_x/\partial x + \partial B_y/\partial y + \partial B_z/\partial z = 0 \tag{8.2}$$

it is seen that only two of the three terms in (8.1a) are independent. Thus, only eight independent parameters are needed for the specification of the spatial variation of the magnetic field.

First, it is desired to examine the effect of the divergence terms as given by (8.1a). In the region of interest, which is the neighborhood of the origin, the magnetic field is assumed to be primarily in the z-direction. It is assumed that the magnetic field has a slow variation in the z-direction, that is $\partial B_z/\partial z$ is nonzero. As a consequence, it is seen from Eq. (8.2) that the terms $\partial B_x/\partial x$ and $\partial B_y/\partial y$ are present. Since the magnetic field has all the components in the close neighborhood of the origin, it is helpful to make use of the concept of magnetic flux lines which are everywhere parallel to **B** and whose density at every point is proportional to the magnitude of **B**. For simplicity, it is assumed that the magnetic flux lines are cylindrically symmetric about the z-axis; this is equivalent to assuming that the perturbations $\partial B_x/\partial x$ and $\partial B_y/\partial y$ are equal. The cylindrical coordinates r, θ, and z are the suitable coordinates to use in view of this simplifying assumption. Then, Eq. (8.2) becomes

$$(1/r)(\partial/\partial r)(rB_r) + \partial B_z/\partial z = 0 \tag{8.3}$$

which shows that only the radial component B_r of the magnetic field is needed to have a variation of B_z with z; therefore, the azimuthal component B_θ can be set equal to zero. The Lorentz force due to B_z is in the perpendicular direction whereas B_r gives rise to a force in the axial direction. The latter force can be evaluated approximately in the following manner. From Eq. (8.3), it can be obtained that

$$\int_0^{r_L} \frac{\partial}{\partial r}(rB_r)\, dr = -\int_0^{r_L} r\frac{\partial B_z}{\partial z}\, dr \tag{8.4}$$

where the integration is carried out over one orbit of Larmor radius r_L. Since $\partial B_z/\partial z$ is a very slowly varying function for $0 < r < r_L$, it can be taken outside the integral sign with the result that Eq. (8.4) yields

$$B_r = -(r_L/2)\partial B_z/\partial z = -(r_L/2)\partial B/\partial z \tag{8.5}$$

All the variations of the magnetic field in the region of interest are very small that it is justifiable to replace B_z by B in Eq. (8.5). The magnetic field is essentially in the axial direction with the result that the orbital velocity $u_\perp$ is in the θ-direction. Therefore, the longitudinal component of the Lorentz force is shown with the help of Eq. (8.5) to be given by

$$F_z = -e(\hat{\boldsymbol{\theta}} u_\perp \times \hat{\mathbf{r}} B_r) \cdot \hat{\mathbf{z}} = eu_\perp B_r = -\frac{eu_\perp r_L}{2}\frac{\partial B}{\partial z} \tag{8.6}$$

For a positive charge $u_\perp$ is in the $-\theta$-direction and also, since $-e$ changes to e, it follows that Eq. (8.6) is the same for negative and positive charges. The substitution of $r_L = u_\perp/\omega_c = u_\perp m/eB$ in Eq. (8.6) gives

$$F_z = m(du_z/dt) = -(w_\perp/B)\partial B/\partial z \tag{8.7}$$

where $w_\perp = \frac{1}{2}mu_\perp^2$ is the transverse kinetic energy of the particle. It is observed from Eq. (8.7) that if the magnetic field changes in the z-direction, an axial force acts on the particle irrespective of whether the charge carried by it is positive or negative and repels it in the direction of decreasing magnetic field. The same result can be brought out in a somewhat different manner.

The multiplication of both sides of Eq. (8.7) by $u_z = dz/dt$ gives the following result:

$$u_z m\frac{du_z}{dt} = \frac{d}{dt}(\tfrac{1}{2}mu_z^2) = -\frac{w_\perp}{B}\frac{\partial B}{\partial z}\frac{dz}{dt} = -\frac{w_\perp}{B}\frac{dB}{dt} \tag{8.8}$$

The total energy $w_{\|} + w_{\perp}$ of a charged particle in a constant magnetic field does not change. Therefore,

$$\frac{d}{dt}(w_{\|}) = \frac{d}{dt}(\tfrac{1}{2}mu_z^2) = -\frac{d}{dt}(w_{\perp}) \tag{8.9}$$

Since

$$\frac{d}{dt}(w_{\perp}) = \frac{d}{dt}\left(\frac{w_{\perp} B}{B}\right) = \frac{w_{\perp}}{B}\frac{dB}{dt} + B\frac{d}{dt}\left(\frac{w_{\perp}}{B}\right)$$

it is found from Eqs. (8.8) and (8.9) that

$$(d/dt)(w_{\perp}/B) = 0 \text{ or } \mathfrak{m} = w_{\perp}/B = \text{constant} \tag{8.10}$$

It is seen from Eq. (8.10) that the orbital magnetic moment is a constant. This constancy holds, rigorously speaking, only in the limit of vanishingly small spatial variations of the magnetic field. Consequently, it is said that the orbital magnetic moment is an *adiabatic invariant.*

Suppose that B increases in the z-direction. Then, in view of the adiabatic invariance of the orbital magnetic moment, $w_{\perp}$ increases and $w_{\|}$ decreases. Therefore, the particle velocity in the direction of increasing magnetic field decreases, may ultimately come to zero and then become reversed. In the reverse direction, B decreases, $w_{\perp}$ decreases and $w_{\|}$ increases. Therefore, the particle is speeded up in the direction of decreasing magnetic field. The phenomenon of reflection of charged particles from a region of sufficiently strongly converging magnetic flux lines is called the *magnetic mirror effect.*

The magnetic flux enclosed by one orbit of the charged particle is obtained as

$$\text{Flux} = \pi r_L^2 B = \frac{\pi u_{\perp}^2 m^2 B}{e^2 B^2} = \frac{2\pi m}{e^2}\frac{w_{\perp}}{B} \tag{8.11}$$

In view of Eq. (8.10), Eq. (8.11) shows that the magnetic flux enclosed by an orbit of the charged particle is also an adiabatic invariant. If the magnetic flux density B increases in the z-direction, the particle is forced to orbit with progressively smaller radius in order to maintain the constancy of the magnetic flux enclosed by an orbit in accordance with Eq. (8.11). In other words, as B increases, the tube of magnetic flux converges and the charged particle orbits around the periphery of this converging tube of magnetic flux.

Consider two coaxial magnetic mirrors facing each other as shown in Fig. 3.5. The magnetic flux lines converge in both directions from the center. For convenience, the magnetic flux lines are assumed to be symmetrical about the $z = 0$ plane. The charged particles are reflected at the two mirrors, travel back and forth in the space between them, and thus are trapped in the so-called magnetic bottle. The magnetic mirror system is used in the laboratory for the confinement of plasmas. Several identical and coaxial magnet coils are wound around a cylindrical tube as

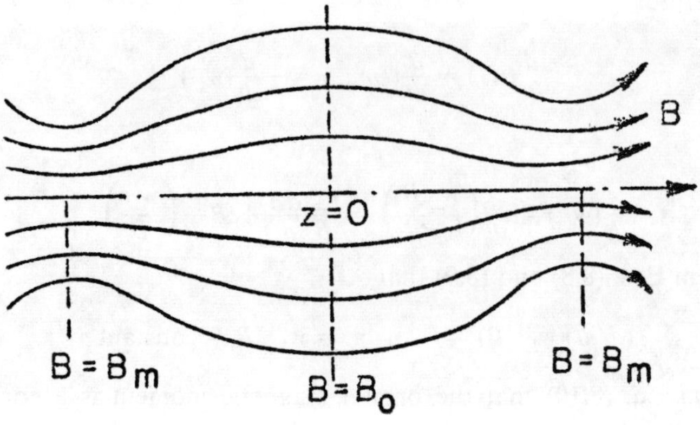

Fig. 3.5. Two coaxial magnetic mirrors facing each other showing that the magnetic flux lines converge in both directions from the center.

shown in Fig. 3.6. These coils are spaced further apart near the center and closer together near the ends. If the current in all the magnet coils encircles the tube in the same direction, the resulting magnetic field is that of the magnetic mirror system as shown in Fig. 3.5.

The magnetic field around the earth constitutes a large magnetic bottle for the trapping of the charged particles of cosmic origin. The magnetic flux lines converge towards the north and the south magnetic poles of the earth. The charged particles traveling along the field lines are reflected at both the poles which thus act as magnetic mirrors. The charged particles trapped in the earth's magnetic field form radiation belts known as *Van Allen belts*.

It is useful to introduce a parameter which is a measure of the effectiveness of a converging magnetic flux lines in reflecting the charged particles. Let u be the speed of a particle and note that in a static magnetic field, u remains a constant. At the center of the magnetic bottle, let B_0 be the value of the magnetic flux density and θ_0 be the angle made by the particle velocity with the axis (z) of the magnetic mirrors. Let B and θ be the corresponding values at a different position along the magnetic bottle. Note that $B > B_0$. Since $u \sin \theta_0$ and $u \sin \theta$ are the perpendicular components of the speed at the two locations, the constancy of the orbital magnetic moment as given by Eq. (8.10) leads to

$$\sin^2 \theta_0 / B_0 = \sin^2 \theta / B \tag{8.12}$$

Since $B > B_0$, it is seen from Eq. (8.12) that $\theta > \theta_0$ and since B increases along the axis of the magnetic bottle, it follows that θ also increases. Let B_m be the value of

B where $\theta = \pi/2$. The locations where $\theta = \pi/2$ are designated as the points of reflection and these locations are arranged to occur at the ends of the magnetic mirror system. From Eq. (8.12), it is obtained that

$$\sin^2\theta_0 = B_0/B_m \quad \text{or} \quad \theta_0 = \sin^{-1}(\sqrt{B_0/B_m}) \tag{8.13}$$

The parameter B_m/B_0 is called the *mirror ratio.* Suppose that the mirror ratio is fixed. Let θ_i be the angle made by the particle velocity at $z = 0$ with the z-axis. If $\pi/2 > \theta_i > \theta_0$, $\sin\theta_i > \sin\theta_0$ and it is found from Eqs. (8.12) and (8.13) that θ becomes equal to $\pi/2$ for values of B less than or equal to B_m; therefore, the corresponding particles are reflected before or at the ends of the magnetic bottle. If $\theta_0 > \theta_i > 0$, $\sin\theta_i < \sin\theta_0$ and then Eqs. (8.12) and (8.13) show that θ never attains the value of $\pi/2$; the particles at the ends have a nonvanishing velocity in the axial direction and therefore escape through the ends of the magnetic mirror system. In view of the reflection symmetry about $z = 0$ and the rotational symmetry about the z-axis, it follows that there is a bi-cone of angle θ_0 and the particles at the center having directions of velocity vectors falling within this cone are not trapped inside the magnetic mirror system. This cone is coaxial with the magnetic mirrors and is called *the loss cone.* The mirror ratio determines the angle of the loss cone.

The ratio of the number of particles reflected from the magnetic mirror in a given time to the total number of particles incident on the mirror in the same time is

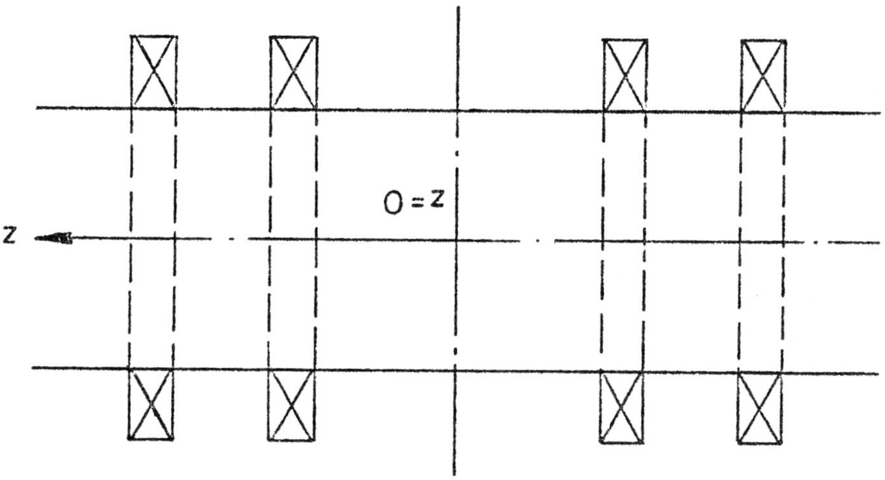

Fig. 3.6. A schematic diagram showing the arrangement of magnet coils for obtaining two coaxial magnetic mirrors facing each other.

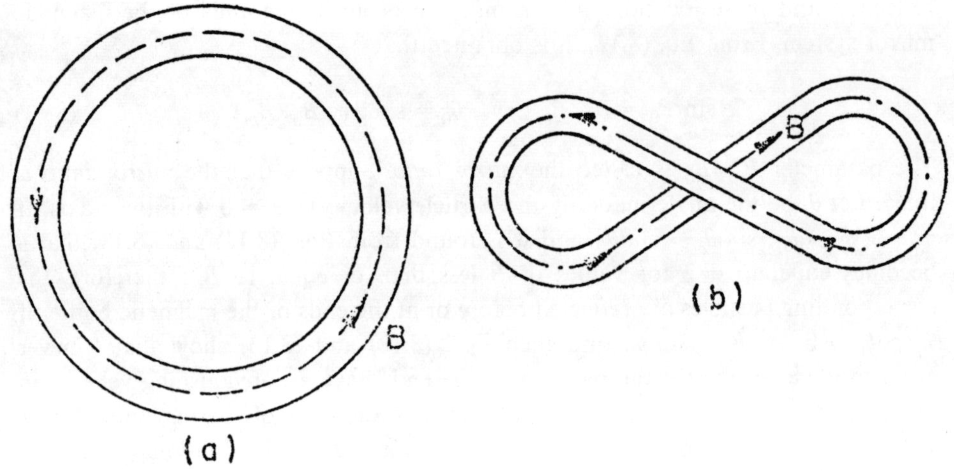

Fig. 3.7. Schematic diagram of (a) torus and (b) figure of eight stellarator showing the magnetic flux lines.

defined as the reflection coefficient R. For particles characterized by an isotropic velocity distribution, R can be evaluated with the following result:

$$R = \cos^2\theta_0 = 1 - B_0/B_m \tag{8.14}$$

The reflection coefficient (8.14) is seen to be applicable for any arbitrary velocity distribution, provided only that the distribution is isotropic. For a mirror ratio of 2, 50% of the particles escape the magnetic mirror system through the ends. One possible device for ensuring total confinement of a plasma is to have a system with no ends. Toroidal geometries (Fig. 3.7a) where the magnetic flux lines close in upon themselves have no ends, but it can be shown that confinement of a plasma within a torus is impossible in view of the radial inhomogeneity of the magnetic flux. *Stellarators* (Fig. 3.7b) have the advantage of endless geometry of the torus and, in their simplest form, the magnetic flux lines are designed to have the shape of a figure of eight in order to overcome the disadvantage of particle drifts present in a torus.

3.9. Fermi acceleration

Consider a charged particle trapped between two magnetic mirrors. If the mirrors move towards each other, the charged particle acquires acceleration which is known as *Fermi acceleration*. Fermi proposed this as a possible mechanism by which two stellar clouds moving towards each other impart acceleration to a cosmic charged particle. If the magnetic field in the clouds is greater than that in the space between them, such a trapping of cosmic charged particles can occur. It is desired to introduce the concept of a *longitudinal adiabatic invariant* and apply it to explain Fermi acceleration.

Consider two magnetic mirrors M_1 and M_2 as shown in Fig. 3.8. The mirrors are coaxial with the z-axis and are a distance L apart. A uniform magnetic field B in the z-direction exists in the region between the mirrors except near the ends where the magnetic field increases to form the mirrors. The mirror M_2 is stationary and the mirror M_1 moves in the z-direction with a uniform speed u_m so that

$$u_m = -dL/dt \tag{9.1}$$

The longitudinal component of the particle velocity is u_z. It is assumed that $u_m \ll u_z$. In other words, in the time taken by the particle to oscillate back and forth between the mirrors, the distance moved by the mirror is small in comparison with the distance L between the mirrors. Near the mirrors, there are end effects and the particle velocity is not equal to u_z. Also, at the moving mirror, u_z changes by an amount which in view of the fact that $u_m \ll u_z$, is negligible. These small end effects can be neglected and the longitudinal particle velocity between the mirrors may be taken to be equal to u_z throughout the space between the mirrors.

Let Δu_z be the change in the particle speed on reflection from the moving mirror. Let u_z^i and u_z^r be the incident and the reflected speeds of the particle at the moving mirror. The corresponding values in the coordinate system in which the mirror is stationary are denoted by $\tilde{u}_z^i$ and $\tilde{u}_z^r$, respectively. The mirror velocity is in the same direction as the reflected velocity but is in the opposite direction to the incident

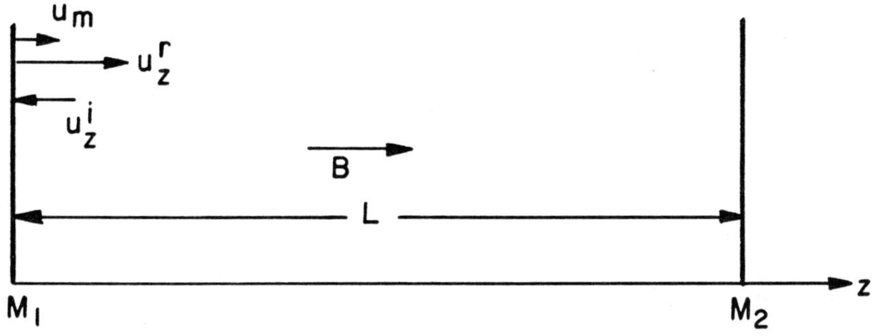

Fig. 3.8. Uniformly moving magnetic mirror system for explaining Fermi's acceleration mechanism.

velocity. Therefore, it follows that

$$\tilde{u}_z^i = u_z^i + u_m \qquad \tilde{u}_z^r = u_z^r - u_m \tag{9.2}$$

In the coordinate system in which the mirror is stationary, the incident and the reflected speeds are equal, $\tilde{u}_z^i = \tilde{u}_z^r$ but only their directions are reversed. Hence, it is obtained from Eqs. (9.2) that

$$\Delta u_z = u_z^r - u_z^i = 2u_m \tag{9.3}$$

The change given by Eq. (9.3) takes place periodically and the period is approximately equal to the time $\Delta t = 2L/u_z$ taken by the particle to oscillate back and forth between the mirrors. Therefore, from Eqs. (9.1) and (9.3), it is found that

$$\frac{du_z}{dt} = \frac{\Delta u_z}{\Delta t} = \frac{u_m}{L} u_z = -\frac{u_z}{L}\frac{dL}{dt} \tag{9.4}$$

Since

$$\frac{du_z}{dt} = \frac{d}{dt}(u_z L \frac{1}{L}) = \frac{1}{L}\frac{d}{dt}(u_z L) - \frac{u_z}{L}\frac{dL}{dt}$$

it follows from Eq. (9.4) that

$$(d/dt)(u_z L) = 0 \quad \text{or} \quad u_z L = \text{constant} \tag{9.5}$$

which is the longitudinal adiabatic invariant referred to previously.

The longitudinal particle speed increases inversely as the distance L between the mirrors. Therefore, the longitudinal kinetic energy $w_\parallel$ of the particle increases inversely as the square of the spacing L between the mirrors. There is a limit on the increase of the longitudinal speed of the particle by Fermi's acceleration mechanism. As u_z increases, the direction of the particle velocity at the center of the magnetic mirror system eventually enters the loss cone and therefore, the particle escapes through the ends of the system. The ratio of the total kinetic energy to the transverse kinetic energy can increase only up to a limit which depends on the reflection coefficient of the magnetic mirrors.

3.10. Gradient and curvature drifts

The gradient and the curvature terms as given by (8.1b) and (8.1c) give rise to drifts of the guiding center and it is desired to deduce the expressions for these drift velocities.

Gradient drift

The gradient terms $\partial B_z/\partial x$ and $\partial B_z/\partial y$ give rise to similar effects and therefore, only $\partial B_z/\partial x$ is assumed to be nonvanishing. The magnetic flux density is essentially in the z-direction but has a small spatial variation in the transverse (x) direction. Let

$\partial B_z/\partial x > 0$. In the absence of spatial variation of B_z, in the xy-plane, a positive charged particle orbits in a circle in the clockwise direction, the Larmor radius being inversely proportional to the magnitude of the B-field. Since B_z increases in the x-direction, it follows that the radius of curvature of the orbit for larger values of x is smaller than that for smaller values of x. The net effect is that the guiding center of the charged particle has a drift velocity in the positive y-direction, as shown in Fig. 3.9. In the x-direction, the orbit has periodic variations and has no drift. It can be shown by similar arguments that a negatively charged particle has a drift in the negative y-direction.

For the purpose of deducing the drift velocity, it should be noted that the variations of the B-field in one orbit are small and therefore Eqs. (3.7)–(3.9) hold. If the time average is taken over one Larmor period, it is found from Eqs. (3.7)–(3.9) that

$$\langle u_x u_y \rangle = 0 \qquad \langle u_x^2 \rangle = u_{\perp 0}^2/2 \tag{10.1}$$

Let $\mathbf{r}_L$ denote the position vector from the instantaneous location of the particle to the guiding center. Then, from the Lorentz force equation

$$m d\mathbf{u}/dt = q\mathbf{u} \times \mathbf{B} \tag{10.2}$$

it can be established that

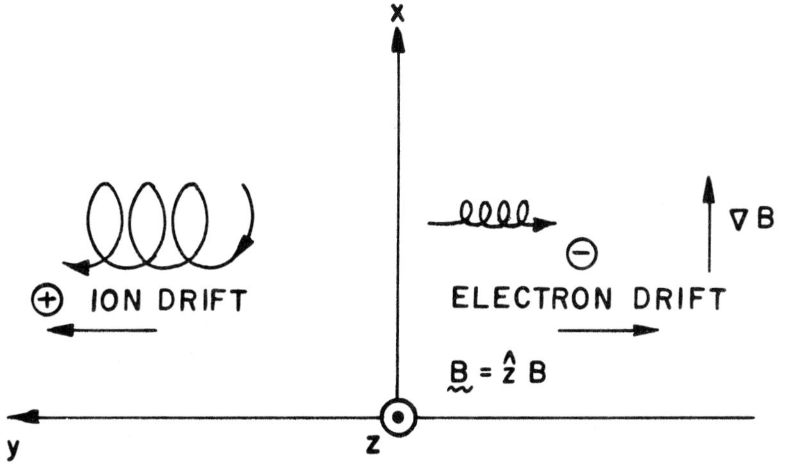

Fig. 3.9. Gradient drift velocities of an electron and an ion in a magnetic field ($\mathbf{B} = \hat{\mathbf{z}}B$) with a spatial variation in the transverse direction ($\partial B_z/\partial x > 0$).

$$\mathbf{r}_L = (m/qB^2)\mathbf{u} \times \mathbf{B} \tag{10.3}$$

The magnitude of Eq. (10.3) is consistent with Eq. (3.14). If the instantaneous position of the guiding center coincides with the origin of the cylindrical coordinates r, θ, and z which denote the position of the particle and if the B-field is in the z-direction, for a positively charged particle $\mathbf{u}$ has been proved in Sec. 3.3 to be in the $-\hat{\boldsymbol{\theta}}$-direction. Hence, Eq. (10.3) shows that r_L is in the $-\hat{\mathbf{r}}$-direction, as it should be. For a negatively charged particle, $\mathbf{u}$ is in the $\hat{\boldsymbol{\theta}}$-direction, but since the sign of the charge is reversed, $\mathbf{r}_L$ is again in the $-\hat{\mathbf{r}}$-direction. Thus, Eq. (10.3) is verified to yield correctly the position vector from the location of the particle to the guiding center.

Let $\mathbf{r}_g$ and $\mathbf{r}$ denote the instantaneous values of the position vectors from an arbitrary origin to the guiding center and the location of the particle, respectively. Then, the following relation holds

$$\mathbf{r}_g = \mathbf{r} + \frac{m}{qB^2}\mathbf{u} \times \mathbf{B} \tag{10.4}$$

Since B has spatial variation, Eq. (10.4) reveals that $\mathbf{r}_g$ also changes in course of time giving rise to the drift velocity referred to previously. The differentiation of Eq. (10.4) with respect to time yields

$$\dot{\mathbf{r}}_g = \dot{\mathbf{r}} + \frac{m}{qB^2}\frac{d\mathbf{u}}{dt} \times \mathbf{B} - \frac{m}{qB^3}\frac{dB}{dt}\mathbf{u} \times \mathbf{B} \tag{10.5}$$

With the help of Eq. (10.2), it can be deduced that

$$\frac{m}{q}\frac{d\mathbf{u}}{dt} \times \mathbf{B} = -\mathbf{B} \times (\mathbf{u} \times \mathbf{B}) = -B^2\left[\mathbf{u} - \left(\mathbf{u} \cdot \frac{\mathbf{B}}{B}\right)\frac{\mathbf{B}}{B}\right] = -\mathbf{u}_\perp B^2 \tag{10.6}$$

Also,

$$\frac{dB}{dt} = \frac{\partial B}{\partial x}\frac{dx}{dt} = u_x \frac{\partial B}{\partial x} \tag{10.7}$$

Together with Eqs. (10.6) and (10.7), Eq. (10.5) yields

$$\dot{\mathbf{r}}_g = \mathbf{u}_\parallel - \frac{m}{qB^2}\frac{\partial B}{\partial x}[\hat{\mathbf{x}}u_x u_y - \hat{\mathbf{y}}u_x^2] \tag{10.8}$$

where $\mathbf{u}_\parallel$ is the velocity of the guiding center parallel to the magnetic field. Note that $\mathbf{u}_\parallel$ is unaffected by the spatial variation of the magnetic field in the transverse direction. If the time average of Eq. (10.8) is taken over one Larmor period, the periodic variations which do not contribute to the drift are eliminated. Hence, such a time average of Eq. (10.8) is taken and in view of Eqs. (10.1), the result is

$$\dot{\mathbf{r}}_g = \mathbf{u}_\parallel + \frac{m}{qB^2}\frac{u_{\perp 0}^2}{2}\hat{\mathbf{y}}\frac{\partial B}{\partial x} = \mathbf{u}_\parallel + \frac{\mathrm{m}}{qB^2}\mathbf{B} \times \nabla B \tag{10.9}$$

In Eq. (10.9) $\mathfrak{m} = \frac{1}{2}mu_{\perp 0}^2/B$ is the orbital magnetic moment. It is observed from Eq. (10.9) that the guiding center, in addition to the parallel velocity $\mathbf{u}_{\parallel}$ which gives rise to the basic helical orbit in a magnetic field, has a drift velocity $\mathbf{u}_G$ which is perpendicular to both the magnetic field and the transverse direction of variation of the magnetic field. The gradient drift velocity $\mathbf{u}_G$ is obtained from Eq. (10.9) to be given by

$$\mathbf{u}_G = (\mathfrak{m}/qB^2)\mathbf{B} \times \nabla B \tag{10.10}$$

It is verified from Eq. (10.10) that the gradient drift is in the positive y-direction for a positively charged particle and in the negative y-direction for a negatively charged particle, as has been argued previously from simple physical considerations. Since it depends on the sign of the charge, it follows that in a neutral plasma the gradient drift of the particles results in an electric current.

A comparison of Eq. (10.10) with Eq. (4.23) shows that the variation of the B-field in the transverse direction is equivalent to an external force $\mathbf{F}_G$ as given by

$$\mathbf{F}_G = -\mathfrak{m}\nabla B \tag{10.11}$$

Curvature drift

The effects of the curvature terms $\partial B_x/\partial z$ and $\partial B_y/\partial z$ are similar and therefore it is sufficient to consider only one of them. It is assumed that only $\partial B_x/\partial z$ is nonvanishing with the result that the magnetic field is two-dimensional. In the zx-plane, the magnetic flux density at the origin is in the z-direction whereas in the neighborhood of the origin, both x- and z-components of the magnetic flux density are present. Consequently, the lines of magnetic flux density are curved near the origin as shown in Fig. 3.10. It is assumed that $\partial B_x/\partial z$ is so small that the radius of curvature of the flux lines near the origin is large compared to the Larmor radius. It is to be proved that the charged particle follows the curved flux lines and the resulting centrifugal force acts like an external force to give rise to a drift of the guiding center of the particle.

The directional tangent dx/dz to the flux lines is equal to the ratio of the x- to the z-component of the magnetic flux density. Hence,

$$dx/dz = B_x/B_z \tag{10.12}$$

It is to be noted that $B_x = 0$ and $B_z = B$ at the origin. If B_x and B_z are expanded into Taylor series about the origin and if only the leading terms in these series are retained, Eq. (10.12) becomes

$$dx/dz = (1/B)(\partial B_x/\partial z)z \tag{10.13}$$

Since dx/dz is small, d^2x/dz^2 is approximately equal to the curvature $1/R$ of the

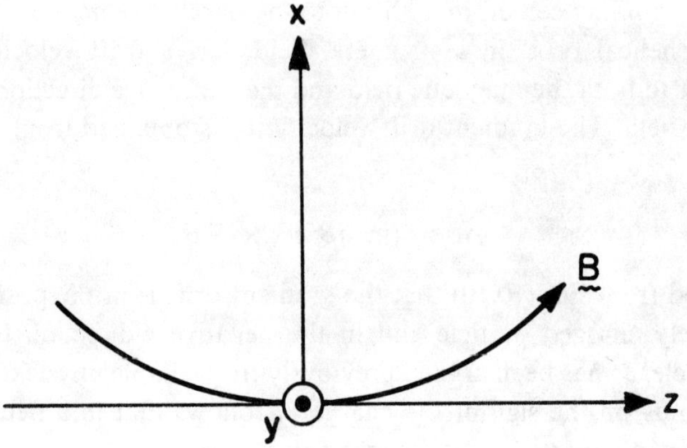

Fig. 3.10. A two-dimensional magnetic field with curved flux lines in the zx-plane ($\partial B_x/\partial z > 0$).

flux lines near the origin. Therefore, Eq. (10.13) gives

$$1/R = (1/B)(\partial B_x/\partial z) \tag{10.14}$$

In view of Eq. (10.14), the magnetic flux density **B** near the origin can be written as

$$\mathbf{B} = B[\hat{\mathbf{x}}z/R + \hat{\mathbf{z}}] \tag{10.15}$$

Consequently, the x-, y-, and z-components of the Lorentz force equation (10.2) become

$$du_x/at = -\omega_c u_y \tag{10.16}$$

$$du_y/dt = -\omega_c(u_z z/R - u_x) \tag{10.17}$$

$$du_z/dt = \omega_c u_y z/R \tag{10.18}$$

Note that Eqs. (10.16)–(10.18) are specialized for an electron for which $q = -e$ and $\omega_c = eB/m$. Since z/R is a small quantity, the terms containing z/R as a factor can be set equal to zero in Eqs. (10.16)–(10.18) as a first approximation, which is identified by the superscript (0). The governing equations of the first approximation are

$$du_x^{(0)}/dt = -\omega_c u_y^{(0)} \tag{10.19}$$

$$du_y^{(0)}/dt = \omega_c u_x^{(0)} \tag{10.20}$$

$$du_z^{(0)}/dt = 0 \tag{10.21}$$

which are the same as those for an electron in a constant and uniform B-field as treated in Sec. 3.3. The solutions of Eqs. (10.19)–(10.21) are

$$x^{(0)} = \frac{u_{\perp 0}}{\omega_c} \sin \omega_c t \qquad u_x^{(0)} = u_{\perp 0} \cos \omega_c t \tag{10.22}$$

$$y^{(0)} = -\frac{u_{\perp 0}}{\omega_c} [\cos \omega_c t - 1] \qquad u_y^{(0)} = u_{\perp 0} \sin \omega_c t \tag{10.23}$$

$$z^{(0)} = u_{z0} t \qquad u_z^{(0)} = u_{z0} \tag{10.24}$$

In obtaining Eqs. (10.22)–(10.24), the particle's initial position is assumed to coincide with the origin and its initial velocity is taken as $\mathbf{u}_0 = \hat{\mathbf{x}} u_{\perp 0} + \hat{\mathbf{z}} u_{z0}$.

Since the terms containing z/R in Eqs. (10.16)–(10.18) are small order quantities, in the next approximation, the coefficients of $1/R$ may be replaced by their first approximations. The governing equations for the second approximation are

$$du_x/dt = -\omega_c u_y \tag{10.25}$$

$$du_y/dt = -\omega_c (u_{z0}^2 t/R - u_x) \tag{10.26}$$

$$du_z/dt = (\omega_c/R) u_{\perp 0} u_{z0} t \sin \omega_c t \tag{10.27}$$

As in Sec. 3.3, Eqs. (10.25) and (10.26) may be combined to yield

$$\frac{d}{dt} U - i\omega_c U = -i\omega_c \frac{u_{z0}^2 t}{R} \qquad U = u_x + iu_y \tag{10.28}$$

or equivalently,

$$\frac{d}{dt}[U e^{-i\omega_c t}] = \frac{u_{z0}^2}{R} t \frac{d}{dt}(e^{-i\omega_c t}) \tag{10.29}$$

If the method of integration by parts is used, Eq. (10.29) can be integrated with the following result:

$$U e^{-i\omega_c t} = \frac{u_{z0}^2}{R}\left(t + \frac{1}{i\omega_c}\right) e^{-i\omega_c t} + u_{\perp 0} \tag{10.30}$$

The constant of integration in Eq. (10.30) is so chosen that if the terms containing $1/R$ are set equal to zero, the solutions obtained from Eq. (10.30) for u_x and u_y are the same as those given in Eqs. (10.22) and (10.23), respectively. From Eq. (10.30), it is found that

$$u_x = (u_{z0}^2/R) t + u_{\perp 0} \cos \omega_c t \tag{10.31}$$

$$u_y = -(u_{z0}^2/R\omega_c) + u_{\perp 0} \sin \omega_c t \tag{10.32}$$

When averaged over a Larmor period, $u_x^{(0)}$ and $u_y^{(0)}$ are seen from Eqs. (10.22) and (10.23) to go to zero but Eqs. (10.24) show that $u_z^{(0)}$ is unchanged by the averaging process. In order to obtain nonvanishing values for u_x and u_y after they are averaged over a Larmor period, it is necessary to evaluate a second approximation to the values of u_x and u_y. For u_z, a second approximation is not necessary. The leading terms in the expressions for u_x, u_y, and u_z after averaging over a Larmor period are determined from Eqs. (10.31), (10.32), and (10.24) as

$$u_x = (u_{z0}^2/R)t \tag{10.33}$$

$$u_y = -u_{z0}^2/R\omega_c \tag{10.34}$$

$$u_z = u_{z0} \tag{10.35}$$

With the help of Eqs. (10.33) and (10.35), it is deduced that

$$x = (u_{z0}^2/R)t^2/2 \qquad z = u_{z0}t \tag{10.36a, b}$$

by taking into account the fact that the initial position of the charged particle is at the origin. The elimination of t from Eqs. (10.36a,b) gives the equation of the path of the guiding center projected onto the zx-plane as

$$x = (1/2R)z^2 \tag{10.37}$$

From Eqs. (10.13) and (10.14), it is seen that

$$dx/dz = z/R \tag{10.38}$$

When Eq. (10.38) is integrated, the equation of the flux line passing through the origin is found to be the same as that given by Eq. (10.37). Hence, it follows that the guiding center moves along the gently curved flux line.

The particle is seen from Eq. (10.33) to have an acceleration u_{z0}^2/R in the x-direction near the origin. It is this acceleration which constrains the particle to move along the curved flux lines. This force per unit mass is balanced by the centrifugal force

$$\mathbf{F}_C = -(mu_{z0}^2/R)\hat{\mathbf{x}} \tag{10.39}$$

This centrifugal force acts like an external force to produce a drift velocity $\mathbf{u}_C$ in accordance with Eq. (4.23) as given by

$$\mathbf{u}_C = -\frac{mu_{z0}^2}{R}\frac{\hat{\mathbf{x}} \times \hat{\mathbf{z}}B}{qB^2} = -\hat{\mathbf{y}}\frac{u_{z0}^2}{R\omega_c} \tag{10.40}$$

which is precisely the same as the drift velocity deduced in Eq. (10.34).

The expressions for the centrifugal force and the resulting drift velocity have been determined for a charged particle in the close neighborhood of the origin where the lines of magnetic flux are curved. These expressions are conveniently recast so as to be applicable to a charged particle at an arbitrary location P along a curved flux line, as depicted in Fig. 3.11. A local coordinate system specified by the orthogonal triplet of unit vectors $\hat{\mathbf{e}}_1$, $\hat{\mathbf{e}}_2$, and $\hat{\mathbf{e}}_3$ is introduced where $\hat{\mathbf{e}}_1$ is along the tangent, $\hat{\mathbf{e}}_2$ along the normal, and $\hat{\mathbf{e}}_3$ along the binormal to the curved magnetic flux line at P. Consider a point Q in the neighborhood of P at a distance ds measured along the curve. Let the arc PQ subtend an angle $d\varphi$ at its center of curvature O and let R be the local radius of curvature. PA and PB denote $\hat{\mathbf{e}}_1$ and $\hat{\mathbf{e}}_1 + d\hat{\mathbf{e}}_1$, respectively, where the latter is the value of $\hat{\mathbf{e}}_1$ at Q. Since $\underline{\angle BPA} = d\varphi$ and $\hat{\mathbf{e}}_1$ is a unit vector, it is seen from the triangle APB that $|d\hat{\mathbf{e}}_1| = d\varphi = ds/R$. Hence, $|d\hat{\mathbf{e}}_1|/ds = 1/R$. In the limit of $d\varphi$ going to zero, it is clear from Fig. 3.11 that $d\hat{\mathbf{e}}_1$ is in the direction of $\hat{\mathbf{e}}_2$. Hence, it is found that

$$\hat{\mathbf{e}}_2/R = d\hat{\mathbf{e}}_1/ds = (\hat{\mathbf{e}}_1 \cdot \nabla)\hat{\mathbf{e}}_1 \tag{10.41}$$

Now, the centrifugal force $\mathbf{F}_C$ at P may be expressed by

$$\mathbf{F}_C = -(mu_z^2/R)\hat{\mathbf{e}}_2 = -2w_{\parallel}(\hat{\mathbf{e}}_1 \cdot \nabla)\hat{\mathbf{e}}_1 \tag{10.42}$$

where $w_{\parallel}$ is the kinetic energy of the particle along the magnetic flux line whose direction is specified by $\hat{\mathbf{e}}_1$. The curvature drift velocity $\mathbf{u}_C$ is deduced from Eqs (4.23) and (10.42) to be

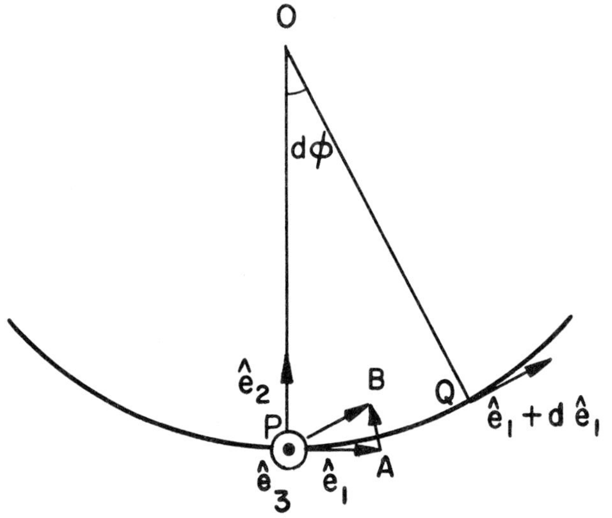

Fig. 3.11. Unit vectors $\hat{\mathbf{e}}_1$, $\hat{\mathbf{e}}_2$, and $\hat{\mathbf{e}}_3$ along the tangent, the normal and the binormal respectively at a location P on a curved line of magnetic flux density.

$$\mathbf{u}_C = 2w_{\parallel}\hat{\mathbf{e}}_1 \times \{(\hat{\mathbf{e}}_1 \cdot \nabla)\hat{\mathbf{e}}_1\}/qB \tag{10.43}$$

Since the curvature drift velocity is in opposite directions for charges of opposite sign, it follows that in a neutral plasma there is an electric current due to the curvature drift.

Shear terms

The shear terms $\partial B_x/\partial y$ and $\partial B_y/\partial x$ as given by (8.1d) enter into the component of $\nabla \times \mathbf{B}$ in the direction of $\mathbf{B}$ and cause twisting of the lines of magnetic flux about each other. As a consequence, the shape of the orbit is changed slightly but no first order drifts are produced. Thus, these terms have no particularly interesting effect on the motion of the charged particles.

3.11. Magnetic pumping

So far consideration has been given only to the effect of spatial variation of the magnetic field on the motion of the charged particle. Now, it is desired to examine the effect of time variation of the magnetic field. According to Maxwell's equations, time-dependent magnetic fields also vary in space. It is assumed that the scale length of the spatial variation of the magnetic field is sufficiently large in comparison with the Larmor radius that in the spatial region of interest, the magnetic field may be legitimately assumed to be uniform in space. Moreover, the magnetic field B is assumed to be in the z-direction with the result that there is cylindrical symmetry. From Maxwell's equations, it is found that the time-varying magnetic field is accompanied by an electric field. In view of the cylindrical symmetry and the axial nature of the magnetic field, only the azimuthal component of the electric field E_θ is present. It is assumed that E_θ is zero along the z-axis.

Let t be the time in which the change in the magnetic field becomes of the order of B. It is then assumed that t is much larger than the Larmor period T_L. In other words, the fractional change of the magnetic field in a time interval of the order of a Larmor period is very small. The azimuthal component of the electric field accelerates the charged particle with the result that the orbit is no longer a circle about the magnetic field lines. Since the time variation of the magnetic field is small, E_θ is small and the orbit of the charged particle is very nearly a circle. There are certain adiabatic invariants associated with the motion of the charged particle and these are now systematically deduced.

From Maxwell's equations, it is seen that

$$(1/r)(\partial/\partial r)(rE_\theta) = -\partial B/\partial t \tag{11.1}$$

which when integrated with respect to r yields

$$\int_0^r (\partial/\partial r)(rE_\theta)\,dr = -\int_0^r r(\partial B/\partial t)\,dr \tag{11.2}$$

Since $\partial B/\partial t$ is a slowly varying function, it can be taken outside the integral sign. Then, the integrations in Eq. (11.2) are performed with the following result:

$$E_\theta = -\frac{r}{2}\frac{\partial B}{\partial t} \quad \text{or} \quad \mathbf{E} = \frac{\mathbf{r}\times\hat{\mathbf{z}}}{2}\frac{\partial B}{\partial t} \tag{11.3}$$

Hence, the Lorentz force equation for an electron becomes

$$m\frac{d\mathbf{u}}{dt} = -e\left[\frac{\mathbf{r}\times\hat{\mathbf{z}}}{2}\frac{\partial B}{\partial t} + \mathbf{u}\times\hat{\mathbf{z}}B\right] \tag{11.4}$$

The components of Eq. (11.4) are given by

$$\dot{u}_x = \ddot{x} = -(y/2)\dot{\omega}_c - \dot{y}\omega_c \tag{11.5}$$

$$\dot{u}_y = \ddot{y} = (x/2)\dot{\omega}_c + \dot{x}\omega_c \tag{11.6}$$

$$\dot{u}_z = \ddot{z} = 0 \tag{11.7}$$

where $\omega_c = eB/m$, $\dot{\omega}_c = e\dot{B}/m$ and the dot denotes differentiation with respect to time. The axial motion of the guiding center is seen from Eq. (11.7) to be unaffected by the time variation in the magnetic field. On setting

$$R = x + iy \tag{11.8}$$

Eqs. (11.5) and (11.6) can be combined to yield

$$\ddot{R} = i\omega_c\dot{R} + i(\dot{\omega}_c/2)R \tag{11.9}$$

The solution of Eq. (11.9) may be sought in the form

$$R = re^{i\theta} \tag{11.10}$$

where r and θ are real but are, in general, functions of time. Moreover,

$$\dot{\theta} = \omega \tag{11.11}$$

is the angular frequency. If Eqs. (11.10) and (11.11) are substituted into Eq. (11.9) and the resulting equation is separated into real and imaginary parts, the following two relations are obtained:

$$(\omega/\omega_c)^2 - \omega/\omega_c - \ddot{r}/\omega_c^2 r = 0 \tag{11.12}$$

and

$$(d/dt)[r^2(\omega - \omega_c/2)] = 0 \tag{11.13}$$

When the magnetic field is a constant, the solution of Eq. (11.9) has been deduced to be given by

$$r = u_{\perp 0}/\omega_c = \text{constant} \qquad \theta = \omega_c t + \theta_0 - \pi/2 \tag{11.14}$$

where $u_{\perp 0}$ and θ_0 depend on the initial conditions. Since the magnetic field varies only very slowly with time, it is to be expected that

$$\ddot{r}/\omega_c^2 r \ll 1 \tag{11.15}$$

Therefore, in the first approximation, the third term on the left side of Eq. (11.12) can be omitted and the following two solutions of ω are obtained:

$$\omega_1 = 0 \qquad \omega_2 = \omega_c \tag{11.16}$$

For each of the two solutions of ω as given by Eqs. (11.16), Eq. (11.13) gives

$$r^2 \omega_c = \text{constant} \tag{11.17}$$

Adiabaticity condition

The implication of the restriction (11.15) can be examined now with the help of Eq. (11.17). According to Eq. (11.17), $r = C\omega_c^{-1/2}$ where C is a constant and by differentiating this expression with respect to time twice, it can be shown that

$$\ddot{r}/\omega_c^2 r = \tfrac{3}{4}(\dot{\omega}_c/\omega_c^2)^2 - \tfrac{1}{2}\ddot{\omega}_c/\omega_c^3 \tag{11.18}$$

Since ω_c is a slowly varying function of time, ω_c at time t can be expanded into a Taylor series around its value for time t_0 with the following result

$$\omega_c(t) = \omega_c(t_0) + (t - t_0)\dot{\omega}_c + \frac{(t - t_0)^2}{2}\ddot{\omega}_c \tag{11.19}$$

Let $\Delta\omega_c = \omega_c(t) - \omega_c(t_0)$ be the change in ω_c in a Larmor period $t - t_0 = T_L = 1/\omega_c$. Then, Eq. (11.19) can be manipulated to yield

$$\Delta B/B = \Delta\omega_c/\omega_c = \dot{\omega}_c/\omega_c^2 + \ddot{\omega}_c/2\omega_c^3 \tag{11.20}$$

It has been assumed that $\Delta B/B \ll 1$; therefore Eq. (11.20) shows that

$$\dot{\omega}_c/\omega_c^2 \ll 1 \qquad \ddot{\omega}_c/2\omega_c^3 \ll 1 \tag{11.21}$$

The restriction (11.15) follows from Eq. (11.18) and the inequalities (11.21). Thus, the restriction (11.15) is equivalent to the adiabaticity assumption that the fractional change in the magnetic field in one Larmor period is negligibly small.

Particle orbit

In view of Eqs. (11.16) and (11.17), the two independent solutions of Eq. (11.9) may be stated in the form

$$R = r_1 e^{i(\omega_1 t + \theta_{10})} \qquad \omega_1 = 0 \qquad r_1^2 \omega_c = \text{constant} \tag{11.22a, b, c}$$

and

$$R = r_2 e^{i(\omega_2 t + \theta_{20})} \qquad \omega_2 = \omega_c \qquad r_2^2 \omega_c = \text{constant} \tag{11.23a, b, c}$$

where r_1, θ_{10}, r_2 and θ_{20} all depend on initial conditions. The orbit of the charged particle can be deduced by linear superposition with the help of Eqs. (11.8), (11.22), and (11.23) to be given in the following vector notation:

$$\mathbf{r} = \mathbf{r}_g + \mathbf{r}_c \tag{11.24}$$

where

$$\mathbf{r}_g = r_1[\hat{\mathbf{x}} \cos(\omega_1 t + \theta_{10}) + \hat{\mathbf{y}} \sin(\omega_1 t + \theta_{10})] \tag{11.25}$$

$$\mathbf{r}_c = r_2[\hat{\mathbf{x}} \cos(\omega_2 t + \theta_{20}) + \hat{\mathbf{y}} \sin(\omega_2 t + \theta_{20})] \tag{11.26}$$

Since $\omega_2 = \omega_c$, it follows that Eq. (11.26) represents the orbital motion of the charged particle at the gyromagnetic frequency about the instantaneous guiding center G as shown in Fig. 3.12. As is to be expected for an electron, the rotation is in the counterclockwise direction. The position vector $\mathbf{r}_g$ as given by Eq. (11.25) represents the location of the guiding center G. In the first approximation, $\omega_1 = 0$ and therefore, Eq. (11.25) shows that the guiding center is stationary and is specified by the cylindrical coordinates r_1 and θ_{10}. Since $\ddot{r}/\omega_c^2 r$ is not equal to zero, it is seen from Eq. (11.12) that in the second approximation ω_1 has a small nonvanishing value. Consequently, Eq. (11.25) reveals that in the next order of approximation, the guiding center precesses slowly about the origin at the angular frequency $\omega = \omega_1$.

Adiabatic invariants

The two adiabatic invariants of the motion of the charged particle are given by Eqs. (11.22c) and (11.23c). From Eq. (11.26), it is found that $|\dot{r}_c| = r_2 \omega_c = u_{\perp 0}$ is the initial value of the speed of the particle in the perpendicular direction. Therefore, Eq. (11.23c) yields

$$r_2^2 \omega_c = \frac{u_{\perp 0}^2}{\omega_c} = \frac{2}{e}\frac{w_\perp}{B} = \frac{2}{e}\mathfrak{m} = \text{constant} \tag{11.27}$$

which establishes the adiabatic invariance of the orbital magnetic moment of the charged particle. When B is increased, Eq. (11.27) shows that the transverse kinetic energy $w_\perp$ of the particle increases linearly with B. If the existence of a large number of particles as in a plasma is taken into account, it is possible to define a transverse temperature $T_\perp$ by the condition

$$\langle w_\perp \rangle = \langle \tfrac{1}{2} m u_{\perp 0}^2 \rangle = KT_\perp \tag{11.28}$$

where $\langle \ \rangle$ indicates the average over all the particles and K is the Boltzmann constant. It is clear from Eq. (11.28) that the transverse temperature $T_\perp$ increases linearly with B.

It is of interest to ascertain the associated change in the number density of the particles as B is increased. This information is contained in the second adiabatic invariant as given by Eq. (11.22c). Since from Eq. (11.22c)

$$r_1^2 \omega_c = (e/\pi m)(\pi r_1^2 B) = \text{constant} \tag{11.29}$$

it follows that the magnetic flux through the circle of radius r_1 is a constant. When the magnetic flux density increases, the magnetic flux tube contracts in accordance with Eq. (11.29) and the guiding center such as G of the charged particles accompany the magnetic field lines in their inward radial displacement. Equivalently, it is said that the charged particles are frozen in the magnetic field lines. Consequently, an increase in the magnetic flux density causes the charged particles

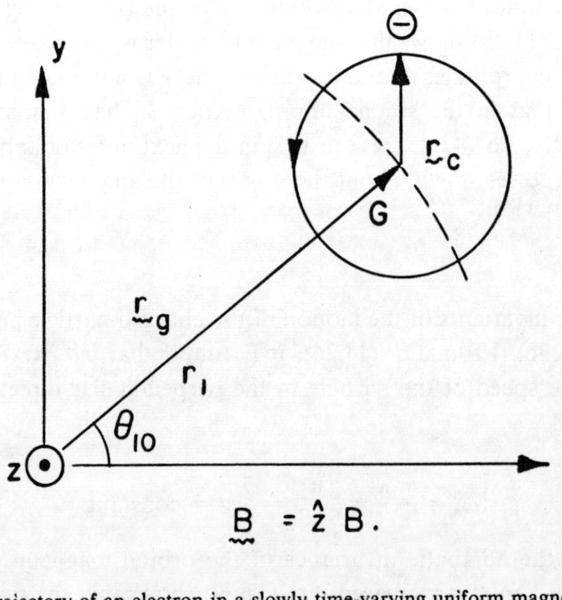

Fig. 3.12. Trajectory of an electron in a slowly time-varying uniform magnetic field ($\mathbf{B} = \hat{z}B$).

to come together resulting in what is known as *magnetic compression*. The number density N of the particles increases due to the magnetic compression and is inversely proportional to the cross-sectional area πr_1^2. From Eq. (11.29), it follows that the magnetic compression causes the number density N to increase linearly with B. In the same manner, if B is decreased, the particles move apart resulting in a magnetic decompression and a decrease in the number density of the charged particles. In the present case of two-dimensional compression, it is found that

$$w_\perp \propto T \propto N \propto B \tag{11.30}$$

Heating of plasma

It is possible to heat a plasma due to collisional effects by periodically compressing and decompressing it magnetically. The compression and the decompression are assumed to take place in a time interval that is very large compared to the Larmor period of the charged particles but is very small compared to the time required for the establishment of the thermal equilibrium. The compression and the decompression is not slow enough for the attainment of equipartition of energy. A change in the velocity and the energy in one direction is unaffected by the other two components. A compression can be one-, two-, or three-dimensional depending on the geometrical arrangement. In the present case of compression by an increase in the axial magnetic field, a two-dimensional compression results since the velocity and the energy in the two directions normal to the magnetic field are changed.

The thermodynamic state of the charged particles subject to adiabatic compressions is governed by the following adiabatic relation

$$PV^\gamma = \text{constant} \tag{11.31}$$

where P is the pressure and V is the volume. Also, the ratio γ of the two specific heats is given by

$$\gamma = (2 + \delta)/\delta \tag{11.32}$$

where δ is the number of degrees of freedom. The relation (11.31) is valid in the limit of very slow compressions. If the plasma is treated as a perfect gas, in addition, the relation

$$PV/T = \text{constant} \tag{11.33}$$

holds. From Eqs. (11.31) and (11.33), it follows that

$$T \propto V^{(1-\gamma)} \propto N^{(\gamma-1)} \tag{11.34}$$

For a two-dimensional compression, $\delta = 2$, $\gamma = 2$, and the relation (11.34) then yields a result which is in conformity with that obtained in the relation (11.30) by the application of orbit theory.

Fermi's acceleration mechanism can be viewed also as an adiabatic compression in one dimension. For $\delta = 1$, $\gamma = 3$ and then according to the relation (11.34), the longitudinal temperature or equivalently the longitudinal kinetic energy is inversely proportional to the square of the volume. Since the longitudinal kinetic energy is proportional to the square of the longitudinal speed u_z and since the volume is proportional to the spacing L between the mirrors, it follows that $u_z L$ = constant, which is in conformity with the result obtained in Eq. (9.5).

For the nonequilibrium state, it is appropriate to use the energy instead of the temperature T. Let w_1 be the initial energy of the plasma. This energy is divided equally among the three degrees of freedom. Note that there are no internal degrees of freedom in a plasma. Let the number of degrees of freedom affected by the magnetic compression be denoted by δ. Then, the energy $w_\perp$ capable of being altered by compression and the energy $w_\parallel$ unaffected by compression are given by

$$w_\perp = (\delta/3)w_1 \qquad w_\parallel = [(3-\delta)/3]w_1 \tag{11.35}$$

Let the volume V_1 be magnetically compressed to the value V_2. Then, $w_\perp$ is seen from the relation (11.34) to be increased by the factor $(V_1/V_2)^{\gamma-1}$ but $w_\parallel$ is unaffected with the result that the energy w_2 after compression is obtained as

$$w_2 = (w_1/3)[(3-\delta) + \delta(V_1/V_2)^{\gamma-1}] \tag{11.36}$$

If immediately after compression, the plasma is decompressed from the volume V_2 back to the original volume V_1, the increased value of $w_\perp$ is decreased by the factor $(V_2/V_1)^{\gamma-1}$ with the result that the energy of the plasma returns to the initial value w_1. Thus, the net effect is that the plasma is not heated. In order to be able to heat the plasma by a finite amount in a complete compression-decompression cycle, it is necessary to transfer part of the energy increase obtained by compression to the degrees of freedom which are unaffected by decompression. This transfer of energy can be brought about by collisional interactions which promote equipartition of energy. If after compression, the plasma is allowed to relax for a time which is sufficiently long compared to that required for the attainment of equipartition of energy, part of the increase in $w_\perp$ achieved during compression is transferred to cause an increase in $w_\parallel$. Then, the plasma is decompressed from the volume V_2 back to the original volume V_1. The decompression, just like the compression, takes place in a time short compared to that required for the establishment of thermal equilibrium. Consequently, the increased value of $w_\parallel$ is unaffected during the decompression but the new value of $w_\perp$ is decreased by the factor $(V_2/V_1)^{\gamma-1}$. In a complete cycle consisting of a compression, a relaxation, and a decompression, the energy of the plasma and therefore, its temperature is increased by a factor which is larger than unity. Thus, by a periodic repetition of these cycles of operation known as *magnetic pumping* a plasma can be heated.

3.12. Drift velocities and current densities

The drift velocities of the charged particles due to the various types of slow variation of the electric and the magnetic fields have been investigated. For the sake of convenience, the expressions for the various drift velocities are collected together as follows:

Electric drift: $$\mathbf{u}_E = \frac{\mathbf{E}_\perp \times \mathbf{B}}{B^2} \tag{4.6}$$

External force drift: $$\mathbf{u}_F = \frac{\mathbf{F}_\perp \times \mathbf{B}}{qB^2} \tag{4.23}$$

Polarization drift: $$\mathbf{u}_P = \frac{m}{qB^2}\dot{\mathbf{E}}_\perp \tag{5.3}$$

Gradient drift: $$\mathbf{u}_G = \frac{\mathfrak{m}}{qB^2}\mathbf{B} \times \nabla B \tag{10.10}$$

Curvature drift: $$\mathbf{u}_C = \frac{2w_\parallel}{qB}\hat{\mathbf{e}}_1 \times \{(\hat{\mathbf{e}}_1 \cdot \nabla)\hat{\mathbf{e}}_1\} \tag{10.43}$$

The charged particles in a magnetic mirror field do not have a drift velocity. Also, in the first order, the charged particles in a slowly time-varying magnetic field do not experience any drift velocity.

All the charged particles in a plasma experience the various drift velocities. Hence, these velocities are averaged over a large number of particles to obtain the average values. The magnetic moment $\mathfrak{m}$ and the parallel kinetic energy $w_\parallel$ are different for the different particles with the result that in the expressions for the average drift velocities, these parameters are assumed to represent their values averaged over a large number of particles. Moreover, some of the physical parameters are different for the ions and the electrons; therefore, additional subscripts i and e are used for distinguishing these quantities. The current density in a plasma is given by

$$\mathbf{J} = N_0 e(\mathbf{v}_i - \mathbf{v}_e) \tag{12.1}$$

where N_0 is the number density and $\mathbf{v}$ is the average velocity. The electric drift does not give rise to a current. The current densities caused by the other drift velocities are tabulated as follows:

Current density due to the external force drift: $$\mathbf{J}_F = \frac{N_0(\mathbf{F}_{\perp i} + \mathbf{F}_{\perp e}) \times \mathbf{B}}{B^2} \tag{12.2}$$

$$\text{Polarization-drift current density: } \mathbf{J}_P = \frac{N_0(m_e + m_i)}{B^2}\dot{\mathbf{E}}_\perp \tag{12.3}$$

$$\text{Gradient-drift current density: } \mathbf{J}_G = \frac{N_0(\mathfrak{m}_i + \mathfrak{m}_e)}{B^2}\mathbf{B} \times \nabla B \tag{12.4}$$

$$\text{Curvature-drift current density: } \mathbf{J}_C = \frac{2N_0}{B}(w_{\|,i} + w_{\|,e})\hat{\mathbf{e}}_1 \times \{(\hat{\mathbf{e}}_1 \cdot \nabla)\hat{\mathbf{e}}_1\} \tag{12.5}$$

In addition to these current densities, there is the magnetization current density $\mathbf{J}_M$ due to the spatial variation of the magnetization $\mathbf{M}$. From Eqs. (3.21) and (3.22), it is found that

$$\text{Magnetization current density: } \mathbf{J}_M = -\nabla \times \left\{N_0(\mathfrak{m}_i + \mathfrak{m}_e)\frac{\mathbf{B}}{B}\right\} \tag{12.6}$$

A knowledge of the various current densities is necessary in several applications of the first-order theory of the orbits of the charged particles.

References

3.1. S. Chandrasekhar, *Plasma Physics*, Chapter 2, The University of Chicago Press, Chicago, 1960.
3.2. J. L. Delcroix, *Plasma Physics*, Chapters 2 and 3, John Wiley, New York, 1965.
3.3. R. Jancel and Th. Kahan, *Electrodynamics of Plasmas*, Chapter 4, John Wiley, New York, 1966.
3.4. M. Kruskal, Elementary orbit and drift theory, in *Plasma Physics*, International Atomic Energy Agency, Vienna, 1965.

Problems

3.1. Consider the motion of an electron in a combination of electric field $\mathbf{E} = \hat{\mathbf{y}}E_y$ and magnetic field $\mathbf{B} = \hat{\mathbf{z}}B$. The fields are constant in time and uniform in space. The initial position and the initial velocity are given by $\mathbf{r}_0 = 0$ and $\mathbf{u}_0 = \hat{\mathbf{x}}u_{x0}$, respectively. Assume that $0 < u_{x0} < E_y/B$. Under the action of the magnetic field alone, the electron moves along a circular arc in the counterclockwise direction. Show that under the action of the combined electric and magnetic fields, initially the electron moves along an arc curved in the clockwise direction. Find the center and the radius of curvature of the arc.

3.2. In the normalized coordinates, the orbit of the electron of Problem 3.1 is given by

$$\tilde{x} = \frac{xB\omega_c}{E_y} = \omega_c t + (A - 1)\sin \omega_c t$$

$$\tilde{y} = \frac{yB\omega_c}{E_y} = (A - 1)(1 - \cos \omega_c t)$$

$$\tilde{z} = \frac{zB\omega_c}{E_y} = 0 \qquad A = u_{x0} B/E_y$$

where ω_c is the gyromagnetic angular frequency of the electron. Show that the electron orbit for $A = (1 - a)$ can be obtained from that for $A = (1 + a)$, by a suitable shift of the coordinates $\tilde{x}$ and $\tilde{y}$. Note that in Fig. 3.4, the electron orbits corresponding to $a = 0$, 0.5, 1.0, and 1.5 are depicted.

3.3. Show that in the limit of low frequencies such that $\omega/\omega_{ce} \ll 1$ and $\omega/\omega_{ci} \ll 1$, the Hall effect conductivity due to the electrons is exactly annulled by that due to the ions. Give a physical explanation for the vanishing of the net Hall effect current in a plasma for the case of very low frequencies. Note that the Hall effect current is perpendicular to both the electric and the magnetic fields.

Prove that in the limit of very low frequencies, the magnetoresistance terms in the conductivity vanish faster than the Hall effect terms.

3.4. Assume that the time-varying electric field is perpendicular to the static magnetic field. Find an approximation to the net current due to the mobility of the electrons and the ions in the limit of very low frequencies and show that the net current is contributed mainly by the ions.

Deduce the equivalent permittivity perpendicular to the static magnetic field and state the reason for it being the same as that given by Eq. (5.8).

3.5. Consider an electron acted upon by a magnetic field $\mathbf{B} = \hat{\mathbf{z}}B$ and an electric field $\mathbf{E} = \hat{\mathbf{y}}E_{y0}\sin \omega t$. The magnetic field is constant and uniform. The electric field is uniform and varies harmonically with time. Assume that the initial conditions of the electron are such that the motion takes place only in the xy-plane. Show that the electron orbit associated with the collective motion of similar particles in a plasma is given by

$$x = \frac{-eE_{y0}}{m(\omega^2 - \omega_c^2)}\left[\frac{\omega}{\omega_c}\cos \omega_c t - \frac{\omega_c}{\omega}\cos \omega t\right]$$

and

$$y = \frac{-eE_{y0}}{m(\omega^2 - \omega_c^2)}\left[\frac{\omega}{\omega_c}\sin \omega_c t - \sin \omega t\right]$$

where ω_c is the electron gyromagnetic angular frequency.

Verify that in the low-frequency limit $\omega/\omega_c \ll 1$, the electron orbits at the angular frequency ω around an ellipse with its major axis perpendicular to the electric field. Find the ratio of the minor to the major axes of the ellipse. Show that in the high-frequency limit $\omega/\omega_c \gg 1$, the electron orbits in a circle at the gyromagnetic angular frequency.

Consider the contribution to the electron orbit which has the same angular frequency as the forcing electric field. Show that for $\omega > \omega_c$, the electron orbit is an ellipse with its major axis along the electric field and the electron motion is in phase with the electric field. Prove that for $\omega < \omega_c$, the orbit is again an ellipse but with its minor axis along the electric field and the electron motion is out of phase with the electric field.

3.6. Consider the electron cyclotron resonance in a plasma. Let a be the dimension of the container transverse to the direction of the magnetostatic field and ν be the average frequency of collision between the electrons. Suppose that $am\omega_c \nu / e|\mathbf{E}_{\perp 0}| \ll 1$ where e is the magnitude of the charge, m the mass and ω_c the gyromagnetic frequency of an electron. Also, $|\mathbf{E}_{\perp 0}|$ is the magnitude of the transverse component of the time-varying electric field. Find whether the finite dimension of the container or the collision is the dominant factor inhibiting the effects of cyclotron resonance.

3.7. Consider a plasma having an isotropic velocity distribution trapped in a magnetic mirror system. Let B_0 and B_m be the magnetic flux densities at the center and at the ends of the magnetic mirror system. Show that the ratio R of the number of particles reflected from the magnetic mirror in a given time to the total number of particles incident on it is given by $R = 1 - B_0/B_m$, as indicated in Eq. (8.14).

3.8. Consider the motion of an electron in the neighborhood of the origin when it is acted upon by a magnetic field which is constant in time but has a small inhomogeneity. The magnetic field is in the z-direction at the origin and has a slow variation along z as given by

$$B_z = B_0(1 + \alpha z)$$

where $\alpha z \ll 1$. Assume that the magnetic field is cylindrically symmetric about the z-axis. Prove that the x-, y-, and z-components of the Lorentz force equation are given by

$$\ddot{x} = -\omega_c[\dot{y} + \alpha(z\dot{y} + \tfrac{1}{2}y\dot{z})]$$

$$\ddot{y} = \omega_c[\dot{x} + \alpha(z\dot{x} + \tfrac{1}{2}x\dot{z})]$$

$$\ddot{z} = -\omega_c(\alpha/2)[x\dot{y} - y\dot{x}]$$

where ω_c is the electron gyromagnetic angular frequency.

The initial conditions of the electron are given by $\mathbf{r}_0 = \hat{\mathbf{x}}\,(x_0 + u_{\perp 0}/\omega_c)$ and $\dot{\mathbf{r}}_0 = \mathbf{u}_0 = \hat{\mathbf{y}}u_{\perp 0} + \hat{\mathbf{z}}u_{z0}$. Find the equation of the magnetic flux line passing through the initial position of the electron.

Solve the Lorentz force equation by a perturbation technique and show that the parallel velocity decreases in the direction of increasing B_z in accordance with the relation

$$\dot{z} = u_z = u_{z0} - \tfrac{1}{2}\alpha u_{\perp 0}^2 t \qquad \text{(a)}$$

With the help of (a), explain the magnetic mirror effect.

Obtain a nonvanishing approximation to the velocities u_x, u_y, and u_z after they are averaged over a Larmor period. Retain only the leading terms of the series in powers of the small parameter α. Show that the average position of the electron follows the magnetic flux line passing through the initial position of the particle.

3.9. Consider the motion of an electron in the neighborhood of the origin under the action of a nonuniform magnetic field which is constant in time. The magnetic flux density is given by

$$\mathbf{B} = B_0[\alpha z\hat{\mathbf{x}} + (1 + \alpha x)\hat{\mathbf{z}}] \tag{a}$$

where $|\alpha z| \ll 1$ and $|\alpha x| \ll 1$. Show that $\mathbf{B}$ as expressed by (a) is consistent with Maxwell's equations. Note that a gradient term and a curvature term are both present in the expression for $\mathbf{B}$.

Obtain the Lorentz force equation in component form, deduce its solution by a perturbation procedure and determine nonvanishing approximations to the velocities u_x, u_y, and u_z after eliminating the terms which have periodic variations by time averaging over a Larmor period. Only the leading terms of the series in powers of the small parameter α need be retained. Assume the following initial conditions for the electron:

$$\mathbf{r} = \hat{\mathbf{x}}(x_0 + u_{\perp 0}/\omega_c) \qquad \text{and} \quad \dot{\mathbf{r}} = \mathbf{u} = \hat{\mathbf{y}}u_{\perp 0} + \hat{\mathbf{z}}u_{z0}$$

Show that the average position of the electron follows the magnetic flux line passing through its initial position. Prove that the resultant drift velocity is given by the sum of the gradient and the curvature drift velocities.

3.10. Show that the magnetic flux density $\mathbf{B}$ in the direction of the axis of a torus is given by $\mathbf{B} = \hat{\boldsymbol{\theta}}B(a/r)$ where B is the magnitude of the flux density at the radius $r = a$. Examine the type of charge separation that occurs due to the spatial variation of the B_θ field in the transverse ($\hat{\mathbf{r}}$) direction. In what direction is the $\mathbf{E} \times \mathbf{B}$ drift where $\mathbf{E}$ is the induced electric field due to the charge separation? Show qualitatively that the gradient and the $\mathbf{E} \times \mathbf{B}$ drifts prevent the confinement of a plasma in the magnetic field of a toroid.

3.11. A plasma confined by an axial magnetic flux density B_z is heated by magnetic pumping using collisional effects. For this purpose, B_z is increased from B_1 to B_2 in a time t_1, maintained at B_2 for a time t_2, decreased from B_2 to B_1 in a time t_1 and maintained at B_1 for a time t_2. This process is repeated periodically. Time t_1 is large compared to the Larmor period but short compared to the time required for the attainment of equipartition of energy. Time t_2 is large compared to that required for the establishment of thermal equilibrium. Show that for each cycle of variation of the magnetic field, the plasma temperature increases by the factor

$$\frac{[2 + 5B_2/B_1 + 2(B_2/B_1)^2]}{9B_2/B_1}$$

3.12. Consider the motion of an electron in a uniform magnetic flux density B_z in the z-direction. The magnetic flux density B_z has a slow time variation as given by $B_z = B(1 - \alpha t)$. Assume the following initial conditions: $\mathbf{r} = (r_L, 0, 0)$ and $\dot{\mathbf{r}} = \mathbf{u} = (0, u_{\perp 0}, 0)$. Obtain the Lorentz force equation, solve it by a perturbation procedure and determine the orbit to the first order in the small parameter α. Calculate the orbital magnetic moment and prove its adiabatic invariance by verifying the absence of linear terms in α.

3.13. Consider a magnetic mirror system whose axis coincides with the z-axis and which is symmetric about the $z = 0$ plane. The mirroring planes are given by $z = -z_m$ and $z = z_m$. The axial magnetic flux density is a function of time and is expressed by

$$B(z, t) = B(0, t)[1 + (|z|/a)^n]$$

where a is a parameter and n is an integer. The adiabatic invariance of the magnetic moment $\mathfrak{m}$ yields

$$\mathfrak{m} = mu_\perp^2/2B(z, t) = mu^2/2B(z_m, t) \qquad u^2 = u_\parallel^2 + u_\perp^2 \tag{A}$$

Note that $u_\parallel$ and $u_\perp$ are respectively the components of the particle velocity parallel and perpendicular to the z-axis and m is the mass of the particle.

With the help of (A), show that the longitudinal adiabatic invariant can be cast into the form:

$$\int_{-z_m}^{z_m} [B(z_m, t) - B(z, t)]^{1/2}\, dz = \text{constant} \tag{B}$$

If the magnetic flux density increases with time, by using (B), prove that the mirroring planes move inwards resulting in an axial compression with the compression factor given by $[B(0, t)/B(0, 0)]^{1/(n+2)}$.

3.14. Consider two infinite, perfectly conducting plates A_1 and A_2 occupying the planes $y = 0$ and $y = d$, respectively. The potential difference between the plates A_1 and A_2 is positive and is given by V. An electron of charge $-e$ and mass m enters the plate A_1 through a small hole and has an initial velocity v in the y-direction. Find the minimum value of the potential difference V necessary to prevent the electron from reaching the plate A_2.

Suppose that the region between the plates is permeated uniformly by a magnetic field of flux density B in the z-direction. Let an ion of charge e and mass m enter with zero initial velocity through a small hole in the plate A_1. As before, the potential difference between the plates A_1 and A_2 is positive and is given by V. Find the minimum value of the magnetic flux density B necessary to prevent the ion from reaching the plate A_2.

3.15. Consider a magnetic mirror system whose axis coincides with the z-axis and which is symmetrical about the $z = 0$ plane. The mirroring planes are given by $z = -z_m$ and $z = z_m$. The axial magnetic flux density is expressed by

$$B(z) = B_0[1 + (z/a)^2]$$

where a is a parameter. Find an expression for $u_{\parallel}$, the component of the particle velocity in the axial direction, in terms of the magnetic moment $\mathfrak{m}$ and $B(z)$; hence prove that the period of small oscillation of a particle of mass m between the mirroring planes is given by

$$T = 2\pi a(m/2\mathfrak{m}B_0)^{1/2}$$

Use Eq. (8.7) to show that the oscillations are of the simple-harmonic type and to verify the expression for the period T.

What is the condition on the total energy and the magnetic moment that the particle be restricted to the region $|z| < z_m$?

3.16. Consider a magnetic mirror system which is cylindrically symmetrical about the z-axis and which has reflection symmetry about the $z = 0$ plane. The axial magnetic flux density B_z increases rapidly with increasing $|z|$ for $|z| > z_0$. The mirroring planes are assumed to occur at $z = \pm z_m$. Suppose that the axial magnetic flux density increases with time. From the results of Problem 3.13, it follows that the mirroring planes $z = \pm z_m$ move inwards with time. Let u_m be the velocity of the mirroring planes. The purpose of this problem is to show that Fermi acceleration of a particle reflected from a moving magnetic mirror is intimately related to betatron acceleration. For this purpose, it is required to deduce the change in the axial component u_z of the particle velocity as given by Eq. (9.3) using an alternative method that brings out this relationship.

In a coordinate system at rest, show that in the mirroring region $|z| > z_0$, the magnetic flux density has a radial component B_r and an axial component B_z, and, that the time variation of the magnetic flux density produces an azimuthal electric field given by

$$E_\theta = u_m B_r$$

A charged particle which enters the mirroring region with a momentum mu_z experiences a force which retards the axial velocity. The particle is said to be reflected if it is thrown out of the mirroring region and its axial velocity is changed from u_z in the positive z-direction to u_z in the negative z-direction or vice versa. Show that the time t_r taken for the particle of charge $-e$ and mass m to be reflected from the mirroring region is specified by the relation:

$$-2mu_z = e \int_0^{t_r} u_\theta B_r \, dt$$

During the time when the particle is in the mirroring region, it is acted upon by an azimuthal force due to an electric field which produces betatron acceleration and changes its energy. Prove that the net change in energy of the particle at each reflection due to betatron acceleration is given by

$$\Delta w = 2u_m mu_z$$

Since the kinetic energy of the particle before reflection is equal to $\frac{1}{2}m(u_z^2 + u_\perp^2)$, verify that the net change in energy of the particle at each reflection can also be written as $\Delta w = mu_z \Delta u_z$ and hence show that the change in the axial component of the velocity is given by

$$\Delta u_z = 2u_m$$

which is identical to that obtained in Eq. (9.3).

Betatron acceleration increases the momentum of the particle in the azimuthal direction. But, when the particle leaves the mirroring region, the increase in momentum appears in the axial component. How does this transfer take place?

3.17. Consider a magnetic mirror system which is cylindrically symmetric about the z-axis. Let u_z and $u_\perp$ be, respectively, the electron velocities in the directions parallel and perpendicular to the axis of the magnetic mirrors. Some authors have introduced the following so-called "loss-cone" distribution function

$$f(\mathbf{u}) = \frac{N}{\pi^{3/2}\alpha_\perp^2 \alpha_z}\left(\frac{u_\perp}{\alpha_\perp}\right)^2 \exp\left[-\left(\frac{u_\perp}{\alpha_\perp}\right)^2 - \left(\frac{u_z}{\alpha_z}\right)^2\right]$$

to describe the electrons inside the magnetic mirror system. Note that α_z^2 and $\alpha_\perp^2$ are proportional to the parallel and the perpendicular temperatures respectively.

Show that N represents correctly the number density of the electrons. Provide arguments for the plausibility of the "loss-cone" distribution function.

CHAPTER 4

Classical Dynamics of Collisions

4.1. Introduction

A number of phenomena in plasmas are governed by the interactions of the particles with the electromagnetic fields; but, certain fundamental processes are determined by the interactions amongst the particles themselves. These interactions are called collisions. Collisional phenomena can be broadly divided into two categories; namely, elastic and inelastic. In elastic collisions, there is conservation of mass, momentum, and energy. The internal states of the particles involved do not change and there is neither creation nor annihilation of particles. For inelastic collisions, either or both of the particles involved have their internal states changed and the number of particles can be either increased or decreased. A charged particle may *recombine* with another to form a neutral atom; it can *attach* itself with a neutral particle to form a heavier charged particle. A collision may cause the energy state of an electron in an atom to be *raised* and electrons can be completely stripped of the atoms resulting in *ionization*. The processes of recombination, attachment, excitation, and ionization are the results of inelastic collisions.

A treatment of collisions between two particles based on classical dynamics is presented in this chapter. The results obtained are valid to a good approximation. More importantly, however, the procedures to be developed are useful whether the mechanics is classical or quantum.

In the classical theory of elastic collisions, the internal structure of the particles is ignored. The particles are represented by spherically symmetric point centers of force. The interaction between two particles is described by a force directed along a straight line connecting the centers of the two particles. The force depends only on the separation distance between the centers of the two particles. It is convenient to describe the collision in the center-of-mass system. The collision between particles is formulated and analyzed in terms of the equivalent one-body problem. The meaning of inverse collision is discussed and the concepts of scattering cross sections are introduced. The relations of the various parameters in the center-of-mass system to those in the laboratory system are determined.

The details of the collision between two perfectly elastic, hard spheres and that between two charged particles interacting through the Coulomb potential are evaluated. For the Coulomb potential, the total scattering and the momentum transfer cross sections become infinite with the result that it becomes necessary to impose a cut-off value on the impact parameter. The physical basis for the introduction of this cut-off value is examined. Finally, the concepts of mean free path and collision frequency are investigated and their relationship to the scattering cross section is pointed out.

4.2. Collision in the center-of-mass system

Consider the collision of two particles 1 and 2 of masses m_1 and m_2, respectively. At time t, particle 1 is located at P and particle 2, at O. As shown in Fig. 4.1, let the vectors CP and CO be denoted by $\mathbf{r}_{1C}$ and $\mathbf{r}_{2C}$ respectively. If the sum of the two mass moments about C is equal to zero, that is, if

$$m_1 \mathbf{r}_{1C} + m_2 \mathbf{r}_{2C} = 0 \tag{2.1}$$

then C is the center of mass (hereinafter abbreviated as c.m.) of the two-particle system. Let $\mathbf{r}_1$, $\mathbf{r}_2$, and $\mathbf{r}_C$ denote, respectively, the positions P, O, and C with respect to an arbitrary origin Q. Then,

$$\mathbf{r}_{1C} = \mathbf{r}_1 - \mathbf{r}_C \qquad \mathbf{r}_{2C} = \mathbf{r}_2 - \mathbf{r}_C \tag{2.2}$$

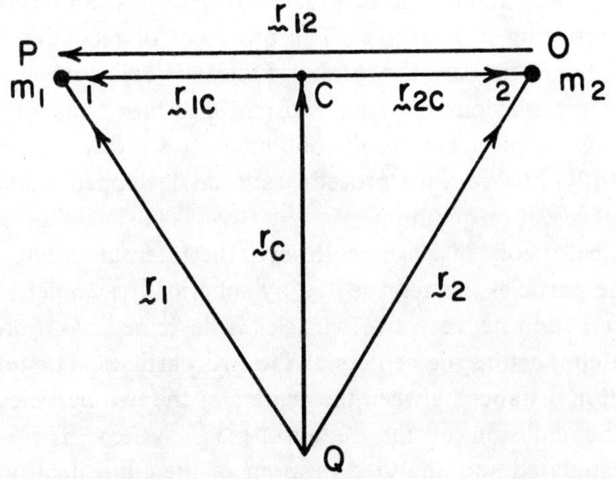

Fig. 4.1. Center of mass of particles 1 and 2.

The substitution of Eqs. (2.2) in Eq. (2.1) yields

$$m_1\mathbf{r}_1 + m_2\mathbf{r}_2 = (m_1 + m_2)\mathbf{r}_C \tag{2.3}$$

It is assumed that the masses of the particles, the total momentum and the total energy are conserved in the collision process. The time derivative of Eq. (2.3), as given by

$$m_1\dot{\mathbf{r}}_1 + m_2\dot{\mathbf{r}}_2 = (m_1 + m_2)\dot{\mathbf{r}}_C \tag{2.4}$$

together with the laws of conservation of mass and momentum shows that

$$\dot{\mathbf{r}}_C = \mathbf{u}_C = \text{constant vector} \tag{2.5}$$

The c.m. *is seen from Eq. (2.5) to move in a straight line with a constant speed* $u_C = |\mathbf{u}_C|$.

Let

$$M_1 = \frac{m_1}{m_1 + m_2} \qquad M_2 = \frac{m_2}{m_1 + m_2} \qquad \text{where } M_1 + M_2 = 1 \tag{2.6}$$

Then, Eq. (2.3) can be recast as

$$M_1\mathbf{r}_1 + M_2\mathbf{r}_2 = \mathbf{r}_C \tag{2.7}$$

The position of particle 1 with respect to that of particle 2 is given by

$$\mathbf{r}_{12} = \mathbf{r} = \mathbf{r}_1 - \mathbf{r}_2 \tag{2.8}$$

The successive elimination of $\mathbf{r}_2$ and $\mathbf{r}_1$ from Eqs. (2.7) and (2.8) together with Eqs. (2.2) and (2.6) gives the following results:

$$\mathbf{r}_1 = \mathbf{r}_C + M_2\mathbf{r} \qquad \mathbf{r}_2 = \mathbf{r}_C - M_1\mathbf{r} \tag{2.9}$$

$$\mathbf{r}_{1C} = M_2\mathbf{r} \qquad \mathbf{r}_{2C} = -M_1\mathbf{r} \tag{2.10}$$

The time derivative of Eq. (2.8) yields the following expression for the instantaneous velocity of particle 1 relative to that of particle 2:

$$\dot{\mathbf{r}}_{12} = \dot{\mathbf{r}} = \dot{\mathbf{r}}_1 - \dot{\mathbf{r}}_2 \tag{2.11}$$

From Eqs. (2.9) and (2.10), the following expressions can be derived:

$$\dot{\mathbf{r}}_1 = \mathbf{u}_C + M_2\dot{\mathbf{r}} \qquad \dot{\mathbf{r}}_2 = \mathbf{u}_C - M_1\dot{\mathbf{r}} \tag{2.12}$$

$$\dot{\mathbf{r}}_{1C} = M_2\dot{\mathbf{r}} \qquad \dot{\mathbf{r}}_{2C} = -M_1\dot{\mathbf{r}} \tag{2.13}$$

In Eqs. (2.13), $\dot{\mathbf{r}}_{1C}$ and $\dot{\mathbf{r}}_{2C}$ are respectively the velocities of particles 1 and 2 relative to the center of mass. From Eqs. (2.13), it follows that in the c.m. system, *the velocities of the two particles are always directed along antiparallel lines and the magnitudes of the velocities of the two particles are inversely proportional to their masses.*

If the particles represent the centers of two hard spheres, collision takes place at the time of contact of the two spheres and is, therefore, a well-defined event in time. Consequently, it is meaningful to identify times as "before" and "after" collision. Suppose that the particles represent two point charges. Then, the force of interaction between the particles depends on the separation distance, becomes smaller as the distance is increased and vanishes only for the limiting case of infinite separation distance. For the case of charged particles, "before" and "after" collision refer, respectively, to times $t = -\infty$ and $t = \infty$ since, only for these times, the force of interaction is zero and the particles move with uniform velocity. The values of $\dot{\mathbf{r}}_1$, $\dot{\mathbf{r}}_2$, and $\dot{\mathbf{r}}$ before collision are denoted by $\mathbf{u}_1$, $\mathbf{u}_2$, and $\mathbf{g}$ respectively. The corresponding values after collision are denoted by $\tilde{\mathbf{u}}_1$, $\tilde{\mathbf{u}}_2$ and $\tilde{\mathbf{g}}$, respectively. Let $|\mathbf{g}| = g$ and $|\tilde{\mathbf{g}}| = \tilde{g}$. In view of Eqs. (2.12), it follows that

$$\text{Before collision:} \quad \mathbf{u}_1 = \mathbf{u}_C + M_2\mathbf{g} \qquad \mathbf{u}_2 = \mathbf{u}_C - M_1\mathbf{g} \tag{2.14}$$

$$\text{After collision:} \quad \tilde{\mathbf{u}}_1 = \mathbf{u}_C + M_2\tilde{\mathbf{g}} \qquad \tilde{\mathbf{u}}_2 = \mathbf{u}_C - M_1\tilde{\mathbf{g}} \tag{2.15}$$

The total energy before collision is obtained from Eqs. (2.6) and (2.14) as

$$W = \tfrac{1}{2}m_1 u_1^2 + \tfrac{1}{2}m_2 u_2^2 = \frac{(m_1 + m_2)}{2}[u_C^2 + M_1 M_2 g^2] \tag{2.16}$$

Note that since the force of interaction is zero, the potential energy is zero before and after collision. In a similar manner, the total energy after collision is evaluated with the help of Eqs. (2.6) and (2.15) as

$$\tilde{W} = \tfrac{1}{2}m_1 \tilde{u}_1^2 + \tfrac{1}{2}m_2 \tilde{u}_2^2 = \frac{(m_1 + m_2)}{2}[u_C^2 + M_1 M_2 \tilde{g}^2] \tag{2.17}$$

The law of conservation of energy together with Eqs. (2.16) and (2.17) yields that $g = \tilde{g}$. *The relative velocity of the two particles is unchanged in magnitude by collision but has its direction altered.*

Let $\mathbf{F}_{12}$ be the force acting on particle 1 due to particle 2. In a similar manner, let $\mathbf{F}_{21}$ be the force acting on particle 2 due to particle 1. These two forces are equal and opposite according to Newton's laws of force. In the laboratory system, the motion of the two particles are governed by the following equations:

$$\mathbf{F}_{12} = m_1\ddot{\mathbf{r}}_1 \qquad \mathbf{F}_{21} = -\mathbf{F}_{12} = m_2\ddot{\mathbf{r}}_2 \tag{2.18a, b}$$

If Eq. (2.18a) is multiplied by m_2, Eq. (2.18b) by $-m_1$ and they are added, the result is

$$\mathbf{F}_{12} = m_r\ddot{\mathbf{r}}_{12} \tag{2.19}$$

where

$$m_r = \frac{m_1 m_2}{m_1 + m_2} \tag{2.20}$$

is known as the *reduced mass*. The force of interaction is assumed to be a central force which acts along the straight line joining the two particles. Hence, from Eq. (2.19)

$$\mathbf{r}_{12} \times \ddot{\mathbf{r}}_{12} = 0 \tag{2.21}$$

Since $\dot{\mathbf{r}}_{12} \times \dot{\mathbf{r}}_{12} = 0$, Eq. (2.21) can be rewritten as

$$\mathbf{r}_{12} \times \ddot{\mathbf{r}}_{12} + \dot{\mathbf{r}}_{12} \times \dot{\mathbf{r}}_{12} = (d/dt)[\mathbf{r}_{12} \times \dot{\mathbf{r}}_{12}] = 0 \tag{2.22}$$

The integration of Eq. (2.22) yields

$$\mathbf{r}_{12} \times \dot{\mathbf{r}}_{12} = \mathbf{K} \tag{2.23}$$

where $\mathbf{K}$ is a constant vector. From Eq. (2.23), the two particles are seen to be always in a plane perpendicular to the constant vector $\mathbf{K}$. The c.m. lying on the straight line joining the two particles is also on the same plane as the two particles. In general, the trajectories of the two particles are curves in the three-dimensional space. It has been shown that the c.m. moves in a straight line with constant speed. Therefore, the plane perpendicular to $\mathbf{K}$ containing the two particles moves in the direction of $\mathbf{u}_C$, as shown in Fig. 4.2. It is easier to represent the trajectories of the particles in a coordinate system in which the c.m. is stationary in the direction parallel to $\mathbf{K}$. In the new coordinate system, the component of the displacement of the c.m. parallel to $\mathbf{K}$ is zero. Therefore, the particles always lie on a fixed plane perpendicular to $\mathbf{K}$. The collision of the two particles is analyzed in this so-called c.m. system since in that system, the trajectories of the particles lie on a fixed plane perpendicular to $\mathbf{K}$ instead of being in the three-dimensional space.

4.3. Equivalent one-body problem

The instantaneous position vectors of particles 1 and 2 relative to the c.m. are given by $\mathbf{r}_{1C}$ and $\mathbf{r}_{2C}$, respectively. Hence, the total instantaneous kinetic energy of the two-particle system is deduced with the help of Eqs. (2.6), (2.13) and (2.20) to be

$$T = \tfrac{1}{2}m_1 \dot{\mathbf{r}}_{1C} \cdot \dot{\mathbf{r}}_{1C} + \tfrac{1}{2}m_2 \dot{\mathbf{r}}_{2C} \cdot \dot{\mathbf{r}}_{2C} = \frac{1}{2} m_r \dot{\mathbf{r}} \cdot \dot{\mathbf{r}} \tag{3.1}$$

In a similar manner, the total instantaneous angular momentum relative to the c.m. is evaluated from Eqs. (2.6), (2.10), (2.13), and (2.20) as

$$\mathbf{\Gamma} = \mathbf{r}_{1C} \times (m_1 \dot{\mathbf{r}}_{1C}) + \mathbf{r}_{2C} \times (m_2 \dot{\mathbf{r}}_{2C}) = \mathbf{r} \times (m_r \dot{\mathbf{r}}) \tag{3.2}$$

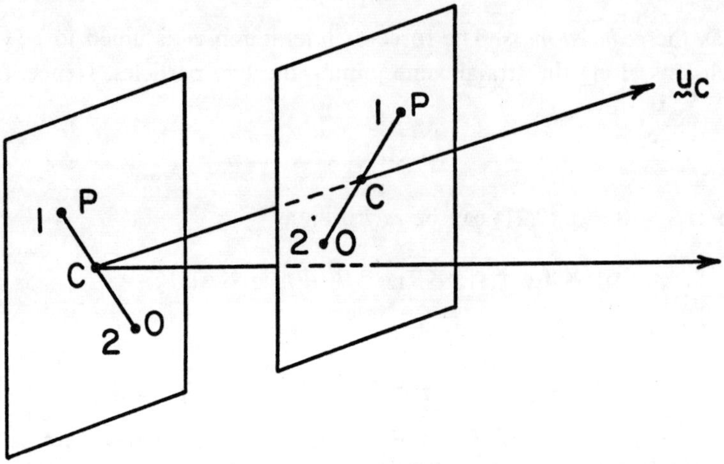

Fig. 4.2. Two sets of positions of particles 1, 2, and their center of mass showing that particles 1 and 2 always lie on a plane perpendicular to **K**.

It is seen from Eqs. (3.1) and (3.2) that the total instantaneous kinetic energy and angular momentum of the two-particle system are equivalent to those of mass m_r whose instantaneous position is denoted by $\mathbf{r}$ relative to a fixed origin. The equations of motion (2.18a) and (2.18b) of the two-particle system lead to Eq. (2.19). It follows from Eq. (2.19) that the collision dynamics of the two-particle system is equivalent to the one-body problem consisting of mass m_r at the position of particle 1 whose location is specified by $\mathbf{r}_{12} = \mathbf{r}$ relative to a fixed origin which coincides with the position of particle 2. This deduction from Eq. (2.19) is consistent with the results expressed in Eqs. (3.1) and (3.2). Thus, the collision dynamics of the two-particle system is conveniently analyzed in terms of this equivalent one-body problem.

Consider a Cartesian coordinate system x, y, and z oriented such that $\mathbf{K}$ is parallel to the z-axis. Therefore, the trajectory of the particle in the equivalent one-body problem lies in the xy-plane. The origin O coincides with the position of particle 2. The x-axis is oriented so as to be antiparallel to the initial velocity $\mathbf{g}$ of particle P, as shown in Fig. 4.3. The particle P is acted upon by a central force which is along the straight line joining the origin O to the instantaneous position of the particle. This force is either attractive or repulsive. In the absence of this force, the particle moves along a straight line with the initial velocity. The distance p of closest approach of the particle to the origin in the absence of the force is called the *impact parameter*.

It is convenient to introduce the polar coordinates r and θ where θ is the angle measured from the x-axis in the counterclockwise direction. Let $\hat{\mathbf{r}}$ and $\hat{\boldsymbol{\theta}}$ be the unit vectors in the directions of increasing r and θ, respectively. Then

$$\mathbf{r} = \hat{\mathbf{r}}r \tag{3.3}$$

where r is equal to the length of OP. It is ascertained that $\hat{\mathbf{r}}$ and $\hat{\boldsymbol{\theta}}$ are independent of r but are functions of θ. It can be established that

$$(d/d\theta)\hat{\mathbf{r}} = \hat{\boldsymbol{\theta}} \qquad (d/d\theta)\hat{\boldsymbol{\theta}} = -\hat{\mathbf{r}} \tag{3.4}$$

From Eqs. (3.3) and (3.4), it is found that

$$\dot{\mathbf{r}} = \hat{\mathbf{r}}\dot{r} + \hat{\boldsymbol{\theta}}r\dot{\theta} \tag{3.5}$$

Therefore, Eqs. (3.1) and (3.2) become

$$T = \tfrac{1}{2}m_r(\dot{r}^2 + r^2\dot{\theta}^2) \tag{3.6}$$

and

$$\boldsymbol{\Gamma} = \hat{\mathbf{z}}m_r r^2\dot{\theta} \tag{3.7}$$

It is evaluated from Eqs. (3.4) and (3.5) that

$$\ddot{\mathbf{r}} = \hat{\mathbf{r}}(\ddot{r} - r\dot{\theta}^2) + \hat{\boldsymbol{\theta}}(r\ddot{\theta} + 2\dot{r}\dot{\theta}) \tag{3.8}$$

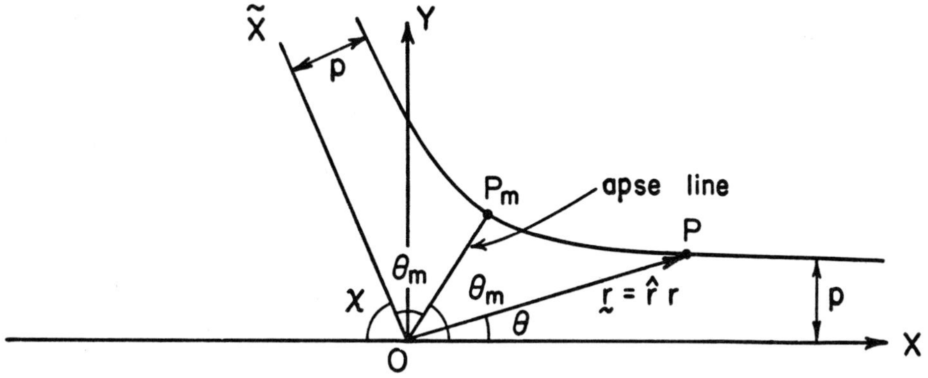

Fig. 4.3. Trajectory of the particle in the equivalent one-body problem showing the symmetry about the apse line.

The r and θ components of the equation of motion (2.19) together with Eq. (3.8) yield

$$m_r(\ddot{r} - r\dot{\theta}^2) = F_r \qquad m_r(r\ddot{\theta} + 2\dot{r}\dot{\theta}) = 0 \tag{3.9a, b}$$

since for a central force the θ-component is zero. On writing Eq. (3.9b) as $(1/r)(d/dt)(m_r r^2\dot{\theta}) = 0$, it follows that the angular momentum as given by Eq. (3.7) is a constant. For $t = -\infty$, it is obtained that

$$\theta = \frac{p}{r} \qquad \dot{\theta} = -\frac{p}{r^2}\dot{r} = \frac{p}{r^2}g \tag{3.10}$$

since initially as time increases r decreases with the result that $\dot{r} = -g$. From Eqs. (3.7) and (3.10), the law of conservation of angular momentum may be stated as

$$m_r r^2\dot{\theta} = m_r pg \tag{3.11}$$

It is assumed that the central force F_r is given in terms of an interaction potential $\phi(r)$ through the relation

$$F_r = -(\partial/\partial r)\phi(r) \tag{3.12}$$

where $\phi(r)$ is the potential energy. Note that the potential energy $\phi(r)$ is a function of only the distance r from the center of force. It is found from Eqs. (3.9a), (3.10), and (3.12) that

$$m_r\ddot{r} - m_r p^2 g^2/r^3 = -(\partial/\partial r)\phi(r) \tag{3.13}$$

If both sides of Eq. (3.13) are multiplied by $\dot{r}$, it follows that

$$\begin{aligned} m_r\dot{r}\ddot{r} - m_r\frac{p^2g^2}{r^3}\dot{r} &= \frac{d}{dt}\left[\frac{1}{2}m_r\dot{r}^2 + \frac{1}{2}m_r\frac{p^2g^2}{r^2}\right] \\ &= -\frac{\partial\phi(r)}{\partial r}\frac{dr}{dt} = -\frac{d\phi(r)}{dt} \end{aligned} \tag{3.14a}$$

The integration of Eq. (3.14a) together with Eqs. (3.6) and (3.11) yields

$$\tfrac{1}{2}m_r[\dot{r}^2 + r^2\dot{\theta}^2] + \phi(r) = T + \phi(r) = \text{constant} \tag{3.14b}$$

which is the statement of the law of conservation of energy. The value of the constant in Eq. (3.14b) can be determined from the initial condition. At $t = -\infty$, $r = \infty$, and $\phi(r) = 0$. From Eq. (3.11), it is seen that $r^2\dot{\theta}^2 = p^2g^2/r^2 = 0$ for $r = \infty$. Since for $t = -\infty$, $\dot{r} = -g$, Eq. (3.14b) becomes

$$\tfrac{1}{2}m_r\dot{r}^2 + \tfrac{1}{2}m_r r^2\dot{\theta}^2 + \phi(r) = \tfrac{1}{2}m_r g^2 \tag{3.15}$$

The laws of conservation of angular momentum and energy as given by Eqs. (3.11) and (3.15), respectively, enable the determination of the trajectory of the particle moving in the field of a central force specified by the interaction potential $\phi(r)$.

From Eqs. (3.11) and (3.15), $\dot{\theta}$ is eliminated and the resulting equation is solved for $\dot{r}$ to yield the following result:

$$\dot{r} = dr/dt = \mp g(1 - p^2/r^2 - \phi(r)/\tfrac{1}{2}m_r g^2)^{1/2} \tag{3.16}$$

The position on the trajectory for which the particle is at the shortest distance from the origin is denoted by P_m. The straight line OP_m is called the apse line. The coordinates of the position of the particle when it is on the apse line are denoted by r_m and θ_m. It is to be noted that P_m is also called the vertex of the trajectory. The trajectory does not necessarily have a vertex for all types of interaction potentials. When the interaction potential is repulsive or when the attractive interaction potential is not a very rapidly varying function of the distance from the center of force, the trajectory possesses a vertex. Only such interaction potentials for which a vertex exists are to be considered. For $\theta < \theta_m$, r is seen from Fig. 4.3 to decrease with time with the result that $\dot{r} < 0$. In a similar manner, it is found that for $\theta > \theta_m$, $\dot{r} > 0$. Hence, in Eq. (3.16) the upper or the lower sign has to be chosen according as $\theta < \theta_m$ or $\theta > \theta_m$.

It is obtained from Eqs. (3.10) that

$$\dot{\theta} = d\theta/dt = pg/r^2 \tag{3.17}$$

From Eqs. (3.16) and (3.17), an expression for $dr/d\theta = \dot{r}/\dot{\theta}$ can be deduced and the result after some rearrangement is

$$d\theta = -\frac{p}{r^2}\left(1 - \frac{p^2}{r^2} - \frac{\phi(r)}{(1/2)m_r g^2}\right)^{-1/2} dr \qquad \text{for } \theta < \theta_m \tag{3.18a}$$

$$= \frac{p}{r^2}\left(1 - \frac{p^2}{r^2} - \frac{\phi(r)}{(1/2)m_r g^2}\right)^{-1/2} dr \qquad \text{for } \theta > \theta_m \tag{3.18b}$$

When Eq. (3.18a) is integrated from $\theta = \theta_1$ to $\theta = \theta_m$ and Eq. (3.18b) from $\theta = \theta_m$ to $\theta = \theta_2$, it follows that

$$\theta_m - \theta_1 = \int_{r_m}^{r} \frac{p}{r^2}\left(1 - \frac{p^2}{r^2} - \frac{\phi(r)}{(1/2)m_r g^2}\right)^{-1/2} dr \qquad \text{for } \theta_1 < \theta_m \tag{3.19a}$$

$$\theta_2 - \theta_m = \int_{r_m}^{r} \frac{p}{r^2}\left(1 - \frac{p^2}{r^2} - \frac{\phi(r)}{(1/2)m_r g^2}\right)^{-1/2} dr \qquad \text{for } \theta_2 > \theta_m \tag{3.19b}$$

It is argued from Eqs. (3.19a) and (3.19b) that the two radius vectors lying on the opposite sides of the apse line and having the same length r are inclined at equal angles to the apse line. In other words, the trajectory is symmetrical about the apse line.

Let $O\tilde{X}$ be parallel to the final velocity of the particle. In view of the symmetry of the trajectory, the distances of the initial and the final positions of the particle from OX and $O\tilde{X}$, respectively, are equal. Also $\angle XOP_m = \angle P_m O\tilde{X} = \theta_m$. The trajectory of the particle is completely specified by Eqs. (3.19) and is dependent on the impact parameter p, the initial velocity $\mathbf{g}$ and the interaction potential $\phi(r)$. The final velocity $\tilde{\mathbf{g}}$ resulting after the collision has the same magnitude as $\mathbf{g}$ but has a different direction. Alternatively, it can be said that the collision process results in the rotation of the velocity vector from $\mathbf{g}$ to $\tilde{\mathbf{g}}$. Thus, the parameter which is significant in assessing the effect of the collision process is *the deflection angle* χ between $\mathbf{g}$ and $\tilde{\mathbf{g}}$, as indicated in Fig. 4.3. It is important to obtain an expression for χ in terms of the impact parameter p, the initial speed g and the interaction potential $\phi(r)$. From Fig. 4.3, it is seen that

$$\chi = \pi - 2\theta_m \tag{3.20}$$

When $r = \infty$, $\theta_1 = 0$, and Eq. (3.19a) then yields the following expression for θ_m:

$$\theta_m = \int_{r_m}^{\infty} \frac{p}{r^2}\left(1 - \frac{p^2}{r^2} - \frac{\phi(r)}{(1/2)m_r g^2}\right)^{-1/2} dr \tag{3.21}$$

The value of the scattering angle χ can be evaluated with the help of Eqs. (3.20) and (3.21) for the specified values of m_r, p, g, and $\phi(r)$.

In Eq. (3.21), r_m is still to be determined; it is the distance of the vertex from the origin. For the vertex $\dot{r} = 0$ and hence Eq. (3.16) yields the following equation for obtaining r_m:

$$1 - \frac{p^2}{r_m^2} - \frac{\phi(r_m)}{(1/2)m_r g^2} = 0 \tag{3.22}$$

Since $\dot{r} = 0$, it follows from Eq. (3.5) that at the vertex, the instantaneous velocity of the particle is tangential to the trajectory.

4.4. Inverse collision

It is convenient to designate the interaction of the particles considered so far as the *direct collision*. It is proposed to give a brief discussion of the *inverse collision*. For this purpose, it is desirable to systematize the equations for obtaining the velocities of the particles before collision from the corresponding values after collision and vice versa. From Eqs. (2.14), it is clear that if the velocity $\mathbf{u}_C$ of c.m. and the velocity $\mathbf{g}$ of particle 1 relative to that of particle 2 are known, the initial velocities of the two particles are determined. It is assumed that the masses of the particles are specified. The velocity $\mathbf{u}_C$ of c.m. is a constant. Therefore, it follows from Eqs. (2.15) that for the evaluation of the final velocities, it is necessary to be able to obtain the final relative velocity $\tilde{\mathbf{g}}$ in terms of its initial value $\mathbf{g}$. Let

$$\tilde{\mathbf{g}} = \mathbf{g} - a\hat{\mathbf{k}} \tag{4.1}$$

where a is a scalar to be determined and $\hat{\mathbf{k}}$ is a unit vector in the direction of the vector obtained by subtracting the final value from the initial value of the relative velocity. Let OA and $O\tilde{A}$ in Fig. 4.4 represent $\mathbf{g}$ and $\tilde{\mathbf{g}}$ respectively. Note that $\mathbf{g}$ and $\tilde{\mathbf{g}}$ are in the directions XO and $O\tilde{X}$, respectively. In view of Eq. (4.1), it follows that $\tilde{A}A = a\hat{\mathbf{k}}$, as indicated in Fig. 4.4. Let OA_m be perpendicular to $\tilde{A}A$. It can be deduced from Eq. (3.20) and the fact that the magnitudes of $\mathbf{g}$ and $\tilde{\mathbf{g}}$ are equal that $\angle O\tilde{A}A_m = \angle OAA_m = \pi/2 - \chi/2 = \theta_m$. Since $\angle XOP_m = \angle P_m O\tilde{X} = \theta_m$, it is found that $\hat{\mathbf{k}}$ is parallel to the apse line OP_m. The relative velocity of the particle at the vertex P_m is perpendicular to the apse line and hence is parallel to OA_m.

The angles AOA_m and $A_m O\tilde{A}$ are each equal to $\chi/2$ and therefore, it is obtained from the triangle AOA_m that

$$\frac{a}{2} = g \sin\left(\frac{\chi}{2}\right) = g \cos\left(\frac{\pi}{2} - \frac{\chi}{2}\right) = g \cos \theta_m = \mathbf{g} \cdot \hat{\mathbf{k}} = -\tilde{\mathbf{g}} \cdot \hat{\mathbf{k}} \tag{4.2}$$

In view of Eq. (4.2), Eq. (4.1) can be rewritten in the following two alternative forms:

$$\tilde{\mathbf{g}} = \mathbf{g} - 2(\mathbf{g} \cdot \hat{\mathbf{k}})\hat{\mathbf{k}} \tag{4.3}$$

and

$$\mathbf{g} = \tilde{\mathbf{g}} - 2(\tilde{\mathbf{g}} \cdot \hat{\mathbf{k}})\hat{\mathbf{k}} \tag{4.4}$$

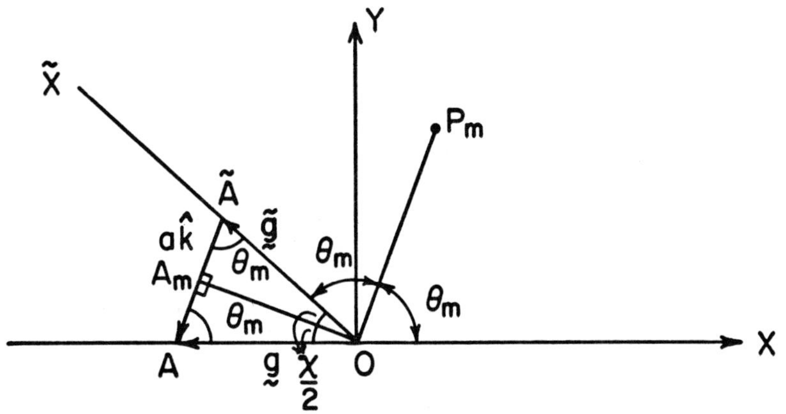

Fig. 4.4. Initial (g) and final (g̃) velocity vectors with reference to the apse line.

If the initial value $\mathbf{g}$ of the relative velocity and the angle χ of deflection in magnitude and direction are known, Eq. (4.3) enables $\tilde{\mathbf{g}}$ to be determined and then Eqs. (2.15) can be used to obtain the final velocities $\tilde{\mathbf{u}}_1$ and $\tilde{\mathbf{u}}_2$ of the two particles.

On the other hand, suppose that the final velocities $\tilde{\mathbf{u}}_1$ and $\tilde{\mathbf{u}}_2$ of the two particles are known. In view of Eqs. (2.15), it follows that $\mathbf{u}_C$ and $\tilde{\mathbf{g}}$ are specified. From the knowledge of $\tilde{\mathbf{g}}$ and the angle χ of deflection in magnitude and direction, Eq. (4.4) enables $\mathbf{g}$ to be obtained. Then, with the help of Eqs. (2.14), the initial velocities $\mathbf{u}_1$ and $\mathbf{u}_2$ of the two particles can be evaluated. Thus, Eqs. (4.3) and (4.4) are helpful in the determination of the final velocities of the particles from the knowledge of their initial values and vice versa. Note that $\hat{\mathbf{k}}$, which is parallel to the apse line, can be determined if the initial relative velocity and the deflection angle χ in magnitude and direction are known.

For the inverse collision, the same particles as in the direct collision are considered, that is, the masses of the two particles in the inverse collision are the same as those in the direct collision. Also, the impact parameter and the interaction potential for the inverse collision are the same as for the direct collision. The relations (2.14), (2.15), (4.2), and (4.3) are reproduced here and are identified by the additional superscript i to indicate that the relations pertain to the inverse collision:

$$\text{Before collision:}\quad \mathbf{u}_1^i = \mathbf{u}_C^i + M_2\mathbf{g}^i \quad \mathbf{u}_2^i = \mathbf{u}_C^i - M_1\mathbf{g}^i \tag{4.5}$$

$$\text{After collision:}\quad \tilde{\mathbf{u}}_1^i = \mathbf{u}_C^i + M_2\tilde{\mathbf{g}}^i \quad \tilde{\mathbf{u}}_2^i = \mathbf{u}_C^i - M_1\tilde{\mathbf{g}}^i \tag{4.6}$$

$$g^i \sin\frac{\chi^i}{2} = \mathbf{g}^i \cdot \hat{\mathbf{k}}^i \tag{4.7}$$

$$\tilde{\mathbf{g}}^i = \mathbf{g}^i - 2(\mathbf{g}^i \cdot \hat{\mathbf{k}}^i)\hat{\mathbf{k}}^i \tag{4.8}$$

For the inverse collision, the initial velocities of the two particles are equal to the corresponding final velocities in the direct collision:

$$\mathbf{u}_1^i = \tilde{\mathbf{u}}_1 \qquad \mathbf{u}_2^i = \tilde{\mathbf{u}}_2 \tag{4.9}$$

In view of the conservation of mass and momentum, and Eqs. (4.9), the velocity $\mathbf{u}_C^i$ of the c.m. in the inverse collision is the same as that in the direct collision:

$$\mathbf{u}_C^i = \mathbf{u}_C \tag{4.10}$$

From Eqs. (2.15), (4.5), (4.9), and (4.10), it is found that

$$\mathbf{g}^i = \tilde{\mathbf{g}} \tag{4.11}$$

For the inverse collision, the deflection angle is the same as that for the direct collision. This requirement together with Eqs. (4.2), (4.7), and (4.11) shows that

$$\hat{\mathbf{k}}^i = -\hat{\mathbf{k}} \tag{4.12}$$

In view of Eq. (4.12), it follows that the vertices of the trajectories in the direct and the inverse collisions lie on opposite sides of the center of force. The substitution of

Eqs. (4.11) and (4.12) in Eq. (4.8) and a comparison of the resulting equation with Eq. (4.4) yields

$$\tilde{\mathbf{g}}^i = \mathbf{g} \tag{4.13}$$

If Eqs. (4.10) and (4.13) are substituted in Eqs. (4.6) and the resulting equations are compared with Eqs. (2.14), it follows that

$$\tilde{\mathbf{u}}_1^i = \mathbf{u}_1 \qquad \tilde{\mathbf{u}}_2^i = \mathbf{u}_2 \tag{4.14}$$

The characteristics of the inverse collision can now be summarized. In the inverse collision, the interacting particles, the impact parameter, the initial (and the final) relative speed, the deflection angle and the interaction potential are the same as in the direct collision. The initial and the final velocities of the particles in the direct collision are interchanged to yield the corresponding values for the inverse collision. The vertex of the trajectory in the inverse collision lies on the side of the center of force opposite to that in which the vertex of the trajectory in the direct collision lies. The trajectories of the direct collision and the corresponding inverse collision are depicted in Fig. 4.5.

4.5. Scattering cross section

So far consideration has been given specifically only to the interaction between two particles. As a first step towards the understanding of the interactions involving many particles that constitute a plasma, it is useful to investigate the interaction of a parallel beam of identical particles with another particle acting as a center of force. As before, it is advantageous to study the phenomenon in the c.m. system. Since they are identical, the interaction potential is the same for all the particles in the incident beam. For simplicity, the beam of particles is assumed to be monoenergetic, that is, all the particles in the beam have the same initial velocity $\mathbf{g}$ with respect to the target particle. Only the impact parameters for the various particles in the beam are different and therefore, the corresponding deflection angles are also different. After interaction, the particles in the incident beam travel in all directions from the center of force. In other words, the incident beam of particles is scattered by the target particle which is, therefore, called the scattering center. It is desirable to introduce a parameter which describes the angular distribution of the number of particles scattered per unit time. The actual number of particles scattered per second depends on the incident particle current density Γ which is the number of particles passing normally the unit cross-sectional area per second. The particle current density Γ is assumed to be independent of the impact parameter, that is, the incident beam is assumed to be homogeneous. Moreover, since the beam is monoenergetic, it follows that the angular distribution of the particles scattered per second is characteristic of the interaction potential.

Let dN_1/dt be the number of particles scattered per second per unit solid angle

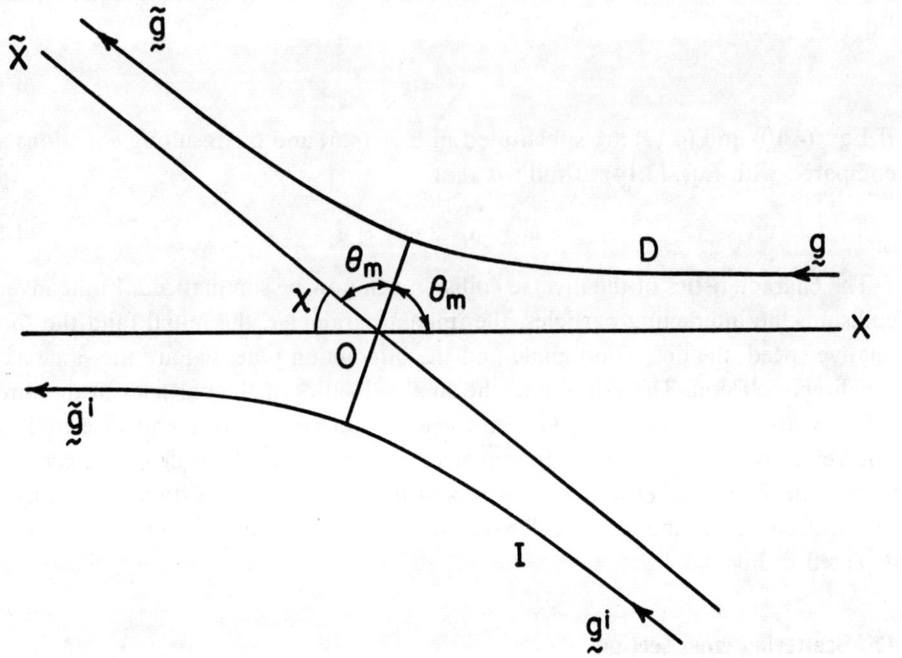

Fig. 4.5. Trajectories of the direct (*D*) and the corresponding inverse (*I*) collision in the center-of-mass system.

around a given direction. The reference direction is taken to be along the *z*-axis and is assumed to coincide with that of the incident beam. The direction of the scattered particles is specified by the spherical coordinates χ and φ, as shown in Fig. 4.6. Note that χ is the deflection or the scattering angle. Since dN_1/dt is proportional to the incident particle current density Γ, it is reasonable to set

$$dN_1/dt = \Gamma\sigma(\chi, \varphi) \tag{5.1}$$

where the proportionality constant $\sigma(\chi, \varphi)$ has the dimensions of an area. In addition, $\sigma(\chi, \varphi)$ gives the angular distribution of the scattered particles and is therefore called the *differential elastic scattering cross section.*

An expression for $\sigma(\chi, \varphi)$ can be obtained in the following manner. The number of particles in the beam lying between the azimuthal angles φ and $\varphi + d\varphi$, and having impact parameters between $p - dp$ and p, incident per second on the target particle is obtained from Fig. 4.6 to be

$$\Gamma p\, dp\, d\varphi \tag{5.2}$$

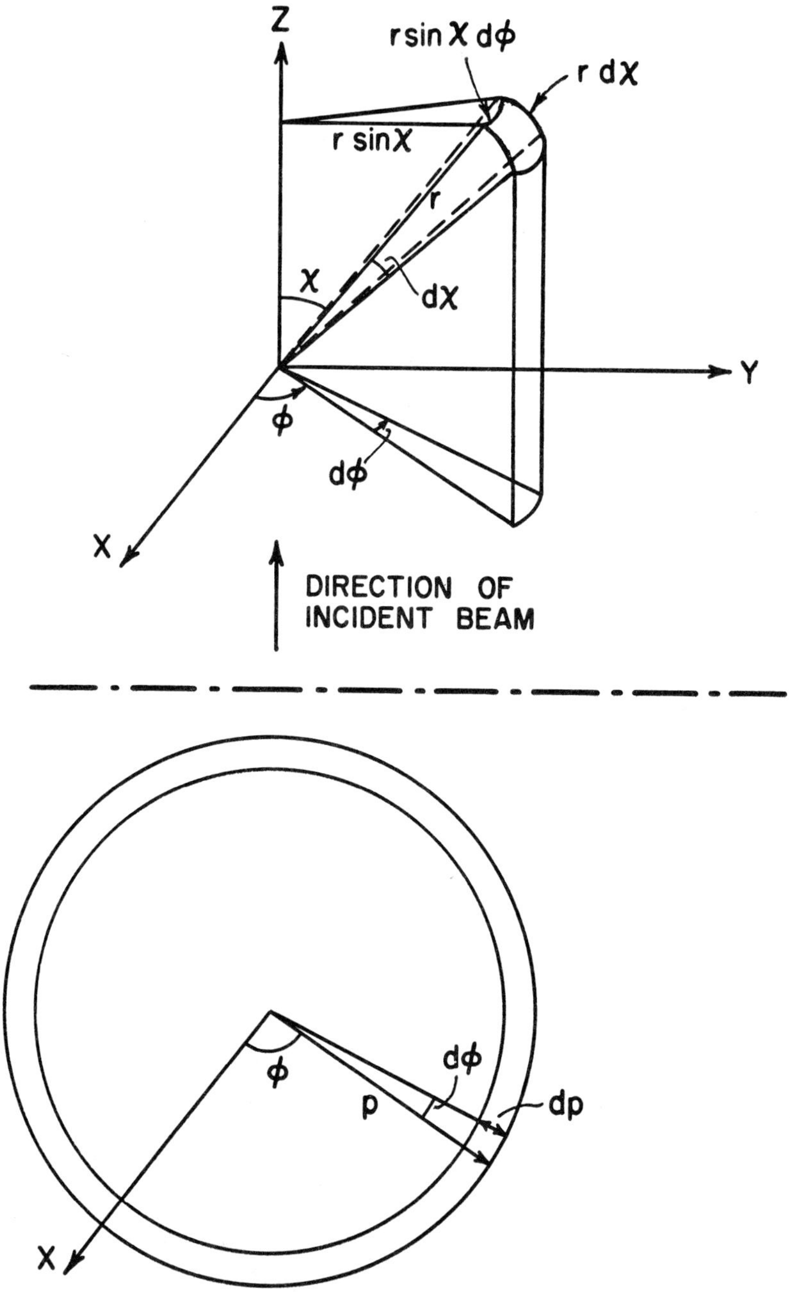

Fig. 4.6. Geometry of the scattering problem in the center-of-mass system for the evaluation of the differential scattering cross section.

The azimuth of the incident particles is not altered in the interaction process. The incident particles are scattered into the solid angle $d\Omega$ contained between φ and $\varphi + d\varphi$, and χ and $\chi + d\chi$. This elemental solid angle is evaluated with the help of Fig. 4.6 as $d\Omega = \sin\chi\, d\chi\, d\varphi$. The number of particles scattered per second into the solid angle $d\Omega$ is determined from Eq. (5.1) as

$$\Gamma\sigma(\chi, \varphi)\sin\chi\, d\chi\, d\varphi \tag{5.3}$$

Since no particles are created or annihilated in this interaction, the expressions (5.2) and (5.3) are equal and hence

$$\sigma(\chi, \varphi) = \frac{p}{\sin\chi}\left|\frac{dp}{d\chi}\right| \tag{5.4}$$

where the magnitude sign has been introduced in Eq. (5.4) to take into account the situations in which the scattering angle χ decreases as the impact parameter p increases. Note that $\sigma(\chi, \varphi)$ is always a positive quantity.

If the interaction potential is isotropic, as has been assumed, the differential scattering cross section is independent of the azimuthal angle. Therefore, hereafter the argument φ in $\sigma(\chi, \varphi)$ is omitted.

The total number of particles scattered per second in all directions from the scattering center is evaluated by integrating Eq. (5.1) throughout the entire solid angle. On noting that the elemental solid angle lying between χ and $\chi + d\chi$ is $d\Omega = 2\pi \sin\chi\, d\chi$, the following result is obtained:

$$(dN_1/dt)_{\text{total}} = \Gamma \int_0^{\pi} \sigma(\chi) 2\pi \sin\chi\, d\chi \tag{5.5}$$

The ratio of the total number of particles scattered per second to the incident particle current density is defined as the *total scattering cross section* σ_t. Then, it follows from Eq. (5.5) that

$$\sigma_t = 2\pi \int_0^{\pi} \sigma(\chi)\sin\chi\, d\chi \tag{5.6}$$

It is important to note that for a single scatterer, the incident particle current density multiplied by the total scattering cross section gives the number of particles scattered per second or equivalently the number of collisions per second.

It is now possible to obtain the average value of any function $F(\chi)$ associated with a single particle and dependent on the scattering angle. The value of this function for the particles scattered per second between χ and $\chi + d\chi$ is $F(\chi)\Gamma\sigma(\chi)2\pi \sin\chi\, d\chi$. The total value of this function for all the particles scattered per second in all directions from the scattering center is obtained by an integration with respect to χ from 0 to π. The result is

$$\int_0^{\pi} F(\chi)\Gamma\sigma(\chi)2\pi \sin\chi\, d\chi \tag{5.7}$$

The average value of $F(\chi)$ is obtained by dividing the expression (5.7) by the total number of particles scattered per second. Therefore, the average value of $F(\chi)$ is deduced from Eqs. (5.5) and (5.6) and the expression (5.7) as

$$\langle F(\chi)\rangle = \frac{1}{\sigma_t}\int_0^\pi F(\chi)\sigma(\chi)2\pi \sin\chi \, d\chi \tag{5.8}$$

It is important to note that a cross section can be defined for various processes of interaction. The total scattering cross section is associated with the number of particles scattered per second. In a similar manner, a cross section can be defined for the rate of transfer of momentum. The momentum of a particle in the beam in the direction of incidence is $m_r g$. Suppose that this particle is deflected through an angle χ; the momentum of this particle in the direction of incidence after the interaction is $m_r g \cos\chi$. Therefore, the momentum transferred by this particle to the scattering center is $m_r g(1 - \cos\chi)$. According to the expression (5.7), the total momentum transferred to the scattering center per second by all the particles scattered in all directions in space is

$$\int_0^\pi m_r g(1 - \cos\chi)\Gamma\sigma(\chi)2\pi \sin\chi \, d\chi \tag{5.9}$$

The incident momentum flux density per second is $m_r g\Gamma$. *The momentum transfer cross section* σ_m is defined as the ratio of the total momentum transferred per second to the scattering center to the incident momentum flux density per second. Hence, it is found from the expression (5.9) that

$$\sigma_m = \int_0^\pi (1 - \cos\chi)\sigma(\chi)2\pi \sin\chi \, d\chi \tag{5.10}$$

Thus, the incident momentum flux density per second when multiplied by the momentum transfer cross section yields the total momentum transferred per second to the scattering center. From Eqs. (5.8) and (5.10), it is deduced that the average value of momentum loss per particle is

$$\langle m_r g(1 - \cos\chi)\rangle = m_r g\sigma_m/\sigma_t \tag{5.11}$$

It is to be noted that as in Eq. (5.10) other transfer cross sections, such as for example the energy transfer cross section, can also be defined.

4.6. Relationship to the laboratory system

Before proceeding to study the details of collision for some important interaction potentials, it is desirable to establish the relationship of the c.m. system to the laboratory system. In experimental observation of collision cross sections, a beam of high velocity particles impinges on a target containing particles which are essentially at rest. The experimental situation, therefore, corresponds approximately to $\mathbf{u}_2 = 0$. For this special and practically important case, it is proposed to establish the relationship between the laboratory and the c.m. systems.

Scattering angle

Since $\mathbf{u}_2 = 0$, it is found from Eqs. (2.14) that

$$\mathbf{u}_1 = \mathbf{g} \qquad \mathbf{u}_C = M_1 \mathbf{g} \tag{6.1}$$

In Fig. 4.7, let $\mathbf{u}_1$ be in the direction of OX. In view of Eqs. (6.1), $\mathbf{g}$ and $\mathbf{u}_C$ are also in the direction of OX. Let OA and AB represent the vectors $M_2 \tilde{\mathbf{g}}$ and $\mathbf{u}_C$, respectively. From Eqs. (2.15), it follows that OB is equal to the final velocity $\tilde{\mathbf{u}}_1$ of particle 1. AA' and BB' in Fig. 4.7 are normal to OX. Since it is equal to the angle between the initial and the final relative velocity vectors, $\mathbf{g}$ and $\tilde{\mathbf{g}}$, the angle $\underline{/AOA'} = \chi$, the deflection angle in the c.m. system. The angle between the initial and the final velocities of particle 1 is defined to be the deflection angle in the laboratory system and is denoted by χ_0. Hence, $\underline{/BOB'} = \chi_0$. It is obtained from Fig. 4.7 that

$$OA' = M_2 g \cos\chi \qquad BB' = AA' = M_2 g \sin\chi \tag{6.2}$$

Again, with the help of Fig. 4.7 and Eqs. (6.2), it can be deduced that

$$\tan\chi_0 = \frac{BB'}{OA' + A'B'} = \frac{M_2 g \sin\chi}{M_2 g \cos\chi + M_1 g} = \frac{\sin\chi}{\cos\chi + m_1/m_2} \tag{6.3}$$

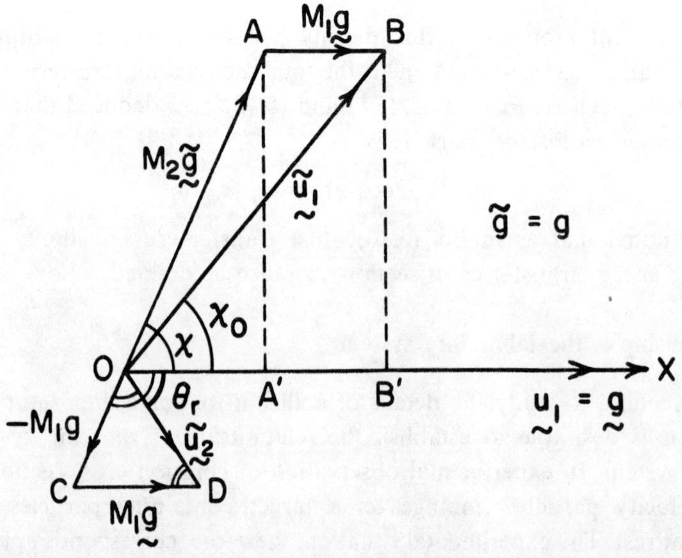

Fig. 4.7. Relationship between the laboratory and the center-of-mass systems.

since $A'B' = AB = M_1 g$. If $m_2 = \infty$, it follows from Eq. (6.3) that $\chi = \chi_0$, that is, the scattering angles in the c.m. and the laboratory systems are identical.

Recoil angle

Let the vector $-M_1 \tilde{\mathbf{g}}$ be represented by OC which is in the direction opposite to that of OA. Let CD represent $\mathbf{u}_C = M_1 \mathbf{g}$. From Eqs. (2.15), it is found that OD is equal to the final velocity $\tilde{\mathbf{u}}_2$ of particle 2. The angle $\underline{\angle A'OD}$ between the initial velocity of particle 1 and the final velocity of particle 2 is called the recoil angle and is denoted by θ_r. Since OC and CD are equal in magnitude and since OA' and CD are parallel, it is seen from Fig. 4.7 that $\underline{\angle DOC} = \underline{\angle CDO} = \underline{\angle A'OD} = \theta_r$ and that

$$\chi = \pi - 2\theta_r \tag{6.4}$$

A comparison of Eq. (6.4) with Eq. (3.20) shows that

$$\theta_r = \theta_m \tag{6.5}$$

where θ_m is defined in Eq. (3.21).

Energy lost on collision

The initial energy W_2 of particle 2 is zero. The final or the recoil energy $\tilde{W}_2$ of particle 2 is equal to its kinetic energy after collision. Hence, it can be deduced from Eqs. (2.15) and (6.1) that

$$\tilde{W}_2 = \tfrac{1}{2} m_2 M_1^2 (\mathbf{g} - \tilde{\mathbf{g}}) \cdot (\mathbf{g} - \tilde{\mathbf{g}}) = m_2 M_1^2 g^2 (1 - \cos\chi) \tag{6.6}$$

where the results $|\mathbf{g}| = |\tilde{\mathbf{g}}| = g$ and $\mathbf{g} \cdot \tilde{\mathbf{g}} = g^2 \cos\chi$ have been used. Let W_1 and $\tilde{W}_1$ denote the initial and the final energies of particle 1. From the law of energy conservation, it is obtained that

$$W_1 - \tilde{W}_1 = \tilde{W}_2 - W_2 = \tilde{W}_2 \tag{6.7}$$

The initial energy of particle 1 is equal to its kinetic energy. Therefore,

$$W_1 = \tfrac{1}{2} m_1 u_1^2 = \tfrac{1}{2} m_1 g^2 \tag{6.8}$$

The fractional loss of energy of the colliding particle is deduced from Eqs. (6.6)–(6.8) as

$$\frac{W_1 - \tilde{W}_1}{W_1} = \frac{2 m_1 m_2}{(m_1 + m_2)^2} (1 - \cos\chi) \tag{6.9}$$

Consider a special case in which the colliding particle is an electron of mass $m_1 = m_e$ and the target is a neutral particle of mass $m_2 = M_n$. Since $m_e \ll M_n$, Eq. (6.9) yields approximately that

$$\frac{W_1 - \tilde{W}_1}{W_1} = \frac{2m_e}{M_n}(1 - \cos\chi) \tag{6.10}$$

It follows from Eq. (6.10) that the fractional energy lost by an electron on collision with a stationary neutral particle such as an atom or molecule is very small and can, therefore, be neglected in a first approximation. Note that $(W_1 - \tilde{W}_1)/W_1$ is identically equal to zero only when the target particle has infinite mass.

Differential scattering cross section

Let $\sigma_0(\chi_0)$ be the differential scattering cross section in the laboratory system. Naturally, the differential scattering cross section depends on the choice of the coordinate system. It is desired to find the relationship between $\sigma(\chi)$ and $\sigma_0(\chi_0)$. Consider the particles in the incident beam with impact parameters lying between $p - dp$ and p. These particles are scattered into the solid angle $d\Omega$ lying between χ and $\chi + d\chi$ in the c.m. system but in the laboratory system, the scattering takes place into the solid angle $d\Omega_0$ lying between χ_0 and $\chi_0 + d\chi_0$. The number of particles scattered per second into the solid angle $d\Omega$ is

$$\Gamma\sigma(\chi)2\pi \sin\chi\, d\chi \tag{6.11}$$

where Γ is the magnitude of the incident particle current density. In a similar manner, the number of particles scattered per second into the solid angle $d\Omega_0$ is obtained to be

$$\Gamma\sigma_0(\chi_0)2\pi \sin\chi_0\, d\chi_0 \tag{6.12}$$

Since the numbers given by the expressions (6.11) and (6.12) are equal, it follows that

$$\sigma_0(\chi_0) = \sigma(\chi)\frac{\sin\chi\, d\chi}{\sin\chi_0\, d\chi_0} = \sigma(\chi)\frac{d(\cos\chi)}{d(\cos\chi_0)} \tag{6.13}$$

It can be shown from Eq. (6.3) that

$$\cos\chi_0 = (1 + \tan^2\chi_0)^{-1/2} = \frac{\cos\chi + m_1/m_2}{[1 + 2(m_1/m_2)\cos\chi + (m_1/m_2)^2]^{1/2}} \tag{6.14}$$

With the help of Eq. (6.14), the value of $d(\cos\chi)/d(\cos\chi_0)$ can be deduced and when this value is substituted into Eq. (6.13), the following result is obtained for the differential scattering cross section in the laboratory system:

$$\sigma_0(\chi_0) = \sigma(\chi)\frac{[1 + 2(m_1/m_2)\cos\chi + (m_1/m_2)^2]^{3/2}}{(1 + (m_1/m_2)\cos\chi)} \tag{6.15}$$

When $m_2 = \infty$, it has been established that $\chi = \chi_0$ and then $\sigma_0(\chi_0)$ should be equal to $\sigma(\chi)$. This result is directly verified to be true from Eq. (6.15).

4.7. Collision between two perfectly elastic, hard spheres.

Let the position of particle 1 represent the center of a hard sphere of radius r_1 and mass m_1. In a similar manner, the position of particle 2 represents the center of another hard sphere of radius r_2 and mass m_2. Let r be the distance between the centers of the two spheres. If $r > r_1 + r_2$, the spheres are not in contact and there is no force of interaction between them. Since the spheres are impenetrable, r can never be less than $r_1 + r_2$. In other words, for $r < r_1 + r_2$ there is infinite repulsive force of interaction between them. For the collision between these two perfectly elastic, hard spheres, the interaction potential may be assumed to be of the form:

$$\phi(r) = \infty \qquad \text{for } r < r_1 + r_2 \tag{7.1a}$$

$$= 0 \qquad \text{for } r \geqslant r_1 + r_2 \tag{7.1b}$$

It is to be noted that the interaction between two molecules of an inert gas can be described approximately in terms of the potential given in Eqs. (7.1).

In the c.m. system in which the sphere of mass m_2 is stationary, the distance r_m from the origin to the vertex of the trajectory of the center of the sphere of mass m_1 is deduced from Eqs. (3.22) and (7.1) as

$$r_m = p \qquad \text{for } r > r_1 + r_2 \tag{7.2}$$

Note that r_m does not exist for $r < r_1 + r_2$. From Eqs. (3.21), (7.1), and (7.2), it is found that

$$\theta_m = \int_p^{\infty} (p/r^2)(1 - p^2/r^2)^{-1/2}\, dr \qquad \text{for } r > r_1 + r_2 \tag{7.3}$$

By changing the variable of integration to $y = p/r$, Eq. (7.3) can be simplified to yield

$$\theta_m = \int_0^1 (1 - y^2)^{-1/2}\, dy = \sin^{-1} y|_0^1 = \pi/2 \qquad \text{for } r > r_1 + r_2 \tag{7.4}$$

For $r = r_1 + r_2$, the distance r_m of closest approach of the centers of the two spheres is clearly equal to $r_1 + r_2$. Therefore,

$$r_m = r_1 + r_2 \qquad \text{for } r = r_1 + r_2 \tag{7.5}$$

The use of Eq. (7.5) in Eq. (3.21) and the evaluation of the resulting integral as in Eq. (7.4) gives

$$\theta_m = \sin^{-1}[p/(r_1 + r_2)] \qquad \text{for } r = r_1 + r_2 \tag{7.6}$$

In view of Eq. (3.20), $\theta_m = \pi/2 - \chi/2$ and therefore, it follows from Eqs. (7.4) and (7.6) that

$$\chi = 0 \qquad \text{for } r > r_1 + r_2 \tag{7.7a}$$

$$\cos(\chi/2) = p/(r_1 + r_2) \qquad \text{for } r = r_1 + r_2 \tag{7.7b}$$

When $r > r_1 + r_2$, the spheres never come into contact and therefore the incident sphere proceeds along the initial path without any deflection, in conformity with the result obtained in Eq. (7.7a). If $r = r_1 + r_2$, the spheres come into contact, as depicted in Fig. 4.8 and there is deflection. The validity of the expression for the deflection angle as given by Eq. (7.7b) can be confirmed from Fig. 4.8.

From Eqs. (7.2) and (7.7a), it is seen that for $p > r_1 + r_2$, $\chi = 0$ and there is no deflection. For $p = r_1 + r_2$, it is obtained from Eq. (7.7b) that $\chi = 0$ and again, there is no deflection. If $0 \leqslant p < r_1 + r_2$, Eq. (7.7b) shows that $0 < \chi \leqslant \pi$ and there is deflection. In other words, there is no deflection of the incident sphere for $p > r_1 + r_2$. The truth of this result follows directly from simple geometrical considerations.

It is found from Eq. (7.7b) that

$$p\left|\frac{dp}{d\chi}\right| = \frac{(r_1 + r_2)^2}{2} \cos\left(\frac{\chi}{2}\right)\sin\left(\frac{\chi}{2}\right) \qquad \text{for } 0 \leqslant \chi \leqslant \pi \tag{7.8}$$

The substitution of Eq. (7.8) into Eq. (5.4) shows that the differential scattering cross section is given by

$$\sigma(\chi) = (r_1 + r_2)^2/4 \qquad \text{for } 0 \leqslant \chi \leqslant \pi \tag{7.9}$$

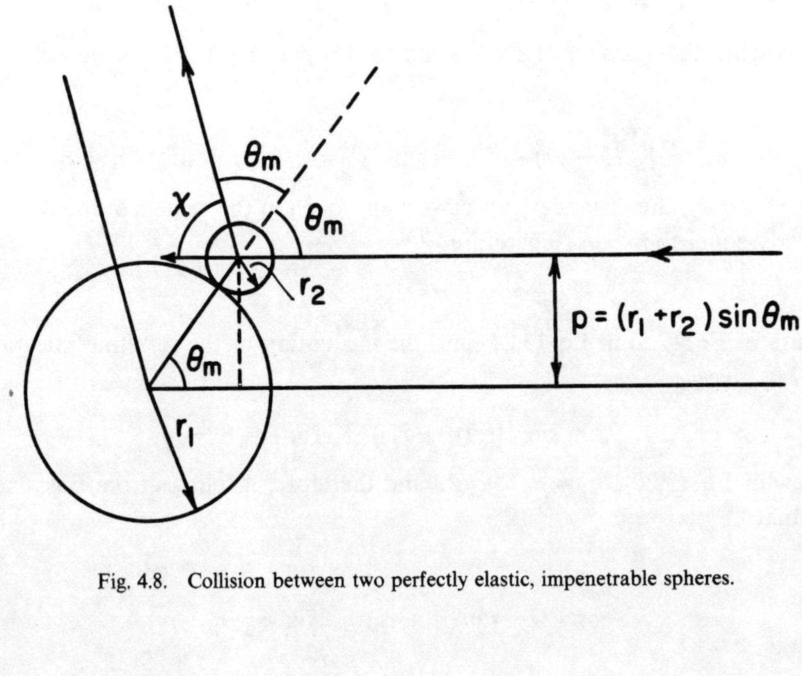

Fig. 4.8. Collision between two perfectly elastic, impenetrable spheres.

It is ascertained from Eq. (7.9) that in the c.m. system, the scattering of a parallel beam of perfectly elastic, hard spheres is *isotropic*.

The total scattering cross section is evaluated from Eqs. (5.6) and (7.9) to be given by

$$\sigma_t = 2\pi \int_0^{\pi} [(r_1 + r_2)^2/4] \sin \chi \, d\chi = \pi(r_1 + r_2)^2 \tag{7.10}$$

It is instructive to deduce the result in Eq. (7.10) by an alternative method and from geometrical considerations. The substitution of Eq. (5.4) into Eq. (5.6) yields

$$\sigma_t = 2\pi \int p dp \tag{7.11}$$

The limits $\chi = 0$ and $\chi = \pi$ transform to $p = r_1 + r_2$ and $p = 0$, respectively. When the integration in Eq. (7.11) is carried out within these limits, it is found that

$$\sigma_t = \pi(r_1 + r_2)^2 \tag{7.12}$$

which is identical to that in Eq. (7.10). It is to be noted that for the impact parameter, there is a cut-off value beyond which collisions do not occur. As is made clear subsequently, it is the existence of the cut-off value for p that leads to a finite value for the total scattering cross section σ_t. A simple geometrical interpretation can be given to Eq. (7.12). If the centers of the incident spheres pass within a distance of $r_1 + r_2$ from the center of the target sphere, the incident spheres come into contact with the target sphere and are therefore scattered. Thus, the current of incident spheres intercepted normally by the area of a circle of radius $(r_1 + r_2)$ are alone scattered. Hence, $\Gamma\pi(r_1 + r_2)^2$ is the number of incident spheres scattered per second and therefore, for σ_t, the expression given in Eq. (7.12) is reproduced.

With the help of Eqs. (5.10) and (7.9), the momentum transfer cross section is evaluated to be given by

$$\sigma_m = \int_0^{\pi} (1 - \cos \chi) \frac{(r_1 + r_2)^2}{4} 2\pi \sin \chi \, d\chi = \pi(r_1 + r_2)^2 \tag{7.13}$$

The average value of momentum loss per particle is found from Eqs. (5.11), (7.12), and (7.13) to be

$$\langle m_r g(1 - \cos \chi) \rangle = m_r g \tag{7.14}$$

Consider a weakly ionized plasma consisting of electrons, ions, and neutral particles. Assume that the motions of the electrons alone are relevant for the phenomenon of interest and those of the ions can be ignored. Since the number density of the neutral particles is a very large fraction of the total number density, so far as electron motions are concerned, only their collisions with the neutral particles are important. As has been mentioned previously, for electron collisions with neutral particles, the mass of the neutral particle can be taken to be infinite in the first approximation. Consequently, the scatterings in the c.m. and the laboratory systems become identical. Also, the reduced mass m_r becomes equal to the electronic mass m_e. According to Eq. (7.14), in the first approximation, it is seen that the entire momentum of an electron is lost in a collision with a neutral particle. If ν is the number of electron collisions with neutral particles per second, that is, if ν is the collision frequency, the rate of loss of momentum of an electron is $\nu m_e \mathbf{u}$ where $\mathbf{u}$ is the electron velocity. Therefore, the Lorentz force equation for an electron in a weakly ionized plasma becomes

$$m_e d\mathbf{u}/dt = -e[\mathbf{E} + \mathbf{u} \times \mathbf{B}] - \nu m_e \mathbf{u} \tag{7.15}$$

where $-e$ is the electronic charge, $\mathbf{E}$ is the electric field, and $\mathbf{B}$ is the magnetic flux density. A weakly ionized plasma wherein the electron motion is governed by Eq. (7.15) is called *a Lorentz gas*. If $\mathbf{v}$ is the average electron velocity, it can be deduced from Eq. (7.15) that

$$m_e d\mathbf{v}/dt = -e[\mathbf{E} + \mathbf{v} \times \mathbf{B}] - \nu m_e \mathbf{v} \tag{7.16}$$

which is called *the Langevin equation*. However, in general, an electron does not lose its entire momentum on collision with a neutral particle. Also, the interaction of an electron with a neutral particle cannot be represented exactly by the model of collision between two perfectly elastic, impenetrable spheres. As a consequence, the term $\nu m_e \mathbf{v}$ representing the average rate of loss of momentum of an electron has to be replaced by the term $\nu_{en} m_e \mathbf{v}$ where ν_{en} is *the effective collision frequency* for momentum transfer between electrons and neutral particles.

4.8. Scattering by Coulomb potential

The nature of interaction between two charged particles is important for the understanding of plasma behavior. Suppose that the interaction between two point charges of mass m_1 and m_2, and, charges q_1 and q_2, is governed by the well-known Coulomb potential:

$$\phi(r) = \frac{q_1 q_2}{4\pi\varepsilon_0 r} \tag{8.1}$$

The medium is assumed to be free space whose permittivity is denoted by ε_0. For convenience, let

$$\frac{q_1 q_2}{4\pi\varepsilon_0 m_r g^2} = p_0 \tag{8.2}$$

From Eqs. (3.22), (8.1), and (8.2), it is found that the distance r_m between the origin and the vertex of the trajectory in the c.m. system is specified by

$$p^2/r_m^2 + 2p_0/r_m - 1 = 0 \tag{8.3}$$

The positive solution of $1/r_m$ is evaluated from Eq. (8.3) as

$$1/r_m = (-p_0 + \sqrt{p_0^2 + p^2})/p^2 \tag{8.4}$$

For simplicity, the scattering phenomenon is treated in the c.m. system. Together with Eq. (8.2), Eqs. (3.19) yield

$$|\theta_m - \theta| = \int_{r_m}^{r} \frac{p}{r^2}\left(1 - \frac{p^2}{r^2} - \frac{2p_0}{r}\right)^{-1/2} dr \tag{8.5}$$

When the variable of integration in Eq. (8.5) is changed to

$$y = (p^2/r + p_0)/\sqrt{p^2 + p_0^2} \tag{8.6}$$

it can be shown with the help of Eq. (8.4) that

$$\begin{aligned} |\theta_m - \theta| &= \int_y^1 (1 - y^2)^{-1/2}\, dy = \sin^{-1} y|_y^1 \\ &= \pi/2 - \sin^{-1}[(p^2/r + p_0)/\sqrt{p^2 + p_0^2}] \end{aligned} \tag{8.7}$$

Therefore,

$$\cos(\theta_m - \theta) = (p^2/r + p_0)/\sqrt{p^2 + p_0^2} \tag{8.8}$$

which can be verified to represent an hyperbola. Thus, in the c.m. system, the trajectory of a point charge acted upon by a Coulomb potential is an hyperbola.

For $r = \infty$, it is seen from Fig. 4.3 that $\theta = 0$ for $\theta < \theta_m$ and $\theta = 2\theta_m$ for $\theta > \theta_m$. In either case, by setting $r = \infty$ in Eq. (8.8) and noting that $\theta_m = \pi/2 - \chi/2$, it is obtained that

$$\cos\theta_m = \sin(\chi/2) = p_0/\sqrt{p^2 + p_0^2} \tag{8.9}$$

It is then a simple matter to show that

$$\tan(\chi/2) = p_0/p \tag{8.10}$$

Several remarks are in order in connection with Eq. (8.10). If $p = p_0$, $\chi = \pi/2$; therefore, p_0 is the value of the impact parameter for which the angle of deflection is 90°. It is noted from Eq. (8.10) that $\chi = \pi$ for $p = 0$, χ decreases as p increases and $\chi = 0$ only for $p = \infty$. There is scattering for all values of the impact parameter which, therefore, has no cut-off value.

The differentiation of Eq. (8.10) with respect to χ yields

$$dp/d\chi = -p_0/2 \sin^2(\chi/2) \tag{8.11}$$

When the values of p and $dp/d\chi$ from Eqs. (8.10) and (8.11), respectively, are substituted into Eq. (5.4), the following expression for the differential scattering cross section is obtained:

$$\sigma(\chi) = p_0^2/4 \sin^4(\chi/2) \tag{8.12}$$

Note that Eq. (8.12) is the well-known Rutherford scattering formula. It is seen from Eq. (8.12) that the scattering is not isotropic. The differential scattering cross section given by Eq. (8.12) is equal to $p_0^2/4$ for the deflection angle $\chi = \pi$, increases monotonically as χ is decreased, becomes very large for very small deflection angles and becomes infinite for $\chi = 0$.

The differential scattering cross section increases so rapidly to infinity as χ goes to zero that both σ_t and σ_m become infinite. The total scattering cross section is obtained from Eqs. (5.6) and (8.12) as

$$\sigma_t = \frac{\pi p_0^2}{2} \int_{\chi_{min}}^{\pi} \frac{\sin \chi \, d\chi}{\sin^4(\chi/2)} = 2\pi p_0^2 \int_{\chi_{min}}^{\pi} \frac{d[\sin(\chi/2)]}{\sin^3(\chi/2)} \tag{8.13}$$

where the lower limit $\chi_{min} = 0$. The reason for writing the lower limit implicitly as χ_{min} is clarified presently. The integral in Eq. (8.13) can be evaluated to yield

$$\sigma_t = \pi p_0^2[\sin^{-2}(\chi_{min}/2) - 1] \tag{8.14}$$

When $\chi_{min} = 0$ is substituted into Eq. (8.14), it is found that $\sigma_t = \infty$. It follows that the particles with very small deflections are responsible for σ_t becoming infinite.

The substitution of Eq. (8.12) into Eq. (5.10) gives the following expression for the momentum transfer cross section:

$$\sigma_m = \frac{\pi p_0^2}{2} \int_{\chi_{min}}^{\pi} \frac{(1 - \cos \chi)\sin \chi \, d\chi}{\sin^4(\chi/2)} = 4\pi p_0^2 \int_{\chi_{min}}^{\pi} \frac{d[\sin(\chi/2)]}{\sin(\chi/2)} \tag{8.15}$$

The evaluation of the integral in Eq. (8.15) is found to yield

$$\sigma_m = 4\pi p_0^2 \ln [1/\sin(\chi_{min}/2)] \tag{8.16}$$

If χ_{min} is set equal to zero in Eq. (8.16), it is seen that $\sigma_m = \infty$. As in the case of σ_t, it is clear that the particles with very small deflections contribute to make σ_m infinite. A comparison of Eqs. (8.14) and (8.16) shows that σ_t tends to infinity more rapidly than σ_m as χ_{min} goes to zero.

In summary, the Coulomb potential gives rise to infinite values for σ_t and σ_m. In order to obtain finite and meaningful values for σ_t and σ_m, it is necessary to modify the basis of treatment of interaction between charged particles.

4.9. Effect of screening

An alternative expression for the total scattering cross section is that given by Eq. (7.11). For the Coulomb potential, the limits of the integral in Eq. (7.11) have to be obtained from Eq. (8.10). For $\chi = \pi$, Eq. (8.10) shows that $p = 0$. Small values of χ correspond to large values of p. Let the large value of p corresponding to $\chi = \chi_{\min}$ be denoted by $p_{\max}$. If $\chi_{\min} = 0$, it is found from Eq. (8.10) that $p_{\max} = \infty$. Therefore, for the Coulomb potential, σ_t in Eq. (7.11) can be written as

$$\sigma_t = 2\pi \int_0^{p_{\max}} p\,dp = \pi p_{\max}^2 \tag{9.1}$$

which goes to infinity for $p_{\max} = \infty$. For the Coulomb potential, Eq. (9.1) shows that σ_t is equal to the area of an infinite plane normal to the beam of incident particles and is therefore infinite. Thus, the infinite value for σ_t can be interpreted as due to the absence of a cut-off value for the impact parameter.

One method of obtaining a finite value for σ_t is to introduce, on some plausible grounds, a cut-off value $p = p_c$ for the impact parameter. With this cut-off, σ_t can be shown from Eq. (9.1) to be given by

$$\sigma_t = \pi p_c^2 \tag{9.2}$$

For $0 < p < p_0$, Eq. (8.10) yields $\pi/2 < \chi < \pi$. It is convenient to designate the deflections corresponding to $\pi/2 < \chi < \pi$ as large-angle deflections. If only the large-angle deflections are taken into account, σ_t becomes

$$\sigma_{t,\text{large}} = \pi p_0^2 \tag{9.3}$$

If the charged particle under consideration is in a plasma it is surrounded by a sphere within which there is a preponderance of particles carrying charges of opposite sign. The scale length of the radius of the sphere is called the Debye length λ_D and the sphere itself is referred to as the Debye sphere. The particles within the Debye sphere carrying charges of opposite sign shield the Coulomb potential due to the charge under consideration and thus significantly reduce its effect on the particles lying external to the Debye sphere. In general, $\lambda_D \gg p_0$. Therefore, for $p_0 < p < \lambda_D$, $\chi < \pi/2$ and the corresponding deflections are designated as small-angle deflections. The contribution to the total scattering cross section from the small-angle deflections is deduced from Eq. (7.11) to be given by

$$\sigma_{t,\text{small}} = 2\pi \int_{p_0}^{\lambda_D} p\,dp = \pi(\lambda_D^2 - p_0^2) \tag{9.4}$$

It follows from Eqs. (9.3) and (9.4) that

$$\sigma_{t,\text{small}}/\sigma_{t,\text{large}} = (\lambda_D/p_0)^2 - 1 \approx (\lambda_D/p_0)^2 \tag{9.5}$$

since $\lambda_D \gg p_0$. It can be argued with the help of Eq. (9.5) that the small number of particles interacting strongly with the target particle and producing large-angle deflections are not nearly as important as the large number of particles interacting mildly with the target particle and therefore, producing only small-angle deflections. For the charged particle in a plasma, it is therefore more legitimate to introduce the cut-off at $p_c = \lambda_D$ than at $p_c = p_0$. If the impact parameter is cut-off at a value equal to the Debye length, Eq. (9.2) yields the following value for σ_t:

$$\sigma_t = \pi\lambda_D^2 \tag{9.6}$$

It is instructive to consider the physical implication of the assumption leading to Eq. (9.6). For the charged particles lying outside the Debye sphere, the shielding of the target particle is assumed to be complete and therefore, there is no interaction with the target particle. But, for the charged particles lying within the Debye sphere, the shielding effect due to the neighboring particles is completely omitted and hence, there is Coulomb type interaction with the target particle.

Let $\chi = \chi_c$ for $p = p_c$. Therefore, Eq. (8.9) gives

$$\sin(\chi_c/2) = [1 + (p_c/p_0)^2]^{-1/2} \tag{9.7}$$

With the help of a procedure similar to that used in deducing Eq. (8.16), and Eq. (9.7), it can be shown that if the impact parameter is cut-off at $p = p_c$, the following expression is obtained for the momentum transfer cross section:

$$\sigma_m = 2\pi p_0^2 \ln [1 + (p_c/p_0)^2] \tag{9.8}$$

As for the case of σ_t, so for the case of σ_m also, it can be proved with the help of Eq. (9.8) that the large number of particles producing small-angle deflections are more important than the small number of particles that produce large-angle deflections. Therefore, for the charged particle in a plasma, it is valid to introduce the cut-off at $p_c = \lambda_D$ and thus include the effect of the small-angle deflections. For this cut-off Eq. (9.8) reduces to

$$\sigma_m = 2\pi p_0^2 \ln [1 + \Lambda^2] \approx 4\pi p_0^2 \ln \Lambda \tag{9.9}$$

where

$$\Lambda = \lambda_D/p_0. \tag{9.10}$$

Note the fact that $\Lambda \gg 1$ has been used in Eq. (9.9).

Since $\Lambda \gg 1$, $\ln \Lambda$ varies very slowly with the variation of the parameters on which Λ depends. Hence, it is usual to make further approximations in the determination of Λ. For this purpose, consider the interaction of an electron gas with a singly charged ion gas. Then $q_2 = -q_1 = e$ where e is the magnitude of the electronic charge. If these charged particles constitute a plasma, the number density of the electrons and the ions are equal; let each be equal to N_0. The two gases are assumed to have no drift velocity. Let the electron and the ion temperatures be each equal to T. Also, the electron and the ion velocities are each assumed to have a Maxwellian distribution. When the square of the relative velocity g is averaged over the velocity distributions of the electrons and the ions, the following result can be shown to be true:

$$\langle g^2 \rangle = 3KT/m_r \tag{9.11}$$

where K is the Boltzmann constant and m_r is the reduced mass corresponding to the masses of an electron and an ion. From the Debye-Hückel theory, it has been established that

$$\lambda_D^2 = \varepsilon_0 KT/N_0 e^2 \tag{9.12}$$

For the value of p_0 in Eq. (9.10), g^2 in Eq. (8.2) is replaced by $\langle g^2 \rangle$. Then, from Eqs. (8.2) and (9.10)–(9.12), it follows that

$$\Lambda = \frac{\lambda_D 4\pi\varepsilon_0 m_r \langle g^2 \rangle}{e^2} = 9\left(\frac{4\pi}{3}\lambda_D^3\right)N_0 = 9N_D \tag{9.13}$$

where N_D is the number of electrons in a Debye sphere. The momentum transfer cross section as given by Eqs. (9.9) and (9.13) is very commonly used in the investigation of plasma behavior. For typical plasmas, $\ln \Lambda$ is of the order of 10. It should be noted that in view of its usefulness in the determination of the transport properties such as diffusion and mobility, the momentum transfer cross section plays a more important role than the total scattering cross section.

From the Debye-Hückel theory, it is known that if the screening effect is taken into account, the interaction potential between two charged particles is of the form

$$\phi(r) = \frac{q_1 q_2}{4\pi\varepsilon_0 r} \exp(-r/\lambda_D) \tag{9.14}$$

If $r \ll \lambda_D$, the Debye potential as given by Eq. (9.14) is very nearly equal to the Coulomb potential as given by Eq. (8.1) but if $r \gg \lambda_D$, the Debye potential is very nearly equal to zero. In deducing the cross sections as given by Eqs. (9.6) and (9.9), it has been assumed that the interaction potential is equal to the Coulomb potential for $r < \lambda_D$ but is equal to zero for $r > \lambda_D$. Such an assumption is a crude approximation to the Debye potential. In spite of the crudeness of the approximation resorted to here for taking into account the screening effects, the results deduced are in good agreement with those evaluated numerically using the Debye potential.*

*E. Everhart, G. Stone, and R.J. Carbone, Classical calculation of differential cross section for scattering from a Coulomb potential with exponential screening, *Phys. Rev.*, **99**, 1287(1955).

4.10. Mean free path and collision frequency

The collision between two particles was investigated first and was then extended to apply to the case of a parallel, homogeneous, and monoenergetic beam of identical particles interacting with a single scattering center. In any practical situation, a beam of particles is incident on a target which, even if small, contains a large number of target particles. Therefore, it is desired to study the phenomenon of interaction of a beam of particles incident on an assembly of scattering centers constituting a target.

If the target particles constitute a periodic structure, the times of arrival of the particles scattered in a given direction can have a definite relationship. This type of scattering is known as *coherent scattering* and occurs, for example, when the target is a crystal which has a periodic arrangement of atoms. In a plasma, the particles are randomly distributed with the result that the times of arrival of the particles scattered in a given direction are also randomly distributed. The latter type of scattering is known as *incoherent scattering*. For incoherent scattering, it is legitimate to assume that the number of particles scattered per second from the incident beam is equal to that due to *one* target particle multiplied by the number of target particles. Moreover, it is assumed that each target particle scatters independently of the others. Further, the scattered particles are assumed to be removed completely from the incident beam with the result that the particle current density diminishes as it penetrates into the target.

Let the target occupy the region $x > 0$, as shown in Fig. 4.9. The number density of the particles in the target is assumed to be uniform and is denoted by N_2. Consider a monoenergetic beam of identical particles traveling in the x-direction and impinging on the target. As before, the incident particle current density $\Gamma(x)$ in the x-direction is assumed to be uniform. Therefore, it is sufficient to consider unit cross-sectional area of the incident beam and the target. Consider two infinitesimally close planes, x and $x + dx$, in the target. The incident particle currents across unit cross-sectional areas at these planes are given by $\Gamma(x)$ and $\Gamma(x + dx)$. The difference $\Gamma(x) - \Gamma(x + dx)$ is equal to the number of particles in the incident beam that are scattered per second by the target particles lying between the planes x and $x + dx$, and within unit cross-sectional area normal to the beam. The number of particles scattered per second, or equivalently, the number of collisions per second is equal to $\Gamma(x)\sigma_t$ per target particle where σ_t is the total scattering cross section. Since there are $N_2\, dx$ target particles, the total number of collisions per second taking place within the specified volume is $\Gamma(x)\sigma_t N_2\, dx$. Hence

$$\Gamma(x) - \Gamma(x + dx) = \Gamma(x)\sigma_t N_2\, dx \tag{10.1}$$

On dividing Eq. (10.1) by dx and taking the limit as dx goes to zero, it is found that

$$-d\Gamma(x)/dx = \Gamma(x)\sigma_t N_2 \tag{10.2}$$

whose solution is given by

$$\Gamma(x) = \Gamma(0)e^{-\sigma_t N_2 x} \tag{10.3}$$

From Eq. (10.3), it is seen that the incident particle current density decays exponentially as it traverses the target. It is to be shown that the attenuation factor $\sigma_t N_2$ is related to the mean distance traveled by the incident particles between collisions.

Consider a unit cross-sectional area as shown in Fig. 4.10. Each of the small shaded circles has an area equal to σ_t, the total scattering cross section. The number of shaded circles is equal to $N_2\, dx$, the number of target particles within the specified volume. The probability of a single incident particle suffering a collision within the distance dx is equal to that of its interception by the shaded regions in Fig. 4.10. This probability is equal to $\sigma_t N_2\, dx$, the ratio of the sum of the areas of the shaded regions to the total area. This definition is valid only if the probability calculated in this manner is much less than unity. If the sum of the areas of the shaded regions is only a small fraction of the total area, it follows that, in general, the shaded circular regions do not overlap. Consequently, the incident particle is intercepted at most by only one of the shaded circles. In other words, the incident particle interacts with only one target particle. Since $\sigma_t N_2\, dx \ll 1$, the average distance between the centers of the shaded circles is clearly much larger than the radius of the circles or, equivalently, the interparticle distance is much larger than the size of the particles in the target. Suppose that $\sigma_t N_2\, dx$ is nearly equal to or greater than unity; then some of the shaded circular areas overlap. In that case, it is possible for an incident particle to be intercepted by two or more of the shaded

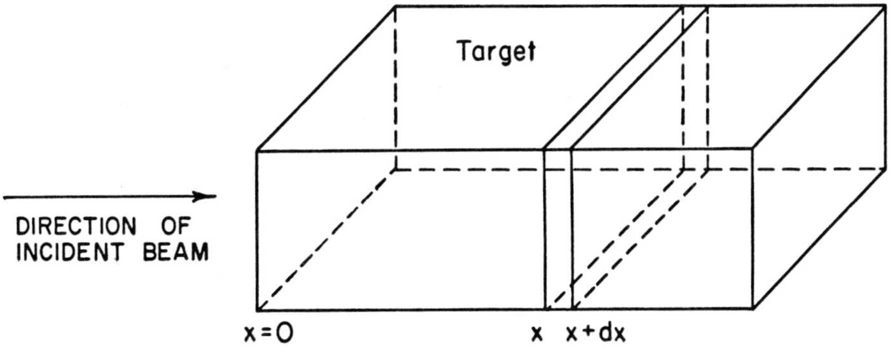

Fig. 4.9. Semiinfinite homogeneous target of unit cross-sectional area with a homogeneous monoenergetic beam of particles incident normally on the surface of the target.

circles resulting in the interaction of the incident particle with more than one target particle. In order for the definition of collision probability to hold good, it is assumed that there are no multiple interactions and that the interparticle distance is much greater than the dimensions of the particles in the target.

Let $P(x)$ be the probability of a particle traveling a distance x without suffering a collision. Thus, x is the *free length* traveled between collisions. The probability $P(x + dx)$ of a particle traveling a distance $x + dx$ without suffering a collision is equal to the probability $P(x)$ of its traveling a distance x without suffering a collision multiplied by the probability of this particle suffering no collision in the length dx. Since $\sigma_t N_2\, dx$ is the probability of a particle suffering a collision in the interval dx, $(1 - \sigma_t N_2\, dx)$ is the probability of its suffering no collision between x and $x + dx$. Therefore, it follows that

$$P(x + dx) = P(x)[1 - \sigma_t N_2\, dx] \tag{10.4}$$

which on rearrangement becomes

$$P(x) - P(x + dx) = P(x)\sigma_t N_2\, dx \tag{10.5}$$

In view of Eqs. (10.1) and (10.3), it is obtained from Eq. (10.5) that

$$P(x) = P(0)e^{-\sigma_t N_2 x} = e^{-\sigma_t N_2 x} = \Gamma(x)/\Gamma(0) \tag{10.6}$$

since the probability $P(0)$ of a particle traveling a zero distance without collision is equal to unity. Thus, x appearing in Eq. (10.3) is seen to be the free length between

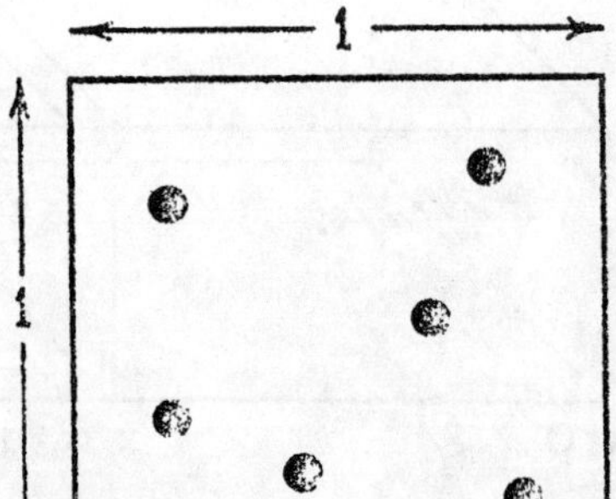

Fig. 4.10. Geometrical determination of the collision probability. Area of each shaded circle is equal to σ_t. Number of shaded circles is equal to $N_2\, dx$.

collisions. The average distance traveled by an incident particle between collisions or the mean free path λ_1 can be deduced in the following manner. The number of particles with a free length x is equal to $\Gamma(x)$. Therefore, $\int_0^\infty x\Gamma(x)\,dx$ gives the total distance traveled by all the incident particles between collisions. Also, $\int_0^\infty \Gamma(x)\,dx$ is the total number of particles having all possible values for their free lengths. Hence, the mean free path λ_1 is defined by

$$\lambda_1 = \frac{\int_0^\infty x\Gamma(x)\,dx}{\int_0^\infty \Gamma(x)\,dx} = \frac{1}{\sigma_t N_2} \tag{10.7}$$

With the help of Eq. (10.3), the integrals in Eq. (10.7) are evaluated to yield for λ_1 the value quoted in Eq. (10.7).

The distance traveled by an incident particle in unit time is equal to g. Since λ_1 is the average distance between collisions, the average number of collisions per second or the collision frequency ν_1 is found from Eq. (10.7) to be

$$\nu_1 = g/\lambda_1 = \sigma_t N_2 g \tag{10.8}$$

It is emphasized that λ_1 and ν_1 are defined with reference to the particle current density and therefore the total scattering cross section σ_t appears in the defining relations. In a similar manner, λ_1 and ν_1 can be defined in terms of the momentum flux density per second. Then, σ_t has to be replaced by the momentum transfer cross section σ_m. The momentum transfer cross section enters into the description of the transport properties such as mobility and diffusion. Hence, with reference to mobility and diffusion, λ_1 and ν_1 defined in terms of σ_m have to be used.

References

4.1. J. L. Delcroix, *Plasma Physics*, Chapters 4 and 5, John Wiley, New York, 1965.
4.2. R. Jancel and Th. Kahan, *Electrodynamics of Plasmas*, Chapter 3, John Wiley, New York, 1966.
4.3. S. Chapman and T. G. Cowling, *The Mathematical Theory of Nonuniform Gases*, Cambridge University Press, London, 1960.

Problems

4.1. Consider particle 1 of mass m colliding with particle 2 of equal mass. Assume that particle 2 is at rest before collision. Express the deflection and the recoil angles in the laboratory system in terms of the deflection angle in the c.m. system and hence show that after collision the two particles travel at right angles to each other.

Find the explicit relationship between the differential scattering cross sections in the c.m. and the laboratory systems.

4.2. Consider particle 1 of mass m_1, initial velocity $\mathbf{u}_1$ and final velocity $\tilde{\mathbf{u}}_1$. Let the corresponding quantities for particle 2 be m_2, $\mathbf{u}_2$, and $\tilde{\mathbf{u}}_2$, respectively. Find the fractional loss of energy of particle 1 due to collision. If the result is specialized for the case of particle 2 at rest before collision, verify that the result given by Eq. (6.9) is reproduced.

4.3. Let the interaction between two charged particles be governed by the Coulomb potential. Show that for head-on collision, the minimum distance of approach of the particle to the center of force is equal to twice the value of the impact parameter for which the deflection angle is equal to 90°. Express this minimum distance in terms of the impact parameter for which the deflection angle is equal to 45°.

4.4. Let the interaction between two charged particles be governed by the so-called *rectangular-well* potential, as defined by

$$\phi(r) = 0 \qquad \text{for } r > a$$

$$\phi(r) = -\phi_0 \qquad \text{for } r \leqslant a$$

Show that the differential scattering cross section is

$$\sigma(\chi) = \frac{n^2 a^2 \{n \cos(\chi/2) - 1\}\{n - \cos(\chi/2)\}}{4 \cos(\chi/2)\{1 - 2n \cos(\chi/2) + n^2\}^2}$$

where

$$n = \{1 + 2\phi_0 / m_r g^2\}^{1/2}$$

Note that m_r is the reduced mass and g is the relative speed before and after collision.

Evaluate the total scattering cross section.

4.5. The central force acting on a particle is attractive and is specified by

$$F_r = -K_{12} r^{-s}$$

where the force constant K_{12} is positive and s is a constant number. Show that

$$v = p/r \qquad v_0 = p(m_r g^2 / K_{12})^{1/(s-1)}$$

are dimensionless parameters. Note that p, m_r, and g are, respectively, the impact parameter, the reduced mass, and the relative speed before and after collision. Express the equation which specifies the distance from the center of force to the vertex of the trajectory in terms of the dimensionless parameters v and v_0. Show that for $s = 2$ there is always a vertex and for $s \geqslant 3$, a vertex exists only if the impact parameter exceeds a critical value. Find an expression for this critical value.

Examine the nature of the trajectory for the case in which the impact parameter is less than the critical value.

4.6. For the central force $F_r = K_{12} r^{-s}$ with $K_{12} > 0$ and $s > 1$, there is always a vertex. Deduce the deflection angle in terms of dimensionless parameters and hence show that the total scattering cross section has the dimensions of $(K_{12}/m_r g^2)^{2/(s-1)}$. Find the form of the force law for which the collision frequency is independent of the relative speed of the particles. Note that this force law is satisfied by the so-called *Maxwellian particles.*

4.7. Find the fraction of the total number of molecules whose free lengths are less than the mean free path.

4.8. Let $P(t)$ be the probability of a particle traveling for a time t without suffering a collision. Show that $P(t) = e^{-\nu_1 t} = e^{-t/\tau_1}$ where ν_1 is the collision frequency and τ_1 is the relaxation time.

Verify that for collisions among like particles of charge $-e$, the relaxation time is given by

$$\tau_1 = 4\pi\varepsilon_0^2 m_r^2 g^3 / N e^4 \ln\Lambda$$

where N is the number density of electrons, m_r the reduced mass, g the relative speed, ε_0 the free space permittivity and Λ is as defined in Eq. (9.10).

4.9. A number of different processes may occur at a collision. A cross section may be defined for each of the processes. Suppose that there are two different processes giving rise to the mean free paths λ_a and λ_b, respectively. Show that the resultant mean free path λ is given by

$$1/\lambda = 1/\lambda_a + 1/\lambda_b$$

4.10. Consider particle 1 of mass m_1, initial velocity $\mathbf{u}_1$ and final velocity $\tilde{\mathbf{u}}_1$. Assume that the initial velocity is directed along the z-axis of a Cartesian coordinate system x, y, and z. Particle 2 of mass m_2 and zero initial velocity acquires the final velocity $\tilde{\mathbf{u}}_2$ after collision with particle 1. Show that the x, y, and z components of the velocity change $\Delta\mathbf{u}_1 = \tilde{\mathbf{u}}_1 - \mathbf{u}_1$ of particle 1 are given by

$$\Delta u_{1x} = M_2 g \sin\chi \cos\varphi$$

$$\Delta u_{1y} = M_2 g \sin\chi \sin\varphi$$

$$\Delta u_{1z} = -M_2 g (1 - \cos\chi)$$

where χ, the deflection angle is the colatitude and φ is the azimuth. Note that $M_2 = m_2/(m_1 + m_2)$ and g is the relative speed before and after collision.

Let an average taken over the entire solid angle in the c.m. system be defined as

$$\langle f \rangle = \int_{\chi=0}^{\pi} \int_{\varphi=0}^{2\pi} f g \sigma(\chi) \sin\chi \, d\chi \, d\varphi$$

where $\sigma(\chi)$ is Rutherford's scattering formula. Establish that

$$\langle \Delta u_{1x} \rangle = \langle \Delta u_{1y} \rangle = 0$$

$$\langle \Delta u_{1z} \rangle = -2\pi M_2 g^2 p_0^2 \ln (1 + \Lambda^2)$$

$$\langle \Delta u_{1x} \Delta u_{1y} \rangle = \langle \Delta u_{1x} \Delta u_{1z} \rangle = \langle \Delta u_{1y} \Delta u_{1z} \rangle = 0$$

$$\langle (\Delta u_{1x})^2 \rangle = \langle (\Delta u_{1y})^2 \rangle = 2\pi M_2^2 g^3 p_0^2 \left[\ln (1 + \Lambda^2) - \frac{\Lambda^2}{1 + \Lambda^2} \right]$$

$$\langle (\Delta u_{1z})^2 \rangle = 4\pi M_2^2 g^3 p_0^2 \frac{\Lambda^2}{1 + \Lambda^2}$$

Note that $\Lambda = \lambda_D / p_0$ where λ_D is the Debye length and p_0 is the impact parameter which yields a deflection angle of 90°. These averages are required in the evaluation of the so-called Fokker-Planck coefficients.

4.11. Let the interaction between particles be governed by the *square-well* potential defined by

$$\phi(r) = \infty \qquad r < a$$

$$\phi(r) = -\phi_0 \qquad a < r < a_0$$

$$\phi(r) = 0 \qquad r > a_0$$

This potential function represents an impenetrable core of radius a surrounded by an attractive well of depth ϕ_0 which extends radially out to a distance a_0 beyond which the interaction potential is zero. Evaluate the differential scattering cross section for the square-well potential.

4.12. Consider the collision between a type 1 particle and a gas of type 2 particles. The type 1 particle moves with a uniform velocity $\mathbf{u}_1$. Type 2 particles have a constant speed u_2 and their velocities are uniformly distributed in all directions. Show that the relative speed g between the type 1 particle and type 2 particles, after averaging over the velocity distribution of type 2 particles, is given by

$$g = u_1 + u_2^2 / 3u_1 \qquad \text{for } u_1 > u_2$$

$$= u_2 + u_1^2 / 3u_2 \qquad \text{for } u_1 < u_2$$

For a homogeneous gas in which $u_1 = u_2 = u$, deduce the following Clausius' formula for the collision frequency:

$$\nu = (4/3) u N \sigma$$

where N is the number density and σ is the scattering cross section.

4.13. Suppose that type 2 particles of Problem 4.12, instead of having a constant speed u_2, are characterized by the Maxwell-Boltzmann distribution function:

$$f_2(\mathbf{u}_2) = N_2(m_2/2\pi KT_2)^{3/2}\exp[-m_2 u_2^2/2KT_2] \qquad \text{(A)}$$

Let σ be the scattering cross section for the case of collision between the type 1 particle and type 2 particles. Using the results of Problem 4.12, show that the collision frequency ν for the interaction between the type 1 particle and type 2 particles is given by

$$\nu = 4\pi\sigma\left[\int_0^{u_1}\left(u_1 + \frac{u_2^2}{3u_1}\right)u_2^2 f_2(\mathbf{u}_2)\,du_2 + \int_{u_1}^{\infty}\left(u_2 + \frac{u_1^2}{3u_2}\right)u_2^2 f_2(\mathbf{u}_2)\,du_2\right] \qquad \text{(B)}$$

Using (A), carry out the integrations in (B) and deduce the following simplified result:

$$\nu = N_2\sigma\left(\frac{2KT_2}{\pi m_2}\right)^{1/2}\left[\exp(-x^2) + \left(2x + \frac{1}{x}\right)\int_0^x \exp(-y^2)\,dy\right]$$

where

$$x = (m_2/2KT_2)^{1/2}u_1$$

4.14. Consider the collisional interaction between type 1 and type 2 particles. Type 2 particles are characterized by the Maxwell-Boltzmann distribution function as given by (A) in Problem 4.13. Type 1 particles are also characterized by a Maxwell-Boltzmann distribution function as given by

$$f_1(\mathbf{u}_1) = N_1(m_1/2\pi KT_1)^{3/2}\exp(-m_1 u_1^2/2KT_1)$$

By averaging the result obtained in Problem 4.13 over the distribution of velocities of type 1 particles, show that the collision frequency for the interaction between type 1 and type 2 particles is expressed by

$$\begin{aligned}\nu = 4\sqrt{\pi}\,N_2\sigma\left(\frac{m_1}{2\pi KT_1}\right)^{3/2}&\int_0^{\infty} u_1^2\exp(-m_1u_1^2/2KT_1)\\ &\times\left[\left(\frac{2KT_2}{m_2}\right)^{1/2}\exp(-m_2u_1^2/2KT_2) + \left\{2u_1\left(\frac{m_2}{2KT_2}\right)^{1/2}\right.\right.\\ &\left.\left.+\left(\frac{2KT_2}{m_2}\right)^{1/2}\frac{1}{u_1}\right\}\int_0^{u_1}\exp(-m_2u_2^2/2KT_2)\,du_2\right]du_1\end{aligned} \qquad \text{(A)}$$

The iterated integral occurring in (A) can be evaluated by inverting the order of integration. Evaluate the integrals occurring in (A) to establish that

$$\nu = N_2 \sigma \sqrt{\frac{8K}{\pi}} \left(\frac{T_1}{m_1} + \frac{T_2}{m_2} \right)^{1/2}$$

If only one kind of particles is present with number density equal to N, average speed $\langle u \rangle = (8KT/\pi m)^{1/2}$ and mutual scattering cross section σ, show that the collision frequency in a homogeneous Maxwellian gas is given by

$$\nu = \sqrt{2}\, N\sigma\langle u \rangle$$

CHAPTER 5

Small Amplitude Waves in a Plasma

5.1. Introduction

The study of waves in a plasma provides information on the plasma properties and hence is useful in plasma diagnostics. It is known that plasma confinement is hampered by instabilities. An investigation of some of these instabilities requires a clear understanding of the nature of wave phenomena in a plasma. In view of these applications, the subject of waves in a plasma assumes importance and merits a detailed consideration even in an introductory treatment of plasma physics. In this chapter, a sufficiently detailed treatment of the theory and some of the applications of the simplest types of waves that exist in an *unbounded, homogeneous* plasma is given. The study is restricted to the *small amplitude* waves; consequently, only the simple linearized forms of the governing equations are used.

It is advantageous to treat wave propagation in a plasma in terms of plane waves partly because of mathematical simplicity and partly because any complex and physically realizable wave motion can be synthesized in terms of plane waves. Therefore, only the theory and applications of plane waves in a plasma are considered in this treatment. Furthermore, since only the general behavior of the plane waves is desired and no detailed comparison between the theoretical predictions and the experimental observations is contemplated, the analysis is still further simplified by the omission of losses due to the collisional interactions.

A cold, homogeneous plasma in a magnetostatic field is called a magnetoionic medium. The electromagnetic theory of magnetoionic medium is known as magnetoionic theory. The governing equations of magnetoionic theory are first deduced. It is found that a cold, homogeneous plasma in a magnetostatic field is equivalent to an anisotropic dielectric characterized by a dyadic relative permittivity whose expression is obtained in terms of the medium parameters both when the ion motions are neglected, such as in the conventional application of magnetoionic theory, and when the effect of heavy ion motions is included, such as in the hydromagnetic extension of magnetoionic theory. In the absence of the magnetostatic field the plasma becomes equivalent to an isotropic dielectric. The characteristics of plane waves in such an isotropic, homogeneous plasma of infinite extent are then discussed.

The characteristics of plane waves propagating along the magnetostatic field in a conventional magnetoionic medium with stationary ions are analyzed. There are two independent modes of propagation; one mode is left circularly polarized and the other is right circularly polarized. The phenomena of atmospheric whistlers and the helicon wave propagation associated with the low frequency branch of the right circularly polarized mode are described. A qualitative discussion of the problem of radio communication black-out at the time of re-entry of a space vehicle into the earth's atmosphere and a simple treatment of the phenomenon of Faraday rotation are given. The plane wave dispersion relations for the case of propagation across the magnetostatic field are investigated.

The theory of plane waves propagating in an arbitrary direction with respect to the magnetostatic field in a magnetoionic medium is known as the Appleton-Hartree theory in honor of E. V. Appleton [1,2] and D. R. Hartree [3] who initiated the study of the characteristics of plane waves in an ionized region. The Appleton-Hartree theory of plane waves propagating in an arbitrary direction with respect to the magnetostatic field is presented in detail. The nonuniform behavior of the dispersion relations that exist as the propagation direction approaches that of the magnetostatic field is emphasized. The polarization of the fields and the Poynting vector in a magnetoionic medium are carefully examined and the difficulties inherent in a plane wave analysis are brought out. As an application of the dispersion relations pertaining to a plane wave propagating in an arbitrary direction with respect to the magnetostatic field, an account is given of the spectrum of Cerenkov radiation due to a point charge moving with uniform velocity in the direction of the magnetostatic field.

The characteristics of plane waves propagating along and across the magnetostatic field in a magnetoionic medium in which the effect of the motion of the heavy ions is included are described. The magnetoionic theory is then systematically extended to hydromagnetic frequencies by incorporating the motion of the heavy ions.

In the conventional magnetoionic theory as well as in its hydromagnetic extension, the plasma is assumed to be cold and the thermal motions of the charged particles are ignored. In the limit of extremely low frequencies and for the case of a fully ionized plasma consisting of only electrons and ions, it is possible to include without much difficulty the effect of the thermal motions of the particles by making use of the magnetohydrodynamic equations in which a scalar kinetic pressure term is retained for the description of the effect of the thermal motions of the particles. This chapter is concluded with an elementary treatment of the so-called magneto-hydrodynamic waves which exist in a fully ionized plasma at extremely low frequencies and which are governed by the well-known magnetohydrodynamic equations.

5.2. Governing equations of magnetoionic theory

There are a number of different methods for the analysis of the behavior of a plasma even from the macroscopic hydrodynamical point of view. These methods are in the increasing order of complexity. As was mentioned previously, the cold and the warm plasma models are two of the most commonly used approximations for the description of the plasma behavior. For the cold plasma model, the plasma becomes equivalent to an anisotropic dielectric which is called a magnetoionic medium. The electromagnetic theory of a magnetoionic medium is called the magnetoionic theory. In this section, the governing equations of magnetoionic theory and its extension to hydromagnetic frequencies are assembled together.

The collision term

A weakly ionized plasma consisting of electrons, ions, and neutral particles in a constant magnetic field constitutes a magnetoionic medium. The conventional magnetoionic theory applies to frequencies which are sufficiently high that the heavy ions are unable to respond and hence are assumed to be immobile. It is usual to assume the medium to be unbounded and uniform, and the magnetic field, uniform and unidirectional. In a weakly ionized gas, only the dynamics of the electrons plays a role in the determination of the plasma characteristics. In the previous chapter, it was argued that collisional interactions of the electrons with the neutral particles alone are important and the Lorentz force equation for an electron becomes

$$N_e m_e d\mathbf{u}/dt = -N_e e[\mathbf{E} + \mathbf{u} \times \mathbf{B}] - \nu(\mathbf{u}) N_e m_e \mathbf{u} \tag{2.1}$$

where $\nu(\mathbf{u})$, the collision frequency, is usually a function of the particle velocity $\mathbf{u}$. When Eq. (2.1) is averaged over the velocity distribution of the electrons, the result is the *Langevin equation*:

$$N_e m_e d\mathbf{v}/dt = -N_e e[\mathbf{E} + \mathbf{v} \times \mathbf{B}] - \nu(\mathbf{v}) N_e m_e \mathbf{v} \tag{2.2}$$

where $\mathbf{v}$ is the average velocity of the electrons. In general, the collision frequency is a function of the average velocity $\mathbf{v}$. It is to be noted that $\mathbf{E}$ and $\mathbf{B}$ occurring in Eqs. (2.1) and (2.2) are the macroscopic electric field and magnetic flux density, respectively. As was pointed out in the previous chapter, Eqs. (2.1) and (2.2) are based on the assumptions that the interaction of an electron with a neutral particle can be represented exactly by the model of collision between two perfectly elastic, impenetrable spheres, and that the entire momentum of an electron is lost on collision with a neutral particle. Since these assumptions are approximations to the actual state of affairs, it is common to represent formally the average rate of loss of momentum density of the electrons as $\nu_{en}(\mathbf{v}) N_e m_e \mathbf{v}$, where $\nu_{en}(\mathbf{v})$ is the effective collision frequency for momentum transfer between the electrons and the neutral particles.

In magnetoionic theory, it is usual to incorporate the effect of collisional interactions of the electrons with the neutral particles phenomenologically as implied in

$$N_e m_e d\mathbf{v}/dt = -N_e e[\mathbf{E} + \mathbf{v} \times \mathbf{B}] - \nu_c N_e m_e \mathbf{v} \tag{2.3}$$

where ν_c, the collision frequency, is assumed to be a constant independent of the average velocity of the electrons. Thus Eq. (2.3) corresponds to a special case in which $\nu_{en}(\mathbf{v})$ is independent of the average velocity. A comparison of Eq. (2.3) with the momentum transport equation (1.11.2) corresponding to the cold plasma model reveals that

$$\mathbf{P}_{\text{coll}} = -\nu_c N_e m_e \mathbf{v} \tag{2.4}$$

Consequently, Eq. (2.4) may be considered to be the magnetoionic approximation to the collisional term $\mathbf{P}_{\text{coll}}$. It is emphasized that $\mathbf{P}_{\text{coll}}$ has not been deduced but only its form has been assumed in Eq. (2.4). It was pointed out that the collision term $(\partial f/\partial t)_{\text{coll}}$ in the Boltzmann equation (1.3.29) is only schematic and that it could be expressed as a multiple integral over the velocity distribution function. It is clear from Eq. (1.9.12) that $\mathbf{P}_{\text{coll}}$ is also a multiple integral of the velocity distribution function. With a complete representation for $(\partial f/\partial t)_{\text{coll}}$, Eq. (1.3.29) becomes an integro-differential equation specifying the distribution function. The solution of the integro-differential equation gives the velocity distribution function which when substituted in Eq. (1.9.12) yields $\mathbf{P}_{\text{coll}}$. This scheme merely indicates the principle for the evaluation of $\mathbf{P}_{\text{coll}}$ but cannot be carried out in practice, at least systematically. In the following chapter, some aspects of the approximate evaluation of $\mathbf{P}_{\text{coll}}$ are discussed. For the treatment of magnetoionic theory, Eq. (2.4) is introduced phenomenologically. Except in special cases such as in *Luxembourg effect* where the nonlinearity introduced by the dependence of the collision frequency on the velocity is crucial for a satisfactory description of the phenomenon, Eq. (2.3) with a constant collision frequency is adequate for the treatment of a variety of problems of wave propagation in a cold plasma. In particular, Eq. (2.3) has proved successful in analyzing the characteristics of wave propagation in the ionosphere.

Anisotropic dielectric

It is proposed to give a treatment of only the small-amplitude waves in a plasma which is stationary as a whole with the result that the electromagnetic fields and the average velocity are first-order, time-dependent perturbations. The magnetostatic flux density $\mathbf{B}_0$ is assumed to be very large in comparison with the time-varying magnetic flux density. The number density N_0 in the unperturbed state is the same for the electrons and the ions. In the perturbed state, the number density of the electrons is equal to $N_0 + N$ where N is also a first-order, time-dependent perturbation. For the present case, the ions are assumed to be stationary and therefore the number density of the ions is not perturbed. If Eq. (2.3) is linearized by the omission of all the second-order terms resulting from the product of two perturbation terms, it is found that

$$m_e \partial \mathbf{v}/\partial t = -e[\mathbf{E} + \mathbf{v} \times \mathbf{B}_0] - \nu_c m_e \mathbf{v} \tag{2.5}$$

The arguments $(\mathbf{r}, t)$ have not been explicitly written down in Eq. (2.5) for $\mathbf{v}$ and $\mathbf{E}$.

Only the time-harmonic case in which all the first-order perturbations have the time dependence of the form $\exp(-i\omega t)$ is to be investigated. The time average of $\mathbf{v}$ is zero with the result that the plasma is stationary, as assumed. If $\mathbf{v}$ had a time-independent term, the time average of $\mathbf{v}$ does not vanish and the plasma as a whole drifts in the direction of the time-independent velocity. Although very interesting, the characteristics of waves in drifting plasmas are not examined in this treatment. For the time-harmonic perturbations, it can be shown from Eq. (2.5) that the phasor amplitudes of the average velocity and the electric field are related as in the following expression:

$$m_e \tilde{\omega} \mathbf{v}(\mathbf{r}) = -ie[\mathbf{E}(\mathbf{r}) + \mathbf{v}(\mathbf{r}) \times \mathbf{B}_0] \tag{2.6}$$

where

$$\tilde{\omega} = \omega + i\nu_c \tag{2.7}$$

Without loss of generality, the static magnetic field may be assumed to be in the z-direction. After some rearrangement, the Cartesian components of Eq. (2.6) are found to be

$$v_x(\mathbf{r}) + \frac{i\omega_{ce}}{\tilde{\omega}} v_y(\mathbf{r}) = -\frac{ie}{\tilde{\omega} m_e} E_x(\mathbf{r}) \tag{2.8a}$$

$$-\frac{i\omega_{ce}}{\tilde{\omega}} v_x(\mathbf{r}) + v_y(\mathbf{r}) = -\frac{ie}{\tilde{\omega} m_e} E_y(\mathbf{r}) \tag{2.8b}$$

$$v_z(\mathbf{r}) = -\frac{ie}{\tilde{\omega} m_e} E_z(\mathbf{r}) \tag{2.8c}$$

where

$$\omega_{ce} = eB_0/m_e \tag{2.9}$$

is the gyromagnetic angular frequency of the electrons. The simultaneous equations (2.8a) and (2.8b) specifying $v_x(\mathbf{r})$ and $v_y(\mathbf{r})$ in terms of $E_x(\mathbf{r})$ and $E_y(\mathbf{r})$ are solved.

Then, the components of $\mathbf{v}(\mathbf{r})$ can be expressed in terms of the components of $\mathbf{E}(\mathbf{r})$ in the following form:

$$\begin{bmatrix} v_x(\mathbf{r}) \\ v_y(\mathbf{r}) \\ v_z(\mathbf{r}) \end{bmatrix} = -\frac{ie}{m_e\tilde{\omega}} \begin{bmatrix} \dfrac{\tilde{\omega}^2}{\tilde{\omega}^2-\omega_{ce}^2} & \dfrac{-i\tilde{\omega}\omega_{ce}}{\tilde{\omega}^2-\omega_{ce}^2} & 0 \\ \dfrac{i\tilde{\omega}\omega_{ce}}{\tilde{\omega}^2-\omega_{ce}^2} & \dfrac{\tilde{\omega}^2}{\tilde{\omega}^2-\omega_{ce}^2} & 0 \\ 0 & 0 & 1 \end{bmatrix} \begin{bmatrix} E_x(\mathbf{r}) \\ E_y(\mathbf{r}) \\ E_z(\mathbf{r}) \end{bmatrix} \tag{2.10}$$

Since only the electrons are mobile, the linearized form of the electric current density is deduced from Eqs. (2.2.2) and (2.10) as

$$\mathbf{J}(\mathbf{r}) = -N_0 e\mathbf{v}(\mathbf{r}) = \tilde{\boldsymbol{\sigma}}_e \cdot \mathbf{E}(\mathbf{r}) \tag{2.11}$$

where the conductivity dyad $\tilde{\boldsymbol{\sigma}}_e$, in the matrix form, is given by

$$\tilde{\boldsymbol{\sigma}}_e = \tilde{\sigma}_{e0} \begin{bmatrix} \dfrac{\tilde{\omega}^2}{\tilde{\omega}^2-\omega_{ce}^2} & \dfrac{-i\tilde{\omega}\omega_{ce}}{\tilde{\omega}^2-\omega_{ce}^2} & 0 \\ \dfrac{i\tilde{\omega}\omega_{ce}}{\tilde{\omega}^2-\omega_{ce}^2} & \dfrac{\tilde{\omega}^2}{\tilde{\omega}^2-\omega_{ce}^2} & 0 \\ 0 & 0 & 1 \end{bmatrix} \tag{2.12}$$

and

$$\tilde{\sigma}_{e0} = iN_0 e^2/m_e\tilde{\omega} \tag{2.13}$$

In accordance with Eqs. (3.6.25) and (3.6.26), the relative permittivity dyad

$$\boldsymbol{\varepsilon}_r = \mathbf{1} + i\tilde{\boldsymbol{\sigma}}_e/\omega\varepsilon_0 \tag{2.14}$$

is evaluated from Eqs. (2.12) and (2.13) as

$$\boldsymbol{\varepsilon}_r = \varepsilon_1(\hat{\mathbf{x}}\hat{\mathbf{x}} + \hat{\mathbf{y}}\hat{\mathbf{y}}) + i\varepsilon_2(\hat{\mathbf{x}}\hat{\mathbf{y}} - \hat{\mathbf{y}}\hat{\mathbf{x}}) + \varepsilon_3\hat{\mathbf{z}}\hat{\mathbf{z}} \tag{2.15}$$

where

$$\varepsilon_1 = 1 - \omega_{pe}^2\tilde{\omega}/\omega(\tilde{\omega}^2 - \omega_{ce}^2) \tag{2.16a}$$

$$\varepsilon_2 = \omega_{pe}^2\omega_{ce}/\omega(\tilde{\omega}^2 - \omega_{ce}^2) \tag{2.16b}$$

$$\varepsilon_3 = 1 - \omega_{pe}^2/\tilde{\omega}\omega \tag{2.16c}$$

and

$$\omega_{pe} = \sqrt{N_0 e^2/m_e\varepsilon_0} \tag{2.17}$$

is the electron plasma angular frequency.

When the collisional interactions of the electrons with the neutral particles are ignored, $\nu_c = 0$ and $\tilde{\omega}$ becomes equal to ω. The values of $\tilde{\boldsymbol{\sigma}}_e$ and $\tilde{\sigma}_{e0}$ for the collisionless case are denoted by $\boldsymbol{\sigma}_e$ and σ_{e0}, respectively. The expressions for $\boldsymbol{\sigma}_e$ and σ_{e0} are the same as those in Eqs. (2.12) and (2.13) with $\tilde{\omega}$ replaced by ω. In the same manner, for the collisionless case, the components of the relative permittivity dyad are obtained from Eqs. (2.16a)–(2.16c) by setting $\tilde{\omega}$ equal to ω. A comparison of Eqs. (2.16a), (2.16b), and (2.16c), respectively, with Eqs. (3.6.28a), (3.6.28b), and (3.6.28c) shows that the collisionless Boltzmann equation and the orbit theory yield the same result for the relative permittivity dyad. As a matter of fact, although not established here, the collisionless Boltzmann equation and the orbit theory are equivalent in all respects.

In the discussion of the characteristics of the small-amplitude waves in an unbounded plasma, the objective is to bring out the general features of these waves and not to make any detailed comparison between the theoretical predictions and the experimental observations. This limited objective can be best fulfilled by the complete omission of the collisional losses. Thus in a lossless uniform plasma with stationary ions and uniform magnetostatic field in the z-direction, the electromagnetic fields satisfy the following time-harmonic Maxwell's equations:

$$\nabla \times \mathbf{E}(\mathbf{r}) = i\omega\mu_0 \mathbf{H}(\mathbf{r}) \tag{2.18}$$

$$\nabla \times \mathbf{H}(\mathbf{r}) = -i\omega\varepsilon_0 \boldsymbol{\varepsilon}_r \cdot \mathbf{E}(\mathbf{r}) \tag{2.19}$$

where the form of the relative permittivity dyad is given by Eq. (2.15). The expressions for the components of the permittivity dyad are the same as those given in Eqs. (2.16a), (2.16b), and (2.16c) with $\tilde{\omega} = \omega$. The electromagnetic fields in a magnetoionic medium are governed by Eqs. (2.18) and (2.19) which reveal that the medium behaves like a dielectric characterized by the permittivity dyad $\varepsilon_0 \boldsymbol{\varepsilon}_r$.

It is convenient to introduce the following normalized parameters:

$$\Omega = \omega/\omega_{pe} \qquad R = \omega_{ce}/\omega_{pe} \tag{2.20}$$

where Ω is the normalized frequency and R is the normalized strength of the magnetostatic field. Together with Eqs. (2.20), Eqs. (2.16a), (2.16b), and (2.16c), with $\tilde{\omega} = \omega$ can be simplified to yield:

$$\varepsilon_1 = 1 - \frac{1}{\Omega^2 - R^2} = \frac{\Omega^2 - 1 - R^2}{\Omega^2 - R^2} \tag{2.21a}$$

$$\varepsilon_2 = \frac{R}{\Omega(\Omega^2 - R^2)} \tag{2.21b}$$

$$\varepsilon_3 = 1 - \frac{1}{\Omega^2} = \frac{\Omega^2 - 1}{\Omega^2} \tag{2.21c}$$

Note that the normalized frequency Ω as defined by Eqs. (2.20) has been used previously in Eq. (2.8.41).

Effect of ion motion

So far the motion of the heavy ions has been omitted. If it is desired to study the wave characteristics at very low frequencies, the effect of the motion of the heavy ions has to be included. Let m_i be the mass and e be the charge of an ion. The linearized, time-harmonic momentum transport equation for the ions is the same as Eq. (2.6) with m_e and $-e$ replaced by m_i and e, respectively. Since only the collisionless case is to be studied, $\tilde{\omega} = \omega$. The gyromagnetic angular frequency of the ions is defined by

$$\omega_{ci} = eB_0/m_i \tag{2.22}$$

Note that by definition the gyromagnetic frequency is positive for both the electrons and the ions. It is ascertained that the expression relating the components of the average velocity of the ions to the components of the electric field is the same as Eq. (2.10) with m_e, $-e$, and ω_{ce} replaced by m_i, e, and $-\omega_{ci}$, respectively. It is now necessary to distinguish the average velocity of the electrons and the ions by the subscripts e and i respectively. If both the electron and the ion motions are taken into account, the linearized form of the electric current density is evaluated with the help of Eqs. (2.2.2) and (2.10) as

$$\mathbf{J} = N_0 e[\mathbf{v}_i(\mathbf{r}) - \mathbf{v}_e(\mathbf{r})] = (\boldsymbol{\sigma}_e + \boldsymbol{\sigma}_i) \cdot \mathbf{E}(\mathbf{r}) \tag{2.23}$$

where $\boldsymbol{\sigma}_e$, the contribution to the electrical conductivity dyad arising from the electron motion, is given by Eqs. (2.12) and (2.13), and, $\boldsymbol{\sigma}_i$, the contribution to the electrical conductivity dyad arising from the ion motion, is also given by Eqs. (2.12) and (2.13) but with m_e, $-e$, and ω_{ce} replaced by m_i, e, and $-\omega_{ci}$ respectively. With the inclusion of the ion motions, the relative permittivity dyad (2.14) has to be modified as

$$\boldsymbol{\varepsilon}_r = \mathbf{1} + i(\boldsymbol{\sigma}_e + \boldsymbol{\sigma}_i)/\omega\varepsilon_0 \tag{2.24}$$

The form of $\boldsymbol{\varepsilon}_r$ as given by Eq. (2.15) remains unaltered. With the help of Eqs. (2.12), (2.13), and (2.24), the components of the relative permittivity dyad are deduced in a straightforward manner to be given by

$$\varepsilon_1 = 1 - \frac{\omega_{pe}^2}{\omega^2 - \omega_{ce}^2} - \frac{\omega_{pi}^2}{\omega^2 - \omega_{ci}^2} \tag{2.25a}$$

$$\varepsilon_2 = \frac{\omega_{pe}^2 \omega_{ce}}{\omega(\omega^2 - \omega_{ce}^2)} - \frac{\omega_{pi}^2 \omega_{ci}}{\omega(\omega^2 - \omega_{ci}^2)} \tag{2.25b}$$

$$\varepsilon_3 = 1 - \frac{\omega_{pe}^2}{\omega^2} - \frac{\omega_{pi}^2}{\omega^2} \tag{2.25c}$$

where

$$\omega_{pi} = \sqrt{N_0 e^2/m_i \varepsilon_0} \tag{2.26}$$

is the ion plasma angular frequency. The expressions for ε_1, ε_2, and ε_3 have become more complicated with the inclusion of the ion motions. The electromagnetic theory of a magnetoionic medium in which the motion of the heavy ions is also taken into account constitutes the hydromagnetic extension to the magnetoionic theory and such an extended magnetoionic theory is governed by Eqs. (2.18), (2.19), (2.15), and (2.25).

It is seen from Eqs. (2.9) and (2.22) that

$$\omega_{ci}/\omega_{ce} = m_e/m_i = m \tag{2.27}$$

In a similar manner, Eqs. (2.17) and (2.26) yield

$$\omega_{pi}/\omega_{pe} = \sqrt{m_e/m_i} = \sqrt{m} \tag{2.28}$$

Since $m = 1/1836$, $\sqrt{m}$ and m can be neglected in comparison with unity in a first approximation. Together with Eqs. (2.20), (2.27) and (2.28), ε_1, ε_2, and ε_3 as given by Eqs. (2.25a), (2.25b), and (2.25c), respectively, may be simplified with the following results:

$$\begin{aligned}\varepsilon_1 &= 1 - \frac{1}{\Omega^2 - R^2} - \frac{m}{\Omega^2 - R^2 m^2} \\ &= \frac{\Omega^4 - \Omega^2\{(1+m) + R^2(1+m^2)\} + R^2 m(1 + m + R^2 m)}{(\Omega^2 - R^2)(\Omega^2 - R^2 m^2)}\end{aligned} \tag{2.29a}$$

$$\varepsilon_2 = \frac{R}{\Omega(\Omega^2 - R^2)} - \frac{Rm^2}{\Omega(\Omega^2 - R^2 m^2)} = \frac{\Omega R(1 - m^2)}{(\Omega^2 - R^2)(\Omega^2 - R^2 m^2)} \tag{2.29b}$$

$$\varepsilon_3 = 1 - \frac{1+m}{\Omega^2} = \frac{\Omega^2 - 1 - m}{\Omega^2} \tag{2.29c}$$

Anisotropy and dispersion

The right side of Eqs. (2.19) and (2.15) enable the components of the electric flux density $\mathbf{D}(\mathbf{r})$ to be written in terms of the components of the electric field as

$$D_x(\mathbf{r}) = \varepsilon_1 E_x(\mathbf{r}) + i\varepsilon_2 E_y(\mathbf{r}) \tag{2.30a}$$

$$D_y(\mathbf{r}) = -i\varepsilon_2 E_x(\mathbf{r}) + \varepsilon_1 E_y(\mathbf{r}) \tag{2.30b}$$

$$D_z(\mathbf{r}) = \varepsilon_3 E_z(\mathbf{r}) \tag{2.30c}$$

The component of $\mathbf{D}(\mathbf{r})$ in the direction of the magnetostatic field is contributed only by the component of the electric field in the same direction. But the components of $\mathbf{D(r)}$ perpendicular to the magnetostatic field are contributed not only by the component of the electric field in the same direction as $\mathbf{D(r)}$ but also by the component which is perpendicular to $\mathbf{D(r)}$ and to the magnetostatic field. This type of relationship between the electric flux density $\mathbf{D(r)}$ and the electric field $\mathbf{E(r)}$ is a manifestation of the *anisotropy* of the magnetoionic medium. If the magnetostatic field is zero, $R = 0$ and $\varepsilon_2 = 0$; therefore $\mathbf{D(r)}$ and $\mathbf{E(r)}$ are in the same direction with the result that the medium becomes isotropic. Thus it is the magnetostatic field which renders the medium anisotropic and therefore the waves progressing in different directions with respect to the magnetostatic field have different characteristics. Moreover, it is seen that ε_1, ε_2, and ε_3 are functions of frequency. As a consequence, the phase and the group velocities of the waves are different for different frequencies, that is, the medium is *dispersive*. It is the anisotropy and the dispersion which make the waves in a magnetoionic medium substantially different from those in an ordinary dielectric medium and therefore the study of waves in a magnetoionic medium is very interesting.

5.3. Plane waves in isotropic plasma

The characteristics of plane waves in an unbounded, uniform plasma without any magnetostatic field are studied first. In the absence of magnetostatic field $R = 0$, and therefore Eqs. (2.21a), (2.21b), and (2.21c) yield $\varepsilon_2 = 0$ and $\varepsilon_1 = \varepsilon_3 = \Omega^{-2}(\Omega^2 - 1)$ for the case of immobile ions. If the ion motion is taken into account, for $R = 0$, it is found from Eqs. (2.29a), (2.29b), and (2.29c) that $\varepsilon_2 = 0$ and $\varepsilon_1 = \varepsilon_3 = \Omega^{-2}(\Omega^2 - 1 - m)$. Since m is negligible in comparison with unity, it follows that for an isotropic plasma, the ion motion does not introduce any significant effect. It is therefore assumed that $\varepsilon_2 = 0$ and $\varepsilon_1 = \varepsilon_3 = \Omega^{-2}(\Omega^2 - 1)$. It is obtained from Eq. (2.15) and the right side of Eq. (2.19) that

$$-i\omega\varepsilon_0 \boldsymbol{\varepsilon}_r \cdot \mathbf{E(r)} = -i\omega\varepsilon_0 \varepsilon_{r0} \mathbf{1} \cdot \mathbf{E(r)} = -i\omega\varepsilon_0 \varepsilon_{r0} \mathbf{E(r)} \tag{3.1}$$

and

$$\varepsilon_{r0} = \Omega^{-2}(\Omega^2 - 1)$$

From Eq. (3.1), it is seen that the medium behaves like an isotropic dielectric characterized by the relative scalar permittivity ε_{r0}.

Consider a plane wave progressing in the x-direction and having the following phase factor:

$$\exp(ik_p kx) \tag{3.2}$$

where $k_p = \omega_p/c = \omega_p\sqrt{\mu_0\varepsilon_0}$ and k is the normalized wavenumber or the propagation coefficient. All the field components depend only on x. Therefore, the components of Eqs. (2.18) and (2.19) are obtained by using Eq. (3.1) as

$$H_x(x) = 0 \tag{3.3a}$$

$$i\omega\mu_0 H_y(x) = -(\partial/\partial x)E_z(x) \tag{3.3b}$$

$$i\omega\mu_0 H_z(x) = (\partial/\partial x)E_y(x) \tag{3.3c}$$

$$E_x(x) = 0 \tag{3.4a}$$

$$-i\omega\varepsilon_0 \varepsilon_{r0} E_y(x) = -(\partial/\partial x)H_z(x) \tag{3.4b}$$

$$-i\omega\varepsilon_0 \varepsilon_{r0} E_z(x) = (\partial/\partial x)H_y(x) \tag{3.4c}$$

It is seen from Eqs. (3.3a) and (3.4a) that there are no components of either the electric or the magnetic fields in the direction x of progression of the plane wave with the result that in an isotropic plasma, a plane wave is a transverse electromagnetic (TEM) wave. It is seen from Eqs. (3.3c) and (3.4b) that $E_y(x)$ and $H_z(x)$ form one set, and, Eqs. (3.3b) and (3.4c) show that $E_z(x)$ and $H_y(x)$ form another set. The second set is obtained from the first set by changing $E_y(x)$ and $H_z(x)$ to $E_z(x)$ and $-H_y(x)$, respectively. This interchange is equivalent to the rotation of the first set of fields about the direction of progression of the plane wave by 90°. The characteristics of these two sets of fields are the same except for their polarizations and it is therefore sufficient to treat only the first set of fields.

In accordance with the expression (3.2), $E_y(x)$ and $H_z(x)$ may be assumed to be of the form

$$E_y(x) = E_{y0}\exp(ik_p kx) \qquad H_z(x) = H_{z0}\exp(ik_p kx) \tag{3.5}$$

where E_{y0} and H_{z0} are the amplitudes of the field components, and, $\exp(ik_p kx)$ is the phase factor. The result of the substitution of Eqs. (3.5) into Eqs. (3.3c) and (3.4b) is the following set of simultaneous equations in E_{y0} and H_{z0}:

$$\omega\mu_0 H_{z0} - k_p kE_{y0} = 0 \tag{3.6a}$$

$$k_p kH_{z0} - \omega\varepsilon_0 \varepsilon_{r0} E_{y0} = 0 \tag{3.6b}$$

A nontrivial solution of the set of homogeneous equations (3.6a) and (3.6b) is possible only if the determinant of their coefficients vanishes. This requirement together with Eq. (3.1) may be shown to lead to the following dispersion relation:

$$k = \sqrt{\Omega^2 - 1} \tag{3.7}$$

From Eq. (3.6a), the relationship between the two amplitudes E_{y0} and H_{z0} is found to be given by

$$H_{z0} = (k/\Omega)\sqrt{\varepsilon_0/\mu_0}\; E_{y0} \tag{3.8}$$

The time-averaged Poynting vector is evaluated by using Eqs. (3.5) and (3.8) as

$$\mathbf{S} = \tfrac{1}{2}\,\mathrm{Re}\,[\hat{\mathbf{y}}E_y(x) \times \hat{\mathbf{z}}H_z^*(x)] = \frac{\hat{\mathbf{x}}}{2}\,\mathrm{Re}\left[|E_{y0}|^2\frac{k^*}{\Omega}\sqrt{\frac{\varepsilon_0}{\mu_0}}\,e^{ik_p(k-k^*)x}\right] \qquad (3.9)$$

Since from Eq. (3.7) k is imaginary or real according as $\Omega < 1$ or $\Omega > 1$, it follows from Eq. (3.9) that

$$\mathbf{S} = 0 \qquad \text{for } \Omega < 1 \qquad (3.10a)$$

$$= \frac{\hat{\mathbf{x}}}{2}|E_{y0}|^2\frac{k}{\Omega}\sqrt{\frac{\varepsilon_0}{\mu_0}} \qquad \text{for } \Omega > 1 \qquad (3.10b)$$

For $\Omega < 1$, k is imaginary, and Eqs. (3.5) show that the fields are exponentially damped. It is seen from Eq. (3.10a) that these fields do not transport any time-averaged powe. Such exponentially damped fields that do not transport time-averaged power are called evanescent waves. For $\Omega > 1$, k is real, and Eqs. (3.5) show that the fields have a linear phase progression in the x-direction. Also, Eq. (3.10b) reveals that these fields transport power in the x-direction. Consequently $\Omega > 1$ is designated as the propagating region and the field components (3.5) are called the propagating fields. It is usual to depict Ω as a function of k in the propagating region ($\Omega > 1$) with the help of Eq. (3.7) as illustrated in Fig. 5.1. Such a portrayal of the dispersion relation is called the Brillouin diagram. For very high frequencies, $\Omega \gg 1$, the propagation coefficient or the wavenumber is deduced from Eqs. (3.5) and (3.7) as $k_p k = \omega_p\sqrt{\mu_0\varepsilon_0}\,\Omega = \omega\sqrt{\mu_0\varepsilon_0}$ which is the same as that for a plane wave in free space. Also, from Eq. (3.8) it can be shown that $E_{y0}/H_{z0} = (\mu_0/\varepsilon_0)^{1/2}$ which is the intrinsic impedance in free space. Thus the plane wave characteristics in a plasma degenerate to those of free space. This result is to be expected from physical considerations since even the electrons in a plasma are unable to respond to the limiting case of an infinite frequency.

The normalized phase velocity v_{ph}/c of the wave where c is the velocity of electromagnetic waves in free space is obtained from Eq. (3.7) as

$$v_{ph}/c = \Omega/k = \Omega(\Omega^2 - 1)^{-1/2} \qquad (3.11)$$

In a similar manner the normalized group velocity v_g/c of the wave can be deduced from Eq. (3.7) to be given by

$$v_g/c = \partial\Omega/\partial k = \Omega^{-1}(\Omega^2 - 1)^{1/2} \qquad (3.12)$$

The frequency dependences of v_{ph}/c and v_g/c are also shown in Fig. 5.1.

5.4. Propagation along the magnetostatic field in an electron plasma

The magnetostatic field is parallel to the z-axis. Let a plane wave be assumed to be propagating parallel to the $\mathbf{B}$-field. All the field components have the phase factor

$$\exp(ik_p k_z z) \tag{4.1}$$

and have no dependence on x or y. The Cartesian components of Eqs. (2.18) and (2.19) together with Eq. (2.15) are found to be as follows:

$$-(\partial/\partial z)E_y(z) = i\omega\mu_0 H_x(z) \tag{4.2a}$$

$$(\partial/\partial z)E_x(z) = i\omega\mu_0 H_y(z) \tag{4.2b}$$

$$H_z(z) = 0 \tag{4.2c}$$

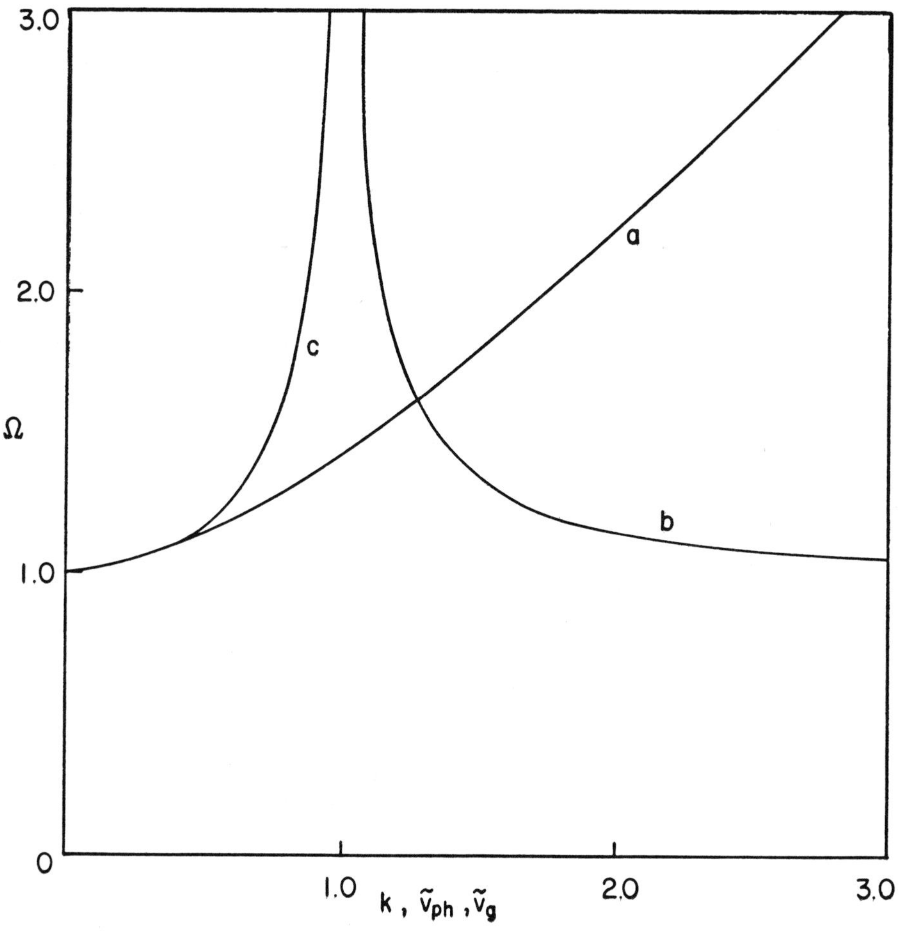

Fig. 5.1. Brillouin diagram for the plane TEM wave in an isotropic plasma: (a) k; (b) $\tilde{v}_{ph} = v_{ph}/c$; (c) $\tilde{v}_g = v_g/c$.

$$-(\partial/\partial z)H_y(z) = -i\omega\varepsilon_0\{\varepsilon_1 E_x(z) + i\varepsilon_2 E_y(z)\} \tag{4.3a}$$

$$(\partial/\partial z)H_x(z) = -i\omega\varepsilon_0\{-i\varepsilon_2 E_x(z) + \varepsilon_1 E_y(z)\} \tag{4.3b}$$

$$E_z(z) = 0 \tag{4.3c}$$

A plane wave propagating along the magnetostatic field in a magnetoionic medium is seen from Eqs. (4.2c) and (4.3c) to be a TEM wave since there is no component of either the electric or the magnetic fields in the direction of propagation.

If Eq. (4.2a) is multiplied by $\mp i/\sqrt{2}$, Eq. (4.2b) by $1/\sqrt{2}$, and they are added together, it can be shown that

$$(\partial/\partial z)E^{\mp}(z) = \pm\omega\mu_0 H^{\mp}(z) \tag{4.4}$$

where

$$E^{\mp}(z) = [E_x(z) \pm iE_y(z)]/\sqrt{2} \tag{4.5}$$

and

$$H^{\mp}(z) = [H_x(z) \pm iH_y(z)]/\sqrt{2} \tag{4.6}$$

Similarly, if Eq. (4.3a) is multiplied by $\mp i/\sqrt{2}$, Eq. (4.3b) by $1/\sqrt{2}$, and they are added together, it is found after some manipulation that

$$(\partial/\partial z)H^{\mp}(z) = \mp\omega\varepsilon_0(\varepsilon_1 \pm \varepsilon_2)E^{\mp}(z) \tag{4.7}$$

The advantage of this manipulation is that the two sets of field components $[E^-(z), H^-(z)]$ and $[E^+(z), H^+(z)]$ are not coupled in Eqs. (4.4) and (4.7). An arbitrary field component perpendicular to the **B**-field can be obtained from a suitable linear superposition of the two uncoupled sets of fields as indicated in the following:

$$\mathbf{E}(z) = \hat{\mathbf{x}}E_x(z) + \hat{\mathbf{y}}E_y(z) = \frac{(\hat{\mathbf{x}} + i\hat{\mathbf{y}})}{\sqrt{2}}E^+(z) + \frac{(\hat{\mathbf{x}} - i\hat{\mathbf{y}})}{\sqrt{2}}E^-(z) \tag{4.8}$$

$$\mathbf{H}(z) = \hat{\mathbf{x}}H_x(z) + \hat{\mathbf{y}}H_y(z) = \frac{(\hat{\mathbf{x}} + i\hat{\mathbf{y}})}{\sqrt{2}}H^+(z) + \frac{(\hat{\mathbf{x}} - i\hat{\mathbf{y}})}{\sqrt{2}}H^-(z) \tag{4.9}$$

It can be verified that

$$\frac{(\hat{\mathbf{x}} \pm i\hat{\mathbf{y}})}{\sqrt{2}} \times \frac{(\hat{\mathbf{x}} \mp i\hat{\mathbf{y}})}{\sqrt{2}} = \mp i\hat{\mathbf{z}} \tag{4.10a}$$

$$\frac{(\hat{\mathbf{x}} \pm i\hat{\mathbf{y}})}{\sqrt{2}} \times \frac{(\hat{\mathbf{x}} \pm i\hat{\mathbf{y}})}{\sqrt{2}} = 0 \tag{4.10b}$$

With the help of Eqs. (4.8)–(4.10), it can be established that

$$\hat{\mathbf{z}} \cdot [\mathbf{E}(z) \times \mathbf{H}^*(z)] = -i[E^+(z)H^{+*}(z) - E^-(z)H^{-*}(z)] \tag{4.11}$$

It follows from Eq. (4.11) that even in the time-averaged Poynting vector, the two sets of fields remain separated. Therefore, the time-averaged power flow per unit area is equal to the sum of the time-averaged power flow per unit area in each of the two sets of fields separately. In view of this, the two sets of fields $[E^-(z), H^-(z)]$ and $[E^+(z), H^+(z)]$ form two independent modes propagating parallel to the magnetostatic field.

Let $E^{\pm}(z) = |E^{\pm}(z)|\exp(i\theta^{\pm})$. If the time factor $\exp(-i\omega t)$ is included in Eq. (4.8) and following the usual convention, the real part of the resulting right side of Eq. (4.8) is taken, the following expression is obtained for the time dependent real electric field:

$$\begin{aligned}\mathbf{E}(z,t) &= 2^{-1/2}|E^+(z)|[\hat{\mathbf{x}}\cos(\omega t-\theta^+)+\hat{\mathbf{y}}\sin(\omega t-\theta^+)]\\ &\quad + 2^{-1/2}|E^-(z)|[\hat{\mathbf{x}}\cos(\omega t-\theta^-)-\hat{\mathbf{y}}\sin(\omega t-\theta^-)]\end{aligned} \tag{4.12}$$

The tip of the electric vector given by the first term on the right side of Eq. (4.12) rotates along a circle in the counterclockwise direction. With the thumb of the right hand in the direction (z) of propagation, the fingers curl in the direction of rotation of the electric vector and therefore the set of fields $[E^+(z), H^+(z)]$ is said to constitute a right circularly polarized wave. In a similar manner, $[E^-(z), H^-(z)]$ constitutes a left circularly polarized wave. The two linearly polarized components of the electric vector which is perpendicular to the direction of propagation are recast in Eq. (4.8) into two circularly polarized components with opposite directions of rotation. The left and the right circularly polarized waves form two independent TEM modes of propagation along the magnetostatic field in an unbounded, uniform plasma.

The characteristics of the two circularly polarized waves are deduced in the following manner. In accordance with the expression (4.1), the field components are assumed to be of the form

$$E^{\mp}(z) = E_0^{\mp}e^{ik_pk_z^{\mp}z} \qquad H^{\mp}(z) = H_0^{\mp}e^{ik_pk_z^{\mp}z} \tag{4.13}$$

where $E_0^{\mp}$ and $H_0^{\mp}$ are the amplitudes and $\exp(ik_pk_z^{\mp}z)$ is the phase factor. The substitution of Eqs. (4.13) into Eqs. (4.4) and (4.7) yields the following set of homogeneous equations specifying $E_0^{\mp}$ and $H_0^{\mp}$:

$$ik_pk_z^{\mp}E_0^{\mp} \mp \omega\mu_0 H_0^{\mp} = 0 \tag{4.14a}$$

$$\pm\omega\varepsilon_0(\varepsilon_1 \pm \varepsilon_2)E_0^{\mp} + ik_pk_z^{\mp}H_0^{\mp} = 0 \tag{4.14b}$$

The condition for the existence of nontrivial solutions of Eqs. (4.14a) and (4.14b) gives the following dispersion relations for the two circularly polarized modes:

$$k_z^- = \Omega\sqrt{\varepsilon_1 + \varepsilon_2} \tag{4.15}$$

and

$$k_z^+ = \Omega\sqrt{\varepsilon_1 - \varepsilon_2} \tag{4.16}$$

Left circularly polarized wave

It can be shown from Eqs. (2.21a) and (2.21b) that

$$\varepsilon_1 \pm \varepsilon_2 = \frac{(\Omega^2 \pm \Omega R - 1)}{\Omega(\Omega \pm R)} = \frac{(\Omega \mp \Omega_1)(\Omega \pm \Omega_2)}{\Omega(\Omega \pm R)} \tag{4.17}$$

where

$$\Omega_{1,2} = \mp R/2 + \sqrt{(R/2)^2 + 1} \tag{4.18}$$

It is verified that $\Omega_1 = 1$ for $R = 0$, decreases as R increases, and becomes equal to R for $R = 1/\sqrt{2}$. As R is further increased, Ω_1 monotonically decreases and attains the value $\Omega_1 = 0$ for $R = \infty$. In a similar manner, it can be ascertained that Ω_2 is greater than 1 and R, increases monotonically as R is increased, and for very large values of R, Ω_2 is approximately equal to R.

The normalized propagation coefficient k_z^- of the left circularly polarized wave is obtained from Eqs. (4.15) and (4.17) as

$$k_z^- = \left[\frac{\Omega(\Omega^2 + \Omega R - 1)}{(\Omega + R)}\right]^{1/2} = \left[\frac{\Omega(\Omega - \Omega_1)(\Omega + \Omega_2)}{(\Omega + R)}\right]^{1/2} \tag{4.19}$$

The propagation of the left circularly polarized wave takes place only when k_z^- is real. Since Ω_1 and Ω_2 are real and positive, it is found from Eq. (4.19) that the left circularly polarized wave propagates only for $\Omega > \Omega_1$. For Ω tending to infinity, k_z^- tends to Ω and the conditions pertaining to free space propagation are attained. The Brillouin diagram for this mode is shown in Fig. 5.2. The frequency $\Omega = \Omega_1$, where $k_z^- = 0$, is called the *cut-off* frequency. The normalized phase velocity is evaluated from Eq. (4.19) to be given by

$$\frac{v_{ph}}{c} = \frac{\Omega}{k_z^-} = \left[\frac{\Omega(\Omega + R)}{(\Omega - \Omega_1)(\Omega + \Omega_2)}\right]^{1/2} \tag{4.20}$$

On rewriting Eq. (4.19) as a polynomial in Ω, $\partial\Omega/\partial k_z^-$ can be determined and expressed with the help of Eq. (4.19) as a function of Ω and R only. The result is the following expression for the normalized group velocity:

$$\frac{v_g}{c} = \frac{\partial\Omega}{\partial k_z^-} = \frac{2(\Omega + R)^2}{[2\Omega(\Omega + R)^2 - R]}\left[\frac{\Omega(\Omega - \Omega_1)(\Omega + \Omega_2)}{(\Omega + R)}\right]^{1/2} \tag{4.21}$$

At the cut-off frequency $\Omega = \Omega_1$, the phase and the group velocities become infinite and zero, respectively. As the frequency is increased beyond $\Omega = \Omega_1$, v_{ph}/c decreases, and v_g/c increases monotonically and approach unity in the limit of Ω tending to infinity. In Fig. 5.2, v_{ph}/c and v_g/c are also depicted as functions of Ω in the propagation region $\Omega > \Omega_1$.

Right circularly polarized wave

The dispersion relation (4.16) for the right circularly polarized wave is seen from Eq. (4.15) to be the same as that for the left circularly polarized wave with ε_2 changed to $-\varepsilon_2$ and this is seen from Eq. (2.21b) to be equivalent to changing R to $-R$. Therefore, from a comparison with Eqs. (4.19)–(4.21), the propagation coefficient k_z^+, the normalized phase velocity v_{ph}/c and the normalized group velocity v_g/c of the right circularly polarized wave are found to be given by

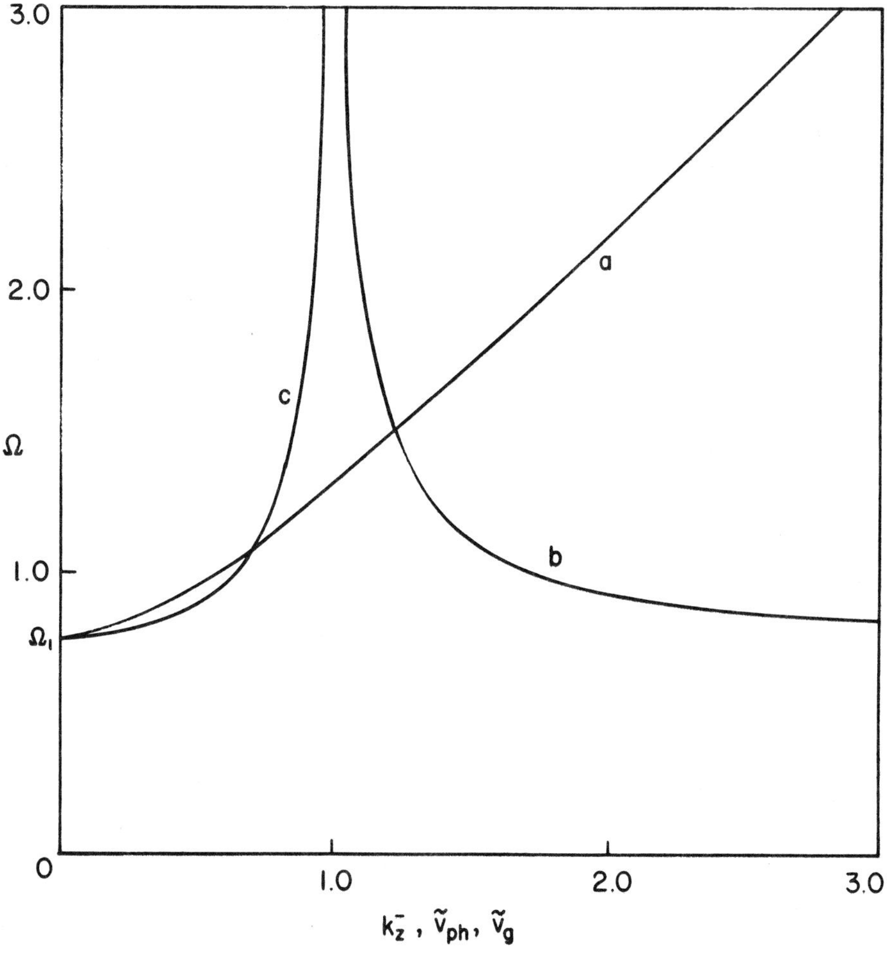

Fig. 5.2. Brillouin diagram for the left circularly polarized wave propagating along the magnetostatic field in a magnetoionic medium: $R = 0.5$; (a) k_z^-; (b) $\tilde{v}_{ph} = v_{ph}/c$; (c) $\tilde{v}_g = v_g/c$.

$$k_z^+ = \left[\frac{\Omega(\Omega^2 - \Omega R - 1)}{(\Omega - R)}\right]^{1/2} = \left[\frac{\Omega(\Omega + \Omega_1)(\Omega - \Omega_2)}{(\Omega - R)}\right]^{1/2} \tag{4.22}$$

$$\frac{v_{ph}}{c} = \frac{\Omega}{k_z^+} = \left[\frac{\Omega(\Omega - R)}{(\Omega + \Omega_1)(\Omega - \Omega_2)}\right]^{1/2} \tag{4.23}$$

$$\frac{v_g}{c} = \frac{\partial\Omega}{\partial k_z^+} = \frac{2(\Omega - R)^2}{[2\Omega(\Omega - R)^2 + R]}\left[\frac{\Omega(\Omega + \Omega_1)(\Omega - \Omega_2)}{(\Omega - R)}\right]^{1/2} \tag{4.24}$$

It was already pointed out that $\Omega_2 > R$. Therefore, it is found from Eq. (4.22) that k_z^+ is real for $0 < \Omega < R$ and $\Omega_2 < \Omega < \infty$. Thus the right circularly polarized wave propagates in the two frequency ranges (i) $0 < \Omega < R$ and (ii) $\Omega_2 < \Omega < \infty$. For the right circularly polarized wave, $\Omega = \Omega_2$, where $k_z^+ = 0$ is the *cut-off* frequency. The frequency for which the propagation coefficient is infinite is called the *resonant* frequency. Such a resonant frequency occurs at $\Omega = R$ for the right circularly polarized wave. When expressed in unnormalized quantities, the resonant frequency is seen to occur at the gyromagnetic frequency of the electrons. The phase and the group velocities become infinite and zero, respectively, at the cut-off frequency $\Omega = \Omega_2$. At the resonant frequency $\Omega = R$, both the velocities become zero and so also at $\Omega = 0$. For Ω tending to infinity, $k_z^+ = \Omega$ and $v_{ph}/c = v_g/c = 1$, and, the propagation characteristics become identical to those of free space. It is to be noted that for $\Omega = R/2$, the phase velocity attains a maximum value $(v_{ph}/c)_{\max} = (1 + 4R^{-2})^{-1}$ which is less than unity. Similarly, the group velocity also has a maximum value in the range $0 < \Omega < R$. In Fig. 5.3, k_z^+, v_{ph}/c, and v_g/c are depicted as functions of Ω in the two propagation ranges $0 < \Omega < R$ and $\Omega_2 < \Omega < \infty$. An important feature of the right circularly polarized wave, in contrast to the left circularly polarized wave, is that one of its propagation ranges extends down to very low frequencies, that is frequencies below the plasma, and the gyromagnetic frequency of the electrons.

Atmospheric whistlers

The very low frequency propagation of the right circularly polarized wave gives rise to two related phenomena which are the naturally occurring *atmospheric whistlers*, and the experimentally observed *helicon* wave propagation.

Let attention be restricted to the low frequency branch of the right circularly polarized wave. Consider a plane wave transient, rich in low frequencies, propagating in the direction of the magnetostatic field. Initially the wave can be decomposed into its various frequency components. The wave packets centered on the various frequencies propagate at their group velocities and thus get dispersed in course of time. At a distant point, in the direction of the magnetostatic field, there is a continuous variation of the times of arrival of the wave packets corresponding to

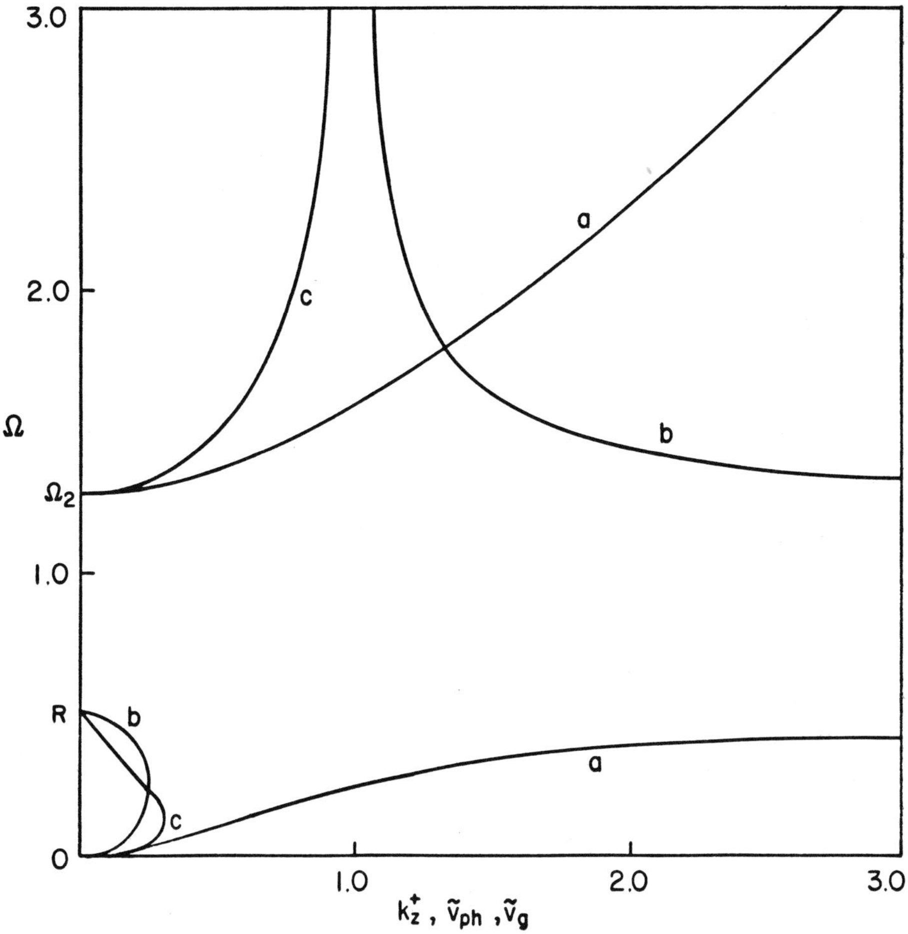

Fig. 5.3. Brillouin diagram for the right circularly polarized wave propagating along the magnetostatic field in a magnetoionic medium: $R = 0.5$; (a) k_z^+; (b) $\tilde{v}_{ph} = v_{ph}/c$; (c) $\tilde{v}_g = v_g/c$.

the various frequencies, the times of arrival being inversely proportional to the corresponding group velocities. From Fig. 5.4 which depicts Ω versus $1/(v_g/c)$, in the very low frequency region, the frequency is seen to decrease continually with time. The conditions envisaged in the foregoing occur to within some approximation in the environment of the earth together with its magnetostatic field and the surrounding plasma of the ionosphere. The transient electromagnetic source is provided by the thunderstorms and lightning, which are rich in very low frequency components. The various frequency constituents of the transient may be idealized to propagate as plane waves in the direction of the earth's magnetostatic field with the result that at a distant point in the direction of the earth's magnetostatic field, the observed frequency slowly decreases with time. The plasma parameters are such that this very low frequency falls in the audio frequency regime, as a consequence of which the frequency versus time spectrum observed on the earth is an audiable tone, with the mean frequency descending slowly with the time of arrival. This is the well-known phenomenon of atmospheric whistlers [4]. The whistlers can be observed in the direct path from the region of the source, in which case they are called fractional hop whistlers or after reflection near the polar regions. In the latter case, the whistlers are called the multiple hop whistlers depending on the number of reflections. As many as ten hop whistlers and even higher, have been observed. The rate of decrease of frequency with the time of arrival decreases with the increase in the number of hops.

The recording of the spectrum of the frequency versus the time is known as an ionogram. Ionograms of whistler activity are continuously recorded at various

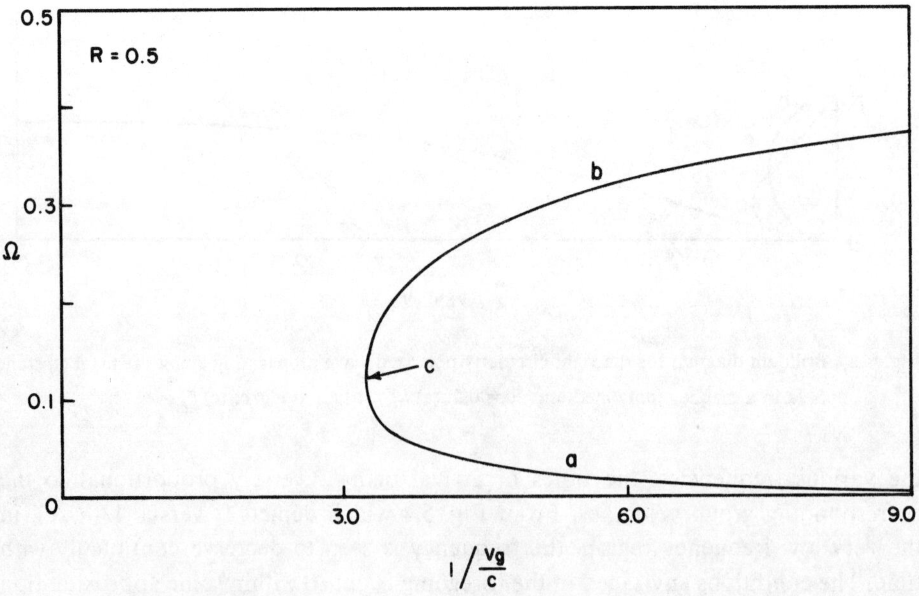

Fig. 5.4. Ω versus $1/(v_g/c)$ for $\Omega \leqslant R$ for the right circularly polarized plane wave propagating along the magnetostatic field in a magnetoionic medium: (a) descending tone whistler region, (b) ascending tone whistler region, (c) nose whistler region.

locations on the earth and are used as an effective diagnostic tool for probing the ionospheric conditions [5].

From Fig. 5.4, it is seen that near the gyromagnetic frequency, it is possible to have the frequency increase with the time of arrival and these are known as the ascending frequency whistlers. The whistlers in the frequency regime where they change from the descending to the ascending tone, are known as the nose whistlers. The ascending frequency and the nose whistlers have also been observed.

Helicons

Consider a slab formed by a plasma medium bounded by two infinite, parallel planes which are d meters apart and are perpendicular to the **B**-field. Let a right circularly polarized wave be launched in the direction of the **B**-field. These waves are successively reflected at the boundaries and form a standing wave whose resonances are defined by the relation

$$n\lambda/2 = d \tag{4.25}$$

approximately, where n is an integer, λ is the wavelength inside the plasma slab, and d is the slab thickness. At very low frequencies, that is, if $\Omega \ll R$, it is found from Eq. (4.22) that

$$k_z^+ = \sqrt{\Omega/R} \ll 1 \tag{4.26}$$

If $k_p k_v$ is the propagation coefficient of the electromagnetic waves in the medium external to the slab and since $k_p k_z^+$ is the propagation coefficient inside the plasma slab, the magnitude of the reflection coefficient at the plasma boundary is equal to $(k_v - k_z^+)/(k_v + k_z^+) \approx 1$ since, $k_z^+ \ll 1$. Therefore, the reflection at the two planes bounding the plasma slab is nearly complete, as a consequence of which Eq. (4.25) is obtained. From Eqs. (4.25) and (4.26), it is deduced that

$$\frac{n}{2}\frac{2\pi}{k_z^+ k_p} = \frac{n\pi c}{\omega_p}\sqrt{\frac{\omega_c}{\omega}} = d \tag{4.27}$$

The condition (4.27) for the resonance of the standing waves is conveniently rewritten in the following two equivalent forms:

$$\omega_n = n^2\pi^2 c^2 \omega_c/\omega_p^2 d^2 \qquad \omega_{cn} = \omega(d\omega_p/n\pi c)^2 \tag{4.28a, b}$$

Note that the subscript n has been added to ω and ω_c to enable the identification of these with the corresponding values of n which gives the number of the standing-wave pattern inside the slab. For given values of d, ω_p, and ω_c, if the frequency of the wave is changed, at $\omega = \omega_n$, there are standing-wave resonances resulting in a large amplitude for the waves inside the slab. The standing wave resonant frequencies ω_n are proportional to the square of the order n of the standing waves.

The conditions discussed in the foregoing can be reproduced approximately in the laboratory [6]. Although only a gaseous medium has been implied for a plasma, a solid such as a metal or a semiconductor can also have the properties of a plasma. In certain metals such as sodium for example, the electrons are extremely mobile whereas the positive ions are stationary, by being locked in position in the crystal lattice. These are the conditions for the validity of the conventional magnetoionic theory. A schematic diagram of the experimental arrangement for obtaining the standing wave resonances in a slab of a solid state plasma is shown in Fig. 5.5. Two small coils, mutually perpendicular and also perpendicular to the large magnetostatic field are wound outside the specimen consisting of a slab of sodium which is cooled to liquid helium temperature, for which case the conductivity of sodium is several thousand times larger than at room temperature. One of the coils is a drive coil which carries an alternating current and excites the right circularly polarized wave in the direction of the magnetostatic field. The second coil is the detector coil which is used to detect the standing-wave or the cavity resonances in the specimen. In the experiment, the voltage in the detector coil is recorded while the frequency of the drive current is varied. The detector shows a voltage maximum at the resonant frequencies and these results are depicted in Fig. 5.6. The frequencies, at

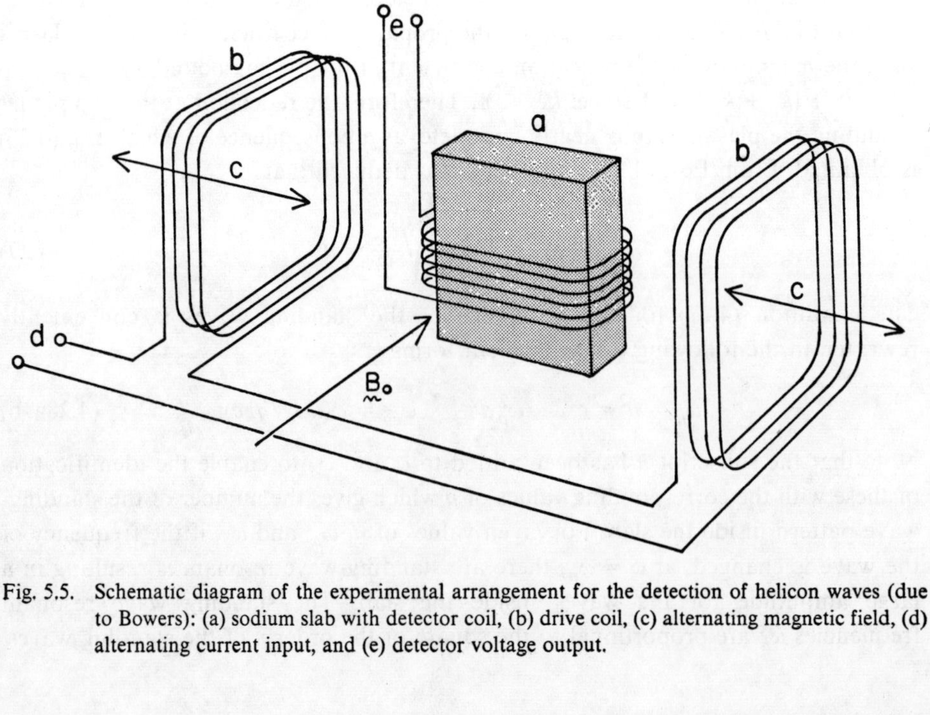

Fig. 5.5. Schematic diagram of the experimental arrangement for the detection of helicon waves (due to Bowers): (a) sodium slab with detector coil, (b) drive coil, (c) alternating magnetic field, (d) alternating current input, and (e) detector voltage output.

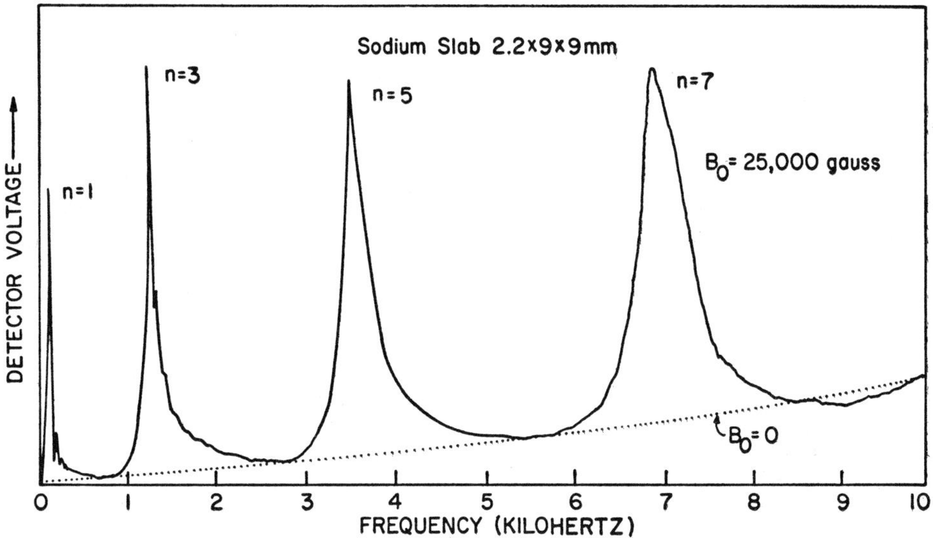

Fig. 5.6. Standing-wave resonances of helicon waves (due to Bowers).

which the voltage maxima occur, were found within the limitations of the theory to be in accordance with that given by Eq. (4.28a). The even order standing-wave effects are cancelled when averaged throughout the thickness of the slab. Consequently, the detector coil which is wound over the entire specimen does not respond to the even order standing waves in this particular experimental arrangement. Therefore, the even order standing-wave resonances do not appear in Fig. 5.6.

The standing-wave resonant frequencies in sodium which contains about 10^{28} electrons per cubic meter are of the order of 100 Hz. It is seen from Eq. (4.28a) that the resonant frequency is inversely proportional to the number density of the electrons. In a semiconductor which contains only 10^{20} electrons per cubic meter, the resonant frequency is of the order of 10^{10} Hz, that is in the microwave range. Aigrain [7] was the first to propose experiments to observe the standing-wave resonances in a semiconductor plasma. He designated this wave motion as a *helicon*. Although the frequency ranges from microwave frequencies in a semiconductor to audio frequencies in a metal, the helicon waves, in general, are the right circularly polarized waves propagating in the direction of the magnetostatic field at "very low frequencies", that is, frequencies much lower than the gyromagnetic frequency of the electrons.

In some experimental investigations of helicon wave propagation, d, ω_p, and ω are fixed and the magnetostatic field is varied. Then the resonances occur for those values of the magnetostatic field for which $\omega_c = \omega_{cn}$ where ω_{cn} is given by Eq. (4.28b). The resonant values of the strength of the magnetostatic field are seen from Eq. (4.28b) to be inversely proportional to the square of the order n of the resonance with the result that the largest magnetostatic field is necessary to excite the fundamental resonance. As the magnetic field strength is increased, the resonances occur one by one in the decreasing order.

Faraday rotation

In the direction of the magnetostatic field, at very low frequencies, only the right circularly polarized wave propagates and certain physical phenomena associated with this propagating wave were discussed. It is now desired to treat a physical phenomenon which occurs only in the range of frequencies where both the right and the left circularly polarized waves propagate. At $z = 0$, without loss of generality, let the electric field be assumed to have only an x-component as given by

$$\mathbf{E}(z=0) = \hat{\mathbf{x}} E_{x0} = \hat{\mathbf{x}} |E_{x0}| e^{i\theta_{E0}} \tag{4.29}$$

As in Eq. (4.8), Eq. (4.29) can be rewritten as

$$\mathbf{E}(z=0) = \frac{(\hat{\mathbf{x}} + i\hat{\mathbf{y}})}{\sqrt{2}} \frac{E_{x0}}{\sqrt{2}} + \frac{(\hat{\mathbf{x}} - i\hat{\mathbf{y}})}{\sqrt{2}} \frac{E_{x0}}{\sqrt{2}} \tag{4.30}$$

The first and the second terms on the right side of Eq. (4.30) are, respectively, the right and the left circularly polarized components. Since these constituents are uncoupled, they propagate independently with the result that the electric vector for any $z > 0$ is obtained as

$$\mathbf{E}(z) = \frac{(\hat{\mathbf{x}} + i\hat{\mathbf{y}})}{\sqrt{2}} \frac{E_{x0}}{\sqrt{2}} e^{ik_p k_z^+ z} + \frac{(\hat{\mathbf{x}} - i\hat{\mathbf{y}})}{\sqrt{2}} \frac{E_{x0}}{\sqrt{2}} e^{ik_p k_z^- z} \tag{4.31}$$

The right side of Eq. (4.31) can be regrouped to yield

$$\begin{aligned} \mathbf{E}(z) &= E_{x0} e^{ik_p(k_z^+ + k_z^-)z/2} \left[\frac{(\hat{\mathbf{x}} + i\hat{\mathbf{y}})}{2} e^{ik_p(k_z^+ - k_z^-)z/2} + \frac{(\hat{\mathbf{x}} - i\hat{\mathbf{y}})}{2} e^{-ik_p(k_z^+ - k_z^-)z/2} \right] \\ &= E_{x0} e^{ik_p(k_z^+ + k_z^-)z/2} [\hat{\mathbf{x}} \cos\{k_p(k_z^+ - k_z^-)z/2\} - \hat{\mathbf{y}} \sin\{k_p(k_z^+ - k_z^-)z/2\}] \end{aligned} \tag{4.32}$$

With the inclusion of the time dependence, Eqs. (4.29) and (4.32) become respectively,

$$\mathbf{E}(z=0,t) = \hat{\mathbf{x}} |E_{x0}| \cos(\omega t - \theta_{E0}) \tag{4.33}$$

$$\begin{aligned} \mathbf{E}(z,t) = |E_{x0}| \cos\left[\omega t - \theta_{E0} - \frac{(k_z^+ + k_z^-)}{2} k_p z\right] \\ \left[\hat{\mathbf{x}} \cos\left\{\frac{k_p(k_z^+ - k_z^-)}{2} z\right\} - \hat{\mathbf{y}} \sin\left\{\frac{k_p(k_z^+ - k_z^-)}{2} z\right\}\right] \end{aligned} \tag{4.34}$$

It is seen that Eq. (4.33) is a linearly polarized wave, polarized in the x-direction and that Eq. (4.34) is also a linearly polarized wave with the polarization direction rotated in the counterclockwise direction by an angle θ_F as given by

$$\theta_F = k_p\left(\frac{k_z^- - k_z^+}{2}\right)z \tag{4.35}$$

Thus, in the range of frequencies where both the modes propagate, the plane of polarization rotates uniformly with distance in the direction of the magnetostatic field. This phenomenon is known as Faraday rotation. The rotation per unit distance θ_F/z is seen from Eq. (4.35) to depend on the difference between the propagation coefficients of the left and the right circularly polarized waves.

It is interesting to find the approximations for θ_F/z for two special cases. For the first case, the plasma is very tenuous and the magnetostatic field very weak so that $\omega \gg \omega_p$ and $\omega \gg \omega_c$. In terms of the normalized parameters Ω and R, the first case corresponds to $\Omega \gg 1$ and $\Omega \gg R$. For a weak magnetostatic field Ω_1 which is only slightly less than unity, is greater than R. Therefore it can be ascertained from Eqs. (4.19) and (4.22) that both the modes propagate only in the range $\Omega > \Omega_2$. For $\Omega \gg 1$ and $\Omega \gg R$, k_z^- given by Eq. (4.19) can be simplified as follows:

$$\begin{aligned} k_z^- &= \Omega\left[1 - \frac{1}{\Omega^2}\left(1 + \frac{R}{\Omega}\right)^{-1}\right]^{1/2} = \Omega\left[1 - \frac{1}{\Omega^2}\left(1 - \frac{R}{\Omega}\right)\right]^{1/2} \\ &= \Omega\left[1 - \frac{1}{2\Omega^2}\left(1 - \frac{R}{\Omega}\right)\right] \qquad \text{for } \Omega \gg 1 \text{ and } \Omega \gg R \end{aligned} \tag{4.36}$$

In a similar manner, it can be obtained from Eq. (4.22) that

$$k_z^+ = \Omega\left[1 - \frac{1}{2\Omega^2}\left(1 + \frac{R}{\Omega}\right)\right] \qquad \text{for } \Omega \gg 1 \text{ and } \Omega \gg R \tag{4.37}$$

With the help of Eqs. (4.35)–(4.37), it can be deduced that for $\Omega \gg 1$ and $\Omega \gg R$:

$$\theta_F/z = \frac{k_p}{2}\frac{R}{\Omega^2} = \frac{\omega_c\omega_p^2}{2c\omega^2} \tag{4.38}$$

The Faraday rotation per unit distance is seen from Eq. (4.38) to be very small. It is interesting to note from Eq. (4.38) that for weak magnetic field strengths, the Faraday rotation is directly proportional to the **B**-field. This special case obtains, for example, in interplanetary space where the plasma is very tenuous and the magnetostatic field very weak. Even though the Faraday rotation per unit distance is very small, since for the propagation in the interplanetary space total distances involved are so enormous, the resultant Faraday rotation is likely to be quite large and has to be contended with.

The second special case corresponds to very large values for the magnetostatic field so that $\omega_c \gg \omega$ and $\omega_c \gg \omega_p$. Equivalently, this case is defined by the inequalities $R \gg \Omega$ and $R \gg 1$. Under these conditions, k_z^- given by Eq. (4.19) can be simplified as follows:

$$k_z^- = \Omega\left[1 - \frac{1}{\Omega R}\left(1 + \frac{\Omega}{R}\right)^{-1}\right]^{1/2} = \Omega\left[1 - \frac{1}{\Omega R}\left(1 - \frac{\Omega}{R}\right)\right]^{1/2}$$
$$= \Omega\left[1 - \frac{1}{2\Omega R}\left(1 - \frac{\Omega}{R}\right)\right] \quad \text{for } R \gg \Omega \text{ and } R \gg 1 \tag{4.39}$$

Note that in deducing Eq. (4.39), it has also been assumed that $\Omega R \gg 1$. Similarly, it can be derived from Eq. (4.22) that

$$k_z^+ = \Omega\left[1 + \frac{1}{2\Omega R}\left(1 + \frac{\Omega}{R}\right)\right] \quad \text{for } R \gg \Omega \text{ and } R \gg 1 \tag{4.40}$$

The substitution of Eqs. (4.39) and (4.40) in Eq. (4.35) yields

$$\theta_F/z = -\frac{k_p}{2}\frac{1}{R} = -\frac{\omega_p^2}{2c\omega_c} \tag{4.41}$$

For $R = 1/\sqrt{2}$, $\Omega_1 = R$ and for $R > 1/\sqrt{2}$, $\Omega_1 < R$; therefore for this special case, both the left and the right circularly polarized waves propagate not only for $\Omega > \Omega_2$ but also for $\Omega_1 < \Omega < R$. It is seen from Eq. (4.41) that for the case of very large **B**-fields, the Faraday rotation is in the clockwise direction in contrast to that for the case of a tenuous plasma and weak **B**-fields for which the Faraday rotation is in the counterclockwise direction. Also, Eq. (4.41) shows that for very large **B**-fields, the Faraday rotation is inversely proportional to the strength of the magnetostatic field. In laboratory experiments with semiconductor plasmas, it is possible to realize the conditions $R \gg \Omega$ and $R \gg 1$ approximately. In such cases, it has been confirmed that the Faraday rotation is inversely proportional to the **B**-field.

The measurement of Faraday rotation is a useful diagnostic tool. The particle density in a plasma can be determined, for example, by the Langmuir probe measurement. This yields the value of the plasma angular frequency ω_{pe} of the electrons. Then, a measurement of the Faraday rotation for a wave of known angular frequency ω enables the determination of the electron gyromagnetic angular frequency ω_{ce} from which the value of the magnetic flux density B *inside* the plasma can be evaluated. If the magnetic flux density *outside* the plasma is known, the parameter β of the plasma device can be calculated and an assessment of the diamagnetic effect of the plasma can be made.

A medium is said to be optically active if it exhibits the phenomenon of Faraday rotation even without any external magnetostatic field. In the absence of the **B**-field, $R = 0$ and $k_z^- = k_z^+ = \sqrt{\Omega^2 - 1}$, and, therefore according to Eq. (4.35) there is no Faraday rotation. Consequently, it is verified that a plasma exhibits Faraday rotation only in the presence of an external magnetostatic field. For this reason, a plasma is said to be magnetoactive.

Radio communication black-out due to re-entry plasma sheath

The theory of wave propagation in a plasma under the influence of a magnetostatic field developed so far enables an approximate explanation for the cause and a qualitative discussion of the possible methods available for the alleviation of the problem of radio communication black-out occurring at the time a space vehicle re-enters the earth's atmosphere. A space vehicle moving at hypersonic velocities, as it re-enters the earth's atmosphere, ionizes the air in contact with it due to the intense heat generated by the friction between the vehicle and the atmosphere with the result that the space vehicle is enveloped by a layer of plasma called the re-entry plasma sheath. The collisional effects in the plasma sheath are quite high but it is possible to explain qualitatively the phenomenon of radio communication black-out during the re-entry of the space vehicle into the earth's atmosphere without taking the collisional effects into account. The frequency of the radio wave is very much less than the plasma frequency of the electrons inside the plasma sheath. In a typical example, $\omega/2\pi = 3 \times 10^9$ Hz and $\omega_p/2\pi = 10^{11}$ Hz so that $\Omega = 0.03$, which is much less than unity. The earth's magnetic field is so small that its effect on the characteristics of the plasma sheath is negligible with the result that the propagation characteristics in a re-entry plasma sheath are essentially the same as in an isotropic plasma. From a treatment of the plane wave characteristics in an isotropic plasma as contained in Sec. 5.3, it is seen that the plasma is opaque to the electromagnetic waves for $\Omega < 1$. Therefore, the electromagnetic waves emanating from the antenna which is usually mounted flush with the surface of the space vehicle are exponentially damped at a high rate as they travel through the plasma sheath. The electromagnetic waves emerging from the plasma sheath are so severely attenuated that they are unable to create a response that can be detected at the distant radio stations located on the earth thus accounting for the radio communication black-out. The severity of the black-out increases naturally as the thickness of the plasma sheath becomes larger.

In principle, a number of methods are available for the alleviation of the problem of radio communication black-out.

1. The thickness of the plasma sheath depends on the local shape of the surface of the space vehicle. By mounting the antenna at a location where the plasma sheath has the least thickness, the attenuation of the electromagnetic waves caused by the plasma sheath can be minimized. But even the smallest thickness of the plasma sheath usually present in a space vehicle produces sufficient attenuation to cause black-out that this method is not expected to be very useful.

2. Another possibility is to increase the frequency used for communication so that ω becomes greater than ω_p and the plasma sheath becomes transparent to the electromagnetic waves. The high radio frequencies which are required to penetrate the plasma sheath correspond to millimeter and submillimeter wavelengths. At

these very short wavelengths the resonant absorption in the intervening atmosphere is very severe. Consequently, the use of very high frequencies such that $\omega > \omega_p$ does not appear to be a practical method of solution of the problem of radio communication black-out.

3. If the wave frequency cannot be increased to become greater than the plasma frequency of the electrons, the natural course is to seek methods for lowering the plasma frequency of the electrons below the wave frequency. The plasma frequency of the electrons can be lowered by decreasing the number density of the electrons. This decrease can be accomplished by injecting the plasma with neutral particles such as water or by seeding the plasma with chemicals which promote recombination of the electrons with the positive ions to form neutral particles. The number density of the electrons should be lowered in a sufficiently large volume of the plasma surrounding the antenna. This method, therefore, requires a large amount of matter to be carried on the space vehicle and the concomitant increase in the load of the space vehicle may create aerodynamical problems.

4. Another possible method is to impose externally a large magnetostatic field on the plasma so as to alter its propagation characteristics in such a manner that the wave frequency falls inside the frequency ranges of propagation. Suppose that the magnetostatic field is perpendicular to the surface of the space vehicle and that the electromagnetic waves passing through the plasma sheath progress in the direction of the magnetostatic field. From Fig. 5.2, it is seen that the cut-off frequency instead of occurring at $\Omega = 1$ as for an isotropic plasma now occurs at a lower frequency $\Omega = \Omega_l$ if the left circularly polarized wave is the mode that contributes to the propagation. By increasing the strength of the magnetostatic field sufficiently, Ω_l can be lowered below the wave frequency so that the plasma becomes transparent to the electromagnetic waves. The required strength of the magnetostatic field is so large that it is impractical to impose the necessary field strength from a space vehicle over a large volume of the plasma surrounding the antenna. Suppose that it is the right circularly polarized wave that contributes to the propagation. For this case, it is seen from Fig. 5.3 that there is a low frequency range $0 < \Omega < R$ of propagation. By increasing the strength of the magnetostatic field the cut-off frequency $\Omega = R$ of the propagation range can be increased such that the wave frequency falls within the pass band $0 < \Omega < R$. As a consequence, the electromagnetic waves propagate through the plasma sheath without suffering the severe attenuation which causes the radio communication black-out. The required value of the magnetic flux density has been estimated to be of the order of a few thousand gauss. It is not impossible to impose this required field strength from a space vehicle over the necessary volume of the plasma surrounding the antenna if compact superconducting magnets are used for producing the magnetic field. However, it is to be pointed out that the feasibility of opening a low frequency "window" for radio communication through a re-entry plasma sheath with the help of an external magnetostatic field has not yet been established by actual experiments.

It appears that a satisfactory and a realizable solution to the problem of radio communication black-out at the time of re-entry of a space vehicle into the earth's atmosphere is yet to be found [8].

5.5. Propagation across the magnetostatic field in an electron plasma

As before, the magnetostatic field is parallel to the z-axis. It is desired to investigate the characteristics of a plane wave propagating perpendicular to the **B**-field. Therefore, it is assumed that the wave is propagating in the x-direction. All the field components have the phase factor

$$\exp(ik_p k_x x) \tag{5.1}$$

and have no dependence on y or z. Therefore the Cartesian components of Eqs. (2.18) and (2.19) together with Eq. (2.15) are given by

$$H_x(x) = 0 \tag{5.2a}$$

$$-(\partial/\partial x)E_z(x) = i\omega\mu_0 H_y(x) \tag{5.2b}$$

$$(\partial/\partial x)E_y(x) = i\omega\mu_0 H_z(x) \tag{5.2c}$$

$$0 = \varepsilon_1 E_x(x) + i\varepsilon_2 E_y(x) \tag{5.3a}$$

$$-(\partial/\partial x)H_z(x) = -i\omega\varepsilon_0[-i\varepsilon_2 E_x(x) + \varepsilon_1 E_y(x)] \tag{5.3b}$$

$$(\partial/\partial x)H_y(x) = -i\omega\varepsilon_0 \varepsilon_3 E_z(x) \tag{5.3c}$$

From Eqs. (5.2a), (5.2b), and (5.3c), it is seen that $[E_z(x), H_y(x)]$ with $H_x(x) \equiv 0$ form one set, and, Eqs. (5.3a), (5.3b), and (5.2c) show that $[E_x(x), E_y(x), H_z(x)]$ form another set. The x-component of the time-averaged Poynting vector

$$\hat{\mathbf{x}} \cdot \tfrac{1}{2}\,\mathrm{Re}\,[\mathbf{E}(x) \times \mathbf{H}^*(x)] = \tfrac{1}{2}\,\mathrm{Re}\,[E_y(x)H_z^*(x) - E_z(x)H_y^*(x)] \tag{5.4}$$

is the time-averaged power transported per unit area in the direction of propagation. Since there are no variations with respect to y or z, Eq. (5.4) may be taken as a measure of the total time-averaged power transported in the direction of propagation. It is found from Eq. (5.4) that the two sets of fields mentioned in the foregoing remain separated even in the expression for the total time-averaged power. In view of this, these two sets of fields form two independent modes of propagation perpendicular to the **B**-field and hence can be treated separately.

TEM mode

The first set of fields forms a TEM mode since it has no component of either the electric or the magnetic field in the direction of propagation. The electric field is polarized linearly in the direction of the **B**-field. Consider this mode in the absence of the **B**-field. Since the electric field is in the z-direction, so also is the average velocity of the electrons. A magnetostatic field imposed in the z-direction, that is the direction of the velocity vector, introduces no additional force with the result that this mode is unaffected. Therefore, the TEM mode with its electric vector linearly polarized in the direction of the **B**-field and propagating perpendicular to the **B**-field should be expected to be identical to that in an isotropic plasma. This result is confirmed presently by the study of the dispersion relation of this mode.

In accordance with the expression (5.1), the field components are assumed to be of the form

$$E_z(x) = E_{z0}\, e^{ik_p k_x x} \qquad H_y(x) = H_{y0}\, e^{ik_p k_x x} \tag{5.5}$$

The substitution of Eqs. (5.5) into Eqs. (5.2b) and (5.3c) yields

$$k_p k_x E_{z0} + \omega\mu_0 H_{y0} = 0 \tag{5.6a}$$

$$\omega\varepsilon_0 \varepsilon_3 E_{z0} + k_p k_x H_{y0} = 0 \tag{5.6b}$$

The propagation coefficient k_x of the TEM mode is deduced from Eqs. (5.6a), (5.6b), and (2.21c) to be given by

$$k_x = \sqrt{\Omega^2 \varepsilon_3} = \sqrt{\Omega^2 - 1} \tag{5.7}$$

which shows that this mode propagates only for $\Omega > 1$. The normalized phase and group velocities are evaluated with the help of Eq. (5.7) to be given by

$$v_{ph}/c = \Omega/k_x = \Omega(\Omega^2 - 1)^{-1/2} \tag{5.8}$$

$$v_g/c = \partial\Omega/\partial k_x = \Omega^{-1}(\Omega^2 - 1)^{1/2} \tag{5.9}$$

Note that Eqs. (5.7), (5.8), and (5.9) are identical with Eqs. (3.7), (3.11), and (3.12), respectively, as anticipated. The cut-off frequency for this mode occurs at $\Omega = 1$. The phase and the group velocities become infinite and zero, respectively, at the cut-off frequency. Also, v_{ph}/c and v_g/c both approach unity in the limit of Ω tending to infinity and the conditions pertaining to the free space propagation are attained. In the range of frequencies where there is propagation, the dependences on Ω of k_x, v_{ph}/c, and v_g/c pertaining to this mode are identical with the corresponding quantities in an isotropic plasma, as illustrated in Fig. 5.1.

TM mode

The second set of fields forms a TM mode since its magnetic field is transverse to the direction of propagation. This mode is also linearly polarized with its magnetic vector lying parallel to the **B**-field. The electric vector in the plane perpendicular to the direction of propagation is also linearly polarized but is perpendicular to the **B**-field.

The propagation characteristics of this mode can be deduced in the following manner. From Eqs. (5.3a) and (5.3b), $E_x(x)$ and $E_y(x)$ can be expressed in terms of $H_z(x)$ with the following result:

$$E_x(x) = \frac{-\varepsilon_2}{\omega\varepsilon_0\varepsilon}\frac{\partial}{\partial x}H_z(x) \qquad E_y(x) = \frac{\varepsilon_1}{i\omega\varepsilon_0\varepsilon}\frac{\partial}{\partial x}H_z(x) \tag{5.10a, b}$$

where

$$\varepsilon = \varepsilon_1^2 - \varepsilon_2^2 \tag{5.11}$$

The substitution of Eq. (5.10b) in Eq. (5.2c) gives the following differential equation satisfied by $H_z(x)$:

$$\frac{\partial^2}{\partial x^2}H_z(x) + \omega^2\mu_0\varepsilon_0\frac{\varepsilon}{\varepsilon_1}H_z(x) = 0 \tag{5.12}$$

In accordance with the expression (5.1), $H_z(x)$ is sought in the form

$$H_z(x) = H_{z0}\exp(ik_p k_x x) \tag{5.13}$$

If Eq. (5.13) is substituted into Eq. (5.12), the requirement of a nontrivial solution of the resulting equation yields the following expression for the propagation coefficient of the TM mode:

$$k_x = \Omega[\varepsilon/\varepsilon_1]^{1/2} \tag{5.14}$$

With the help of Eqs. (2.21a), (4.17), and (5.11), k_x given by Eq. (5.14) can be expressed in terms of Ω and R as follows:

$$\begin{aligned} k_x &= \Omega\left[\frac{(\varepsilon_1 - \varepsilon_2)(\varepsilon_1 + \varepsilon_2)}{\varepsilon_1}\right]^{1/2} \\ &= \left[\frac{(\Omega - \Omega_1)(\Omega + \Omega_1)(\Omega - \Omega_2)(\Omega + \Omega_2)}{(\Omega - \Omega_u)(\Omega + \Omega_u)}\right]^{1/2} \end{aligned} \tag{5.15}$$

where

$$\Omega_u = \sqrt{1 + R^2} \tag{5.16}$$

It can be shown with the help of Eqs. (4.18) and (5.16) that $\Omega_1 < \Omega_u < \Omega_2$. Consequently, it follows from Eq. (5.15) that the TM mode propagates for $\Omega_1 < \Omega < \Omega_u$ and $\Omega_2 < \Omega < \infty$. It is verified that $k_x = 1$ and k_x tends to Ω for $\Omega = 1$ and Ω tending to infinity respectively. From Fig. 5.7, where the Brillouin diagram for this mode is illustrated, it is seen that $\Omega = \Omega_1$ and $\Omega = \Omega_2$ are the two cut-off frequencies and these same cut-off frequencies were obtained for the case of propagation parallel to the magnetostatic field. The resonant frequency $\Omega = \Omega_u$ is known as the normalized *upper hybrid resonant* frequency. In terms of the unnormalized quantities the upper hybrid resonant frequency is given by $\omega_u = \sqrt{\omega_{pe}^2 + \omega_{ce}^2}$. The normalized phase and the normalized group velocities can be derived with the help of Eq. (5.15) to be given by

$$v_{ph}/c = \frac{\Omega}{k_x} = \left[\frac{\Omega^2(\Omega^2 - \Omega_u^2)}{(\Omega^2 - \Omega_1^2)(\Omega^2 - \Omega_2^2)}\right]^{1/2} \tag{5.17}$$

$$\begin{aligned} v_g/c &= \frac{\partial \Omega}{\partial k_x} \\ &= \frac{(\Omega^2 - \Omega_u^2)^2}{\Omega[\Omega^4 - 2\Omega^2(R^2+1) + R^4 + 3R^2 + 1]}\left[\frac{(\Omega^2 - \Omega_1^2)(\Omega^2 - \Omega_2^2)}{(\Omega^2 - \Omega_u^2)}\right]^{1/2} \end{aligned} \tag{5.18}$$

The phase and the group velocities become infinite and zero, repectively, at the two cut-off frequencies $\Omega = \Omega_1$ and $\Omega = \Omega_2$. At the resonant frequency $\Omega = \Omega_u$, both the velocities become zero. For Ω tending to infinity, both v_{ph}/c and v_g/c tend to unity. The group velocity has a maximum value in the range $\Omega_1 < \Omega < \Omega_u$. The normalized phase and the normalized group velocities are also depicted in Fig. 5.7 as functions of Ω in the frequency ranges where there is propagation of the TM mode.

5.6. Propagation in an arbitrary direction with respect to the magnetostatic field in an electron plasma

The magnetostatic field is in the z-direction. The spatial dependence of the field quantities associated with a plane wave propagating in an arbitrary direction with respect to the **B**-field is of the form

$$E_n(\mathbf{r}) = E_{n0}\exp[ik_p\{\hat{\mathbf{x}}k_x + \hat{\mathbf{y}}k_y + \hat{\mathbf{z}}k_z\}\cdot\mathbf{r}] \qquad n = x, y, z \tag{6.1a}$$

$$H_n(\mathbf{r}) = H_{n0}\exp[ik_p\{\hat{\mathbf{x}}k_x + \hat{\mathbf{y}}k_y + \hat{\mathbf{z}}k_z\}\cdot\mathbf{r}] \qquad n = x, y, z \tag{6.1b}$$

With the help of Eqs. (6.1), the Cartesian components of Eqs. (2.18) and (2.19) together with Eq. (2.15) are found to be given by

$$k_y E_{z0} - k_z E_{y0} = \frac{\omega\mu_0}{k_p} H_{x0} \tag{6.2a}$$

$$k_z E_{x0} - k_x E_{z0} = \frac{\omega\mu_0}{k_p} H_{y0} \tag{6.2b}$$

$$k_x E_{y0} - k_y E_{x0} = \frac{\omega\mu_0}{k_p} H_{z0} \tag{6.2c}$$

$$k_y H_{z0} - k_z H_{y0} = -\frac{\omega\varepsilon_0}{k_p}[\varepsilon_1 E_{x0} + i\varepsilon_2 E_{y0}] \tag{6.3a}$$

$$k_z H_{x0} - k_x H_{z0} = -\frac{\omega\varepsilon_0}{k_p}[-i\varepsilon_2 E_{x0} + \varepsilon_1 E_{y0}] \tag{6.3b}$$

$$k_x H_{y0} - k_y H_{x0} = -\frac{\omega\varepsilon_0 \varepsilon_3}{k_p} E_{z0} \tag{6.3c}$$

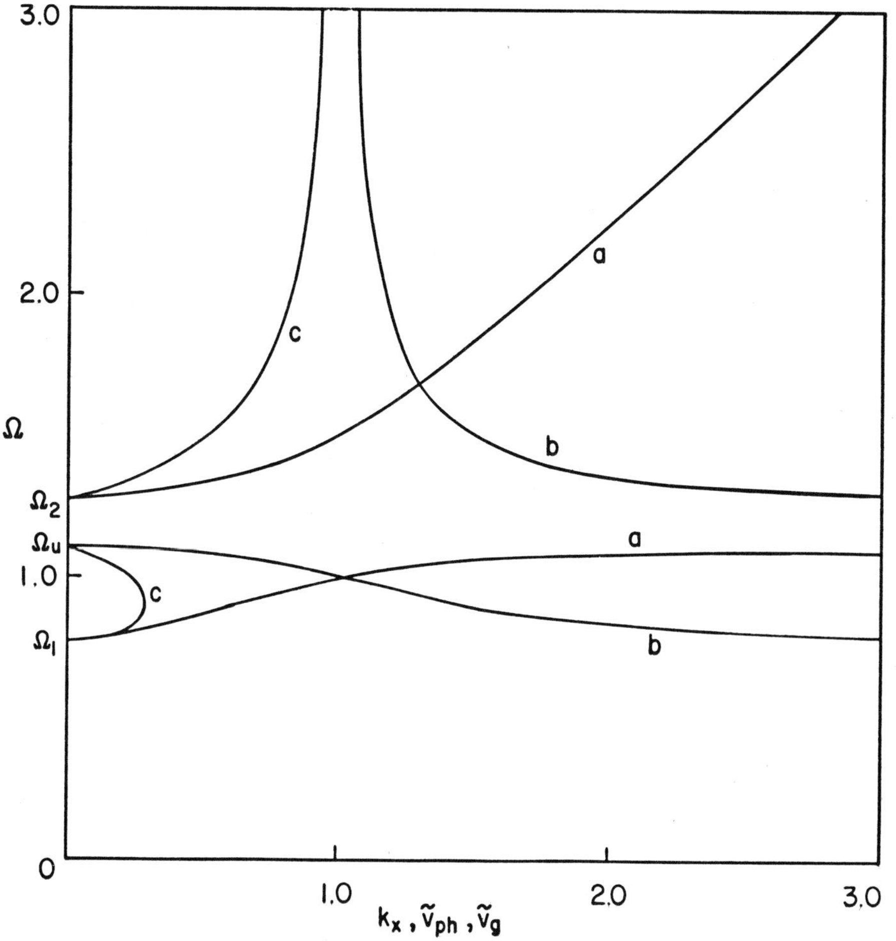

Fig. 5.7. Brillouin diagram for the TM mode propagating across the magnetostatic field in a magnetoionic medium: $R = 0.5$; (a) k_x; (b) $\tilde{v}_{ph} = v_{ph}/c$; (c) $\tilde{v}_g = v_g/c$.

If H_{x0} and H_{y0} from Eqs. (6.2a) and (6.2b) are substituted in Eqs. (6.3b) and (6.3a), respectively, two simultaneous equations in E_{x0} and E_{y0} are obtained. The solution of these equations gives E_{x0} and E_{y0} in terms of E_{z0} and H_{z0}. The substitution of these results for E_{x0} and E_{y0} back in Eqs. (6.2b) and (6.2a), respectively, enables H_{x0} and H_{y0} to be expressed in terms of E_{z0} and H_{z0}. The final results are given by

$$E_{x0} = -\frac{k_z}{\Omega^2 \bar{\varepsilon}}(\bar{\varepsilon}_1 k_x - i\varepsilon_2 k_y)E_{z0} - \frac{k_p}{\omega\varepsilon_0 \bar{\varepsilon}}(\bar{\varepsilon}_1 k_y + i\varepsilon_2 k_x)H_{z0} \tag{6.4}$$

$$E_{y0} = -\frac{k_z}{\Omega^2 \bar{\varepsilon}}(\bar{\varepsilon}_1 k_y + i\varepsilon_2 k_x)E_{z0} + \frac{k_p}{\omega\varepsilon_0 \bar{\varepsilon}}(\bar{\varepsilon}_1 k_x - i\varepsilon_2 k_y)H_{z0} \tag{6.5}$$

$$H_{x0} = \frac{k_p}{\omega\mu_0}\left[k_y + \frac{k_z^2}{\Omega^2 \bar{\varepsilon}}(\bar{\varepsilon}_1 k_y + i\varepsilon_2 k_x)\right]E_{z0} - \frac{k_z}{\Omega^2 \bar{\varepsilon}}(\bar{\varepsilon}_1 k_x - i\varepsilon_2 k_y)H_{z0} \tag{6.6}$$

$$H_{y0} = -\frac{k_p}{\omega\mu_0}\left[k_x + \frac{k_z^2}{\Omega^2 \bar{\varepsilon}}(\bar{\varepsilon}_1 k_x - i\varepsilon_2 k_y)\right]E_{z0} - \frac{k_z}{\Omega^2 \bar{\varepsilon}}(\bar{\varepsilon}_1 k_y + i\varepsilon_2 k_x)H_{z0} \tag{6.7}$$

where

$$\bar{\varepsilon}_1 = \varepsilon_1 - k_z^2/\Omega^2 \qquad \bar{\varepsilon} = \bar{\varepsilon}_1^2 - \varepsilon_2^2 \tag{6.8}$$

The substitution of Eqs. (6.4)–(6.7) in Eqs. (6.2c) and (6.3c) yields the following pair of simultaneous equations in E_{z0} and H_{z0}:

$$\frac{k_z i\varepsilon_2}{\Omega^2 \bar{\varepsilon}}(k_x^2 + k_y^2)E_{z0} + \frac{k_p}{\omega\varepsilon_0}\left[\Omega^2 - \frac{\bar{\varepsilon}_1}{\bar{\varepsilon}}(k_x^2 + k_y^2)\right]H_{z0} = 0 \tag{6.9}$$

$$\frac{k_p}{\omega\mu_0}\left[\frac{(k_x^2 + k_y^2)}{\bar{\varepsilon}}\left(\bar{\varepsilon} + \bar{\varepsilon}_1 \frac{k_z^2}{\Omega^2}\right) - \Omega^2 \varepsilon_3\right]E_{z0} + \frac{k_z i\varepsilon_2}{\Omega^2 \bar{\varepsilon}}(k_x^2 + k_y^2)H_{z0} = 0 \tag{6.10}$$

A nontrivial solution of Eqs. (6.9) and (6.10) exists only if the determinant of the coefficients of E_{z0} and H_{z0} vanishes. This condition gives a functional relationship between **k** and ω known as the dispersion equation. It is worthwhile to note that in Eqs. (6.9) and (6.10) and therefore in the dispersion equation, the wavenumbers k_x and k_y associated with the directions perpendicular to the magnetostatic field appear only in the combination

$$k_p^2 = k_x^2 + k_y^2 \tag{6.11}$$

This is a consequence of the cylindrical symmetry that exists about the direction of the magnetostatic field.

Dispersion relation

In view of this cylindrical symmetry, without loss of generality, the normalized propagation vector **k** may be assumed to lie in the *xz*-plane with the result that

$\hat{\mathbf{z}} \times \mathbf{k}$ coincides with the y-axis. The angle between $\mathbf{k}$ and the direction of the magnetostatic field is denoted by θ_{ph} so that

$$k_x = k \sin \theta_{ph} \qquad k_y = 0 \qquad k_z = k \cos \theta_{ph} = kl \tag{6.12}$$

Together with Eqs. (6.12), Eqs. (6.1) describe the spatial dependence of the field quantities. It is convenient to rewrite Eqs. (6.9) and (6.10) as follows:

$$\frac{k_z}{\Omega}\frac{\varepsilon_2}{\bar{\varepsilon}}k_x^2 i\sqrt{\varepsilon_0}\, E_{z0} + \left\{\Omega^2 - \frac{\bar{\varepsilon}_1}{\bar{\varepsilon}}k_x^2\right\}\sqrt{\mu_0}\, H_{z0} = 0 \tag{6.13}$$

$$-\left[k_x^2\left(1 + \frac{\bar{\varepsilon}_1}{\bar{\varepsilon}}\frac{k_z^2}{\Omega^2}\right) - \Omega^2\varepsilon_3\right] i\sqrt{\varepsilon_0}\, E_{z0} + \frac{k_z}{\Omega}\frac{\varepsilon_2}{\bar{\varepsilon}}k_x^2\sqrt{\mu_0}\, H_{z0} = 0 \tag{6.14}$$

The determinant of the coefficients of $i\sqrt{\varepsilon_0}\, E_{z0}$ and $\sqrt{\mu_0}\, H_{z0}$ in Eqs. (6.13) and (6.14) is obtained as

$$\frac{k_z^2}{\Omega^2}\frac{\varepsilon_2^2}{\bar{\varepsilon}^2}k_x^4 + \Omega^2 k_x^2\left(1 + \frac{\bar{\varepsilon}_1}{\bar{\varepsilon}}\frac{k_z^2}{\Omega^2}\right) - \Omega^4\varepsilon_3 + \frac{\bar{\varepsilon}_1}{\bar{\varepsilon}}k_x^2(\Omega^2\varepsilon_3 - k_x^2) - \frac{k_z^2}{\Omega^2}\frac{\bar{\varepsilon}_1^2}{\bar{\varepsilon}^2}k_x^4 \tag{6.15}$$

On combining the first and the last terms in the expression (6.15) with the help of Eqs. (6.8) and rearranging the resultant expression as a polynomial in k_x^2, it is found that

$$-\frac{k_x^4}{\bar{\varepsilon}}\left(\frac{k_z^2}{\Omega^2} + \bar{\varepsilon}_1\right) + \frac{\Omega^2 k_x^2}{\bar{\varepsilon}}\left(\bar{\varepsilon} + \bar{\varepsilon}_1\frac{k_z^2}{\Omega^2} + \bar{\varepsilon}_1\varepsilon_3\right) - \Omega^4\varepsilon_3 \tag{6.16}$$

A nontrivial solution of Eqs. (6.13) and (6.14) exists only if the determinant of the coefficients of $i\sqrt{\varepsilon_0}\, E_{z0}$ and $\sqrt{\mu_0}\, H_{z0}$ vanishes. This condition leads to the dispersion equation which with the help of Eqs. (6.8) and the expression (6.16) can be deduced to be given by

$$k_x^4 + \frac{\Omega^2 k_x^2}{\varepsilon_1}\left[(\varepsilon_1 + \varepsilon_3)\frac{k_z^2}{\Omega^2} - (\varepsilon_1^2 - \varepsilon_2^2 + \varepsilon_1\varepsilon_3)\right] + \frac{\Omega^4\varepsilon_3}{\varepsilon_1}\left[\left(\varepsilon_1 - \frac{k_z^2}{\Omega^2}\right)^2 - \varepsilon_2^2\right] = 0 \tag{6.17}$$

Since in Eqs. (6.9) and (6.10), k_x and k_y appear only in the combination given by Eq. (6.11), it follows that Eq. (6.17) is valid if k_x^2 is replaced by k_ρ^2. This observation is useful for purposes of comparison later on.

The substitution of Eqs. (6.12) in Eq. (6.17) and a suitable regrouping of the terms yields the following quadratic equation in k^2:

$$A_2 k^4 + A_1 k^2 + A_0 = 0 \tag{6.18}$$

where

$$A_2 = \varepsilon_1 - (\varepsilon_1 - \varepsilon_3)l^2 \tag{6.19a}$$

$$A_1 = -\Omega^2[\varepsilon_1^2 - \varepsilon_2^2 + \varepsilon_1\varepsilon_3 - (\varepsilon_1^2 - \varepsilon_2^2 - \varepsilon_1\varepsilon_3)l^2] \tag{6.19b}$$

$$A_0 = \Omega^4\varepsilon_3(\varepsilon_1^2 - \varepsilon_2^2) = \Omega^4\varepsilon_3\varepsilon \tag{6.19c}$$

In Eqs. (6.18) and (6.19), k is the magnitude of the propagation vector and $\theta_{ph} = \cos^{-1} l$ is its direction with respect to that of the magnetostatic field. Let k_1^2 and k_2^2 be the two possible solutions of Eq. (6.18) for k^2. The field components associated with these two different values of k^2 form two different sets whose linear combination is the most general solution for the plane wave fields propagating in a direction making an arbitrary angle with respect to the **B**-field. It is to be established that these two different sets of fields are independent and consequently, constitute the two plane wave *modes* in a magnetoionic medium.

Orthogonality

From Eq. (6.13), it is obtained that

$$\sqrt{\mu_0}\, H_{z0} = -\frac{k_z \varepsilon_2 k_x^2}{\Omega(\Omega^2 \bar{\varepsilon} - \bar{\varepsilon}_1 k_x^2)} i\sqrt{\varepsilon_0}\, E_{z0} \tag{6.20}$$

With the help of Eqs. (5.11), (6.8), (6.12), and (6.20), the following expressions can be deduced from Eqs. (6.4)–(6.7):

$$i\sqrt{\varepsilon_0}\, E_{x0} = -\frac{k_z k_x(\Omega^2 \bar{\varepsilon}_1 - k_x^2)}{\Omega^2(\Omega^2 \bar{\varepsilon} - \bar{\varepsilon}_1 k_x^2)} i\sqrt{\varepsilon_0}\, E_{z0} \tag{6.21}$$

$$\sqrt{\varepsilon_0}\, E_{y0} = -\frac{k_x k_z \varepsilon_2}{(\Omega^2 \bar{\varepsilon} - \bar{\varepsilon}_1 k_x^2)} i\sqrt{\varepsilon_0}\, E_{z0} \tag{6.22}$$

$$\sqrt{\mu_0}\, H_{x0} = \frac{k_z^2 k_x \varepsilon_2}{\Omega(\Omega^2 \bar{\varepsilon} - \bar{\varepsilon}_1 k_x^2)} i\sqrt{\varepsilon_0}\, E_{z0} \tag{6.23}$$

$$i\sqrt{\mu_0}\, H_{y0} = -\frac{k_x[\Omega^2 \varepsilon - \varepsilon_1 k^2]}{\Omega(\Omega^2 \bar{\varepsilon} - \bar{\varepsilon}_1 k_x^2)} i\sqrt{\varepsilon_0}\, E_{z0} \tag{6.24}$$

It is convenient to introduce a new set of coordinates x', y, and z' obtained by the rotation of the original coordinates about the y-axis such that z' is in the direction of propagation (Fig. 5.8). Then, from Eqs. (6.20)–(6.23) together with Eqs. (6.8) and (6.12), the expressions for the field components in the new coordinate system are evaluated to be given by

$$\begin{aligned} i\sqrt{\varepsilon_0}\, E_{x'0} &= i\sqrt{\varepsilon_0}\, [\cos\theta_{ph} E_{x0} - \sin\theta_{ph} E_{z0}] \\ &= -\sin\theta_{ph} \frac{(\Omega^2 \varepsilon - \varepsilon_1 k^2)}{(\Omega^2 \bar{\varepsilon} - \bar{\varepsilon}_1 k_x^2)} i\sqrt{\varepsilon_0}\, E_{z0} \end{aligned} \tag{6.25}$$

$$\begin{aligned} i\sqrt{\varepsilon_0}\, E_{z'0} &= i\sqrt{\varepsilon_0}\, [\sin\theta_{ph} E_{x0} + \cos\theta_{ph} E_{z0}] \\ &= \cos\theta_{ph} \frac{(\Omega^4 \varepsilon - 2\varepsilon_1 \Omega^2 k^2 + k^4)}{\Omega^2(\Omega^2 \bar{\varepsilon} - \bar{\varepsilon}_1 k_x^2)} i\sqrt{\varepsilon_0}\, E_{z0} \end{aligned} \tag{6.26}$$

$$\sqrt{\mu_0}\, H_{x'0} = \sqrt{\mu_0}\, [\cos\theta_{ph} H_{x0} - \sin\theta_{ph} H_{z0}] = \frac{k^3 \varepsilon_2 \sin\theta_{ph}\cos\theta_{ph}}{\Omega(\Omega^2\bar{\varepsilon} - \bar{\varepsilon}_1 k_x^2)} i\sqrt{\varepsilon_0}\, E_{z0} \quad (6.27)$$

$$\sqrt{\mu_0}\, H_{z'0} = \sqrt{\mu_0}\, [\sin\theta_{ph} H_{x0} + \cos\theta_{ph} H_{z0}] = 0 \quad (6.28)$$

The total time-averaged power radiated can be obtained by integrating the component of the time-averaged Poynting vector in the direction of the wavevector, throughout an infinite plane surface parallel to the phase front. This total time-averaged power is seen to become infinite. However, since the density of the time-averaged power flow is uniform throughout the plane surface of integration, it is legitimate to consider only the power flow density. In order to establish the orthogonality between the two sets of plane wave fields, it is necessary to show that

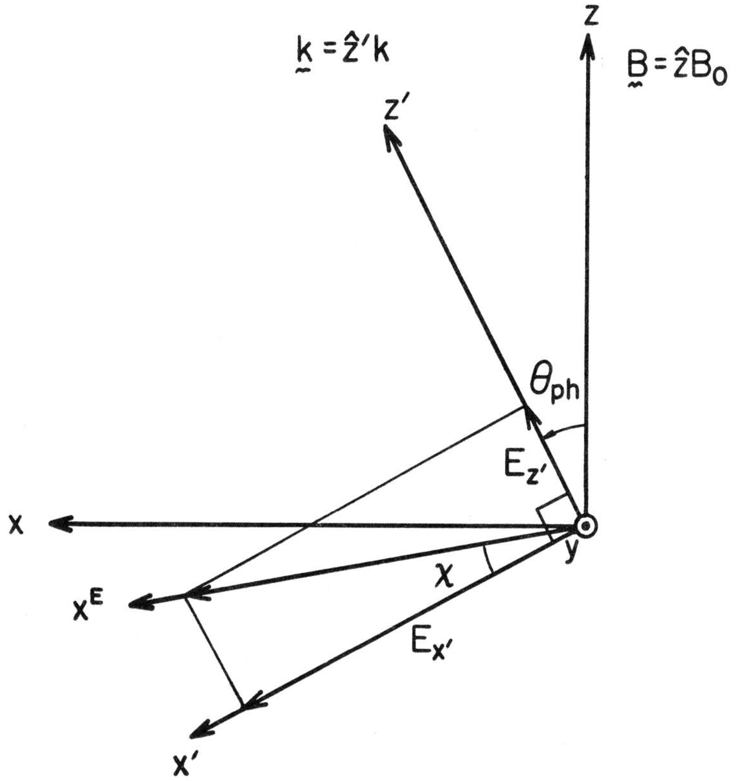

Fig. 5.8. Geometry for the plane wave propagation in an arbitrary direction with respect to the **B**-field: $x'y$: **H**-plane, $x^E y$: **E**-plane.

the time-averaged power flow density can be obtained from the superposition of those due to the two sets of fields separately. This is the same procedure that was used in obtaining Eqs. (4.11) and (5.4) which are respectively the statements of orthogonality for the plane wave propagation parallel and perpendicular to the **B**-field.

For the present case, the time-averaged power flow density is given by

$$\begin{aligned}\hat{\mathbf{z}}' \cdot \mathbf{S} = S_{z'} &= \tfrac{1}{2}\,\mathrm{Re}\,(E^*_{x'} H_y - E^*_y H_{x'}) \\ &= \frac{1}{2\sqrt{\mu_0 \varepsilon_0}}\,\mathrm{Re}\,[(i\sqrt{\varepsilon_0}\, E_{x'})^* i\sqrt{\mu_0}\, H_y - (\sqrt{\varepsilon_0}\, E_y)^* \sqrt{\mu_0}\, H_{x'}]\end{aligned} \tag{6.29}$$

Let the fields associated with the propagation coefficients k_1 and k_2 be denoted by the superscripts 1 and 2, respectively. Thus, the z-component of the electric field is given by

$$E_z = E_z^{(1)} + E_z^{(2)} \tag{6.30}$$

where the first and the second terms on the right side of Eq. (6.30) belong to the first and the second set of fields respectively. It is to be noted that $E_z^{(1)}$ and $E_z^{(2)}$ contain in addition to the amplitude terms, the phase factors $\exp(ik_p k_1 z')$ and $\exp(ik_p k_2 z')$, respectively. In a similar manner, it is found from Eq. (6.25) that

$$\begin{aligned} i\sqrt{\varepsilon_0}\, E_{x'} &= \sum_{n=1,2} i\sqrt{\varepsilon_0}\, E_{x'}{}^{(n)} \\ &= \sum_{n=1,2} -\sin\theta_{ph} \frac{(\Omega^2 \varepsilon - \varepsilon_1 k_n^2)}{d_n} i\sqrt{\varepsilon_0}\, E_z^{(n)}\end{aligned} \tag{6.31}$$

where d_n is the value of $(\Omega^2 \bar{\varepsilon} - \bar{\varepsilon}_1 k_x^2)$ with k replaced by k_n. Similarly, the other field components can be written as the sum of the individual sets of fields. The substitution of Eq. (6.30) and the terms similar to Eq. (6.31) into Eq. (6.29) may be shown to yield

$$\begin{aligned} S_{z'} = {} & \frac{1}{2\sqrt{\mu_0 \varepsilon_0}}\,\mathrm{Re} \sum_{n=1,2} [(i\sqrt{\varepsilon_0}\, E_{x'}{}^{(n)})^* i\sqrt{\mu_0}\, H_y^{(n)} - (\sqrt{\varepsilon_0}\, E_y^{(n)})^* \sqrt{\mu_0}\, H_{x'}{}^{(n)}] \\ & + S_{z'}^c \end{aligned} \tag{6.32}$$

where

$$\begin{aligned} S_{z'}^c = \frac{1}{2\sqrt{\mu_0 \varepsilon_0}}\,\mathrm{Re}\,[& (i\sqrt{\varepsilon_0}\, E_{x'}{}^{(1)})^* i\sqrt{\mu_0}\, H_y^{(2)} + (i\sqrt{\varepsilon_0}\, E_{x'}{}^{(2)})^* i\sqrt{\mu_0}\, H_y^{(1)} \\ & - (\sqrt{\varepsilon_0}\, E_y^{(1)})^* \sqrt{\mu_0}\, H_{x'}{}^{(2)} - (\sqrt{\varepsilon_0}\, E_y^{(2)})^* \sqrt{\mu_0}\, H_{x'}{}^{(1)}] \end{aligned} \tag{6.33}$$

Together with Eqs. (6.12), (6.22), (6.24), (6.25), and (6.27), Eq. (6.33) can be manipulated to give

$$S_{z'}^{c} = S_{z'}^{(1),(2)} + S_{z'}^{(2),(1)} \tag{6.34}$$

and

$$S_{z'}^{(1),(2)} = \frac{k_2 \sin^2\theta_{ph}}{\Omega d_1 d_2 2\sqrt{\mu_0 \varepsilon_0}} \text{Re}\,[(i\sqrt{\varepsilon_0}\,E_z^{(1)})^*(i\sqrt{\varepsilon_0}\,E_z^{(2)})]Q^{(1),(2)} \tag{6.35}$$

where

$$Q^{(1),(2)} = (\Omega^2\varepsilon - \varepsilon_1 k_1^2)(\Omega^2\varepsilon - \varepsilon_1 k_2^2) + k_1^2 k_2^2 l^2 \varepsilon_2^2 \tag{6.36}$$

Note that $S_{z'}^{(2),(1)}$ is obtained from $S_{z'}^{(1),(2)}$ by interchanging k_1 and k_2. From Eq. (6.36), it is found that

$$Q^{(1),(2)} = \Omega^4\varepsilon^2 - \Omega^2\varepsilon\varepsilon_1(k_1^2 + k_2^2) + k_1^2 k_2^2(\varepsilon_1^2 + l^2\varepsilon_2^2) \tag{6.37}$$

It follows from Eqs. (6.18) and (6.19) that

$$A_2(k_1^2 + k_2^2) = \Omega^2[\varepsilon_1^2 - \varepsilon_2^2 + \varepsilon_1\varepsilon_3 - (\varepsilon_1^2 - \varepsilon_2^2 - \varepsilon_1\varepsilon_3)l^2] \tag{6.38}$$

and

$$A_2 k_1^2 k_2^2 = \Omega^4 \varepsilon_3 \varepsilon \tag{6.39}$$

The substitution of Eqs. (6.38) and (6.39) in Eq. (6.37) and the use of Eq. (6.19a) may be shown to yield

$$\begin{aligned} A_2 Q^{(1),(2)} = {} & \Omega^4\varepsilon^2\{\varepsilon_1 - (\varepsilon_1 - \varepsilon_3)l^2\} - \Omega^4\varepsilon\varepsilon_1\{\varepsilon + \varepsilon_1\varepsilon_3 - (\varepsilon_1^2 - \varepsilon_2^2 - \varepsilon_1\varepsilon_3)l^2\} \\ & + \Omega^4\varepsilon_3\varepsilon(\varepsilon_1^2 + \varepsilon_2^2 l^2) = 0 \end{aligned} \tag{6.40}$$

In view of Eq. (6.40), $S_{z'}^{(1),(2)}$ and therefore $S_{z'}^{(2),(1)}$ vanish. Consequently $S_{z'}^{c} = 0$, with the result that Eq. (6.32) shows that the resultant time-averaged power flow density is equal to the sum of the time-averaged power flow densities due to each of the two sets of fields separately. Therefore, the two sets of fields associated with the propagation coefficients k_1 and k_2, respectively, constitute the two independent plane wave modes propagating in an arbitrary direction with respect to the **B**-field.

Frequency ranges of propagation

With the help of Eqs. (2.21a,b,c), Eqs. (6.19a,b,c) can be recast in terms of Ω and R as follows:

$$A_2 = \frac{B_2}{\Omega^2(\Omega^2 - R^2)} \qquad A_1 = \frac{B_1}{\Omega^2(\Omega^2 - R^2)} \qquad A_0 = \frac{B_0}{\Omega^2(\Omega^2 - R^2)} \tag{6.41}$$

where

$$B_2 = \Omega^4 - \Omega^2(1 + R^2) + R^2 l^2 \tag{6.42a}$$

$$B_1 = -\Omega^2[2\Omega^4 - 2\Omega^2(R^2 + 2) + 2 + R^2(1 + l^2)] \tag{6.42b}$$

$$B_0 = \Omega^2(\Omega^2 - 1)\{\Omega^4 - \Omega^2(R^2 + 2) + 1\} \tag{6.42c}$$

From Eqs. (6.41), it is seen that Eq. (6.18) reduces to

$$B_2 k^4 + B_1 k^2 + B_0 = 0 \tag{6.43}$$

The two solutions of Eq. (6.43) are given by

$$k_{1,2}^2 = -\frac{B_1}{2B_2} \pm \frac{1}{2B_2}\sqrt{B_1^2 - 4B_0 B_2} \tag{6.44}$$

The upper and the lower signs in Eq. (6.44) correspond to k_1^2 and k_2^2, respectively. The discriminant $(B_1^2 - 4B_0 B_2)$ can be simplified by using Eqs. (6.42 a,b,c) and the result is

$$B_1^2 - 4B_0 B_2 = 4\Omega^2 R^2[(\Omega^2 - 1)^2 l^2 + \frac{\Omega^2 R^2}{4}(1 - l^2)^2] \tag{6.45}$$

Since Eq. (6.45) is positive, it follows that k_1^2 and k_2^2 are always real. If

$$B_0 B_2 < 0 \qquad \left|\frac{1}{2B_2}\sqrt{B_1^2 - 4B_0 B_2}\right| > \left|\frac{B_1}{2B_2}\right|$$

and hence the solution given by Eq. (6.44) which corresponds to the positive sign in front of the radical yields a positive value for k^2 resulting in the propagation of the corresponding mode. If

$$B_0 B_2 > 0 \qquad \left|\frac{1}{2B_2}\sqrt{B_1^2 - 4B_0 B_2}\right| < \left|\frac{B_1}{2B_2}\right|$$

and hence the signs of k_1^2 and k_2^2 are the same as that of $-B_1/2B_2$. Therefore, both the modes propagate if $B_1/2B_2 < 0$.

Once the propagation direction of the plane wave is specified, l is known, and, then Ω and R are the only remaining parameters in Eqs. (6.42) and (6.43). In the Ω^2 - R^2 space, the regions which correspond to k_1 and k_2 real, that is, which correspond to the propagation regions of modes 1 and 2 respectively, are obtained in the following manner. The parametric equations for $B_2 = 0$, $B_1 = 0$, and $B_0 = 0$ are seen from Eqs. (6.42a)–(6.42c) to be given by the following curves in the Ω^2 - R^2 space:

(a) $R^2 = \Omega^2(\Omega^2 - 1)/(\Omega^2 - l^2)$ for $B_2 = 0$ (6.46a)

(b) $R^2 = (\Omega^2 - 1)^2/[\Omega^2 - (1 + l^2)/2)]$ for $B_1 = 0$ (6.46b)

(c) $\Omega^2 = 1$ for $B_0 = 0$ (6.46c)

(d) $R^2 = \Omega^{-2}(\Omega^2 - 1)^2$ for $B_0 = 0$ (6.46d)

(e) $\Omega^2 = 0$ for $B_1 = 0$ and $B_0 = 0$ (6.46e)

The curves (a)–(e) are shown in Fig. 5.9, and these curves are seen to divide the Ω^2 - R^2 space into several regions inside each of which the signs of B_2, B_1, and B_0 do not change. B_2 changes sign only on crossing the curve corresponding to $B_2 = 0$, that is, the curve (a). In a similar manner, B_1 and B_0 change sign only on crossing the curves corresponding to $B_1 = 0$ [i.e., the curve (b)] and $B_0 = 0$ [i.e., the curves (c) and (d)], respectively. If the signs of B_2, B_1, and B_0 are found for any one region, their signs in all the other regions are obtained by inspection.

For convenience, the important characteristics of the various curves are summarized here. The curves (b) and (d) are both tangential to the Ω^2-axis at $\Omega^2 = 1$. The curve (b) is always above the curve (d). For $\Omega^2 > 1$, the curve (a) lies above the curve (b). Note that $l^2 < (1 + l^2)/2 < 1$. Also, the curves (a) and (d) intersect at $\Omega^2 = l^2/(1 + l^2)$ and $R^2 = 1/[l^2(1 + l^2)]$.

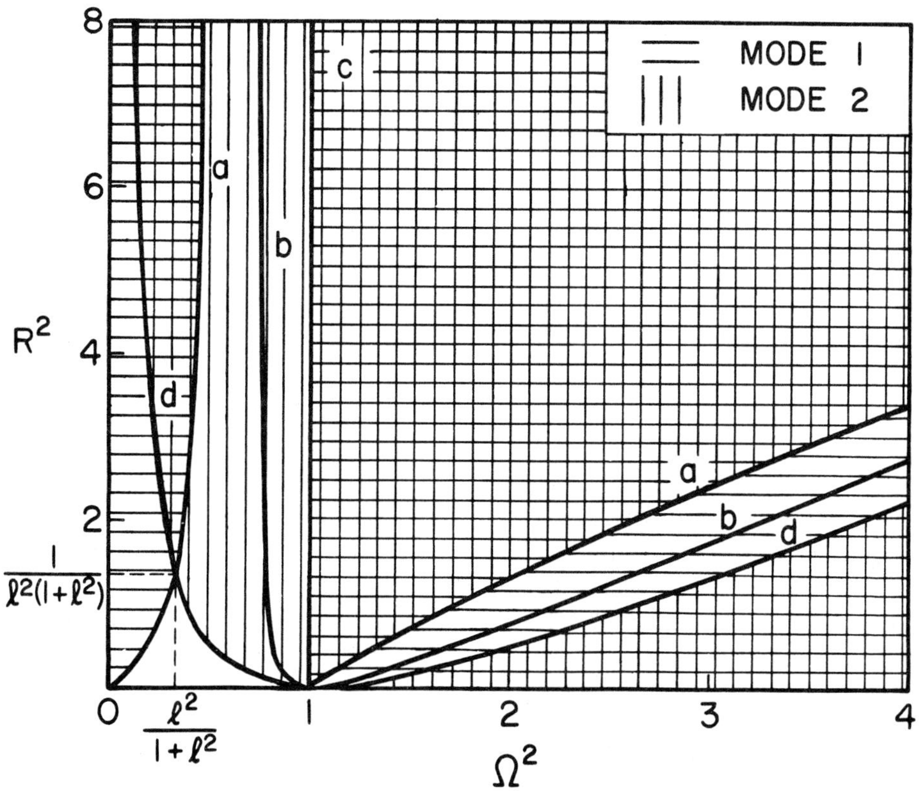

Fig. 5.9. Propagation regions of the plane wave modes 1 and 2 for the arbitrary direction of propagation with respect to the magnetostatic field.

For Ω^2 very large and $R^2 = 0$, it is seen from Eqs. (6.42a) and (6.42c) that both B_2 and B_0 are positive. The signs of B_2 and B_0 in all the other regions in the Ω^2 - R^2 space are found by inspection, as indicated earlier. In the regions where $B_2 B_0 < 0$, either k_1 or k_2 is real, i.e., one of the two modes propagates. If in these regions, $B_2 > 0$, it follows from Eq. (6.44) that k_1 is real and therefore, mode 1 propagates. The region in Fig. 5.9 where mode 1 alone propagates is indicated by shading with horizontal lines. On the other hand, if $B_2 < 0$, Eq. (6.44) shows that k_2 is real and therefore, in the corresponding region mode 2 propagates. The region in Fig. 5.9 where mode 2 alone propagates is indicated by shading with vertical lines.

In the Ω^2 - R^2 space, the regions where $B_2 B_0 > 0$ are determined and in these regions the signs of $B_1/2B_2$ are also ascertained. If in the regions where $B_2 B_0 > 0$, $B_1/2B_2 < 0$, both k_1 and k_2 are real. Therefore, both the modes propagate in these regions which are indicated in Fig. 5.9 by shading with both horizontal and vertical lines.

An examination of Fig. 5.9 reveals the frequency ranges of propagation of the two modes. Let the two values of Ω^2 corresponding to the intersection points of the line R^2 = constant with the curve (a) be denoted by Ω_g^2 and Ω_p^2 and these are evaluated from Eq. (6.46a) to be given by

$$\Omega_{g,p}^2 = \frac{1 + R^2}{2} \mp \left\{\left(\frac{1 + R^2}{2}\right)^2 - R^2 l^2\right\}^{1/2} \tag{6.47}$$

In Eq. (6.47), Ω_g^2 and Ω_p^2 correspond to the upper and the lower signs, respectively, and are always positive real. Also, the line R^2 = constant intersects the curve (d) for two values of Ω^2, namely Ω_1^2 and Ω_2^2. The expressions for Ω_1 and Ω_2, which are always positive real are the same as those given by Eq. (4.18). The frequency ranges of propagation of the two modes are found from Fig. 5.9 to be given by

$$\text{Mode 1} \qquad 0 < \Omega < \Omega_g \qquad 1 < \Omega < \infty \tag{6.48a}$$

$$\text{Mode 2} \qquad \Omega_1 < \Omega < \Omega_p \qquad \Omega_2 < \Omega < \infty \tag{6.48b}$$

With the help of Fig. 5.9, at least one mode is seen to propagate in any frequency range, for $R^2 > 1/[l^2(1 + l^2)]$. If $R^2 < 1/[l^2(1 + l^2)]$, there is a frequency band which lies entirely below $\Omega^2 = 1$ and in which there is no propagation. As R^2 is decreased from the value $1/[l^2(1 + l^2)]$, the width of this cut-off band increases progressively until finally for the isotropic ($R = 0$) plasma, this cut-off band of frequencies extends from $\Omega^2 = 0$ to $\Omega^2 = 1$.

Cut-off and resonance frequencies

The cut-off frequencies are those for which $k = 0$. It follows from Eqs. (6.42c) and (6.43) that $k = 0$ for $\Omega = 0$, $\Omega = 1$, $\Omega = \Omega_1$, and $\Omega = \Omega_2$. It is customary not to designate $\Omega = 0$ as a cut-off frequency. For Ω tending to zero, it is found from Fig. 5.9 that only k_1 is real and from Eqs. (6.42), (6.44) and (6.45), it is verified that $k_1 \approx (\Omega/Rl)^{1/2}$ for Ω tending to zero. As a consequence the normalized phase velocity in this limit is given by $v_{ph}/c = \sqrt{\Omega Rl}$, which also goes to zero for Ω tending to 0. Consequently, if the cut-off frequencies are redefined to be those for which the phase velocity is infinite, the frequency $\Omega = 0$ is naturally eliminated. It is noted that the cut-off frequencies are independent of l and therefore, of the direction of propagation of the plane wave with respect to the **B**-field. This statement is true only subject to a proviso to be introduced subsequently.

The resonance frequencies are defined to be those for which k is infinite. It is found from Eqs. (6.42) and (6.43) that k is infinite for values of Ω for which $B_2 = 0$ and these frequencies are verified to be $\Omega = \Omega_g$ and $\Omega = \Omega_p$ as given by Eq. (6.47). These resonant frequencies are functions of the direction of propagation of the plane wave with respect to the **B**-field. As the direction of propagation of the plane wave is changed so as to approach the direction normal to the **B**-field, l becomes smaller and as a consequence Ω_g becomes smaller and Ω_p becomes larger. In the limiting case of propagation across the **B**-field, $l = 0$ and therefore, $\Omega_g = 0$ and $\Omega_p = \sqrt{1 + R^2} = \Omega_u$, the upper hybrid resonant frequency. Since Ω_1 and Ω_2 are independent of the direction of propagation, it is seen from the relations (6.48a,b) that the ranges of propagation of the higher frequency branches of modes 1 and 2 are unaffected as the propagation direction is changed. However, the frequency ranges of the lower frequency branches of modes 1 and 2 are, respectively, diminished and increased, as the propagation direction approaches that perpendicular to the **B**-field. In the limiting case of propagation across the **B**-field, the lower frequency branch of mode 1 disappears completely and the propagation range of the lower frequency branch of mode 2 becomes a maximum and is given by $\Omega_1 < \Omega < \Omega_u$.

As the propagation direction is changed so as to approach that of the **B**-field, l becomes progressively larger and reaches the limiting value of unity. As l becomes larger, Ω_g increases for all values of R in such a way as to approach $\min(R, 1)$, as shown in Fig. 5.10, where $\min(R, 1)$ denotes the minimum of R and 1. Also, Ω_p decreases continuously for all values of R so as to approach $\max(R, 1)$, where $\max(R, 1)$ stands for the maximum of R and 1. From Eq. (6.43), it follows that the cut-off and the resonant frequencies are given by the values of Ω which make B_0/B_2 zero and infinity, respectively. For $\theta_{ph} \neq 0$, irrespective of how small θ_{ph} is, $l \neq 1$ and the factor $(\Omega^2 - 1)$ in B_0 is not cancelled by either of the factors $(\Omega^2 - \Omega_g^2)$ and $(\Omega^2 - \Omega_p^2)$ of B_2 with the result that $\Omega = 1$ is a cut-off frequency, and $\Omega = \Omega_g$ and $\Omega = \Omega_p$ are the two resonant frequencies. But for $\theta_{ph} = 0$, $l = 1$ and the factor $(\Omega^2 - 1)$ in B_0 is cancelled by one of the two factors, $(\Omega^2 - \Omega_g^2)$ or $(\Omega^2 - \Omega_p^2)$ of B_2. The factor of B_2 that is cancelled is seen from Fig. 5.10 to be equal to $(\Omega^2 - \Omega_p^2)$ if $R < 1$ or $(\Omega^2 - \Omega_g^2)$ if $R > 1$. As a consequence, the cut-off frequency $\Omega = 1$ disappears in a nonuniform manner for the case of propagation in the direction of

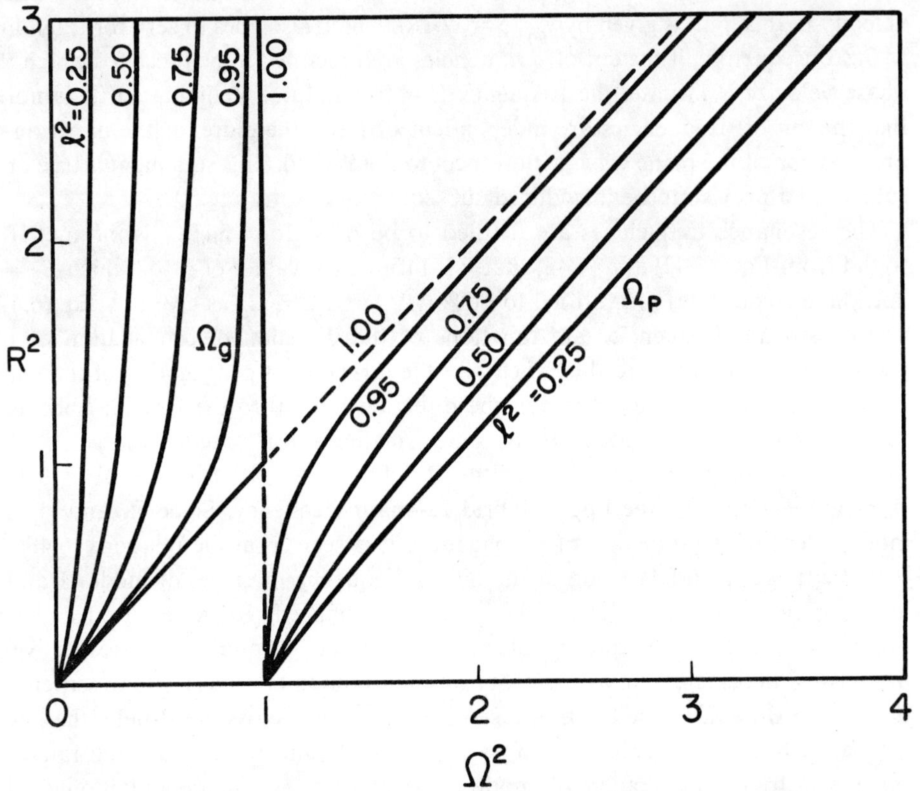

Fig. 5.10. Variation of the normalized resonant frequencies Ω_g and Ω_p with $l = \cos\theta_{ph}$.

the **B**-field. Subject to this condition, alluded to earlier, the cut-off frequencies remain independent of the direction of propagation. The resonant frequencies $\Omega = \Omega_p$ for $R < 1$ and $\Omega = \Omega_g$ for $R > 1$ disappear also for the case of propagation parallel to the **B**-field. The resonant frequencies $\Omega = \Omega_g$ for $R < 1$ and $\Omega = \Omega_p$ for $R > 1$ join together to yield for all values of R the resonant frequency $\Omega = R$ for the case of propagation in the direction of the **B**-field.

In conformity with the foregoing discussion, the resonant frequency lines in Fig. 5.10 for the limiting case of $l = 1$ should be interpreted in the following manner. The resonant frequency line corresponding to $\Omega^2 = \Omega_g^2$ for $R^2 \leqslant 1$ and that corresponding to $\Omega^2 = \Omega_p^2$ for $R^2 \geqslant 1$ join to form the resonant frequency line $\Omega^2 = R^2$ whereas the resonant frequency line $\Omega^2 = 1$ corresponding to that of Ω_p^2 for $R^2 \leqslant 1$ and of Ω_g^2 for $R^2 \geqslant 1$ disappears. For the special case of propagation in the direction of the **B**-field, the regions of propagation of modes 1 and 2 are depicted in Fig. 5.11 in the Ω^2 - R^2 parameter space. It is found with the help of Eqs. (6.42a) and (6.44) that when the resonant frequency line $\Omega^2 = 1$ disappears, the sign in front of the radical in Eq. (6.44) is reversed for $\Omega^2 < 1$ with the result that for the limiting case of propagation in the direction of the **B**-field, the mode designations 1 and 2 get interchanged for $\Omega^2 < 1$. Consequently, for $l = 1$, the frequency ranges of propagation of the two modes are as follows:

$$\text{Mode 1} \qquad \Omega_1 < \Omega < \infty \tag{6.49a}$$

$$\text{Mode 2} \qquad 0 < \Omega < R \qquad \Omega_2 < \Omega < \infty \tag{6.49b}$$

It is not difficult to identify modes 1 and 2 referred to in the relations (6.49a,b) with the left and the right circularly polarized modes introduced in Sec. 5.4.

As the propagation direction is changed from the perpendicular to the parallel

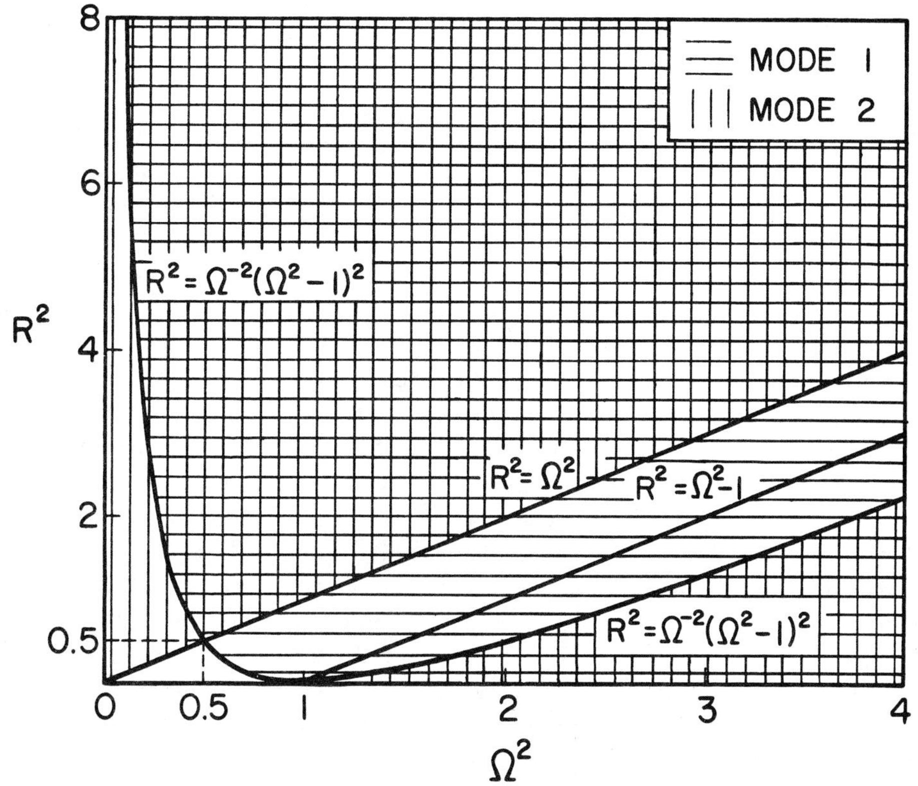

Fig. 5.11. Propagation regions of the plane wave modes 1 and 2 for the special case of propagation in the direction of the **B**-field.

direction with respect to the **B**-field, the ranges of Ω_g and Ω_p are found from Eq. (6.47) to be given by

$$0 < \Omega_g < \min(R, 1) \qquad \Omega_u > \Omega_p > \max(R, 1) \tag{6.50}$$

For propagation in the direction of the **B**-field and for $R < 1$, Ω_g becomes equal to the gyromagnetic frequency of the electrons. Consequently, Ω_g is called the *gyromagnetic resonance frequency*. Similarly, for propagation very close to the direction of the **B**-field and for $R < 1$, Ω_p becomes very close to the plasma frequency of the electrons with the result that Ω_p is called the *plasma resonance frequency*. The reason for the choice of the subscripts g and p in Eq. (6.47) is now clarified. For any frequency in the range,

$$\text{Gyromagnetic resonance region: } 0 < \Omega < \min(R, 1) \tag{6.51a}$$

the gyromagnetic resonance occurs in some direction with respect to the **B**-field. As a result, the frequency range (6.51a) is called the gyromagnetic resonance region. In a similar manner, for any frequency in the range,

$$\text{Plasma resonance region: } \max(R, 1) < \Omega < \Omega_u \tag{6.51b}$$

the plasma resonance occurs for the plane wave propagation in some direction with respect to the **B**-field, as a consequence of which the frequency range (6.51b) is called the plasma resonance region.

It is emphasized that the direction of propagation parallel to the **B**-field is a very special case. For propagation in the direction of the **B**-field, $\Omega = 1$ is not a cut-off frequency. But for propagation directions arbitrarily close to the **B**-field, $\Omega = 1$ is a cut-off frequency. There is no plasma resonance for propagation in the direction of the **B**-field. But, as soon as the direction of propagation is changed ever so little from that of the **B**-field, the plasma resonance frequency Ω_p comes into existence. For $R < 1$, this value of Ω_p is slightly greater than $\Omega = 1$, increases continuously as the propagation direction approaches that perpendicular to the **B**-field and becomes equal to the upper hybrid resonance frequency for the limiting case of propagation across the **B**-field. Similarly, for $R < 1$, the gyromagnetic resonance frequency Ω_g which is equal to R for propagation in the direction of the **B**-field, decreases continuously and vanishes in the limiting case of propagation across the **B**-field. However, for $R > 1$ and for propagation directions slightly different from that of the **B**-field, Ω_p is slightly greater than $\Omega = R$, increases continuously as the propagation direction moves away from that of the **B**-field and attains the value $\Omega = \Omega_u$ in the limiting case of propagation across the **B**-field. But, for $R > 1$, the gyromagnetic resonance frequency Ω_g, which is equal to $\Omega = R$ for propagation in the direction of the **B**-field, *discontinuously* jumps to a value which is slightly less than $\Omega = 1$ as soon as the propagation direction is changed by an arbitrarily small amount away from that of the **B**-field and for further changes of the direction of propagation away from that of the **B**-field, Ω_g continuously decreases to attain the value zero in the limiting case of propagation across the **B**-field.

A better understanding of the characteristics of the plane wave propagation in the directions very close to that of the **B**-field can be obtained by using the warm plasma model which is an extension of the magnetoionic theory and in which a scalar pressure term is introduced to account for the thermal motions of the electrons [9]. Another mode of propagation is brought into existence by this extension of the magnetoionic theory. It is found that as the propagation direction approaches very close to that of the **B**-field, the characteristics of plane wave propagation undergo very rapid changes as a result of the decoupling of the longitudinal electron plasma wave which has its electric vector in the direction of propagation. As the finite temperature of the plasma is decreased so as to vanish in the limit, the extended treatment approaches that of the magnetoionic theory, and there is a continual increase in the rapidity of the changes near the direction of the **B**-field since these changes are confined to progressively smaller and smaller ranges of direction close to that of the **B**-field. Finally, in the limit of the magnetoionic theory, there is a nonuniform behavior of the plane wave characteristics for the case of propagation approaching the direction of the **B**-field.

Brillouin diagrams, phase and group velocities

From Eqs. (6.42), (6.44), and (6.45), it can be shown that

$$\begin{aligned} k_{1,2}^2 = {} & [\Omega^4 - \Omega^2(1 + R^2) + R^2 l^2]^{-1} \\ & [\Omega^2\{\Omega^4 - \Omega^2(R^2 + 2) + 1 + \frac{R^2}{2}(1 + l^2)\} \\ & \pm \Omega R\{(\Omega^2 - 1)^2 l^2 + \frac{\Omega^2 R^2}{4}(1 - l^2)^2\}^{1/2}] \end{aligned} \tag{6.52}$$

The normalized phase velocity v_{ph}/c is found from Eq. (6.52) with the help of the relation

$$v_{ph}/c = \Omega/k_{1,2} \tag{6.53}$$

To obtain the group velocity, Eq. (6.43) together with Eqs. (6.42) is rearranged as a polynomial in Ω^2 as follows:

$$\Omega^8 + a_6\Omega^6 + a_4\Omega^4 + a_2\Omega^2 + a_0 = 0 \tag{6.54}$$

where

$$a_6 = -(2k^2 + R^2 + 3) \tag{6.55a}$$

$$a_4 = k^4 + 2(2 + R^2)k^2 + R^2 + 3 \tag{6.55b}$$

$$a_2 = -[(1 + R^2)k^4 + \{2 + R^2(1 + l^2)\}k^2 + 1] \tag{6.55c}$$

$$a_0 = R^2 l^2 k^4 \tag{6.55d}$$

With the help of Eqs. (6.54) and (6.55) the normalized group velocity v_g/c is evaluated to be given by

$$v_g/c = \partial\Omega/\partial k = (k/\Omega)P_N/P_D \tag{6.56}$$

where

$$P_N = 2\Omega^6 - 2\Omega^4(k^2 + R^2 + 2) + \Omega^2\{2k^2(1 + R^2) + 2 + R^2(1 + l^2)\} \\ - 2k^2R^2l^2 \tag{6.57a}$$

$$P_D = 4\Omega^6 + 3\Omega^4 a_6 + 2\Omega^2 a_4 + a_2 \tag{6.57b}$$

Note that v_g/c exists only for the values of Ω for which the corresponding value of k is real. The normalized group velocities of the two modes are determined from Eqs. (6.56) and (6.57) by using the corresponding value of k from Eq. (6.52).

From the previous discussion, it is found that $k = 0$ for $\Omega = 0$, $\Omega = 1$, $\Omega = \Omega_1$ and $\Omega = \Omega_2$, and, $k = \infty$ for $\Omega = \Omega_g$ and $\Omega = \Omega_p$. Also, $v_{ph}/c = 0$, for $\Omega = 0$, $\Omega = \Omega_g$ and $\Omega = \Omega_p$ and $v_{ph}/c = \infty$ for $\Omega = 1$, $\Omega = \Omega_1$ and $\Omega = \Omega_2$. It follows from Eqs. (6.56) and (6.57) that $v_g/c = 0$ when $k = 0$, that is, for $\Omega = 0$, $\Omega = 1$, $\Omega = \Omega_1$ and $\Omega = \Omega_2$. For k tending to infinity, it can be shown from Eqs. (6.54)–(6.57) that

$$\frac{v_g}{c} \to -\frac{1}{\Omega k}\frac{(\Omega^2 - \Omega_g^2)(\Omega^2 - \Omega_p^2)}{\left(\Omega^2 - \frac{1 + R^2}{2}\right)} \quad \text{for } k \to \infty \tag{6.58}$$

Note that $k = \infty$, for $\Omega = \Omega_g$ and $\Omega = \Omega_p$. It is therefore evident from the relation (6.58) that $v_g/c = 0$, for $k = \infty$, i.e., for $\Omega = \Omega_g$ and $\Omega = \Omega_p$.

The propagation coefficient k, the normalized phase velocity v_{ph}/c and the normalized group velocity v_g/c are depicted as functions of Ω in Figs. 5.12 and 5.13 for modes 1 and 2, respectively. The figures correspond to the propagation direction specified by $\theta_{ph} = 45°$, that is, $l^2 = 0.5$.

5.7. Polarization and Poynting vector in magnetoionic theory

The polarization of a wave gives information on the behavior with respect to time of the various field vectors at a fixed location in space. It is usual to describe the polarization by the nature of the locus traced by the terminal point of the field vector in one time period, at a fixed location in space. For the case of a plane wave propagating parallel to the **B**-field, it was shown that the terminal point of the electric vector traced a circle in a plane perpendicular to the direction of propagation and the circles were traced in opposite directions for the two modes of propagation. Consequently, these modes were called the right and the left circularly polarized waves depending upon the sense of rotation of the electric vector with respect to the direction of propagation. For the case of propagation perpendicular to the **B**-field, the terminal point of the electric vector for the TEM mode and that of the magnetic vector for the TM mode traced straight lines in the direction of the **B**-field, that is in a plane perpendicular to the direction of propagation. Therefore, these modes are linearly polarized waves. For the case of propagation in an arbitrary direction, it is shown in the following treatment, that the electric vector does not, in general, lie in a plane perpendicular to the direction of propagation and that the terminal points of the field vectors trace ellipses of which the circles and the straight lines obtained previously are special cases.

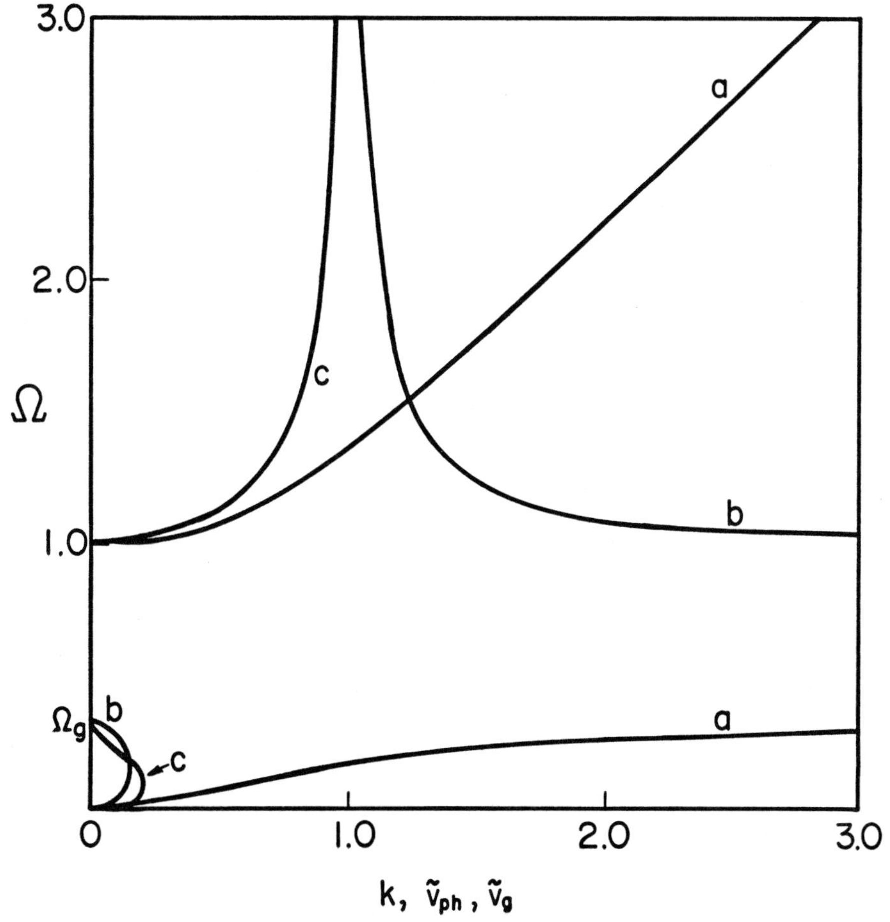

Fig. 5.12. Brillouin diagram for mode 1 propagating at an angle $\theta_{ph} = 45°$ to the magnetostatic field in a magnetoionic medium: $R = 0.5$; (a) k_1; (b) $\tilde{v}_{ph} = v_{ph}/c$; (c) $\tilde{v}_g = v_g/c$.

Polarization equation

Let Eqs. (6.1a) and (6.1b) be written in a vector form as

$$\mathbf{E}(\mathbf{r}) = \mathbf{E}_0 \exp(ik_p \mathbf{k} \cdot \mathbf{r}) \qquad \mathbf{H}(\mathbf{r}) = \mathbf{H}_0 \exp(ik_p \mathbf{k} \cdot \mathbf{r}) \tag{7.1}$$

where $\mathbf{k}$ is the normalized wavevector. Since the operator $\nabla = ik_p\mathbf{k}$, it is obtained from Eqs. (2.18), (2.19), and (7.1) that

$$\mathbf{k} \times \mathbf{E}_0 = (\omega\mu_0/k_p)\mathbf{H}_0 \tag{7.2}$$

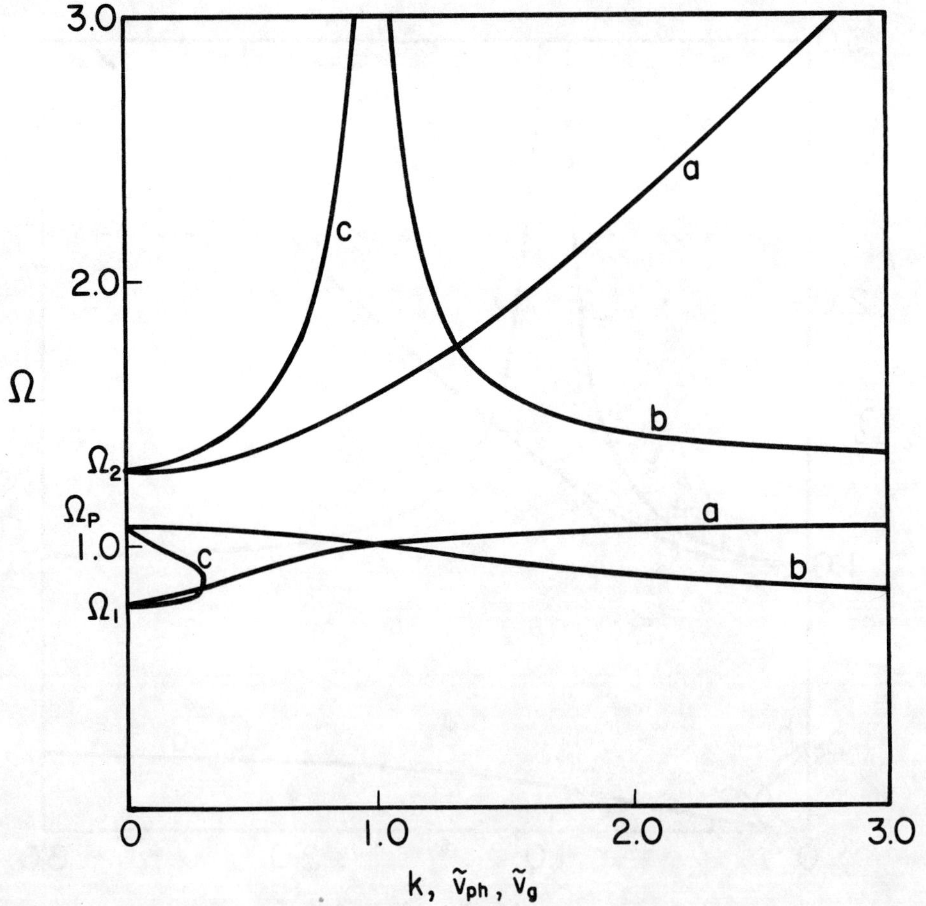

Fig. 5.13. Brillouin diagram for mode 2 propagating at an angle $\theta_{ph} = 45°$ to the magnetostatic field in a magnetoionic medium: $R = 0.5$; (a) k_2; (b) $\tilde{v}_{ph} = v_{ph}/c$; (c) $\tilde{v}_g = v_g/c$.

$$\mathbf{k} \times \mathbf{H}_0 = -(\omega\varepsilon_0/k_p)\varepsilon_r \cdot \mathbf{E}_0 \tag{7.3}$$

From Eq. (7.2), it is found that

$$\mathbf{E}_0 \cdot \mathbf{H}_0 = 0 \qquad \mathbf{k} \cdot \mathbf{H}_0 = 0 \tag{7.4}$$

It is seen from Eqs. (7.4) that the electric and the magnetic vectors are always perpendicular to each other and that the magnetic vector, which has no component in the direction of propagation, always lies in the phase front, that is, in a plane perpendicular to the direction of propagation. This latter result is consistent with that deduced previously as given by Eq. (6.28). In the coordinate system x', y, and z' where z' coincides with the direction of propagation,

$$\mathbf{k} = \hat{\mathbf{z}}'k \tag{7.5}$$

The components of Eq. (7.2) are given by

$$(\omega\mu_0/k_p)H_{x'0} = -kE_{y0} \tag{7.6}$$

$$(\omega\mu_0/k_p)H_{y0} = kE_{x'0} \tag{7.7}$$

From Eqs. (7.6) and (7.7), it is found that

$$E_{y0}/E_{x'0} = -H_{x'0}/H_{y0} = \rho_r \tag{7.8}$$

where ρ_r is called the *polarization ratio*. It follows from Eqs. (7.6) and (7.7) that for a propagating wave for which k is real, $E_{y0}/H_{x'0}$ and $E_{x'0}/H_{y0}$ are both real. Hence, for example, the phase angles of $E_{x'0}$ and H_{y0} are the same.

From Eqs. (6.12), (6.22), (6.25), and (7.8), it is deduced that

$$\rho_r = \frac{ik^2 \cos\theta_{ph}\varepsilon_2}{(\Omega^2\varepsilon - \varepsilon_1 k^2)} \tag{7.9}$$

It can be verified that, as indicated in Eq. (7.8), the same expression for ρ_r as given by Eq. (7.9) is also obtained from Eqs. (6.24) and (6.27). Let ρ_{r1} and ρ_{r2} be the two polarization ratios associated with the two modes of propagation corresponding to the propagation coefficients k_1 and k_2. Then, it follows from Eqs. (6.36), (6.40), and (7.9) that

$$\rho_{r1}\rho_{r2} = \frac{ik_1^2\cos\theta_{ph}\varepsilon_2}{(\Omega^2\varepsilon - \varepsilon_1 k_1^2)} \cdot \frac{ik_2^2\cos\theta_{ph}\varepsilon_2}{(\Omega^2\varepsilon - \varepsilon_1 k_2^2)} = 1 \tag{7.10}$$

In view of Eq. (7.10), it is obtained from Eq. (7.9) that

$$\rho_{r1} + \rho_{r2} = \frac{1}{\rho_{r1}} + \frac{1}{\rho_{r2}} = \frac{-i}{\varepsilon_2\cos\theta_{ph}}\left[\Omega^2\varepsilon\frac{(k_1^2 + k_2^2)}{k_1^2 k_2^2} - 2\varepsilon_1\right] \tag{7.11}$$

If Eqs. (6.38) and (6.39) are substituted in Eq. (7.11), the result after some simplification is

$$\rho_{r1} + \rho_{r2} = \frac{-i(\varepsilon_1^2 - \varepsilon_2^2 - \varepsilon_1 \varepsilon_3)(1 - l^2)}{\varepsilon_2 \varepsilon_3 \cos \theta_{ph}} \tag{7.12}$$

With the help of Eqs. (2.21a,b,c) it can be established that

$$\varepsilon_1^2 - \varepsilon_2^2 - \varepsilon_1 \varepsilon_3 = \varepsilon_1 - \varepsilon_3 = -\frac{R^2}{\Omega^2(\Omega^2 - R^2)} \tag{7.13}$$

Further use of Eqs. (2.21b,c) and (7.13) in Eq. (7.12) yields

$$\rho_{r1} + \rho_{r2} = iR\Omega(1 - l^2)/l(\Omega^2 - 1) \tag{7.14}$$

From Eqs. (7.10) and (7.14), the polarization ratio can be shown to satisfy the following quadratic equation known as the *polarization equation*:

$$\rho_r^2 - \frac{iR\Omega(1 - l^2)}{l(\Omega^2 - 1)}\rho_r + 1 = 0 \tag{7.15}$$

Polarization in the magnetic meridian plane

It is seen from Eq. (7.9) that the polarization ratio for a loss-free magnetoionic medium is always imaginary. In view of the foregoing, it is assumed that

$$\rho_r = |\rho_r|e^{\pm i\pi/2} \qquad E_{x'0} = |E_{x'0}|e^{i\theta} \qquad H_{y0} = |H_{y0}|e^{i\theta} \tag{7.16}$$

In the plane of the phase front, that is, in the $x'y$-plane, the time-dependent electric vector is obtained with the help of Eqs. (7.8) and (7.16) to be given by

$$\begin{aligned} \mathbf{E}(t) &= \text{Re}\,[\{\hat{\mathbf{x}}' E_{x'0} + \hat{\mathbf{y}}\rho_r E_{x'0}\}e^{-i\omega t}] \\ &= [\hat{\mathbf{x}}' \cos(\omega t - \theta) \pm \hat{\mathbf{y}}|\rho_r|\sin(\omega t - \theta)]|E_{x'0}| \end{aligned} \tag{7.17}$$

where the upper and the lower signs correspond to those in Eqs. (7.16). It is found from Eq. (7.17) that the terminal point of the electric vector in the $x'y$-plane traces an ellipse with its principal axes along $\hat{\mathbf{x}}'$ and $\hat{\mathbf{y}}$. The sense of rotation of the electric vector in the $x'y$-plane with reference to the direction of propagation is positive for the upper sign in Eq. (7.17) and negative for the lower sign. If $|\rho_r| < 1$, the major axis of the ellipse is along the x'-axis which is in the *magnetic meridian plane* that contains the direction of the magnetostatic field and the propagation direction. For $|\rho_r| > 1$, the minor axis is in the magnetic meridian plane. For the sake of definiteness, it is assumed that $|\rho_r| < 1$ or $|\rho_{r1}| < 1$.

The magnetic vector is seen from Eqs. (7.4) and (7.5) to lie entirely in the $x'y$-

plane. With the help of Eqs. (7.8) and (7.16), the time-dependent magnetic vector is obtained to be given by

$$\begin{aligned} \mathbf{H}(t) &= \mathrm{Re}\,[\{-\hat{\mathbf{x}}'\rho_r H_{y0} + \hat{\mathbf{y}} H_{y0}\} e^{-i\omega t}] \\ &= [\mp\hat{\mathbf{x}}'|\rho_r|\sin(\omega t - \theta) + \hat{\mathbf{y}}\cos(\omega t - \theta)]|H_{y0}| \end{aligned} \tag{7.18}$$

where the upper and the lower signs correspond to those in Eqs. (7.16). The terminal point of the magnetic vector is seen from Eq. (7.18) to trace an ellipse in the $x'y$-plane. The sense of rotation of the magnetic vector is the same as that of the electric vector discussed previously. The principal axes of the ellipse are along the x'- and y-axes; but the major axis of the ellipse corresponding to the magnetic vector is at right angles to that of the ellipse traced by the terminal point of the electric vector in the $x'y$-plane. In Fig. 5.14, the polarizations of the electric and the magnetic vectors in the $x'y$-plane are depicted.

There are two modes of propagation of plane waves in any direction with respect to the **B**-field. Let the polarizations of the electric and the magnetic vectors

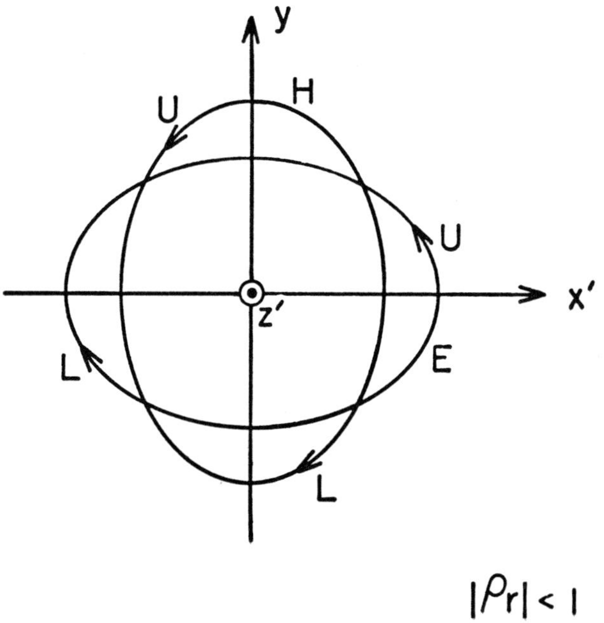

Fig. 5.14. Polarizations of the electric and the magnetic vectors in the **H**-plane. *U*: for the upper sign in Eqs. (7.17) and (7.18); *L*: for the lower sign in Eqs. (7.17) and (7.18); *z'*: direction of propagation.

discussed in the foregoing correspond to mode 1. When referring to Eqs. (7.16)–(7.18), let an additional subscript 1 be added to ρ_r so as to be able to distinguish it as being associated with mode 1. In view of Eqs. (7.10) and (7.16), the polarization ratio associated with mode 2 is of the form

$$\rho_{r2} = |\rho_{r2}| e^{\mp i\pi/2} \tag{7.19}$$

Note that the phase angle of ρ_{r2} given in Eq. (7.19) is negative of the phase angle of ρ_{r1} as given in Eqs. (7.16). The time-dependent electric vector associated with mode 2 is deduced from Eqs. (7.8) and (7.19) to be given by

$$\mathbf{E}(t) = [\hat{\mathbf{x}}' \cos(\omega t - \theta_2) \mp \hat{\mathbf{y}} |\rho_{r2}| \sin(\omega t - \theta_2)] |E_{x'0,2}| \tag{7.20}$$

where θ_2 and $|E_{x'0,2}|$ are the phase angle and the magnitude of $E_{x'0,2}$; the additional subscript 2 is used to indicate that these quantities correspond to mode 2. The terminal point of the electric vector given in Eq. (7.20) traces an ellipse with its principal axes along the x'- and y-axes. If $|\rho_{r1}| \lesseqgtr 1$, then $|\rho_{r2}| \gtreqless 1$ with the result that the major axis of the ellipse associated with mode 2 is *at right angles* to that associated with mode 1. The sense of rotation of the ellipse associated with mode 2 is seen from a comparison of Eqs. (7.17) and (7.20), to be *opposite* to that of the ellipse associated with mode 1. However, since $|\rho_{r1}| = 1/|\rho_{r2}|$, the ratio of the minor to the major axes of the ellipses associated with the two modes remain the same. In Fig. 5.15, the polarizations of the electric vector in the $x'y$-plane for the two modes of propagation are shown together so as to bring out clearly the differences between them.

Polarization in the E-plane

In contrast to the magnetic vector, the electric vector is seen from Eq. (6.26) to have a component in the direction of propagation. From Eqs. (6.25) and (6.26), it is clear that the two components of the electric vector in the magnetic meridian plane are in time phase. Let the resultant electric vector in the magnetic meridian plane make an angle χ with the x'-axis, where $-\pi/2 < \chi < \pi/2$. It can be deduced with the help of Fig. 5.8, Eqs. (6.25) and (6.26) that

$$\tan \chi = \frac{E_{z'0}}{E_{x'0}} = -\cot \theta_{ph} \frac{(k^4 - 2\varepsilon_1 k^2 \Omega^2 + \Omega^4 \varepsilon)}{\Omega^2(\Omega^2 \varepsilon - \varepsilon_1 k^2)} \tag{7.21}$$

Since E_{y0} is in time quadrature with both $E_{x'0}$ and $E_{z'0}$, it follows that the total electric vector rotates in a plane containing the y-axis and the direction of the resultant of $E_{x'0}$ and $E_{z'0}$. The plane in which the electric vector rotates is called the **E**-plane and this forms an angle χ with the **H**-plane which is the wave-normal ($x'y$) plane. Let the resultant of $E_{x'0}$ and $E_{z'0}$ be in the direction of $\hat{\mathbf{x}}^E$. The time-dependent total electric vector is obtained with the help of Eqs. (7.8), (7.16), and (7.21) to be given by

$$\begin{aligned}\mathbf{E}(t) &= \text{Re}\,[\{\hat{\mathbf{x}}'E_{x'0} + \hat{\mathbf{y}}\rho_r E_{x'0} + \hat{\mathbf{z}}' \tan \chi E_{x'0}\}e^{-i\omega t}] \\ &= \left[\frac{\hat{\mathbf{x}}^E}{\cos\chi}\cos(\omega t - \theta) \pm \hat{\mathbf{y}}|\rho_r|\sin(\omega t - \theta)\right]|E_{x'0}|\end{aligned} \tag{7.22}$$

where the upper and the lower signs correspond to those in Eqs. (7.16). It is found from Eq. (7.22) that the total electric vector is in the $x^E y$-plane and its terminal point traces an ellipse with its principal axes along $\hat{\mathbf{x}}^E$ and $\hat{\mathbf{y}}$. Since $\cos\chi$ is always positive, the sense of rotation of the electric vector in the $x^E y$-plane with reference to the direction of propagation is positive for the upper sign and negative for the

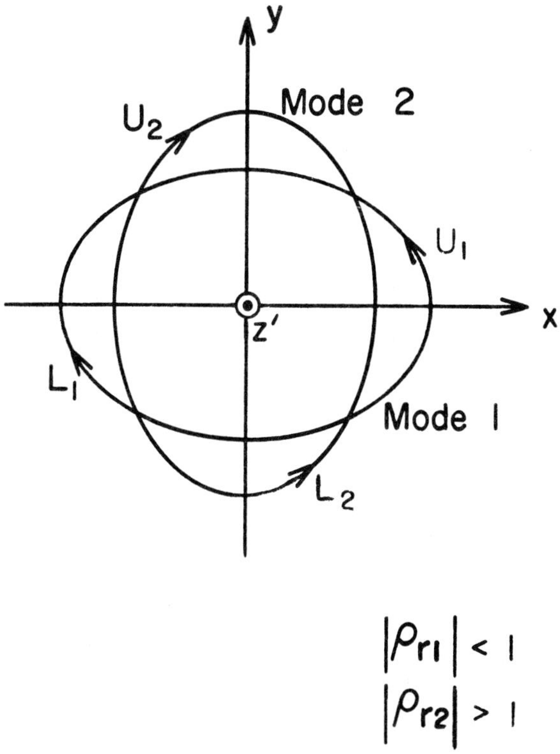

Fig. 5.15. Polarizations of the electric vector in the **H**-plane for modes 1 and 2. U_1: for the upper sign in Eq. (7.17); L_1: for the lower sign in Eq. (7.17); U_2: for the upper sign in Eq. (7.20); L_2: for the lower sign in Eq. (7.20); z' : direction of propagation.

lower sign. Since the polarization of the electric vector in the $x'y$-plane is the projection of the corresponding quantity in the $x^E y$-plane, it is natural to expect that the senses of rotation of the electric vector in these two planes to be identical as is obtained from Eq. (7.22). If $|\rho_r| < \sec \chi$, the major axis of the ellipse is along the x^E-axis which is in the magnetic meridian plane. On the other hand, if $|\rho_r| > \sec \chi$, the minor axis of the ellipse is in the magnetic meridian plane.

From a rearrangement of Eq. (6.18) together with Eqs. (6.19) and (7.13) it can be deduced that

$$l^2 = \frac{\varepsilon_1 k^4 - \Omega^2 k^2(\varepsilon_1^2 - \varepsilon_2^2 + \varepsilon_1 \varepsilon_3) + \Omega^4 \varepsilon_3(\varepsilon_1^2 - \varepsilon_2^2)}{(\varepsilon_1 - \varepsilon_3)(k^4 - k^2\Omega^2)} \tag{7.23}$$

With the help of Eqs. (7.13) and (7.23), it is obtained that

$$1 - l^2 = -\varepsilon_3 \frac{[k^4 - 2\varepsilon_1 k^2\Omega^2 + \Omega^4 \varepsilon]}{(\varepsilon_1 - \varepsilon_3)(k^4 - k^2\Omega^2)} \tag{7.24}$$

Hence, it follows from Eqs. (7.23) and (7.24) that

$$\begin{aligned} \tan^2 \theta_{ph} &= \frac{1 - l^2}{l^2} = -\frac{\varepsilon_3(k^4 - 2\varepsilon_1 k^2\Omega^2 + \Omega^4 \varepsilon)}{(\varepsilon_1 k^2 - \varepsilon\Omega^2)(k^2 - \Omega^2 \varepsilon_3)} \\ &= -\varepsilon_3 \frac{[k^2 - \Omega^2(\varepsilon_1 - \varepsilon_2)][k^2 - \Omega^2(\varepsilon_1 + \varepsilon_2)]}{[\varepsilon_1 k^2 - \varepsilon\Omega^2][k^2 - \Omega^2 \varepsilon_3]} \end{aligned} \tag{7.25}$$

The dispersion equation (6.18) expressed in the form (7.25) is due to Astrom [10]. The use of Eq. (7.25) enables Eq. (7.21) to be written in the following equivalent form

$$\tan \chi = -\tan \theta_{ph} \frac{(k^2 - \Omega^2 \varepsilon_3)}{\Omega^2 \varepsilon_3} \tag{7.26}$$

Special directions of propagation

It is instructive to examine the values of the polarization ratio ρ_r and the position of the **E**-plane specified by $\tan \chi$ under various special situations. For propagation in the direction of the **B**-field, that is, for $\theta_{ph} = 0$, the two possible values of k are obtained from Eqs. (4.15) and (4.16) as

$$k^2 = (k_z^-)^2 = \Omega^2(\varepsilon_1 + \varepsilon_2) \qquad k^2 = (k_z^+)^2 = \Omega^2(\varepsilon_1 - \varepsilon_2) \tag{7.27}$$

Then, from Eq. (7.9) it is found that

$$\rho_r = e^{-i\pi/2} \quad \text{for } k = k_z^- \qquad \rho_r = e^{i\pi/2} \quad \text{for } k = k_z^+ \tag{7.28}$$

For $k = k_z^-$ and $k = k_z^+$ as given by Eqs. (7.27), it is seen from Eqs. (7.25) and (7.26) that $\chi = 0$. Moreover, Eq. (7.22) then shows that the electric vectors of the waves associated with $k = k_z^-$ and $k = k_z^+$ are both in the wave-normal plane and are left and right circularly polarized, respectively. These results are in conformity with those obtained in Sec. 5.4.

For the TEM mode of propagation perpendicular ($\theta_{ph} = \pi/2$) to the **B**-field, $k^2 = \Omega^2 \varepsilon_3$ and Eq. (7.9) gives $\rho_r = 0$. Also, from Eqs. (7.25) and (7.26), it is found that $\chi = 0$. With the help of Eq. (7.22), it is ascertained that the electric vector is in the wave-normal plane and is linearly polarized in the z-direction, that is, the direction of the **B**-field. For the TM mode of propagation perpendicular to the **B**-field, $k^2 = (\Omega^2 \varepsilon/\varepsilon_1)$ and it is found from Eq. (7.9) together with Eq. (7.23) that $\rho_r = \infty$. Therefore, it is clear from Eq. (7.8) that $E_{x'0} = H_{y0} = 0$. Also, from Eqs. (7.25) and (7.26), it is found that $\chi = \pi/2$ and hence the **E**-plane contains the propagation direction and is perpendicular to the wave-normal (or the **H**-) plane. Since $H_{z'0} = 0$ always, it follows that for the TM mode of propagation perpendicular to the **B**-field, the magnetic vector is linearly polarized in the x'-direction or equivalently in the direction of the **B**-field. These results for the case of propagation perpendicular to the **B**-field are in agreement with those obtained in Sec. 5.5. The polarization of the electric field for the case of propagation perpendicular to the **B**-field was not previously treated and a discussion of the same is included here for the sake of completeness. Since $\rho_r = \infty$, $\chi = \pi/2$ and $|E_{x'0}| = 0$, Eq. (7.22) is not useful in determining the polarization of the electric vector for this special case. It can be obtained from Eqs. (6.22) and (6.26) that

$$\frac{E_{y0}}{E_{z'0}} = -i \frac{k^2 \sin \theta_{ph} \varepsilon_2 \Omega^2}{[\Omega^4 \varepsilon - 2\varepsilon_1 \Omega^2 k^2 + k^4]} = \rho_r' = |\rho_r'| e^{\pm i\pi/2} \tag{7.29}$$

where ρ_r' is imaginary. Then, the time-dependent electric vector is obtained with the help of Eq. (7.29) to be given by

$$\begin{aligned} \mathbf{E}(t) &= \text{Re}\,[\{\hat{\mathbf{y}}\rho_r' E_{z'0} + \hat{\mathbf{z}}' E_{z'0}\} e^{-i\omega t}] \\ &= [\pm \hat{\mathbf{y}} |\rho_r'| \sin(\omega t - \theta') + \hat{\mathbf{z}}' \cos(\omega t - \theta')] |E_{z'0}| \end{aligned} \tag{7.30}$$

where $E_{z'0} = |E_{z'0}| e^{i\theta'}$. The polarization of the electric vector is seen from Eq. (7.30) to be an ellipse with its principal axes along $\hat{\mathbf{y}}$ and $\hat{\mathbf{z}}'$. Noting that for $\theta_{ph} = \pi/2$, $\hat{\mathbf{x}}' = -\hat{\mathbf{z}}$, the sense of rotation of the electric vector in the yz'-plane with reference to the direction of the **B**-field is seen to be positive for the upper and negative for the lower signs. The substitution of $\theta_{ph} = \pi/2$ and $k^2 = \Omega^2 \varepsilon/\varepsilon_1$ in Eq. (7.29) yields

$$\rho_r' = i\varepsilon_1/\varepsilon_2 \qquad \text{for } \theta_{ph} = \pi/2 \tag{7.31}$$

Hence, if $|\varepsilon_1/\varepsilon_2| < 1$, the major axis of the ellipse is in the propagation direction.

From Eqs. (6.22), (6.25), and (6.26), it is found that

$$E_{x'0} : E_{y0} : E_{z'0} = i \sin\theta_{ph}(\Omega^2\varepsilon - \varepsilon_1 k^2)$$
$$: -k^2 \sin\theta_{ph}\cos\theta_{ph}\varepsilon_2 : -i\frac{\cos\theta_{ph}}{\Omega^2}(\Omega^4\varepsilon - 2\varepsilon_1\Omega^2 k^2 + k^4) \tag{7.32}$$

With the help of Eqs. (7.13) and (7.24), it can be shown that

$$\Omega^2\varepsilon_3 - k^2(1 - l^2) = \varepsilon_3\frac{(k^2 - \Omega^2\varepsilon_1)(k^2 - \Omega^2\varepsilon_3)}{(\varepsilon_1 - \varepsilon_3)(k^2 - \Omega^2)} \tag{7.33}$$

On multiplying each term on the right-hand side of Eq. (7.32) by $\cot\theta_{ph}\,\varepsilon_3 (k^2 - \Omega^2\varepsilon_1)/(\Omega^2\varepsilon - \varepsilon_1 k^2)$ and making use of Eqs. (7.23), (7.25), and (7.33), Eq. (7.32) may be recast in the following form

$$E_{x'0} : E_{y0} : E_{z'0} = i\cos\theta_{ph}\varepsilon_3(k^2 - \Omega^2\varepsilon_1) : \varepsilon_2(\Omega^2\varepsilon_3 - k^2\sin^2\theta_{ph})$$
$$: -i\frac{\sin\theta_{ph}}{\Omega^2}(k^2 - \Omega^2\varepsilon_1)(k^2 - \Omega^2\varepsilon_3) \tag{7.34}$$

Special frequencies

At the resonant frequencies, $\Omega = \Omega_g$ and $\Omega = \Omega_p$, $k = \infty$ and it follows from Eq. (7.34) that the electric vector is in the direction of propagation except when $\theta_{ph} = 0$. For this special case of propagation parallel to the **B**-field, it has been shown in Sec. 5.4 that there is no component of the electric field in the propagation direction for any frequency. The right circularly polarized wave propagating parallel to the **B**-field has a resonance at the gyromagnetic frequency of the electrons $\Omega = R$ at which frequency the electric vector is in the wave-normal plane. Except for the resonance frequency $\Omega = R$ occurring in the direction of the **B**-field, in general, at a resonance frequency, the electric vector is in the direction of propagation.

For $\Omega = 1$, Eq. (7.32) should be used and not Eq. (7.34) since the latter has been obtained from Eq. (7.32) by the multiplication of a factor that included ε_3 which is equal to zero for $\Omega = 1$. At the cut-off ($k = 0$) frequency $\Omega = 1$, it can be deduced from Eq. (7.32) and Fig. 5.8 that $E_{x0} = E_{y0} = 0$; therefore the electric field is linearly polarized in the direction of the **B**-field. At the other cut-off frequencies $\Omega = \Omega_1$ and $\Omega = \Omega_2$, it can be shown from Eq. (7.34) and Fig. 5.8 that

$$E_{x0} : E_{y0} : E_{z0} = -i\varepsilon_1 : \varepsilon_2 : 0 \tag{7.35}$$

Therefore, the electric field is in a plane perpendicular to the **B**-field. Since for $\Omega = \Omega_1$, $\varepsilon_1 = -\varepsilon_2$, it follows from Eq. (7.35) that the electric field is left circularly polarized with reference to the direction of the **B**-field. Similarly for $\Omega = \Omega_2$, $\varepsilon_1 = \varepsilon_2$; and therefore, it follows from Eq. (7.35) that the electric field is right circularly polarized with reference to the **B**-field.

As has been pointed out by Allis, Buchsbaum and Bers [11], an interesting situation occurs when $k^2 = \Omega^2 \varepsilon_1$. It can be shown from Eq. (7.21) that

$$\tan \chi = -\cot \theta_{ph} \qquad \chi = \pi/2 + \theta_{ph} \tag{7.36}$$

Therefore, the **E**-plane is parallel to the z-axis, that is, the direction of the **B**-field. Hence, $E_{x0} = 0$. From Eq. (7.9), it is found that for this case $\rho_r = -i(\varepsilon_1/\varepsilon_2)\cos\theta_{ph}$ and Eq. (7.22) then shows that the electric field is elliptically polarized in the yz-plane. As the magnetoionic parameters are varied so that the **E**-plane passes through the yz-plane, the sense of rotation of the electric vector in the **E**-plane does not change, but the sense of rotation of the projection of the electric vector on the xy-plane is changed. This phenomenon occurs at very low frequencies and for directions of propagation other than that perpendicular to the **B**-field.

The polarization ratio ρ_r and the angle χ between the **E**- and the **H**-planes are shown, respectively, in Figs. 5.16 and 5.17, as functions of frequency for both modes 1 and 2. The parameters used are $R = 0.5$ and $\theta_{ph} = 45°$. From Eq. (7.9), it can be proved that for $\Omega = 0$, $\rho_r = i$, which is the same as that obtained in Fig. 5.16. At the cut-off frequency $\Omega = 1$, it is seen from Fig. 5.16 that $\rho_{r1} = 0$. In view of Eq. (7.10), at this frequency ρ_{r2} should be infinite as is confirmed in Fig. 5.16. At the cut-off frequencies $\Omega = \Omega_1$ and $\Omega = \Omega_2$, since the waves are, respectively, left and right circularly polarized in the xy-plane, it follows that $\rho_r = -i\sqrt{2}$ for $\Omega = \Omega_1$ and $\rho_r = i\sqrt{2}$ for $\Omega = \Omega_2$ and these values of ρ_r at the cut-off frequencies are also obtained in Fig. 5.16. It is seen from Fig. 5.16 that $|\rho_{r1}| < 1$ and $|\rho_{r2}| > 1$. Consequently, it follows that the major axes of the polarization ellipses for the two modes remain in their respective planes throughout the entire frequency ranges of propagation.

From Eq. (7.26), it can be deduced that $\chi = \theta_{ph}$ for $\Omega = 0$. At the resonance frequencies $\Omega = \Omega_g$ and $\Omega = \Omega_p$, since the **E**-plane has been shown to contain the propagation direction, it follows that $\chi = \pi/2$. In a similar manner, at the cut-off frequencies $\Omega = \Omega_1$ and $\Omega = \Omega_2$, since the **E**-plane has been shown to be perpendicular to the **B**-field, it is obtained that $\chi = \theta_{ph}$. At the cut-off frequency $\Omega = 1$, the **E**-plane has been shown to be parallel to the **B**-field with the result that χ should be equal to $-[\pi/2 - \theta_{ph}]$. For $\Omega = 1$, $\varepsilon_3 = 0$, and $\varepsilon = \varepsilon_1$; also, it can be verified from Eq. (6.52) that $k_2 = 1$. Therefore, it follows either from Eq. (7.21) or (7.26) that $\chi = -\pi/2$ for $\Omega = 1$ for mode 2. For these special values of frequency, the anticipated values for the angle χ are obtained numerically as may be seen from Fig. 5.17. At very high frequencies, the **E**-plane is seen to approach (from the opposite directions for the two modes) the **H**-plane; this is to be expected since free-space characteristics are attained in the limit of very high frequencies.

In the preceding discussion of the polarization of the fields, the medium has been assumed to be loss free with the result that the principal axes of the ellipses lie in the magnetic meridian plane and in a direction perpendicular to this plane. The present discussion can be extended to include the effect of the collisional losses [12, 13]. The net result is that the principal axes are rotated as well as the axis ratio is changed.

The Poynting vector

It is interesting to consider the nature of the locus traced in one time period by the terminal point of the time-dependent Poynting vector **S**(t) which is deduced from

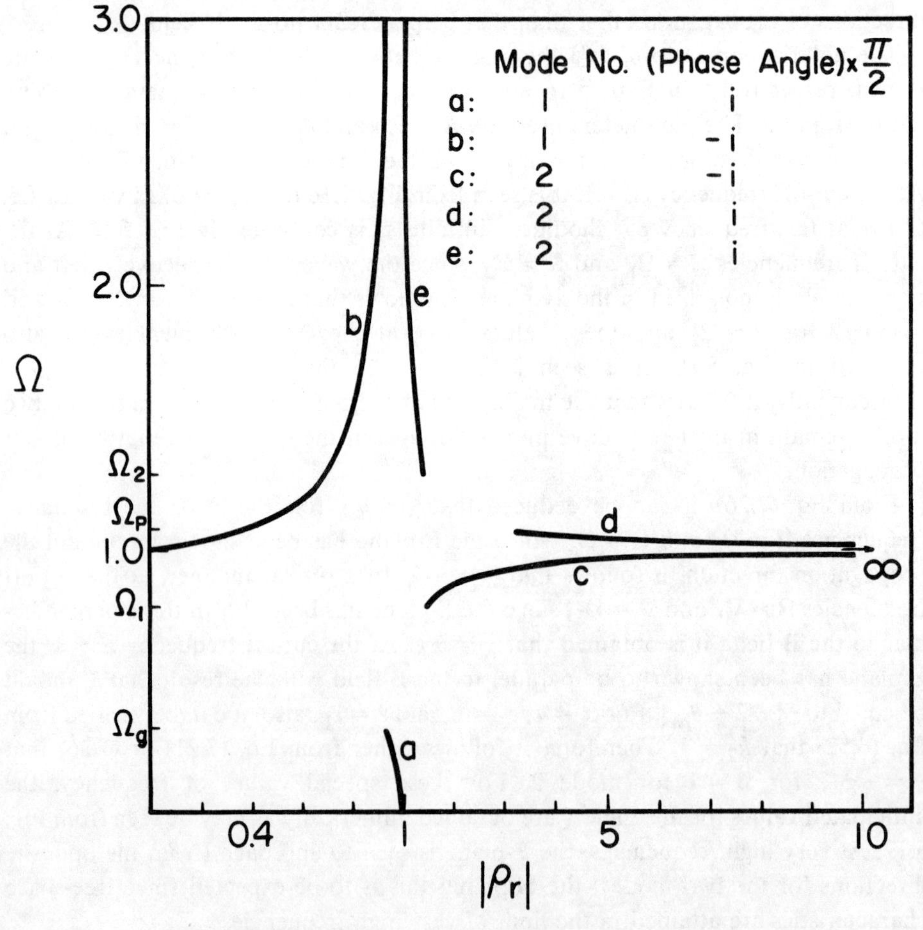

Fig. 5.16. Polarization ratio ρ_r for modes 1 and 2 as a function of frequency: $R = 0.5$; $\theta_{ph} = 45°$.

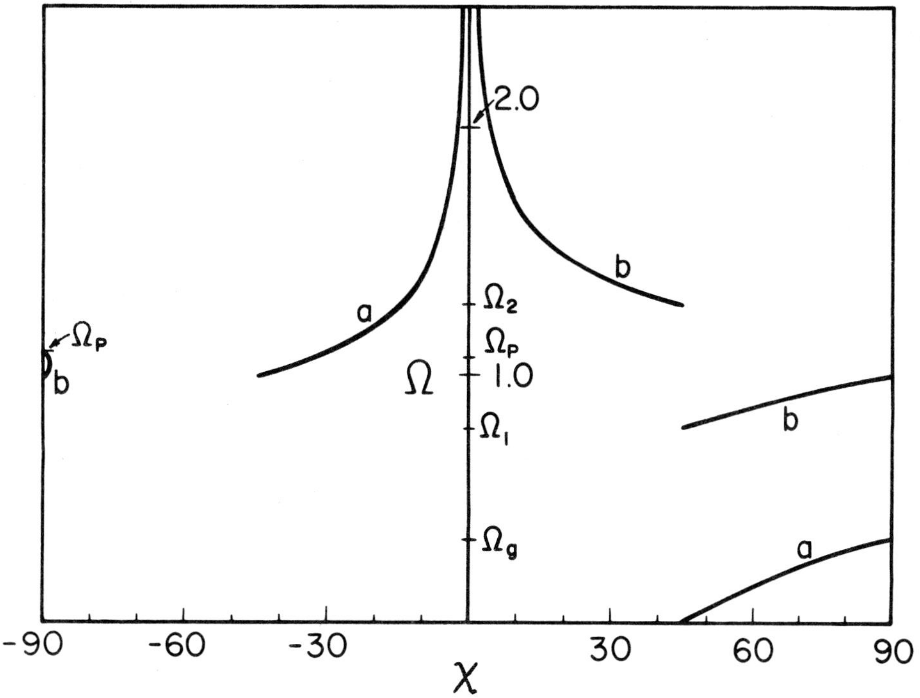

Fig. 5.17. Angle χ between the **E**- and the **H**-planes for modes 1 and 2 as a function of frequency: $R = 0.5$; $\theta_{ph} = 45°$; (a) mode 1; (b) mode 2.

Eqs. (7.18) and (7.22) to be given by

$$\mathbf{S}(t) = \mathbf{E}(t) \times \mathbf{H}(t)$$
$$= [-\hat{\mathbf{x}}' \tan\chi \cos^2(\omega t - \theta) \mp \hat{\mathbf{y}}|\rho_r|\tan\chi \cos(\omega t - \theta)\sin(\omega t - \theta) \qquad (7.37)$$
$$+ \hat{\mathbf{z}}'\{\cos^2(\omega t - \theta) + |\rho_r|^2\sin^2(\omega t - \theta)\}]|E_{x'0}||H_{y0}|$$

where the upper and the lower signs correspond to those in Eqs. (7.16). The time average of $\mathbf{S}(t)$ is obtained as

$$\mathbf{S} = \frac{\omega}{2\pi}\int_0^{2\pi/\omega} \mathbf{S}(t)\,dt = \tfrac{1}{2}|E_{x'0}||H_{y0}|[-\hat{\mathbf{x}}' \tan\chi + \hat{\mathbf{z}}'(1 + |\rho_r|^2)] \qquad (7.38)$$

A number of remarks are in order in connection with Eqs. (7.37) and (7.38). Whereas the time averages of $\mathbf{E}(t)$ and $\mathbf{H}(t)$ vanish, the time-averaged Poynting vector is a nonvanishing quantity which lies in the magnetic meridian plane. The time-averaged Poynting vector has components parallel as well as perpendicular to the wavevector. When $\omega t - \theta = 0$, since from Eq. (7.37) and Fig. 5.18, $\mathbf{S}(t) = |E_{x'0}||H_{y0}|\hat{\mathbf{z}}^E/\cos\chi$ where $\hat{\mathbf{z}}^E = \hat{\mathbf{x}}^E \times \hat{\mathbf{y}}$, it follows that $\mathbf{S}(t)$ is normal to the $x^E y$-plane. When $\omega t - \theta = \pi/2$, Eq. (7.37) yields $\mathbf{S}(t) = |E_{x'0}||H_{y0}||\rho_r|^2\hat{\mathbf{z}}'$ and therefore $\mathbf{S}(t)$ is normal to the $x'y$-plane. It can be ascertained from Eq. (7.37) that $\mathbf{S}(t)$ moves uniformly on the surface of a cone and passes through twice in a time period the normals to the $x'y$- and the $x^E y$-planes. Consequently, $\mathbf{S}$ in Eq. (7.38) makes with $\hat{\mathbf{z}}'$ an angle which is always less than χ. Therefore, the important result that the time-averaged Poynting vector $\mathbf{S}$ makes an acute angle with the direction of the wave-vector follows. If δ_1 is the angle made by $\mathbf{S}$ with $\hat{\mathbf{z}}'$, it follows from Eq. (7.38) and Fig. 5.18 that

$$\tan\delta_1 = \tan\chi/(1 + |\rho_r|^2) \qquad |\delta_1| < \pi/2 \tag{7.39}$$

and

$$\mathbf{S} = |E_{x'0}||H_{y0}|(1 + |\rho_r|^2)\hat{\mathbf{z}}_1^S/2\cos\delta_1 \tag{7.40}$$

From Eqs. (7.37) and (7.38), it is obtained that

$$\begin{aligned}\mathbf{S}(t) - \mathbf{S} = \tfrac{1}{2}|E_{x'0}||H_{y0}|[\{-\hat{\mathbf{x}}'\tan\chi + (1 - |\rho_r|^2)\hat{\mathbf{z}}'\}\cos\{2(\omega t - \theta)\}\\ \mp \hat{\mathbf{y}}|\rho_r|\tan\chi\sin 2(\omega t - \theta)]\end{aligned} \tag{7.41}$$

It is convenient to introduce (Fig. 5.18) another unit vector $\hat{\mathbf{z}}_2^S$ which makes an angle δ_2 with $\hat{\mathbf{z}}'$ such that

$$\tan\delta_2 = -\tan\chi/(1 - |\rho_r|^2) \qquad |\delta_2| < \pi/2 \tag{7.42}$$

Then, Eq. (7.41) can be rewritten in the form

$$\mathbf{S}(t) - \mathbf{S} = -\tfrac{1}{2}|E_{x'0}||H_{y0}|\tan\chi\left[\hat{\mathbf{z}}_2^S\frac{\cos 2(\omega t - \theta)}{\sin\delta_2} \pm \hat{\mathbf{y}}|\rho_r|\sin 2(\omega t - \theta)\right] \tag{7.43}$$

It is clear from Eq. (7.43) that the terminal point of the vector $[\mathbf{S}(t) - \mathbf{S}]$ traces an ellipse in $z_2^S y$-plane. The principal axes of the ellipse are along the y-axis and the z_2^S-axis which is in the magnetic meridian plane. The sense of rotation of $[\mathbf{S}(t) - \mathbf{S}]$ in the $z_2^S y$-plane is such that the sense of rotation of the projection of $[\mathbf{S}(t) - \mathbf{S}]$ onto the $x'y$-plane is the same as that of the magnetic vector. Therefore the senses of rotation of $[\mathbf{S}(t) - \mathbf{S}]$ with time are opposite for the two plane wave modes. The angular velocity with which the terminal point of $[\mathbf{S}(t) - \mathbf{S}]$ traces the ellipse is twice that which the terminal points of $\mathbf{E}(t)$ and $\mathbf{H}(t)$ trace their respective ellipses. If $\rho_r < \operatorname{cosec}\delta_2$, the major axis of the ellipse is along the z_2^S-axis which is in the magnetic meridian plane. On the other hand, if $|\rho_r| > |\operatorname{cosec}\delta_2|$, the minor axis of the ellipse is in the magnetic meridian plane. For $R = 0.5$ and $\theta_{ph} = 45°$, it was shown that $|\rho_{r1}| < 1$ and hence the major axis of the ellipse associated with $[\mathbf{S}(t) - \mathbf{S}]$ corresponding to mode 1 is in the magnetic meridian plane. The $z_2^S y$-plane in which $[\mathbf{S}(t) - \mathbf{S}]$ lies is conveniently designated as the Poynting vector plane or simply the **S**-plane. It is interesting to note that in all cases whether it is the electric, the magnetic or the Poynting vector, the difference between the vector and its time average always traces an ellipse with one of its principal axes in the magnetic meridian plane.

For the special case of $\theta_{ph} = 0$, corresponding to the plane wave propagating parallel to the **B**-field, $\chi = 0$ and $|\rho_r| = 1$ with the result that it follows from either Eqs. (7.18) and (7.22) or Eq. (7.37) that $\mathbf{S}(t) = |E_{x'0}||H_{y0}|\hat{\mathbf{z}}'$. Therefore, $[\mathbf{S}(t) - \mathbf{S}]$ vanishes. The Poynting vector does not vary with time and is parallel to the wave-vector **k**.

The angles δ_1 and δ_2 made by the direction $\hat{\mathbf{z}}'$ of the normalized wavevector **k** respectively with the time-averaged Poynting vector **S** and the projection of $[\mathbf{S}(t) - \mathbf{S}]$ onto the magnetic meridian plane are depicted as functions of frequency in Figs. 5.19 and 5.20 for both the modes in their respective frequency ranges of existence. As before, the parameters used are $R = 0.5$ and $\theta_{ph} = 45°$.

So far the two plane wave modes have been treated separately. It is now instructive to inquire further in what sense these two modes are independent. It has

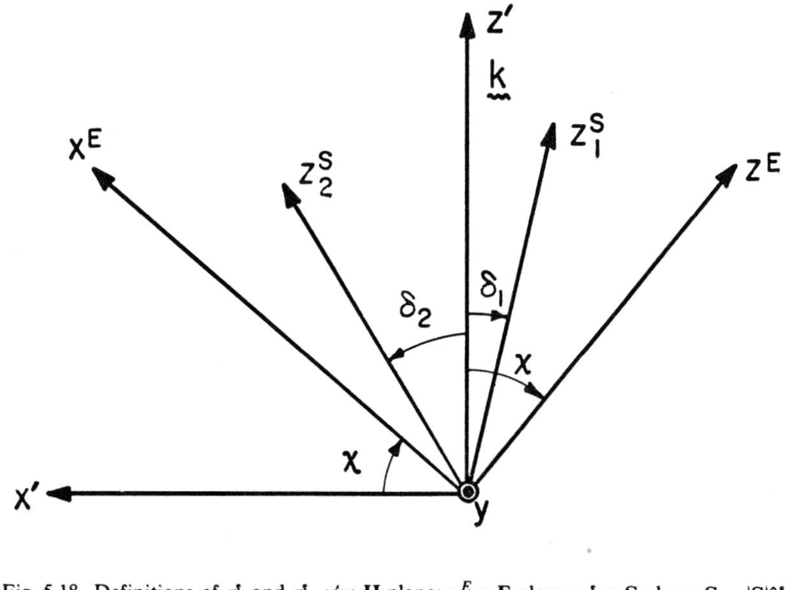

Fig. 5.18. Definitions of z_1^s and z_2^s. $x'y$: **H**-plane; $x^E y$: **E**-plane; $z_2^s y$: **S**-plane, $\mathbf{S} = |\mathbf{S}|\hat{\mathbf{z}}_1^s$.

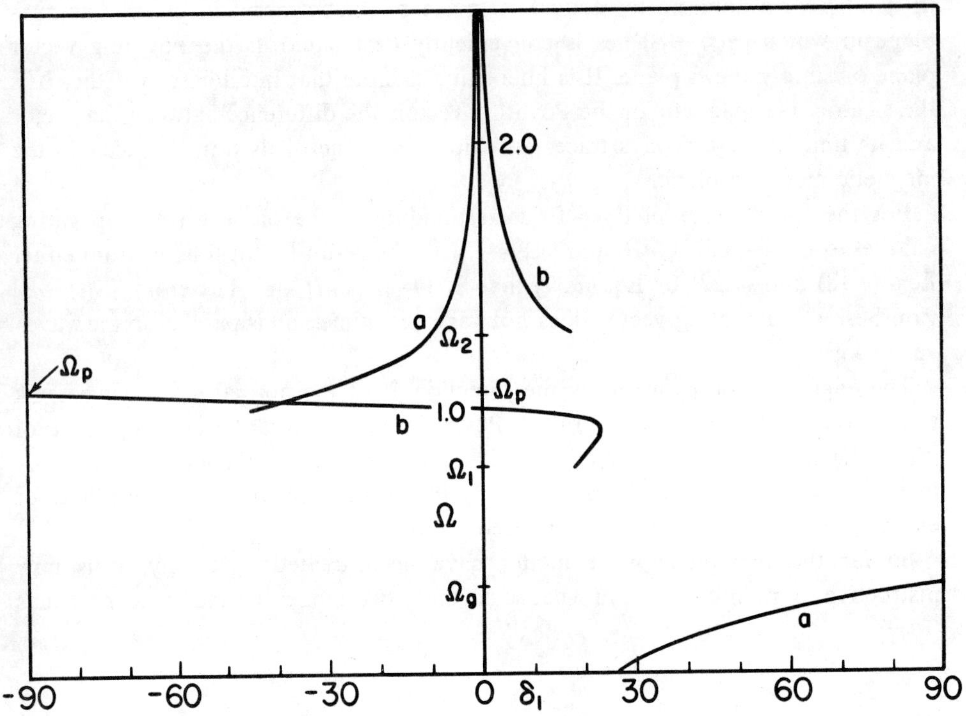

Fig. 5.19. Angle δ_1 between the time averaged Poynting vector **S** and the wavevector **k** for modes 1 and 2 as a function of frequency: $R = 0.5$; $\theta_{ph} = 45°$; (a) mode 1; (b) mode 2.

been established previously that the resultant time-averaged Poynting vector $S_{z'}$ parallel to the direction of the wavevector **k** due to the simultaneous presence of the two sets of fields is equal to the sum of the z'-components of the time-averaged Poynting vector associated with the two sets of fields separately. This result assures in some sense that the total power is equal to the sum of the powers associated with the two sets of fields independently. However, the components of the time-averaged Poynting vector **S** normal to the wavevector **k** do not appear to satisfy orthogonality in the conventional sense, as is established presently.

The y-component of **S** is given by

$$S_y = \tfrac{1}{2}\,\mathrm{Re}\,(E_{z'}^{*} H_{x'}) = \frac{-1}{2\sqrt{\mu_0 \varepsilon_0}}\,\mathrm{Im}\,[(i\sqrt{\varepsilon_0}\,E_{z'})^{*}(\sqrt{\mu_0}\,H_{x'})] \tag{7.44}$$

It is helpful to rewrite Eq. (6.30) so as to exhibit the phase factor $P(k_n) = \exp[i\{k_p k_n z' - \omega t\}]$ with $n = 1, 2$ explicitly as follows:

$$E_{z}^{(n)} = E_{z,n} P(k_n) \qquad n = 1, 2 \tag{7.45}$$

Note that in Eq. (7.45), $E_{z,1}$ and $E_{z,2}$ give the amplitude terms alone. The substitution of Eq. (7.45) and the terms similar to Eq. (6.31) in Eq. (7.44) yields

$$S_y = \sum_{n=1,2} S_y^{(n)} + y_{12} \tag{7.46}$$

where

$$S_y^{(n)} = \frac{-1}{2\sqrt{\mu_0 \varepsilon_0}} \operatorname{Im}\,[(i\sqrt{\varepsilon_0}\, E_{z'}^{(n)})^*(\sqrt{\mu_0}\, H_{x'}^{(n)})] \tag{7.47}$$

and

$$y_{12} = \frac{-1}{2\sqrt{\mu_0 \varepsilon_0}} \operatorname{Im}\,[(i\sqrt{\varepsilon_0}\, E_{z'}^{(1)})^*(\sqrt{\mu_0}\, H_{x'}^{(2)}) + (i\sqrt{\varepsilon_0}\, E_{z'}^{(2)})^*(\sqrt{\mu_0}\, H_{x'}^{(1)})] \tag{7.48}$$

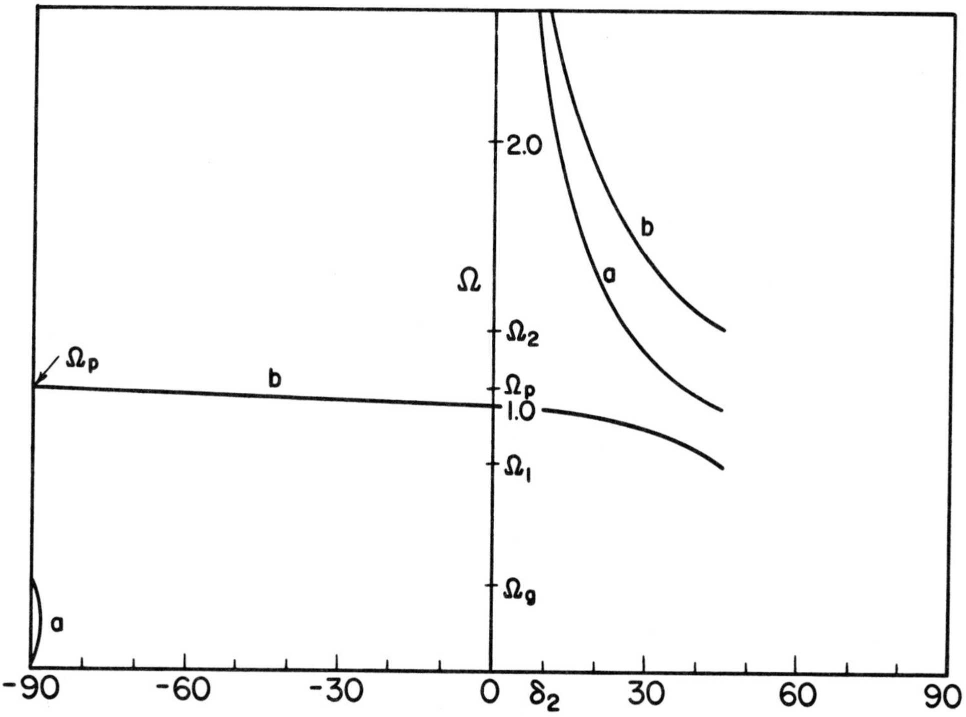

Fig. 5.20. Angle δ_2 between the projection of [**S**(t) − **S**] on the magnetic meridian plane and the wavevector **k** for modes 1 and 2 as a function of frequency: $R = 0.5$, $\theta_{ph} = 45°$; (a) mode 1; (b) mode 2.

Together with Eqs. (6.26), (6.27), and (7.45), Eqs. (7.47) and (7.48) may be simplified to yield

$$S_y^{(n)} = 0 \qquad n = 1, 2 \tag{7.49}$$

$$y_{12} = \frac{-l^2 \sin\theta_{ph}\,\varepsilon_2}{2\sqrt{\mu_0\varepsilon_0}\,\Omega^3 d_1 d_2} D_1 \,\mathrm{Im}\,[(i\sqrt{\varepsilon_0}\,E_{z,1})^*(i\sqrt{\varepsilon_0}\,E_{z,2})P^*(k_1)P(k_2)] \tag{7.50}$$

where

$$D_1 = k_2^3(k_1^4 - 2\varepsilon_1\Omega^2 k_1^2 + \Omega^4\varepsilon) - k_1^3(k_2^4 - 2\varepsilon_1\Omega^2 k_2^2 + \Omega^4\varepsilon) \tag{7.51}$$

With the help of Eqs. (6.19a), (6.38), (6.39), and (7.13), Eq. (7.51) may be simplified to yield

$$D_1 = \Omega^4\varepsilon(k_1 - k_2)(\varepsilon_3 - \varepsilon_1)\sin^2\theta_{ph}\,(k_1 k_2 + \Omega^2)/A_2 \tag{7.52}$$

Since k_1 and k_2 are never equal, it follows from Eq. (7.52) that $D_1 \neq 0$ except for the special case of $\theta_{ph} = 0$. When the two sets of fields are present simultaneously, owing to D_1 not being equal to zero identically, the time-averaged Poynting vector has a component perpendicular to the magnetic meridian plane. The presence of the factor $P^*(k_1)P(k_2) = \exp\{-ik_p(k_1 - k_2)z'\}$ in Eq. (7.50) shows that y_{12} and hence S_y have spatial variations in the direction of **k**. It is a common practice to treat each set of fields separately and obtain the result that the time-averaged Poynting vector has no component perpendicular to the magnetic meridian plane [11–13]. This conventional result is not rigorously correct when both the sets of fields are present simultaneously. However, the conventional result can be obtained and therefore justified only by taking a spatial average in the direction of progression of the phase fronts over the space period $2\pi/k_p(k_1 - k_2)$.

The x'-component of **S** is given by

$$S_{x'} = -\tfrac{1}{2}\,\mathrm{Re}\,E_{z'}^* H_y = \frac{-1}{2\sqrt{\mu_0\varepsilon_0}}\,\mathrm{Re}\,[(i\sqrt{\varepsilon_0}\,E_{z'})^*(i\sqrt{\mu_0}\,H_y)] \tag{7.53}$$

The substitution of Eq. (7.45) and terms similar to Eq. (6.31) in Eq. (7.53) yields

$$S_{x'} = \sum_{n=1,2} S_{x'}^{(n)} + x_{12} \tag{7.54}$$

where

$$S_{x'}^{(n)} = \frac{-1}{2\sqrt{\mu_0\varepsilon_0}}\,\mathrm{Re}\,[(i\sqrt{\varepsilon_0}\,E_{z'}^{(n)})^*(i\sqrt{\mu_0}\,H_y^{(n)})] \tag{7.55}$$

and

$$x_{12} = \frac{-l\sin\theta_{ph}}{2\sqrt{\mu_0\varepsilon_0}\,\Omega^3 d_1 d_2} D_2\,\mathrm{Re}\,[(i\sqrt{\varepsilon_0}\,E_{z,1})^*(i\sqrt{\varepsilon_0}\,E_{z,2})P^*(k_1)P(k_2)] \tag{7.56}$$

$$D_2 = k_2(k_1^4 - 2\varepsilon_1 \Omega^2 k_1^2 + \Omega^4 \varepsilon)(k_2^2 \varepsilon_1 - \Omega^2 \varepsilon) + k_1(k_2^4 - 2\varepsilon_1 \Omega^2 k_2^2 + \Omega^4 \varepsilon)(k_1^2 \varepsilon_1 - \Omega^2 \varepsilon) \tag{7.57}$$

With the help of Eq. (7.25), Eq. (7.57) may be simplified to yield

$$D_2 = \frac{-\sin^2\theta_{ph}}{\varepsilon_3 l^2}(k_1^2\varepsilon_1 - \Omega^2\varepsilon)(k_2^2\varepsilon_1 - \Omega^2\varepsilon)(k_1 + k_2)(k_1 k_2 - \Omega^2\varepsilon_3) \tag{7.58}$$

Except for the special case of $\theta_{ph} = 0$, $D_2 \neq 0$ and therefore, it follows that the resultant $S_{x'}$ due to the simultaneous presence of the two sets of fields is not equal to the sum of the x'-components of the time-averaged Poynting vector associated with the two sets of fields separately. However, by performing a spatial average as before, the x'-components of **S** also satisfy the orthogonality relation as is normally assumed to be the case.

Difficulties inherent in the plane wave analysis

The treatment of wave propagation in a magnetoionic medium in terms of homogeneous plane waves meets with a number of difficulties. The total power associated with the plane wave can be obtained by integrating the component $S_{z'}$ of the time-averaged Poynting vector throughout an infinite plane surface parallel to the phase fronts. From Eqs. (6.22), (6.24), (6.25), (6.27), and (6.32), this power can be shown to be infinite and therefore such a plane wave cannot be physically realized. This difficulty is inherent in a homogeneous plane wave irrespective of the medium in which it exists. Therefore, the homogeneous plane wave has to be abandoned as an entity and has to be used only as a constituent to build up other types of waves which have associated with them finite power and which are therefore physically realizable.

The components of the time-averaged Poynting vector parallel to the phase fronts do not strictly satisfy orthogonality. Consequently, the two possible sets of fields form independent modes of plane wave propagation in a magnetoionic medium only in a limited sense. The special case of propagation parallel to the direction of the **B**-field is an exception.

Another difficulty associated with the homogeneous plane waves is that the kinematic and the dynamic approaches to the concept of group velocity do not yield identical results. The kinematic approach is based on the linear superposition of harmonic components to form a suitable spatially localized quasimonochromatic disturbance. A plane wave field spatially localized in the direction of propagation ($\hat{\mathbf{z}}'$) can be constructed on the basis of the Fourier integral representation by choosing the amplitude function properly to form a wave packet. The wave packet propagates with the group velocity whose normalized value is obtained to be given by $\mathbf{v}_g/c = \hat{\mathbf{z}}'(\partial\Omega/\partial k)$, which is in the direction of the wavevector **k**. The dynamic approach, which is based on the energy principle, identifies the group velocity as the ratio of the time-averaged power flux density to the time-averaged total stored energy density. The direction of the group velocity as obtained from the dynamic approach is the same as that of the time-averaged Poynting vector which has an additional component normal to the wavevector. Thus, the two approaches yield different directions for the group velocity.

In view of the fact that there is a component of the time-averaged Poynting vector perpendicular to the wavevector, it follows that energy supply is needed not only in the direction of the wavevector but also parallel to the phase fronts in order to sustain a plane wave in an anisotropic region such as a magnetoionic medium. As a result of this, Lighthill [14] had concluded previously that it is unrealistic to treat wave propagation in magnetoionic media in terms of homogeneous plane waves alone. Additional arguments are now provided to strengthen Lighthill's original conclusion. Notwithstanding the fact that the magnetoionic theory, that is the theory of homogeneous plane waves in a magnetoionic medium, has had significant success in explaining a variety of wave phenomenon in the ionosphere, the difficulties mentioned in the foregoing advocate caution against losing sight of the fact that an infinite plane wave is only a part of any physically realizable wave field, and drawing conclusions by treating it as a whole entity.

5.8. Spectrum of Cerenkov radiation in a magnetoionic medium

Synchronism condition

Cerenkov radiation is a coherent phenomenon and it refers generally to the emission from a point charge moving in a straight line at a constant speed. A qualitative explanation of this effect in terms of Huygens' spherical wavelets is possible. For this purpose, it is simple to consider an ordinary dielectric medium characterized by a relative permittivity ε_r which is independent of frequency. Let v be the velocity of electromagnetic waves in the dielectric medium. A point charge is assumed to move in a straight line with a constant velocity u. The charge continuously emits spherical wavelets as it moves along the straight line and these wavelets spread radially outwards at the velocity v. If $u < v$, it is seen from Fig. 5.21a that the spherical wavelets do not interfere constructively in any direction to give a coherent radiation. But if $u > v$, as shown in Fig. 5.21b, the spherical wavelets interfere constructively on the surface of a cone. The conical surface moves normally to itself in a direction which makes an angle θ_c with the direction of motion of the charge. Thus there is a coherent radiation with a conical wavefront. In the other directions, the spherical wavelets interfere destructively and hence do not yield a coherent radiation. From Fig. 5.21b, the direction of the wave normal is found to be given by

$$\cos \theta_c = v/u \tag{8.1}$$

which is called the *coherence condition.*

In free space, $v = c$ where c is the velocity of light in vacuum. Since the charged particle cannot travel at a velocity greater than that of light in vacuum, it is obtained that $u < c$. Therefore, there is no Cerenkov radiation in free space. In the dielectric medium, the velocity of electromagnetic waves is $v = c/\sqrt{\varepsilon_r}$ where ε_r is greater than unity. Hence, $v < c$. Therefore, in a dielectric medium, the velocity u of the point charge can be greater than v and yet be less than the velocity of light in vacuum with the result that Cerenkov radiation is possible in an ordinary dielectric medium.

The time-dependent phenomenon in an ordinary dielectric with a constant relative permittivity can be Fourier analyzed to yield the behavior of the various frequency components. For each frequency, the point charge emits time-harmonic spherical wavelets with their phases progressing radially outwards at a phase velocity v_{ph} which, being independent of frequency, is equal to v. The various spherical wavelets are in phase on a conical surface whose normal makes an angle θ_{ph} with the direction of motion of the charge, where

$$\cos\theta_{ph} = v_{ph}/u \tag{8.2}$$

If $u < v_{ph}$, Eq. (8.2) reveals that the spherical wavelets do not have the same phase on any surface and there is no Cerenkov radiation. If $u > v_{ph}$, there is

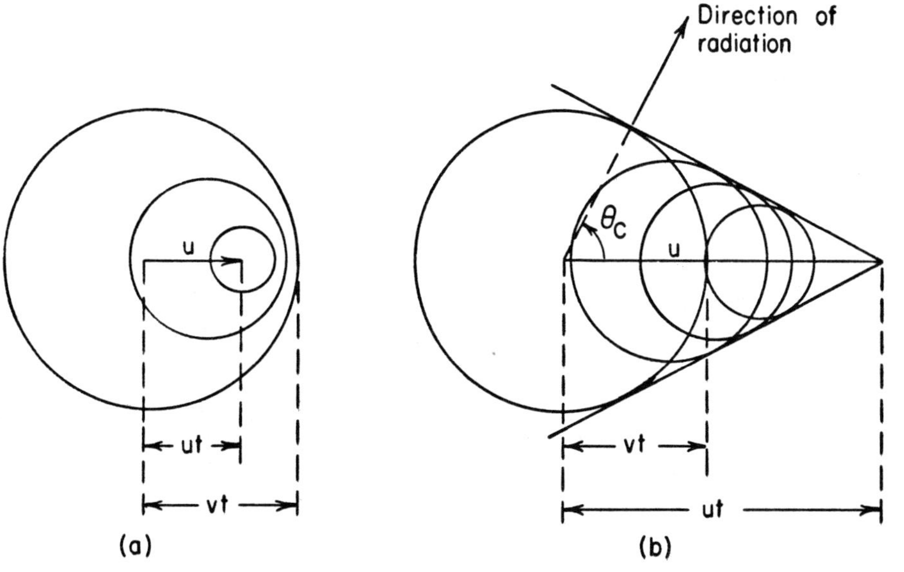

Fig. 5.21. Spherical wavelets produced by a point charge moving with uniform velocity in a straight line: (a) $u < v$, (b) $u > v$.

Cerenkov radiation in the form of a conical wave. For an ordinary dielectric, v_{ph} is independent of frequency and Eq. (8.2) shows that for all the frequency components, the surfaces of constant phase coincide with the same conical surface as specified by Eq. (8.2). All the frequency components have identical behavior; therefore, it is an easy matter to sum up their contributions and determine the nature of the phenomenon in the time domain. As was discussed previously, the conical phase front with its apex coinciding with the position of the charged particle becomes the wavefront in the time domain. The wavefront is a surface of discontinuity such that the radiation is confined to the regions behind it.

In a dispersive medium in which the phase velocity is a function of frequency, it is seen from Eq. (8.2) that different conical phase fronts are obtained for the different frequency components. When the responses due to the various frequency components are synthesized it is found that in the time domain the wavefront is very complicated. As a matter of fact, the wavefront cannot even be explicitly determined for most cases. This type of behavior is obtained in a magnetoionic medium which is dispersive. Even in a dispersive medium the frequency components that are emitted by the Cerenkov mechanism and the directions of the wave normals of the conical phase fronts pertaining to each frequency can be determined with the help of Eq. (8.2). Such a study for a magnetoionic medium is very useful in gaining some understanding of the naturally occurring electromagnetic emissions through the Cerenkov mechanism by the charged solar particles traversing the ionized regions of the exosphere in the direction of the terrestrial magnetic field. For this reason and as an application of the dispersion relation of a plane wave progressing in an arbitrary direction with respect to the magnetostatic field in a magnetoionic medium, a treatment of the spectrum of Cerenkov radiation in a magnetoionic medium is given in this section for the simplest case in which the point charge is moving with uniform velocity in the direction of the magnetostatic field. It should be noted that the presence of anisotropy further complicates the study of Cerenkov radiation in a magnetoionic medium.

The magnetostatic field is in the z-direction. The point charge moves with uniform velocity u in the direction of the magnetostatic field. The propagation coefficient of the wave progressing in a direction making an angle θ_{ph} with that of the magnetostatic field is ω/v_{ph}. The component of the propagation coefficient in the direction of the magnetostatic field is $(\omega/v_{ph})\cos\theta_{ph}$. Therefore, the phase velocity $v_{ph,z}$ of the wave in the direction of motion of the charge is $v_{ph}/\cos\theta_{ph}$ which according to Eq. (8.2) is equal to the particle velocity u. Therefore, the coherence condition (8.2) can be stated as

$$v_{ph,z} = u \tag{8.3}$$

which is also called the synchronism condition. The coherence condition as stated in Eq. (8.3) is in accordance with the previously discussed requirement for the interaction between the wave and the particle.

Dispersion relation for Cerenkov radiation

The propagation coefficient of the wave along and across the magnetostatic field are given by $k_p k_z$ and $k_p k_\rho$, respectively, where k_z and k_ρ are the normalized values and $k_p = \omega_p / c = \omega_p \sqrt{\mu_0 \varepsilon_0}$ is the normalizing factor. From Eq. (6.17), k_z and k_ρ are known to be related by the following dispersion relation

$$k_\rho^4 + B k_\rho^2 + C = 0 \tag{8.4}$$

where

$$B = \frac{\Omega^2}{\varepsilon_1}\left[(\varepsilon_1 + \varepsilon_3)\frac{k_z^2}{\Omega^2} - (\varepsilon_1^2 - \varepsilon_2^2 + \varepsilon_1 \varepsilon_3)\right] \tag{8.5a}$$

and

$$C = \frac{\Omega^4 \varepsilon_3}{\varepsilon_1}\left[\left(\varepsilon_1 - \frac{k_z^2}{\Omega^2}\right)^2 - \varepsilon_2^2\right] \tag{8.5b}$$

If the wave is generated by the point charge moving at a constant velocity u along the z-axis, i.e., in the direction of the magnetostatic field, then Eq. (8.3) shows that the component $k_p k_z$ of the propagation coefficient in the direction of motion of the charge is fixed by the particle velocity as given by

$$k_p k_z = \omega / u \qquad \text{or} \qquad k_z = \Omega c / u = \Omega \gamma \tag{8.6}$$

where $\gamma = c/u$ is greater than unity and γ^{-1} gives the particle velocity normalized by the velocity c of light in vacuum. In view of Eq. (8.6), Eqs. (8.5a) and (8.5b) become

$$B = \frac{\Omega^2}{\varepsilon_1}[(\varepsilon_1 + \varepsilon_3)\gamma^2 - (\varepsilon_1^2 - \varepsilon_2^2 + \varepsilon_1 \varepsilon_3)] \tag{8.7a}$$

$$C = \frac{\Omega^4 \varepsilon_3}{\varepsilon_1}[(\varepsilon_1 - \gamma^2)^2 - \varepsilon_2^2] \tag{8.7b}$$

It is necessary to evaluate the discriminant $(B/2)^2 - C$ of Eq. (8.4). For this purpose, with the help of Eq. (7.13), B given by Eq. (8.7a) is rewritten in the form

$$B = \frac{\Omega^2}{\varepsilon_1}[(\varepsilon_1 - \varepsilon_3)(\gamma^2 - 1) + 2\varepsilon_3(\gamma^2 - \varepsilon_1)] \tag{8.8}$$

From Eqs. (8.7b) and (8.8), it is obtained that

$$\begin{aligned}\left(\frac{B}{2}\right)^2 - C = \frac{\Omega^4}{4\varepsilon_1^2}[&(\varepsilon_1 - \varepsilon_3)^2(\gamma^2 - 1)^2 + 4\varepsilon_3(\varepsilon_1 - \varepsilon_3)(\gamma^2 - 1)(\gamma^2 - \varepsilon_1) \\ &+ 4\varepsilon_3^2(\gamma^2 - \varepsilon_1)^2 - 4\varepsilon_3\varepsilon_1(\gamma^2 - \varepsilon_1)^2 + 4\varepsilon_3\varepsilon_1\varepsilon_2^2]\end{aligned} \tag{8.9}$$

The third and the fourth terms within the square brackets in Eq. (8.9) are equal to

$-4\varepsilon_3(\varepsilon_1-\varepsilon_3)(\gamma^2-\varepsilon_1)^2$ which, when combined with the second term, yields $4\varepsilon_3(\varepsilon_1-\varepsilon_3)(\varepsilon_1-1)(\gamma^2-\varepsilon_1)$. It is possible to rewrite Eq. (7.13) as

$$(\varepsilon_1-\varepsilon_3)(\varepsilon_1-1)=\varepsilon_2^2 \tag{8.10}$$

Therefore, the second, the third, and the fourth terms together equal $4\varepsilon_3\varepsilon_2^2(\gamma^2-\varepsilon_1)$ which, when combined with the fifth term, gives $4\varepsilon_3\varepsilon_2^2\gamma^2$. Hence, Eq. (8.9) simplifies as

$$\left(\frac{B}{2}\right)^2-C=\frac{\Omega^4}{4\varepsilon_1^2}[(\varepsilon_1-\varepsilon_3)^2(\gamma^2-1)^2+4\varepsilon_3\varepsilon_2^2\gamma^2] \tag{8.11}$$

With the help of Eqs. (2.21 a, b, and c) and (7.13), B, C and $(B/2)^2-C$ as given by Eqs. (8.7a), (8.7b), and (8.11) can be expressed in terms of Ω and R as follows:

$$\begin{aligned}B=-(\Omega^2-1-R^2)^{-1}[2\Omega^4(1-\gamma^2)\\-2\Omega^2\{2-\gamma^2+R^2(1-\gamma^2)\}+2+R^2(1-\gamma^2)]\end{aligned} \tag{8.12}$$

$$\begin{aligned}C=(\Omega^2-1)(\Omega^2-1-R^2)^{-1}\\ [\Omega^4(1-\gamma^2)^2-\Omega^2(1-\gamma^2)\{2+R^2(1-\gamma^2)\}+1]\end{aligned} \tag{8.13}$$

$$(B/2)^2-C=R^2(\Omega^2-1-R^2)^{-2}\Delta \qquad \Delta=(\Omega^2-1)\gamma^2+R^2(\gamma^2-1)^2/4 \tag{8.14}$$

If k is the normalized propagation coefficient, then

$$k_\rho=k\sin\theta_{ph} \qquad k_z=k\cos\theta_{ph} \tag{8.15a, b}$$

Since θ_{ph} has to be real, it follows from Eqs. (8.6) and (8.15b) that k is real; therefore k_ρ is real. Equivalently, k_ρ^2 is real and positive. In order for the point charge moving with uniform velocity in the direction of the magnetostatic field to emit electromagnetic waves, it is required that k_ρ^2 be real and positive. This requirement together with Eq. (8.4) imposes certain restrictions on B and C for the existence of Cerenkov radiation. Since B and C are functions of γ, R, and Ω, the implication is that for a given value of the normalized particle speed γ^{-1} and a specified value of the normalized strength R of the magnetostatic field, there are only certain ranges of Ω for which there is radiation of electromagnetic waves by the Cerenkov mechanism. Thus the frequency spectrum of Cerenkov radiation can be obtained.

Frequency spectrum

For a specified value of γ or the speed of the point charge, in the Ω^2-R^2 space, it is proposed to determine the various regions which correspond to the existence of Cerenkov radiation. Therefore, for any given value of the normalized strength R of the magnetostatic field, the frequency spectrum of Cerenkov radiation can be obtained. The two solutions of k_ρ^2 are found from Eq. (8.4) as

$$k_\rho^2 = -B/2 \pm \sqrt{[(B/2)^2 - C]} \tag{8.16}$$

It is convenient to use Eqs. (8.14), rewrite the two solutions in Eq. (8.16) and designate them as follows:

$$k_{\rho 1}^2 = -B/2 + R\sqrt{\Delta}\,/(\Omega^2 - 1 - R^2) \tag{8.17a}$$

$$k_{\rho 2}^2 = -B/2 - R\sqrt{\Delta}\,/(\Omega^2 - 1 - R^2) \tag{8.17b}$$

Also, the electromagnetic fields associated with $k_{\rho 1}$ and $k_{\rho 2}$ are designated as modes 1 and 2, respectively. It is convenient to distinguish two cases, namely (i) $C < 0$ and (ii) $C > 0$. For the case (i), $(B/2)^2 - C > 0$ and $|\sqrt{(B/2)^2 - C}| > |B/2|$; therefore, that solution in Eq. (8.16) which has a positive sign in front of the radical is real and positive. With reference to Eqs. (8.17a) and (8.17b), it follows that if $C < 0$, $k_{\rho 1}^2$ or $k_{\rho 2}^2$ is real and positive according as $\Omega^2 > \Omega_u^2 = 1 + R^2$ or $\Omega^2 < \Omega_u^2$. For the case (ii), if $(B/2)^2 - C < 0$, both $k_{\rho 1}^2$ and $k_{\rho 2}^2$ are complex. But if $C > 0$ and $(B/2)^2 - C > 0$, $|\sqrt{(B/2)^2 - C}| < |B/2|$ and therefore, both the solutions in Eq. (8.16) are real and have the same sign as $-B$. Hence, if $B < 0$, both $k_{\rho 1}^2$ and $k_{\rho 2}^2$ are real and positive. Thus from a knowledge of the signs of B, C, and $\sqrt{(B/2)^2 - C}$, the various regions in the Ω^2-R^2 space where $k_{\rho 1}^2$ and $k_{\rho 2}^2$ are real and positive can be determined.

The signs of B, C, and $(B/2)^2 - C$ in the various regions of the Ω^2-R^2 space can be obtained in the following manner. The parametric equations corresponding to $B = 0$, $B = \infty$, $C = 0$, $C = \infty$ and $(B/2)^2 - C = 0$ are deduced from Eqs. (8.12), (8.13), and (8.14) to be given by the following curves in the Ω^2-R^2 space:

(a) $\Omega^2 = 1$ for $C = 0$

(b) $R^2 = \Omega^{-2}[\Omega^2 + 1/(\gamma^2 - 1)]^2$ for $C = 0$

(c) $\Omega^2 = 1 + R^2$ for $B = C = \infty$

(d) $R^2 = [\Omega^2 + 1/(\gamma^2 - 1)][\Omega^2 - 1][\Omega^2 - \tfrac{1}{2}]^{-1}$ for $B = 0$

(e) $\Omega^2 = 1 - R^2(\gamma^2 - 1)^2/4\gamma^2$ for $(B/2)^2 = C$ (8.18)

The curves (*a*)–(*e*) are sketched in Fig. 5.22 for $\gamma^2 = 5$ and these curves are seen to divide the Ω^2-R^2 space into several regions. For the solar particles traversing the ionized regions of the exosphere γ is of the order of 100 and therefore $\gamma^2 = 10^4$. But the value $\gamma^2 = 5$ has been chosen for Fig. 5.22 in order to exhibit clearly the various regions in the Ω^2-R^2 space. It is to be noted that the relative positions of the various curves are unaltered for any other value of γ^2 greater than five.

A brief description of some of the relevant properties of the various parametric curves (*a*)–(*e*) is useful in explicitly writing down the frequency ranges of excitation of the two modes. The curve (*e*) is a straight line; it intersects the $\Omega^2 = 0$ axis at $R^2 = R_1^2 = 4\gamma^2(\gamma^2-1)^{-2}$ and is tangent to the curve (*b*) at $\Omega^2 = \Omega_2^2 = (\gamma^2+1)^{-1}$, $R^2 = R_2^2 = 4\gamma^4[(\gamma^2+1)(\gamma^2-1)^2]^{-1}$. The curve (*b*) has a minimum at $\Omega^2 = \Omega_3^2 = (\gamma^2-1)^{-1}$, $R^2 = R_3^2 = 4(\gamma^2-1)^{-1}$ and intersects the curve (*a*) which is a straight line at $\Omega^2 = \Omega_0^2 = 1$, $R^2 = R_0^2 = \gamma^4(\gamma^2-1)^{-2}$. The curve (*d*) may be shown to pass through the tangent point (Ω_2^2, R_2^2) and to lie, for $\Omega^2 > 1$, below the curve (*b*) and above the curve (*c*), which is a straight line. Once the particle velocity, that is γ^2, is known, the parametric curves can be drawn in the Ω^2 - R^2 space.

It is seen from Eq. (8.12) that B changes sign on crossing the curves corresponding to $B = 0$ and $B = \infty$. Similarly, Eqs. (8.13) and (8.14) show that C changes sign on

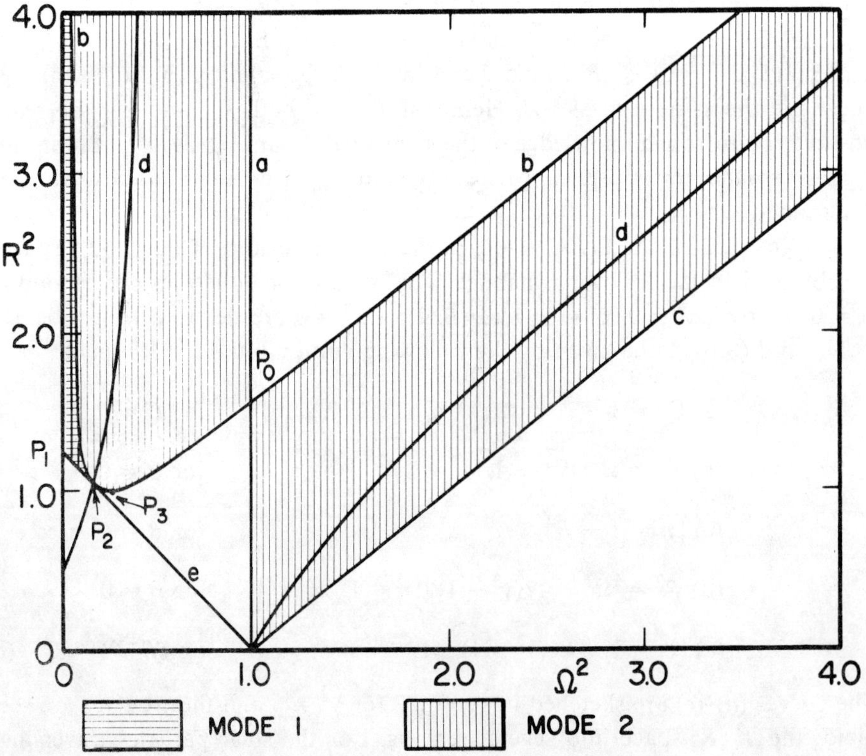

Fig. 5.22. Regions of emission of modes 1 and 2 in a magnetoionic medium by the Cerenkov mechanism. $P_0 : (1, R_0^2)$; $P_1 : (0, R_1^2)$; $P_2 : (\Omega_2^2, R_2^2)$; $P_3 : (\Omega_3^2, R_3^2)$.

crossing the curves corresponding to $C = 0$ and $C = \infty$, and, $(B/2)^2 - C$ changes sign on crossing the curve corresponding to $(B/2)^2 = C$. Therefore, the various parametric curves divide the Ω^2-R^2 space into several regions inside which the signs of B, C and $(B/2)^2 - C$ remain the same. If these signs are determined for the various regions in the Ω^2-R^2 space, the characteristics of the roots $k_{\rho 1}^2$ and $k_{\rho 2}^2$ can also be obtained for all values of Ω and R.

The signs of B, C and $(B/2)^2 - C$ are determined for any one region in the Ω^2-R^2 space and their signs for the other regions are obtained by inspection. From Eqs. (8.14) and the equation for the curve (*e*), it is found that for the region lying below the curve (*e*), $(B/2)^2 - C$ is negative with the result that $k_{\rho 1}^2$ and $k_{\rho 2}^2$ are both complex. On crossing the curve (*e*), the sign of $(B/2)^2 - C$ changes. Hence $(B/2)^2 - C$ is positive for all the regions lying above and to the right of the curve (*e*); therefore, in these regions $k_{\rho 1}^2$ and $k_{\rho 2}^2$ are both real. For R^2 not too large and Ω^2 very large, it is found from Eqs. (8.12) and (8.13) that $B > 0$ and $C > 0$. On crossing the curve (*c*) corresponding to $B = C = \infty$, the signs of both B and C change. Thus the signs of B and C can be obtained for all the regions. There are two regions in which $C < 0$ and these regions are marked by shading with vertical lines in Fig. 5.22. For both the regions, $\Omega^2 < 1 + R^2$. Therefore, in these two regions $k_{\rho 2}^2$ is real and positive, and, corresponds to the excitation of mode 2. There is one region corresponding to small values of Ω^2 for which $C > 0$, $(B/2)^2 - C > 0$ and $B < 0$. This region is indicated in Fig. 5.22 by shading with both horizontal and verticle lines. Both $k_{\rho 1}^2$ and $k_{\rho 2}^2$ are real and positive in this region, and corresponds to the excitation of both modes 1 and 2. Consequently in Fig. 5.22, in the region marked by shading with horizontal lines, $k_{\rho 1}^2$ is real and positive corresponding to the excitation of mode 1 and in the regions indicated by shading with vertical lines, $k_{\rho 2}^2$ is real and positive corresponding to the excitation of mode 2. The operating line is determined by the value of R and is a straight line parallel to the Ω^2-axis. The frequency spectrum of Cerenkov radiation is given by the range of frequencies associated with those portions of the operating line lying in the regions where k_ρ^2 is real and positive.

With the help of Fig. 5.22, it is possible to determine explicit analytical expressions for the frequency spectrum of Cerenkov radiation. For this purpose, the relative magnitudes of R_0^2, R_1^2, R_2^2, and R_3^2 are required. It can be established that for any value of γ^2, the following inequalities are satisfied:

$$R_1^2 > R_2^2 > R_3^2 \tag{8.19}$$

Also, it can be proved that

$$R_0^2 > R_1^2 \qquad \text{for } \gamma^2 > 4 \tag{8.20}$$

It was already pointed out that for the application under consideration γ^2 is of the order of 10^4 and is greater than 4. Therefore for any value of γ^2 greater than 4, the relative magnitudes of R_0^2, R_1^2, R_2^2, and R_3^2 remain the same and is given by

$$R_0^2 > R_1^2 > R_2^2 > R_3^2 \tag{8.21}$$

Note that in Fig. 5.22 which corresponds to $\gamma^2 = 5$, the inequalities (8.21) hold good. It is found from Eq. (8.18e) that any operating line $R^2 = R_{op}^2$ where $0 \leqslant R_{op}^2 < R_1^2$ intersects the curve (*e*) at a value of $\Omega^2 = \Omega_{60}^2$ where Ω_{60}^2 is given by

$$\Omega_{60}^2 = 1 - R_{op}^2(\gamma^2 - 1)^2/4\gamma^2 \tag{8.22}$$

With the help of Eq. (8.18b), it can be shown that any operating line $R^2 = R_{op}^2$ where $R_3^2 < R_{op}^2 < \infty$ intersects the curve (*b*) at two values of Ω^2, namely Ω_{2S}^2 and Ω_{2L}^2 whose values are specified by

$$\Omega_{2S,2L} = R/2 \mp [(R/2)^2 - (\gamma^2 - 1)^{-1}]^{1/2} \tag{8.23}$$

where Ω_{2S} is the smaller of the two roots. The expressions (8.22) and (8.23) are necessary to define the limits of the frequency spectrum. From an examination of Fig. 5.22 together with the inequalities (8.21), the frequency spectrum of Cerenkov radiation can be explicitly determined and the results are given in the following table:

	Mode 1		Mode 2
1. $0 < R^2 < R_3^2$	—	—	$1 < \Omega < \Omega_u$
2. $R_3^2 < R^2 < R_2^2$	—	$\Omega_{2S} < \Omega < \Omega_{2L}$	$1 < \Omega < \Omega_u$
3. $R_2^2 < R^2 < R_1^2$	$\Omega_{60} < \Omega < \Omega_{2S}$	$\Omega_{60} < \Omega < \Omega_{2L}$	$1 < \Omega < \Omega_u$
4. $R_1^2 < R^2 < R_0^2$	$0 < \Omega < \Omega_{2S}$	$0 < \Omega < \Omega_{2L}$	$1 < \Omega < \Omega_u$
5. $R_0^2 < R^2 < \infty$	$0 < \Omega < \Omega_{2S}$	$0 < \Omega < 1$	$\Omega_{2L} < \Omega < \Omega_u$

It is emphasized that the frequency spectrum of Cerenkov radiation consists of one or more bands of *continuous* frequencies and in certain frequency ranges there is simultaneous emission of two sets of fields or modes having different characteristics.

Characteristics and application of Cerenkov radiation in an electron plasma

The characteristics of the wave at a particular frequency emitted by the Cerenkov mechanism can be evaluated in the following manner. If the particle speed and the medium parameters are specified, γ and R are known, and then the frequency spectrum of Cerenkov radiation can be deduced. For a particular normalized frequency lying within the emitted spectrum, the normalized radial propagation coefficient $k_{\rho n}$ is determined from Eqs. (8.17) where the subscript n (1 or 2) indicates the order of the mode. The normalized axial propagation coefficient k_z is given by Eq. (8.6). The magnitude and the direction of the normalized wavevector are determined from Eqs. (8.15) to be given by

$$k_n = \sqrt{k_{\rho n}^2 + k_z^2} \qquad \theta_{ph,n} = \tan^{-1}(k_{\rho n}/k_z) \tag{8.24}$$

The normalized group velocity $\mathbf{v}_{gn}/c$ can be determined from Eq. (8.4). Let $A_{\rho n}$ and A_{zn} be the radial and the axial components of $\mathbf{v}_{gn}/c$. Then, the magnitude and the direction of the normalized group velocity are obtained as

$$v_{gn}/c = \sqrt{A_{\rho n}^2 + A_{zn}^2} \qquad \theta_{g,n} = \tan^{-1}(|A_{\rho n}|/A_{zn}) \tag{8.25}$$

The direction of the group velocity is the same as that of the Cerenkov *ray*. In general, the wave normal and the ray are in different directions and this is a manifestation of the anisotropy of the medium.

There is an interesting phenomenon associated with mode 1 whose spectrum extends down to very low frequencies [15]. The characteristics of radiation due to streams of solar charged particles moving along the magnetostatic field of the earth, may be obtained approximately by treating the process as a Cerenkov radiation due to a point charge moving with uniform velocity in the direction of the magnetostatic field in an unbounded magnetoionic medium. The wave-normal direction θ_{ph}, the direction θ_g of the group velocity, and the magnitude of the normalized group velocity v_g/c are depicted in Fig. 5.23 as functions of frequency for mode 1 for the appropriate values of the parameters γ and R. It is seen that the emitted radiation is confined to directions very close ($\leqslant 20°$) to that of the magnetostatic field, and has a group velocity whose magnitude increases with the frequency. The frequency spectrum of mode 1 is usually in the audiofrequency range for the medium parameters corresponding to the exosphere. The emission associated with mode 1 is propagated essentially along the direction of the magnetostatic field, and has a frequency versus time spectrum observed on the earth as an audible tone with mean frequency descending slowly with the time of arrival, giving rise to the well-known phenomenon of whistlers. This phenomenon was treated previously in terms of the plane waves propagating along the magnetostatic field in a magnetoionic medium. Here that treatment is extended to include the propagation in directions other than that of the magnetostatic field as well as a possible excitation mechanism. From Fig. 5.22, it is noted that mode 2 is emitted simultaneously with mode 1. But the emission associated with mode 2 cannot account for the observed phenomenon, since (i) the emission is not confined to the directions close to that of the magnetostatic field, and (ii) the magnitude of the group velocity decreases with the frequency.

5.9. Propagation along and across the magnetostatic field in an electron-ion plasma

The conventional magnetoionic theory treated in Secs. 5.6 and 5.7 is concerned with a study of the characteristics of a plane wave propagating in an arbitrary direction with respect to the magnetostatic field in a cold, homogeneous, and unbounded electron plasma. This theory, which is an approximation in which the motion of the electrons is taken into account but those of the heavy ions are neglected, describes accurately the characteristics of a plane wave for frequencies which are sufficiently greater than the gyromagnetic frequency of the ions. In order to obtain the plane wave dispersion relations for low frequencies which are of the order of the gyromagnetic frequency of the ions or lower, it is necessary to include the motion of the ions. For convenience, a plasma in which the motion of the electrons and the ions are both taken into account is called an electron-ion plasma. It was pointed out in Sec. 5.2 that the form of the relative permittivity dyad $\boldsymbol{\varepsilon}_r$ as given by Eq. (2.15) for the magnetostatic field in the z-direction is unchanged by the inclusion of the

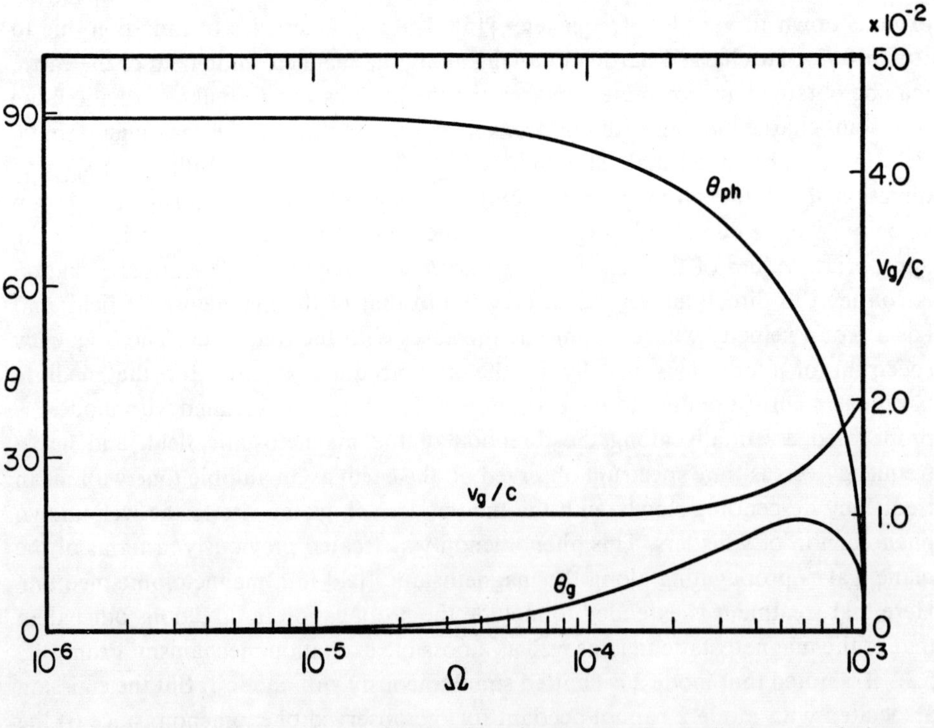

Fig. 5.23. Wave-normal direction θ_{ph}, ray direction θ_g, and magnitude of normalized group velocity v_g/c as a function of normalized frequency Ω for mode 1: $\gamma = 100$; $R = 0.1$.

motion of the heavy ions but the dependence of the components ε_1, ε_2, and ε_3 of the relative permittivity dyad on the medium parameters becomes more complicated. If the ion motions are included, ε_1, ε_2, and ε_3, instead of being given by Eqs. (2.2la), (2.2lb), and (2.2lc), respectively, as for an electron plasma, are now given by Eqs. (2.29a), (2.29b), and (2.29c), respectively. The treatment of the dispersion relations for a plane wave in an electron plasma remains valid even for an electron-ion plasma up to the stage in which the analysis is in terms of ε_1, ε_2, and ε_3.

Propagation along the magnetostatic field

The two independent modes of propagation along the magnetostatic field in an electron-ion plasma are still the left and right circularly polarized waves with the normalized propagation coefficients given by Eqs. (4.15) and (4.16) respectively. Let v_{ph}^{-}/c and v_{ph}^{+}/c denote the normalized phase velocities of the left and right circularly polarized waves propagating along the magnetostatic field. Then, it follows from Eqs. (4.15) and (4.16) that

$$v_{ph}^{-}/c = \Omega/k_z^{-} = (\varepsilon_1 + \varepsilon_2)^{-1/2} \tag{9.1}$$

$$v_{ph}^{+}/c = \Omega/k_z^{+} = (\varepsilon_1 - \varepsilon_2)^{-1/2} \tag{9.2}$$

For an electron-ion plasma, it is obtained from Eqs. (2.29a) and (2.29b) that

$$\begin{aligned} \varepsilon_1 + \varepsilon_2 &= 1 - \frac{1}{\Omega^2 - R^2} + \frac{R}{\Omega(\Omega^2 - R^2)} - \frac{m}{\Omega^2 - R^2 m^2} - \frac{Rm^2}{\Omega(\Omega^2 - R^2 m^2)} \\ &= 1 - \frac{1}{\Omega(\Omega + R)} - \frac{m}{\Omega(\Omega - Rm)} = 1 - \frac{1 + m}{(\Omega + R)(\Omega - Rm)} \\ &= \frac{\Omega^2 + \Omega R(1 - m) - (1 + m + R^2 m)}{(\Omega + R)(\Omega - Rm)} \end{aligned} \tag{9.3}$$

In a similar manner, it is found from Eqs. (2.29a) and (2.29b) that

$$\varepsilon_1 - \varepsilon_2 = \frac{\Omega^2 - \Omega R(1 - m) - (1 + m + R^2 m)}{(\Omega - R)(\Omega + Rm)} \tag{9.4}$$

Together with Eqs. (9.3) and (9.4), Eqs. (9.1) and (9.2) become

$$v_{ph}^{-}/c = \left[\frac{(\Omega + R)(\Omega - Rm)}{\Omega^2 + \Omega R(1 - m) - (1 + m + R^2 m)}\right]^{1/2} \tag{9.5a}$$

$$= \left[\frac{(\Omega + R)(\Omega - Rm)}{(\Omega - \Omega_1)(\Omega + \Omega_2)}\right]^{1/2} \tag{9.5b}$$

$$v_{ph}^{+}/c = \left[\frac{(\Omega - R)(\Omega + Rm)}{\Omega^2 - \Omega R(1 - m) - (1 + m + R^2 m)}\right]^{1/2} \tag{9.6a}$$

$$= \left[\frac{(\Omega - R)(\Omega + Rm)}{(\Omega + \Omega_1)(\Omega - \Omega_2)}\right]^{1/2} \tag{9.6b}$$

Note that in unnormalized quantities, just as $\Omega = R$ corresponds to the gyromagnetic frequency of the electrons, $\Omega = Rm$ corresponds to the gyromagnetic frequency of the ions.

In this investigation, m is taken to be equal to 1/1836 which corresponds to a hydrogen plasma. In general, $m \ll 1$ and is therefore negligible in comparison with unity. For $\Omega = 0$, the phase velocities of the left and the right circularly polarized modes are seen from Eqs. (9.5a) and (9.6a) to become equal, as given by

$$v_{ph}^{-} = v_{ph}^{+} = c\left[\frac{R^2 m}{1 + m + R^2 m}\right]^{1/2} = v_\alpha\left[1 + \left(\frac{v_\alpha}{c}\right)^2\right]^{-1/2} \tag{9.7}$$

where

$$v_\alpha = c\left[\frac{R^2 m}{1 + m}\right]^{1/2} = \left[\frac{B_0^2/\mu_0}{N_0(m_e + m_i)}\right]^{1/2} \tag{9.8}$$

is known as the Alfvén wave velocity. It was shown that in a perfectly conducting fluid, the magnetic flux lines are frozen into the fluid. Thus the tubes of flux may be thought of as possessing inertial mass of linear mass density equal to $N_0(m_e + m_i)$ of the fluid. Also, it was proved in the matrix (2.9.27) that the stress caused by the magnetic flux density B_0 is equivalent to an isotropic magnetic pressure $B_0^2/2\mu_0$ and a tension B_0^2/μ_0 along the magnetic flux lines. The former is balanced by a decrease in the fluid pressure with the result that only the tension along the flux lines remains. Hence, if the fluid is incompressible, the tubes of flux behave like strings of linear mass density $N_0(m_e + m_i)$ under a tension B_0^2/μ_0. By analogy with the transverse vibrations of stretched strings, it is to be expected that, when the fluid is disturbed from rest, the tubes of flux perform transverse vibrations, the phase velocity of the waves generated being given by (tension/linear mass density)$^{1/2}$ = $[(B_0^2/\mu_0)/N_0(m_e + m_i)]^{1/2}$. The possibility of this type of wave was pointed out first by Alfvén and hence v_α given by Eq. (9.8) is called the Alfvén wave velocity. For most practical cases of propagation in the ionosphere with which the theory developed in this and the following section is mainly concerned, the Alfvén wave velocity v_α is much less than the free space electromagnetic wave velocity such that $v_\alpha^2 \ll c^2$ which is equivalent to the condition $R^2 m \ll 1$. Hence $R^2 m$ is also negligible in comparison with unity. Therefore, in this treatment m and $R^2 m$ are ignored in comparison with unity.

It is therefore reasonable to approximate the denominators in Eqs. (9.5a) and (9.6a) as $\Omega^2 \pm \Omega R - 1 = (\Omega \mp \Omega_1)(\Omega \pm \Omega_2)$ where Ω_1 and Ω_2 are the same as those defined by Eq. (4.18). Hence Eqs. (9.5b) and (9.6b) follow from Eqs. (9.5a) and (9.6a) respectively. It can be established that $\Omega_1 > Rm$ and $\Omega_2 > R$. Since the two modes propagate only if their corresponding phase velocities are real, it follows from Eqs. (9.5b) and (9.6b) that the left circularly polarized mode propagates in the two frequency ranges $0 < \Omega < Rm$ and $\Omega_1 < \Omega < \infty$, and, the right circularly polarized mode propagates in the two frequency ranges $0 < \Omega < R$ and $\Omega_2 < \Omega < \infty$. The phase velocities given by Eqs. (9.5b) and (9.6b) are shown in Fig. 5.24 as functions of Ω for $R = 1/\sqrt{2}$.

To facilitate reference, the left and the right circularly polarized modes are denoted by L and R, respectively. Also, the higher and the lower frequency branches of the two modes are distinguished by the superscripts h and l, respectively. It is seen from Fig. 5.24 that the phase velocities of both the L^h and the R^h modes are always greater than the velocity of electromagnetic waves in free space. The phase velocities of the L^h and the R^h modes start with infinite values at their cut-off frequencies $\Omega = \Omega_1$ and $\Omega = \Omega_2$, respectively, continuously decrease as Ω is increased and asymptotically approach the value c in the limit of infinite frequency. The phase velocity of the L^l mode starts with a value equal to $v_\alpha[1 + (v_\alpha/c)^2]^{-1/2}$

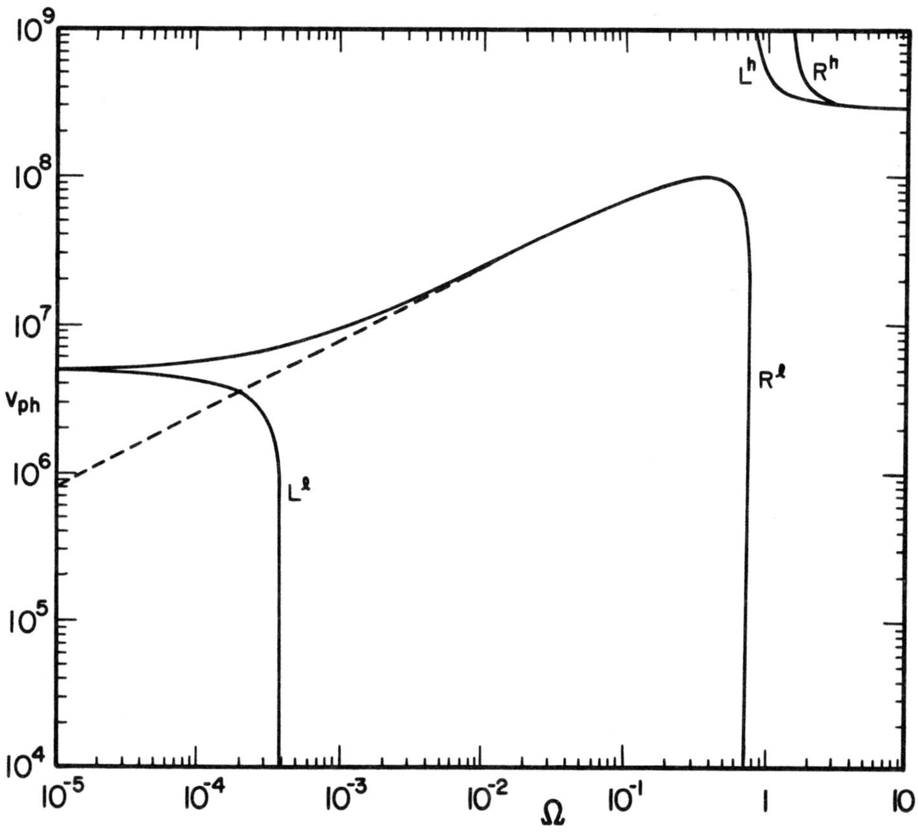

Fig. 5.24. Phase velocity v_{ph} as a function of normalized frequency Ω for propagation along the magnetostatic field in an electron-ion plasma: $R = 1/\sqrt{2}$; $c = 3 \times 10^8$ m/sec.

for zero frequency, continuously decreases as Ω is increased and goes to zero at the resonant frequency $\Omega = Rm$. The phase velocity of the R^l mode also starts with the same value as the L^l mode for zero frequency, increases continuously as Ω is increased and reaches a maximum value equal to $c[1 + 4/(1 + m)R^2]^{-1/2}$ for $\Omega = R(1 - m)/2$. For further increase in Ω, the phase velocity of the R^l mode decreases and goes to zero at the resonant frequency $\Omega = R$.

In Fig. 5.24, the dashed line corresponds to the conventional magnetoionic approximation in which the motion of the heavy ions is neglected. It is seen that the effect of the inclusion of the motion of the ions changes the dispersion curve only in the lower end of the range of the so-called "whistler" frequencies. The dispersion curves of the L^h and R^h modes are practically unaffected by the inclusion of the motion of the ions. The L^l mode is absent in the conventional magnetoionic theory. It is therefore concluded that the extension of the conventional magnetoionic theory by the inclusion of the motion of the ions introduces changes only for the extremely low frequencies of the order of the ion gyromagnetic frequency and lower, and these changes become insignificant above the lower end of the range of whistler frequencies.

Propagation across the magnetostatic field

For propagation across the magnetostatic field even in an electron-ion plasma, the two independent modes of propagation are the TEM and the TM modes which are linearly polarized respectively with their electric and magnetic vectors parallel to the magnetostatic field. The normalized propagation coefficients of the TEM and the TM modes are given by Eqs. (5.7) and (5.15), respectively. Let v_{ph}^o/c and v_{ph}^e/c denote, respectively, the normalized phase velocities of the TEM and the TM modes propagating across the magnetostatic field. Then, it follows from Eqs. (5.7) and (5.15) that

$$v_{ph}^o/c = (\varepsilon_3)^{-1/2} \tag{9.9}$$

$$v_{ph}^e/c = \left[\frac{\varepsilon_1}{(\varepsilon_1 - \varepsilon_2)(\varepsilon_1 + \varepsilon_2)}\right]^{1/2} \tag{9.10}$$

Since m can be neglected in comparison with unity, it is found from Eqs. (2.29c) and (9.9) that

$$v_{ph}^o/c = \Omega(\Omega^2 - 1 - m)^{-1/2} = \Omega(\Omega^2 - 1)^{-1/2} \tag{9.11}$$

It follows from Eqs. (2.29a), (9.3), (9.4), and (9.10) that

$$\begin{aligned} v_{ph}^e/c = {} & [\Omega^4 - \Omega^2\{1 + m + R^2(1 + m^2)\} + R^2 m(1 + m + R^2 m)]^{1/2} \\ & \times [\Omega^2 + \Omega R(1 - m) - (1 + m + R^2 m)]^{-1/2} \\ & \times [\Omega^2 - \Omega R(1 - m) - (1 + m + R^2 m)]^{-1/2} \end{aligned} \tag{9.12}$$

Since m and R^2m can be neglected in comparison with unity, Eq. (9.12) can be simplified as

$$v_{ph}^e/c = \left[\frac{\Omega^4 - \Omega^2(1 + R^2) + R^2m}{\{\Omega^2 + \Omega R - 1\}\{\Omega^2 - \Omega R - 1\}}\right]^{1/2} \tag{9.13a}$$

$$= \left[\frac{(\Omega^2 - \Omega_u^2)(\Omega^2 - \Omega_L^2)}{(\Omega^2 - \Omega_1^2)(\Omega^2 - \Omega_2^2)}\right]^{1/2} \tag{9.13b}$$

where Ω_1 and Ω_2 are the same as those defined by Eq. (4.18); also Ω_u and Ω_L are the zeros of ε_1 and are given approximately by

$$\Omega_u = \sqrt{1 + R^2} \qquad \Omega_L = [R^2m/(1 + R^2)]^{1/2} \tag{9.14a, b}$$

Note that previously Ω_u has been designated as the normalized upper hybrid resonant frequency and Ω_L given by Eq. (9.14b) is called the normalized *lower hybrid resonant* frequency.

It can be proved that $\Omega_1 < \Omega_u < \Omega_2$ and $\Omega_L < \Omega_1$. Then, it follows from Eqs. (9.11) and (9.13b) that the TEM mode propagates approximately for $\Omega > 1$ and that the TM mode propagates in the following three frequency ranges: $0 < \Omega < \Omega_L$, $\Omega_1 < \Omega < \Omega_u$, and $\Omega_2 < \Omega < \infty$. It is found from Eqs. (9.7) and (9.12) that for $\Omega = 0$, the phase velocity of the TM mode is the same as those of the two modes propagating along the magnetostatic field. The phase velocities of the TEM and the TM modes as given respectively by Eqs. (9.11) and (9.13) are depicted in Fig. 5.25 as functions of Ω for $R = 1/\sqrt{2}$.

It is seen from Fig. 5.25 that the phase velocity of the TEM mode is always greater than c. It starts with a value of infinity at the cut-off frequency $\Omega = 1$, continuously decreases as Ω is increased and asymptotically approaches the value c in the limit of infinite frequency. For the sake of convenience, the higher, the middle, and the lower frequency branches of the TM mode are distinguished by the superscripts h, m, and l, respectively. The phase velocity of the TM^h mode is always greater than c. It starts with a value of infinity at the cut-off frequency $\Omega = \Omega_2$, decreases continuously as Ω is increased and asymptotically attains the value c in the limit of infinite frequency. The phase velocity of the TM^m mode starts with a value of infinity at the cut-off frequency $\Omega = \Omega_1$ (which is always less than unity), decreases as Ω is increased and attains the value c approximately for $\Omega = 1$. For further increase in Ω, the phase velocity decreases and rapidly goes to zero at the upper hybrid resonant frequency $\Omega = \Omega_u$. The phase velocity of the TM^l mode starts with a value equal to $v_\alpha[1 + (v_\alpha/c)^2]^{-1/2}$ for zero frequency, continuously decreases as Ω is increased and rapidly goes to zero at the lower hybrid resonant frequency $\Omega = \Omega_L$. When $R^2 \ll 1$, which corresponds to the case in which the gyromagnetic frequency of the electrons is considerably less than the plasma frequency of the electrons, it is found from Eq. (9.14b) that $\Omega_L = \sqrt{R \times Rm}$. In unnormalized quantities $\omega_L = \sqrt{\omega_{ce}\omega_{ci}}$, that is, the lower hybrid resonant frequency becomes equal to the geometric mean of the gyromagnetic frequencies of the electrons and the ions. This geometric mean frequency is often referred to as the limit frequency up to which the propagation range of the low frequency branch of the TM mode extends for the case of propagation across the magnetostatic field in an electron-ion plasma.

It is usual to designate the TEM and the TM modes propagating across the magnetostatic field as the ordinary and the extraordinary modes, respectively. The reason for naming the TEM mode as the ordinary mode is that it is independent of the presence of the magnetostatic field. The justification for the use of the superscript o in Eq. (9.9) pertaining to the TEM mode and the superscript e in Eq. (9.10) pertaining to the TM mode is provided now.

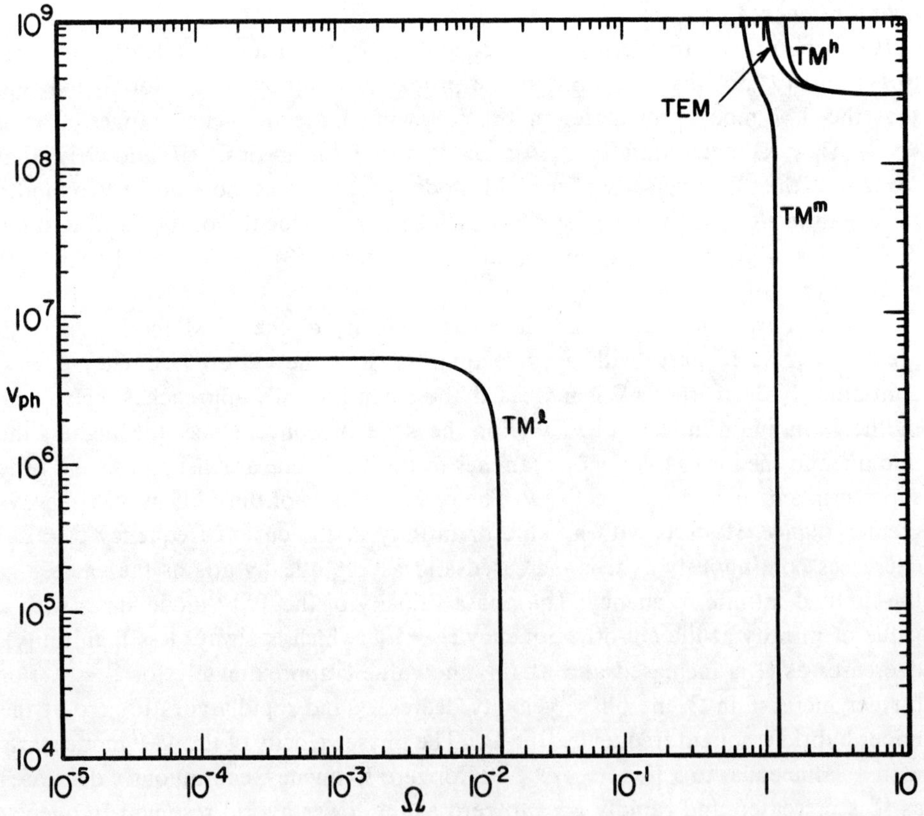

Fig. 5.25. Phase velocity v_{ph} as a function of normalized frequency Ω for propagation across the magnetostatic field in an electron-ion plasma: $R = 1/\sqrt{2}$; $c = 3 \times 10^8$ m/sec.

5.10. Magnetoionic theory at hydromagnetic frequencies

Astrom [10, 16] was perhaps the first to study the characteristics of plane wave propagation in a multicomponent plasma with emphasis on the dispersion relations for the two hydromagnetic waves which propagate at extremely low frequencies. These two hydromagnetic waves are the generalizations of the Alfvén waves [17] to arbitrary directions of propagation with respect to the magnetostatic field. Since the works of Alfvén and Astrom, a number of investigators [18–23] have extended in various respects the understanding of the plane wave dispersion relations in a cold, homogeneous electron-ion plasma. The analysis as given here of the characteristics of a plane wave propagating in an arbitrary direction with respect to the magnetostatic field in an electron-ion plasma follows closely a recent treatment [24] by the author.

Dispersion equation

The normalized propagation coefficient k of a plane wave propagating at an angle $\theta_{ph} = \cos^{-1} l$ with respect to the magnetostatic field satisfies the dispersion equation as given by Eqs. (6.18) and (6.19). With the help of Eqs. (2.29 a,b,c), Eqs. (6.19 a,b,c) can be expressed in terms of Ω, R, and m. If the resulting dispersion equation is simplified by the removal of a common denominator, it can be deduced that Eqs. (6.18) and (6.19) become

$$B_2 k^4 + B_1 k^2 + B_0 = 0 \tag{10.1}$$

where

$$\begin{aligned} B_2 = {} & \Omega^6 - \Omega^4\{1 + m + R^2(1 + m^2)\} \\ & + \Omega^2 R^2(l^2 + m + m^2 + m^3 l^2 + R^2 m^2) - R^4 m^2(1 + m)l^2 \end{aligned} \tag{10.2a}$$

$$\begin{aligned} B_1 = {} & -\Omega^2[2\Omega^6 - 2\Omega^4\{2 + 2m + R^2(1 + m^2)\} \\ & + \Omega^2\{2(1 + m)^2 + R^2(1 + l^2 + 4m + 4m^2 + m^3(1 + l^2) + 2R^2 m^2)\} \\ & - R^2 m\{(1 + m)^2 + R^2 m(1 + m)\}(1 + l^2)] \end{aligned} \tag{10.2b}$$

$$\begin{aligned} B_0 = {} & \Omega^4(\Omega^2 - 1 - m)[\Omega^4 - \Omega^2\{2 + 2m + R^2(1 + m^2)\} \\ & + \{(1 + m)^2 + R^2 m(2 + 2m + R^2 m)\}] \end{aligned} \tag{10.2c}$$

So far, no approximations have been made in the expressions for B_2, B_1, and B_0. As stated previously, both m and $R^2 m$ can be neglected in comparison with unity with the result that the expressions for the coefficients can be simplified to yield

$$B_2 = \Omega^6 - \Omega^4(1 + R^2) + \Omega^2 R^2(l^2 + m) - R^4 m^2 l^2$$

$$= [\Omega^4 - \Omega^2(1 + R^2) + R^2(l^2 + m)]\left[\Omega^2 - \frac{R^2 m^2 l^2}{l^2 + m}\right] \tag{10.3a}$$

$$B_1 = -\Omega^2[2\Omega^6 - 2\Omega^4(2 + R^2) + \Omega^2\{2 + R^2(1 + l^2)\} - R^2 m(1 + l^2)]$$

$$= -\Omega^2[2\Omega^4 - 2\Omega^2(2 + R^2) + 2 + R^2(1 + l^2)] \tag{10.3b}$$

$$\left[\Omega^2 - \frac{R^2 m(1 + l^2)}{\{2 + R^2(1 + l^2)\}}\right]$$

$$B_0 = \Omega^4(\Omega^2 - 1)[\Omega^4 - \Omega^2(2 + R^2) + 1] \tag{10.3c}$$

The two solutions of Eq. (10.1) are given by

$$k_{1,2}^2 = -B_1/2B_2 \pm (1/2B_2)\sqrt{B_1^2 - 4B_0 B_2} \tag{10.4}$$

The upper and the lower signs in Eq. (10.4) correspond to k_1^2 and k_2^2 respectively. The discriminant $(B_1^2/4 - B_0 B_2)$ can be simplified by using Eqs. (6.19) and (10.2) with the following result.

$$\frac{B_1^2}{4} - B_0 B_2 = \Omega^4(\Omega^2 - R^2)^2(\Omega^2 - R^2 m^2)^2[\varepsilon_2^2 \varepsilon_3^2 l^2 + \frac{1}{4}(\varepsilon - \varepsilon_1 \varepsilon_3)^2(1 - l^2)^2] \tag{10.5}$$

Since Eq. (10.5) is positive, it follows that k_1^2 and k_2^2 are always real. If

$$B_0 B_2 < 0 \qquad \text{then } |(1/2B_2)\sqrt{B_1^2 - 4B_0 B_2}| > |B_1/2B_2|$$

and hence the solution given by Eq. (10.4) which corresponds to a positive factor in front of the radical gives a positive value for k^2 resulting in the propagation of the corresponding mode. If

$$B_0 B_2 > 0 \qquad \text{then } |(1/2B_2)\sqrt{B_1^2 - 4B_0 B_2}| < |B_1/2B_2|$$

and therefore the signs of k_1^2 and k_2^2 are the same as that of $-B_1/2B_2$. Consequently, both the modes propagate if $B_1/2B_2 < 0$.

Determination of the propagation regions

If the propagation direction is specified, l is known, and, since m is also known, Ω and R are the only remaining parameters in Eqs. (10.3) and (10.4). In the Ω^2 - R^2 space, the regions which correspond to k_1 and k_2 real can be determined by the procedure outlined in Sec. 5.6. For that purpose, the parametric equations for $B_2 = 0$, $B_1 = 0$ and $B_0 = 0$ are required and these are found from Eqs. (10.3a)–(10.3c) to be given by the following curves in the Ω^2 - R^2 space:

$$(a) \quad R^2 = \Omega^2(\Omega^2 - 1)/(\Omega^2 - l^2 - m) \qquad \text{for } B_2 = 0 \qquad (10.6a)$$

$$(b) \quad \Omega^2 = R^2 m^2 l^2/(l^2 + m) \qquad \text{for } B_2 = 0 \qquad (10.6b)$$

$$(c) \quad R^2 = (\Omega^2 - 1)^2 \Big/ \left(\Omega^2 - \frac{1 + l^2}{2}\right) \qquad \text{for } B_1 = 0 \qquad (10.6c)$$

$$(d) \quad R^2 = 2\Omega^2/(1 + l^2)(m - \Omega^2) \qquad \text{for } B_1 = 0 \qquad (10.6d)$$

$$(e) \quad \Omega^2 = 1 \qquad \text{for } B_0 = 0 \qquad (10.6e)$$

$$(f) \quad R^2 = \Omega^{-2}(\Omega^2 - 1)^2 \qquad \text{for } B_0 = 0 \qquad (10.6f)$$

The curves (a)–(f) are depicted in Fig. 5.26 and these curves are seen to divide the Ω^2 - R^2 space into several regions. It is found from Eqs. (10.3a), (10.3b), and (10.3c) that B_2, B_1, and B_0 change sign only on crossing the curves corresponding to $B_2 = 0$, $B_1 = 0$, and $B_0 = 0$, respectively. Consequently, inside the several regions into which the Ω^2 - R^2 space is divided by the parametric curves (a)–(f), the signs of B_2, B_1, and B_0 remain the same. If the signs of B_2, B_1, and B_0 are determined for any one region, their signs in all the other regions are obtained by inspection.

A knowledge of the important characteristics of the various curves is necessary and these characteristics are summarized here. The curves (c) and (f) are tangential to the Ω^2-axis at $\Omega^2 = 1$. The curve (c) is always above the curve (f). For $\Omega^2 > 1$, the curve (a) lies above the curve (c). The curve (b) always lies above the curve (d). For $R^2 m \ll 1$, the curves (b) and (d) do not intersect with (f). If the directions which are very nearly parallel to the magnetostatic field are excluded, it can be shown that $l^2 + m < (1 + l^2)/2 < 1$; therefore for $\Omega^2 < 1$, the curves (a) and (c) do not intersect.

For Ω^2 very large and $R^2 = 0$, it is found from Eqs. (10.3a), (10.3b), and (10.3c) that $B_2 > 0$, $B_1 < 0$, and $B_0 > 0$. On crossing the curve (f), B_0 changes sign and becomes negative. In a similar manner, the signs of B_2, B_1, and B_0 are obtained for all the regions in the Ω^2 - R^2 space. It is found that $B_2 B_0 < 0$ in the regions marked 1, 2, and 3 in Fig. 5.26 and therefore in these three regions, either k_1 or k_2 is real, that is, one of the two modes propagates. In the regions 1 and 3, $B_2 > 0$ and therefore it follows from Eq. (10.4) that k_1 is real and corresponds to the propagation of mode 1. In Fig. 5.26, the regions 1 and 3 where mode 1 alone propagates are indicated by shading with horizontal lines. In the region 2, $B_2 < 0$ and Eq. (10.4) shows that k_2 is real leading to the propagation of mode 2. In Fig. 5.26, the region 2 where mode 2 alone propagates is indicated by shading with vertical lines.

In the Ω^2 - R^2 space, the regions where $B_2 B_0 > 0$ are determined and in these regions the signs of $B_1/2B_2$ are also ascertained. In the region 8, $B_1/2B_2 > 0$, and hence neither of the two modes propagates in this region. In the regions 4, 5, 6, and

7, $B_1/2B_2 < 0$ and hence both k_1 and k_2 are real. The regions 4, 5, 6, and 7 where both the modes propagate are indicated by shading with both horizontal and vertical lines.

The frequency ranges of propagation of the two modes are obtained from an examination of Fig. 5.26. The results are valid for only such values of R which satisfy the criterion $R^2m \ll 1$. Let the two values of Ω^2 corresponding to the intersection points of the operating line R^2 = constant with the curve (a) be denoted by Ω_{ge}^2 and Ω_p^2. From Eq. (10.6a), Ω_{ge}^2 and Ω_p^2 are evaluated to be given by

$$\Omega_{ge}^2, \Omega_p^2 = \frac{1+R^2}{2} \mp \left[\left(\frac{1+R^2}{2}\right)^2 - R^2(l^2+m)\right]^{1/2} \tag{10.7}$$

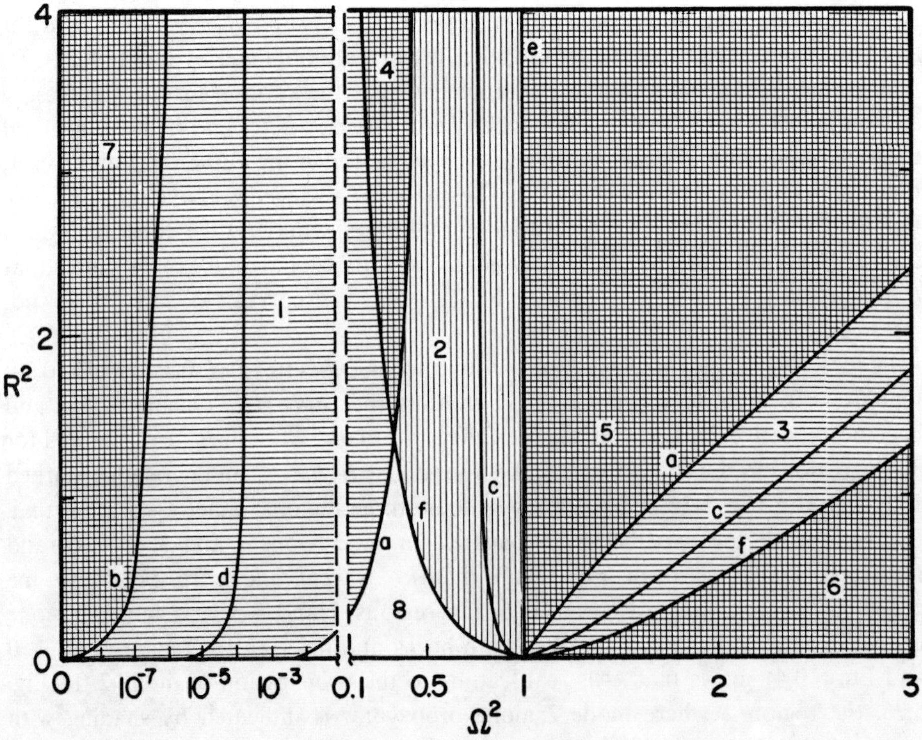

Fig. 5.26. Regions of propagation of ▤ mode 1; ▥ mode 2 in the $\Omega^2 - R^2$ space. 1 and 3: $B_0 < 0$; $B_2 > 0$; 2: $B_0 > 0$; $B_2 < 0$; 4 and 6: $B_0 > 0$; $B_1 < 0$; $B_2 > 0$; 5 and 7: $B_0 < 0$; $B_1 > 0$; $B_2 < 0$; 8: $B_0 < 0$; $B_1 < 0$; $B_2 < 0$.

where Ω_{ge}^2 and Ω_p^2 correspond to the upper and the lower signs respectively and are always positive real. In addition, let

$$\Omega_{gi}^2 = R^2 m^2 l^2/(l^2 + m) \tag{10.8}$$

Also, the line R^2 = constant intersects the curve (f) at two values of Ω^2, namely Ω_1^2 and Ω_2^2. The expressions for Ω_1 and Ω_2, which are always positive real, are the same as those given by Eq. (4.18). From Fig. 5.26, modes 1 and 2 are found to propagate in the following frequency ranges:

$$\text{Mode 1: } 0 < \Omega < \Omega_{ge} \qquad 1 < \Omega < \infty \tag{10.9a}$$

$$\text{Mode 2: } 0 < \Omega < \Omega_{gi} \qquad \Omega_1 < \Omega < \Omega_p \qquad \Omega_2 < \Omega < \infty \tag{10.9b}$$

The normalized phase velocity is deduced from Eq. (10.4) through the relation

$$v_{ph}/c = \Omega/k \tag{10.10}$$

The phase velocities of the two modes are evaluated with the help of Eqs. (10.4) and (10.10), and, are shown in Fig. 5.27 as functions of the normalized frequency Ω for $R = 1/\sqrt{2}$ and $l = 1/\sqrt{2}$ which corresponds to the propagation vector making an angle of 45° with the magnetostatic field. The low and the high frequency branches of mode 1 are distinguished by the superscripts l and h respectively. Similarly, the low, the middle, and the high frequency branches of mode 2 are distinguished by the superscripts l, m, and h, respectively. Since most of the features of the dispersion curves are evident from Fig. 5.27, it is sufficient to only emphasize the fact that for mode 1^l, the phase velocity is seen to increase with the frequency for a range of frequencies above approximately the ion gyromagnetic frequency. This special feature of the dispersion of mode 1^l is known to account for the whistler phenomenon.

Dispersion relations near $\theta_{ph} = \pi/2$ and $\theta_{ph} = 0$

It is instructive to examine how the dispersion for the arbitrary direction of propagation approaches the limiting values corresponding to the propagation along and across the magnetostatic field. As the propagation direction approaches the direction normal to the magnetostatic field, l^2 becomes smaller and as a consequence both Ω_{ge}^2 and Ω_{gi}^2 become smaller but Ω_p^2 becomes larger. In the limiting case of propagation across the magnetostatic field, $l^2 = 0$ and therefore $\Omega_{gi}^2 = 0$. Also, Ω_{ge}^2 and Ω_p^2 attain the limiting values $R^2 m/(1 + R^2)$ and $1 + R^2$, respectively. The cut-off frequencies $\Omega = \Omega_1$, $\Omega = 1$, and $\Omega = \Omega_2$ do not depend on l and, therefore, on the propagation direction. Hence, as the direction of propagation approaches that perpendicular to the magnetostatic field, the propagation range of mode 1^h and mode 2^h are unchanged and that of mode 2^m increases to the limiting range $\Omega_1 < \Omega < \sqrt{1 + R^2}$ The propagation range of mode 2^l continuously decreases and finally vanishes, and that of mode 1^l also continuously decreases but attains the limiting range $0 < \Omega < \{R^2 m/(1 + R^2)\}^{1/2}$ for propagation across the magnetostatic field. Note that mode 1^l propagating across the magnetostatic field becomes the TM^l mode. Also, modes 1^h, 2^m, and 2^h become, respectively, the TEM, the TM^m, and the TM^h modes.

As the propagation direction approaches that of the magnetostatic field, l^2 becomes progressively larger and reaches the limiting value of unity. Consequently, Ω_{gi} increases and attains for $l^2 = 1$ the limiting value $\Omega_{gi} = Rm$, which is the normalized ion gyromagnetic frequency. For this reason, Ω_{gi} for an arbitrary direction of propagation is also called the normalized ion gyromagnetic resonant

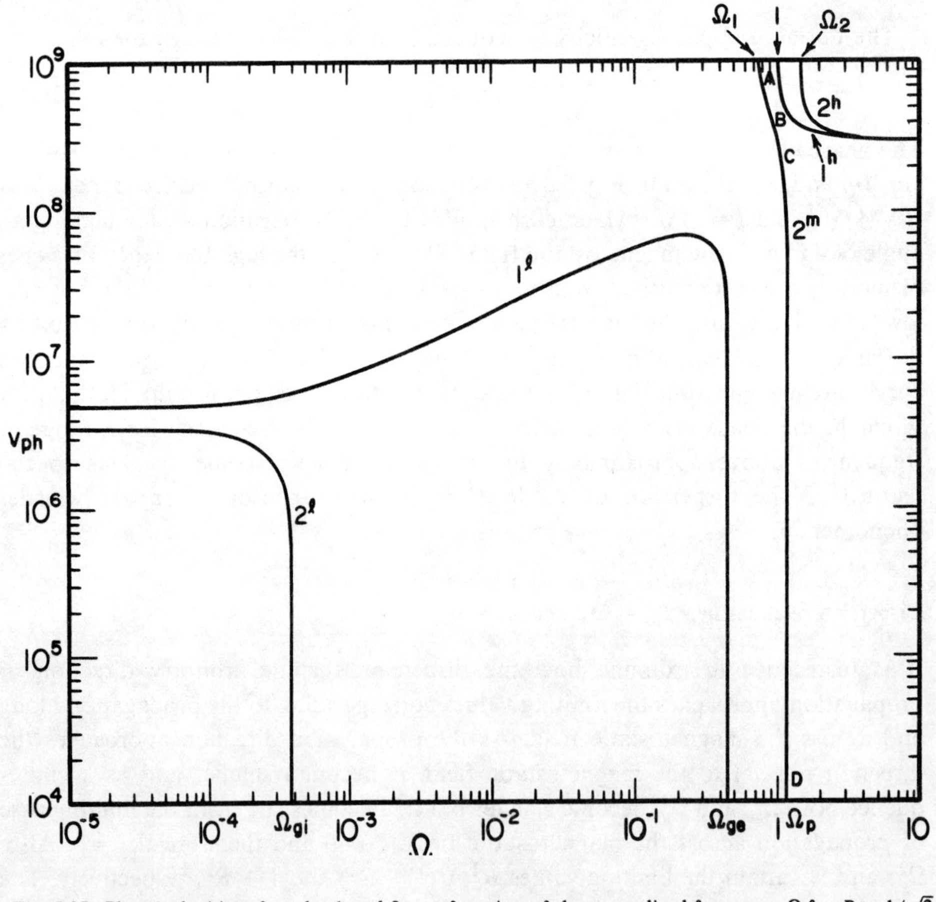

Fig. 5.27. Phase velocities of modes 1 and 2 as a function of the normalized frequency Ω for $R = 1/\sqrt{2}$ and $l = 1/\sqrt{2}$.

frequency. Furthermore, Ω_{ge}^2 increases, and Ω_p^2 decreases, in such a manner as to approach respectively the smaller or the larger of the two values, 1 and R^2. In the limiting case of propagation along the magnetostatic field, the resonant frequency line corresponding to Ω_{ge}^2 for $R^2 \leqslant 1$ and that corresponding to Ω_p^2 for $R^2 \geqslant 1$ join to form the resonant frequency line $\Omega^2 = R^2$, whereas the resonant frequency line $\Omega^2 = 1$ corresponding to Ω_p^2 for $R^2 \leqslant 1$, and that corresponding to Ω_{ge}^2 for $R^2 \geqslant 1$ disappears. This nonuniform behavior of the resonant frequency as the propagation direction approaches that of the magnetostatic field is identical to that in an electron plasma and has been treated in detail in Sec. 5.6. There are two changes associated with the disappearance of the resonant frequency line $\Omega^2 = 1$. First, it is found from Eqs. (10.3 a, b, c) that $(\Omega^2 - 1)$ is a factor in B_2, B_1, and B_0. Therefore, this factor is cancelled from B_2 appearing in front of the radical in Eq. (10.4). As a consequence, for $\Omega^2 < 1$, the sign of the term in front of the radical in Eq. (10.4) is reversed with the result that the nomenclature of modes 1 and 2 get interchanged. For propagation along the magnetostatic field modes 1^l and 2^l become modes R^l, and L^l, respectively. Mode 2^m disappears and the cut-off frequency of mode 1^h extends down to $\Omega = \Omega_1$. Finally for the case of propagation along the magnetostatic field, one of the modes propagates in the two frequency ranges, $0 < \Omega < Rm$ and $\Omega_1 < \Omega < \infty$, and the other mode propagates in the two frequency ranges $0 < \Omega < R$ and $\Omega_2 < \Omega < \infty$. These results are in agreement with those deduced in the previous section.

A simple physical explanation can be provided for the disappearance of the middle branch of mode 2 and the discontinuous lowering of the cut-off frequency of the high frequency branch of mode 1, as the propagation direction coincides with that of the magnetostatic field. If the polarization is examined in accordance with the theory given in Sec. 5.7, it is found that the electric vector corresponding to the region marked AB in Fig. 5.27 for mode 1^h and that marked CD for mode 2^m tends to become parallel to the direction of propagation as it approaches that of the magnetostatic field. For the limiting case of propagation along the magnetostatic field, the wave motions corresponding to the regions marked AB and CD become entirely longitudinal. Such a longitudinal type of wave motion cannot be sustained in a magnetoionic medium and hence it disappears. The portion of mode 1^h for $\Omega > 1$ merges with that of mode 2^m for $\Omega < 1$ to form one continuous dispersion curve which is associated with mode L^h.

Resonant frequencies

A study of the rapidity of change of the resonant frequencies Ω_{ge}, Ω_p, and Ω_{gi} with the change of the direction of propagation reveals an interesting feature of the dispersion relations. In Fig. 5.28, the resonant frequencies Ω_{ge}, Ω_p, and Ω_{gi} are plotted as functions of θ_{ph}, the angle between the direction of propagation and the magnetostatic field for $R = 2/\sqrt{3}$. It is found from Fig. 5.28b that the frequency Ω_{gi}

of ion gyromagnetic resonance does not change appreciably from its value *Rm* pertaining to the propagation along the magnetostatic field, as the direction of propagation is changed away from that of the magnetostatic field, till the propagation direction comes very close to that perpendicular to the magnetostatic field. Then, Ω_{gi} starts to decrease rapidly and goes to zero for the case of propagation across the magnetostatic field. This rapid variation of the resonant frequency Ω_{gi} of mode 2^l causes the dispersion in the close neighborhood of the direction across the magnetostatic field to differ considerably from those in the other directions of propagation.

From the development contained in Sec. 5.6, it is found that Ω_{ge} and Ω_p are the frequencies of electron gyromagnetic and electron plasma resonance respectively. For $R < 1$, the frequency of electron gyromagnetic resonance Ω_{ge} starts with a value equal to the lower hybrid resonance frequency Ω_L for $\theta_{ph} = \pi/2$, continuously increases as θ_{ph} is reduced and attains the value R for $\theta_{ph} = 0$. Note that R is the normalized electron gyromagnetic frequency. It is this fact which is responsible for the name acquired by Ω_{ge}. But for $R > 1$, the limiting value attained by the frequency of electron gyromagnetic resonance Ω_{ge} for $\theta_{ph} = 0$ is the electron plasma frequency $\Omega = 1$ rather than the electron gyromagnetic frequency $\Omega = R$. Similarly, for $R < 1$ the frequency of plasma resonance Ω_p starts with a value equal to the upper hybrid resonant frequency Ω_u for $\theta_{ph} = \pi/2$, continuously decreases as θ_{ph} is reduced and attains the value of the electron plasma frequency $\Omega = 1$ for $\theta_{ph} = 0$.

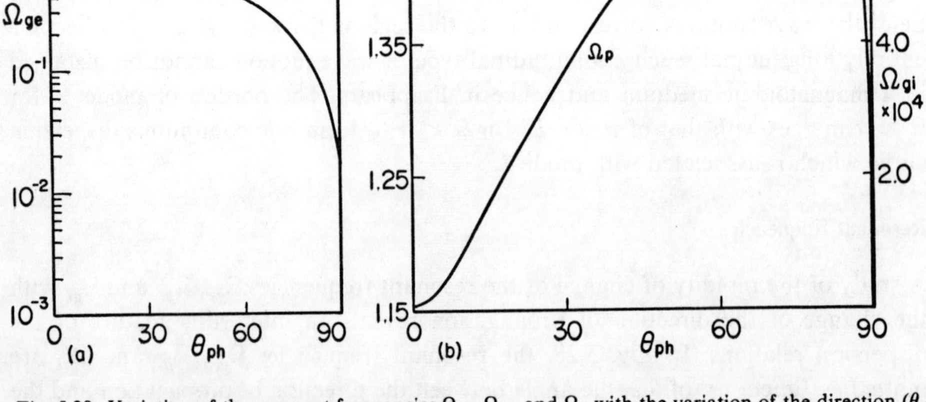

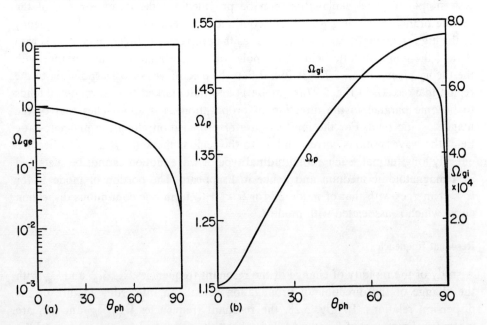

Fig. 5.28. Variation of the resonant frequencies Ω_{ge}, Ω_{gi}, and Ω_p with the variation of the direction (θ_{ph}) of propagation for $R = 2/\sqrt{3}$.

But for $R > 1$, the limiting value attained by Ω_p for $\theta_{ph} = 0$ is the electron gyromagnetic frequency $\Omega = R$ rather than the electron plasma frequency $\Omega = 1$. Therefore, it follows that for $R > 1$ or $\omega_{ce} > \omega_{pe}$ the frequency Ω_{ge} of the electron gyromagnetic resonance undergoes discontinuous change to attain the proper limiting value equal to the electron gyromagnetic frequency $\Omega = R$ for propagation along the magnetostatic field. This *discontinuous* change in the frequency Ω_{ge} of the electron gyromagnetic resonance for $R > 1$, and the *discontinuous* lowering of the cut-off frequency of the higher frequency branch of mode 1 from the value $\Omega = 1$ to the value $\Omega = \Omega_1$ account for the rapid variation of the dispersion relations as the propagation direction approaches that of the magnetostatic field.

Phase velocity diagrams

An examination of the dependence of the dispersion relations on the angular separation between the propagation direction and the magnetostatic field over the entire frequency range reveals an interesting feature of the variation of the location of occurrence of the various resonances as a function of frequency. The phase velocities of the two modes normalized to the electromagnetic wave velocity c are plotted in Fig. 5.29 as functions of the angle θ_{ph} in the range $0 \leqslant \theta_{ph} < \pi/2$ for $R = 1/\sqrt{2}$ and for several values of Ω.

In the limit of zero frequency, Eqs. (10.3) and (10.1) can be shown to yield the following dispersion equation:

$$R^4 m^2 l^2 k^4 - R^2 m(1 + l^2)\Omega^2 k^2 + \Omega^4 = (R^2 m l^2 k^2 - \Omega^2)(R^2 m k^2 - \Omega^2) = 0 \quad (10.11)$$

The solution of Eq. (10.11) is seen to give the following phase velocities for the two modes:

$$v_{ph,1} = c\Omega/k_1 = R\sqrt{m}\, c \qquad v_{ph,2} = c\Omega/k_2 = R\sqrt{m}\, lc \qquad (10.12a, b)$$

It is found from Eqs. (10.12) that the phase velocity of mode 1 is greater than that of mode 2. Hence modes 1 and 2 are designated as the fast and the slow hydromagnetic waves respectively. For extremely low frequencies, the fast hydromagnetic wave has a phase velocity that does not change with the propagation direction and the slow hydromagnetic wave has a phase velocity that has a $\cos \theta_{ph}$ dependence and hence goes to zero for propagation across the magnetostatic field. The phase velocity curves for $\Omega = 0$ are depicted in Fig. 5.29a. It is to be noted that the fast and the slow hydromagnetic waves are parts of modes $1'$ and $2'$, respectively.

As the frequency is increased, the resonance of the slow hydromagnetic wave (mode $2'$) occurs at a progressively smaller angle θ_{ph} and finally for the frequency equal to or greater than the gyromagnetic frequency ω_{ci} of the ions, this phase velocity curve disappears. This transition is depicted in Figs. 5.29b and 5.29c. From Fig. 5.28b, it is found that as the frequency Ω is increased from zero, the angle at which the ion gyromagnetic resonance occurs decreases very slowly until the frequency comes very close to the gyromagnetic frequency of the ions and then it decreases rapidly to zero.

When Ω is greater than ω_{ci}/ω_{pe} but less than $\{R^2 m/(1 + R^2)\}^{1/2}$, only mode 1^l propagates. As shown in Fig. 5.29d, the phase velocity of this mode decreases as θ_{ph} is increased. For $\Omega = \{R^2 m/(1 + R^2)\}^{1/2}$, this mode has a resonance at $\theta_{ph} = \pi/2$. For further increase in the frequency, it is seen from Fig. 5.29e that the phase velocity near $\theta_{ph} = 0$ increases and at the same time the resonant angle, that is the angle at which the phase velocity goes to zero, decreases with the result that mode 1^l now does not propagate for θ_{ph} greater than the resonant angle. When the frequency becomes equal to or greater than the electron gyromagnetic frequency $\Omega = R$, this mode ceases to propagate for any angle.

For $\Omega > \Omega_1$, mode 2^m starts propagating and it is seen from Fig. 5.29f, that its phase velocity is greater than c for $\Omega < 1$ and decreases slightly as θ_{ph} is increased. When the frequency exceeds the electron plasma frequency, mode 2^m has a resonance at an angle $\theta_{ph} = 0$. Also, mode 1^h starts propagating. As depicted in Figs.

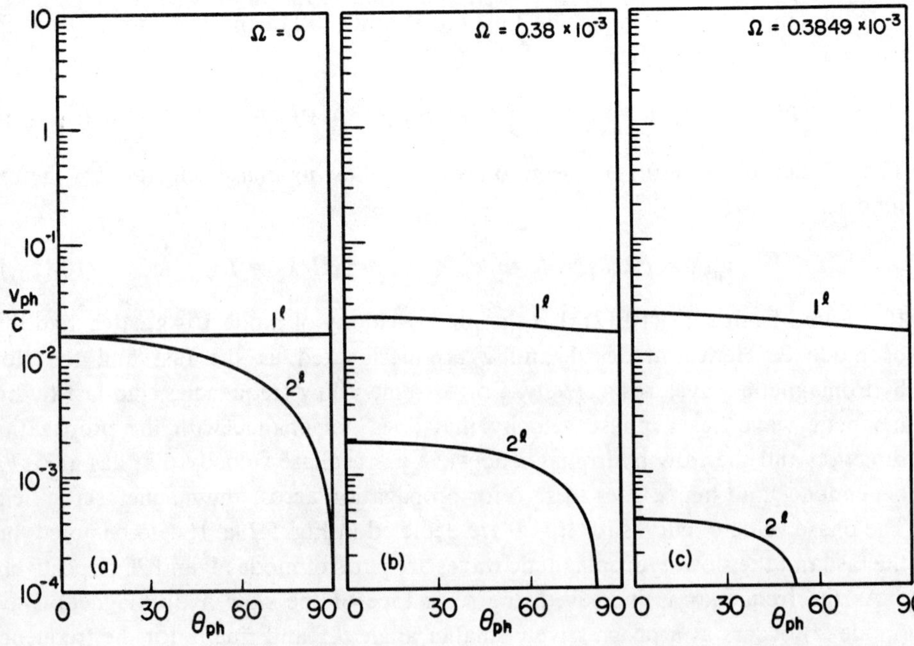

Fig. 5.29. Phase velocity diagrams as a function of frequency for $R = 1/\sqrt{2}$.

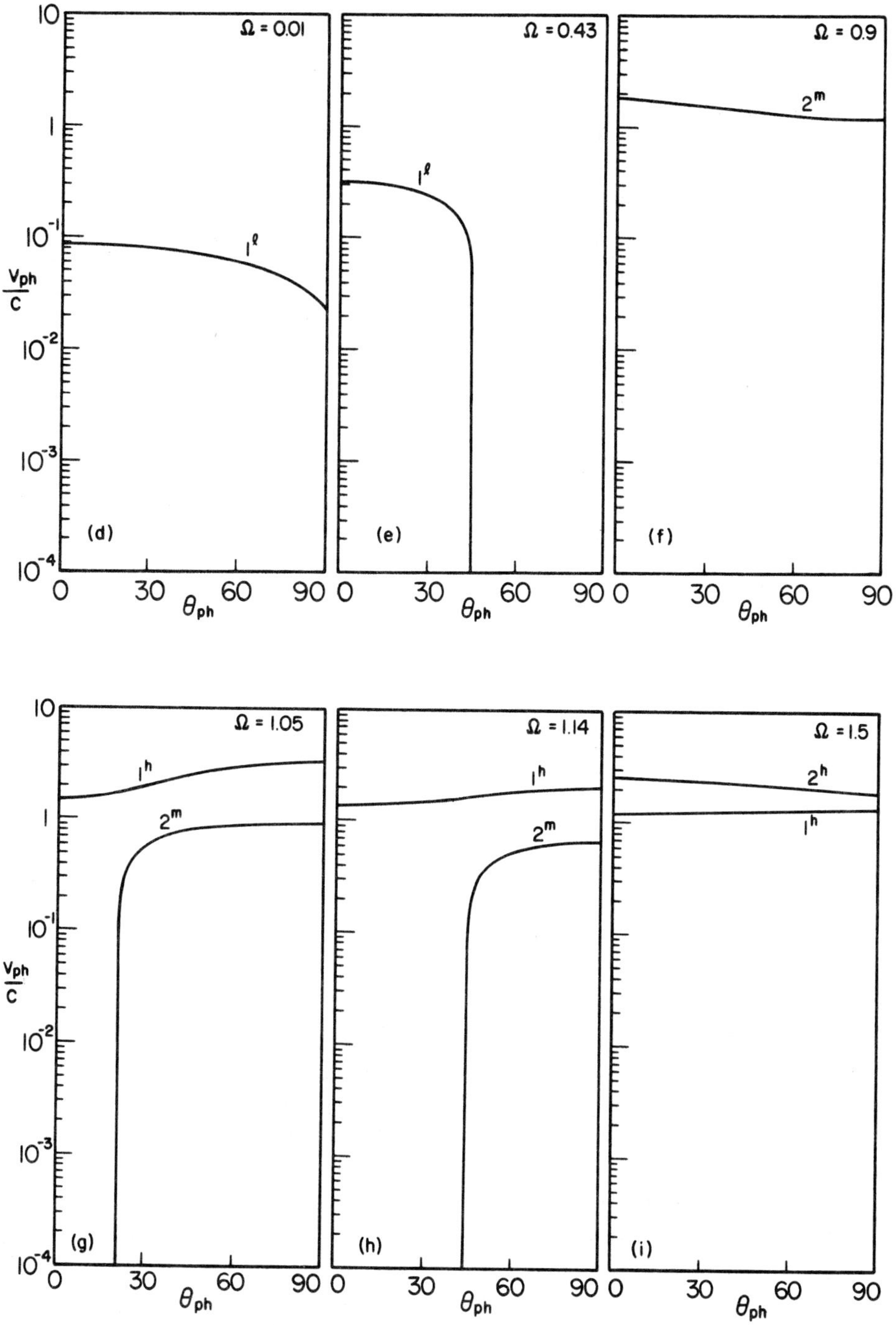
Ω = 0.01
Ω = 0.43
Ω = 0.9
Ω = 1.05
Ω = 1.14
Ω = 1.5
V_{ph}/c
θ_{ph}
(d)
(e)
(f)
(g)
(h)
(i)
1^{ℓ}
2^{m}
1^{h}
2^{h}
10
1
10^{-1}
10^{-2}
10^{-3}
10^{-4}
0
30
60
90

5.29g and 5.29h, the phase velocity of mode 1^h is always greater than c and increases slightly as θ_{ph} is increased. As the frequency is still further increased, the resonant angle of mode 2^m increases with the result that mode 2^m propagates only in the range of directions which are more and more nearly perpendicular to the magnetostatic field. Finally for $\Omega \geqslant \sqrt{1 + R^2}$, mode 2^m ceases to propagate in any direction. At the same time, the phase velocity of mode 1^h becomes closer and closer to the free space electromagnetic wave velocity c. For $\Omega > \Omega_2$, mode 2^h starts propagating. As shown in Fig. 5.29i, the phase velocity of mode 2^h is greater than c, and that of mode 1^h for all directions of propagation. For any further increase in the frequency, the variation with respect to direction of both modes 1^h and 2^h is progressively reduced until in the limit of infinite frequency these two modes are characterized by the free space electromagnetic wave velocity for all directions of propagation.

An interesting fact has emerged from the study of the variation of the phase velocity with propagation direction over the entire frequency range. It is found from Fig. 5.29 that the ion and the electron gyromagnetic resonances first occur in the direction perpendicular to the magnetostatic field and move toward the direction of the magnetostatic field as the frequency is continuously increased but the plasma resonance first occurs along the magnetostatic field and moves towards the direction perpendicular to the magnetostatic field as the frequency is increased. In the three frequency ranges $Rm < \Omega < \Omega_L$, $\min(R, 1) < \Omega < \max(R, 1)$ and $\sqrt{1 + R^2} < \Omega < \infty$, there is no resonance in any direction with respect to the magnetostatic field. For the other frequency ranges, either mode 1 or mode 2 has a resonance in some direction.

The phase velocity curve has a reflection symmetry about the plane perpendicular to the magnetostatic field and a rotational symmetry about the direction of the magnetostatic field. If the phase velocity curves are reflected about the plane perpendicular to the magnetostatic field and rotated about the direction of the magnetostatic field, the so-called phase velocity surfaces are obtained. The length of the radius vector from the origin to any point on the phase velocity surface gives the phase velocity of the wave propagating in the direction of the radius vector. In general, the phase velocity surface is a two-sheeted surface corresponding to the two possible modes of propagation in any direction. The phase velocity surfaces have been found to be useful in the analysis of wave propagation problems in an anisotropic medium such as a plasma in a magnetostatic field.

In summary, it is pointed out that there is a rapid variation of the dispersion in the neighborhood of $\theta_{ph} = 0$ and $\theta_{ph} = \pi/2$. The former is due to the discontinuous change of the electron gyromagnetic resonance frequency, the discontinuous lowering of the cut-off frequency of the higher frequency branch of mode 1 and the disappearance of the middle branch of mode 2 in the close neighborhood of $\theta_{ph} = 0$. The latter is due to the rapid variation of the frequency of the ion gyromagnetic resonance in the close neighborhood of $\theta_{ph} = \pi/2$. A physical explanation of the disappearance of the middle branch of mode 2 is given based on the polarization studies. The variation of the dispersion caused by the variation in the propagation direction is examined over the entire frequency range, and this study has brought out an interesting feature in the variation of the location of occurrence of the various resonances as a function of frequency.

5.11. Magnetohydrodynamic waves

In magnetoionic theory and in its hydromagnetic extension, the thermal motions of the charged particles are ignored. In the limit of extremely small frequencies and for a fully ionized plasma, it is possible to include the effect of the thermal motions of the particles without much difficulty. For this purpose, it is convenient to use the magnetohydrodynamic equations in which a scalar kinetic pressure term is retained in order to take into account in an approximate manner the effect of the thermal motions of the particles. The linearized magnetohydrodynamic equations for a perfectly conducting fluid as given by Eqs. (2.9.1), (2.9.2), (2.9.5), and (2.9.7)–(2.9.11) are collected together here for convenience:

$$(\partial/\partial t)\rho_m + \rho_{m0}(\nabla \cdot \mathbf{V}) = 0 \tag{11.1}$$

$$\rho_{m0}(\partial/\partial t)\mathbf{V} = -a^2 \nabla \rho_m + \mathbf{J} \times \mathbf{B}_0 \tag{11.2}$$

$$\mathbf{E} + \mathbf{V} \times \mathbf{B}_0 = 0 \tag{11.3}$$

$$\nabla \times \mathbf{E} = -(\partial/\partial t)\mathbf{B} \tag{11.4}$$

$$\nabla \times \mathbf{B} = \mu_0 \mathbf{J} \tag{11.5}$$

$$\nabla \cdot \mathbf{B} = 0 \tag{11.6}$$

$$\nabla \cdot \mathbf{E} = 0 \tag{11.7}$$

Note that the equation of motion (2.9.2) has been combined with the equation of state (2.9.7) to obtain Eq. (11.2) wherein a is the sound speed in a perfect gas consisting of a mixture of two kinds of neutral particles characterized by the electronic and the ionic masses respectively. The magnetostatic field $\mathbf{B}_0$ is assumed to be in the z-direction. Only the plane wave characteristics are to be investigated. As before, it is helpful to analyze first the special cases of propagation along and across the magnetostatic field before proceeding to the general case of arbitrary direction of propagation.

Propagation along the magnetostatic field

Consider a plane wave propagating in the z-direction. All the field quantities depend only on z and t. Hence, Eq. (11.1), the components of Eqs. (11.2)–(11.5), Eqs. (11.6), and (11.7) become

$$(\partial/\partial t)\rho_m = -\rho_{m0}(\partial/\partial z)V_z \tag{11.8}$$

$$\rho_{m0}(\partial/\partial t)V_x = J_y B_0 \tag{11.9a}$$

$$\rho_{m0}(\partial/\partial t)V_y = -J_x B_0 \tag{11.9b}$$

$$\rho_{m0}(\partial/\partial t)V_z = -a^2(\partial/\partial z)\rho_m \tag{11.9c}$$

$$E_x = -V_y B_0 \tag{11.10a}$$

$$E_y = V_x B_0 \tag{11.10b}$$

$$E_z = 0 \tag{11.10c}$$

$$(\partial/\partial t)B_x = (\partial/\partial z)E_y \tag{11.11a}$$

$$(\partial/\partial t)B_y = -(\partial/\partial z)E_x \tag{11.11b}$$

$$(\partial/\partial t)B_z = 0 \tag{11.11c}$$

$$\mu_0 J_x = -(\partial/\partial z)B_y \tag{11.12a}$$

$$\mu_0 J_y = (\partial/\partial z)B_x \tag{11.12b}$$

$$J_z = 0 \tag{11.12c}$$

$$(\partial B_z/\partial z) = 0 \tag{11.13}$$

$$(\partial E_z/\partial z) = 0 \tag{11.14}$$

It follows from Eqs. (11.10c), (11.14), (11.11c), (11.13), and (11.12c) that

$$E_z = B_z = J_z = 0 \tag{11.15}$$

It is also found that Eqs. (11.8) and (11.9c), containing ρ_m and V_z, form one set, and Eqs. (11.9a,b), (11.10a,b), (11.11a,b), and (11.12a,b), containing V_x, V_y, E_x, E_y, B_x, B_y, J_x, and J_y, form another set.

The second set can be manipulated into a compact form. For that purpose, it is necessary to define

$$V^{\mp} = 2^{-1/2}[V_x \pm iV_y] \tag{11.16}$$

In a similar manner, $E^{\mp}$, $B^{\mp}$, and $J^{\mp}$ are defined. If Eq. (11.9a) is multiplied by $1/\sqrt{2}$, Eq. (11.9b) by $\pm i/\sqrt{2}$, and they are added together, the result is

$$\rho_{m0}(\partial/\partial t)V^{\mp} = \mp iB_0 J^{\mp} \tag{11.17}$$

In a similar manner, the following results are deduced from Eqs. (11.10a,b), (11.11a,b) and (11.12a,b):

$$E^{\mp} = \pm iB_0 V^{\mp} \tag{11.18}$$

$$(\partial/\partial t)B^{\mp} = \mp i(\partial/\partial z)E^{\mp} \tag{11.19}$$

$$\mu_0 J^{\mp} = \pm i(\partial/\partial z)B^{\mp} \tag{11.20}$$

By combining Eq. (11.20) with Eq. (11.17), and Eq. (11.18) with Eq. (11.19), it is found that

$$\rho_{m0}(\partial/\partial t)V^{\mp} = (B_0/\mu_0)(\partial/\partial z)B^{\mp} \tag{11.21a}$$

$$(\partial/\partial t)B^{\mp} = B_0(\partial/\partial z)V^{\mp} \tag{11.21b}$$

The elimination of $B^{\mp}$ from Eqs. (11.21a) and (11.21b) yields

$$(\partial^2/\partial t^2)V^{\mp} = v_\alpha^2(\partial^2/\partial z^2)V^{\mp} \tag{11.22}$$

where

$$v_\alpha = [B_0^2/\mu_0 \rho_{m0}]^{1/2} \tag{11.23}$$

The general solution of Eq. (11.22) is of the form

$$V^{\mp} = f_t^{\mp}(z - v_\alpha t) + g_t^{\mp}(z + v_\alpha t) \tag{11.24}$$

where $f_t^{\mp}$ and $g_t^{\mp}$ are any functions of their respective arguments. It is verified that $f_t^{\mp}$ represents a plane wave traveling in the positive z-direction with a velocity v_α and without any attenuation or distortion. Similarly, $g_t^{\mp}$ represents a plane wave with the same properties but traveling in the negative z-direction. These are transverse waves since the electric field, the electric current density, the magnetic flux density, and the velocity of mass flow are all perpendicular to the direction of propagation. There are two types of waves propagating in the same direction with the same velocity v_α. In the time harmonic case, the fields distinguished by the superscript $-$ are left circularly polarized and those distinguished by the superscript $+$ are right circularly polarized. Suppose that time harmonic plane waves having the phase factor $\exp\{i(kz - \omega t)\}$ are sought. The propagation coefficient and the phase velocity are then obtained from Eq. (11.22) to be given by

$$k = \omega/v_\alpha \qquad v_{ph} = \omega/k = v_\alpha \tag{11.25}$$

It is found from Eqs. (11.23) and (11.25) that the phase velocity is independent of the frequency, that is, there is no dispersion. It is this absence of dispersion in the frequency domain that leads to waves without attenuation and distortion in the time domain. Another important property of these waves is the absence of any density fluctuation ρ_m or perturbations of kinetic pressure. These waves are called the *Alfvén waves* and v_α is known as the Alfvén wave velocity. Previously, the Alfvén wave velocity v_α as given by Eq. (11.23) was deduced in a different manner by using the analogy with the transverse vibrations of stretched strings. Thus for propagation along the magnetostatic field, two types of Alfvén waves, one with left and the other right circular polarizations exist and these waves are independent of the compressibility of the plasma.

It can be deduced from Eqs. (11.8) and (11.9c) that

$$(\partial^2/\partial t^2)\rho_m = a^2(\partial^2/\partial z^2)\rho_m \tag{11.26}$$

whose general solution is given by

$$\rho_m = f_l(z - at) + g_l(z + at) \tag{11.27}$$

As before, f_l represents a plane wave propagating in the positive z-direction with a velocity a and without any distortion or attenuation. Similarly g_l represents a plane wave with the same properties but traveling in the negative z-direction. The wave represented by Eq. (11.27) is a longitudinal wave since the velocity of mass flow associated with this wave is in the direction of propagation. There is no electric field, electric current density, and magnetic flux density associated with this wave. Also, this wave contains all the density fluctuations. Therefore, Eq. (11.27) represents a *longitudinal sound* wave. For the time-harmonic, plane longitudinal sound waves, the propagation coefficient and the phase velocity can be deduced from Eq. (11.26) to be given by

$$k = \omega/a \qquad v_{ph} = \omega/k = a \tag{11.28}$$

The phase velocity of the sound wave is also independent of the frequency and consequently, there is no dispersion.

Propagation across the magnetostatic field

In order to determine the characteristics of a plane wave propagating perpendicular to the magnetostatic field, all the field quantities are assumed to depend on x and t only. The propagation direction is parallel to the x-axis which is perpendicular to the direction (z) of the magnetostatic field. For this case, the following results are obtained:

$$(\partial/\partial t)\rho_m + \rho_{m0}(\partial/\partial x)V_x = 0 \tag{11.29}$$

$$\rho_{m0}(\partial/\partial t)V_x = -a^2(\partial/\partial x)\rho_m + J_y B_0 \tag{11.30a}$$

$$\rho_{m0}(\partial/\partial t)V_y = -J_x B_0 \tag{11.30b}$$

$$\rho_{m0}(\partial/\partial t)V_z = 0 \tag{11.30c}$$

$$E_x + V_y B_0 = 0 \tag{11.31a}$$

$$E_y - V_x B_0 = 0 \tag{11.31b}$$

$$E_z = 0 \tag{11.31c}$$

$$(\partial/\partial t)B_x = 0 \tag{11.32a}$$

$$(\partial/\partial t)B_y = (\partial/\partial x)E_z \tag{11.32b}$$

$$(\partial/\partial t)B_z = -(\partial/\partial x)E_y \tag{11.32c}$$

$$\mu_0 J_x = 0 \tag{11.33a}$$

$$\mu_0 J_y = -(\partial/\partial x)B_z \tag{11.33b}$$

$$\mu_0 J_z = (\partial/\partial x)B_y \tag{11.33c}$$

$$(\partial/\partial x)B_x = 0 \tag{11.34}$$

$$(\partial/\partial x)E_x = 0 \tag{11.35}$$

With the help of Eqs. (11.30b), (11.30c), (11.3la), (11.3lc), (11.32a), (11.32b), (11.33a), (11.33c), (11.34), and (11.35), it can be proved that

$$J_x = J_z = E_x = E_z = B_x = B_y = V_y = V_z = 0 \tag{11.36}$$

The only nonvanishing quantities are J_y, E_y, B_z, and V_x, and these quantities are governed by Eqs. (11.29), (11.30a), (11.31b), (11.32c), and (11.33b). In Fig. 5.30, the directions of these nonvanishing field quantities as well as those of the magnetostatic field and the wavevector are indicated. The substitution of J_y from Eq. (11.33b) into Eq. (11.30a) yields

$$\rho_{m0}(\partial/\partial t)V_x = -a^2(\partial/\partial x)\rho_m - (B_0/\mu_0)(\partial/\partial x)B_z \tag{11.37}$$

In a similar manner, the substitution of E_y from Eq. (11.31b) into Eq. (11.32c) gives

$$(\partial/\partial t)B_z = -B_0(\partial/\partial x)V_x \tag{11.38}$$

Let

$$a^2\rho_m + (B_0/\mu_0)B_z = C_z \tag{11.39}$$

Then, it can be deduced from Eqs. (11.29), (11.38), and (11.39) that

$$(\partial/\partial t)C_z = -v_{ms}^2\rho_{m0}(\partial/\partial x)V_x \tag{11.40}$$

where

$$v_{ms}^2 = B_0^2/\mu_0\rho_{m0} + a^2 = v_\alpha^2 + a^2 \tag{11.41}$$

From Eqs. (11.37) and (11.39), it is found that

$$\rho_{m0}(\partial/\partial t)V_x = -(\partial/\partial x)C_z \tag{11.42}$$

The elimination of C_z from Eqs. (11.40) and (11.42) yields

$$(\partial^2/\partial t^2)V_x = v_{ms}^2(\partial^2/\partial x^2)V_x \tag{11.43}$$

whose general solution is given by

$$V_x = f_{ms}(x - v_{ms} t) + g_{ms}(x + v_{ms} t) \tag{11.44}$$

In Eq. (11.44), f_{ms} represents a plane wave propagating in the positive x-direction with a velocity v_{ms}, and, without any attenuation or distortion. In a similar manner, g_{ms} represents a plane wave with the same properties but traveling in the negative x-direction. Just like the longitudinal wave, the wave represented by Eq. (11.43) has a velocity of mass flow in the direction of propagation as well as a fluctuating mass density associated with it. Moreover, the wave specified by Eq. (11.43) is similar to an electromagnetic wave since associated with it there are time varying electric and magnetic fields lying in a plane perpendicular to the direction of propagation. For this reason, this wave is called the *magnetosonic wave.* The time varying magnetic field is perpendicular to the direction of propagation but parallel to the magnetostatic field but the time varying electric field is perpendicular to both the magnetostatic field and the direction of propagation. For the time-harmonic, plane magnetosonic wave, the propagation coefficient and the phase velocity can be determined from Eq. (11.43) to be given by

$$k = \omega/v_{ms} \qquad v_{ph} = \omega/k = v_{ms} = \sqrt{v_\alpha^2 + a^2} \tag{11.45}$$

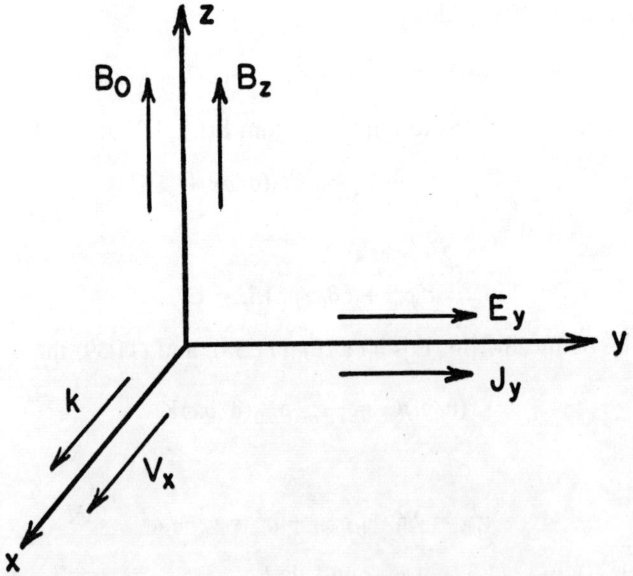

Fig. 5.30. Relative directions of the various field components associated with a magnetosonic wave propagating across the magnetostatic field in a compressible plasma.

The subscripts *ms* are used to denote that v_{ms} is the magnetosonic wave velocity. As expected, the phase velocity of the magnetosonic wave is independent of frequency and therefore the magnetosonic wave is nondispersive.

In the absence of collisions, the time-varying magnetic flux affects the particle velocities in the two directions perpendicular to itself and the compression is two-dimensional. Hence in the evaluation of the sound speed a appearing in Eq. (11.44), the ratio of the specific heats may be set equal to 2 both for the electrons and for the ions. For the magnetosonic wave, there are two kinds of restoring forces, namely the gradient of the kinetic pressure and the gradient of the compressional stresses between the magnetic flux lines. It is seen from Eq. (11.39) that the perturbation of the kinetic pressure, $P = a^2 \rho_m$, and the perturbation of the magnetic pressure, $B_0 B_z / \mu_0$, act in phase for the magnetosonic wave. If the kinetic pressure is much greater than the magnetic pressure, $v_{ms} \approx a$ and the magnetosonic wave becomes essentially an acoustic wave similar to the previously discussed ion plasma mode with a phase velocity as given by Eq. (2.8.49). On the other hand, if the magnetic pressure is very large compared to the kinetic pressure, $v_{ms} \approx v_\alpha$, that is, the magnetosonic wave velocity becomes equal to the Alfvén wave velocity. However, it is to be noted that the Alfvén and the magnetosonic waves involve quite different types of magnetic stresses.

Propagation in an arbitrary direction

It is now desired to investigate the characteristics of a plane wave propagating in an arbitrary direction with respect to the magnetostatic field. Without loss of generality, the direction $\hat{\mathbf{k}}$ of propagation is assumed to lie in the *zx*-plane such that $\hat{\mathbf{z}} \times \hat{\mathbf{k}}$ is in the *y*-direction. All the field components are dependent on *x*, *z*, and *t* but are independent of *y*. Hence Eq. (11.1), the components of Eqs. (11.2)–(11.5), Eqs. (11.6), and (11.7) become

$$(\partial \rho_m / \partial t) + \rho_{m0}\{(\partial V_x / \partial x) + (\partial V_z / \partial z)\} = 0 \tag{11.46}$$

$$\rho_{m0}(\partial V_x / \partial t) = -a^2(\partial \rho_m / \partial x) + J_y B_0 \tag{11.47a}$$

$$\rho_{m0}(\partial V_y / \partial t) = -J_x B_0 \tag{11.47b}$$

$$\rho_{m0}(\partial V_z / \partial t) = -a^2(\partial \rho_m / \partial z) \tag{11.47c}$$

$$E_x + V_y B_0 = 0 \tag{11.48a}$$

$$E_y - V_x B_0 = 0 \tag{11.48b}$$

$$E_z = 0 \tag{11.48c}$$

$$(\partial B_x / \partial t) = (\partial E_y / \partial z) \tag{11.49a}$$

$$-(\partial B_y / \partial t) = (\partial E_x / \partial z) - (\partial E_z / \partial x) \tag{11.49b}$$

$$-(\partial B_z / \partial t) = (\partial E_y / \partial x) \tag{11.49c}$$

$$\mu_0 J_x = -(\partial B_y / \partial z) \tag{11.50a}$$

$$\mu_0 J_y = (\partial B_x / \partial z) - (\partial B_z / \partial x) \tag{11.50b}$$

$$\mu_0 J_z = (\partial B_y / \partial x) \tag{11.50c}$$

$$(\partial B_x / \partial x) + (\partial B_z / \partial z) = 0 \tag{11.51}$$

$$(\partial E_x / \partial x) + (\partial E_z / \partial z) = 0 \tag{11.52}$$

It follows from Eqs. (11.47b), (11.48a), (11.48c), (11.49b), (11.50a), (11.50c), and (11.52) that V_y, J_x, E_x, E_z, B_y, and J_z form one set. Moreover, Eqs. (11.48c) and (11.52) show that $E_z = 0$ and E_x is independent of x. The substitution of J_x from Eq. (11.50a) into Eq. (11.47b) yields

$$\rho_{m0}(\partial V_y / \partial t) = (B_0 / \mu_0)(\partial B_y / \partial z) \tag{11.53}$$

In a similar manner, it is obtained from Eqs. (11.48a), (11.48c), and (11.49b) that

$$(\partial B_y / \partial t) = B_0(\partial V_y / \partial z) \tag{11.54}$$

Since $E_z = 0$ and E_x is independent of x, it is seen from Eqs. (11.49b), (11.54), and (11.50a) that B_y, V_y, and J_x are also independent of x. Also, it is found from Eq. (11.50c) that

$$J_z = 0 \tag{11.55}$$

Thus, this set actually represents a wave propagating along the magnetostatic field in the z-direction. The elimination of B_y from Eqs. (11.53) and (11.54) gives

$$(\partial^2 V_y / \partial t^2) = v_\alpha^2(\partial^2 V_y / \partial z^2) \tag{11.56}$$

which represents a linearly polarized Alfvén wave propagating in the z-direction. The field components associated with this wave are B_y, V_y, E_x, and J_x. Therefore, this is a transverse Alfvén wave. It is deduced from Eq. (11.16) that

$$2^{-1/2}(V^- + V^+) = V_x \qquad -i2^{-1/2}(V^- - V^+) = V_y \tag{11.57}$$

In view of Eqs. (11.57), for the time-harmonic fields, the two different circularly polarized Alfvén waves can also be thought of as two different linearly polarized Alfvén waves. When the propagation direction is changed from that of the magnetostatic field to an arbitrary direction which lies in the zx-plane, that Alfvén wave with the time varying magnetic flux density linearly polarized normal to the zx-plane, that is, in the y-direction, is unaffected [25] and the other Alfvén wave interacts with the longitudinal sound wave to form two kinds of magnetohydrodynamic waves whose characteristics are investigated presently [26]. This Alfvén wave which propagates only along the magnetostatic field is mistakenly considered as propagating in a direction making an angle θ_{ph} with that of the magnetostatic field and with a velocity $v_\alpha \cos\theta_{ph}$ and is called the oblique Alfvén wave.

If J_y from Eq. (11.50b) is substituted into Eq. (11.47a), the result is

$$\rho_{m0}\frac{\partial V_x}{\partial t} = -a^2\frac{\partial \rho_m}{\partial x} + \frac{B_0}{\mu_0}\left\{\frac{\partial B_x}{\partial z} - \frac{\partial B_z}{\partial x}\right\} \tag{11.58}$$

Similarly, if E_y from Eq. (11.48b) is substituted into Eqs. (11.49a) and (11.49c), it is obtained that

$$(\partial B_x/\partial t) = B_0(\partial V_x/\partial z) \tag{11.59}$$

$$-(\partial B_z/\partial t) = B_0(\partial V_x/\partial x) \tag{11.60}$$

If Eq. (11.58) is differentiated with respect to time and, if $\partial\rho_m/\partial t$, $\partial B_x/\partial t$, and $\partial B_z/\partial t$ are substituted from Eqs. (11.46), (11.59), and (11.60), it is found that

$$\frac{\partial^2 V_x}{\partial t^2} = a^2\left\{\frac{\partial^2 V_x}{\partial x^2} + \frac{\partial^2 V_z}{\partial x\partial z}\right\} + v_\alpha^2\left\{\frac{\partial^2 V_x}{\partial z^2} + \frac{\partial^2 V_x}{\partial x^2}\right\} \tag{11.61}$$

Similarly, it can be deduced from Eqs. (11.47c) and (11.46) that

$$\partial^2 V_z/\partial t^2 = a^2\{\partial^2 V_x/\partial x\partial z + \partial^2 V_z/\partial z^2\} \tag{11.62}$$

In the investigation of the properties of the two magnetohydrodynamic waves as specified by Eqs. (11.61) and (11.62), it is convenient to proceed to the time-harmonic case first, determine the phase velocities of the two waves and then deduce the characteristics of the waves in the time domain.

Let all the field quantities have the phase factor of the form

$$f(x,z,t) = f_0\exp[i\{k(x\sin\theta_{ph} + z\cos\theta_{ph}) - \omega t\}] \tag{11.63}$$

Together with Eq. (11.63), Eqs. (11.61) and (11.62) may be shown to yield the following pair of simultaneous equations for the amplitudes of V_x and V_z:

$$[\omega^2 - k^2\{a^2\sin^2\theta_{ph} + v_\alpha^2\}]V_x - k^2a^2\cos\theta_{ph}\sin\theta_{ph}V_z = 0 \tag{11.64}$$

$$-k^2a^2\cos\theta_{ph}\sin\theta_{ph}V_x + [\omega^2 - k^2a^2\cos^2\theta_{ph}]V_z = 0 \tag{11.65}$$

If the determinant of the coefficients of the system of equations (11.64) and (11.65) is set equal to zero, the following simple dispersion equation satisfied by the phase velocity $v_{ph} = \omega/k$ is obtained:

$$v_{ph}^4 - v_{ph}^2(v_\alpha^2 + a^2) + a^2v_\alpha^2\cos^2\theta_{ph} = 0 \tag{11.66}$$

It is seen from Eq. (11.66) that the phase velocity is independent of the frequency and hence the waves are nondispersive. Consequently, in the time domain, the two plane waves propagate in the direction $\hat{\mathbf{k}} = \hat{\mathbf{x}}\sin\theta_{ph} + \hat{\mathbf{z}}\cos\theta_{ph}$ without attenuation and distortion. The velocities of the two waves are different and are specified by Eq. (11.66).

For propagation along the magnetostatic field, $\theta_{ph} = 0$ and Eq. (11.66) gives the following two solutions: $v_{ph} = v_\alpha$ and $v_{ph} = a$ which pertain to the Alfvén and the sound waves, respectively, and are in accordance with the previous results. In a similar manner, for the special case of propagation across the magnetostatic field, $\theta_{ph} = \pi/2$ and Eq. (11.66) yields one nontrivial solution $v_{ph} = \sqrt{v_\alpha^2 + a^2}$ which corresponds to the magnetosonic wave and which also is in conformity with the previous results. The two solutions of Eq. (11.66) are evaluated to be given by

$$v_{ph,f},\, v_{ph,s} = \left(\tfrac{1}{2}[v_\alpha^2 + a^2 \pm \{(v_\alpha^2 - a^2)^2 + 4v_\alpha^2 a^2 \sin^2\theta_{ph}\}^{1/2}]\right)^{1/2} \qquad (11.67)$$

where the upper and the lower signs correspond to $v_{ph,f}$ and $v_{ph,s}$, respectively. Both the solutions in Eq. (11.67) are real and positive, and $v_{ph,f} > v_{ph,s}$. Therefore, the wave corresponding to $v_{ph,f}$ is called a *fast wave* and that corresponding to $v_{ph,s}$ is called a *slow wave*.

Consider the case in which the Alfvén wave velocity is greater than the sound velocity, that is, $v_\alpha > a$. Then, Eq. (11.67) yields

$$v_{ph,f} = v_\alpha \qquad v_{ph,s} = a \qquad \text{for } \theta_{ph} = 0$$

$$v_{ph,f} = \sqrt{v_\alpha^2 + a^2} \qquad v_{ph,s} = 0 \qquad \text{for } \theta_{ph} = \pi/2.$$

Thus for $v_\alpha > a$, the fast wave becomes the Alfvén wave for $\theta_{ph} = 0$ and the magnetosonic wave for $\theta_{ph} = \pi/2$; the slow wave becomes the sound wave for $\theta_{ph} = 0$ and vanishes for $\theta_{ph} = \pi/2$. The phase velocity curves of the fast and the slow waves as evaluated from Eq. (11.67) are shown in Fig. 5.31 for the case of $v_\alpha > a$.

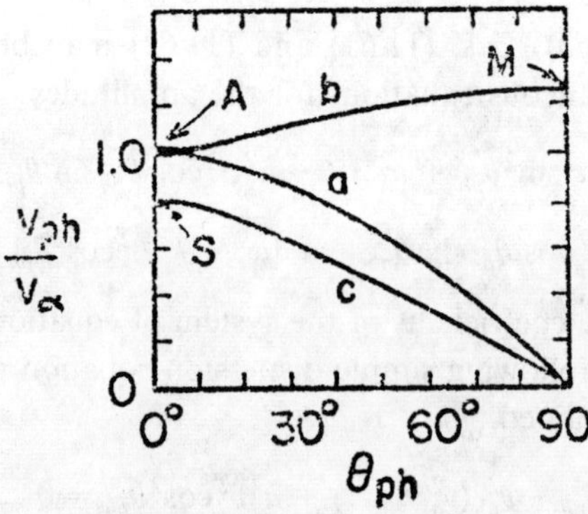

Fig. 5.31. Phase velocity curves for the oblique Alfvén, the fast and the slow waves for $a/v_\alpha = 0.8$: (a) the oblique Alfvén wave; (b) the fast wave; (c) the slow wave; *A*: Alfvén wave; *S*: sound wave; *M*: magnetosonic wave.

It is interesting to consider two limiting cases. If the magnetostatic field B_0 is very weak so that $v_\alpha \ll a$. It is then obtained from Eq. (11.67) that $v_{ph,f} = a$ and $v_{ph,s} = v_\alpha \cos\theta_{ph}$ with the result that the fast and the slow waves become equivalent to an acoustic and the oblique Alfvén waves, respectively. If B_0 is so large that $v_\alpha \gg a$, Eq. (11.67) gives $v_{ph,f} = v_\alpha$ and $v_{ph,s} = a\cos\theta_{ph}$. Consequently, the fast and the slow waves become an Alfvén wave and an acoustic wave, respectively.

References

5.1. E. V. Appleton, Geophysical influences on the transmission of wireless waves, *Proc. Phys. Soc.* **37**, 16D–22D (1925).

5.2. E. V. Appleton, Wireless studies of the ionosphere, *J. Inst. Elec. Engr.* **71**, 642–650 (1932).

5.3. D. R. Hartree, The propagation of electromagnetic waves in a refracting medium in a magnetic field, *Proc. Camb. Phil. Soc.* **27**, 143–162 (1931).

5.4. L. R. O. Storey, An investigation of whistling atmospherics, *Phil. Trans. Roy. Soc. London* **A246**, 113–141, (1953).

5.5. R. A. Helliwell, *Whistlers and related ionospheric phenomenon*, Stanford University Press, Stanford, Calif., 1965.

5.6. R. Bowers, Plasmas in solids, *Scientific American* **209**, No.5, 46–53, (1963).

5.7. P. Aigrain, Les Hélicons dans les semiconducteurs, *Proceedings of the International Conference on Semiconductor Physics, Prague*, 1960, 224–226, Czechoslovak Academy of Science, Prague, 1961.

5.8. M. P. Bachynski, Electromagnetic wave penetration of reentry plasma sheaths, *Radio Science J. Res. NBS/USNC-URSI* **69D**, No.2, 147–154 (1965).

5.9. S. R. Seshadri, Wave propagation in a compressible ionosphere, I and II, *Radio Science J. Res. NBS/USNC-URSI* **68D**, No.12, 1285–1307 (1964).

5.10. E. Astrom, On waves in an ionized gas, *Arkiv. Fysik* **2**, No.42, 443–457 (1951).

5.11. W. P. Allis, S. J. Buchsbaum, and A. Bers, *Waves in Anisotropic Plasmas*, The M.I.T. Press, Cambridge, Mass., 1963.

5.12. J. A. Ratcliffe, *The Magnetoionic Theory and its Applications to the Ionosphere*, Cambridge University Press, Cambridge, England, 1959.

5.13. K. G. Budden, *Radio Waves in the Ionosphere*, Cambridge University Press, Cambridge, England, 1961.

5.14. M. J. Lighthill, Studies on magnetohydrodynamic waves and other anisotropic wave motions, *Phil. Trans. Roy. Soc.* **A252**, 397–430 (1960).

5.15. S. R. Seshadri, Cerenkov radiation in a magnetoionic medium, *Electronics Lett.* **3**, No.6, 271–274 (1967).

5.16. E. Astrom, Magnetohydrodynamic waves in a plasma, *Nature* **165**, 1019–1020 (1950).

5.17. H. Alfvén, Existence of electromagnetic-hydrodynamic waves, *Nature* **150**, 405–406 (1942).

5.18. C. O. Hines, Generalized magneto-hydrodynamic formulae, *Proc. Camb. Phil. Soc.* **49**, 299–307 (1953).

5.19. T. H. Stix, Oscillations of a cylindrical plasma, *Phys. Rev.* **106**, No.6, 1146–1150 (1957).

5.20. L. R. O. Storey, A method to detect the presence of ionized hydrogen in the outer atmosphere, *Canad. J. Phys.* **34**, 1153–1163 (1956).

5.21. C. O. Hines, Heavy-ion effects in audio-frequency radio propagation, *J. Atmosph. Terr. Phys.* **11**, 36–42 (1957).

5.22. J. A. Fejer, Hydromagnetic wave propagation in the ionosphere, *J. Atmosph. Terr. Phys.* **18**, 135–146 (1960).

5.23. C. O. Hines, The relation between hydromagnetic waves and the magnetoionic theory, in *Electromagnetic Theory and Antennas* (E. C. Jordan, Ed.), Pergamon Press, New York, 287–299, 1963.

5.24. S. R. Seshadri, The magnetoionic theory at hydromagnetic frequencies, *J. Atmosph. Terr. Phys.* **27**, No.5, 617–634 (1965).

5.25. L. Spitzer, *Physics of Fully Ionized Gases*, 2nd ed., Chapter 3, Interscience Publishers, New York, 1962.
5.26. V. C. A. Ferraro and C. Plumpton, *An Introduction to Magnetofluid Mechanics*, Chapter 3, Oxford University Press, London, England, 1961.

Problems

5.1. Consider a homogeneous plasma of infinite extent with stationary ions. The electrons are assumed to have a time-independent drift velocity $\mathbf{v} = \hat{\mathbf{x}}v_d$. It is desired to find the characteristics of a longitudinal plane wave with the electric vector lying in the direction of propagation. If the wave propagates in the direction of the electron stream, the electric field associated with the wave may be assumed to be of the form

$$E_x(x) = E_{x0}\exp[i\{k_p kx - \omega t\}]$$

Show that there exists two types of longitudinal electron plasma waves having the following dispersion relations:

$$k = (\Omega - 1)/\tilde{v}_d \qquad k = (\Omega + 1)/\tilde{v}_d$$

where $\tilde{v}_d = v_d\sqrt{\mu_0\varepsilon_0}$ is the normalized drift velocity. Sketch the Brillouin diagram for the two waves in the region of positive Ω and k. These two waves are designated as the *fast* and the *slow* longitudinal space-charge waves in a "cold" collisionless electron stream. Identify these two waves on the Brillouin diagram.

5.2. The characteristics of a transverse electromagnetic plane wave were analyzed for the case in which the electron gas had no drift velocity. Suppose that the electrons are assumed to have a time-independent drift velocity $\mathbf{v} = \hat{\mathbf{x}}v_d$ in the direction of propagation of the plane wave. Examine how the dispersion relation (3.7) is changed for a "cold" collisionless electron stream.

5.3. Deduce the expression (4.21) for the normalized group velocity of the left circularly polarized plane wave propagating along the magnetostatic field in a magnetoionic medium.

5.4. For the right circularly polarized plane wave propagating along the magnetostatic field in a magnetoionic medium, establish that the time-averaged power transported vanishes for those values of the physical parameters for which the normalized propagation coefficient k_z^+ is imaginary.

Find the expression for the time-averaged Poynting vector for the case in which k_z^+ is real and note that, as is to be expected intuitively, the time-averaged Poynting vector becomes infinite at the resonant frequency $\Omega = R$.

5.5. Consider a homogeneous plasma of infinite extent with stationary ions. The electrons are assumed to have a time-independent drift velocity $\mathbf{v} = \hat{\mathbf{z}}v_d$ in the direction of the magnetostatic field which permeates uniformly throughout the plasma. Show that the normalized propagation coefficient k_z of a transverse electromagnetic plane wave propagating along the magnetostatic field is given by

$$k_z = \left[\Omega^2 - \frac{(\Omega - k_z \tilde{v}_d)}{(\Omega - k_z \tilde{v}_d \pm R)}\right]^{1/2}$$

where $\tilde{v}_d = v_d\sqrt{\mu_0 \varepsilon_0}$ is the normalized drift velocity of the electron stream. Verify that in the absence of the drift, k_z becomes equal to the normalized propagation coefficient of the left and the right circularly polarized plane electromagnetic waves propagating along the magnetostatic field in a magnetoionic medium.

5.6. Deduce the expression (5.18) for the normalized group velocity of the plane TM wave propagating across the magnetostatic field in a magnetoionic medium.

5.7. Show that the electric field **E(r)** in a magnetoionic medium satisfies the differential equation:

$$\nabla \times \nabla \times \mathbf{E}(\mathbf{r}) - k_p^2 \Omega^2 \varepsilon_r \cdot \mathbf{E}(\mathbf{r}) = 0 \tag{A}$$

The spatial dependence of the electric field is of the form $\mathbf{E}(\mathbf{r}) = \mathbf{E}_0 \exp[ik_p k(mx + lz)]$, where $l = \cos\theta_{ph}$, $m = \sin\theta_{ph}$, and θ_{ph} is the angle between the direction of propagation and the magnetostatic field which is in the z-direction. If the amplitude of the electric field is written as a column matrix $[E_0]$ as in (B), show that the curl operation $\nabla \times$ can be written as a 3×3 matrix [M] as in (C):

$$[E_0] = \begin{bmatrix} E_{x0} \\ E_{y0} \\ E_{z0} \end{bmatrix} \qquad [M] = ik_p k \begin{bmatrix} 0 & -l & 0 \\ l & 0 & -m \\ 0 & m & 0 \end{bmatrix} \tag{B, C}$$

The differential equation (A) can be reduced to the following matrix form:

$$[M][M][E_0] - k_p^2 \Omega^2 \begin{bmatrix} \varepsilon_1 & i\varepsilon_2 & 0 \\ -i\varepsilon_2 & \varepsilon_1 & 0 \\ 0 & 0 & \varepsilon_3 \end{bmatrix} [E_0] = 0 \tag{D}$$

Find, in the matrix form, the three homogeneous equations satisfied by E_{x0}, E_{y0}, and E_{z0}. Hence, deduce the dispersion relation and verify that it is identical to that given by Eqs. (6.18) and (6.19).

5.8. Deduce Eqs. (6.4)–(6.7) with the help of Eqs. (6.2a), (6.2b), (6.3a), and (6.3b).

5.9. Deduce the expression for $i\sqrt{\mu_0}\, H_{y0}$ as given by Eq. (6.24) from Eqs. (6.7) and (6.20).

5.10. Deduce the expression for $\sqrt{\mu_0}\, H_{x'0}$ as given by Eq. (6.27) from Eqs. (6.20) and (6.23).

5.11. For the special case of propagation in the direction of the magnetostatic field, construct Fig. 5.11 and deduce the regions of propagation of modes 1 and 2, starting from the dispersion equation as given by Eqs. (6.42) and (6.43).

5.12. For the case of propagation across the magnetostatic field, determine the regions of propagation of the two modes with the help of a parametric diagram in the Ω^2 - R^2 space. Start from the dispersion equation as given by Eqs. (6.42) and (6.43).

5.13. The two modes propagating along the magnetostatic field in a magnetoionic medium are circularly polarized in opposite directions. For some applications, it is convenient to represent the two modes propagating in a direction inclined to the magnetostatic field by a very small angle in terms of the so-called quasicircularly polarized modes which are the perturbations of the two circularly polarized modes pertaining to the case of propagation along the magnetostatic field. In view of the nonuniform convergence that exists as the propagation direction approaches that of the magnetostatic field, find the frequency range in which such a representation in terms of the quasicircularly polarized modes is not valid.

5.14. Deduce the normalized group velocity as given by Eqs. (6.56) and (6.57) starting from the dispersion relation as stated in Eqs. (6.42) and (6.43).

5.15. Find the form of the relative permittivity dyad $\boldsymbol{\varepsilon}_r$ when the strength of the magnetostatic field is infinite. This type of medium is called *uniaxially* anisotropic. Determine the characteristics of the two possible plane wave modes that can propagate in an uniaxially anisotropic medium starting from Eqs. (2.18) and (2.19).

5.16. Show that no Cerenkov radiation is possible in a cold isotropic plasma.

5.17. Show that for a charged particle moving with a uniform velocity u in a magnetoionic medium along the direction of the magnetostatic field which is assumed to be in the z-direction, the normalized group velocity of the electromagnetic wave generated is given by

$$\mathbf{v}_g/c = \hat{\boldsymbol{\rho}} A_{\rho n} + \hat{\mathbf{z}} A_{zn}$$

where

$$A_{\rho n} = \frac{k_{\rho n}}{\Omega} \frac{P_{\rho n}}{Q_n} \qquad A_{zn} = \frac{k_z}{\Omega} \frac{P_{zn}}{Q_n}$$

$$\begin{aligned} Q_n = {} & 4\Omega^6 - 3\Omega^4\{2k_{\rho n}^2 + 2k_z^2 + 3 + R^2\} \\ & + 2\Omega^2\{k_{\rho n}^4 + k_z^4 + 2k_{\rho n}^2 k_z^2 + 2k_{\rho n}^2(2 + R^2) + 2k_z^2(2 + R^2) + 3 + R^2\} \\ & - \{k_{\rho n}^4(1 + R^2) + k_z^4(1 + R^2) + 2k_{\rho n}^2 k_z^2(1 + R^2) \\ & + k_{\rho n}^2(2 + R^2) + 2k_z^2(1 + R^2) + 1\} \end{aligned}$$

$$\begin{aligned} P_{\rho n} = {} & 2\Omega^6 - 2\Omega^4\{k_{\rho n}^2 + k_z^2 + 2 + R^2\} \\ & + \Omega^2\{2k_{\rho n}^2(1 + R^2) + 2k_z^2(1 + R^2) + 2 + R^2\} - R^2 k_z^2 \end{aligned}$$

$$\begin{aligned} P_{zn} = {} & 2\Omega^6 - 2\Omega^4\{k_{\rho n}^2 + k_z^2 + 2 + R^2\} + 2\Omega^2(1 + R^2)\{k_{\rho n}^2 + k_z^2 + 1\} \\ & - R^2(k_{\rho n}^2 + 2k_z^2) \end{aligned}$$

5.18. Consider the radiation in a magnetoionic medium due to a point charge moving with a uniform velocity u in the direction of the magnetostatic field. Show that for the normalized upper hybrid resonant frequency $\Omega = \sqrt{1 + R^2}$, $k_{\rho n}$ is infinite. Find the direction of the wave normal. From the results of the previous problem, prove that the direction of the Cerenkov ray is given by $\theta_g = \pi$. Note the interesting result that the energy due to Cerenkov radiation at the upper hybrid resonant frequency is transported in a direction *opposite* to that of the motion of the charge.

5.19. Prove that a point charge moving with a uniform velocity u in an isotropic, warm electron plasma excites the longitudinal electron plasma wave. Find the frequency spectrum of the emitted radiation. Show that the wave normal and the ray are in the same direction.

Examine if the transverse electromagnetic wave is excited.

5.20. Prove that a point charge moving with a uniform velocity u in the direction of the anisotropic axis in an uniaxially anisotropic plasma excites one of the two possible modes. Find the frequency spectrum of the emitted radiation. Start from Eqs. (2.18) and (2.19) or the results of Problem 5.15.

5.21. For the special case of propagation in the direction of the magnetostatic field, construct Fig. 5.26 and deduce the regions of propagation of modes 1 and 2, starting from the dispersion equation for a plane wave in an electron-ion plasma, as given by Eqs. (10.1) and (10.2).

5.22. For the case of propagation of a plane wave across the magnetostatic field in an electron-ion plasma, determine the regions of propagation of the two modes with the help of a parametric diagram in the Ω^2 - R^2 space. Start from the dispersion equation as given by Eqs. (10.1) and (10.2).

5.23. Show that the electric vector corresponding to the region marked AB in Fig. 5.27 for mode 1^h and that marked CD for mode 2^m tends to become parallel to the propagation direction as the propagation direction approaches that of the magnetostatic field.

5.24. The characteristics of the left and the right circularly polarized transverse Alfvén waves propagating in the direction of the magnetostatic field in a compressible, fully ionized plasma are independent of the compressibility of the medium and the phase velocities of both these modes are equal to the Alfvén wave velocity v_α. Since these waves are unaffected by the compressibility of the medium, the phase velocities of these two modes should be identical with those of the corresponding waves in a cold, electron-ion plasma in the limit of zero frequency. The phase velocities of the two modes propagating along the magnetostatic field in a cold, electron-ion plasma in the limit of zero frequency were deduced in Sec. 5.9 to be given by $v_\alpha[1 + (v_\alpha/c)^2]^{1/2}$ where c is the free space electromagnetic wave velocity. Find the reason for the difference in the results obtained in the two cases.

5.25. For the transverse Alfvén waves propagating along the magnetostatic field in a compressible plasma, show that the magnetic and the kinetic energy densities of the wave motion are equal.

5.26. Sketch the lines of total magnetic flux due to the magnetostatic and the time-varying magnetic fields associated with the Alfvén wave propagating along the magnetostatic field and the magnetosonic wave propagating across the magnetostatic field. Show that the types of stress caused by the magnetic flux for these two waves are different.

5.27. Sketch the phase velocity curves for the fast and the slow magnetohydrodynamic waves for the case in which the sound velocity is greater than the Alfvén wave velocity and show that it is the sound wave that transforms into the magnetosonic wave for propagation across the magnetostatic field.

5.28. Let $V_{\parallel}$ and $V_{\perp}$ be the components of the velocity of mass flow that are, respectively, parallel and perpendicular to the direction of propagation. For the time-harmonic waves, show that $V_{\parallel}$ and $V_{\perp}$ are in phase for the fast wave and 180° out of phase for the slow wave. Deduce that the perturbations of the kinetic and the magnetic pressures are in phase for the fast wave and 180° out of phase for the slow wave.

5.29. Show that for $v_\alpha \gg a$, the velocity $\mathbf{V}$ is essentially parallel to the magnetostatic field for one wave but for the other wave, it is perpendicular to the direction of propagation and is in the plane containing the directions of propagation and the magnetostatic field.

5.30. Calculate the Alfvén wave velocity for mercury in a magnetic flux of density 1000 G.

5.31. Consider a homogeneous, unbounded magnetohydrodynamic medium in which the sound velocity a is less than the Alfvén wave velocity v_α. A point charge moves with a uniform velocity u in the direction of the magnetostatic field such that $v_\alpha > u > a$. Show that the oblique Alfvén wave is not excited.

Let k_ρ be the propagation coefficient of the waves excited by the moving charge in the direction normal to the magnetostatic field. Find the dispersion equation satisfied by k_ρ in terms of a, v_α, u, and the angular frequency ω. Solve the dispersion equation to show that no wave is generated.

CHAPTER 6

Applications of the Boltzmann Equation

6.1. Introduction

In this concluding chapter, some examples of the kinetic-theory treatment of plasmas are given. The kinetic-theory treatment of plasmas is based on the Boltzmann equation and Maxwell's equations. As was pointed out previously, the solution of the Boltzmann equation is difficult and therefore approximation procedures are necessary. The approximations of the Boltzmann equation resorted to in the following treatment can be grouped into three categories. In the first type of approximation, the Boltzmann collision term is set equal to zero resulting in what is known as the Boltzmann-Vlasov equation. An introductory treatment of the theory of plasma oscillations is developed using the Boltzmann-Vlasov equation. Only the small-amplitude waves in unbounded plasmas which are close to the equilibrium are considered.

First, the discussion is limited to the special case in which the anisotropy due to an external magnetostatic field is absent. The theory of the plasma oscillations is formulated as an initial-value problem. The plasma oscillations separate into three independent groups. The first group is the longitudinal plasma wave, and the second and the third groups are the transverse electromagnetic waves with two different polarizations. The dispersion equations are deduced and analyzed for all the three cases. For the longitudinal wave in the limit of zero temperature, the kinetic theory predicts stationary oscillations at the plasma frequency, and thus confirms the results based on the cold plasma model. In the long-wavelength limit, the kinetic theory leads to undamped progressive waves having a dispersion relation which is identical to that derived using the warm plasma model. In addition to reproducing the results of the cold and the warm plasma models, as well as establishing the regions of their validity, the kinetic theory confirms the possibility of the wave-particle interaction which was qualitatively brought out earlier in Sec. 2.12. If the equilibrium state is characterized by the isotropic Maxwell-Boltzmann distribution function, the wave-particle interaction results in the temporal damping of the wave and the so-called Landau damping constant is derived explicitly.

The dispersion relation of the transverse electromagnetic wave also reduces to the result of the cold plasma model in the limit of long wavelength. Although temporal damping exists in principle, it is insignificant for the transverse electromagnetic wave since the phase velocity of the wave is never in the neighborhood of the thermal velocity of the electrons with the result that the wave-particle interaction is ineffective.

The effect of the magnetostatic field is incorporated and the characteristics of the plasma oscillations having its propagation coefficient parallel to the magnetostatic field are then studied. Again the plasma oscillations separate into three independent groups. The first group is the longitudinal plasma wave; the second and the third groups are the left and the right circularly polarized transverse electromagnetic waves, respectively. The characteristics of the longitudinal plasma wave are unaffected for this orientation of the magnetostatic field. The dispersion relations of the two transverse electromagnetic waves reduce to the results of the cold plasma model in the limit of long wavelength. The temporal damping is insignificant for the left circularly polarized wave and for the right circularly polarized wave except in the neighborhood of the gyromagnetic frequency of the electrons. Near the gyromagnetic frequency, the phase velocity of the right circularly polarized wave is close to the thermal velocity of the electrons with the result that there is efficient wave-particle interaction. Consequently, there is the so-called cyclotron damping whose coefficient is deduced.

Next the characteristics of the plasma oscillations with the propagation coefficient perpendicular to the magnetostatic field are studied. The longitudinal plasma wave does not exist independently for any orientation of the magnetostatic field other than parallel to the propagation coefficient of the wave. For the perpendicular orientation of the magnetostatic field, again the plasma oscillations separate into two groups which are designated as the TM and the TEM modes. The dispersion relations for these modes are quite complicated and only numerical work can reveal their characteristics. For the TM mode, only the resonance characteristics are studied. For the zero-temperature limit, in addition to the upper hybrid resonant frequency as predicted by the cold plasma model, there are resonances at all the harmonics of the gyromagnetic frequency of the electrons other than the fundamental. For a finite temperature, the resonances occur at all the harmonics of the gyromagnetic frequency including the fundamental. These cyclotron-harmonic resonances are also present for the TEM mode. One important feature of the plasma oscillations with the propagation coefficient perpendicular to the magnetostatic field is the absence of the Landau damping.

In the second type of approximation of the Boltzmann equation, the collision term of the Boltzmann equation is replaced by a relaxation model. After explaining the plausability of the relaxation model as well as pointing out its drawbacks, it is shown that if the relaxation collision frequency is independent of the particle velocity, it becomes identical to the constant collision frequency introduced phenomenologically in magnetoionic theory. For the special case of a constant collision frequency, the simple transport processes such as the diffusion, the mobility and the electrical conduction are investigated for the case of a weakly ionized plasma permeated by a magnetostatic field. An approximate theory of the ambipolar diffusion is then presented for the simple case in which the external magnetostatic field is absent. The electrical conduction and the diffusion are then analyzed for the general case in which the relaxation collision frequency is a function of the particle velocity, and the expressions for the dyadic electrical conductivity and the dyadic diffusion coefficient are deduced. In particular, the use of the magnetostatic field for inhibiting the diffusion in the perpendicular direction is emphasized.

In the third type of approximation of the Boltzmann equation, an integral expression is derived for the collision term making use of certain assumptions, in particular those of binary encounters and molecular chaos. For the case of a plasma which has no spatial variation and which is free from the action of external forces, it is shown that there is a detailed balance between the effect of a collision and that of the corresponding inverse collision for the existence of the equilibrium state. With the help of the principle of the detailed balance, the equilibrium or the steady state is proved to be characterized by the Maxwell-Boltzmann distribution function. Finally, the results of the collision between an electron and a neutral particle are used to establish that the Boltzmann collision term reduces to the relaxation model for a weakly ionized plasma, and an expression of the relaxation collision frequency in terms of the momentum transfer cross section is obtained.

6.2. Longitudinal plane plasma wave in a hot, isotropic plasma

Consider a homogeneous plasma of infinite extent. The frequencies of interest are sufficiently high with the result that the heavy positive ions are legitimately assumed to be immobile but they provide the necessary positive neutralizing background for the mobile electrons. Let N_0 be the uniform number density of the positive ions. Initially, the plasma is in equilibrium with the electrons distributed throughout all of physical space with the same uniform number density N_0 as that of the ions. The electrons are characterized by the velocity distribution function $f_0(\mathbf{u})$. In view of Eq. (1.2.1), it follows that

$$N_0 = \int_{-\infty}^{\infty} f_0(\mathbf{u})\, d\mathbf{u} \tag{2.1}$$

Initially, the electric charge and current densities are zero throughout the plasma. Since the sources for the macroscopic electric and magnetic fields are these charge and current densities, it follows that the electric field is zero. In this and the following section, it is assumed that there is no external magnetostatic field with the result that the magnetic field is also zero initially.

At time $t = 0$, the plasma is perturbed slightly from its state of equilibrium so that for times subsequent to $t = 0$, the electrons are characterized by the velocity distribution function $f(\mathbf{r}, \mathbf{u}, t)$ as given by

$$f(\mathbf{r}, \mathbf{u}, t) = f_0(\mathbf{u}) + g(\mathbf{r}, \mathbf{u}, t) \tag{2.2}$$

where $g(\mathbf{r}, \mathbf{u}, t)$ is always small compared to $f_0(\mathbf{u})$ and

$$g(\mathbf{r}, \mathbf{u}, t) = 0 \qquad \text{for } t < 0 \tag{2.3}$$

On account of the perturbation, there are nonvanishing electric charge and current densities which in turn give rise to macroscopic electric and magnetic fields. There are two distinct possibilities. The perturbation $g(\mathbf{r}, \mathbf{u}, t)$ of the velocity distribution function and the resulting electromagnetic fields may decay in time with characteristic frequencies and decay constants with the result that the plasma returns to the original state of equilibrium in course of time. Therefore, the original state of equilibrium is said to be *stable*. On the other hand, it is also possible for the perturbation $g(\mathbf{r}, \mathbf{u}, t)$ of the velocity distribution function and the resulting electromagnetic fields to increase in time with characteristic growth rates; either a new state of equilibrium is reached or the growth may continue indefinitely in time. In this case, the original state of equilibrium is said to be *unstable*.

For the sake of definiteness and simplicity, the details of the analysis are carried out for the case where initially the electrons are characterized by the Maxwell-Boltzmann distribution function:

$$f_0(\mathbf{u}) = N_0\left(\frac{m_e}{2\pi K T_e}\right)^{3/2} e^{-m_e u^2/2KT_e} \qquad u^2 = u_x^2 + u_y^2 + u_z^2 \tag{2.4}$$

where m_e is the electron mass, T_e the electron temperature, and K is the Boltzmann constant. The distribution function given by Eqs. (2.4) is homogeneous and isotropic, and these features of the distribution function simplify the analysis. For other initial velocity distribution functions for the electrons, the analysis proceeds along similar lines with possible complexities in the details and different results.

Solution of the linearized Boltzmann-Vlasov equation

The velocity distribution function $f(\mathbf{r}, \mathbf{u}, t)$ satisfies the Boltzmann equation (1.3.29) where the term $(\partial f/\partial t)_{\text{coll}}$ on the right-hand side arises due to the collisional interactions of the electrons with the ions. It is assumed that these collisional interactions are negligible and therefore $(\partial f/\partial t)_{\text{coll}}$ is set equal to zero. In other words, the treatment is restricted to the special case of a collisonless plasma. The Boltzmann equation with $(\partial f/\partial t)_{\text{coll}} = 0$ is commonly called *the Boltzmann-Vlasov equation*. Thus the velocity distribution function $f(\mathbf{r}, \mathbf{u}, t)$ satisfies the Boltzmann-Vlasov equation:

$$\frac{\partial}{\partial t}f(\mathbf{r},\mathbf{u},t) + (\mathbf{u}\cdot\nabla_r)f(\mathbf{r},\mathbf{u},t) - \frac{e}{m_e}[\mathbf{E}(\mathbf{r},t) + \mathbf{u}\times\mathbf{B}(\mathbf{r},t)]\cdot\nabla_u f(\mathbf{r},\mathbf{u},t) = 0 \quad (2.5)$$

where $-e$ is the charge on an electron. Since the macroscopic electric and magnetic fields vanish in the equilibrium state, it follows that $\mathbf{E}(\mathbf{r},t)$ and $\mathbf{B}(\mathbf{r},t)$ arise due to the perturbation $g(\mathbf{r},\mathbf{u},t)$ in the velocity distribution function. Therefore $\mathbf{E}(\mathbf{r},t)$ and $\mathbf{B}(\mathbf{r},t)$ are first-order small quantities and

$$\mathbf{E}(\mathbf{r},t) = 0 \qquad \mathbf{B}(\mathbf{r},t) = \mu_0\mathbf{H}(\mathbf{r},t) = 0 \qquad \text{for } t < 0 \tag{2.6}$$

The result of substituting Eq. (2.2) into Eq. (2.5) is

$$\begin{aligned}\frac{\partial}{\partial t}g(\mathbf{r},\mathbf{u},t) + (\mathbf{u}\cdot\nabla_r)g(\mathbf{r},\mathbf{u},t) - \frac{e}{m_e}[\mathbf{E}(\mathbf{r},t) + \mathbf{u}\times\mathbf{B}(\mathbf{r},t)]\cdot\nabla_u f_0(\mathbf{u})\\ - \frac{e}{m_e}[\mathbf{E}(\mathbf{r},t) + \mathbf{u}\times\mathbf{B}(\mathbf{r},t)]\cdot\nabla_u g(\mathbf{r},\mathbf{u},t) = 0\end{aligned} \tag{2.7}$$

The fourth term on the left-hand side of Eq. (2.7) is a product of two first-order small quantities and hence is a second-order small quantity. Consequently, the fourth term on the left-hand side of Eq. (2.7) can be omitted in comparison with the three remaining terms which are first-order small quantities. Thus the *linearized* Boltzmann-Vlasov equation becomes

$$\frac{\partial}{\partial t}g(\mathbf{r},\mathbf{u},t) + (\mathbf{u}\cdot\nabla_r)g(\mathbf{r},\mathbf{u},t) - \frac{e}{m_e}[\mathbf{E}(\mathbf{r},t) + \mathbf{u}\times\mathbf{B}(\mathbf{r},t)]\cdot\nabla_u f_0(\mathbf{u}) = 0 \tag{2.8}$$

A special choice of $g(\mathbf{r},\mathbf{u},t)$ introduces further simplification in the analysis without the loss of the essentials of the plasma behavior under consideration. The perturbation $g(\mathbf{r},\mathbf{u},t)$ of the velocity distribution function, instead of being localized in the configuration space, is assumed to be periodic in space as given by

$$g(\mathbf{r},\mathbf{u},t) = g(\mathbf{u},t)e^{ikx} \tag{2.9}$$

where the propagation coefficient k is real. Note that Eq. (2.9) is in phasor notation and $g(\mathbf{u},t)$ is, in general, complex.

The perturbation charge density is obtained from Eq. (2.2.1) as

$$\rho(\mathbf{r},t) = eN_0 - e\int_{-\infty}^{\infty} f(\mathbf{r},\mathbf{u},t)\,d\mathbf{u} \tag{2.10}$$

With the help of Eqs. (2.1), (2.2) and (2.9), Eq. (2.10) can be simplified to yield

$$\rho(\mathbf{r},t) = -e\int_{-\infty}^{\infty} g(\mathbf{r},\mathbf{u},t)\,d\mathbf{u} = \rho(t)e^{ikx} \tag{2.11}$$

Since the ions are immobile, the perturbation current density is found from Eq. (2.2.2) as

$$\mathbf{J}(\mathbf{r},t) = -e\int_{-\infty}^{\infty} \mathbf{u}f(\mathbf{r},\mathbf{u},t)\,d\mathbf{u} \tag{2.12}$$

The substitution of Eq. (2.2) into Eq. (2.12) may be shown to yield

$$\mathbf{J}(\mathbf{r}, t) = -e \int_{-\infty}^{\infty} \mathbf{u} f_0(\mathbf{u})\, d\mathbf{u} - e \int_{-\infty}^{\infty} \mathbf{u} g(\mathbf{r}, \mathbf{u}, t)\, d\mathbf{u} = -e \int_{-\infty}^{\infty} \mathbf{u} g(\mathbf{r}, \mathbf{u}, t)\, d\mathbf{u} \qquad (2.13)$$

where the current density $-e \int_{-\infty}^{\infty} \mathbf{u} f_0(\mathbf{u})\, d\mathbf{u}$ in the equilibrium state has been assumed to be equal to zero. For $f_0(\mathbf{u})$ given by Eqs. (2.4), $\mathbf{u} f_0(\mathbf{u})$ is an odd function of $\mathbf{u}$ and therefore $-e \int_{-\infty}^{\infty} \mathbf{u} f_0(\mathbf{u})\, d\mathbf{u} = 0$. Thus the equilibrium distribution function as given by Eqs. (2.4) is consistent with the assumption that the current density in the equilibrium state is equal to zero. In view of Eq. (2.9), Eq. (2.13) shows that the spatial dependence of $\mathbf{J}(\mathbf{r}, t)$ is of the form:

$$\mathbf{J}(\mathbf{r}, t) = \mathbf{J}(t) e^{ikx} \qquad (2.14)$$

Since the macroscopic electric and magnetic fields are caused by the perturbation charge and current densities, it follows that $\mathbf{E}(\mathbf{r}, t)$ and $\mathbf{B}(\mathbf{r}, t)$ have the same spatial dependence as $\rho(\mathbf{r}, t)$ and $\mathbf{J}(\mathbf{r}, t)$. Therefore from Eqs. (2.11) and (2.14), $\mathbf{E}(\mathbf{r}, t)$ and $\mathbf{B}(\mathbf{r}, t)$ are found to be of the form

$$\mathbf{E}(\mathbf{r}, t) = \mathbf{E}(t) e^{ikx} \qquad \mathbf{B}(\mathbf{r}, t) = \mathbf{B}(t) e^{ikx} \qquad (2.15)$$

From Eqs. (2.9), (2.11), (2.14), and (2.15), it is clear that the initial disturbance is in the form of a plane wave with the phase fronts perpendicular to the x-axis. The problem then is to determine the characteristics of this plane wave for times subsequent to $t = 0$ by using the Boltzmann-Vlasov equation. In view of Eqs. (2.3) and (2.6), it follows from Eqs. (2.9), (2.11), and (2.13)–(2.15) that

$$\begin{aligned} g(\mathbf{u}, t) &= 0 \qquad \rho(t) = 0 \qquad \mathbf{J}(t) = 0 \\ \mathbf{E}(t) &= 0 \qquad \mathbf{B}(t) = 0 \qquad \text{for } t < 0 \end{aligned} \qquad (2.16)$$

The equilibrium distribution function $f_0(\mathbf{u})$ as given by Eqs. (2.4) is isotropic. Therefore it can be deduced that

$$\frac{\partial}{\partial u_i} f_0(\mathbf{u}) = \frac{d}{du} f_0(\mathbf{u}) \frac{\partial u}{\partial u_i} = \frac{u_i}{u} \frac{d}{du} f_0(\mathbf{u}) \qquad i = x, y, z \qquad (2.17)$$

and

$$\nabla_u f_0(\mathbf{u}) = (\mathbf{u}/u)(d/du) f_0(\mathbf{u}) \qquad (2.18)$$

It is obtained from Eq. (2.18) that

$$[\mathbf{u} \times \mathbf{B}(\mathbf{r}, t)] \cdot \nabla_u f_0(\mathbf{u}) = 0 \qquad (2.19)$$

The result of substituting Eqs. (2.9), (2.11), (2.14), (2.15), (2.18), and (2.19) into Eq. (2.8) is

$$\frac{\partial}{\partial t} g(\mathbf{u}, t) + iku_x g(\mathbf{u}, t) - \frac{e}{m_e} \mathbf{E}(t) \cdot \frac{\mathbf{u}}{u} f_0'(u) = 0 \qquad (2.20)$$

In Eq. (2.20), since it depends only on the particle speed u, $f_0(\mathbf{u})$ given by Eqs. (2.4) has been set equal to $f_0(u)$ and therefore $(d/du)f_0(u)$ has been denoted by $f_0'(u)$ where prime denotes differentiation with respect to the argument.

The differential equation (2.20) can be solved by the method of the Fourier transform in the time domain. For this purpose, the following two basic relations connecting a function of time, $f(t)$ and its Fourier transform $\bar{f}(\omega)$ are needed:

$$f(t) = \int_{-\infty}^{\infty} \bar{f}(\omega)e^{-i\omega t}\, d\omega \tag{2.21a}$$

$$\bar{f}(\omega) = \frac{1}{2\pi}\int_{-\infty}^{\infty} f(t)e^{i\omega t}\, dt \tag{2.21b}$$

It can be shown by an integration by parts that

$$\frac{1}{2\pi}\int_{-\infty}^{\infty} e^{i\omega t}\frac{\partial}{\partial t}g(\mathbf{u},t)\, dt = -\frac{1}{2\pi}g(\mathbf{u},0) - i\omega\bar{g}(\mathbf{u},\omega) \tag{2.22}$$

since $g(\mathbf{u},t) = 0$ for $t < 0$. If Eq. (2.20) is multiplied by $(1/2\pi)e^{i\omega t}$ and integrated with respect to t from $-\infty$ to ∞, it can be shown on using Eq. (2.22) that

$$-\frac{1}{2\pi}g(\mathbf{u},0) + i(ku_x - \omega)\bar{g}(\mathbf{u},\omega) - \frac{e}{m_e}\bar{\mathbf{E}}(\omega)\cdot\frac{\mathbf{u}}{u}f_0'(u) = 0 \tag{2.23}$$

The solution of Eq. (2.23) for $\bar{g}(\mathbf{u},\omega)$ gives

$$\bar{g}(\mathbf{u},\omega) = \frac{1}{i(\omega - ku_x)}\left[-\frac{1}{2\pi}g(\mathbf{u},0) - \frac{e}{m_e}\bar{\mathbf{E}}(\omega)\cdot\frac{\mathbf{u}}{u}f_0'(u)\right] \tag{2.24}$$

Current density

It can be deduced from Eqs. (2.9), (2.13), and (2.14) that

$$\mathbf{J}(t) = -e\int_{-\infty}^{\infty} \mathbf{u}g(\mathbf{u},t)\, d\mathbf{u} \tag{2.25}$$

If the Fourier transform of both sides of Eq. (2.25) is taken with the help of Eq. (2.21b), the result is

$$\bar{\mathbf{J}}(\omega) = -e\int_{-\infty}^{\infty} \mathbf{u}\bar{g}(\mathbf{u},\omega)\, d\mathbf{u} \tag{2.26}$$

From Eqs. (2.24) and (2.26), it is obtained that

$$\bar{\mathbf{J}}(\omega) = \int_{-\infty}^{\infty} d\mathbf{u}\,\mathbf{u}\frac{ie}{(\omega - ku_x)}\left[-\frac{1}{2\pi}g(\mathbf{u},0) - \frac{e}{m_e}\bar{\mathbf{E}}(\omega)\cdot\frac{\mathbf{u}}{u}f_0'(u)\right] \tag{2.27}$$

The integral in Eq. (2.27) is actually a triple integral with respect to the variables u_x, u_y, and u_z; the integration ranges for all the three variables are from $-\infty$ to ∞. The x-component of Eq. (2.27) is given by

$$\bar{J}_x(\omega) = \int_{-\infty}^{\infty} d\mathbf{u} \frac{ieu_x}{(\omega - ku_x)} \left[-\frac{1}{2\pi} g(\mathbf{u}, 0) - \frac{e}{m_e} \bar{\mathbf{E}}(\omega) \cdot \frac{\mathbf{u}}{u} f_0'(u) \right] \tag{2.28}$$

Note that

$$\int_{-\infty}^{\infty} d\mathbf{u} \frac{u_x}{(\omega - ku_x)} \frac{u_i}{u} f_0'(u) = 0 \qquad \text{for } i = y, z \tag{2.29}$$

since the integrand is an odd function of u_i. Together with Eqs. (2.17) and (2.29), Eq. (2.28) can be simplified to yield

$$\bar{J}_x(\omega) = -\int_{-\infty}^{\infty} d\mathbf{u} \frac{ieu_x}{2\pi(\omega - ku_x)} g(\mathbf{u}, 0) - \frac{ie^2}{m_e} \bar{E}_x(\omega) \int_{-\infty}^{\infty} d\mathbf{u} \frac{u_x[\partial f_0(u)/\partial u_x]}{(\omega - ku_x)} \tag{2.30}$$

In a similar manner, the y- and the z-components of Eq. (2.27) can be deduced to be given by

$$\bar{J}_y(\omega) = -\int_{-\infty}^{\infty} d\mathbf{u} \frac{ieu_y}{2\pi(\omega - ku_x)} g(\mathbf{u}, 0) - \frac{ie^2}{m_e} \bar{E}_y(\omega) \int_{-\infty}^{\infty} d\mathbf{u} \frac{u_y[\partial f_0(u)/\partial u_y]}{(\omega - ku_x)} \tag{2.31}$$

$$\bar{J}_z(\omega) = -\int_{-\infty}^{\infty} d\mathbf{u} \frac{ieu_z}{2\pi(\omega - ku_x)} g(\mathbf{u}, 0) - \frac{ie^2}{m_e} \bar{E}_z(\omega) \int_{-\infty}^{\infty} d\mathbf{u} \frac{u_z[\partial f_0(u)/\partial u_z]}{(\omega - ku_x)} \tag{2.32}$$

It is important to note from Eqs. (2.30)- (2.32) that $\bar{J}_x(\omega)$, $\bar{J}_y(\omega)$ and $\bar{J}_z(\omega)$ depend only on $\bar{E}_x(\omega)$, $\bar{E}_y(\omega)$, and $\bar{E}_z(\omega)$, respectively. This feature is a manifestation of the isotropy of the medium insofar as the macroscopic electromagnetic properties are concerned and is to be expected in a stationary plasma in the absence of external magnetostatic field.

Separation into the various modes

The macroscopic electric and magnetic fields satisfy the following Maxwell's equations:

$$\nabla \times \mathbf{E}(\mathbf{r}, t) = -\mu_0 \frac{\partial}{\partial t} \mathbf{H}(\mathbf{r}, t) \tag{2.33}$$

$$\nabla \times \mathbf{H}(\mathbf{r}, t) = \varepsilon_0 \frac{\partial}{\partial t} \mathbf{E}(\mathbf{r}, t) + \mathbf{J}(\mathbf{r}, t) \tag{2.34}$$

Since the electric current density and the electromagnetic fields depend only on x as given by Eqs. (2.14) and (2.15), it can be shown that $\nabla = ik\hat{\mathbf{x}}$. Hence Eqs. (2.33) and (2.34) simplify as

$$ik\hat{\mathbf{x}} \times \mathbf{E}(t) = -\mu_0 \frac{\partial}{\partial t} \mathbf{H}(t) \tag{2.35}$$

$$ik\hat{\mathbf{x}} \times \mathbf{H}(t) = \varepsilon_0 \frac{\partial}{\partial t} \mathbf{E}(t) + \mathbf{J}(t) \tag{2.36}$$

The Fourier transforms of Eqs. (2.35) and (2.36) can be deduced with the help of Eqs. (2.21b) and (2.22). The results are

$$k\hat{\mathbf{x}} \times \bar{\mathbf{E}}(\omega) = \omega\mu_0 \bar{\mathbf{H}}(\omega) - (i\mu_0/2\pi)\mathbf{H}(0) \tag{2.37}$$

$$k\hat{\mathbf{x}} \times \bar{\mathbf{H}}(\omega) = -\omega\varepsilon_0 \bar{\mathbf{E}}(\omega) - i\bar{\mathbf{J}}(\omega) + (i\varepsilon_0/2\pi)\mathbf{E}(0) \tag{2.38}$$

where $\mathbf{E}(0)$ and $\mathbf{H}(0)$ are the values of $\mathbf{E}(t)$ and $\mathbf{H}(t)$ for $t = 0$. Note that $\mathbf{E}(0)$ and $\mathbf{H}(0)$ correspond to the initial value of the perturbation of the velocity distribution function.

In component form Eqs. (2.37) and (2.38) become

$$0 = \omega\mu_0 \bar{H}_x(\omega) - (i\mu_0/2\pi)H_x(0) \tag{2.39a}$$

$$-k\bar{E}_z(\omega) = \omega\mu_0 \bar{H}_y(\omega) - (i\mu_0/2\pi)H_y(0) \tag{2.39b}$$

$$k\bar{E}_y(\omega) = \omega\mu_0 \bar{H}_z(\omega) - (i\mu_0/2\pi)H_z(0) \tag{2.39c}$$

$$0 = -\omega\varepsilon_0 \bar{E}_x(\omega) - i\bar{J}_x(\omega) + (i\varepsilon_0/2\pi)E_x(0) \tag{2.40a}$$

$$-k\bar{H}_z(\omega) = -\omega\varepsilon_0 \bar{E}_y(\omega) - i\bar{J}_y(\omega) + (i\varepsilon_0/2\pi)E_y(0) \tag{2.40b}$$

$$k\bar{H}_y(\omega) = -\omega\varepsilon_0 \bar{E}_z(\omega) - i\bar{J}_z(\omega) + (i\varepsilon_0/2\pi)E_z(0) \tag{2.40c}$$

An examination of Eqs. (2.30)–(2.32), (2.39) and (2.40) shows that the electromagnetic fields can be separated into the following four independent groups:

1. $\bar{E}_x(\omega)$, $\bar{J}_x(\omega)$
2. $\bar{H}_x(\omega)$
3. $\bar{E}_y(\omega)$, $\bar{H}_z(\omega)$, $\bar{J}_y(\omega)$
4. $\bar{E}_z(\omega)$, $\bar{H}_y(\omega)$, $\bar{J}_z(\omega)$

The first group contains an electric field and a current density in the direction of the wave normal of the initial plane wave disturbance created in the plasma. Like the current density, the average particle velocity is also in the wave-normal direction. Also, this group contains no magnetic field. Therefore the first group gives the longitudinal plane plasma wave whose characteristics are investigated in this section. The third and the fourth groups contain electric and magnetic fields which are perpendicular to the wave normal of the initial plane wave disturbance set up in the plasma. The electric field, the magnetic field and the wave normal form a mutually perpendicular triad. The electric current densities and the average particle velocities are also perpendicular to the wave-normal direction. Thus the third and the fourth groups give the two different polarizations of the transverse plane electromagnetic wave whose characteristics are discussed in the following section. The second group contains only the magnetic field in the wave-normal direction. The second group has no current density associated with it and hence is not influenced by the collective motion of the electrons. Therefore the second group does not form a natural mode of oscillation of the plasma caused by the cumulative contribution arising from the motion of the electrons and hence can be discarded.

Determination of the contour of integration in the ω-plane

The substitution of $\bar{J}_x(\omega)$ from Eq. (2.40a) into Eq. (2.30) and the solution of the resulting equation for $\bar{E}_x(\omega)$ may be shown to give the following result:

$$\bar{E}_x(\omega) = G_x(\omega)/D_l(k_p, \omega) \tag{2.41}$$

where

$$G_x(\omega) = \frac{i}{2\pi\omega} E_x(0) - \frac{1}{\omega\varepsilon_0} \int_{-\infty}^{\infty} d\mathbf{u} \frac{eu_x}{2\pi(\omega - k_p u_x)} g(\mathbf{u}, 0) \tag{2.42}$$

and

$$D_l(k_p, \omega) = 1 + \frac{\omega_{pe}^2}{N_0 \omega} \int_{-\infty}^{\infty} d\mathbf{u} \frac{u_x(\partial f_0(u)/\partial u_x)}{(\omega - k_p u_x)} \tag{2.43}$$

The subscript p is added to k to indicate that it is the wavenumber corresponding to the longitudinal plasma wave. From Eqs. (2.21a) and (2.15), the longitudinal electric field in the time domain is obtained as

$$E_x(\mathbf{r}, t) = e^{ik_p x} \int_{-\infty}^{\infty} \bar{E}_x(\omega) e^{-i\omega t}\, d\omega \tag{2.44}$$

where the contour is essentially along the real axis in the ω-plane.

In order that $E_x(\mathbf{r}, t)$ may have a unique value, it is necessary to specify the contour of integration in Eq. (2.44) unambiguously. Let $\omega = \omega_r + i\omega_i$ where ω_r and ω_i are both real; therefore ω_r and ω_i are the real and the imaginary parts of complex ω. For $t < 0$ and $\omega_i > 0$, the term $(-i\omega t)$ in Eq. (2.44) has a negative real part which increases as ω_i increases. Therefore, for $t < 0$, there is no contribution to the integral in Eq. (2.44) for an integration along a semicircular path in the upper half of the complex ω-plane in the limiting case where the radius of the semicircle is infinite. This limiting case of the semicircular path is called for convenience the infinite semicircular path. The contour which is essentially a straight line along the real axis in the complex ω-plane for the integration in Eq. (2.44), can be transformed into a closed contour by joining the terminals of the real axis with the infinite semicircular path in the upper half of the complex ω-plane without altering the value of $E_x(\mathbf{r}, t)$ for $t < 0$. The perturbation of the velocity distribution function which causes $E_x(\mathbf{r}, t)$ is initiated at $t = 0$. The principle of causality demands that $E_x(\mathbf{r}, t) = 0$ for $t < 0$; there should be no fields before the starting of the source. According to a well-known theorem in complex variables, the value of an integral with a closed contour of integration is identically zero if there are no singularities of the integrand within the closed path. In order that $E_x(\mathbf{r}, t)$ given by Eq. (2.44) may conform to the causality requirement, there should be no singularities of the integrand within the closed contour formed by the real axis and the infinite semicircular path in the upper half of the complex ω-plane.

Suppose that the integrand in Eq. (2.44) has singularities on the real axis as well as in both halves of the complex ω-plane. Then the contour of integration along the real axis should be indented as shown in Fig. 6.1 so that all the singularities are

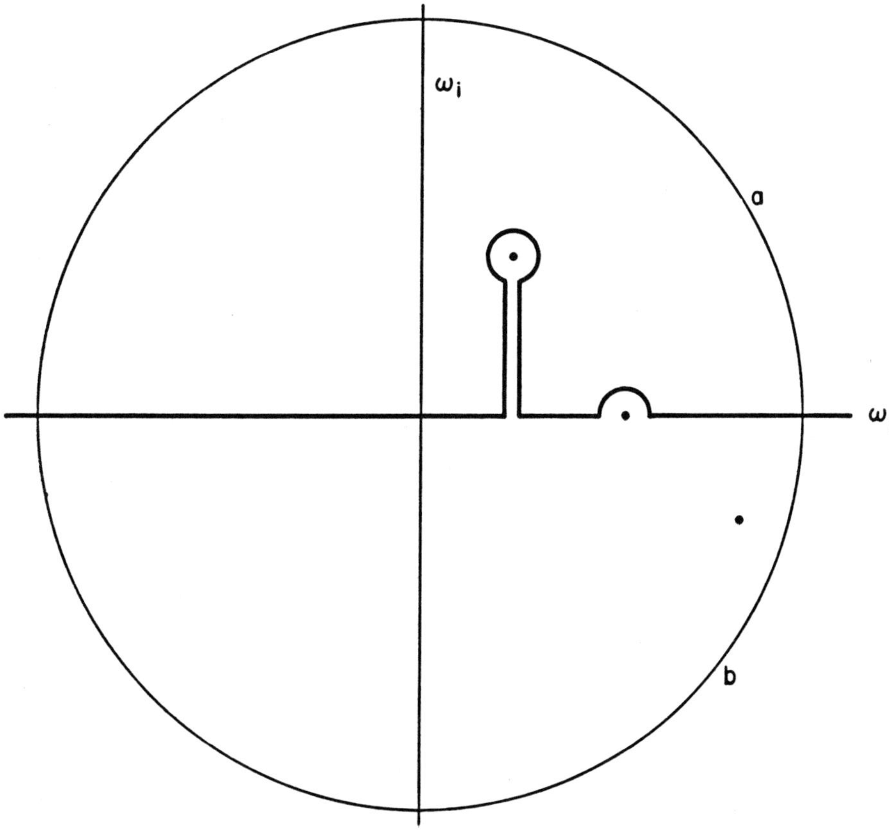

Fig. 6.1. Contour of integration in the complex ω-plane for the integral in Eq. (2.44): (a) infinite semicircular path for time $t < 0$; (b) infinite semicircular path for time $t > 0$.

below the indented contour along the real axis. It is in anticipation of these required indentations that the contour of integration of Eq. (2.44) was referred to previously as lying *essentially* along the real axis. If the contour along the real axis is indented as indicated, there are no singularities within the closed contour formed by the indented path along the real axis and the infinite semicircular path in the upper half of the complex ω-plane with the result that $E_x(\mathbf{r}, t) = 0$ for $t < 0$ in accordance with the principle of causality. Thus, without any ambiguity, the contour of integration in Eq. (2.44) is along the real axis with indentations such that all the singularities are below the path of integration. This prescription enables the removal of any ambiguity in the integration path near the singularities.

For example, for the integration with respect to u_x in Eq. (2.43), the path of integration is along the real axis in the u_x-plane and there is a singularity on the real axis at $u_x = \omega/k_p$; therefore, it is necessary to determine the path of integration in the neighborhood of the singularity. In the ω-plane, the integration path is indented from above in the neighborhood of the singularity, that is, real ω is altered to become complex with a sufficiently large imaginary part so that all the singularities lie below the integration path. Therefore real ω should be treated as the limiting case of a complex ω with a positive imaginary part as the imaginary part goes to zero. Consequently, the singularity in the u_x-plane is in the upper half of the complex u_x-plane whereas the integration path is along the real axis. In the limiting case of real ω, the path of integration in the u_x-plane is along the real axis but indented from below the singularity at $u_x = \omega/k_p$ as shown in Fig. 6.2.

Returning to the integral in Eq. (2.44), it can be shown with the help of the arguments used before that for $t > 0$, the indented contour along the real axis can

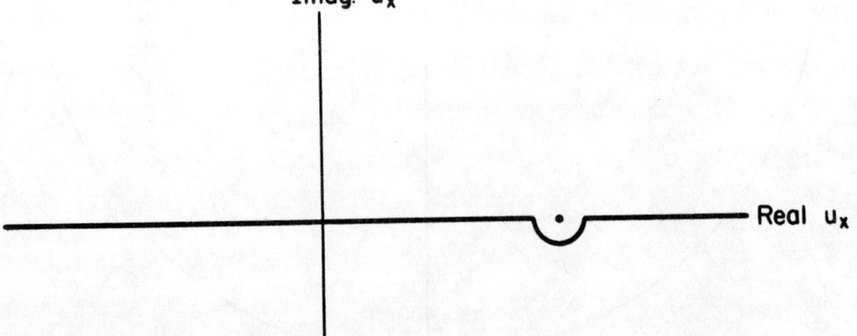

Fig. 6.2. Contour of integration in the complex u_x-plane for the integral in Eq. (2.43) showing the integration path in the neighborhood of the pole $u_x = \omega/k_p$.

be closed by an infinite semicircular path in the lower half of the complex ω-plane without changing the value of $E_x(\mathbf{r}, t)$. The closed contour for the evaluation of $E_x(\mathbf{r}, t)$ for $t > 0$ contains all the singularities of the integrand in Eq. (2.44). The value of an integral with a closed contour of integration is equal to the sum of the contributions arising from the neighborhood of all the singularities enclosed by the path. Thus the nature of the singularities of the integrand in Eq. (2.44) determines the behavior of $E_x(\mathbf{r}, t)$ for times subsequent to $t = 0$.

Contribution from a simple pole

The integrand in Eq. (2.44) can have, in general, several types of singularities. For the purposes of the present treatment, it is sufficient to consider only the simplest possible singularity. Let $\bar{E}_x(\omega)$ be written as $\bar{E}_x(\omega) = N(\omega)/D(\omega)$ where neither $N(\omega)$ nor $D(\omega)$ becomes infinite for any value of ω. Suppose that $D(\omega)$ has a simple zero at $\omega = \omega_s$ where ω_s is either real or complex. For $\omega = \omega_s$, the integrand in Eq. (2.44) becomes infinite and this type of singularity is known as a *simple pole*. In other words, the integrand in Eq. (2.44) is said to have a simple pole at $\omega = \omega_s$. Since it is necessary for the present analysis, the contribution to the integral arising from the neighborhood of a simple pole is deduced, as follows.

For definiteness, consider a pole P in the upper half of the complex ω-plane as shown in Fig. 6.3. The contour of integration in the neighborhood of the pole P is assumed to be a circle BCD of radius δ. At a suitable stage in the analysis, the radius δ is allowed to proceed to the limiting value of zero. It is to be noted that the net contribution to the integral arising from integration along the paths AB and DE connecting the circle BCD to the real axis is zero. For carrying out the integration along the circular path BCD, only the value of the integrand in the close neighborhood of $\omega = \omega_s$ is needed. Therefore $D(\omega)$ can be expanded into a Taylor series and only the leading term need be retained. The result is

$$D(\omega) = (\omega - \omega_s)D'(\omega_s) \qquad D'(\omega_s) = [\partial D(\omega)/\partial\omega]_{\omega=\omega_s} \tag{2.45}$$

For the sake of convenience, let

$$\tilde{N}(\omega) = N(\omega)e^{-i\omega t} \qquad \omega - \omega_s = \delta e^{i\theta} \tag{2.46}$$

For the integration path BCD, δ is a constant and θ varies from $3\pi/2$ to $-\pi/2$. With the help of Eqs. (2.45) and (2.46), the contribution to $E_x(\mathbf{r}, t)$ given by Eq. (2.44) arising from the integration along the path BCD may be expressed as

$$E_x(\mathbf{r}, t) = e^{ik_p x} \int_{3\pi/2}^{-\pi/2} \frac{\tilde{N}(\omega_s + \delta e^{i\theta})}{\delta e^{i\theta} D'(\omega_s)} \delta e^{i\theta} i\, d\theta \tag{2.47}$$

On proceeding to the limiting value of $\delta = 0$, the integral in Eq. (2.47) can be evaluated with the following result:

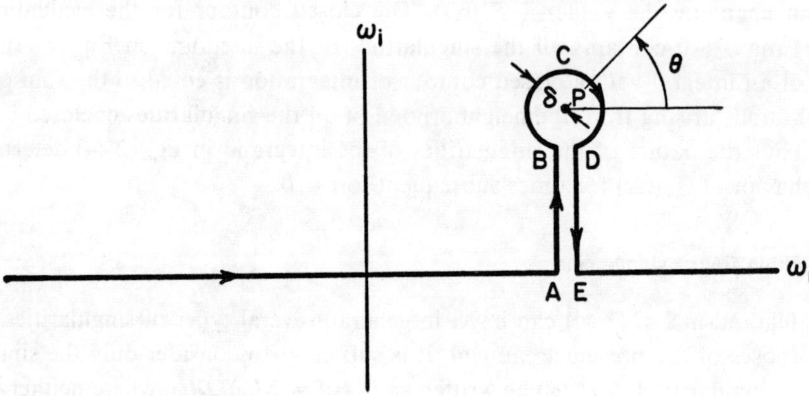

Fig. 6.3. Contour of integration in the complex ω-plane for the evaluation of the contribution arising from the neighborhood of a complex pole $\omega = \omega_s$.

$$E_x(\mathbf{r}, t) = e^{ik_p x}\left[-2\pi i \frac{\tilde{N}(\omega_s)}{D'(\omega_s)}\right] \qquad \omega_s \text{ complex} \tag{2.48}$$

Note that $\tilde{N}(\omega_s)/D'(\omega_s)$ is called the residue of the integrand in Eq. (2.44) at the pole $\omega = \omega_s$. Thus the value of the integral in Eq. (2.44) is equal to $-2\pi i$ times the residue. Instead of being in the negative (clockwise) direction, if the integration path around the pole is in the positive (counterclockwise) direction, the value of the integral can be shown to become $+2\pi i$ times the residue.

It can be verified that the result given by Eq. (2.48) is valid even for a pole in the lower half of the complex ω-plane. For a pole *on* the real axis, the path of integration in the neighborhood of the pole is a semicircle with the result that it can be shown by the procedure used before that the value of the integral in Eq. (2.44) is equal to $-\pi i$ times the residue. If $\tilde{N}(\omega_s)$ is expressed in terms of $N(\omega_s)$, Eq. (2.48) and the corresponding result for ω_s real become

$$E_x(\mathbf{r}, t) = e^{ik_p x}\left[-a\pi i \frac{N(\omega_s)}{D'(\omega_s)} e^{-i\omega_s t}\right] \tag{2.49}$$

where $a = 2$ or 1 according as ω_s is complex or real.

If ω_s is complex with a positive imaginary part, Eq. (2.49) shows that $E_x(\mathbf{r}, t)$ which is equal to zero for $t < 0$ grows with time for times subsequent to $t = 0$ showing that the equilibrium state is unstable. On the other hand if ω_s is complex with a negative imaginary part, $E_x(\mathbf{r}, t)$ is seen from Eq. (2.49) to decay with time for $t > 0$ showing that the original equilibrium state is stable. If ω_s is real, $E_x(\mathbf{r}, t)$ becomes an oscillatory function of time with a constant amplitude. The time average of $E_x(\mathbf{r}, t)$ is equal to zero and hence reduces to the equilibrium value. For this reason, the equilibrium state corresponding to ω_s real is also considered to be stable. The frequency of oscillation and the growth or the decay rates are completely specified by the values of ω_s. It was stated previously that the objective is to determine the temporal behavior of the plane wave set up in the plasma at time $t = 0$. This behavior has been shown now to be completely determined by the values of ω_s. Thus the determination of the singularities of the integrand in Eq. (2.44) enables the time behavior of the plane wave to be obtained completely.

Dispersion equation for the longitudinal plasma wave

The singularities of the integrand in Eq. (2.44) is seen from Eq. (2.41) to be contributed by the infinities of $G_x(\omega)$ and the zeros of $D_l(k_p, \omega)$. The infinities, if any, of $G_x(\omega)$ depend on the details of the initial perturbation and are not characteristic of the plasma. Therefore, the infinities of $G_x(\omega)$ give no information on the natural behavior of the plasma and are not given any further consideration. On the other hand, the zeros of $D_l(k_p, \omega)$ depend on the intrinsic behavior of the plasma and therefore determine the natural modes of the system. Thus the solutions of

$$D_l(k_p, \omega) = 0 \tag{2.50}$$

determine the time behavior of the longitudinal plasma wave as contributed by the intrinsic properties of the plasma. Note that Eq. (2.50) is known as the *dispersion equation.* The subscript l denotes that the dispersion equation pertains to the *l*ongitudinal plasma wave.

Before proceeding with the analysis of the dispersion relation (2.50) for the equilibrium state characterized by the Maxwell-Boltzmann distribution function, it is useful to examine the results for the limiting case of a cold plasma. In the equilibrium state, since the electrons are also at rest, the velocity distribution function is given by

$$f_{0C}(u) = N_0 \delta(u_x)\delta(u_y)\delta(u_z) \tag{2.51}$$

where $\delta(x)$ is Dirac's delta function and is defined as follows:

$$\delta(x) = 0 \text{ for } x \neq 0 \text{ and } \int_{-\infty}^{\infty} \delta(x)\, dx = 1 \tag{2.52}$$

With the help of Eqs. (2.52), it is seen that $f_{0C}(u)$ given by Eq. (2.51) is consistent with Eq. (2.1). Before substituting Eq. (2.51) into Eq. (2.43), it is convenient to simplify the integral in Eq. (2.43) as follows:

$$\begin{aligned}\int_{-\infty}^{\infty} d\mathbf{u} \frac{u_x(\partial f_0(u)/\partial u_x)}{(\omega - k_p u_x)} &= -\frac{1}{k_p}\int_{-\infty}^{\infty} d\mathbf{u} \frac{\partial f_0(u)}{\partial u_x}\left[1 - \frac{\omega}{\omega - k_p u_x}\right] \\ &= \frac{\omega}{k_p}\int_{-\infty}^{\infty} d\mathbf{u} \frac{(\partial f_0(u)/\partial u_x)}{(\omega - k_p u_x)}\end{aligned} \tag{2.53}$$

since

$$\int_{-\infty}^{\infty} d\mathbf{u}\frac{\partial f_0(u)}{\partial u_x} = \int_{-\infty}^{\infty} du_y \int_{-\infty}^{\infty} du_z [f_0(u)]\Big|_{u_x=-\infty}^{u_x=\infty} = 0$$

In Eq. (2.53) the integral with respect to u_x is further simplified by an integration by parts. The result is

$$\begin{aligned}\frac{\omega}{k_p}\int_{-\infty}^{\infty} du_y \int_{-\infty}^{\infty} du_z \left[\frac{f_0(u)}{\omega - k_p u_x}\Big|_{u_x=-\infty}^{u_x=\infty} - k_p \int_{-\infty}^{\infty} \frac{f_0(u)du_x}{(\omega - k_p u_x)^2}\right] \\ = -\omega \int_{-\infty}^{\infty} d\mathbf{u}\frac{f_0(u)}{(\omega - k_p u_x)^2}\end{aligned} \tag{2.54}$$

Together with Eqs. (2.43), (2.53), and (2.54), the dispersion equation (2.50) becomes

$$1 = \frac{\omega_{pe}^2}{N_0}\int_{-\infty}^{\infty} d\mathbf{u}\frac{f_0(u)}{(\omega - k_p u_x)^2} \tag{2.55}$$

In view of the following property of Dirac's delta function

$$\int_{-\infty}^{\infty} g(x')\delta(x' - x)\,dx' = g(x) \tag{2.56}$$

the substitution of Eq. (2.51) into Eq. (2.55) yields

$$\omega^2 = \omega_{pe}^2 \tag{2.57}$$

Since the dispersion relation (2.57) is independent of the propagation coefficient k_p, it follows that the oscillations at $\omega = \omega_{pe}$ are stationary in character. If it is disturbed, a cold plasma breaks into undamped, stationary oscillations at the electron plasma frequency. These are the electron plasma oscillations which have been previously treated on the basis of the fluid dynamical description of the plasma. It was pointed out that the particle velocity and the electric field associated with the plasma oscillations are in the same direction. There is no magnetic field associated with these oscillations. Thus the electron plasma oscillations are longitudinal, electrostatic, and stationary and these results are in accordance with those deduced previously in Sec. 2.2.

Long-wavelength limit

In the equilibrium state, let the electrons be characterized by the Maxwell-Boltzmann distribution function (2.4), as has been assumed throughout except for the derivation of Eq. (2.57). If Eqs. (2.4) are substituted into Eq. (2.55) and the integrations with respect to u_y and u_z are carried out with the help of Eqs. (1.4.7), the result is

$$1 = \frac{\omega_{pe}^2}{\omega^2}\left(\frac{m_e}{2\pi KT_e}\right)^{1/2} \int_{-\infty}^{\infty} du_x \left(1 - \frac{k_p u_x}{\omega}\right)^{-2} e^{-m_e u_x^2/2KT_e} \tag{2.58}$$

which shows that the velocity distribution in the direction of the propagation coefficient of the initial plane wave disturbance is alone relevant. It is therefore clear that the problem of the electron plasma oscillations is essentially one-dimensional in nature and identical results would have been obtained by assuming a one-dimensional velocity distribution function at the outset.

By evaluating the integral in Eq. (2.58) for the limiting case in which the phase velocity ω/k_p of the wave is very large compared to the velocity of almost all of the particles, the results pertaining to the warm plasma model obtained in Sec. 2.7 can be reproduced. In this high phase velocity or long-wavelength limit, it is reasonable to assume that $k_p u_x/\omega \ll 1$ and hence $(1 - k_p u_x/\omega)^{-2}$ can be expanded into a binomial series. If only the first three terms of the series are retained, Eq. (2.58) becomes

$$1 = \frac{\omega_{pe}^2}{\omega^2}\left(\frac{m_e}{2\pi KT_e}\right)^{1/2} \int_{-\infty}^{\infty} du_x \left(1 + 2\frac{k_p u_x}{\omega} + 3\frac{k_p^2 u_x^2}{\omega^2}\right) e^{-m_e u_x^2/2KT_e} \tag{2.59}$$

The second term within the brackets in Eq. (2.59) gives an integrand which is an odd function of u_x and hence the corresponding integral vanishes. The remaining integral can be evaluated with the help of Eqs. (1.4.7) to yield

$$\omega^2 = \omega_{pe}^2\left(1 + 3\frac{KT_e}{m_e}\frac{k_p^2}{\omega^2}\right) \tag{2.60}$$

In the long-wavelength limit, the second term is small compared to the first term and hence can be omitted in the first approximation reproducing Eq. (2.57). In the small second term, ω can be replaced by ω_{pe} in accordance with the first approximation and the following second approximation is obtained:

$$\omega^2 = \omega_{pe}^2\left(1 + 3\frac{KT_e}{m_e}\frac{k_p^2}{\omega_{pe}^2}\right) \tag{2.61}$$

which is known as the *Bohm-Gross dispersion relation.* It was already demonstrated that the first approximation is related to the cold plasma model. A comparison of Eq. (2.7.20) with Eq. (2.61) shows that the second approximation given in Eq. (2.61) is related to the warm plasma model.

Analysis of the dispersion equation

By retaining additional terms in the binomial series expansion of $(1 - k_p u_x/\omega)^{-2}$, additional approximations to the value of ω can be obtained. In all these approximations, ω continues to remain real and therefore the plasma oscillations have a constant amplitude in time, that is, the oscillations neither grow nor decay with time. It is usual to terminate the approximations to ω at the stage given by Eq. (2.61) which with the help of Eq. (2.3.10) can be rewritten as

$$\omega^2 = \omega_{pe}^2(1 + 3\lambda_{De}^2 k_p^2) \tag{2.62}$$

where λ_{De} is the Debye length for the electrons. It can be deduced by a careful analysis of the dispersion equation for $f_0(u)$ given by Eqs. (2.4) that ω has a negative imaginary part leading to the damping of the oscillations with time. It is now proposed to deduce this damping constant. If $f_0(u)$ as given by Eqs. (2.4) is substituted into Eq. (2.53), the integrals with respect to u_y and u_z are evaluated with the help of Eqs. (1.4.7) and the differentiation with respect to u_x is carried out, it can be shown that

$$\int_{-\infty}^{\infty} d\mathbf{u} \frac{u_x(\partial f_0(u)/\partial u_x)}{(\omega - k_p u_x)}$$
$$= \frac{\omega}{k_p^2}\frac{N_0}{\sqrt{2\pi}}\left(\frac{m_e}{KT_e}\right)^{3/2} \int_{-\infty}^{\infty} du_x \left[1 + \frac{v_{ph}}{u_x - v_{ph}}\right] e^{-m_e u_x^2/2KT_e} \tag{2.63}$$

where $v_{ph} = \omega/k_p$ is the phase velocity of the wave. The integral corresponding to the first term within the square brackets in Eq. (2.63) can be evaluated by using Eqs. (1.4.7).

It is convenient to introduce the following dimensionless parameter:

$$\tilde{v}_{ph} = v_{ph}(m_e/2KT_e)^{1/2} = (\omega/k_p)(m_e/2KT_e)^{1/2} \tag{2.64}$$

Also the dummy variable of integration is changed as

$$\tilde{u}_x = u_x(m_e/2KT_e)^{1/2}$$

If the result of these operations on Eq. (2.63) is substituted into Eq. (2.43), the dispersion equation (2.50) may be shown to become

$$-\frac{k_p^2}{\omega_{pe}^2}\frac{KT_e}{m_e} = 1 + \tilde{v}_{ph} I(\tilde{v}_{ph}) \tag{2.65}$$

where

$$I(\tilde{v}_{ph}) = \pi^{-1/2} \int_{-\infty}^{\infty} \frac{d\tilde{u}_x}{(\tilde{u}_x - \tilde{v}_{ph})} e^{-\tilde{u}_x^2} \tag{2.66}$$

As shown in Fig. 6.2, the contour of integration for the integral in Eq. (2.66) is along the real axis and indented from below the pole at $\tilde{u}_x = \tilde{v}_{ph}$. The contribution to the integral due to the pole is equal to πi times the residue of the integrand at the pole. The remaining contribution to the integral is the principal value of the integral which is identified by prefixing P to the integral sign and is defined as follows:

$$P\int_{-\infty}^{\infty}\frac{d\tilde{u}_x}{(\tilde{u}_x-\tilde{v}_{ph})}e^{-\tilde{u}_x^2}=\lim_{\delta\to 0}\left(\int_{\tilde{v}_{ph}+\delta}^{\infty}+\int_{-\infty}^{\tilde{v}_{ph}-\delta}\right)\frac{d\tilde{u}_x}{(\tilde{u}_x-\tilde{v}_{ph})}e^{-\tilde{u}_x^2} \tag{2.67}$$

Therefore the dispersion equation (2.65) can be written as

$$-\frac{k_p^2}{\omega_{pe}^2}\frac{KT_e}{m_e}=1+i\sqrt{\pi}\,\tilde{v}_{ph}e^{-\tilde{v}_{ph}^2}+\tilde{v}_{ph}G(\tilde{v}_{ph},1) \tag{2.68}$$

where

$$G(\tilde{v}_{ph},s)=\pi^{-1/2}P\int_{-\infty}^{\infty}\frac{d\tilde{u}_x}{(\tilde{u}_x-\tilde{v}_{ph})}e^{-s\tilde{u}_x^2} \tag{2.69}$$

Dispersion function

The integral in Eq. (2.69) cannot be evaluated explicitly but can be recast into a more convenient form for numerical evaluation. On multiplying the numerator and the denominator of the integrand by $(\tilde{u}_x+\tilde{v}_{ph})$, it is noted that the integrand corresponding to $\tilde{u}_x$ in the numerator is an odd function of $\tilde{u}_x$ and hence its integral vanishes. Therefore Eq. (2.69) becomes

$$G(\tilde{v}_{ph},s)=\tilde{v}_{ph}\pi^{-1/2}P\int_{-\infty}^{\infty}\frac{d\tilde{u}_x}{(\tilde{u}_x^2-\tilde{v}_{ph}^2)}e^{-s\tilde{u}_x^2} \tag{2.70}$$

From Eq. (2.70), it can be deduced that

$$\frac{d}{ds}G(\tilde{v}_{ph},s)+\tilde{v}_{ph}^2G(\tilde{v}_{ph},s)=-\tilde{v}_{ph}\pi^{-1/2}\int_{-\infty}^{\infty}d\tilde{u}_x\,e^{-s\tilde{u}_x^2}=-\tilde{v}_{ph}s^{-1/2} \tag{2.71}$$

where the integral in Eq. (2.71) has been evaluated with the help of Eqs. (1.4.7). It is possible to rewrite Eq. (2.71) as

$$\frac{d}{ds}[G(\tilde{v}_{ph},s)e^{s\tilde{v}_{ph}^2}]=-\tilde{v}_{ph}s^{-1/2}e^{s\tilde{v}_{ph}^2} \tag{2.72}$$

In order to be able to integrate Eq. (2.72), it is necessary to obtain the value of the integral

$$\begin{aligned}G(\tilde{v}_{ph},0)&=\pi^{-1/2}P\int_{-\infty}^{\infty}\frac{d\tilde{u}_x}{(\tilde{u}_x-\tilde{v}_{ph})}\\&=\pi^{-1/2}\lim_{\delta\to 0}\left(\int_{\tilde{v}_{ph}+\delta}^{\infty}+\int_{-\infty}^{\tilde{v}_{ph}-\delta}\right)\frac{d\tilde{u}_x}{(\tilde{u}_x-\tilde{v}_{ph})}\end{aligned} \tag{2.73}$$

For the second integral in Eq. (2.73), the integration variable is changed from $\tilde{u}_x$ to $-\tilde{u}_x$; therefore, Eq. (2.73) becomes

$$
\begin{aligned}
G(\tilde{v}_{ph}, 0) &= \pi^{-1/2} \lim_{\delta \to 0} \left[\int_{\tilde{v}_{ph}+\delta}^{\infty} \frac{d\tilde{u}_x}{(\tilde{u}_x - \tilde{v}_{ph})} - \int_{-\tilde{v}_{ph}+\delta}^{\infty} \frac{d\tilde{u}_x}{(\tilde{u}_x + \tilde{v}_{ph})} \right] \\
&= \pi^{-1/2} \lim_{\delta \to 0} [\ln(\tilde{u}_x - \tilde{v}_{ph})|_{\tilde{v}_{ph}+\delta}^{\infty} - \ln(\tilde{u}_x + \tilde{v}_{ph})|_{-\tilde{v}_{ph}+\delta}^{\infty}] \\
&= \pi^{-1/2} \ln\left(\frac{\tilde{u}_x - \tilde{v}_{ph}}{\tilde{u}_x + \tilde{v}_{ph}} \right)_{\tilde{u}_x = \infty} = 0
\end{aligned}
\tag{2.74}
$$

Both sides of Eq. (2.72) are integrated from $s = 0$ to $s = 1$ and Eq. (2.74) is used to obtain

$$
G(\tilde{v}_{ph}, 1) = -\tilde{v}_{ph} e^{-\tilde{v}_{ph}^2} \int_0^1 s^{-1/2} e^{s\tilde{v}_{ph}^2} \, ds \tag{2.75}
$$

If the integration variable is changed to $w = \sqrt{s}\, \tilde{v}_{ph}$, Eq. (2.75) can be simplified as

$$
G(\tilde{v}_{ph}, 1) = -2 \int_0^{\tilde{v}_{ph}} e^{w^2 - \tilde{v}_{ph}^2} \, dw \tag{2.76}
$$

Together with Eq. (2.76), the dispersion equation (2.68) can be expressed as

$$
-\frac{k_p^2}{\omega_{pe}^2} \frac{KT_e}{m_e} = 1 - 2\tilde{v}_{ph} \int_0^{\tilde{v}_{ph}} e^{w^2 - \tilde{v}_{ph}^2} \, dw + i\sqrt{\pi}\, \tilde{v}_{ph} e^{-\tilde{v}_{ph}^2} \tag{2.77}
$$

The integral appearing in Eq. (2.77) is known as *the dispersion function* and its values are extensively tabulated [1]. The functional relationship between ω and k_p can be evaluated with the help of Eq. (2.77) which is therefore the dispersion equation for the longitudinal plasma wave. A value of $\tilde{v}_{ph}$ is chosen and the corresponding value of the dispersion function is obtained from the tables. Then Eq. (2.77) can be used to evaluate the value of k_p. A knowledge of $\tilde{v}_{ph}$ and k_p enables ω to be determined from Eq. (2.64). Thus the dispersion relation, that is, the relationship between ω and k_p can be determined.

Landau damping

It can be shown that Eq. (2.77) predicts the damping of the longitudinal plasma waves with time. For this purpose an approximate evaluation of the dispersion function is necessary. Such an approximation can be obtained in a straightforward manner for the long-wavelength limit. Since the real part of ω has already been found for the long-wavelength limit, the evaluation of the dispersion relation for the long-wavelength limit from Eq. (2.77) provides, on the one hand, a partial check on the accuracy of Eq. (2.77) and, on the other, enables an explicit expression to be obtained for the imaginary part of ω.

Consider the dispersion function

$$
I = 2\tilde{v}_{ph} \int_0^{\tilde{v}_{ph}} e^{w^2 - \tilde{v}_{ph}^2} \, dw \tag{2.78}
$$

In order to find an approximation to Eq. (2.78) for the limiting case of $\bar{v}_{ph}$ tending to infinity, Eq. (2.78) is rewritten by transforming the variable of integration to $z = \bar{v}_{ph}^2 - w^2$ with the following result:

$$I = \int_0^{\bar{v}_{ph}^2} e^{-z}\left(1 - \frac{z}{\bar{v}_{ph}^2}\right)^{-1/2} dz \tag{2.79}$$

Since z is less than $\bar{v}_{ph}^2$ over the entire range of integration, for the purpose of finding an asymptotic expansion for I in inverse powers of $\bar{v}_{ph}$ for the case of $\bar{v}_{ph}$ tending to infinity, it is reasonable to expand $[1 - (z/\bar{v}_{ph}^2)]^{-1/2}$ into a binomial series. On retaining only the first three terms in this series, it is found that

$$I = \int_0^{\bar{v}_{ph}^2} e^{-z}\left(1 + \frac{z}{2\bar{v}_{ph}^2} + \frac{3z^2}{8\bar{v}_{ph}^4}\right) dz \tag{2.80}$$

The contribution to the integral in Eq. (2.80) arising from the upper limit is exponentially small and can be ignored. This fact gives justification for the binomial expansion of $[1 - (z/\bar{v}_{ph}^2)]^{-1/2}$ even though the range of z extends up to $\bar{v}_{ph}^2$. Since

$$\int e^{-z}\, dz = -e^{-z} \qquad \int ze^{-z}\, dz = -ze^{-z} - e^{-z}$$

$$\int z^2 e^{-z}\, dz = -z^2 e^{-z} - 2ze^{-z} - 2e^{-z}$$

the evaluation of the contribution to the integral in Eq. (2.80) arising from the lower limit alone gives

$$I = 1 + (1/2\bar{v}_{ph}^2) + (3/4\bar{v}_{ph}^4) \tag{2.81}$$

By substituting Eq. (2.81) into Eq. (2.77), it can be shown that

$$\frac{k_p^2}{\omega_{pe}^2}\frac{KT_e}{m_e} 2\bar{v}_{ph}^2 = \left(1 + \frac{3}{2\bar{v}_{ph}^2}\right) - i\sqrt{\pi}\, 2\bar{v}_{ph}^3 \exp(-\bar{v}_{ph}^2) \tag{2.82}$$

With the help of Eqs. (2.64) and (2.3.10), Eq. (2.82) can be simplified as

$$\begin{aligned}\frac{\omega^2}{\omega_{pe}^2} &= 1 + 3k_p^2\lambda_{De}^2\frac{\omega_{pe}^2}{\omega^2} \\ &\quad - i\sqrt{\frac{\pi}{8}}\frac{2}{(k_p\lambda_{De})^3}\left(\frac{\omega}{\omega_{pe}}\right)^3 \exp\left[-\frac{1}{2(k_p\lambda_{De})^2}\left(\frac{\omega}{\omega_{pe}}\right)^2\right]\end{aligned} \tag{2.83}$$

In the long-wavelength limit, the second term is small and the third term is exponentially small compared to the first term and hence the square root of Eq. (2.83) is obtained as

$$\frac{\omega}{\omega_{pe}} = 1 + \tfrac{3}{2}k_p^2\lambda_{\text{De}}^2\frac{\omega_{pe}^2}{\omega^2}$$
$$- i\sqrt{\frac{\pi}{8}}\frac{1}{(k_p\lambda_{\text{De}})^3}\left(\frac{\omega}{\omega_{pe}}\right)^3\exp\left[-\frac{1}{2(k_p\lambda_{\text{De}})^2}\left(\frac{\omega}{\omega_{pe}}\right)^2\right] \tag{2.84}$$

As before, in the small second term, the first approximation $\omega = \omega_{pe}$ is used and in the exponentially small third term, the first approximation $\omega = \omega_{pe}$ is substituted for the factor in front of the exponent and the second approximation $\omega = \omega_{pe}(1 + (3/2)k_p^2\lambda_{\text{De}}^2)$ is substituted for the argument of the exponential function. Then Eq. (2.84) gives

$$\omega = \omega_{\text{real}} + i\omega_{\text{imag}} \tag{2.85}$$

where

$$\omega_{\text{real}} = \omega_{pe}(1 + (3/2)k_p^2\lambda_{\text{De}}^2) \tag{2.86}$$

and

$$\omega_{\text{imag}} = -\sqrt{\frac{\pi}{8}}\,\omega_{pe}(k_p\lambda_{\text{De}})^{-3}\exp\left[-\frac{1}{2(k_p\lambda_{\text{De}})^2} - \frac{3}{2}\right] \tag{2.87}$$

The real part of ω given by Eq. (2.86) is identical to that deduced previously in Eq. (2.62). As anticipated, the imaginary part of ω as given by Eq. (2.87) is negative showing that the longitudinal plasma waves decrease in amplitude as a function of time. This damping of the longitudinal plasma waves was first pointed out in an important paper by L. D. Landau [2] and ω_{imag} given by Eq. (2.87) is usually known as *the Landau damping constant.*

It is important to note that the damping of the longitudinal plasma waves arises inspite of the lack of dissipative mechanisms, like the collisional interactions between the electrons and the ions. This damping is due to the wave-particle interaction whose qualitative treatment was already presented in Sec. 2.12. The electrons with initial velocities quite close to the phase velocity of the wave are trapped inside the moving potential wells of the wave and this trapping process results in a net interchange of energy between the particles and the wave. For the Maxwell-Boltzmann velocity distribution function, the electrons gain energy and the amplitude of the wave decays in time. It is to be noted that the Landau damping constant ω_{imag} is contributed by the pole of the integrand in Eq. (2.66) and this pole occurs at $u_x = v_{ph} = \omega/k_p$, that is at a value of the electron velocity in the direction of the wave normal of the plane wave equal to the phase velocity of the wave. This is a mathematical manifestation of the fact that the wave-particle interaction is effective when the initial velocities of the electrons are quite close to the phase velocity of the wave. For an appropriate initial velocity distribution of the electrons,

ω_{imag} can be positive. In that case the wave amplitude grows in time indicating an unstable equilibrium.

In summary, the theory of the longitudinal plasma waves based on the Boltzmann-Vlasov equation has not only reproduced the results of the cold and the warm plasma models but has also lead to new results. First, the result based on the warm plasma model has been clarified to be true only in the long-wavelength limit. More importantly, the theory based on the Boltzmann-Vlasov equation has established the possibility of temporally decaying or even growing longitudinal plasma waves.

6.3. Transverse plane electromagnetic wave in a hot, isotropic plasma

The third group of fields consists of $\bar{E}_y(\omega)$, $\bar{H}_z(\omega)$, and $\bar{J}_y(\omega)$ which are governed by Eqs. (2.31), (2.39c), and (2.40b). The dispersion equation for the transverse plane electromagnetic wave can be deduced from Eqs. (2.31), (2.39c), and (2.40b). If $\bar{H}_z(\omega)$ is eliminated from Eqs. (2.39c) and (2.40b), it is found that

$$\bar{J}_y(\omega) = \frac{\varepsilon_0}{i\omega}(k^2c^2 - \omega^2)\bar{E}_y(\omega) + \frac{\varepsilon_0}{2\pi}E_y(0) + \frac{k}{2\pi\omega}H_z(0) \tag{3.1}$$

The substitution of Eq. (3.1) into Eq. (2.31) and the solution of the resulting equation for $\bar{E}_y(\omega)$ may be shown to give

$$\bar{E}_y(\omega) = G_y(\omega)/D_t(k_{em}, \omega) \tag{3.2}$$

where

$$G_y(\omega) = -\frac{i\omega}{2\pi}E_y(0) - \frac{ik_{em}}{2\pi\varepsilon_0}H_z(0) + \frac{\omega}{\varepsilon_0}\int_{-\infty}^{\infty} d\mathbf{u}\, \frac{eu_y}{2\pi(\omega - k_{em}u_x)}\, g(\mathbf{u}, 0) \tag{3.3}$$

and

$$D_t(k_{em}, \omega) = k_{em}^2c^2 - \omega^2 - \frac{\omega_{pe}^2\,\omega}{N_0}\int_{-\infty}^{\infty} d\mathbf{u}\, \frac{u_y(\partial f_0(u)/\partial u_y)}{(\omega - k_{em}u_x)} \tag{3.4}$$

The subscript *em* is added to k to indicate that it is the wavenumber corresponding to the transverse electromagnetic wave. Also $c = (\mu_0\varepsilon_0)^{-1/2}$ is the velocity of electromagnetic waves in free space. It can be shown by a procedure used in the previous section that

$$D_t(k_{em}, \omega) = 0 \tag{3.5}$$

is the *dispersion equation* which determines the time behavior of the transverse electromagnetic wave as contributed by the intrinsic properties of the plasma. The subscript t denotes that the dispersion equation pertains to the *t*ransverse electromagnetic wave.

It is useful to examine first the solution of the dispersion equation (3.5) for the limiting case of a cold plasma characterized by the velocity distribution function as given by Eq. (2.51). The integral with respect to u_y in Eq. (3.4) is simplified by an integration by parts and in carrying out this integration by parts, it is noted that $f_0(u)$ vanishes at $u_y = \pm\infty$ more rapidly than that of any growth due to an algebraic function of u_y. Therefore Eq. (3.4) together with Eq. (3.5) gives

$$k_{em}^2 c^2 - \omega^2 + \frac{\omega_{pe}^2 \omega}{N_0} \int_{-\infty}^{\infty} d\mathbf{u} \frac{f_0(u)}{(\omega - k_{em} u_x)} = 0 \tag{3.6}$$

In view of Eq. (2.56), the substitution of Eq. (2.51) into Eq. (3.6) gives the following result

$$k_{em}^2 c^2 - \omega^2 + \omega_{pe}^2 = 0 \tag{3.7}$$

which is identical to that obtained in Eq. (2.7.27). It is clear from Eq. (3.7) that even in the cold plasma approximation, the transverse electromagnetic wave is a propagating disturbance for $\omega > \omega_{pe}$. The value of ω given by Eq. (3.7) is real showing that the amplitude of the transverse electromagnetic wave remains constant in time.

Returning to the equilibrium state characterized by the Maxwell-Boltzmann distribution function (2.4) for the electrons, $f_0(u)$ given by Eqs. (2.4) is substituted into Eq. (3.6) and the integrals with respect to u_y and u_z are evaluated with the help of Eqs. (1.4.7). For simplifying the resulting integral, the following new variable of integration is used:

$$\tilde{u}_x = u_x(m_e/2KT_e)$$

If as before the dimensionless parameter

$$\tilde{v}_{ph} = v_{ph}(m_e/2KT_e)^{1/2} = (\omega/k_{em})(m_e/2KT_e)^{1/2} \tag{3.8}$$

is introduced where now v_{ph} is the phase velocity of the transverse electromagnetic wave, the dispersion equation (3.6) may be shown to become

$$k_{em}^2 c^2 - \omega^2 = \omega_{pe}^2 \tilde{v}_{ph} I(\tilde{v}_{ph}) \tag{3.9}$$

where $I(\tilde{v}_{ph})$ is defined by Eq. (2.66). As pointed out before, the contour of integration for the integral $I(\tilde{v}_{ph})$ is along the real axis and indented from below the pole at $\tilde{u}_x = \tilde{v}_{ph}$. It is convenient to write $I(\tilde{v}_{ph})$ as the sum of the contribution arising from the pole and the principal value of the integral. Then Eq. (3.9) together with Eqs. (2.65), (2.68), and (2.76) can be expressed as

$$k_{em}^2 c^2 - \omega^2 = i\sqrt{\pi}\, \omega_{pe}^2 \tilde{v}_{ph} e^{-\tilde{v}_{ph}^2} - \omega_{pe}^2 2\tilde{v}_{ph} \int_0^{\tilde{v}_{ph}} e^{w^2 - \tilde{v}_{ph}^2} dw \tag{3.10}$$

The integral appearing in Eq. (3.10) is the dispersion function whose values are tabulated [1]. From Eq. (3.10), the functional relationship between ω and k_{em}, that is the dispersion relationship for the transverse electromagnetic wave, can be evaluated.

Landau damping of transverse electromagnetic waves

It follows from Eq. (3.10) that ω has a small negative imaginary part showing that the transverse electromagnetic wave is also damped in time. It can be established that compared to the Landau damping of the longitudinal plasma waves, the Landau damping of the transverse electromagnetic wave is negligibly small. For this purpose, it is desirable to evaluate the dispersion function approximately. Such an approximation is deduced for the long-wavelength limit and is contained in Eqs. (2.78) and (2.81). For obtaining the first approximation to the real part of ω, it is sufficient to retain only the leading term in Eq. (2.81). Then in the long-wavelength limit Eq. (3.10) simplifies to

$$k_{em}^2 c^2 - \omega^2 + \omega_{pe}^2 = i\sqrt{\pi}\, \omega_{pe}^2 \tilde{v}_{ph} e^{-\tilde{v}_{ph}^2} \tag{3.11}$$

For the long-wavelength limit, the small term on the right-hand side of Eq. (3.11) can be omitted in the first approximation with the result that ω is specified by Eq. (3.7). When the phase velocity is computed from Eq. (3.7), it is found that $v_{ph} > c$ for $\omega > \omega_{pe}$. Therefore $\tilde{v}_{ph}$ becomes of the order of the ratio of the velocity of electromagnetic waves in free space to the average thermal velocity of the electrons and is a very large number. Since $\tilde{v}_{ph}$ is very large, the term on the right-hand side of Eq. (3.11) is very small and hence the Landau damping of the transverse electromagnetic wave becomes negligible. On the other hand, the phase velocity of the longitudinal plasma wave becomes of the order of the thermal velocity of the electrons for the higher frequencies where $\tilde{v}_{ph}$ is not a large number with the result that the Landau damping constant becomes significant.

Simple physical explanation can be given for the difference in the importance of the Landau damping between the longitudinal plasma and the transverse electromagnetic waves. It was pointed out that the Landau damping is due to the wave-particle interaction in which the wave loses energy and this interaction takes place efficiently when the phase velocity of the wave is in the neighborhood of the thermal velocity of the electrons. For the longitudinal plasma wave, there are frequencies in which the phase velocity of the wave is of the order of the thermal velocities of the electrons; consequently the wave-particle interaction proceeds efficiently and the Landau damping becomes significant. For the transverse electromagnetic wave, throughout the frequency range of propagation, the phase velocity is greater than the velocity of electromagnetic waves in free space; therefore the conditions for the efficient wave-particle interaction do not obtain resulting in a negligible damping of the wave.

It can be proved that the dispersion equation for the fourth group of fields which consist of $\bar{E}_z(\omega)$, $\bar{H}_y(\omega)$, and $\bar{J}_z(\omega)$ and which are governed by Eqs. (2.32), (2.39b), and (2.40c) is the same as that given by Eq. (3.5). Therefore the characteristics of the fourth and the third groups of fields are identical. Indeed the fourth group of fields can be obtained from the third group by a rotation through 90° about the wave-normal direction. Thus the third and the fourth groups of fields form two independent transverse electromagnetic plane waves having the same characteristics but with their electric vectors polarized linearly at right angles to each other.

Suppose that the initial perturbations of the velocity distribution function, the electric and the magnetic fields are all zero. For this case, Eq. (3.3) shows that $G_y(\omega) = 0$. In order for $\bar{E}_y(\omega)$ given by Eq. (3.2) to have a nontrivial solution, it is then necessary that $D_t(k_{em}, \omega) = 0$. Therefore, in order to obtain the dispersion equation (3.5), it is not necessary to solve an initial-value problem; it is sufficient to deduce the requirement for a nontrivial solution of Maxwell's equations and the Boltzmann-Vlasov equation. It is an easy matter to verify that the dispersion equation (2.50) for the longitudinal plasma wave can be obtained by this simplified method. In the following two sections where the characteristics of waves in a plasma under the influence of an external magnetostatic field are analyzed on the basis of the Boltzmann-Vlasov equation, the required dispersion equations are deduced using this simplified method and without posing an initial-value problem.

6.4. Propagation along the magnetostatic field in a hot plasma

Consider a homogeneous, unbounded plasma consisting of mobile electrons in a neutralizing background of stationary ions of number density N_0. A uniform magnetostatic field $\mathbf{B}_0 = \hat{\mathbf{z}}B_0$ is assumed to be impressed throughout the plasma in the z-direction. The electrons whose equilibrium number density is also equal to N_0 is characterized by the velocity distribution function $f(\mathbf{r}, \mathbf{u}, t)$. As before, the velocity distribution function is assumed to consist of a small perturbation $g(\mathbf{r}, \mathbf{u}, t)$ superimposed on the equilibrium distribution function $f_0(u)$ which for the sake of simplicity is taken to be the Maxwellian distribution specified by Eqs. (2.4). Thus

$$f(\mathbf{r}, \mathbf{u}, t) = f_0(u) + g(\mathbf{r}, \mathbf{u}, t) \tag{4.1}$$

where $g(\mathbf{r}, \mathbf{u}, t) \ll f_0(u)$. In the equilibrium state, the electric charge and current densities are zero throughout the plasma; therefore the electric field is also zero but the magnetic field is equal to the external magnetostatic field $\mathbf{B}_0$.

Let $\mathbf{E}(\mathbf{r}, t)$ and $\mathbf{B}(\mathbf{r}, t)$ be the electric field and the magnetic field respectively associated with the charge and the current densities caused by the perturbation $g(\mathbf{r}, \mathbf{u}, t)$ in the velocity distribution function. Therefore like $g(\mathbf{r}, \mathbf{u}, t)$, $\mathbf{E}(\mathbf{r}, t)$ and $\mathbf{B}(\mathbf{r}, t)$ are also first-order small quantities. Whereas $\mathbf{E}(\mathbf{r}, t)$ is the total electric field, the total magnetic field $\mathbf{B}_t(\mathbf{r}, t)$ is given by

$$\mathbf{B}_t(\mathbf{r}, t) = \mathbf{B}_0 + \mathbf{B}(\mathbf{r}, t). \tag{4.2}$$

Noting that the equilibrium distribution function is homogeneous, the linearized Boltzmann-Vlasov equation for the case with an external magnetostatic field can be deduced from Eq. (2.7) with the help of Eqs. (4.1) and (4.2) as

$$\begin{aligned} &\frac{\partial}{\partial t} g(\mathbf{r}, \mathbf{u}, t) + (\mathbf{u} \cdot \nabla_r) g(\mathbf{r}, \mathbf{u}, t) - \frac{e}{m_e}[\mathbf{E}(\mathbf{r}, t) + \mathbf{u} \times \mathbf{B}_t(\mathbf{r}, t)] \cdot \nabla_u f_0(u) \\ &\quad - \frac{e}{m_e}(\mathbf{u} \times \mathbf{B}_0) \cdot \nabla_u g(\mathbf{r}, \mathbf{u}, t) = 0 \end{aligned} \tag{4.3}$$

Perturbation of the equilibrium distribution function

For the investigation of the characteristics of plane waves propagating along the magnetostatic field, all the field quantities may be assumed to have the phase factor $\exp[i(k_z z - \omega t)]$. Therefore

$$\mathbf{E}(\mathbf{r}, t) = \mathbf{E} e^{i(k_z z - \omega t)} \qquad \mathbf{B}(\mathbf{r}, t) = \mathbf{B} e^{i(k_z z - \omega t)} \qquad g(\mathbf{r}, \mathbf{u}, t) = g(\mathbf{u}) e^{i(k_z z - \omega t)} \tag{4.4}$$

In Eqs. (4.4) **E**, **B**, and $g(\mathbf{u})$ are independent of space and time; hence they are the phasor amplitudes. Since the equilibrium distribution function is isotropic, Eqs. (2.17)–(2.19) are applicable. With the help of Eqs. (2.19) and (4.4), Eq. (4.3) can be simplified as

$$-i(\omega - k_z u_z) g(\mathbf{u}) - \frac{e}{m_e} \mathbf{E} \cdot \nabla_u f_0(u) - \frac{e B_0}{m_e} (\mathbf{u} \times \hat{\mathbf{z}}) \cdot \nabla_u g(\mathbf{u}) = 0 \tag{4.5}$$

Note that

$$\omega_{ce} = e B_0 / m_e \tag{4.6}$$

is the gyromagnetic angular frequency of the electrons and is positive.

It is convenient to introduce the cylindrical coordinates (u_ρ, φ, u_z) in the velocity space so that

$$u_x = u_\rho \cos \varphi \qquad u_y = u_\rho \sin \varphi \qquad u_z = u_z \tag{4.7}$$

Hence

$$\begin{aligned} \frac{dg(\mathbf{u})}{d\varphi} &= \frac{\partial g(\mathbf{u})}{\partial u_x} \frac{du_x}{d\varphi} + \frac{\partial g(\mathbf{u})}{\partial u_y} \frac{du_y}{d\varphi} + \frac{\partial g(\mathbf{u})}{\partial u_z} \frac{du_z}{d\varphi} \\ &= -u_y \frac{\partial g(\mathbf{u})}{\partial u_x} + u_x \frac{\partial g(\mathbf{u})}{\partial u_y} \\ &= -(\mathbf{u} \times \hat{\mathbf{z}}) \cdot \nabla_u g(\mathbf{u}) \end{aligned} \tag{4.8}$$

The substitution of Eq. (4.8) into Eq. (4.5) together with Eq. (4.6) gives the following ordinary differential equation for the determination of $g(\mathbf{u})$:

$$\frac{dg(\mathbf{u})}{d\varphi} - \frac{i(\omega - k_z u_z)}{\omega_{ce}} g(\mathbf{u}) = \frac{e}{m_e \omega_{ce}} \mathbf{E} \cdot \nabla_u f_0(u) \tag{4.9}$$

It follows from Eq. (2.17) that

$$\mathbf{E} \cdot \nabla_u f_0(u) = \frac{1}{u} f_0'(u)[E_x u_x + E_y u_y + E_z u_z] = \frac{1}{u} f_0'(u)\mathbf{E} \cdot \mathbf{u} \tag{4.10}$$

It is convenient to express the field vector in a plane perpendicular to the magnetostatic field as a linear superposition of two oppositely directed circularly polarized components. Therefore, with the help of Eqs. (3.6.2) and (3.6.3), **E** and **u** can be expressed as follows:

$$\mathbf{E} = E_+ \frac{(\hat{\mathbf{x}} + i\hat{\mathbf{y}})}{\sqrt{2}} + E_- \frac{(\hat{\mathbf{x}} - i\hat{\mathbf{y}})}{\sqrt{2}} + \hat{\mathbf{z}} E_z \tag{4.11}$$

$$\mathbf{u} = u_+ \frac{(\hat{\mathbf{x}} + i\hat{\mathbf{y}})}{\sqrt{2}} + u_- \frac{(\hat{\mathbf{x}} - i\hat{\mathbf{y}})}{\sqrt{2}} + \hat{\mathbf{z}} u_z \tag{4.12}$$

where

$$E_\pm = (E_x \mp iE_y)/\sqrt{2} \tag{4.13}$$

and

$$u_\pm = (u_x \mp iu_y)/\sqrt{2} \tag{4.14}$$

As explained in Sec. 3.6, E_+ and E_- are the positive and the negative circularly polarized electric fields, respectively. Similar meanings apply to u_+ and u_- also. With the help of Eqs. (3.6.6), (3.6.7), (4.11), and (4.12), Eq. (4.10) can be simplified as

$$\mathbf{E} \cdot \nabla_u f_0(u) = \frac{1}{u} f_0'(u)[E_+ u_- + E_- u_+ + E_z u_z] \tag{4.15}$$

From Eqs. (4.7) and (4.14), it is found that

$$u_+ = u_\rho e^{-i\varphi}/\sqrt{2} \qquad u_- = u_\rho e^{i\varphi}/\sqrt{2} \tag{4.16}$$

With the help of Eqs. (2.17), (4.15), and (4.16), the right-hand side of Eq. (4.9) can be expressed as

$$\frac{e}{m_e \omega_{ce}} \mathbf{E} \cdot \nabla_u f_0(u) = G_+(\mathbf{u}) + G_-(\mathbf{u}) + G_z(\mathbf{u}) \tag{4.17}$$

where

$$G_+(\mathbf{u}) = G_+(u_\rho, u_z)e^{i\varphi} \qquad G_-(\mathbf{u}) = G_-(u_\rho, u_z)e^{-i\varphi} \qquad G_z(\mathbf{u}) = G_z(u_\rho, u_z) \tag{4.18}$$

$$G_+(u_\rho, u_z) = \frac{e}{m_e \omega_{ce}} \frac{\partial f_0(u)}{\partial u_\rho} \frac{E_+}{\sqrt{2}}$$
$$G_-(u_\rho, u_z) = \frac{e}{m_e \omega_{ce}} \frac{\partial f_0(u)}{\partial u_\rho} \frac{E_-}{\sqrt{2}} \tag{4.19}$$
$$G_z(u_\rho, u_z) = \frac{e}{m_e \omega_{ce}} \frac{\partial f_0(u)}{\partial u_z} E_z$$

Let

$$g(\mathbf{u}) = g_+(\mathbf{u}) + g_-(\mathbf{u}) + g_z(\mathbf{u}) \tag{4.20}$$

where $g_+(\mathbf{u})$, $g_-(\mathbf{u})$, and $g_z(\mathbf{u})$ are, respectively, the solutions of the differential equation (4.9) corresponding to $G_+(\mathbf{u})$, $G_-(\mathbf{u})$, and $G_z(\mathbf{u})$ on the right-hand side of Eq. (4.9).

It is proposed to present the details of the procedure for deducing the expression for $g_+(\mathbf{u})$. First the differential equation (4.9) satisfied by $g_+(\mathbf{u})$ can be rewritten as

$$\begin{aligned} \frac{d}{d\varphi}\left[g_+(\mathbf{u})\exp\left\{\frac{-i(\omega - k_z u_z)}{\omega_{ce}}\varphi\right\}\right] \\ = G_+(u_\rho, u_z)\exp\left[i\left\{\frac{-(\omega - k_z u_z)}{\omega_{ce}}\varphi + \varphi\right\}\right] \end{aligned} \tag{4.21}$$

The exponential term vanishes at $\varphi = -\infty$ since ω has a vanishingly small positive imaginary part. If both sides of Eq. (4.21) are integrated with respect to φ from $\varphi = -\infty$ to $\varphi = \varphi$, the result is

$$g_+(\mathbf{u}) = \frac{i\omega_{ce} G_+(u_\rho, u_z) e^{i\varphi}}{(\omega - \omega_{ce} - k_z u_z)} + C_+ \exp\left\{i\frac{(\omega - k_z u_z)}{\omega_{ce}}\varphi\right\} \tag{4.22}$$

The constant of integration C_+ can be shown to be zero in the following manner. Physical considerations demand that $g_+(\mathbf{u})$ be a unique function of $\mathbf{u}$. Therefore, the value of $g_+(\mathbf{u})$ should not change if φ is increased or decreased by integral multiples of 2π. Such a result can be obtained only if $C_+ = 0$. Hence Eq. (4.22) simplifies to

$$g_+(\mathbf{u}) = g_+(u_\rho, u_z) e^{i\varphi} \tag{4.23}$$

where

$$g_+(u_\rho, u_z) = \frac{i\omega_{ce} G_+(u_\rho, u_z)}{(\omega - \omega_{ce} - k_z u_z)} \tag{4.24}$$

In a similar manner, it can be deduced that

$$g_-(\mathbf{u}) = g_-(u_\rho, u_z) e^{-i\varphi} \tag{4.25}$$

where

$$g_-(u_\rho, u_z) = \frac{i\omega_{ce} G_-(u_\rho, u_z)}{(\omega + \omega_{ce} - k_z u_z)} \tag{4.26}$$

and

$$g_z(\mathbf{u}) = g_z(u_\rho, u_z) = \frac{i\omega_{ce} G_z(u_\rho, u_z)}{(\omega - k_z u_z)} \tag{4.27}$$

Thus the phasor amplitude $g(\mathbf{u})$ of the perturbation of the velocity distribution function of the electrons has been obtained explicitly in terms of the equilibrium distribution function.

Current density

Let the phasor amplitude of the current density **J** be also separated into the two oppositely directed circularly polarized components and the longitudinal component as given in Eq. (4.12) for the particle velocity. Then from Eqs. (2.25) and (4.12), it is found that

$$J_+ = -e \int_0^\infty du_\rho \int_0^{2\pi} d\varphi \int_{-\infty}^{\infty} du_z u_\rho u_+ g(\mathbf{u}) \tag{4.28}$$

$$J_- = -e \int_0^\infty du_\rho \int_0^{2\pi} d\varphi \int_{-\infty}^{\infty} du_z u_\rho u_- g(\mathbf{u}) \tag{4.29}$$

and

$$J_z = -e \int_0^\infty du_\rho \int_0^{2\pi} d\varphi \int_{-\infty}^{\infty} du_z u_\rho u_z g(\mathbf{u}) \tag{4.30}$$

If u_+ and u_- as given by Eqs. (4.16) and $g(\mathbf{u})$ as given by Eqs. (4.20), (4.23), (4.25), and (4.27) are substituted into Eqs. (4.28)–(4.30) and the integrals with respect to φ are evaluated, the following simple results are obtained:

$$J_+ = -\pi e\sqrt{2} \int_0^\infty du_\rho \int_{-\infty}^{\infty} du_z u_\rho^2 g_+(u_\rho, u_z) \tag{4.31}$$

$$J_- = -\pi e\sqrt{2} \int_0^\infty du_\rho \int_{-\infty}^{\infty} du_z u_\rho^2 g_-(u_\rho, u_z) \tag{4.32}$$

and

$$J_z = -2\pi e \int_0^\infty du_\rho \int_{-\infty}^{\infty} du_z u_\rho u_z g_z(u_\rho, u_z) \tag{4.33}$$

It is clear from Eqs. (4.19), (4.24), (4.26), (4.27), and (4.31) – (4.33) that J_+, J_-, and J_z depend, respectively, on only E_+, E_-, and E_z. It is in anticipation of this advantage that every field vector was decomposed into the sum of the two oppositely directed circularly polarized components and the longitudinal component.

Separation into the various modes

For the fields with the phase factor as given by Eqs. (4.4), Maxwell's equations (2.33) and (2.34) reduce to

$$k_z \hat{\mathbf{z}} \times \mathbf{E} = \omega\mu_0 \mathbf{H} \tag{4.34}$$

$$k_z \hat{\mathbf{z}} \times \mathbf{H} = -\omega\varepsilon_0 \mathbf{E} - i\mathbf{J} \tag{4.35}$$

In component form Eqs. (4.34) and (4.35) become

$$-k_z E_y = \omega\mu_0 H_x \tag{4.36a}$$

$$k_z E_x = \omega\mu_0 H_y \tag{4.36b}$$

$$0 = \omega\mu_0 H_z \tag{4.36c}$$

$$-k_z H_y = -\omega\varepsilon_0 E_x - iJ_x \tag{4.37a}$$

$$k_z H_x = -\omega\varepsilon_0 E_y - iJ_y \tag{4.37b}$$

$$0 = -\omega\varepsilon_0 E_z - iJ_z \tag{4.37c}$$

If Eq. (4.36a) is multiplied by $1/\sqrt{2}$, Eq. (4.36b) by $\mp i/\sqrt{2}$ and they are added, it is found that

$$\mp ik_z E_\pm = \omega\mu_0 H_\pm \tag{4.38}$$

In a similar manner, it can be deduced from Eqs. (4.37a) and (4.37b) that

$$\mp ik_z H_\pm = -\omega\varepsilon_0 E_\pm - iJ_\pm \tag{4.39}$$

Note that $J_\pm$ and $H_\pm$ are defined as in Eq. (4.13).

For a wave with its propagation coefficient in the z-direction, the time-varying magnetic field in the longitudinal (z) direction is seen from Eq. (4.36c) to be zero. Since J_+, J_-, and J_z depend, respectively, on only E_+, E_-, and E_z, it follows from Eqs. (4.37c), (4.38), and (4.39) that the total electromagnetic field separates into three independent groups as given by

1. E_z, J_z
2. E_-, H_-, J_-
3. E_+, H_+, J_+

The first group represents the longitudinal plasma wave and has no magnetic field associated with it. The second group represents the left circularly polarized transverse electromagnetic wave. Similarly, the third group represents the right circularly polarized transverse electromagnetic wave. The advantage of decomposing each transverse field vector into the sum of the two oppositely directed circularly polarized components is that the two different circular polarizations separate into two independent sets of fields. The characteristics of the longitudinal and the two transverse waves can be analyzed separately.

Longitudinal plasma wave

First it is convenient to recast Eq. (4.33) as follows:

$$J_z = -e \int_{-\infty}^{\infty} d\mathbf{u}\, u_z g_z(u_\rho, u_z) \tag{4.40}$$

If the value of J_z from Eq. (4.37c) and the expression for $g_z(u_\rho, u_z)$ from Eqs. (4.27) and (4.19) are substituted into Eq. (4.40), it can be shown that

$$i\omega\varepsilon_0 E_z \left[1 + \frac{\omega_{pe}^2}{N_0 \omega} \int_{-\infty}^{\infty} d\mathbf{u} \frac{u_z}{(\omega - k_p u_z)} \frac{\partial f_0(u)}{\partial u_z}\right] = 0 \tag{4.41}$$

Since k_z now refers to the longitudinal plasma wave, it has been changed to k_p. Except for the fact that the directions x and z are interchanged, the term within the square brackets in Eq. (4.41) is identically equal to that given by Eq. (2.43). In order for the existence of a nontrivial solution for E_z, Eq. (4.41) requires that

$$D_l(k_p, \omega) = 0 \tag{4.42}$$

which is the dispersion equation for the longitudinal plasma wave. For the wave normal in the direction of the magnetostatic field, the longitudinal plasma wave separates out as an independent mode of oscillation and its dispersion equation is found, by comparing Eqs. (2.50) and (4.42), to be identical to the case of the plasma with no external magnetostatic field.

A simple physical explanation can be given for the magnetostatic field having no influence on the longitudinal plasma wave when impressed in the longitudinal direction. As was pointed out before, only the perturbation in the distribution of the velocities of the electrons in the longitudinal direction has a role in the determination of the characteristics of the longitudinal plasma wave. A magnetostatic field in the longitudinal direction exerts no force in the longitudinal direction and therefore does not influence the distribution of velocities in the longitudinal direction accounting for the magnetostatic field in the longitudinal direction having no effect on the characteristics of the longitudinal plasma wave.

Transverse electromagnetic wave

The two circularly polarized waves can be treated together. By eliminating $H_\pm$ from Eqs. (4.38) and (4.39), $J_\pm$ can be expressed in terms of $E_\pm$ as

$$J_\pm = -\frac{i\varepsilon_0}{\omega}[k_z^2 c^2 - \omega^2]E_\pm \tag{4.43}$$

With the help of Eqs. (4.19), (4.24), (4.26), (4.31), and (4.32), it can be deduced that

$$J_{\pm} = -\frac{i\pi e^2}{m_e} E_{\pm} \int_0^{\infty} du_\rho \int_{-\infty}^{\infty} du_z \frac{u_\rho^2}{(\omega \mp \omega_{ce} - k_z u_z)} \frac{\partial f_0(u)}{\partial u_\rho} \tag{4.44}$$

The elimination of $J_{\pm}$ from Eqs. (4.43) and (4.44) may be shown to yield

$$E_{\pm}\left[k_z^2 c^2 - \omega^2 - \frac{\omega_{pe}^2 \omega \pi}{N_0} \int_0^{\infty} du_\rho \int_{-\infty}^{\infty} du_z \frac{u_\rho^2}{(\omega \mp \omega_{ce} - k_z u_z)} \frac{\partial f_0(u)}{\partial u_\rho}\right] = 0 \tag{4.45}$$

For the existence of a nontrivial solution for $E_{\pm}$, it is required that the term within the square brackets in Eq. (4.45) vanish and this requirement yields the dispersion equation for the transverse electromagnetic wave. If $f_0(u)$ is substituted from Eqs. (2.4), the differentiation with respect to u_ρ is carried out explicitly and the integral with respect to u_ρ is evaluated with the help of Eqs. (1.4.7), the dispersion equation deduced from Eq. (4.45) can be shown to become

$$k_z^2 c^2 - \omega^2 + \omega_{pe}^2 \omega \left(\frac{m_e}{2\pi K T_e}\right)^{1/2} \int_{-\infty}^{\infty} du_z \frac{e^{-m_e u_z^2/2KT_e}}{(\omega \mp \omega_{ce} - k_z u_z)} = 0 \tag{4.46}$$

where the upper and the lower signs correspond to the right and the left circularly polarized waves respectively.

The most probable speed of the electrons is seen from Eq. (1.4.20) to be given by

$$a_e = \sqrt{\frac{2KT_e}{m_e}} \tag{4.47}$$

The superscript + (−) is used on k_z to denote that its value corresponds to the right (left) circularly polarized wave. Let

$$\beta_{\pm} = \frac{\omega}{k_z^{\pm} a_e} \tag{4.48}$$

denote the phase velocity of the wave normalized to the most probable thermal speed a_e of the electrons. Also, let

$$\alpha_{\pm} = (\omega \mp \omega_{ce})/k_z^{\pm} a_e \tag{4.49}$$

By setting $\tilde{u}_z = u_z/a_e$, the dispersion equation (4.46) can be recast with the help of Eqs. (4.47)–(4.49) into the form

$$(k_z^{\pm} c)^2 - \omega^2 - \omega_{pe}^2 \beta_{\pm} I(\alpha_{\pm}) = 0 \tag{4.50}$$

where the integral I is the same as that defined by Eq. (2.66).

The contour of integration for the integral $I(\alpha_{\pm})$ is along the real axis in the $\tilde{u}_z$-plane and indented from below the pole at $\tilde{u}_z = \alpha_{\pm}$. Therefore $I(\alpha_{\pm})$ can be written as the sum of the contribution from the pole and the principle value of the integral.

Hence, with the help of Eqs. (2.65), (2.68), and (2.76), Eq. (4.50) can be recast as

$$(k_z^{\mp} c)^2 - \omega^2 - i\sqrt{\pi}\, \beta_{\pm} \omega_{pe}^2 e^{-\alpha_{\pm}^2} + 2\omega_{pe}^2 \beta_{\pm} \int_0^{\alpha_{\pm}} \exp(w^2 - \alpha_{\pm}^2)\, dw = 0 \qquad (4.51)$$

which is the dispersion equation for the right and the left circularly polarized transverse electromagnetic wave propagating along the magnetostatic field in a hot plasma whose equilibrium state is characterized by the Maxwell-Boltzmann distribution function.

For $k_z^{\mp}$ real, an examination of Eq. (4.51) reveals that ω has a negative imaginary part. Therefore, it follows that if the waves are set up at time $t = 0$, there is a damping of their amplitudes with time for times subsequent to $t = 0$. It is important to establish if this temporal damping is significant or not.

Suppose that $|\alpha_{\pm}| \gg 1$. Then the dispersion function appearing in Eq. (4.51) can be expanded into an asymptotic series in inverse powers of $\alpha_{\pm}$. For the purpose of obtaining the first approximation to the value of ω, it is sufficient to retain only the leading term. From Eqs. (2.78) and (2.81), it is found that

$$\int_0^{\alpha_{\pm}} \exp(w^2 - \alpha_{\pm}^2)\, dw = \frac{1}{2\alpha_{\pm}} \qquad (4.52)$$

which together with Eqs. (4.48) and (4.49) can be used to simplify Eq. (4.51) with the following result:

$$(k_z^{\mp} c)^2 - \omega^2 + \omega_{pe}^2 \frac{\omega}{(\omega \mp \omega_{ce})} = i\sqrt{\pi}\, \beta_{\pm} \omega_{pe}^2 e^{-\alpha_{\pm}^2} \qquad (4.53)$$

For $|\alpha_{\pm}| \gg 1$, the exponentially small term on the right-hand side of Eq. (4.53) can be omitted in the first approximation and the resulting equation can be manipulated into the form

$$(k_z^{\mp} c)^2 / \omega_{pe}^2 = \Omega \frac{[\Omega^2 \mp \Omega R - 1]}{(\Omega \mp R)} \qquad (4.54)$$

where

$$\Omega = \frac{\omega}{\omega_{pe}} \quad \text{and} \quad R = \frac{\omega_{ce}}{\omega_{pe}} \qquad (4.55)$$

Note that $k_z^{\mp} c/\omega_{pe}$ is the normalized propagation coefficient as used in Sec. 5.4. The right-hand side of Eq. (4.54) is verified to be equal to the square of the normalized propagation coefficient pertaining to the cold plasma model on a comparison with Eq. (5.4.19) for the left circularly polarized wave, and Eq. (5.4.22) for the right circularly polarized wave. The dispersion equation (4.51) gives correctly the results of the cold plasma model in the limit of $|\alpha_{\pm}| \gg 1$. Alternatively, it follows that the results of the cold plasma model are valid only if $|\alpha_{\pm}| \gg 1$.

The dispersion relations for the long-wavelength limit as obtained in Eq. (4.54) are portrayed in terms of normalized quantities in Figs. 5.2 and 5.3 for the left and the right circularly polarized waves, respectively. For a specified real propagation coefficient, the frequency of oscillation of the left circularly polarized wave is real and is such that

$$\omega > -\frac{\omega_{ce}}{2} + \sqrt{\left(\frac{\omega_{ce}}{2}\right)^2 + \omega_{pe}^2} \tag{4.56}$$

Also the phase velocity ω/k_z^- of the left circularly polarized wave is greater than c for all k_z^- and hence β_- is a large number of the order of the ratio of the velocity of electromagnetic waves in free space to the thermal velocity of the electrons. Since $\alpha_-/\beta_- = (\omega + \omega_{ce})/\omega$ is positive and greater than unity, it follows that $\alpha_- \gg 1$ for all k_z^-. Since always $\alpha_- \gg 1$, the imaginary term on the right-hand side of Eq. (4.53) is exponentially small and can be ignored; therefore, as in the case of the transverse electromagnetic waves in a hot plasma without a magnetostatic field, the Landau damping of the left circularly polarized wave propagating along the magnetostatic field in a hot plasma is always negligibly small. Also, for all real propagation coefficients, the cold plasma model is a very good approximation insofar as the determination of the characteristics of the left circularly polarized waves is concerned.

From Fig. 5.3 for a specified real propagation coefficient, it is seen that the frequency of oscillation of the right circularly polarized wave is real and is such that

$$a)\quad 0 < \omega < \omega_{ce} \qquad b)\quad \omega > \frac{\omega_{ce}}{2} + \sqrt{\left(\frac{\omega_{ce}}{2}\right)^2 + \omega_{pe}^2} \tag{4.57}$$

However, the results do not hold for frequencies of the order of the ion plasma frequency and lower because it is then not valid to ignore the ion motions. Therefore the very low frequency region is omitted from further consideration. It is convenient to examine first the frequency ranges (4.57a) and (4.57b) except the frequencies in the close neighborhood of $\omega = \omega_{ce}$. Then, it is found from Fig. 5.3 that the phase velocity ω/k_z^+ of the right circularly polarized wave is greater than c in the frequency range (4.57b) and is less than but of the order of c in the frequency range (4.57a) except, as mentioned before, in the close neighborhood of $\omega = \omega_{ce}$. Therefore, β_+ is a large number. Since $|\alpha_+/\beta_+| = |(\omega - \omega_{ce})/\omega|$ is of the order unity, it follows that $|\alpha_+| \gg 1$ for ω not close to ω_{ce}. The arguments advanced before for the left circularly polarized wave show that the temporal damping is negligibly small and the cold plasma model is a very good approximation even for the right circularly polarized wave if ω is not close to ω_{ce}.

Cyclotron damping

For ω in the close neighborhood of ω_{ce}, the right circularly polarized wave exhibits an interesting feature. For $\omega \approx \omega_{ce}$, it is seen from Fig. 5.3 that the phase velocity is of the order of the thermal velocity of the electrons and lower with the result that $\beta_+ \lesssim 1$. Also, since $|\alpha_+/\beta_+| = |(\omega - \omega_{ce})/\omega| \ll 1$, it follows that $|\alpha_+| \ll 1$. Therefore Eq. (4.53) which is applicable for $|\alpha_+| \gg 1$ and the cold plasma approximation as given by Eq. (4.54) are not valid for ω approximately equal to ω_{ce}. For the purpose of obtaining results which are valid for $\omega \approx \omega_{ce}$, it is necessary to deduce an approximation of the dispersion equation (4.51) for the limiting case of $|\alpha_+| \ll 1$.

As a first approximation, α_+ can be set equal to zero in Eq. (4.51). Then with the help of Eq. (4.48), Eq. (4.51) can be simplified to yield

$$\left(\frac{k_z^+ c}{\omega}\right)^3 - \left(\frac{k_z^+ c}{\omega}\right) = i\sqrt{\pi}\,\frac{c}{a_e}\left(\frac{\omega_{pe}}{\omega_{ce}}\right)^2 \tag{4.58}$$

Since $\omega/k_z^+ c \ll 1$, the second term in Eq. (4.58) can be omitted in comparison to the first term with the result that

$$\left(\frac{\omega}{k_z^+ c}\right)^3 = \frac{-i}{\sqrt{\pi}}\frac{a_e}{c}\left(\frac{\omega_{ce}}{\omega_{pe}}\right)^2 \tag{4.59}$$

The solution of Eq. (4.59) gives the following real and imaginary values for ω:

$$\omega_{\text{real}} = \frac{\sqrt{3}}{2}k_z^+\left(\frac{a_e}{\sqrt{\pi}}c^2\frac{\omega_{ce}^2}{\omega_{pe}^2}\right)^{1/3} \tag{4.60a}$$

and

$$\omega_{\text{imag}} = -\frac{k_z^+}{2}\left(\frac{a_e}{\sqrt{\pi}}c^2\frac{\omega_{ce}^2}{\omega_{pe}^2}\right)^{1/3} \tag{4.60b}$$

Since the imaginary part of ω satisfying the dispersion equation is negative, it follows that if the right circularly polarized wave is set up at time $t = 0$ to propagate along the magnetostatic field in a hot plasma, the waves are temporally damped for ω_{real} in the close neighborhood of ω_{ce}. This damping corresponds to the Landau damping of the longitudinal plasma wave and is called the *cyclotron damping*. There are some differences between the cyclotron damping and the Landau damping. For the cyclotron damping, the acceleration is essentially perpendicular to the drift motion of the particle and has a negligible influence on the longitudinal drift velocity as a consequence of which trapping becomes insignificant.

If the magnetostatic field is absent, $\alpha_\pm = \beta_\pm = \tilde{v}_{ph}$ and Eq. (4.51) becomes identical to Eq. (3.10). Thus in the limiting case of zero magnetostatic field, the dispersion equation (4.51) reduces correctly to that of the transverse electromagnetic waves in an isotropic plasma. In this limiting case, the dispersion relations of the two circularly polarized waves become identical. Therefore the two circularly polarized waves can be linearly superposed to obtain the two different linearly polarized, transverse electromagnetic waves discussed in the previous section.

There is another interesting feature associated with the right circularly polarized wave. For a real propagation coefficient k_z^+, Fig. 5.3 shows that there are two natural frequencies of oscillation. But for the left circularly polarized wave, there is only one natural frequency of oscillation corresponding to a real propagation coefficient k_z^-.

6.5. Propagation across the magnetostatic field in a hot plasma

As before, the magnetostatic field is assumed to be in the z-direction. For the investigation of the characteristics of plane waves propagating across the magnetostatic field, all the field quantities are assumed to have the phase factor $\exp[i(k_x x - \omega t)]$. Therefore

$$\mathbf{E}(\mathbf{r}, t) = \mathbf{E}e^{i(k_x x - \omega t)} \qquad \mathbf{B}(\mathbf{r}, t) = \mathbf{B}e^{i(k_x x - \omega t)} \qquad g(\mathbf{r}, \mathbf{u}, t) = g(\mathbf{u})e^{i(k_x x - \omega t)} \tag{5.1}$$

where **E**, **B**, and $g(\mathbf{u})$ are the phasor amplitudes. The propagation coefficient k_x is assumed to be real. As in the previous cases, it is proposed to deduce the dispersion equation giving the functional relationship between k_x and ω, and from an analysis of the dispersion equation, determine the intrinsic time behavior for the case in which the equilibrium state is characterized by the Maxwell-Boltzmann distribution function.

Perturbation of the equilibrium distribution function

The linearized Boltzmann-Vlasov equation is given by Eq. (4.3) which can be simplified with the help of Eqs. (2.19) and (5.1) into a form expressed by Eq. (4.5) with $k_x u_x$ substituted for $k_z u_z$. If Eqs. (4.6)–(4.8) are used, the equation corresponding to Eq. (4.9) and applicable for the present case can be shown to be given by

$$\frac{dg(\mathbf{u})}{d\varphi} - i\frac{(\omega - k_x u_x)}{\omega_{ce}} g(\mathbf{u}) = \frac{e}{m_e \omega_{ce}} \mathbf{E} \cdot \nabla_u f_0(u) \tag{5.2}$$

It is found from Eqs. (2.17), (4.7), and (4.10) that

$$\mathbf{E} \cdot \nabla_u f_0(u) = E_x \cos\varphi \frac{\partial f_0(u)}{\partial u_\rho} + E_y \sin\varphi \frac{\partial f_0(u)}{\partial u_\rho} + E_z \frac{\partial f_0(u)}{\partial u_z} \tag{5.3}$$

For the first-order differential equation (5.2), the integrating factor is obtained as

$$\begin{aligned} h(\varphi) &= \exp\left[-\int_0^\varphi i\frac{(\omega - k_x u_\rho \cos\varphi)}{\omega_{ce}}\, d\varphi\right] \\ &= \exp\left[-i\frac{\omega}{\omega_{ce}}\varphi + \frac{ik_x u_\rho}{\omega_{ce}} \sin\varphi\right] \end{aligned} \tag{5.4}$$

Note that $h(\varphi)h(-\varphi) = 1$. With the help of Eqs. (5.3) and (5.4), the solution of Eq. (5.2) can be deduced as

$$g(\mathbf{u}) = \frac{e}{m_e\omega_{ce}} h(-\varphi) \int_{-\infty}^{\varphi} \left[\{E_x \cos\varphi'' + E_y \sin\varphi''\} \frac{\partial f_0(u)}{\partial u_\rho} + E_z \frac{\partial f_0(u)}{\partial u_z} \right] h(\varphi'')\, d\varphi'' \tag{5.5}$$

Since ω has a vanishingly small positive imaginary part, the integral in Eq. (5.5) becomes zero at the lower limit. If the variable of integration is changed to $\varphi'' = \varphi - \varphi'$, Eq. (5.5) becomes

$$g(\mathbf{u}) = \frac{e}{m_e\omega_{ce}} \exp\left(-\frac{ik_x u_\rho}{\omega_{ce}} \sin\varphi\right) \int_0^{\infty} \left[\{E_x \cos(\varphi - \varphi') + E_y \sin(\varphi - \varphi')\} \frac{\partial f_0(u)}{\partial u_\rho} + E_z \frac{\partial f_0(u)}{\partial u_z} \right] \exp\left[i\frac{\omega}{\omega_{ce}}\varphi' + \frac{ik_x u_\rho}{\omega_{ce}} \sin(\varphi - \varphi') \right] d\varphi' \tag{5.6}$$

Since φ occurs only as the argument of periodic functions of period 2π, it follows, in accordance with the physical requirement, that $g(\mathbf{u})$ is a single-valued function of φ.

Conductivity and relative permittivity dyads

The knowledge of $g(\mathbf{u})$ enables the phasor amplitude of the current density to be obtained from Eq. (2.26) as

$$\mathbf{J} = -e \int_0^{\infty} u_\rho\, du_\rho \int_0^{2\pi} d\varphi \int_{-\infty}^{\infty} du_z\, \mathbf{u} g(\mathbf{u}) \tag{5.7}$$

It is convenient to express $\mathbf{J}$ as

$$\mathbf{J} = \boldsymbol{\sigma} \cdot \mathbf{E} \tag{5.8}$$

where $\boldsymbol{\sigma}$ is the conductivity dyad. From Eqs. (5.6)–(5.8), the component σ_{xx} of the conductivity dyad is seen to be expressed by the following integral:

$$\sigma_{xx} = -\frac{e^2}{m_e\omega_{ce}} \int_0^{\infty} du_\rho \int_0^{2\pi} d\varphi \int_{-\infty}^{\infty} du_z\, u_\rho^2 \cos\varphi \exp\left(-\frac{ik_x u_\rho \sin\varphi}{\omega_{ce}}\right) \times \int_0^{\infty} d\varphi' \cos(\varphi - \varphi') \frac{\partial f_0(u)}{\partial u_\rho} \exp\left[i\frac{\omega}{\omega_{ce}}\varphi' + \frac{ik_x u_\rho}{\omega_{ce}} \sin(\varphi - \varphi') \right] \tag{5.9}$$

In simplifying the expression for σ_{xx}, it is advantageous to evaluate first the integral with respect to φ' which is obtained from Eq. (5.9) as

$$I_1 = \int_0^\infty d\varphi' \cos(\varphi - \varphi')\exp\{g_1(\varphi')\} \tag{5.10}$$

where

$$g_1(\varphi') = i\frac{\omega}{\omega_{ce}}\varphi' + \frac{ik_x u_\rho}{\omega_{ce}} \sin(\varphi - \varphi') \tag{5.11}$$

From Eq. (5.11), it can be shown that

$$\cos(\varphi - \varphi') = \frac{\omega}{k_x u_\rho} + \frac{i\omega_{ce}}{k_x u_\rho}\frac{dg_1(\varphi')}{d\varphi'} \tag{5.12}$$

Hence, Eq. (5.10) can be rewritten as

$$\begin{aligned} I_1 &= \frac{\omega}{k_x u_\rho}\int_0^\infty d\varphi' \exp\{g_1(\varphi')\} + \frac{i\omega_{ce}}{k_x u_\rho}\int_0^\infty d[e^{g_1(\varphi')}] \\ &= \frac{\omega}{k_x u_\rho}\int_0^\infty d\varphi' \exp\{g_1(\varphi')\} - \frac{i\omega_{ce}}{k_x u_\rho}\exp\left(\frac{ik_x u_\rho}{\omega_{ce}} \sin \varphi\right) \end{aligned} \tag{5.13}$$

Let

$$\zeta = \frac{k_x u_\rho}{\omega_{ce}} \tag{5.14}$$

It is known that

$$\exp\{i\zeta \sin(\varphi - \varphi')\} = \sum_{n=-\infty}^{\infty} J_n(\zeta)e^{in(\varphi - \varphi')} \tag{5.15}$$

where J_n is the Bessel function of the first kind of order n. The substitution of Eq. (5.15) into Eq. (5.13) yields

$$\begin{aligned} I_1 &= -\frac{i}{\zeta}\exp(i\zeta \sin \varphi) + \frac{\omega}{k_x u_\rho}\sum_{n=-\infty}^{\infty} J_n(\zeta)e^{in\varphi}\int_0^\infty d\varphi' e^{i(\frac{\omega}{\omega_{ce}} - n)\varphi'} \\ &= -\frac{i}{\zeta}\exp(i\zeta \sin \varphi) + \frac{i\omega}{k_x u_\rho}\sum_{n=-\infty}^{\infty} J_n(\zeta)e^{in\varphi}\Big/\left(\frac{\omega}{\omega_{ce}} - n\right) \end{aligned} \tag{5.16}$$

The integral with respect to φ is then evaluated. From Eqs. (5.9), (5.10), and (5.16), the integral with respect to φ is found as

$$I_2 = \int_0^{2\pi} d\varphi \cos \varphi\left[-\frac{i}{\zeta} + \frac{i\omega}{k_x u_\rho}\sum_{n=-\infty}^{\infty}\frac{J_n(\zeta)\exp\{g_2(\varphi)\}}{(\omega/\omega_{ce} - n)}\right] \tag{5.17}$$

where

$$g_2(\varphi) = in\varphi - i\zeta \sin \varphi \tag{5.18}$$

The first term within the square brackets in Eq. (5.17) integrates to zero. It can be deduced from Eq. (5.18) that

$$\cos\varphi = \frac{n}{\zeta} + \frac{i}{\zeta}\frac{dg_2(\varphi)}{d\varphi} \tag{5.19}$$

Together with Eq. (5.19), Eq. (5.17) becomes

$$I_2 = \frac{i\omega}{k_x u_\rho} \sum_{n=-\infty}^{\infty} \frac{J_n(\zeta)}{(\omega/\omega_{ce} - n)} \left[\frac{n}{\zeta} \int_0^{2\pi} d\varphi \exp\{g_2(\varphi)\} + \frac{i}{\zeta} \int_0^{2\pi} d(\exp\{g_2(\varphi)\}) \right] \tag{5.20}$$

The second integral within the square brackets in Eq. (5.20) vanishes and the first integral can be written as a Bessel function in view of the relation

$$\int_0^{2\pi} e^{i(n\varphi - \zeta\sin\varphi)} d\varphi = 2\pi J_n(\zeta) \tag{5.21}$$

Hence, I_2 given by Eq. (5.20) simplifies to

$$I_2 = \frac{i2\pi\omega}{\zeta^2\omega_{ce}} \sum_{n=-\infty}^{\infty} \frac{nJ_n^2(\zeta)}{(\omega/\omega_{ce} - n)} \tag{5.22}$$

which can be rewritten in a slightly different form as

$$I_2 = \frac{i2\pi}{\zeta^2} \sum_{n=-\infty}^{\infty} J_n^2(\zeta) \frac{n(\omega/\omega_{ce} - n + n)}{(\omega/\omega_{ce} - n)} = \frac{i2\pi}{\zeta^2} \sum_{n=-\infty}^{\infty} \frac{n^2 J_n^2(\zeta)}{(\omega/\omega_{ce} - n)} \tag{5.23}$$

since

$$\sum_{n=-\infty}^{\infty} nJ_n^2(\zeta) = 0 \tag{5.24}$$

The result given by Eq. (5.24) follows by changing n to $-n$ and noting that $J_{-n}(\zeta) = (-1)^n J_n(\zeta)$.

With the use of Eq. (5.23) in Eq. (5.9), σ_{xx} can be expressed as

$$\sigma_{xx} = -\frac{i2\pi e^2}{m_e\omega_{ce}} \int_0^{\infty} du_\rho \int_{-\infty}^{\infty} du_z\, u_\rho^2 \zeta^{-2} \frac{\partial f_0(u)}{\partial u_\rho} \sum_{n=-\infty}^{\infty} \frac{n^2 J_n^2(\zeta)}{(\omega/\omega_{ce} - n)} \tag{5.25}$$

The expression for σ_{xx} given by Eq. (5.25) is valid for any equilibrium isotropic, distribution function $f_0(u)$. However, as in the previous cases, further details are restricted to the case in which the equilibrium is characterized by the Maxwell-Boltzmann distribution function (2.4) which is rewritten here for convenience:

$$f_0(u) = N_0 \left(\frac{m_e}{2\pi KT_e}\right)^{3/2} e^{-m_e(u_\rho^2 + u_z^2)/2KT_e} \tag{5.26}$$

If Eq. (5.26) is substituted into Eq. (5.25), the integral with respect to u_z can be carried out using Eqs. (1.4.7). For the simplification of the remaining integral with respect to u_ρ, the following parameter is introduced:

$$\tilde{\nu} = \frac{KT_e k_x^2}{m_e \omega_{ce}^2} \tag{5.27}$$

The differentiation with respect to u_ρ appearing in Eq. (5.25) is explicitly carried out, and, Eqs. (5.14) and (5.27) are used. Then Eq. (5.25) can be simplified and recast into the form:

$$\sigma_{xx} = \frac{iN_0 e^2}{m_e \omega_{ce} \tilde{\nu}^2} \sum_{n=-\infty}^{\infty} \frac{n^2}{(\omega/\omega_{ce} - n)} \int_0^{\infty} d\zeta\, \zeta J_n^2(\zeta) e^{-\zeta^2/2\tilde{\nu}} \tag{5.28}$$

Watson [3] has given the following *Weber's second exponential integral*:

$$\int_0^{\infty} \exp(-p^2 t^2) J_n(at) J_n(bt) t\, dt = \frac{1}{2p^2} \exp\left(-\frac{a^2 + b^2}{4p^2}\right) I_n\left(\frac{ab}{2p^2}\right) \tag{5.29}$$

where $I_n(x)$ is the Bessel function of the second kind and is related to the ordinary Bessel function with an imaginary argument as given by

$$I_n(x) = (-i)^n J_n(ix) \tag{5.30}$$

When Eq. (5.29) is used in Eq. (5.28), the result is

$$\sigma_{xx} = \frac{iN_0 e^2}{m_e \omega_{ce}} \frac{e^{-\tilde{\nu}}}{\tilde{\nu}} \sum_{n=-\infty}^{\infty} \frac{n^2 I_n(\tilde{\nu})}{(\omega/\omega_{ce} - n)} \tag{5.31}$$

The component σ_{zz} of the conductivity dyad is seen from Eqs. (5.6)–(5.8) to be given by

$$\begin{aligned} \sigma_{zz} = -\frac{e^2}{m_e \omega_{ce}} \int_0^{\infty} du_\rho \int_0^{2\pi} d\varphi \int_{-\infty}^{\infty} du_z\, u_\rho u_z \exp(-i\zeta \sin\varphi) \\ \times \int_0^{\infty} d\varphi' \frac{\partial f_0(u)}{\partial u_z} \exp\{g_1(\varphi')\} \end{aligned} \tag{5.32}$$

The integrals occurring in Eq. (5.32) can be evaluated as before leading to the following result:

$$\sigma_{zz} = \frac{iN_0 e^2}{m_e \omega_{ce}} e^{-\tilde{\nu}} \sum_{n=-\infty}^{\infty} \frac{I_n(\tilde{\nu})}{(\omega/\omega_{ce} - n)} \tag{5.33}$$

Four components of the conductivity dyad can be shown to vanish. If the integral expressions for σ_{xz}, σ_{yz}, σ_{zx}, and σ_{zy} are written by using Eqs. (5.6)–(5.8), the integrands are found to be odd functions of u_z and if the integrations with respect to u_z are performed first, it is ascertained that

$$\sigma_{xz} = \sigma_{yz} = \sigma_{zx} = \sigma_{zy} = 0 \tag{5.34}$$

For those aspects of the characteristics of waves propagating across the magnetostatic field in a hot plasma that are treated in this section, the components σ_{xy}, σ_{yx}, and σ_{yy} of the conductivity dyad are not required and therefore detailed expressions for these components are not deduced here.

If the current density given by Eq. (5.8) is combined with the displacement current density, an equivalent relative permittivity dyad $\boldsymbol{\varepsilon}_r$ can be defined. As in Eq. (3.6.26), the expression for $\boldsymbol{\varepsilon}_r$ is found to be

$$\boldsymbol{\varepsilon}_r = \mathbf{1} + \frac{i\boldsymbol{\sigma}}{\omega\varepsilon_0} \tag{5.35}$$

where $\mathbf{1}$ is the unit dyad. It follows from Eqs. (5.31) and (5.33)–(5.35) that

$$\varepsilon_{xx} = 1 - \frac{\omega_{pe}^2}{\omega\omega_{ce}}\frac{e^{-\tilde{\nu}}}{\tilde{\nu}}\sum_{n=-\infty}^{\infty}\frac{n^2 I_n(\tilde{\nu})}{(\omega/\omega_{ce} - n)} \tag{5.36}$$

$$\varepsilon_{zz} = 1 - \frac{\omega_{pe}^2}{\omega\omega_{ce}}e^{-\tilde{\nu}}\sum_{n=-\infty}^{\infty}\frac{I_n(\tilde{\nu})}{(\omega/\omega_{ce} - n)} \tag{5.37}$$

and

$$\varepsilon_{xz} = \varepsilon_{yz} = \varepsilon_{zx} = \varepsilon_{zy} = 0 \tag{5.38}$$

As mentioned before, the expressions for the remaining components of $\boldsymbol{\varepsilon}_r$ are omitted here.

Separation into TM and TEM modes

For the fields with the phase factor as given by Eqs. (5.1), Maxwell's equations (2.33) and (2.34) reduce to

$$k_x\hat{\mathbf{x}} \times \mathbf{E} = \omega\mu_0\mathbf{H} \tag{5.39}$$

$$k_x\hat{\mathbf{x}} \times \mathbf{H} = -\omega\varepsilon_0\boldsymbol{\varepsilon}_r \cdot \mathbf{E} \tag{5.40}$$

In component form, Eqs. (5.39) and (5.40) become

$$0 = \omega\mu_0 H_x \tag{5.41a}$$

$$-k_x E_z = \omega\mu_0 H_y \tag{5.41b}$$

$$k_x E_y = \omega\mu_0 H_z \tag{5.41c}$$

$$0 = -\omega\varepsilon_0(\varepsilon_{xx}E_x + \varepsilon_{xy}E_y) \tag{5.42a}$$

$$-k_x H_z = -\omega\varepsilon_0(\varepsilon_{yx}E_x + \varepsilon_{yy}E_y) \tag{5.42b}$$

$$k_x H_y = -\omega\varepsilon_0\varepsilon_{zz}E_z \tag{5.42c}$$

It follows from Eq. (5.41a) that $H_x = 0$ and the waves are transverse magnetic (TM) with respect to the direction (x) of propagation. The remaining field components decouple into the following two sets:

TM mode:	1.	H_z, E_x, E_y
TEM mode:	2.	E_z, H_y

The first set represents the TM mode and is governed by Eqs. (5.41c), (5.42a), and (5.42b). The second set represents the TEM mode which is a degenerate case of the TM mode and is governed by Eqs. (5.41b) and (5.42c).

The dispersion equations for these two modes can be deduced as follows: The expression for E_x obtained from Eq. (5.42a) is substituted into Eq. (5.42b) with the following result:

$$k_x H_z = \omega\varepsilon_0(\varepsilon_{yy} - \varepsilon_{yx}\varepsilon_{xy}/\varepsilon_{xx})E_y \tag{5.43}$$

If the expression for E_y from Eq. (5.41c) is substituted into Eq. (5.43), it is found that

$$H_z\left[\frac{k_x^2 c^2}{\omega^2} - \frac{(\varepsilon_{xx}\varepsilon_{yy} - \varepsilon_{xy}\varepsilon_{yx})}{\varepsilon_{xx}}\right] = 0 \tag{5.44}$$

For a nontrivial solution for H_z and therefore for E_x and E_y also, it is required that

$$\frac{k_x^2 c^2}{\omega^2} - \frac{(\varepsilon_{xx}\varepsilon_{yy} - \varepsilon_{xy}\varepsilon_{yx})}{\varepsilon_{xx}} = 0 \tag{5.45}$$

which is the dispersion equation for the TM mode. If the expression for H_y from Eq. (5.42c) is used in Eq. (5.41b), it is obtained that

$$E_z\left(\frac{k_x^2 c^2}{\omega^2} - \varepsilon_{zz}\right) = 0$$

As before, a nontrivial solution exists for E_z and H_y only if

$$\frac{k_x^2 c^2}{\omega^2} - \varepsilon_{zz} = 0 \tag{5.46}$$

which is the dispersion equation for the TEM mode.

Quasistatic mode

The dispersion equation (5.45) is very complicated and it has not yet been analyzed in detail. Only a special case of Eq. (5.45) has been extensively investigated [4,5]. Suppose that $k_x c/\omega$ is much greater than ε_{yx} and ε_{yy}. For finite values of E_x and E_y, it is seen from Eq. (5.42b) that H_z is negligibly small and is equal to zero in the limiting case of $k_x c/\omega$ equal to infinity. Then, it follows from Eq. (5.41c) that E_y is also negligibly small and vanishes in the limit of $k_x c/\omega = \infty$. From Eq. (5.42a), it is ascertained that a nontrivial solution for E_x can exist only if

$$\varepsilon_{xx} = 0 \tag{5.47}$$

which is the dispersion equation for the limiting case in which $k_x c/\omega \gg 1$. For this limiting case, the magnetic field is negligible and the electric field is essentially in the direction of propagation. Thus Eq. (5.47) is the dispersion equation for the so-called *quasistatic* wave propagating across the magnetostatic field. In fact, Eq. (5.47) can be derived by omitting the magnetic field at the outset and using the laws of electrostatics instead of Maxwell's equations. Therefore, Eq. (5.47) is also called the dispersion equation for the *electrostatic* wave. For $k_x c/\omega = \infty$, it can be shown that the components of the permittivity dyad are independent of $k_x c/\omega$. As a consequence, the limiting case in which $k_x c/\omega = \infty$ is seen to be obtained from Eq. (5.45) if the dispersion equation (5.47) is satisfied. Although it is strictly correct only if $k_x c/\omega = \infty$, Eq. (5.47) can be considered to be a reasonably good approximation for $k_x c/\omega \gg 1$. It is the special case given by Eq. (5.47) that has been extensively investigated and is proposed to be treated here. Since $k_x c/\omega = \infty$ defines the resonance condition, it is clear that the solution of Eq. (5.47) gives the resonant frequencies for the TM mode propagating across the magnetostatic field in a hot plasma.

Absence of complex solutions for ω

The dispersion equation for the quasistatic wave is found from Eqs. (5.36) and (5.47) to be given by

$$1 = \frac{\omega_{pe}^2}{\omega\omega_{ce}} \frac{e^{-\tilde{\nu}}}{\tilde{\nu}} \sum_{n=-\infty}^{\infty} \frac{n^2 I_n(\tilde{\nu})}{(\omega/\omega_{ce} - n)} \tag{5.48}$$

Since $I_{-n}(\tilde{\nu}) = I_n(\tilde{\nu})$, it follows that

$$0 = \sum_{n=-\infty}^{\infty} n I_n(\tilde{\nu}) \tag{5.49}$$

If Eq. (5.49) is multiplied by $(\omega_{pe}^2/\omega\omega_{ce})(e^{-\tilde{\nu}}/\tilde{\nu})$ and added to Eq. (5.48), it is obtained that

$$\tilde{\nu}\frac{\omega_{ce}^2}{\omega_{pe}^2} = e^{-\tilde{\nu}} \sum_{n=-\infty}^{\infty} \frac{n I_n(\tilde{\nu})}{(\omega/\omega_{ce} - n)} \tag{5.50}$$

It can be shown by setting $y = 0$ in the following relation

$$e^{\tilde{\nu}\cos y} = \sum_{n=-\infty}^{\infty} I_n(\tilde{\nu}) e^{iny} \tag{5.51}$$

that

$$1 = e^{-\tilde{\nu}} \sum_{n=-\infty}^{\infty} I_n(\tilde{\nu}) \tag{5.52}$$

The addition of Eqs. (5.50) and (5.52) yields

$$1 + \tilde{\nu}\frac{\omega_{ce}^2}{\omega_{pe}^2} = \omega e^{-\tilde{\nu}} \sum_{n=-\infty}^{\infty} \frac{I_n(\tilde{\nu})}{(\omega - \omega_{ce} n)} \tag{5.53}$$

Let $\omega = \omega_r + i\omega_i$. From Eq. (5.27), it is ascertained that $\tilde{\nu}$ is real and positive; therefore $I_n(\tilde{\nu})$ is also real and positive. The real and the imaginary parts of Eq. (5.53) are deduced to be given by

$$1 + \tilde{\nu}\frac{\omega_{ce}^2}{\omega_{pe}^2} = e^{-\tilde{\nu}} \sum_{n=-\infty}^{\infty} I_n(\tilde{\nu})\left[1 + \frac{\omega_{ce} n(\omega_r - \omega_{ce} n)}{\{(\omega_r - \omega_{ce} n)^2 + \omega_i^2\}}\right] \tag{5.54}$$

$$0 = -\omega_i e^{-\tilde{\nu}} \sum_{n=-\infty}^{\infty} I_n(\tilde{\nu}) \frac{\omega_{ce} n}{\{(\omega_r - \omega_{ce} n)^2 + \omega_i^2\}} \tag{5.55}$$

If Eq. (5.55) is multiplied by ω_r/ω_i and is added to Eq. (5.54), the result, on using Eq. (5.52), is

$$\tilde{\nu}\frac{\omega_{ce}^2}{\omega_{pe}^2} = -e^{-\tilde{\nu}} \sum_{n=-\infty}^{\infty} I_n(\tilde{\nu}) \frac{(\omega_{ce} n)^2}{\{(\omega_r - \omega_{ce} n)^2 + \omega_i^2\}} \tag{5.56}$$

The left-hand side of Eq. (5.56) is positive and the right-hand side is negative. Hence Eq. (5.56) can never be fulfilled. Therefore, the operation of multiplication by ω_r/ω_i is invalid. If $\omega_r = 0$, Eq. (5.54) can never be satisfied. Therefore, ω_i should vanish. In other words, the dispersion equation (5.53) for the quasistatic wave has only real solutions for ω. Thus, there is neither instability nor temporal damping of the waves.

The real solutions for ω of the dispersion equation (5.50) can be obtained explicitly for two limiting cases. Suppose that $\tilde{\nu} \ll 1$. It is seen from Eq. (5.27) that this case corresponds to the zero-temperature limit. For $\tilde{\nu} \ll 1$, $I_1(\tilde{\nu}) = I_{-1}(\tilde{\nu}) = \tilde{\nu}/2$ and $I_n(\tilde{\nu}) = I_{-n}(\tilde{\nu}) = o(\tilde{\nu}^n)$. If ω/ω_{ce} is not close to n, for the infinite series on the right-hand side of Eq. (5.50), only the terms corresponding to $n = \pm 1$ are significant; the other terms are small and can be neglected. Then Eq. (5.50) can be simplified to yield the following solution for ω:

$$\omega = \omega_u = \sqrt{\omega_{pe}^2 + \omega_{ce}^2} \tag{5.57}$$

Note that ω_u is known as the upper hybrid resonant frequency. The treatment of the characteristics of waves propagating across the magnetostatic field in a cold plasma as contained in Sec. 5.5 gives also the resonance at $\omega = \omega_u$. Hence the present theory of a hot plasma reproduces correctly the results of the cold plasma model.

Harmonic resonances

In addition, the present theory gives other resonant frequencies not predicted by the cold plasma model. If $\omega/\omega_{ce} - n = o(\tilde{\nu}^{n-1})$, for the infinite series on the right-hand side of Eq. (5.50), only the terms corresponding to $\pm n$ are important and the other terms can be neglected. Also, the nth terms can be arranged to satisfy the dispersion equation (5.50). This argument is not applicable for $n = 0$ and the first-order terms reproduce Eq. (5.57). Hence in the limit of $\tilde{\nu} \ll 1$, the hot plasma theory gives also the following resonant frequencies:

$$\omega = n\omega_{ce} \qquad n \geqslant 2 \qquad \text{for } \tilde{\nu} \ll 1 \tag{5.58}$$

The resonances occurring at the second and the higher harmonics of the electron gyromagnetic frequency in the limit of zero temperature are not predicted by the cold plasma model.

Since $k_x c/\omega \gg 1$, it follows from Eq. (5.27) that $\tilde{\nu} \gg 1$ for a sufficiently high temperature. For this case $e^{-\tilde{\nu}} I_n(\tilde{\nu}) = o(\tilde{\nu}^{-1/2})$. Then, the dispersion equation (5.50) yields the following solutions

$$\omega = n\omega_{ce} \qquad n \geqslant 1 \qquad \text{for } \tilde{\nu} \gg 1 \tag{5.59}$$

Thus, in the high temperature limit, the resonances occur at the fundamental as well as at all the harmonics of the electron gyromagnetic frequency.

For intermediate values of $\tilde{\nu}$, the dispersion equation can be solved by a numerical procedure. For this purpose, Eq. (5.50) is rewritten as

$$\tilde{\nu}\frac{\omega_{ce}^2}{\omega_{pe}^2} = F(\omega/\omega_{ce}, \tilde{\nu}) \qquad F(\omega/\omega_{ce}, \tilde{\nu}) = 2e^{-\tilde{\nu}} \sum_{n=1}^{\infty} \frac{n^2 I_n(\tilde{\nu})}{[(\omega/\omega_{ce})^2 - n^2]} \tag{5.60}$$

For a specified value of $\tilde{\nu}$, $F(\omega/\omega_{ce}, \tilde{\nu})$ is plotted in terms of ω/ω_{ce} as shown in Fig. 6.4 for $\tilde{\nu} = 0.1$. The intersection points of this curve with the horizontal line corresponding to $\tilde{\nu}\omega_{ce}^2/\omega_{pe}^2$ give the resonant frequencies in the normalized form ω/ω_{ce}. It is verified from Fig. 6.4 that for $\tilde{\nu} \ll 1$, the first harmonic of the electron gyromagnetic frequency is not a solution of the dispersion equation (5.50).

The variation of the normalized resonant frequencies ω/ω_{ce} as a function of $\tilde{\nu}$ has also been investigated numerically [6]. In Fig. 6.5, the normalized resonant frequency ω/ω_{ce} is depicted as a function of $\sqrt{\tilde{\nu}}$. It is seen from Fig. 6.5 that the resonant frequency curve corresponding to $\omega = n\omega_{ce}$ for $\tilde{\nu} \ll 1$, as $\tilde{\nu}$ becomes large, goes over to the curve corresponding to $\omega = n\omega_{ce}$ for $\tilde{\nu} \gg 1$, provided $n\omega_{ce} > \omega_u$. These curves all have one maximum value in ω/ω_{ce} and the maximum value of ω becomes closer to $n\omega_{ce}$ as n is increased. The resonant frequency curve corresponding to $\omega = \omega_u$ for $\tilde{\nu} \ll 1$ becomes the curve corresponding to $\omega = n\omega_{ce}$, where $n\omega_{ce}$ is next below ω_u. The remaining curves starting at $\omega = n\omega_{ce}$ for $\tilde{\nu} = 0$ go over to the curves corresponding to $\omega = (n-1)\omega_{ce}$ for $\tilde{\nu} \gg 1$, where $n \geqslant 2$. It is observed that

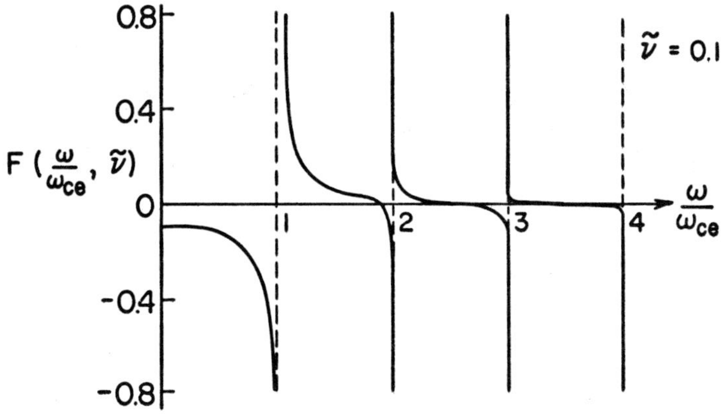

Fig. 6.4. Plot of $F(\omega/\omega_{ce}, \tilde{\nu})$ versus ω/ω_{ce} for $\tilde{\nu} = 0.1$.

below each resonant frequency curve corresponding to frequencies greater than the upper hybrid resonant frequency, there is a range of ω in which resonance does not occur for any value of $\tilde{\nu}$. These gaps in the resonant frequencies were first deduced by Gross [7].

There is no Landau damping for the quasistatic waves propagating across the magnetostatic field. But Landau damping is present for the longitudinal waves in an isotropic plasma and the longitudinal waves propagating along the magnetostatic field in an anisotropic plasma. This is an important difference between the quasistatic waves treated here and the longitudinal waves analyzed previously.

The present analysis of the characteristics of the quasistatic waves can be generalized to an arbitrary direction of propagation and a treatment of this subject exists in the available literature [4,5].

TEM mode

The dispersion equation for the TEM mode propagating across the magnetostatic field in a hot plasma is seen from Eqs. (5.37) and (5.46) to be given by

$$\frac{k_x^2 c^2}{\omega^2} = 1 - \frac{\omega_{pe}^2}{\omega\omega_{ce}} e^{-\tilde{\nu}} \sum_{n=-\infty}^{\infty} \frac{I_n(\tilde{\nu})}{(\omega/\omega_{ce} - n)} \tag{5.61}$$

As before, Eq. (5.61) can be analyzed numerically. However, even without a numerical work, some useful results can be obtained directly from Eq. (5.61). For $\tilde{\nu} \ll 1$, only the term corresponding to $n = 0$ is significant; all the other terms are small and can be neglected with the result that Eq. (5.61) becomes

$$\frac{k_x^2 c^2}{\omega^2} = 1 - \frac{\omega_{pe}^2}{\omega^2} \tag{5.62}$$

where the relation $I_0(0) = 1$ is used. The characteristics of the TEM mode propagating across the magnetostatic field in a cold plasma were analyzed in Sec. 5.5 and the result of the cold plasma model, when written in the present unnormalized form, is identical to Eq. (5.62). Thus, the hot-plasma theory reproduces correctly the result of the cold plasma model in the limit of zero temperature.

For $\tilde{\nu} \gg 1$, since $e^{-\tilde{\nu}} I_n(\tilde{\nu}) = o(\tilde{\nu}^{-1/2})$, Eq. (5.61) reduces to

$$\frac{k_x c}{\omega} = 1 \tag{5.63}$$

indicating that the wave characteristics become identical to those in free space. The condition $\tilde{\nu} \gg 1$ together with Eqs. (5.27) and (5.63) is equivalent to the requirement

$$\omega \gg \omega_{ce} \left(\frac{m_e c^2}{K T_e} \right)^{1/2} \tag{5.64}$$

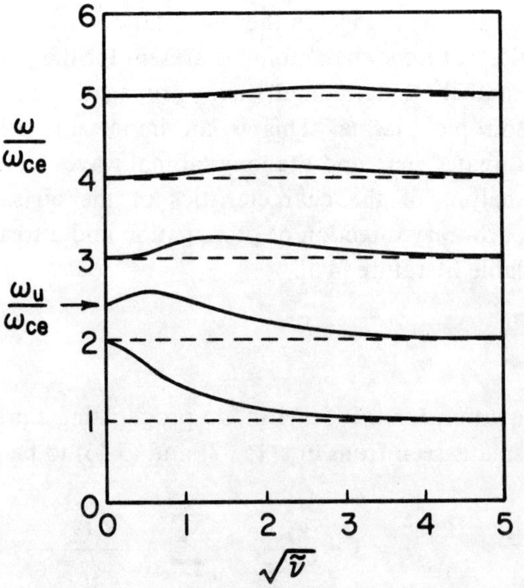

Fig. 6.5. Resonant frequency curves for the quasistatic waves propagating across the magnetostatic field for $(\omega_{ce}/\omega_{pe})^2 = 0.2$.

which shows that the frequency is very high. For very high frequencies, Eq. (5.62) reduces to Eq. (5.63). Therefore, even for $\tilde{\nu} \gg 1$, the hot-plasma theory gives a result which is in accordance with that predicted by the cold plasma model.

For $\omega = n\omega_{ce}$, where n is an integer, Eq. (5.61) shows that $k_x c/\omega = \infty$. Therefore, according to the hot-plasma theory, the TEM mode has resonances at the fundamental and all the harmonics of the electron gyromagnetic frequency. These harmonic resonances are not predicted by the cold plasma model.

The foregoing discussion shows that the analysis of the characteristics of waves propagating across the magnetostatic field in a hot plasma is difficult. For the general case of propagation in an arbitrary direction with respect to the magnetostatic field, the analysis becomes more complicated, insofar as the necessary details are involved. Certain interesting aspects of waves in a hot plasma with a magnetostatic field are treated by Stix [8], Montgomery and Tidman [9], Akhiezer [10], and, Clemmow and Dougherty [11].

6.6. Relaxation model for the collision term

In the preceding four sections, an introductory treatment of the theory of plasma oscillations was given based on the Boltzmann equation in which the collision term was set equal to zero. The resulting equation is known as the Boltzmann-Vlasov equation and this equation is not very restrictive, as it might appear since a part of the interaction among the particles is included in the Lorentz force term. In this and the following four sections, the transport processes in a plasma are investigated using another approximation for the Boltzmann equation. In this approximation, the collision term is replaced by a simple relaxation model. The Boltzmann equation with the relaxation model for the collision term has proved useful, particularly for the weakly ionized plasma.

Transport processes

Consider a plasma with spatial inhomogeneity. For example, the plasma density may be assumed to vary in space. On account of their random thermal motions, the electrons have collisions with the other particles. Even though individually the electrons move at random, the collisional interactions create a tendency for the electrons to drift from the high-density to the low-density regions. In a similar manner, if there is a variation of the plasma temperature in space, the electrons tend to drift from the high-temperature to the low-temperature regions and this drift is promoted by the collisional interactions of the electrons with the other particles. The drifting of the particles in a spatially inhomogeneous plasma caused by the collisional interactions is called *diffusion*. In view of the presence of the drift velocity, it follows that the velocity distribution function is *anisotropic*. In course of time, an equilibrium is reached and the average velocity of the electrons vanishes. Therefore, the velocity distribution function is *isotropic* in the equilibrium state.

The drifting of the electrons can occur even in a spatially homogeneous plasma if an external force is present. As an example, consider a spatially homogeneous plasma under the action of an electric field. As before, individually the electrons move at random because of their finite temperature but on the average, the electrons have a velocity in the direction opposite to that of the electric field. The drifting of the particles in a plasma induced by an external force is called *mobility*. Since the electrons have mass and carry charge, the drifting of the electrons is accompanied by the transport of mass and the conduction of electricity. Previously the electrical conductivity in a plasma was studied using the macroscopic description of the plasma.

In a spatially inhomogeneous plasma acted on by an external force, the particle current is due to both the diffusion and the mobility. These two transport processes can be separated only if the average velocity due to the external force is small compared to the average thermal velocity. As mentioned before, the particle current is accompanied by the transport of mass and electrical conduction. In addition, there are other transport processes. Since the electrons have kinetic energy, their drift is accompanied by the transport of energy and heat conduction. In this treatment, attention is restricted to the diffusion and the electrical conduction only. Moreover, these transport processes are proposed to be studied using the Boltzmann equation with a relaxation model for the collision term.

Reasonableness of the relaxation model

Let $f(\mathbf{r}, \mathbf{u}, t)$ be the anisotropic velocity distribution function characterizing the electrons in the nonequilibrium state and let $f_0(\mathbf{r}, u)$ be the isotropic velocity distribution function describing the equilibrium state. In the relaxation model, the collision term is approximated as follows:

$$\left(\frac{\partial f}{\partial t}\right)_{\text{coll}} = -\nu_r(u)[f(\mathbf{r}, \mathbf{u}, t) - f_0(\mathbf{r}, u)] \tag{6.1}$$

where $1/\nu_r(u) = \tau(u)$ has the dimension of time and is known as the relaxation time. The approximation (6.1) is particularly applicable for a very weakly ionized gas in which the neutral particles are uniformly distributed in space. Since the ions are so few in number compared to the neutral particles, the electron-ion collisions are unimportant and only the collisions of the electrons with the neutral particles are significant. In view of the fact that the neutral particles are uniformly distributed, it follows that the frequency of collision between the electrons and the neutral particles is independent of the position coordinate $\mathbf{r}$ and depends only on the relative speed between the electrons and the neutral particles. The heavy neutral particles are assumed to be at rest and therefore the collision frequency depends only on the speed u of the electrons.

It is possible to argue the plausibility of the approximation (6.1) in the following manner. Suppose that the electrons are distributed uniformly in space. Let $\mathbf{F}$ be the external force acting on the plasma and let $f(\mathbf{u}, t)$ be the velocity distribution function of the electrons. The external force $\mathbf{F}$ is removed at time $t = 0$. For times subsequent to $t = 0$, it is seen from Eqs. (1.3.29) and (6.1) that $f(\mathbf{u}, t)$ is governed by the following equation:

$$\frac{\partial}{\partial t} f(\mathbf{u}, t) = -\nu_r(u)[f(\mathbf{u}, t) - f_0(u)] \tag{6.2}$$

where $f_0(u)$ is the isotropic distribution function characterizing the equilibrium state. Since

$$\frac{\partial}{\partial t} f(\mathbf{u}, t) = \frac{\partial}{\partial t}[f(\mathbf{u}, t) - f_0(u)]$$

the solution of Eq. (6.2) is deduced to be given by

$$f(\mathbf{u}, t) - f_0(u) = [f(\mathbf{u}, 0) - f_0(u)]e^{-\nu_r(u)t} \tag{6.3}$$

As time t increases, the right-hand side of Eq. (6.3) goes to zero and the anisotropic distribution function $f(\mathbf{u}, t)$ approaches the isotropic distribution function $f_0(u)$, the rate of approach being dependent on the collision frequency $\nu_r(u)$. Therefore, $\nu_r(u)$ is called the velocity-dependent relaxation collision frequency and $\tau(u)$ is called the relaxation time. Thus the relaxation model (6.1) for the collision term predicts correctly the relaxation of the anisotropic distribution function to the isotropic distribution function when the external force causing the deviation from the equilibrium state is removed.

Another argument can be advanced to emphasize the reasonableness of the relaxation model (6.1). Consider the special case in which $\nu_r(u) = \nu_r$, that is, the relaxation collision frequency is independent of the speed of the electrons. Then, the collision term $\mathbf{P}_{\text{coll}}$ appearing in the momentum transport equation is found from Eqs. (1.9.12) and (6.1) to be given by

$$\begin{aligned} \mathbf{P}_{\text{coll}} &= -\int_{-\infty}^{\infty} m_e \mathbf{u}\nu_r[f(\mathbf{r}, \mathbf{u}, t) - f_0(\mathbf{r}, u)]\, d\mathbf{u} \\ &= -m_e \nu_r \int_{-\infty}^{\infty} \mathbf{u} f(\mathbf{r}, \mathbf{u}, t)\, d\mathbf{u} \\ &= -m_e \nu_r N_e \mathbf{v} \end{aligned} \tag{6.4}$$

where N_e is the number density of the electrons. A comparison of Eq. (6.4) with Eq. (5.2.4) shows that if the relaxation collision frequency is independent of the speed of the electrons, it becomes identical to the *constant collision frequency* ν_c which was introduced phenomenologically in the momentum transport equation as given by Eq. (5.2.3) to take into account the effect of the collisional interactions of the electrons with the neutral particles in a weakly ionized plasma. Thus the relaxation model (6.1) leads to results which for a special case are identical to those obtained with the use of the constant collision frequency as introduced previously in connection with magnetoionic theory.

Drawback of the relaxation model

Although it appears reasonable, the relaxation model for the collision term as given by Eq. (6.1) oversimplifies the entire relaxation phenomena. In order to establish this result, again suppose that the electrons are distributed uniformly in space with the result that Eq. (6.3) is applicable. Let $g(\mathbf{u})$ be any property of the electrons which depends on the electron velocity $\mathbf{u}$. It is convenient to introduce the following notations:

$$\int_{-\infty}^{\infty} f(\mathbf{u}, t)\, d\mathbf{u} = N(t) \qquad \int_{-\infty}^{\infty} f_0(u)\, d\mathbf{u} = N_0 \tag{6.5}$$

$$\int_{-\infty}^{\infty} g(\mathbf{u}) f(\mathbf{u}, t)\, d\mathbf{u} = N(t)Q(t) \qquad \int_{-\infty}^{\infty} g(\mathbf{u}) f_0(u)\, d\mathbf{u} = N_0 Q_0 \tag{6.6}$$

It follows from Eqs. (6.5) that N_0 is the equilibrium number density of the electrons and $N(t)$ is the corresponding quantity pertaining to the nonequilibrium state. In a similar manner, Eqs. (6.6) show that $Q(t)$ and Q_0 are the averages of the function $g(\mathbf{u})$ over the distribution of velocities in the nonequilibrium and the equilibrium states respectively.

Let the external force $\mathbf{F}$ which perturbs the equilibrium state be removed at time $t = 0$. Then the relaxation of the velocity distribution function $f(\mathbf{u}, t)$ is given by Eq. (6.3). Since $f(\mathbf{u}, t)$ approaches $f_0(u)$, it follows that $N(t)$ approaches N_0. Consider the special case in which $\nu_r(u)$ is independent of the speed u of the electrons. If both sides of Eq. (6.3) are multiplied by $g(\mathbf{u})$ and integrated over the entire velocity distribution, it is found with the help of Eqs. (6.6) that

$$N(t)Q(t) - N_0 Q_0 = [N(0)Q(0) - N_0 Q_0]e^{-\nu_r t} \tag{6.7}$$

The first difficulty with the relaxation model given by Eq. (6.1) appears from Eqs. (6.5) which show that the number density is not conserved. It is possible to modify the model for the collision term in such a way that the number density is preserved [12]. Since $N(t)$ approaches N_0, it follows from Eq. (6.7) that $Q(t)$ approaches Q_0. However, more important is the fact that every average property $Q(t)$ of the electrons approaches the corresponding equilibrium value Q_0 at a rate which depends only on ν_r. Thus the relaxation model as given by Eq. (6.1) predicts the *same* relaxation time for the approach of the average velocity and the kinetic energy to their respective equilibrium values. A detailed analysis of the collision process between the electrons and the neutral particles does not bear out this prediction.

As explained in Sec. 4.5, a transfer cross section can be defined for each physical property of the electrons. The momentum transfer cross section σ_m as defined in Eq. (4.5.10) is given by

$$\sigma_m = \int_0^{\pi} [1 - \cos(\chi)]\sigma(\chi)2\pi \sin(\chi)\, d\chi \tag{6.8}$$

where $\sigma(\chi)$ is the differential scattering cross section and χ is the scattering angle. The fractional energy lost by an electron on collision with a stationary neutral particle is given by Eq. (4.6.10). The expression for the energy transfer cross section σ_e can be deduced by using Eq. (4.6.10) in the same manner as Eq. (4.5.10) was obtained. The result is

$$\sigma_e = \frac{2m_e}{M_n} \int_0^{\pi} [1 - \cos(\chi)]\sigma(\chi)2\pi \sin(\chi)\, d\chi = \frac{2m_e}{M_n}\sigma_m \tag{6.9}$$

In Sec. 4.10, the collision frequency is shown to be proportional to the transfer cross section. Hence it follows from Eqs. (6.8) and (6.9) that if $1/\nu_r$ is the relaxation time for the momentum, the relaxation time for the energy is given by $M_n/2m_e\nu_r$. For nonrelativistic velocities, the relaxation time for the momentum is the same as that for the average velocity. Since the mass M_n of a neutral particle is much larger than the mass m_e of an electron, it follows that the relaxation time for the energy is very long compared to that for the average velocity.

A simple physical explanation can be given for the existence of the difference in the relaxation times for the average velocity and the kinetic energy in a weakly ionized gas. When an electron collides with a massive neutral particle, the recoil of the neutral particle is negligibly small with the result that there is an extremely small fractional change in the kinetic energy or the speed of the electron. This change is of the order of the ratio of the mass of an electron to that of a neutral particle. Even though its speed is changed only very slightly, depending on the impact parameter, the electron can get scattered through all angles from $\chi = 0$ to $\chi = \pi$. Therefore, the fractional change in the electron velocity is of the order of unity. After averaging over all the scattering angles and the velocity distribution of the electrons, the rate of change of the kinetic energy of the electrons is of the order of the ratio of the mass of an electron to that of a neutral particle whereas the corresponding rate of change of the average velocity of the electrons is of the order of unity. Hence the relaxation time for the kinetic energy of the electrons is greater than that for the average velocity by a factor which is of the order of the ratio of the mass of a neutral particle to that of an electron.

The relaxation model for the collision term as given by Eq. (6.1) does not correctly predict the different relaxation rates for the various physical quantities associated with the electrons. In spite of this drawback, the relaxation model is used for the derivation of the various transport coefficients partly because of its simplicity and partly because it gives the needed first approximation to these transport coefficients.

6.7. Conductivity and diffusion for a constant collision frequency

It is instructive to evaluate the conductivity and the diffusion first for the special case in which the relaxation collision frequency is independent of the speed of the electrons. For this special case, the collision term $\mathbf{P}_{\text{coll}}$ appearing in the momentum transport equation (1.9.14) was computed in Eq. (6.4) and was found to be the same as that for the constant collision frequency introduced previously in Sec. 5.2. Consequently, the conductivity and the diffusion for the constant collision frequency ν_c can be deduced from the momentum transport equation (1.9.14) and Eq. (6.4). For this purpose, it is assumed that the deviations from the equilibrium state caused by the spatial inhomogeneity and the external forces are very small. In other words, the spatial inhomogeneity and the external forces are first-order small quantities. As a result, the average velocity $\mathbf{v}_e$ is also a first-order small quantity. Since the velocity distribution is approximately isotropic, the pressure dyad $\mathbf{\Psi}$ can be approximated by the scalar pressure P as given by Eq. (1.9.18) and the second-order small quantity $(\mathbf{v}_e \cdot \nabla)\mathbf{v}_e$ can be neglected.

Under the approximations mentioned in the foregoing, the momentum transport equation (1.9.14) together with Eqs. (1.9.18) and (6.4) becomes for the electrons characterized by the mass m_e and the charge $-e$:

$$N_e m_e \frac{\partial}{\partial t}\mathbf{v}_e = -\nabla(N_e K T_e) - N_e e(\mathbf{E} + \mathbf{v}_e \times \mathbf{B}_0) - m_e N_e \nu_c \mathbf{v}_e \tag{7.1}$$

The magnetostatic field is assumed to be so large in comparison with any time-varying magnetic field that in the Lorentz force term $-N_e e\mathbf{v}_e \times \mathbf{B}_0$, only the magnetostatic field $\mathbf{B}_0$ is used. The average velocity $\mathbf{v}_e$ is contributed partly by the electric field and partly by the gradient of the kinetic pressure. It is convenient to evaluate these two contributions separately.

Mobility and conductivity

For the purpose of deducing the mobility and the conductivity, the contribution to the average velocity due to the electric field alone is required. Therefore the pressure gradient term in Eq. (7.1) can be set equal to zero with the result that

$$m_e \frac{\partial}{\partial t}\mathbf{v}_e = -e(\mathbf{E} + \mathbf{v}_e \times \mathbf{B}_0) - m_e \nu_c \mathbf{v}_e \tag{7.2}$$

Only the time-harmonic case in which all the first-order perturbations have the time dependence of the form $\exp(-i\omega t)$ is to be investigated. Then Eq. (7.2) becomes equal to that given by Eqs. (5.2.6) and (5.2.7). As before, the magnetostatic field $\mathbf{B}_0$ is assumed to be in the z-direction. The rectangular components of the average velocity are the same as those given in Eq. (5.2.10). The linearized particle current density $\mathbf{\Gamma}_e$ for the electrons can be expressed as

$$\mathbf{\Gamma}_e = N_0 \mathbf{v}_e = N_0 \boldsymbol{\mu} \cdot \mathbf{E} \tag{7.3}$$

where N_0 is the equilibrium number density of the electrons. Note that the difference $(N_e - N_0)$ is a first-order small quantity. The mobility dyad $\boldsymbol{\mu}$ is defined by Eq. (7.3). In matrix form, the mobility dyad can be shown from Eqs. (7.3) and (5.2.10) to be given by

$$\boldsymbol{\mu} = \begin{bmatrix} \mu_\perp & -\mu_H & 0 \\ \mu_H & \mu_\perp & 0 \\ 0 & 0 & \mu_\parallel \end{bmatrix} \tag{7.4}$$

where

$$\mu_\perp = -ie\tilde{\omega}/m_e(\tilde{\omega}^2 - \omega_{ce}^2) \qquad \mu_H = e\omega_{ce}/m_e(\tilde{\omega}^2 - \omega_{ce}^2) \tag{7.5a}$$

$$\mu_\parallel = -ie/m_e\tilde{\omega} \tag{7.5b}$$

The subscripts $\parallel$ and $\perp$ stand for parallel and perpendicular to the magnetostatic field, respectively, and the subscript H denotes the Hall effect. It is to be noted that the Hall effect current which is in a direction perpendicular to both the magnetostatic field and the electric field has already been mentioned in Sec. 3.6.

The electric current density $\mathbf{J}_e$ is obtained as

$$\mathbf{J}_e = -e\boldsymbol{\Gamma}_e = -eN_0\boldsymbol{\mu}\cdot\mathbf{E} = \tilde{\boldsymbol{\sigma}}_e\cdot\mathbf{E} \tag{7.6}$$

with the result that the electrical conductivity $\tilde{\boldsymbol{\sigma}}_e$ is related to the mobility as follows:

$$\tilde{\boldsymbol{\sigma}}_e = -N_0 e\boldsymbol{\mu} \tag{7.7}$$

As before, the tilde on $\tilde{\boldsymbol{\sigma}}_e$ is used to indicate that the collisional interactions have been taken into account. The conductivity dyad can be deduced from Eqs. (7.4), (7.5), and (7.7) to be given by

$$\tilde{\boldsymbol{\sigma}}_e = \begin{bmatrix} \sigma_\perp & -\sigma_H & 0 \\ \sigma_H & \sigma_\perp & 0 \\ 0 & 0 & \sigma_\parallel \end{bmatrix} \tag{7.8}$$

where

$$\sigma_\perp = iN_0e^2\tilde{\omega}/m_e(\tilde{\omega}^2 - \omega_{ce}^2) \qquad \sigma_H = -N_0e^2\omega_{ce}/m_e(\tilde{\omega}^2 - \omega_{ce}^2) \tag{7.9a}$$

$$\sigma_\parallel = iN_0e^2/m_e\tilde{\omega} \tag{7.9b}$$

In the absence of the magnetostatic field $\omega_{ce} = 0$. It follows therefore from Eqs. (7.5) and (7.9) that

$$\mu_H = 0 \qquad \mu_\perp = \mu_\parallel \tag{7.10}$$

and

$$\sigma_H = 0 \qquad \sigma_\perp = \sigma_\parallel \tag{7.11}$$

It is seen from Eqs. (7.10) and (7.11) that for $\mathbf{B}_0 = 0$, the mobility and the conductivity dyads reduce to scalar quantities equal to $\mu_\parallel$ and $\sigma_\parallel$, respectively.

The steady state in which no physical property of the system varies with time corresponds to $\omega = 0$ and $\tilde{\omega} = i\nu_c$. For this steady state case, the components of the mobility and the conductivity dyads, as given by Eqs. (7.5) and (7.9), respectively, simplify to yield the following expressions:

$$\mu_\perp = -e\nu_c/m_e(\nu_c^2 + \omega_{ce}^2) \qquad \mu_H = -e\omega_{ce}/m_e(\nu_c^2 + \omega_{ce}^2) \tag{7.12a}$$

$$\mu_\parallel = -e/m_e\nu_c \tag{7.12b}$$

$$\sigma_\perp = N_0 e^2 \nu_c/m_e(\nu_c^2 + \omega_{ce}^2) \qquad \sigma_H = N_0 e^2 \omega_{ce}/m_e(\nu_c^2 + \omega_{ce}^2) \tag{7.13a}$$

$$\sigma_\parallel = N_0 e^2/m_e\nu_c \tag{7.13b}$$

Diffusion

In order to investigate the diffusion, it is sufficient to take into account the contribution to the average velocity arising from the pressure gradient term alone. Also attention is restricted to the steady state case corresponding to the absence of time variation in all the physical parameters of the system. Hence Eq. (7.1) when expressed in terms of the particle current density $\mathbf{\Gamma}_e$ simplifies to the form

$$\nu_c\mathbf{\Gamma}_e + \omega_{ce}\mathbf{\Gamma}_e \times \hat{\mathbf{z}} = -\frac{1}{m_e}\nabla(N_e KT_e) \tag{7.14}$$

If Eq. (7.14) is written in component form, a set of simultaneous equations in the components of $\mathbf{\Gamma}_e$ is obtained. When these simultaneous equations are solved, $\mathbf{\Gamma}_e$ can be expressed as follows:

$$\mathbf{\Gamma}_e = -\nabla \cdot [\mathbf{D}_f N_e] \tag{7.15}$$

The dyadic free diffusion coefficient $\mathbf{D}_f$ is defined by Eq. (7.15). In the matrix form, the expression for $\mathbf{D}_f$ is found to be given by

$$\mathbf{D}_f = \begin{bmatrix} D_\perp & D_H & 0 \\ -D_H & D_\perp & 0 \\ 0 & 0 & D_\parallel \end{bmatrix} \tag{7.16}$$

where

$$D_\perp = \frac{KT_e}{m_e}\frac{\nu_c}{(\nu_c^2 + \omega_{ce}^2)} \qquad D_\mathrm{H} = \frac{KT_e}{m_e}\frac{\omega_{ce}}{(\nu_c^2 + \omega_{ce}^2)} \qquad D_\parallel = \frac{KT_e}{m_e}\frac{1}{\nu_c} \tag{7.17}$$

It has been assumed that the constant collision frequency ν_c and the gyromagnetic angular frequency ω_{ce} of the electrons are independent of the position coordinates. The particle current density given in Eq. (7.15) is due to the spatial variation of the kinetic pressure and is called the *diffusion current density*.

If the magnetostatic field is absent, the Hall diffusion coefficient D_H vanishes and the perpendicular diffusion coefficient $D_\perp$ becomes equal to the parallel diffusion coefficient $D_\parallel$. Therefore, the dyadic diffusion coefficient $\mathbf{D}_f$ reduces to a scalar diffusion coefficient $D_\parallel$ and the expression corresponding to Eq. (7.15) becomes

$$\boldsymbol{\Gamma}_e = -\nabla(D_\parallel N_e) \tag{7.18}$$

The ratio of the scalar diffusion coefficient $D_\parallel$ to the scalar mobility $\mu_\parallel$ is found to be

$$D_\parallel/\mu_\parallel = -KT_e/e \tag{7.19}$$

The relation given by Eq. (7.19) is independent of the collision frequency and is called *the Einstein relation.* It is also found from Eqs. (7.12) and (7.17) that

$$\mathbf{D}_f = -\frac{KT_e}{e}\boldsymbol{\mu}_f \tag{7.20}$$

Thus the dyadic diffusion coefficient is seen to be proportional to the dyadic mobility, the proportionality factor being given by the Einstein relation.

In the discussion of the diffusion phenomenon Eq. (1.9.18) has been used and Eq. (1.9.18) implies that the electrons are characterized by the Maxwell-Boltzmann distribution function. Therefore the diffusion coefficient and the Einstein relation are valid only for the Maxwellian distribution of velocities.

It is instructive to compare the magnitudes of the diffusion currents for the electrons with the corresponding quantities for the ions in the various directions with respect to the magnetostatic field. The directions (*a*) parallel to the magnetostatic field, (*b*) perpendicular to the magnetostatic field but parallel to the gradient of the pressure, and (*c*) perpendicular to both the magnetostatic field and the pressure gradient are the three important directions to consider. The ions are characterized by the mass m_i and the charge e. Since the gyromagnetic angular frequency of the ions is given by $\omega_{ci} = eB_0/m_i$, it follows that the replacing of m_e and $-e$ with m_i and e respectively is equivalent to changing ω_{ce} to $-\omega_{ci}$. Hence the diffusion coefficients for the ions can be obtained from the corresponding coefficients for the electrons as given by Eqs. (7.17) by changing m_e and ω_{ce} to m_i and $-\omega_{ci}$, respectively. The number densities of the electrons and the ions are approximately equal and therefore the diffusion currents for the ions are proportional to the corresponding diffusion coefficients.

For the same temperature and the collision frequency, the diffusion current along the magnetostatic field is seen from the expression for $D_{\parallel}$ in Eqs. (7.17) to be inversely proportional to the mass of the particle and hence is larger for the electrons than for the ions.

For the sake of convenience, the direction perpendicular to the magnetostatic field but parallel to the pressure gradient is called the perpendicular direction and the direction parallel to the magnetostatic field is designated as the parallel direction. Since $D_{\perp} < D_{\parallel}$, the diffusion current in the perpendicular direction is always less than that in the parallel direction both for the electrons and for the ions. For $\omega_{ce} \ll \nu_c$, $D_{\perp} = KT_e/m_e\nu_c = D_{\parallel}$. Therefore, for a very weak magnetostatic field, the diffusion current for the electrons is larger than that for the ions in the perpendicular direction provided the temperature and the collision frequency are approximately equal for the two kinds of particles. For $\omega_{ce} \gg \nu_c$, $D_{\perp} = KT_e m_e \nu_c/e^2 B_0^2$. If the temperature and the collision frequency are approximately equal for both types of particles, the diffusion current for the ions in the perpendicular direction is very large compared to that for the electrons since the ion mass is much larger than the electron mass. In the parallel and the perpendicular directions, the electron and the ion diffusion particle currents are in the same direction.

The direction perpendicular to both the magnetostatic field and the pressure gradient is that of the Hall current. The Hall diffusion particle currents for the electrons and the ions are in the opposite directions. For $\omega_{ce} \ll \nu_c$, $D_{\mathrm{H}} = KT_e eB_0/m_e^2\nu_c^2$. Therefore for a weak magnetostatic field, and, for the same temperature and the collision frequency, the Hall diffusion current for the electrons is much larger than that of the ions. For $\omega_{ce} \gg \nu_c$, $D_{\mathrm{H}} = KT_e/eB_0$. Hence for a very strong magnetostatic field, the Hall diffusion coefficient is independent of the mass of the particle and the collision frequency; it is directly proportional to the particle energy and is inversely proportional to the strength of the magnetostatic field.

The mobility is proportional to the diffusion coefficient and the proportionality factor has a different sign for the electrons and the ions. Hence the foregoing remarks excepting those which pertain to the current directions are applicable also to the relative magnitudes of the electron and the ion mobilities. Since the particle temperature does not appear in the mobility, the stated requirement on the temperatures of the particles has to be removed when applying the foregoing remarks to the relative magnitudes of the electron and the ion mobilities under various situations.

6.8. Ambipolar diffusion

In the previous section, the diffusion of the electrons and that of the ions in a weakly ionized gas were treated separately without taking into account their interaction. A simple treatment of the diffusion phenomenon taking into account the mutual interaction between the electrons and the ions is proposed to be presented in this section.

Suppose that the magnetostatic field is absent and that the particle temperature is independent of position. For the stationary state in which the physical parameters do not vary with time, the diffusion equation for the electrons is obtained from Eq. (7.1) or directly from Eqs. (7.17) and (7.18) to be given by

$$\boldsymbol{\Gamma}_e = -D_e \nabla N_e \qquad D_e = KT_e/m_e \nu_{ce} \tag{8.1a, b}$$

The subscript e has been added to ν_c to indicate that the constant collision frequency ν_{ce} pertains to the collisional interactions between the electrons and the neutral particles. The diffusion equation for the ions is given by

$$\boldsymbol{\Gamma}_i = -D_i \nabla N_i \qquad D_i = KT_i/m_i \nu_{ci} \tag{8.2a, b}$$

where ν_{ci} is the constant collision frequency for the interactions between the ions and the neutral particles. Since the diffusion coefficient is inversely proportional to the particle mass as seen from Eqs. (8.1b) and (8.2b), the electrons diffuse faster leaving an excess of positive charge behind. The resulting charge density gives rise to an electric field $\mathbf{E}_s$ whose effect is not included in the diffusion equations (8.1a) and (8.2a). The diffusion in which the effect of the space charge is not included is known as the *free diffusion* and therefore D_e and D_i as given by Eqs. (8.1b) and (8.2b) are called the free diffusion coefficients.

The electric field $\mathbf{E}_s$ caused by the space charge is in the direction of the diffusion currents. Therefore, this electric field retards the electrons and accelerates the ions with the result that the two kinds of charged particles diffuse at a rate which is intermediate in value to their free diffusion rates. This combined diffusion of the electrons and the ions forced by the space charge electric field is called *the ambipolar diffusion*.

In order to investigate the characteristics of the ambipolar diffusion, it is necessary to retain the electric field due to the space charge in the diffusion equation. Thus the diffusion equation for the electrons is deduced from Eq. (7.1) as

$$\boldsymbol{\Gamma}_e = -D_e \nabla N_e + \mu_e N_e \mathbf{E}_s \tag{8.3}$$

where

$$\mu_e = -e/m_e \nu_{ce} \tag{8.4}$$

is the mobility of the electrons. In a similar manner, the diffusion equation for the ions is given by

$$\boldsymbol{\Gamma}_i = -D_i \nabla N_i + \mu_i N_i \mathbf{E}_s \tag{8.5}$$

where

$$\mu_i = e/m_i \nu_{ci} \tag{8.6}$$

is the mobility of the ions. Since μ_e is negative and μ_i is positive, it is verified from Eqs. (8.3) and (8.5) that the electric field due to the space charge reduces the particle current density of the electrons and augments that of the ions, as anticipated from physical considerations.

The space charge electric field $\mathbf{E}_s$ is governed by Poisson's equation:

$$\nabla \cdot \mathbf{E}_s = (N_i - N_e)e/\varepsilon_0 \tag{8.7}$$

Also, for the stationary state, the following equations of continuity for the electrons and the ions are obtained:

$$\nabla \cdot \boldsymbol{\Gamma}_e = I \tag{8.8}$$

$$\nabla \cdot \boldsymbol{\Gamma}_i = I \tag{8.9}$$

where I, the ionization rate, is the number of particles of each kind created per unit volume in unit time. The system of equations consisting of Eqs. (8.3), (8.5), and (8.7)–(8.9) should be solved in order to determine the characteristics of the ambipolar diffusion.

In view of the presence of the terms like $N_e \mathbf{E}_s$ and $N_i \mathbf{E}_s$, this system of equations is nonlinear and cannot be solved analytically without making some simplifying approximations. From Eqs. (8.8) and (8.9), it is found that

$$\nabla \cdot \boldsymbol{\Gamma}_e = \nabla \cdot \boldsymbol{\Gamma}_i \tag{8.10}$$

which suggests the following first approximation:

$$\boldsymbol{\Gamma}_e = \boldsymbol{\Gamma}_i \tag{8.11}$$

The approximation (8.11) is consistent with but not implied by Eq. (8.10) and is more stringent than Eq. (8.10). For one-dimensional diffusion, Eq. (8.10) can be integrated and if the constant of integration is set equal to zero, Eq. (8.11) is obtained. But for the general case Eq. (8.11) does not follow from Eq. (8.10).

Following the treatment by Allis [13], the second simplifying approximation is taken to be given by

$$N_i = CN_e \qquad C = \text{constant} \tag{8.12}$$

For two limiting cases, the validity of the approximation (8.12) can be established. For low density plasmas, that is for N_e and N_i very small, the electric field due to the space charge is negligible and the two types of particles diffuse separately. Hence, Eqs. (8.1a) and (8.2a) are applicable. From Eqs. (8.1a), (8.2a), and (8.10), it is found that

$$\nabla^2 N_i = \frac{D_e}{D_i} \nabla^2 N_e \tag{8.13}$$

which shows that Eq. (8.12) is valid with $C = D_e/D_i$. If N_e and N_i are sufficiently large but with both of them being still very small compared to the number density N_n of the neutral particles, the coupling between the two types of particles is complete resulting in what is known as the *perfect ambipolar diffusion*. For this limiting case, $N_e = N_i$ and therefore Eq. (8.12) is valid with $C = 1$. Since it is valid for the two limiting cases of free diffusion and perfect ambipolar diffusion, Eq. (8.12) may be expected to apply reasonably accurately for the intermediate cases.

Effective and ambipolar diffusion coefficients

With the help of the approximations (8.11) and (8.12), it is possible to manipulate the system of equations (8.3) and (8.5) into such a form as to deduce the effective value of the diffusion coefficient which incorporates the coupling due to the space charge. If Eq. (8.5) is multiplied by $\mu_e N_e$ and is subtracted from Eq. (8.3) after it is multiplied by $\mu_i N_i$, the result on using Eqs. (8.11) and (8.12) is

$$(\mu_i N_i - \mu_e N_e)\mathbf{\Gamma} = -(\mu_i D_e - \mu_e D_i) N_i \nabla N_e \tag{8.14}$$

The space charge density ρ and the conductivity σ of the plasma are given by

$$\rho = (N_i - N_e)e \tag{8.15}$$

and

$$\sigma = (\mu_i N_i - \mu_e N_e)e \tag{8.16}$$

Since

$$1 - \frac{\mu_e \rho}{\sigma} = \frac{N_i(\mu_i - \mu_e)}{(\mu_i N_i - \mu_e N_e)}$$

it follows that

$$D_s = D_a\left(1 - \frac{\mu_e \rho}{\sigma}\right) = \frac{(\mu_i D_e - \mu_e D_i)N_i}{(\mu_i N_i - \mu_e N_e)} \tag{8.17}$$

where

$$D_a = \frac{(\mu_i D_e - \mu_e D_i)}{(\mu_i - \mu_e)} \tag{8.18}$$

By definition, D_a is the *ambipolar diffusion coefficient*. The substitution of Eq. (8.17) into Eq. (8.14) yields

$$\Gamma = -D_s \nabla N_e \tag{8.19}$$

A comparison of Eq. (8.19) with Eq. (8.1a) shows that D_s is *the effective diffusion coefficient* of the electrons in the presence of the space charge and it is not a function of position.

The mobility μ_e of the electrons is negative and the conductivity σ is always positive. Also, from Eqs. (8.12) and (8.15), it is found that $\rho = (C-1)N_e e$. The constant $C = D_e/D_i \gg 1$ in the limit of low density and $C = 1$ in the limit of high density. Thus the constant C is greater than unity and ρ is positive. The minimum value of D_s is then seen from Eq. (8.17) to be equal to D_a and it occurs when $\rho = 0$, that is, in the limit of high density. This limiting case corresponds to the perfect ambipolar diffusion.

The substitution of Eqs. (8.19) and (8.11) into Eq. (8.3) may be shown to give the electric field due to the space charge as

$$\mathbf{E}_s = -\frac{(D_s - D_e)}{\mu_e}\frac{\nabla N_e}{N_e} \tag{8.20}$$

So far Poisson's equation (8.7) has not been used. Now the charge density ρ can be obtained from Eqs. (8.7) and (8.20) as

$$\rho = -\varepsilon_0 \frac{(D_s - D_e)}{\mu_e}\nabla \cdot \left(\frac{\nabla N_e}{N_e}\right) = \varepsilon_0 \frac{(D_s - D_e)}{\mu_e}\left[\left(\frac{\nabla N_e}{N_e}\right)^2 - \frac{\nabla^2 N_e}{N_e}\right] \tag{8.21}$$

In view of the approximation (8.12), in Eq. (8.21) N_e can be replaced by N_i and in such cases where the expressions are valid both for N_e and N_i the subscript on N is conveniently dropped.

Effect of the relative magnitudes of the diffusion and the Debye lengths

The distance over which the diffusion takes place can be characterized by a length parameter Λ or diffusion length as defined by

$$\Lambda^2 = -\frac{N}{\nabla^2 N} \tag{8.22}$$

From Eqs. (8.8), (8.11), (8.12), and (8.19), it is found that

$$-D_s \nabla^2 N_e = -\frac{D_s}{C}\nabla^2 N_i = I$$

Therefore Eq. (8.22) is positive and has the dimensions of the square of a length. Since it is linear, Eq. (8.22) does not depend on the value of N but depends primarily only on the form of the container and its dimensions. In general, Λ is of the order of the dimensions of the container.

The point in the plasma where $\nabla N = 0$ is identified as the "center". For example, if the container has an axis of symmetry, $\nabla N = 0$ at the geometrical center. Usually

Eq. (8.22) is evaluated at the center. The space charge density ρ_0 at the center is found from Eqs. (8.21) and (8.22) as

$$\rho_0 = \varepsilon_0 \frac{(D_s - D_e)}{\mu_e} \frac{1}{\Lambda^2} \tag{8.23}$$

Note that μ_e is negative. If $\rho_0 = 0$, $D_s = D_e$. As the space charge density at the center increases, D_s decreases and when D_s attains the minimum value D_a corresponding to the perfect ambipolar diffusion, the space charge must be saturated.

It is interesting to compare the diffusion length Λ with the Debye length of the electrons as given by

$$\lambda_{\mathrm{D}e}^2 = \frac{\varepsilon_0 KT_e}{N_e e^2} = -\frac{\varepsilon_0 D_e}{N_e e \mu_e} \tag{8.24}$$

where the Einstein relation (7.19) has been used. If Eq. (8.24) is substituted into Eq. (8.23) and Eq. (8.15) is used, the following relation pertaining to the densities at the center is obtained:

$$\frac{(N_i - N_e)}{N_e} = \frac{(D_e - D_s)}{D_e} \frac{\lambda_{\mathrm{D}e}^2}{\Lambda^2} \tag{8.25}$$

Two cases can be distinguished according to the order of magnitude of the ratio of the Debye length to the diffusion length. Previously these two cases were identified according to the magnitude of the density of the charged particles. Suppose that the Debye length is very large compared to the diffusion length. From Eqs. (8.24) and (8.25), this case is seen to correspond to the low-density limit. For this case, the dimensions of the container are not sufficiently large to allow the Debye shielding of the ions by the electrons with the result that the electrons and the ions diffuse independently.

On the other hand, suppose that the Debye length is very small compared to the diffusion length. This case is seen from Eqs. (8.24) and (8.25) to correspond to the high-density limit. Also Eq. (8.25) shows that for this case, $N_i \approx N_e$. Since the Debye length is very small compared to the diffusion length, the screening of the ions by the electrons is complete and the deviation from the state of electrical neutrality is negligible. This is the case of perfect ambipolar diffusion in which the electrons and the ions drift together. The electric field due to the space charge becomes a negligible perturbation.

In the limit of perfect ambipolar diffusion, the effective diffusion coefficient D_s becomes equal to the ambipolar diffusion coefficient D_a. From Eq. (8.18), D_a can be expressed as

$$D_a = D_i \left(\frac{\mu_i D_e}{\mu_e D_i} - 1 \right) \Big/ \left(\frac{\mu_i}{\mu_e} - 1 \right) \tag{8.26}$$

Since $|\mu_i/\mu_e| \ll 1$, it can be neglected in comparison with unity. When the free transport coefficients are substituted from Eqs. (8.1b), (8.2b), (8.4), and (8.6), D_a given by Eq. (8.26) simplifies to

$$D_a = D_i\left(1 + \frac{T_e}{T_i}\right) \tag{8.27}$$

Thus the ambipolar diffusion coefficient is of the order of magnitude of the free diffusion coefficient for the ions.

Comparison between the approximate and the exact theories

The effective diffusion coefficient D_s can be expressed in terms of the conductivity σ_0 at the center. If the value of $\mu_e \rho_0$ from Eq. (8.23) is substituted into Eq. (8.17) and the resulting equation solved for D_s, it can be shown that

$$\frac{D_s}{D_e} = \frac{D_a}{D_e}\left[1 + \frac{\sigma_0 \Lambda^2}{\varepsilon_0 D_e}\right] \Big/ \left[\frac{D_a}{D_e} + \frac{\sigma_0 \Lambda^2}{\varepsilon_0 D_e}\right] \tag{8.28}$$

The two limits D_e and D_a of D_s respectively corresponding to $\sigma_0 = 0$ and $\sigma_0 = \infty$ are reproduced by Eq. (8.28). In Fig. 6.6, D_s/D_e is depicted by dashed lines as a function of the parameter $\sigma_0 \Lambda^2/\varepsilon_0 D_e$.

The validity of the present theory of ambipolar diffusion rests on the reasonableness of the approximation of congruence (8.11) and the approximation of propor-

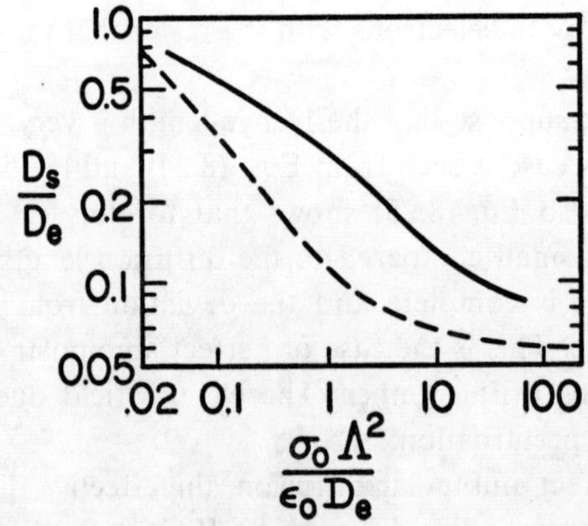

Fig. 6.6. Normalized effective diffusion coefficient D_s/D_e as a function of the parameter $\sigma_0 \Lambda^2/\varepsilon_0 D_e$. The dashed line corresponds to the approximate theory and the solid line corresponds to the exact theory.

tionality (8.12). Allis [13] has evaluated the solution of the exact system of equations (8.3), (8.5), and (8.7)–(8.9) by numerical techniques. In Fig. 6.6, D_s/D_e evaluated from the exact theory is also shown (solid line) as a function of the same parameter $\sigma_0 \Lambda^2/\varepsilon_0 D_e$. The effective diffusion coefficient obtained from the present approximate theory is less than that deduced from the exact theory for the cases intermediate between the limits of free diffusion and perfect ambipolar diffusion.

The charged particles have been assumed to be created in a uniform manner throughout the plasma and to recombine on the walls after having diffused towards them. The characteristics of ambipolar diffusion depend on the nature of the assumptions made concerning the recombination of the charged particles as well as on the geometry of the container. Without examining any specific problem, only the general features of the ambipolar diffusion have been presented here.

6.9. Conductivity for a velocity-dependent collision frequency

The conductivity of a weakly ionized plasma immersed in a magnetostatic field was deduced in Sec. 6.7 for the special case in which the relaxation collision frequency $\nu_r(u)$ is independent of the speed of the electrons. In this section, it is proposed to extend the results for the conductivity to the general case in which the collision frequency $\nu_r(u)$ is dependent on the speed of the electrons. For this purpose, consider a weakly ionized plasma of infinite extent, permeated throughout by a uniform magnetostatic field $\mathbf{B}_0$ in the z-direction. The charged particles are assumed to be uniformly distributed with the result that the equilibrium velocity distribution function is homogeneous.

The total electric and magnetic fields are given by $\mathbf{E}(\mathbf{r}, t)$ and $\mathbf{B}_0 + \mathbf{B}(\mathbf{r}, t)$, respectively, where $\mathbf{E}(\mathbf{r}, t)$ and $\mathbf{B}(\mathbf{r}, t)$ are first-order small quantities. As a result of the perturbing electromagnetic field, the velocity distribution function $f(\mathbf{r}, \mathbf{u}, t)$ deviates slightly from the equilibrium value. As before, it is convenient to express $f(\mathbf{r}, \mathbf{u}, t)$ as

$$f(\mathbf{r}, \mathbf{u}, t) = f_0(u) + g(\mathbf{r}, \mathbf{u}, t) \tag{9.1}$$

The equilibrium distribution function $f_0(u)$ is assumed to be isotropic as in Eq. (6.1). Attention is restricted only to the case in which the homogeneous, isotropic distribution function $f_0(u)$ is given by the Maxwell-Boltzmann distribution function. Note that $g(\mathbf{r}, \mathbf{u}, t)$ is very small compared to $f_0(u)$. For the relaxation model, the collision term in the Boltzmann equation now becomes

$$\left(\frac{\partial f}{\partial t}\right)_{\text{coll}} = -\nu_r(u) g(\mathbf{r}, \mathbf{u}, t) \tag{9.2}$$

When Eqs. (9.1) and (9.2) are substituted into the Boltzmann equation and the second-order small quantities are neglected, it is found that

$$\frac{\partial}{\partial t}g(\mathbf{r},\mathbf{u},t) + (\mathbf{u}\cdot\nabla_r)g(\mathbf{r},\mathbf{u},t) - \frac{e}{m_e}[\mathbf{E}(\mathbf{r},t) + \mathbf{u}\times\mathbf{B}_t(\mathbf{r},t)]\cdot\nabla_u f_0(u)$$
$$-\frac{e}{m_e}(\mathbf{u}\times\mathbf{B}_0)\cdot\nabla_u g(\mathbf{r},\mathbf{u},t) = -\nu_r(u)g(\mathbf{r},\mathbf{u},t) \tag{9.3}$$

Since the perturbation $g(\mathbf{r},\mathbf{u},t)$ in the velocity distribution function depends on the position coordinate, it follows that the density of the particles has a spatial variation. Hence the particle current density is partly due to the diffusion and partly induced by the electromagnetic fields. For the purpose of evaluating the conductivity, the particle current density due to the diffusion need not be taken into account. Therefore, the perturbation $g(\mathbf{r},\mathbf{u},t)$ in the velocity distribution function can be assumed to be essentially independent of the position coordinate $\mathbf{r}$, that is, it can be assumed to be of the form $g(\mathbf{u},t)$. Under this approximation, the second term on the left side of Eq. (9.3) vanishes.

If it is evaluated by using the foregoing approximation, the electric current density at a given position in the plasma depends only on the value of the macroscopic electric field at the same position in the plasma. Hence a *local* relation is obtained for the conductivity. In other words, the procedure that is to be used here for deducing the expression for the conductivity neglects the existence of any *nonlocal* effects which would cause the current density at a given position to depend on the electric field not only at the same position but at other positions as well.

As before, all the field quantities are assumed to have the harmonic time dependence of the form $\exp(-i\omega t)$. Then for the phasor amplitudes, Eq. (9.3) simplifies to the form:

$$-i[\omega + i\nu_r(u)]g(\mathbf{u}) - \frac{e}{m_e}[\mathbf{E}(\mathbf{r}) + \mathbf{u}\times\mathbf{B}_t(\mathbf{r})]\cdot\nabla_u f_0(u)$$
$$-\frac{e}{m_e}(\mathbf{u}\times\mathbf{B}_0)\cdot\nabla_u g(\mathbf{u}) = 0 \tag{9.4}$$

Since $f_0(u)$ is isotropic, Eq. (2.19) holds good and therefore Eq. (9.4) can be written as

$$-i[\omega + i\nu_r(u)]g(\mathbf{u}) - \frac{eB_0}{m_e}(\mathbf{u}\times\hat{\mathbf{z}})\cdot\nabla_u g(\mathbf{u}) = \frac{e}{m_e}\mathbf{E}(\mathbf{r})\cdot\nabla_u f_0(u) \tag{9.5}$$

It is helpful to note that Eq. (9.5) is identical to Eq. (4.5) with $-k_z u_z$ replaced by $i\nu_r(u)$. Hence the solution of Eq. (9.5) as well as the expressions for the current density can be obtained by inspection of the corresponding results contained in Sec. 6.4.

If the field components are decomposed into the right circularly polarized, the left circularly polarized and the longitudinal components, the two circularly polarized components of the current density are found from Eq. (4.44) to be given by

$$J_{\pm}(\mathbf{r}) = -\frac{i\pi e^2}{m_e}E_{\pm}(\mathbf{r})\int_0^{\infty}du_\rho\int_{-\infty}^{\infty}du_z\frac{u_\rho^2}{[\omega \mp \omega_{ce} + i\nu_r(u)]}\frac{\partial f_0(u)}{\partial u_\rho} \tag{9.6}$$

The longitudinal component of the current density is obtained from Eqs. (4.19), (4.27) and (4.40) as

$$J_z(\mathbf{r}) = -\frac{ie^2}{m_e} E_z(\mathbf{r}) \int_{-\infty}^{\infty} d\mathbf{u} \frac{u_z}{[\omega + i\nu_r(u)]} \frac{\partial f_0(u)}{\partial u_z} \tag{9.7}$$

Now it is possible to define the following three components of the conductivity as

$$J_{\pm}(\mathbf{r}) = \sigma_{\pm} E_{\pm}(\mathbf{r}) \qquad J_z(\mathbf{r}) = \sigma_z E_z(\mathbf{r}) \tag{9.8}$$

It can be deduced with the help of Eqs. (9.6)–(9.8) that

$$\sigma_{\pm} = -\frac{i\pi e^2}{m_e} \int_0^{\infty} du_\rho \int_{-\infty}^{\infty} du_z \frac{1}{[\omega \mp \omega_{ce} + i\nu_r(u)]} u_\rho^2 \frac{\partial f_0(u)}{\partial u_\rho} \tag{9.9}$$

and

$$\sigma_z = -\frac{ie^2}{m_e} \int_{-\infty}^{\infty} d\mathbf{u} \frac{1}{[\omega + i\nu_r(u)]} u_z \frac{\partial f_0(u)}{\partial u_z} \tag{9.10}$$

It is convenient to transform to the spherical coordinates (u, θ, φ) in the velocity space. The following relations

$$u_\rho \, du_\rho \, du_z = u^2 \sin\theta \, d\theta \, du \qquad d\mathbf{u} = u_\rho \, du_\rho \, du_z \tag{9.11a}$$

$$u_\rho = u \sin\theta \qquad u_z = u \cos\theta \tag{9.11b}$$

and

$$\frac{\partial f_0(u)}{\partial u_\rho} = \sin\theta \frac{\partial f_0(u)}{\partial u} \qquad \frac{\partial f_0(u)}{\partial u_z} = \cos\theta \frac{\partial f_0(u)}{\partial u} \tag{9.11c}$$

are then valid With the help of Eqs. (9.11), Eqs. (9.9) and (9.10) can be recast as

$$\sigma_{\pm} = -\frac{i\pi e^2}{m_e} \int_0^{\infty} du \int_0^{\pi} d\theta \frac{\sin^3\theta}{[\omega \mp \omega_{ce} + i\nu_r(u)]} u^3 \frac{\partial f_0(u)}{\partial u} \tag{9.12}$$

and

$$\sigma_z = -\frac{ie^2}{m_e} \int_0^{\infty} du \int_0^{\pi} d\theta \int_0^{2\pi} d\varphi \frac{\cos^2\theta \sin\theta}{[\omega + i\nu_r(u)]} u^3 \frac{\partial f_0(u)}{\partial u} \tag{9.13}$$

If the integrations with respect to the angular variables are carried out, Eqs. (9.12) and (9.13) become

$$\sigma_{\pm} = -\frac{ie^2}{m_e} \frac{4\pi}{3} \int_0^{\infty} du \, [\omega \mp \omega_{ce} + i\nu_r(u)]^{-1} u^3 \frac{\partial f_0(u)}{\partial u} \tag{9.14}$$

and

$$\sigma_z = -\frac{ie^2}{m_e} \frac{4\pi}{3} \int_0^{\infty} du \, [\omega + i\nu_r(u)]^{-1} u^3 \frac{\partial f_0(u)}{\partial u} \tag{9.15}$$

For the Maxwell-Boltzmann distribution function $f_0(u)$, the integrals in Eqs. (9.14) and (9.15) cannot be evaluated explicitly, and, the conductivities $\sigma_\pm$ and σ_z have to be determined only by a numerical procedure, especially when the collision frequency is an arbitrary function of the speed u of the electrons. If the collision frequency can be expressed as a polynomial in u, for the two limiting cases of very high and very low collision frequencies, simple expressions can be deduced for $\sigma_\pm$ and σ_z from Eqs. (9.14) and (9.15), respectively. For the special case in which $\nu_r(u)$ is independent of u, Eqs. (9.14) and (9.15) can be simplified to yield

$$\sigma_\pm = \frac{iN_0 e^2}{m_e}\frac{1}{[\omega \mp \omega_{ce} + i\nu_c]} \tag{9.16}$$

and

$$\sigma_z = \frac{iN_0 e^2}{m_e}\frac{1}{[\omega + i\nu_c]} \tag{9.17}$$

where ν_c stands for the constant value of $\nu_r(u)$. In matrix form, Eqs. (9.8) can be expressed as

$$\begin{bmatrix} J_+ \\ J_- \\ J_z \end{bmatrix} = \begin{bmatrix} \sigma_+ & 0 & 0 \\ 0 & \sigma_- & 0 \\ 0 & 0 & \sigma_z \end{bmatrix}\begin{bmatrix} E_+ \\ E_- \\ E_z \end{bmatrix} \tag{9.18}$$

From Eq. (4.13), it is found that

$$\begin{bmatrix} E_+ \\ E_- \\ E_z \end{bmatrix} = \begin{bmatrix} \frac{1}{\sqrt{2}} & \frac{-i}{\sqrt{2}} & 0 \\ \frac{1}{\sqrt{2}} & \frac{i}{\sqrt{2}} & 0 \\ 0 & 0 & 1 \end{bmatrix}\begin{bmatrix} E_x \\ E_y \\ E_z \end{bmatrix} \tag{9.19}$$

Inversion of a relation similar to Eq. (9.19) for the components of the current density yields

$$\begin{bmatrix} J_x \\ J_y \\ J_z \end{bmatrix} = \begin{bmatrix} \frac{1}{\sqrt{2}} & \frac{1}{\sqrt{2}} & 0 \\ \frac{i}{\sqrt{2}} & \frac{-i}{\sqrt{2}} & 0 \\ 0 & 0 & 1 \end{bmatrix}\begin{bmatrix} J_+ \\ J_- \\ J_z \end{bmatrix} \tag{9.20}$$

With the help of Eqs. (9.18)–(9.20), it can be proved that

$$\begin{bmatrix} J_x \\ J_y \\ J_z \end{bmatrix} = \begin{bmatrix} \sigma_\perp & -\sigma_H & 0 \\ \sigma_H & \sigma_\perp & 0 \\ 0 & 0 & \sigma_\| \end{bmatrix}\begin{bmatrix} E_x \\ E_y \\ E_z \end{bmatrix} \tag{9.21}$$

where

$$\sigma_\perp = \frac{1}{2}(\sigma_+ + \sigma_-) \qquad \sigma_H = \frac{i}{2}(\sigma_+ - \sigma_-) \qquad \sigma_\| = \sigma_z \tag{9.22}$$

The Cartesian components of the conductivity dyad can be determined from Eqs. (9.14), (9.15), and (9.22). For the special case of a constant collision frequency ν_c, with the help of Eqs. (9.16), (9.17), and (9.22), it can be shown that

$$\sigma_\perp = \frac{iN_0 e^2}{m_e}\frac{\tilde{\omega}}{(\tilde{\omega}^2 - \omega_{ce}^2)} \qquad \sigma_H = -\frac{N_0 e^2}{m_e}\frac{\omega_{ce}}{(\tilde{\omega}^2 - \omega_{ce}^2)} \qquad \sigma_\| = \frac{iN_0 e^2}{m_e \tilde{\omega}} \tag{9.23}$$

where

$$\tilde{\omega} = \omega + i\nu_{ce} \tag{9.24}$$

The results deduced in Eqs. (9.23) and (9.24) are in agreement with those in Eqs. (7.9a) and (7.9b). Thus for the special case of a constant collision frequency Eqs. (9.14) and (9.15) lead to the same results as those evaluated from the momentum transport equation.

6.10. Diffusion for a velocity-dependent collision frequency

In Sec. 6.7 the diffusion coefficient for a weakly ionized plasma in a magnetostatic field was derived for the special case in which the relaxation collision frequency $\nu_r(u)$ is independent of the speed of the electrons. In this section, the dyadic diffusion coefficient is to be deduced for the general case in which the collision frequency $\nu_r(u)$ is a function of the speed of the electrons. It was pointed out that the diffusion is caused by the spatial inhomogeneity of the electrons. Therefore, in order to study the diffusion phenomenon, the variation of the density of the electrons in space has to be specifically taken into account. Hence, the equilibrium velocity distribution function of the electrons is assumed to be inhomogeneous but isotropic. In particular, it is taken to be of the form

$$f_0(\mathbf{r}, u) = N(\mathbf{r})\left(\frac{m_e}{2\pi KT_e}\right)^{3/2} \exp(-m_e u^2/2KT_e) \tag{10.1}$$

where the spatial variation of the electron density is exhibited. The electron temperature can also have a spatial variation and the following treatment is general enough to include the spatial variation in the electron temperature.

Since the particle current caused by the diffusion alone is of interest, the external electromagnetic fields can be set equal to zero. The electric field induced by the effects of the space charge is also neglected and attention is restricted to the phenomenon of free diffusion. As is usual, the diffusion is studied only for the stationary case in which all the physical parameters are independent of time. The diffusion causes the velocity distribution function $f(\mathbf{r}, \mathbf{u})$ of the electrons to deviate from the equilibrium value $f_0(\mathbf{r}, u)$. Hence $f(\mathbf{r}, \mathbf{u})$ is assumed to be of the form

$$f(\mathbf{r}, \mathbf{u}) = f_0(\mathbf{r}, u) + g(\mathbf{r}, \mathbf{u}) \tag{10.2}$$

where $g(\mathbf{r}, \mathbf{u})$ is a first-order small quantity. The collision term in the Boltzmann equation is approximated on the basis of the relaxation model as

$$\left(\frac{\partial f}{\partial t}\right)_{\text{coll}} = -\nu_r(u) g(\mathbf{r}, \mathbf{u}) \tag{10.3}$$

The spatial inhomogeneities are assumed to be a small perturbation of the equilibrium values with the result that the gradient in the configuration space is a first-order small quantity. Hence if Eqs. (10.2) and (10.3) are substituted into the Boltzmann equation, and only the first-order small quantities are retained, it is seen that

$$(\mathbf{u} \cdot \nabla_r) f_0(\mathbf{r}, u) - \frac{e}{m_e}(\mathbf{u} \times \mathbf{B}_0) \cdot \nabla_u g(\mathbf{r}, \mathbf{u}) = -\nu_r(u) g(\mathbf{r}, \mathbf{u}) \tag{10.4}$$

In deducing Eq. (10.4), it is noted that $f_0(\mathbf{r}, u)$ is isotropic and Eq. (2.19) has been used. As before, the magnetostatic field $\mathbf{B}_0$ is assumed to be in the z-direction. If the cylindrical coordinates (u_ρ, φ, u_z) in the velocity space are used, Eqs. (4.7) and (4.8) enable Eq. (10.4) to be reduced to the form

$$\begin{aligned} &\frac{d}{d\varphi} g(\mathbf{r}, \mathbf{u}) + \frac{\nu_r(u)}{\omega_{ce}} g(\mathbf{r}, \mathbf{u}) \\ &\qquad = -\frac{1}{\omega_{ce}}\left[u_\rho \cos\varphi \frac{\partial}{\partial x} + u_\rho \sin\varphi \frac{\partial}{\partial y} + u_z \frac{\partial}{\partial z}\right] f_0(\mathbf{r}, u) \end{aligned} \tag{10.5}$$

Let

$$g(\mathbf{r}, \mathbf{u}) = g_x(\mathbf{r}, \mathbf{u}) + g_y(\mathbf{r}, \mathbf{u}) + g_z(\mathbf{r}, \mathbf{u}) \tag{10.6}$$

where $g_x(\mathbf{r}, \mathbf{u})$, $g_y(\mathbf{r}, \mathbf{u})$ and $g_z(\mathbf{r}, \mathbf{u})$ are the solutions of the differential equation (10.5) corresponding to the first, the second and the third terms within the square brackets on the right-hand side of Eq. (10.5). Thus

$$\begin{aligned} \left[\frac{d}{d\varphi} + \frac{\nu_r(u)}{\omega_{ce}}\right] g_x(\mathbf{r}, \mathbf{u}) &= \exp\left(-\frac{\nu_r(u)}{\omega_{ce}}\varphi\right) \frac{d}{d\varphi}\left[g_x(\mathbf{r}, \mathbf{u}) \exp\left(\frac{\nu_r(u)}{\omega_{ce}}\varphi\right)\right] \\ &= -\frac{1}{\omega_{ce}} u_\rho \cos\varphi \frac{\partial}{\partial x} f_0(\mathbf{r}, u) \end{aligned} \tag{10.7}$$

The solution of Eq. (10.7) for $g_x(\mathbf{r}, \mathbf{u})$ which is a periodic function of φ with a period of 2π is found to be given by

$$\begin{aligned} g_x(\mathbf{r}, \mathbf{u}) &= -\frac{1}{\omega_{ce}} u_\rho \frac{\partial}{\partial x} f_0(\mathbf{r}, u) \exp\left(-\frac{\nu_r(u)}{\omega_{ce}}\varphi\right) \int_{-\infty}^{\varphi} \cos\varphi' \exp\left(\frac{\nu_r(u)}{\omega_{ce}}\varphi'\right) d\varphi' \\ &= -u_\rho \frac{\partial}{\partial x} f_0(\mathbf{r}, u) [\nu_r(u) \cos\varphi + \omega_{ce} \sin\varphi] / (\nu_r^2(u) + \omega_{ce}^2) \end{aligned} \tag{10.8}$$

In a similar manner, the following results can be derived:

$$g_y(\mathbf{r}, \mathbf{u}) = -u_\rho \frac{\partial}{\partial y} f_0(\mathbf{r}, u)[\nu_r(u)\sin\varphi - \omega_{ce}\cos\varphi]/(\nu_r^2(u) + \omega_{ce}^2) \tag{10.9}$$

and

$$g_z(\mathbf{r}, \mathbf{u}) = -u_z \frac{\partial}{\partial z} f_0(\mathbf{r}, u)/\nu_r(u) \tag{10.10}$$

Thus $g(\mathbf{r}, \mathbf{u})$ and hence the velocity distribution function $f(\mathbf{r}, \mathbf{u})$ are known in terms of the equilibrium distribution function $f_0(\mathbf{r}, u)$.

The particle current density is then deduced as

$$\Gamma_e = \int_{-\infty}^{\infty} d\mathbf{u}\, \mathbf{u}[f_0(\mathbf{r}, u) + g(\mathbf{r}, \mathbf{u})] = \int_{-\infty}^{\infty} d\mathbf{u}\, \mathbf{u} g(\mathbf{r}, \mathbf{u}) \tag{10.11}$$

Only the anisotropic part $g(\mathbf{r}, \mathbf{u})$ in the velocity distribution function is seen from Eq. (10.11) to contribute to the particle current density. The x-component of the particle current density is found from Eq. (10.11) to be

$$\Gamma_{ex} = \int_0^{\infty} du_\rho \int_0^{2\pi} d\varphi \int_{-\infty}^{\infty} du_z\, u_\rho^2 \cos\varphi\, g(\mathbf{r}, \mathbf{u}) \tag{10.12}$$

If Eqs. (10.6) and (10.8)–(10.10) are used in Eq. (10.12) and the integration with respect to φ is then carried out, the following is the result:

$$\Gamma_{ex} = -\pi \int_0^{\infty} du_\rho \int_{-\infty}^{\infty} du_z\, u_\rho^3 [\nu_r(u) \frac{\partial}{\partial x} f_0(\mathbf{r}, u) - \omega_{ce} \frac{\partial}{\partial y} f_0(\mathbf{r}, u)]/(\nu_r^2(u) + \omega_{ce}^2) \tag{10.13}$$

With the help of Eqs. (9.11), Eq. (10.13) can be expressed in terms of the spherical coordinates (u, θ, φ) in the velocity space as

$$\Gamma_{ex} = -\pi \int_0^{\infty} du \int_0^{\pi} d\theta\, u^4 \sin^3\theta [\nu_r(u) \frac{\partial}{\partial x} f_0(\mathbf{r}, u) - \omega_{ce} \frac{\partial}{\partial y} f_0(\mathbf{r}, u)]/(\nu_r^2(u) + \omega_{ce}^2) \tag{10.14}$$

The integration with respect to the angular variable can be carried out and Γ_{ex} can be written as

$$\Gamma_{ex} = -\frac{\partial}{\partial x}(D_\perp N(r)) - \frac{\partial}{\partial y}(-D_H N(r)) \tag{10.15}$$

where

$$D_\perp = \frac{4\pi}{3N(r)} \int_0^{\infty} du \frac{u^4}{[\nu_r^2(u) + \omega_{ce}^2]} \nu_r(u) f_0(\mathbf{r}, u) \tag{10.16}$$

and

$$D_{\mathrm{H}} = \frac{4\pi}{3N(r)} \int_0^{\infty} du \frac{u^4}{[\nu_r^2(u) + \omega_{ce}^2]} \omega_{ce} f_0(\mathbf{r}, u) \tag{10.17}$$

In a similar manner, it can be proved from Eq. (10.11) that

$$\Gamma_{ey} = -\frac{\partial}{\partial x}(D_{\mathrm{H}} N(r)) - \frac{\partial}{\partial y}(D_{\perp} N(r)) \tag{10.18}$$

and

$$\Gamma_{ez} = -\frac{\partial}{\partial z}(D_{\parallel} N(r)) \tag{10.19}$$

where $D_{\perp}$ and D_{H} are the same as those given by Eqs. (10.16) and (10.17), respectively, and

$$D_{\parallel} = \frac{4\pi}{3N(r)} \int_0^{\infty} du \frac{u^4}{\nu_r(u)} f_0(\mathbf{r}, u) \tag{10.20}$$

With the help of Eqs. (10.15), (10.18), and (10.19), the particle current density can be succinctly written as follows:

$$\boldsymbol{\Gamma}_e = -\boldsymbol{\nabla} \cdot (\mathbf{D}_f N(r)) \tag{10.21}$$

where, in the matrix form, the dyadic $\mathbf{D}_f$ is given by

$$\mathbf{D}_f = \begin{bmatrix} D_{\perp} & D_{\mathrm{H}} & 0 \\ -D_{\mathrm{H}} & D_{\perp} & 0 \\ 0 & 0 & D_{\parallel} \end{bmatrix} \tag{10.22}$$

A comparison of Eq. (10.21) with Eq. (7.15) shows that $\mathbf{D}_f$ is the dyadic diffusion coefficient and its components are obtained in Eqs. (10.16), (10.17), and (10.20) for the general case in which the relaxation collision frequency is dependent on the speed of the electrons.

For the special case in which $\nu_r(u)$ is a constant ν_c independent of u, Eqs. (10.16), (10.17), and (10.20) can be simplified with the help of Eq. (10.1) to yield

$$D_{\perp} = \frac{KT_e}{m_e} \frac{\nu_c}{(\nu_c^2 + \omega_{ce}^2)} \qquad D_{\mathrm{H}} = \frac{KT_e}{m_e} \frac{\omega_{ce}}{(\nu_c^2 + \omega_{ce}^2)} \qquad D_{\parallel} = \frac{KT_e}{m_e} \frac{1}{\nu_c} \tag{10.23}$$

The results deduced in Eqs. (10.22) and (10.23) are in agreement with those in Eqs. (7.16) and (7.17). Therefore, for the special case of a constant collision frequency, Eqs. (10.16), (10.17), (10.20), and (10.22) lead to the same results as those evaluated from the momentum transport equation.

In the absence of the magnetostatic field, the dyadic diffusion constant reduces to a scalar constant equal to $D_{\parallel}$. Therefore, the application of a magnetostatic field does not alter the diffusion in the parallel direction. It can be verified from Eqs. (10.16) and (10.20) that $D_{\perp} < D_{\parallel}$. Thus the magnetostatic field inhibits the diffusion in the perpendicular direction and hence has found application in the confinement of plasmas.

6.11. Integral expression for the collision term

So far two approximations of the Boltzmann equation have been studied. In the first approximation, the collision term was set equal to zero and the resulting Boltzmann-Vlasov equation was used for an introductory treatment of the oscillations in an unbounded plasma. In the second approximation, the collision term was replaced by a simple relaxation model. The Boltzmann equation with a relaxation model for the collision term was used for an elementary treatment of the transport processes in a plasma. Now a third application of the Boltzmann equation is to be presented. For this purpose, a detailed representation for the collision term in terms of the velocity distribution function is required. In this section, an integral representation for the collision term is deduced based on certain simplifying assumptions.

The collision term $(\partial f/\partial t)_{\text{coll}}$ represents the time rate of change of the distribution function $f(\mathbf{r}, \mathbf{u}, t)$ as a result of the collisional interactions with the other particles. The number density of the particles having velocities in the range from $\mathbf{u}_1$ to $\mathbf{u}_1 + d\mathbf{u}_1$ is given by $f(\mathbf{r}, \mathbf{u}_1, t) d\mathbf{u}_1$. For convenience, these particles are referred to as type 1 and are said to be for short in the velocity range $d\mathbf{u}_1$. Also for notational simplicity, $f(\mathbf{r}, \mathbf{u}_1, t)$ is denoted by f^1. Thus the number density of the particles with the velocity range $d\mathbf{u}_1$ at time t is given by $f^1 d\mathbf{u}_1$. Due to the collisional interactions, the number density of the particles with the velocity range $d\mathbf{u}_1$ at time $t + dt$ is different from $f^1 d\mathbf{u}_1$ and is given by

$$\left[f^1 + \left(\frac{\partial f^1}{\partial t}\right)_{\text{coll}} dt\right] d\mathbf{u}_1$$

Hence the change in the number density of the type 1 particles in the time interval dt is given by

$$\Delta N = \left(\frac{\partial f^1}{\partial t}\right)_{\text{coll}} dt\, d\mathbf{u}_1 \tag{11.1}$$

As a result of the collisional interactions, some of the particles which were originally in the velocity range $d\mathbf{u}_1$ leave that range and other particles which were originally outside the velocity range $d\mathbf{u}_1$ enter that range. Hence it is convenient to split ΔN given by Eq. (11.1) into two parts as

$$\Delta N = \Delta N_{\text{in}} - \Delta N_{\text{out}} \tag{11.2}$$

where ΔN_{in} and ΔN_{out} are the contributions to ΔN as a result of the particles entering and leaving respectively the velocity range $d\mathbf{u}_1$ in the time interval dt. If expressions for ΔN_{out} and ΔN_{in} can be found in terms of the velocity distribution function, Eqs. (11.1) and (11.2) can be used to deduce $(\partial f^1/\partial t)_{\text{coll}}$ or in general $(\partial f/\partial t)_{\text{coll}}$.

For the purpose of obtaining an expression for ΔN_{out}, consider the particles having velocities in the range from $\mathbf{u}_2$ to $\mathbf{u}_2 + d\mathbf{u}_2$. These particles are referred to as type 2 and are said to be in the velocity range $d\mathbf{u}_2$. The number density of the particles in the velocity range $d\mathbf{u}_2$ at time t is $f(\mathbf{r}, \mathbf{u}_2, t)d\mathbf{u}_2 = f^2 d\mathbf{u}_2$. First consider the interaction between the type 1 and the type 2 particles. As a result of the interaction some of the type 1 particles change from the velocity range $d\mathbf{u}_1$ to $d\tilde{\mathbf{u}}_1$ and some of the type 2 particles change from the velocity range $d\mathbf{u}_2$ to $d\tilde{\mathbf{u}}_2$. Each collision between the type 1 and the type 2 particles causes one of the type 1 particles to leave the velocity range $d\mathbf{u}_1$. By calculating the number of collisions between the type 1 and the type 2 particles in the time interval dt, the number of particles leaving the velocity range $d\mathbf{u}_1$ in the time interval dt due to the collisional interactions between the type 1 and the type 2 particles can be obtained. If this number is integrated for all possible velocity ranges $d\mathbf{u}_2$ extending from $\mathbf{u}_2 = -\infty$ to $\mathbf{u}_2 = \infty$, an expression for ΔN_{out} can be found.

It is helpful to examine the geometry of the collision between one of the type 1 and another of the type 2 particles in a coordinate system in which the type 1 particle is stationary. As shown in Fig. 6.7, the type 1 particle is at the origin and the impact parameter is p. The z-axis passes through the origin and is parallel to the initial direction of motion of the type 2 particle. The plane containing the z-axis and the initial direction of motion of the type 2 particle is called the collision plane. It is convenient to introduce a reference plane, namely the xz-plane which contains the z-axis. Clearly, the xy-plane passes through the origin and is perpendicular to the reference plane. The collision plane makes an angle φ with the reference plane. Let $\mathbf{g}$ be the initial relative velocity of the type 2 particle. In the absence of

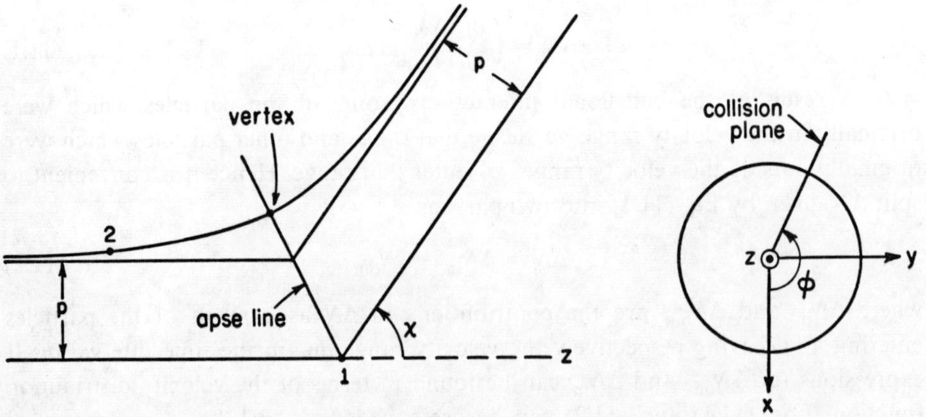

Fig. 6.7. Geometry of the collision between a type 1 particle and a type 2 particle in a coordinate system in which the type 1 particle is at rest showing the impact parameter p and the angle φ of the collision plane.

interaction between the two particles, the type 2 particle passes through the xy-plane at a distance p from the origin or the type 1 particle.

The elastic collision between the two particles has been treated in Secs. 4.2 and 4.3. There it has been shown that the type 2 particle always lies in the collision plane specified by the angle φ. The relative speed g of the type 2 particle is unchanged in the collision process and only its direction is altered. These features of the collision between the two particles enable the number of collisions between the type 1 and the type 2 particles in the time interval dt to be deduced in a straightforward manner.

In the collision process, even though the relative speed of the type 2 particle is unchanged, its direction of motion is deflected through an angle χ. It is assumed that this deflection takes place in a time interval that is very short compared to the time interval dt in which the distribution function has a significant variation. Also, it is assumed that the path length in which the deflection is completed is very small compared to the distance in which the distribution function has a significant variation. In view of these assumptions, for the purpose of the following analysis, it is legitimate to consider the collision between the two particles to take place *at* the time the type 2 particle passes the apse line. Then the number of collisions can be determined by counting the number of particles passing the vertices of their respective trajectories. Also, since the relative speed is unchanged in the collision process, the number of particles passing the vertices of their respective trajectories is also equal to the number of particles passing the xy-plane in the same time interval in the absence of any force of interaction between the two particles. The constancy of the relative speed g is valid only when the external forces are absent. Thus, a third assumption is involved, namely that the external forces are negligibly small compared to the short-range force of interaction between the two particles.

Consider the type 2 particles having impact parameters between p and $p + dp$ and with their collision planes lying between the angles φ and $\varphi + d\varphi$. The number of collisions of this part of the type 2 particles with the type 1 particle in the time interval dt is equal to the number of the type 2 particles lying within an elementary cylindrical volume of height $g\,dt$ and of sides lying between p and $p + dp$, and, φ and $\varphi + d\varphi$. With the help of Fig. 6.8, this number is obtained as

$$f^2\,d\mathbf{u}_2\,p\,dp\,d\varphi\,g\,dt \tag{11.3}$$

The number of collisions of the indicated part of the type 2 particles with all the type 1 particles in the time interval dt is determined by multiplying the number given by the expression (11.3) by the number $f^1\,d\mathbf{u}_1$ of the type 1 particles with the following result:

$$f^1\,d\mathbf{u}_1\,f^2\,d\mathbf{u}_2\,p\,dp\,d\varphi\,g\,dt \tag{11.4}$$

In deducing the expression (11.4), it has been assumed that the number density of particles having the velocity range $d\mathbf{u}_1$ is *independent* of the number density of particles having the velocity range $d\mathbf{u}_2$. In the dynamics of a system of interacting particles, the existence of a particle with a particular velocity in a volume element $d\mathbf{r}$ at time t influences the probability that another particle in the same volume element $d\mathbf{r}$ at the same time t has a specified velocity. The existence of this correlation has been omitted or the *molecular-chaos* [14] assumption has been used in the derivation of the expression (11.4).

If the number given by the expression (11.4) is integrated for all angles of the collision plane from $\varphi = 0$ to $\varphi = 2\pi$ and for all impact parameters from $p = 0$ to $p = \infty$, the result gives the number of collisions in the time interval dt between the type 1 and the type 2 particles. If the result is further integrated for all possible velocities of the type 2 particles ranging from $\mathbf{u}_2 = -\infty$ to $\mathbf{u}_2 = \infty$, the number of collisions in the time interval dt between the type 1 particles and all other particles is obtained. This number gives ΔN_{out} since each collision results in one of the type 1 particles leaving the velocity range $d\mathbf{u}_1$. Thus it is found that

$$\Delta N_{\text{out}} = d\mathbf{u}_1\, dt \int_{-\infty}^{\infty} d\mathbf{u}_2 \int_0^{\infty} dp \int_0^{2\pi} d\varphi f^1 f^2\, pg \tag{11.5}$$

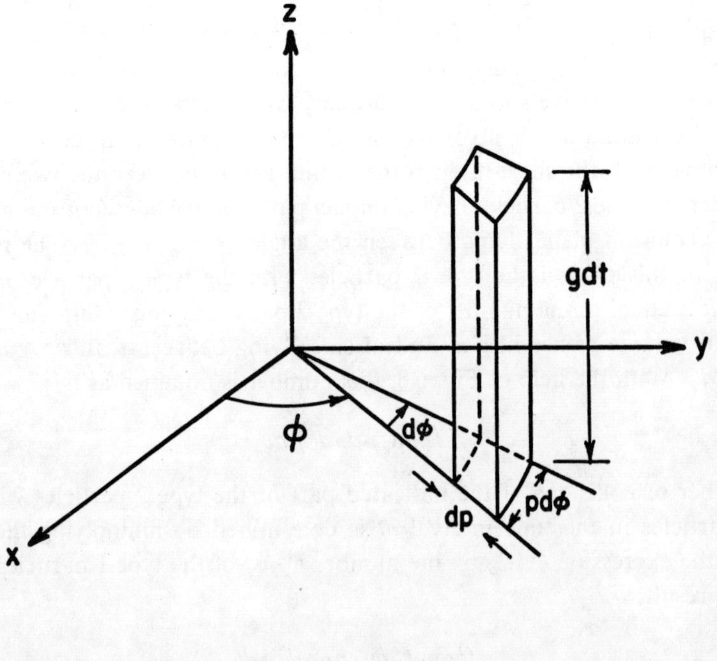

Fig. 6.8. Volume element of height $g\,dt$ and base area bounded by the curves corresponding to p, $p + dp$, φ, and $\varphi + d\varphi$.

As a result of the collision, the velocity of the type 1 particle changes from $\mathbf{u}_1$ to, say $\tilde{\mathbf{u}}_1$. In a similar manner, the velocity of the interacting type 2 particle changes from $\mathbf{u}_2$ to $\tilde{\mathbf{u}}_2$. In other words, the velocity range of the colliding type 1 particle changes from $d\mathbf{u}_1$ to $d\tilde{\mathbf{u}}_1$ and that of the participating type 2 particle changes from $d\mathbf{u}_2$ to $d\tilde{\mathbf{u}}_2$. For convenience, the particles in the velocity ranges $d\tilde{\mathbf{u}}_1$ and $d\tilde{\mathbf{u}}_2$ are referred to as type $\tilde{1}$ and type $\tilde{2}$ particles respectively. The theory of the inverse collision treated in Sec. 4.4 shows that if the type $\tilde{1}$ and the type $\tilde{2}$ particles interact, the velocity of the type $\tilde{1}$ particle changes from $\tilde{\mathbf{u}}_1$ to $\mathbf{u}_1$ and that of the type $\tilde{2}$ particle changes from $\tilde{\mathbf{u}}_2$ to $\mathbf{u}_2$. Therefore, for the purpose of finding the number density of particles entering the velocity range $d\mathbf{u}_1$ in the time interval dt, it is sufficient to consider the interaction of the type $\tilde{1}$ with all the other particles.

For notational simplicity, let $f(\mathbf{r}, \tilde{\mathbf{u}}_1, t)$ and $f(\mathbf{r}, \tilde{\mathbf{u}}_2, t)$ be denoted by $\tilde{f}^1$ and $\tilde{f}^2$, respectively. In the same manner as the expression (11.4) was deduced, it can be shown that the number of collisions of the type $\tilde{2}$ particles having impact parameters between p and $p + dp$ and with their collision planes lying between the angles φ and $\varphi + d\varphi$ with all the type $\tilde{1}$ particles in the time interval dt is given by

$$\tilde{f}^1 d\tilde{\mathbf{u}}_1 \tilde{f}^2 d\tilde{\mathbf{u}}_2 p\, dp\, d\varphi g\, dt \tag{11.6}$$

The relative speed g is the same for the direct and the inverse encounters. Note that each and every one of the collisions given by the expression (11.6) results in a particle entering the velocity range $d\mathbf{u}_1$.

If the number given by the expression (11.6) is integrated for all angles of the collision plane from $\varphi = 0$ to $\varphi = 2\pi$, for all impact parameters from $p = 0$ to $p = \infty$ and for all possible velocities of the type $\tilde{2}$ particles ranging from $\tilde{\mathbf{u}}_2 = -\infty$ to $\tilde{\mathbf{u}}_2 = \infty$, the number of collisions in the time interval dt between the type $\tilde{1}$ particles and all the other particles is found with the following result

$$\Delta N_{\text{in}} = d\tilde{\mathbf{u}}_1\, dt \int_{-\infty}^{\infty} d\tilde{\mathbf{u}}_2 \int_0^{\infty} dp \int_0^{2\pi} d\varphi \tilde{f}^1 \tilde{f}^2 pg \tag{11.7}$$

As indicated in Eq. (11.7), this number gives the number density of particles entering the velocity range $d\mathbf{u}_1$ in the time interval dt.

Jacobian of the transformation

In order to be able to combine Eqs. (11.5) and (11.7) as required in Eq. (11.2), it is advantageous to change the variable of integration in Eq. (11.7) from $d\tilde{\mathbf{u}}_2$ to $d\mathbf{u}_2$. For this purpose, the relationship between $d\mathbf{u}_1 d\mathbf{u}_2$ and $d\tilde{\mathbf{u}}_1 d\tilde{\mathbf{u}}_2$ is used. This relationship is given by the Jacobian transformation

$$d\mathbf{u}_1\, d\mathbf{u}_2 = |J| d\tilde{\mathbf{u}}_1\, d\tilde{\mathbf{u}}_2 \tag{11.8}$$

where

$$J = \frac{\partial(\mathbf{u}_1, \mathbf{u}_2)}{\partial(\tilde{\mathbf{u}}_1, \tilde{\mathbf{u}}_2)} = \begin{vmatrix} \dfrac{\partial \mathbf{u}_1}{\partial \tilde{\mathbf{u}}_1} & \dfrac{\partial \mathbf{u}_1}{\partial \tilde{\mathbf{u}}_2} \\ \dfrac{\partial \mathbf{u}_2}{\partial \tilde{\mathbf{u}}_1} & \dfrac{\partial \mathbf{u}_2}{\partial \tilde{\mathbf{u}}_2} \end{vmatrix} \tag{11.9}$$

Each of the terms in the determinant given by Eq. (11.9) is itself a 3×3 subdeterminant. For example

$$\frac{\partial \mathbf{u}_1}{\partial \tilde{\mathbf{u}}_1} = \begin{vmatrix} \dfrac{\partial u_{1x}}{\partial \tilde{u}_{1x}} & \dfrac{\partial u_{1x}}{\partial \tilde{u}_{1y}} & \dfrac{\partial u_{1x}}{\partial \tilde{u}_{1z}} \\ \dfrac{\partial u_{1y}}{\partial \tilde{u}_{1x}} & \dfrac{\partial u_{1y}}{\partial \tilde{u}_{1y}} & \dfrac{\partial u_{1y}}{\partial \tilde{u}_{1z}} \\ \dfrac{\partial u_{1z}}{\partial \tilde{u}_{1x}} & \dfrac{\partial u_{1z}}{\partial \tilde{u}_{1y}} & \dfrac{\partial u_{1z}}{\partial \tilde{u}_{1z}} \end{vmatrix} \tag{11.10}$$

It is now proposed to establish that

$$|J| = 1 \tag{11.11}$$

In the same manner as Eq. (11.8), the following transformation is also obtained:

$$d\tilde{\mathbf{u}}_1\, d\tilde{\mathbf{u}}_2 = |\tilde{J}|\, d\mathbf{u}_1\, d\mathbf{u}_2 \tag{11.12}$$

where

$$\tilde{J} = \frac{\partial(\tilde{\mathbf{u}}_1, \tilde{\mathbf{u}}_2)}{\partial(\mathbf{u}_1, \mathbf{u}_2)} \tag{11.13}$$

Note that $\tilde{J}$ is the same as J with the quantities without and with tildes interchanged. From Eqs. (11.8) and (11.12), it is found that

$$d\mathbf{u}_1\, d\mathbf{u}_2 = |J|\, d\tilde{\mathbf{u}}_1\, d\tilde{\mathbf{u}}_2 = |J||\tilde{J}|\, d\mathbf{u}_1\, d\mathbf{u}_2$$

or alternatively that

$$|J||\tilde{J}| = 1 \tag{11.14}$$

With the help of Eqs. (4.2.14), (4.2.15), (4.4.3), and (4.4.4), $\mathbf{u}_1$ and $\mathbf{u}_2$ can be expressed in terms of either $\tilde{\mathbf{u}}_1$ or $\tilde{\mathbf{u}}_2$. In a similar manner, $\tilde{\mathbf{u}}_1$ and $\tilde{\mathbf{u}}_2$ can also be expressed in terms of either $\mathbf{u}_1$ or $\mathbf{u}_2$. The final quantities for the direct encounter as given by Eqs. (4.2.15) and (4.4.4) are obtained from the initial quantities as given by Eqs. (4.2.14) and (4.4.3) respectively by interchanging the quantities without and with the tildes. Hence it follows that the relations for $\tilde{\mathbf{u}}_1$ and $\tilde{\mathbf{u}}_2$ can be deduced from the corresponding relations for $\mathbf{u}_1$ and $\mathbf{u}_2$ by interchanging the quantities without and with the tildes. Consequently the partial derivative of any component of $\mathbf{u}_1$ or $\mathbf{u}_2$ with respect to any component of $\tilde{\mathbf{u}}_1$ or $\tilde{\mathbf{u}}_2$ equals the same partial derivative but for the interchange of the quantities without and with the tildes.

The following example is illustrative of the general result. From Eqs. (4.2.14) and (4.2.15), it is found that

$$\mathbf{u}_1 - \tilde{\mathbf{u}}_1 = M_2(\mathbf{g} - \tilde{\mathbf{g}}) \tag{11.15}$$

If Eqs. (4.4.4) and (4.2.15) are used to express the right-hand side of Eq. (11.15) in terms of $\tilde{\mathbf{u}}_1$, the result is

$$\mathbf{u}_1 = \tilde{\mathbf{u}}_1 - 2\hat{\mathbf{k}}[(\tilde{\mathbf{u}}_1 - \mathbf{u}_C) \cdot \hat{\mathbf{k}}] \tag{11.16}$$

If Eqs. (4.4.3) and (4.2.14) are used to express the right-hand side of Eq. (11.15) in terms of $\mathbf{u}_1$, it is obtained that

$$\tilde{\mathbf{u}}_1 = \mathbf{u}_1 - 2\hat{\mathbf{k}}[(\mathbf{u}_1 - \mathbf{u}_C) \cdot \hat{\mathbf{k}}] \tag{11.17}$$

It is clear from Eqs. (11.16) and (11.17) that

$$\frac{\partial \mathbf{u}_1}{\partial \tilde{\mathbf{u}}_1} = \frac{\partial \tilde{\mathbf{u}}_1}{\partial \mathbf{u}_1} \tag{11.18}$$

In a similar manner the corresponding terms in the expressions for the two Jacobians J and $\tilde{J}$ can be proved to be identical. Hence

$$J = \tilde{J} \tag{11.19}$$

Together with Eq. (11.19), Eq. (11.14) leads to Eq. (11.11) and thus the following transformation is established:

$$d\mathbf{u}_1\, d\mathbf{u}_2 = d\tilde{\mathbf{u}}_1\, d\tilde{\mathbf{u}}_2 \tag{11.20}$$

In view of Eq. (11.20), Eq. (11.7) can be rewritten as

$$\Delta N_{\text{in}} = d\mathbf{u}_1\, dt \int_{-\infty}^{\infty} d\mathbf{u}_2 \int_0^{\infty} dp \int_0^{2\pi} d\varphi \tilde{f}^1 \tilde{f}^2 pg \tag{11.21}$$

From Eqs. (11.1), (11.2), (11.5), and (11.21), it follows that

$$\left(\frac{\partial f^1}{\partial t}\right)_{\text{coll}} = \int_{-\infty}^{\infty} d\mathbf{u}_2 \int_0^{\infty} dp \int_0^{2\pi} d\varphi (\tilde{f}^1 \tilde{f}^2 - f^1 f^2) pg \tag{11.22}$$

If $\mathbf{u}_1$ is changed to $\mathbf{u}$ and hence f^1 to f, Eq. (11.22) yields the collision term in the Boltzmann equation as

$$\left(\frac{\partial f}{\partial t}\right)_{\text{coll}} = \int_{-\infty}^{\infty} d\mathbf{u}_2 \int_0^{\infty} dp \int_0^{2\pi} d\varphi (\tilde{f}\tilde{f}^2 - ff^2) pg \tag{11.23}$$

Note that $\tilde{\mathbf{u}}_2$ is known in terms of $\mathbf{u}_2$, the velocity of the center of mass $\mathbf{u}_C$ and the direction $\hat{\mathbf{k}}$ of the apse line. Similarly $\tilde{\mathbf{u}}$ is known in terms of $\mathbf{u}$, $\mathbf{u}_C$ and $\hat{\mathbf{k}}$. If the masses of the two types of particles are known, $\mathbf{u}_C$ can be expressed in terms of $\mathbf{u}_2$ and $\mathbf{u}$. If the impact parameter p and the interaction potential are known, the direction $\hat{\mathbf{k}}$ of the apse line can also be expressed in terms of $\mathbf{u}_2$ and $\mathbf{u}$. Therefore the integrand of Eq. (11.23) is a function of $\mathbf{r}$, $\mathbf{u}$, t, $\mathbf{u}_2$, p, and φ. The masses of the two particles and any constant occurring in the interaction potential appear as parameters. After integration Eq. (11.23) is a function of **r**, **u** and t only. It is usual to denote the fivefold integral appearing in Eq. (11.23) by a single integral sign and hence express the collision term in the Boltzmann equation as

$$\left(\frac{\partial f}{\partial t}\right)_{\text{coll}} = \int d\mathbf{u}_2 \, dp \, d\varphi (\bar{f}\bar{f}^2 - ff^2) pg \tag{11.24}$$

It is common to refer to Eq. (11.24) as the Boltzmann collision term.

Assumptions involved in the derivation of the collision term

Some of the assumptions involved in the derivation of the Boltzmann collision term were indicated in the course of deducing Eq. (11.23). These assumptions are summarized here for the sake of convenience:

(a) The interaction length is much less than the scale length of the spatial variation of the distribution function.
(b) The interaction time is much less than the scale length of the temporal variation of the distribution function.
(c) The collisions are considered to be binary in nature.
(d) The molecular-chaos assumption is used and the interacting particles are assumed to be uncorrelated.

The first two assumptions are equivalent to the assumption that the external force is negligibly small compared to the force of interaction between the particles. In a plasma, the Debye shielding imposes a limit on the interaction length and time. Therefore the first two assumptions appear to be reasonable provided very strong or rapidly fluctuating forces are not present.

Since Coulomb force of interaction between two charged particles is a long-range force, in a plasma each charged particle interacts with every other charged particle lying within at least a Debye length. When two particles are interacting with each other, they are also interacting with the other particles and are therefore correlated via the other particles. Consequently the assumptions of binary collision and molecular chaos are not, strictly speaking, valid for a plasma. However, under certain conditions, the existence of the Debye shielding of the particles carrying a charge of one kind by those carrying the opposite kind of charge enables the encounters to be treated as if they were binary. It was pointed out in Sec. 4.9 that the small number of short-range interactions are not nearly as important as the large number of long-range interactions. But each long-range interaction produces only a small deflection in the path of the particle. Even though each particle interacts with all the particles lying within at least its Debye sphere, since the individual effects are very weak, the cumulative effect of the large number of simultaneous interactions on the path of a particle can be considered as due to a succession of individual binary encounters. Nevertheless, the validity of the Boltzmann collision term is of suspect for a plasma and the results obtained for a plasma with the Boltzmann collision term have to be interpreted cautiously.

6.12. Boltzmann's *H*-theorem

Consider the particles to be uniformly distributed in space and having no density gradients. Then, the velocity distribution function $f(\mathbf{u}, t)$ is independent of the position vector $\mathbf{r}$. Suppose that the plasma is completely free from the action of external forces. Then the Boltzmann equation (1.3.29) together with the collision term given by Eq. (11.24) becomes

$$\frac{\partial}{\partial t}f(\mathbf{u}, t) = \int d\mathbf{u}_2\, dp\, d\varphi(\tilde{f}\tilde{f}^2 - ff^2)pg \tag{12.1}$$

It was pointed out previously that the distribution function tends towards the Maxwell-Boltzmann distribution function characterizing the equilibrium state. It is proposed to examine how the Boltzmann equation, in the equilibrium state, leads to the Maxwell-Boltzmann distribution function.

For this purpose, it is convenient to introduce Boltzmann's function $H(t)$ defined by

$$H(t) = \int f(\mathbf{u}, t) \ln f(\mathbf{u}, t)\, d\mathbf{u} \tag{12.2}$$

which is a function of only t. From Eqs. (12.2) and (12.1), it is deduced that

$$\begin{aligned}\frac{\partial}{\partial t}H(t) &= \int [1 + \ln f(\mathbf{u}, t)]\frac{\partial}{\partial t}f(\mathbf{u}, t)\, d\mathbf{u} \\ &= \int [1 + \ln f](\tilde{f}\tilde{f}^2 - ff^2)pg\, d\mathbf{u}\, d\mathbf{u}_2\, dp\, d\varphi\end{aligned} \tag{12.3}$$

where the notation $f = f(\mathbf{u}, t)$ has been used. The interchange of the dummy variables of integration $\mathbf{u}$ and $\mathbf{u}_2$ in Eq. (12.3) yields

$$\frac{\partial}{\partial t}H(t) = \int [1 + \ln f^2](\tilde{f}\tilde{f}^2 - ff^2)pg\, d\mathbf{u}\, d\mathbf{u}_2\, dp\, d\varphi \tag{12.4}$$

It is found on combining Eqs. (12.3) and (12.4) that

$$\frac{\partial}{\partial t}H(t) = \tfrac{1}{2}\int [2 + \ln ff^2](\tilde{f}\tilde{f}^2 - ff^2)pg\, d\mathbf{u}\, d\mathbf{u}_2\, dp\, d\varphi \tag{12.5}$$

In Eq. (12.5) let the variables of integration be changed from $\mathbf{u}$ and $\mathbf{u}_2$ to $\tilde{\mathbf{u}}$ and $\tilde{\mathbf{u}}_2$, respectively. In view of the transformation relation similar to Eq. (11.20), it follows that

$$\frac{\partial}{\partial t}H(t) = \tfrac{1}{2}\int [2 + \ln \tilde{f}\tilde{f}^2](ff^2 - \tilde{f}\tilde{f}^2)pg\, d\mathbf{u}\, d\mathbf{u}_2\, dp\, d\varphi \tag{12.6}$$

Addition of Eqs. (12.5) and (12.6) may be shown to give

$$\frac{\partial}{\partial t} H(t) = \frac{1}{4} \int \ln(ff^2/\tilde{f}\tilde{f}^2)[\tilde{f}\tilde{f}^2 - ff^2] pg \, d\mathbf{u} \, d\mathbf{u}_2 \, dp \, d\varphi \tag{12.7}$$

Note that in all the indicated transformations of variables, the limits of integration are unchanged.

If $\tilde{f}\tilde{f}^2 > ff^2$, $\ln(ff^2/\tilde{f}\tilde{f}^2)$ is negative and $[\tilde{f}\tilde{f}^2 - ff^2]$ is positive and vice versa if $\tilde{f}\tilde{f}^2 < ff^2$. Both the factors are zero for $\tilde{f}\tilde{f}^2 = ff^2$. Since all the other factors appearing in the integrand of Eq. (12.7) are positive, it follows that

$$\frac{\partial}{\partial t} H(t) \leqslant 0. \tag{12.8}$$

In words, $H(t)$ can never grow; that is Boltzmann's H-theorem. According to the condition (12.8), $H(t)$ monotonically decreases with time to attain a limiting value.

When the distribution function is independent of time, $(\partial/\partial t)H(t) = 0$. Since the integrand of Eq. (12.7) is negative or zero for all values of the independent variables, the steady state corresponding to $(\partial/\partial t)H(t) = 0$ is reached *only* if the integrand of Eq. (12.7) is zero, i.e., if

$$ff^2 = \tilde{f}\tilde{f}^2 \tag{12.9}$$

for all values of the independent variables. It is verified from Eq. (12.1) that the steady state is reached if Eq. (12.9) is satisfied. Thus Eq. (12.9) is the unique condition describing the steady state. In the following section, it is proved that Eq. (12.9) leads to the Maxwellian state.

6.13. Equilibrium velocity distribution function

Law of summation invariants

It is best to begin by introducing the concept of summation invariants. If a function $g(\mathbf{u})$ of the particle velocity is associated with each particle and if the sum of these functions for two colliding particles is conserved during a collision, then $g(\mathbf{u})$ is called a summation invariant. Thus

$$g(\mathbf{u}) + g^{(2)}(\mathbf{u}_2) = \tilde{g}(\tilde{\mathbf{u}}) + \tilde{g}^{(2)}(\tilde{\mathbf{u}}_2) \tag{13.1}$$

for the collision between a particle having the initial velocity $\mathbf{u}$ and the final velocity $\tilde{\mathbf{u}}$, and particle 2 having the initial and the final velocities given by $\mathbf{u}_2$ and $\tilde{\mathbf{u}}_2$, respectively, provided the function g is a summation invariant. The summation invariant has been taken to be a scalar function of the particle velocity but it could also be a vector function.

Let m and m_2 be the mass of the first particle and particle 2, respectively. Then, the laws of conservation of mass, momentum and energy can be stated as follows:

$$m + m_2 = m + m_2 \tag{13.2}$$

$$m\mathbf{u} + m_2\mathbf{u}_2 = m\tilde{\mathbf{u}} + m_2\tilde{\mathbf{u}}_2 \tag{13.3}$$

$$\tfrac{1}{2}mu^2 + \tfrac{1}{2}m_2u_2^2 = \tfrac{1}{2}m\tilde{u}^2 + \tfrac{1}{2}m_2\tilde{u}_2^2 \tag{13.4}$$

From Eqs. (13.2)–(13.4), it is clear that

$$g_1(\mathbf{u}) = m \qquad \mathbf{g}_2(\mathbf{u}) = m\mathbf{u} \qquad \text{and} \quad g_3(\mathbf{u}) = \tfrac{1}{2}mu^2 \tag{13.5}$$

are summation invariants.

The mass conservation law (13.2) is trivial and does not give any information. Together with the equations specifying the impact parameter p and the angle φ of the collision plane, Eqs. (13.3) and (13.4) give altogether six equations. Thus the six "unknowns" consisting of the three components of the final velocity $\tilde{\mathbf{u}}$ and the three components of the final velocity $\tilde{\mathbf{u}}_2$ can be completely determined in terms of the six "knowns" consisting of the three components each of the initial velocities $\mathbf{u}$ and $\mathbf{u}_2$. Suppose that there is another summation invariant and it leads to an equation similar to Eq. (13.4). Since the new equation gives no additional information, it follows that the new equation is a linear combination of the four equations, the three components of Eq. (13.3) and Eq. (13.4). Consequently, the new summation invariant is also the same linear combination of the three summation invariants given by Eqs. (13.5). Thus, it is established that any summation invariant in a collision process is expressible as a linear combination of the three summation invariants given by Eqs. (13.5). For the sake of convenience in referring to it, this is called the law of summation invariants.

Maxwell-Boltzmann distribution function

The equilibrium velocity distribution function can be deduced starting from Eq. (12.9) and using the law of summation invariants. Before doing so, it is helpful to emphasize the physical implication of Eq. (12.9). In view of the relations similar to Eqs. (11.5) and (11.7), it follows that in the time interval dt, the number density of the particles leaving the velocity range $d\mathbf{u}$ and that of the particles leaving the velocity range $d\mathbf{u}_2$ due to the collisional interactions between the particles in these two velocity ranges is exactly balanced by the number densities of the particles entering these two velocity ranges due to the inverse collisions between the particles in the velocity ranges $d\tilde{\mathbf{u}}$ and $d\tilde{\mathbf{u}}_2$. This is an example of the general principle of detailed balance in statistical mechanics. This principle states that in the equilibrium state of a gas, the effect of each type of collision is exactly compensated by the effect of the corresponding inverse collision.

From Eq. (12.9), it is found that

$$\ln f + \ln f^2 = \ln \tilde{f} + \ln \tilde{f}^2 \tag{13.6}$$

with the result that $\ln f$ is a summational invariant. The law of summational invariants shows that

$$\ln f = m[a_1 + a_{2x}u_x + a_{2y}u_y + a_{2z}u_z - \tfrac{1}{2}a_3(u_x^2 + u_y^2 + u_z^2)] \tag{13.7}$$

where a_1, a_{2x}, a_{2y}, a_{2z}, and a_3 are undetermined coefficients. The minus sign in front of a_3 is used for latter convenience. By completing the squares, the right-hand side of Eq. (13.7) can be written as follows:

$$\ln f = \ln a_0 - \tfrac{1}{2}ma_3(\mathbf{u} - \mathbf{u}_0)^2 \tag{13.8}$$

where

$$\ln a_0 = m\left[a_1 + \frac{1}{2a_3}(a_{2x}^2 + a_{2y}^2 + a_{2z}^2)\right] \tag{13.9}$$

and

$$\mathbf{u}_0 = (\hat{\mathbf{x}}a_{2x} + \hat{\mathbf{y}}a_{2y} + \hat{\mathbf{z}}a_{2z})/a_3 \tag{13.10}$$

It is obtained from Eq. (13.8) that

$$f = a_0\exp[-\tfrac{1}{2}ma_3(\mathbf{u} - \mathbf{u}_0)^2] \tag{13.11}$$

Note that there are five undetermined coefficients in Eq. (13.11)—exactly the same number as in Eq. (13.7).

The five constants in Eq. (13.11) can be expressed in terms of the number density N, the average velocity $\mathbf{v}$ and the kinetic temperature T whose definitions are as follows:

$$N = \int f d\mathbf{u} \qquad \mathbf{v} = (1/N)\int \mathbf{u} f d\mathbf{u} \qquad (1/N)\int \tfrac{1}{2}mu^2 f d\mathbf{u} - \tfrac{1}{2}mv^2 = \tfrac{3}{2}KT \tag{13.12}$$

where K is the Boltzmann constant. By the procedure outlined in Sec. 1.4, it can be shown with the help of Eqs. (13.11) and (13.12) that

$$f = N\left(\frac{m}{2\pi KT}\right)^{3/2} \exp[-m(\mathbf{u} - \mathbf{v})^2/2KT] \tag{13.13}$$

which is the Maxwell-Boltzmann distribution function. Thus the equilibrium or the steady state velocity distribution function of a system of particles distributed uniformly in space and free from the action of any external forces is the Maxwell-Boltzmann distribution function as given by Eq. (13.13). Any distribution function which differs from that given by Eq. (13.13) approaches it in course of time.

If the gas is contained in a vessel, in the steady state, there can be no drift velocity and hence $\mathbf{v} = 0$ resulting in the following isotropic distribution function:

$$f_0 = N\left(\frac{m}{2\pi KT}\right)^{3/2} \exp[-mu^2/2KT] \tag{13.14}$$

Note that N and T are independent of the position coordinate $\mathbf{r}$. As a matter of fact, Eq. (13.14) can be proved to be the steady state solution of the Boltzmann equation for a system of particles enclosed in a container with smooth and perfectly reflecting walls and free from any external forces even without resorting to the assumption that the particles are uniformly distributed within the container [15].

6.14. The Boltzmann collision term for a weakly ionized plasma

For a weakly ionized plasma, as mentioned previously, only the collisional interactions of the electrons with the neutral particles are important. It is assumed that the neutral particles are uniformly distributed in space. The spatial inhomogeneity of, and the external forces on, the electrons are assumed to be very small. In other words, the state of the electrons is very close to equilibrium. For simplicity, the electrons are assumed to have no drift in the equilibrium state. Therefore the velocity distribution function of the electrons in the equilibrium state is isotropic and is similar to that in Eq. (13.14).

In the nonequilibrium state caused by the spatial inhomogeneity and the external forces, the velocity distribution function is anisotropic. Since the nonequilibrium is only a slight perturbation of the equilibrium state, it is clear that the anisotropy in the distribution function is very small. Let u, θ and φ be the spherical coordinates in the velocity space where θ is the angle measured from the u_z-axis. Then the equilibrium distribution function depends only on the speed u and the nonequilibrium distribution function depends in addition on θ and φ.

Series expansion of the velocity distribution function

In view of the fact that the anisotropy in the distribution function is small, it follows that the dependence of the velocity distribution function on θ and φ is very weak. This suggests the expansion of the velocity distribution function in terms of the angular variables θ and φ. Since φ varies from 0 to 2π, the velocity distribution function $f(\mathbf{r}, \mathbf{u}, t)$ can be expanded into a Fourier series in φ. Also, since θ varies from 0 to π, $\cos\theta$ varies from -1 to 1. A function defined in the range from -1 to 1 can be expanded into a series of Legendre polynomials. Hence $f(\mathbf{r}, \mathbf{u}, t)$ can be expanded into a series of Legendre polynomials in $\cos\theta$. Consequently $f(\mathbf{r}, \mathbf{u}, t)$ can be expanded in terms of the angular variables θ and φ as follows:

$$f(\mathbf{r}, \mathbf{u}, t) = \sum_{m=0}^{\infty} \sum_{n=0}^{\infty} P_n^m(\cos\theta)[f_{mn}(\mathbf{r}, u, t)\cos m\varphi + g_{mn}(\mathbf{r}, u, t)\sin m\varphi] \tag{14.1}$$

where $P_n^m(\cos\theta)$ is the associated Legendre polynomials.

The leading term in Eq. (14.1) corresponds to $m = 0$ and $n = 0$. Since $P_0^0(\cos\theta) = 1$, the leading term of $f(\mathbf{r}, \mathbf{u}, t)$ is equal to $f_{00}(\mathbf{r}, u, t)$. The next higher order term in Eq. (14.1) corresponds to $m = 0$ and $n = 1$ since $P_0^1(\cos\theta) = 0$. In view of the fact that $P_1^0(\cos\theta) = \cos\theta$, the second term in the series (14.1) is equal to $\cos\theta f_{01}(\mathbf{r}, u, t)$. Since the anisotropy is small, it is sufficient to retain only the first two terms in the series with the result that

$$f(\mathbf{r}, \mathbf{u}, t) = f_{00}(\mathbf{r}, u, t) + \frac{\mathbf{u} \cdot \hat{\mathbf{u}}_z}{u} f_{01}(\mathbf{r}, u, t) \tag{14.2}$$

The first term in Eq. (14.2) is the isotropic distribution function corresponding to the equilibrium state. The second term in Eq. (14.2) is the anisotropic part caused by the spatial inhomogeneity and the external forces.

Approximation for the collision integral

Let $\mathbf{u}_2$ and $\tilde{\mathbf{u}}_2$ be, respectively, the velocity of the neutral particles before and after the collision with the electrons. Therefore, f^2 and $\tilde{f}^2$ represent the velocity distribution functions of the neutral particles before and after the collision, respectively. Since the mass of a neutral particle is very large compared to that of an electron, in the first approximation, the neutral particles may be assumed to be stationary and unaffected by the collisional interactions with the electrons. Therefore,

$$\mathbf{u}_2 = \tilde{\mathbf{u}}_2 = 0 \qquad f^2 = \tilde{f}^2 \tag{14.3}$$

Since the number density N_n of the neutral particles is given by

$$N_n = \int f^2 \, d\mathbf{u}_2$$

it follows from Eqs. (11.23) and (14.3) that

$$\left(\frac{\partial f}{\partial t}\right)_{\text{coll}} = N_n \int_0^\infty dp \int_0^\infty d\varphi(\tilde{f} - f)pg \tag{14.4}$$

Since the neutral particles are at rest, the electron velocity is equal to the relative velocity of the electrons with respect to the neutral particles. Therefore, $\mathbf{u} = \mathbf{g}$ may be taken as the velocity of the electrons before the collision and $\tilde{\mathbf{u}} = \tilde{\mathbf{g}}$ the velocity after the collision. Then f and $\tilde{f}$ give the velocity distribution functions of the electrons before and after the collision, respectively. In the first approximation, the electrons do not lose any kinetic energy during the collisional interaction with the neutral particles. The speed of the electrons is, therefore, unchanged in the collision process. Hence it is found that

$$u = \tilde{u} \qquad f_{00}(\mathbf{r}, u, t) = \tilde{f}_{00}(\mathbf{r}, \tilde{u}, t) \qquad f_{01}(\mathbf{r}, u, t) = \tilde{f}_{01}(\mathbf{r}, \tilde{u}, t) \tag{14.5}$$

Together with Eqs. (14.5), Eq. (14.2) gives

$$f = f_{00}(\mathbf{r}, u, t) + \frac{\mathbf{u} \cdot \hat{\mathbf{u}}_z}{u} f_{01}(\mathbf{r}, u, t) \tag{14.6a}$$

and

$$\tilde{f} = f_{00}(\mathbf{r}, u, t) + \frac{\tilde{\mathbf{u}} \cdot \hat{\mathbf{u}}_z}{u} f_{01}(\mathbf{r}, u, t) \tag{14.6b}$$

From Eqs. (14.6a) and (14.6b), it is obtained that

$$\tilde{f} - f = \frac{(\tilde{\mathbf{u}} - \mathbf{u}) \cdot \hat{\mathbf{u}}_z}{u} f_{01}(\mathbf{r}, u, t) \tag{14.7}$$

Even though the speed of the electrons is unchanged, since they can be scattered through any angle χ from 0 to π, it follows that their velocities can be significantly affected by the collisions. Without loss of generality, the u_z-axis can be taken to be parallel to the initial relative velocity of the electron. Hence Eq. (14.7) reduces to

$$\tilde{f} - f = -(1 - \cos \chi) f_{01}(\mathbf{r}, u, t) \tag{14.8}$$

If Eq. (14.8) is substituted into Eq. (14.4), the result is

$$\left(\frac{\partial f}{\partial t}\right)_{\text{coll}} = -N_n u f_{01}(\mathbf{r}, u, t)\sigma_m \tag{14.9}$$

where

$$\sigma_m = \int_0^\infty \int_0^{2\pi} (1 - \cos \chi) p \, dp \, d\varphi \tag{14.10}$$

With the help of Eqs. (4.5.4) and (4.5.10), it is seen that σ_m given by Eq. (14.10) is the momentum transfer cross section for the collision of the electrons with the neutral particles. If $f_{01}(\mathbf{r}, u, t)$ in Eq. (14.9) is eliminated by using Eq. (14.2), it is found that

$$\left(\frac{\partial f}{\partial t}\right)_{\text{coll}} = -\nu_r(u)[f - f_0] \tag{14.11}$$

and

$$\nu_r(u) = N_n u \sigma_m \tag{14.12}$$

where in accordance with the previous notation, the isotropic distribution function f_{00} characterizing the equilibrium state of the electrons has been replaced by f_0.

It is interesting to note that Eq. (14.11) is identical to the relaxation model for the collision term introduced previously in Sec. 6.6. Thus the foregoing development provides a justification for the relaxation model for the Boltzmann collision term for a weakly ionized plasma. Also, the relaxation collision frequency is expressed in terms of the collision cross section in Eq. (14.12) which is seen to be consistent with a relation similar to Eq. (4.10.8) using the momentum transfer cross section. For a specified force of interaction between the electrons and the neutral particles, Eq. (14.12) enables the relaxation collision frequency $\nu_r(u)$ to be obtained as an explicit function of the velocity of the electrons.

References

6.1. B. D. Fried and S. D. Conte, *The Plasma Dispersion Function*, Academic Press, New York, 1961.
6.2. L. D. Landau, On the vibrations of the electronic plasma, *J. Phys. U.S.S.R.* **10**, 25–34 (1946); *Collected Papers*, Pergamon, Oxford, 1965, 445–460.
6.3. G. N. Watson, *Theory of Bessel Functions*, Cambridge University Press, London, 1966, 395–396.
6.4. A. G. Sitenko and K. N. Stepanov, On the oscillations of an electron plasma in a magnetic field, *Sov. Phys. JETP* **4**, No. 4, 512–520 (1957).
6.5. I. B. Bernstein, Waves in a plasma in a magnetic field, *Phys. Rev.* **109**, No. 1, 10–21 (1958).
6.6. F. W. Crawford, Cyclotron harmonic waves in warm plasmas, *Radio Science J. Res. NBS/USNC-URSI* **69D**, No. 6, 789–805 (1965).
6.7. E. P. Gross, Plasma oscillations in a static magnetic field, *Phys. Rev.* **82**, No. 2, 232–242 (1951).
6.8. T. H. Stix, *The Theory of Plasma Waves*, McGraw-Hill, New York, 1962.
6.9. D. C. Montgomery and D. A. Tidman, *Plasma Kinetic Theory*, McGraw-Hill, New York, 1964.
6.10. A. I. Akhiezer, I. A. Akhiezer, R. V. Polovin, A. G. Sitenko, and K. N. Stepanov, *Collective Oscillations in a Plasma*, The M.I.T. Press, Cambridge, Mass., 1967.

6.11. P. C. Clemmow and J. P. Dougherty, *Electrodynamics of Particles and Plasmas*, Addison-Wesley, Reading, Mass., 1969.
6.12. P. L. Bhatnagar, E. P. Gross, and M. Krook, A model for collision processes in gases, Part I. Small amplitude processes in charged and neutral one-component systems, *Phys. Rev.* **94**, No. 3, 511–525 (1954).
6.13. W. P. Allis, Motions of ions and electrons, in *Handbuch der Physik*, **21**, Springer-Verlag OHG, Berlin, 1956, 397–400.
6.14. J. Jeans, *The Dynamical Theory of Gases*, Cambridge University Press, London, 1954.
6.15. S. Chapman and T. G. Cowling, *The Mathematical Theory of Nonuniform Gases*, Cambridge University Press, London, 1960.

Problems

6.1. If the equilibrium state is characterized by the Maxwell-Boltzmann velocity distribution function, show that the following dispersion equation for the longitudinal plasma wave

$$1 = -\frac{\omega_{pe}^2}{N_0\,\omega} \int_{-\infty}^{\infty} d\mathbf{u} \frac{u_x}{(\omega - k_p u_x)} \frac{\partial f_0(u)}{\partial u_x}$$

can be recast into the form

$$1 + \frac{k_p^2}{\omega_{pe}^2} \frac{KT_e}{m_e} = -i\omega \int_0^{\infty} dt \exp\{i\omega t - k_p^2 t^2 KT_e/2m_e\}$$

Note that the integral converges irrespective of whether the imaginary part of ω is positive or negative. Hence deduce the new form of the dispersion equation both when the imaginary part of ω is positive and when it is negative.

6.2. Suppose that the equilibrium state is characterized by the Maxwell-Boltzmann velocity distribution function. Show that the dispersion equation for the longitudinal plasma wave as stated in Problem 6.1 can be expressed as

$$H(k_p, z) = \frac{k_p^2}{\omega_{pe}^2} - \int_{-\infty}^{\infty} \frac{G(u_x)}{(u_x - z)} du_x = 0$$

where

$$z = \omega/k_p$$

and

$$G(u_x) = -\frac{1}{\sqrt{2\pi}} \left(\frac{m_e}{KT_e}\right)^{3/2} u_x \exp(-m_e u_x^2/2KT_e)$$

Note that the contour of integration in the complex ω-plane is essentially along the real axis but with indentations such that the contour is above all the singularities.

Show that the domain of definition of $H(k_p, z)$ can be extended to the upper half of the ω-plane for which the imaginary part of ω is positive.

Let x and y be the real and the imaginary parts of z. Similarly, let Re $\{H(k_p, z)\}$ and Im $\{H(k_p, z)\}$ denote the real and the imaginary parts of $H(k_p, z)$. Choose x and a positive y such that they are the solutions of the imaginary part of the dispersion equation. By adding (x/y) Im $\{H(k_p, z)\}$ to the real part of the dispersion equation, establish that the real part of the dispersion equation can never be satisfied. Hence show that the longitudinal plasma wave is not unstable.

Find the analytic continuation of $H(k_p, z)$ valid in the lower half of the complex ω-plane.

6.3. Consider a plasma whose equilibrium state is characterized by the following so-called square distribution of velocities in the direction of the wave normal of the longitudinal plasma wave set up initially in the plasma:

$$f_0(\mathbf{u}) = N_0\,\delta(u_y)\delta(u_z)/2a_s \qquad \text{for } |u_x| < a_s$$
$$= 0 \qquad \text{for } |u_x| > a_s$$

Show that there is neither instability nor Landau damping. Find the frequency of oscillation.

6.4. Consider a plasma whose equilibrium state is characterized by the following so-called *resonance distribution* of velocities in the direction of the wave normal of the longitudinal plasma wave set up initially in the plasma:

$$f_0(\mathbf{u}) = N_0\,\delta(u_y)\delta(u_z)\frac{a_r}{\pi}(u_x^2 + a_r^2)^{-1}$$

Show that the longitudinal plasma wave is not unstable. Find the frequency of oscillation. Evaluate the Landau damping constant and compare it with the corresponding value obtained for the Maxwell-Boltzmann distribution of velocities.

6.5. The longitudinal plasma wave is an electrostatic oscillation. Hence it should be possible to deduce its dispersion relation by using the laws of electrostatics instead of Maxwell's equations. Derive the following dispersion equation for the longitudinal plasma wave

$$1 = \frac{\omega_{pe}^2}{N_0}\int_{-\infty}^{\infty} d\mathbf{u}\,\frac{f_0(\mathbf{u})}{(\omega - k_p u_x)^2}$$

starting from the Boltzmann-Vlasov equation and using the following laws of electrostatics:

$$\mathbf{E}(\mathbf{r}, t) = -\nabla\Phi(\mathbf{r}, t) \qquad \nabla^2\Phi(\mathbf{r}, t) = -\rho(\mathbf{r}, t)/\varepsilon_0$$

where $\Phi(\mathbf{r}, t)$ is the scalar potential.

6.6. Consider a homogeneous plasma of infinite extent with stationary ions. The electrons are assumed to have a drift velocity $\mathbf{v} = \hat{\mathbf{x}}v_d$. Assume that the magnetic field resulting from the electron current is negligible. The results of Problem 6.5 show that the dispersion equation of the longitudinal plasma wave in terms of the equilibrium distribution function $f_0(\mathbf{u})$ is unchanged by the drift of the electrons.

Using the cold plasma model, find the appropriate velocity distribution function. For a longitudinal plasma wave with the wave normal in the direction of the electron stream, deduce the dispersion equation. Find the frequency of oscillation and verify the absence of instability for the case of the single electron stream.

6.7. Consider a homogeneous plasma of infinite extent consisting of a neutralizing background of stationary ions of number density N_0. The electrons consist of two streams of equal number density. The first stream has a drift velocity $\mathbf{v} = \hat{\mathbf{x}}v_d$ and the second stream the drift velocity $\mathbf{v} = -\hat{\mathbf{x}}v_d$. Thus the electrons consist of two identical streams drifting in opposite directions with the same speed. As in Problem 6.6 neglect the magnetic field due to the electron currents.

Using the cold plasma model, find the appropriate velocity distribution function. For a longitudinal plasma wave with the wave normal in the direction of the first electron stream, deduce the dispersion equation and express it as a polynomial in $\tilde{\omega} = \omega/\omega_{pe}$.

Assume that the propagation coefficient k_p of the longitudinal plasma wave is real. Show that the frequency of oscillation is specified by the following relations:

$$\tilde{\omega}^2 = \tilde{\omega}_1^2 = \frac{B}{2} + \sqrt{\left(\frac{B}{2}\right)^2 - C}$$

and

$$\tilde{\omega}^2 = \tilde{\omega}_2^2 = \frac{B}{2} - \sqrt{\left(\frac{B}{2}\right)^2 - C}$$

where

$$B = 2\alpha^2 + 1 \qquad C = \alpha^2(\alpha^2 - 1) \qquad \alpha = k_p v_d/\omega_{pe}$$

Note that α is the normalized speed of the two electron streams.

Show that instability can arise if $\alpha < 1$. Find the value of α for which the growth rate is a maximum. What is the maximum value of this growth rate? This instability of the longitudinal plasma wave which arises in the presence of the two oppositely drifting electron streams is called *the two-stream instability*.

6.8. Consider an unbounded, homogeneous plasma in which the effect of the motion of the ions is taken into account. Show that the dispersion equation for the longitudinal plasma wave is given by

$$1 = \frac{\omega_{pe}^2}{N_0 k_p^2} \int_{-\infty}^{\infty} d\mathbf{u} \frac{1}{(u_x - \omega/k_p)} \frac{\partial f_0^e(\mathbf{u})}{\partial u_x} + \frac{\omega_{pi}^2}{N_0 k_p^2} \int_{-\infty}^{\infty} d\mathbf{u} \frac{1}{(u_x - \omega/k_p)} \frac{\partial f_0^i(\mathbf{u})}{\partial u_x}$$

where $f_0^e(\mathbf{u})$ and $f_0^i(\mathbf{u})$ are the equilibrium distribution functions of the electrons and the ions, respectively. The plasma angular frequency of the ions is given by $\omega_{pi} = (N_0 e^2/m_i \varepsilon_0)^{1/2}$ where m_i is the mass of an ion.

Prove that for the cold plasma model, the result of the solution of the dispersion equation is the same as that for the case of the stationary ions except that the electron mass is replaced by the reduced mass of an electron and an ion, $m_e m_i/(m_e + m_i)$.

6.9. In the electron-ion plasma introduced in Problem 6.8, suppose that the electrons have a drift velocity given by $\mathbf{v} = \hat{\mathbf{x}} v_d$ in the direction of the wave normal of the longitudinal plasma wave. Using a cold plasma model, deduce the dispersion equation and express it as a polynomial in $\tilde{\omega} = \omega/\omega_{pe}$. Use $\alpha = k_p v_d/\omega_{pe}$ and $m = m_e/m_i$ as parameters. This dispersion equation has been shown to yield solutions which correspond to the instability in the longitudinal plasma waves. [See I. B. Bernstein and S. K. Trehan, Plasma oscillations, *Nuclear Fusion* **1**, 3–41 (1960)].

6.10. Show that the dispersion equation for the longitudinal plasma wave in an electron-ion plasma as stated in Problem 6.8 can be recast into the form

$$1 + k_p^{-2}\lambda_{\mathrm{D}e}^{-2}\left(1 + i\sqrt{\pi}\, z_e e^{-z_e^2} - 2z_e \int_0^{z_e} e^{w^2 - z_e^2}\, dw\right) \qquad \text{(A)}$$
$$+ k_p^{-2}\lambda_{\mathrm{D}i}^{-2}\left(1 + i\sqrt{\pi}\, z_i e^{-z_i^2} - 2z_i \int_0^{z_i} e^{w^2 - z_i^2}\, dw\right) = 0$$

where $\lambda_{\mathrm{D}e}$ and $\lambda_{\mathrm{D}i}$ are the Debye lengths corresponding to the electrons and the ions, respectively, and,

$$z_e = \omega/k_p a_e \qquad z_i = \omega/k_p a_i \qquad a_e = \sqrt{2KT_e/m_e} \qquad a_i = \sqrt{2KT_i/m_i}$$

Under the conditions

$$a_i \ll \omega_{\mathrm{real}}/k_p \ll a_e \qquad \omega_{\mathrm{imag}} \ll \omega_{\mathrm{real}} \qquad \text{(B)}$$

prove that the dispersion equation can be reduced to the form:

$$1 + k_p^{-2}\lambda_{\mathrm{D}e}^{-2}\left(1 - \frac{m_e}{2m_i} z_e^{-2} + i\sqrt{\pi}\, z_e\right) = 0$$

Show that for the weakly damped oscillations in the low frequency and the low phase velocity range as defined by the conditions in (B), the frequency of oscillation and the Landau damping constant are given by

$$\omega_{\mathrm{real}} = a_s k_p (1 + k_p^2 \lambda_{\mathrm{D}e}^2)^{-1/2} \qquad \omega_{\mathrm{imag}} = -\sqrt{\frac{\pi m_e}{8m_i}}\, a_s k_p (1 + k_p^2\lambda_{\mathrm{D}e}^2)^{-2} \qquad \text{(C)}$$

where $a_s = \sqrt{KT_e/m_i}$. Note that the second condition in (B) is fulfilled by ω_{imag} as obtained in (C) and the first condition in (B) is satisfied only if $T_e/T_i \gg 1 + k_p^2\lambda_{\mathrm{D}e}^2$, i.e., only if the plasma is strongly nonisothermal, with hot electrons and cold ions.

Establish that in the long-wave range, these low-frequency oscillations are essentially the ion acoustic waves discussed in Sec. 2.8.

6.11. Consider a homogeneous plasma of infinite extent with stationary ions of number density N_0. The electrons are characterized by the following velocity distribution function

$$f_0(\mathbf{u}) = N_0 \delta(u_y)\delta(u_z)\frac{a_r}{2\pi}\left[\{(u_x - v_d)^2 + a_r^2\}^{-1} + \{(u_x + v_d)^2 + a_r^2\}^{-1}\right]$$

This distribution function corresponds to the presence of two identical electron streams drifting in the positive and in the negative x-directions with the same speed v_d and a_r is a measure of the thermal spreading of the velocities.

Show that the velocity distribution function has two peaks only for $v_d/a_r > 1/\sqrt{3}$.

Consider a longitudinal plasma wave set up in the plasma with the wave normal in the direction of one of the streams. Let k_p be the propagation coefficient. Find the condition on v_d/a_r for the absence of instability.

Establish the requirements on v_d/a_r and $k_p v_d/a_r$ for the existence of unstable longitudinal plasma waves.

6.12. The dispersion equation for the transverse electromagnetic wave in a hot, isotropic plasma can be expressed as

$$H(\omega) = k_{em}^2 c^2 - \omega^2 + \omega_{pe}^2 \omega \int_{-\infty}^{\infty} \frac{du_x g(u_x)}{(\omega - k_{em} u_x)} = 0$$

where

$$g(u_x) = (m_e/2\pi KT_e)^{1/2} \exp(-m_e u_x^2/2KT_e)$$

Noting that $H(\omega)$ is valid in the upper half of the complex ω-plane, establish the absence of instability of the transverse electromagnetic wave.

6.13. Evaluate the cyclotron damping constant for the right circularly polarized wave propagating along the magnetostatic field in an unbounded, homogeneous plasma with stationary ions of number density N_0 and hot electrons characterized by the following velocity distribution function:

$$f_0(\mathbf{u}) = N_0 \frac{a_r}{\pi^2} (u^2 + a_r^2)^{-2}$$

The magnetostatic field is in the z-direction.

6.14. Consider a homogeneous plasma of infinite extent with stationary ions. A magnetostatic field is assumed to permeate uniformly throughout the plasma in the z-direction. In the equilibrium state, the electrons are characterized by a velocity distribution function which is cylindrically symmetrical about the direction of the magnetostatic field, that is, $f_0(\mathbf{u}) = f_0(u_\rho, u_z)$. Noting that for this case $[\mathbf{u} \times \mathbf{B}(\mathbf{r}, t)] \cdot \nabla_u f_0(\mathbf{u}) \neq 0$, deduce that the dispersion equation for the right circularly polarized, transverse electromagnetic wave propagating along the magnetostatic field is given by

$$k_z^2 c^2 = \omega^2 + \frac{\omega_{pe}^2 \pi}{N_0} \int_0^{\infty} du_\rho \int_{-\infty}^{\infty} du_z \frac{u_\rho^2}{(\omega - \omega_{ce} - k_z u_z)} \left[(\omega - k_z u_z) \frac{\partial f_0(\mathbf{u})}{\partial u_\rho} + k_z u_\rho \frac{\partial f_0(\mathbf{u})}{\partial u_z} \right]$$

Suppose that in the equilibrium state, the velocity distribution function of the electrons is given as

$$f_0(\mathbf{u}) = N_0 \left(\frac{m_e}{2\pi KT_e} \right)^{3/2} \exp\left[-\frac{m_e}{2KT_e} \{u_\rho^2 + (u_z - v_d)^2\} \right]$$

This distribution function corresponds to the electrons streaming in the direction of the magnetostatic field with the speed v_d. Show that the dispersion equation can be written now in the following simplified form:

$$k_z^2 c^2 = \omega^2 - \frac{\omega_{pe}^2}{N_0} \int_{-\infty}^{\infty} d\mathbf{u} \frac{(\omega - k_z v_d) f_0(\mathbf{u})}{(\omega - \omega_{ce} - k_z u_z)}$$

Find the form of the distribution function $f_0(\mathbf{u})$ and the corresponding dispersion equation for the limiting case of $T_e = 0$.

6.15. Deduce the dispersion equation for the quasistatic wave propagating across the magnetostatic field in a hot plasma starting from the Boltzmann-Vlasov equation and the laws of electrostatics as stated in Problem 6.5.

6.16. Show that the Boltzmann-Vlasov equation describing the equilibrium state of the plasma under the influence of a uniform external magnetostatic field is satisfied by any homogeneous velocity distribution function which is cylindrically symmetrical with respect to the magnetostatic field.

6.17. Deduce the dispersion equation for the quasistatic wave propagating in an arbitrary direction with respect to the magnetostatic field in a hot plasma, starting from the Boltzmann-Vlasov equation and the laws of electrostatics. Carry through the derivation as far as possible for an arbitrary value for the strength of the magnetostatic field. Then, specialize for the case of a very weak magnetostatic field. In particular, show that the frequency of oscillation is specified by

$$\omega^2 = \omega_{pe}^2 + 3\frac{KT_e}{m_e}k^2 + \omega_{ce}^2 \sin^2\theta$$

where θ is the angle between the propagation direction and the magnetostatic field; all the other parameters have the usual meaning.

6.18. Obtain the dispersion relation for the longitudinal plane plasma wave in an isotropic plasma using the laws of electrostatics and the Boltzmann equation with a relaxation model for the collision term. Show that in the limit of the relaxation frequency tending to zero, the dispersion relation becomes identical to that derived in Sec. 6.2 using the causality considerations. Hence note that any ambiguity in the integration path in the neighborhood of the singularities can also be removed by using the Boltzmann equation with a relaxation model for the collision term instead of the Boltzmann-Vlasov equation together with the causality requirement.

6.19. Consider a weakly ionized plasma in the absence of any external electromagnetic fields. Let N_{e0} and N_e be the equilibrium number density and the perturbation of the number density of the electrons respectively. Let m_e be the electron mass. The electron temperature T_e and the relaxation collision frequency ν_{ce} for the interactions between the electrons and the neutral particles are assumed to be constants. Write the continuity and the momentum transport equations and combine them to obtain the following diffusion equation

$$\frac{\partial}{\partial t}N_e = D_e \nabla^2 N_e \qquad D_e = KT_e/m_e\nu_{ce}$$

under the assumption that the changes in the number density are negligibly small in the average period between two successive collisions of the electrons with the neutral particles.

6.20. For a weakly ionized plasma, the diffusion equation for the electrons in the presence of the electric field due to the space charge is given by

$$\boldsymbol{\Gamma}_e = -D_e \nabla N_e + \mu_e N_e \mathbf{E}_s$$

where

$$D_e = KT_e/m_e \nu_{ce} \qquad \mu_e = -e/m_e \nu_{ce}$$

By the application of simple dimensional considerations show that the foregoing equation reduces to the equation of free diffusion if the Debye length λ_{De} of the electrons is very large compared to the diffusion length Λ.

6.21. Consider the case of a perfect ambipolar diffusion in a weakly ionized plasma immersed in a uniform magnetostatic field $\mathbf{B}_0$ in the z-direction. Note that the perfect ambipolar diffusion is characterized by the conditions (a) $\boldsymbol{\Gamma}_e = \boldsymbol{\Gamma}_i$ and (b) $N_e = N_i$.

Show that the diffusion equation for the electrons in the presence of the space charge electric field $\mathbf{E}_s$ is given by

$$\boldsymbol{\Gamma}_e = -\nabla \cdot (\mathbf{D}_{fe} N_e) + N_e \boldsymbol{\mu}_e \cdot \mathbf{E}_s$$

where

$$\mathbf{D}_{fe} = \begin{bmatrix} D_{e\perp} & D_{eH} & 0 \\ -D_{eH} & D_{e\perp} & 0 \\ 0 & 0 & D_{e\parallel} \end{bmatrix} \qquad \begin{aligned} D_{e\perp} &= \frac{KT_e}{m_e} \frac{\nu_{ce}}{(\nu_{ce}^2 + \omega_{ce}^2)} \\ D_{eH} &= \frac{KT_e}{m_e} \frac{\omega_{ce}}{(\nu_{ce}^2 + \omega_{ce}^2)} \\ D_{e\parallel} &= \frac{KT_e}{m_e} \frac{1}{\nu_{ce}} \end{aligned}$$

and

$$\boldsymbol{\mu}_e = \begin{bmatrix} \mu_{e\perp} & -\mu_{eH} & 0 \\ \mu_{eH} & \mu_{e\perp} & 0 \\ 0 & 0 & \mu_{e\parallel} \end{bmatrix} \qquad \begin{aligned} \mu_{e\perp} &= \frac{-e\nu_{ce}}{m_e(\nu_{ce}^2 + \omega_{ce}^2)} \\ \mu_{eH} &= \frac{-e\omega_{ce}}{m_e(\nu_{ce}^2 + \omega_{ce}^2)} \\ \mu_{e\parallel} &= \frac{-e}{m_e \nu_{ce}} \end{aligned}$$

Deduce the corresponding diffusion equation for the ions in the presence of the space charge electric field $\mathbf{E}_s$.

Eliminate the space charge electric field from the diffusion equations for the electrons and the ions. Then determine the dyadic ambipolar diffusion coefficient $\mathbf{D}_{fa}$.

Verify that the application of the magnetostatic field does not alter the ambipolar diffusion coefficient.

6.22. The *thermal conductivity* $\mathcal{K}$ is usually defined for a constant kinetic pressure $P_e = N_e(\mathbf{r})KT_e(\mathbf{r})$; the number density $N_e(\mathbf{r})$ and the temperature $T_e(\mathbf{r})$ of the electrons can have spatial variation. Therefore, the equilibrium velocity distribution function of the electrons can be assumed to be of the form:

$$f_0(\mathbf{r}, u) = N_e(\mathbf{r})\left[\frac{m_e}{2\pi K T_e(\mathbf{r})}\right]^{3/2} \exp(-m_e u^2/2KT_e(\mathbf{r}))$$

Neglect the presence of the electromagnetic field and assume that the relaxation collision frequency is a constant independent of the electron speed u. For the stationary state in which the physical parameters do not vary with time, find an expression for the nonequilibrium distribution function $f(\mathbf{r}, \mathbf{u})$ by applying a perturbation technique to the Boltzmann equation with a relaxation model for the collision term.

Evaluate the vector flux density of heat $\mathbf{q}$ as defined by

$$\mathbf{q} = \int_{-\infty}^{\infty} d\mathbf{u}\, \tfrac{1}{2} m_e u^2\, \mathbf{u} f(\mathbf{r}, \mathbf{u})$$

and show that it is given by

$$\mathbf{q} = -\mathcal{K}\,\nabla T_e(\mathbf{r}) \qquad \mathcal{K} = \frac{5KP_e}{2m_e \nu_c}$$

The parameter $\mathcal{K}$ which determines the vector flux density of heat $\mathbf{q}$ caused by the temperature gradient is known as the thermal conductivity.

6.23. In Problem 6.22, include the presence of an external magnetostatic field $\mathbf{B}_0 = \hat{\mathbf{z}} B_0$ in the z-direction and deduce an expression for the nonequilibrium distribution function $f(\mathbf{r}, \mathbf{u})$. Show that the vector flux density of heat $\mathbf{q}$ can be expressed as

$$\mathbf{q} = -\boldsymbol{\mathcal{K}} \cdot \nabla T_e(\mathbf{r}),$$

where $\boldsymbol{\mathcal{K}}$ is the dyadic thermal conductivity. Prove that in the matrix form $\boldsymbol{\mathcal{K}}$ is given by

$$\boldsymbol{\mathcal{K}} = \begin{bmatrix} \mathcal{K}_\perp & -\mathcal{K}_\mathrm{H} & 0 \\ \mathcal{K}_\mathrm{H} & \mathcal{K}_\perp & 0 \\ 0 & 0 & \mathcal{K}_\parallel \end{bmatrix}$$

Obtain the expressions for the components of $\boldsymbol{\mathcal{K}}$.

6.24. The *coefficient of viscosity* η is defined as the shear stress produced by unit velocity gradient. For this purpose, the electrons may be assumed to have an average velocity $v_x(z)$ in the x-direction such that the gradient of $v_x(z)$ is in the z-direction. Therefore the equilibrium velocity distribution function of the electrons may be taken to be of the form

$$f_0(\mathbf{r}, \mathbf{u}) = N_e\left(\frac{m_e}{2\pi K T_e}\right)^{3/2} \exp\left[-\frac{m_e}{2KT_e}\{(u_x - v_x(z))^2 + u_y^2 + u_z^2\}\right]$$

Neglect the presence of the electromagnetic field and assume the relaxation collision frequency to be a constant independent of the electron speed.

The coefficient of viscosity is defined by the relation

$$\eta \frac{d}{dz} v_x(z) = -P_{xz}$$

where P_{xz} is the appropriate component of the kinetic pressure dyad. Deduce the expression for η.

6.25. If the relaxation collision frequency is independent of the speed of the electrons, the electrical conductivity in the parallel direction is obtained as

$$\sigma_z = \frac{iN_0 e^2}{m_e} \frac{1}{[\omega + i\nu_c]}$$

This suggests that even when the relaxation collision frequency is a function of the speed of the electrons, an effective collision frequency $\nu_{eff}(\omega)$ can be introduced such that

$$\sigma_z = \frac{iN_0 e^2}{m_e} \frac{1}{[\omega + i\nu_{eff}(\omega)]}$$

Show that in the limits of low frequency ($\omega \ll \nu_{eff}(\omega)$) and high frequency ($\omega \gg \nu_{eff}(\omega)$), the effective collision frequency $\nu_{eff}(\omega)$ is independent of the angular frequency ω.

6.26. Suppose that the electrons are characterized by the Maxwell-Boltzmann distribution function and that the relaxation collision frequency is given by $\nu_r(u) = \nu_c u^n$ where ν_c is a constant. For the high-frequency limit, deduce the expression for the effective collision frequency $\nu_{eff}(\omega)$.

What is the value of the average collision frequency $< \nu_r(u) >$? Compare the values of $< \nu_r(u) >$ and $\nu_{eff}(\omega)$.

6.27. Assuming that the dependence of the collision frequency on the mass of the charged particle is the same as that found in Problem 6.26, verify that the viscosity of the plasma is contributed principally by the ions whereas the other transport processes like the diffusion, the electrical conductivity and the thermal conductivity are dominated by the electrons.

6.28. Show by using the laws of conservation of momentum and energy in a collision that the Maxwell-Boltzmann distribution function given by

$$f = N\left(\frac{m}{2\pi KT}\right)^{3/2} \exp[-m(\mathbf{u} - \mathbf{v})^2/2KT]$$

satisfies the equation of detailed balance: $\bar{f}\bar{f}^2 = ff^2$.

6.29. Consider the collisional interaction between the electrons and the ions which are characterized respectively by the following distribution functions:

$$f_e = N\left(\frac{m_e}{2\pi KT_e}\right)^{3/2} \exp[-m_e(\mathbf{u}_e - \mathbf{v}_e)^2/2KT_e]$$

and

$$f_i = N\left(\frac{m_i}{2\pi KT_i}\right)^{3/2} \exp[-m_i(\mathbf{u}_i - \mathbf{v}_i)^2/2KT_i]$$

Show that in the absence of the external forces, the mixture of electrons and ions reach an equilibrium state only if $T_e = T_i$ and $\mathbf{v}_e = \mathbf{v}_i$.

6.30. Let $G(\mathbf{u})$ be any function of the particle velocity. Use the integral expression for the Boltzmann collision term to show that

$$\int G(\mathbf{u})\left(\frac{\partial f}{\partial t}\right)_{\text{coll}} d\mathbf{u} = \int [\tilde{G}(\tilde{\mathbf{u}}) - G(\mathbf{u})] ff^2 gp\, dp\, d\varphi\, d\mathbf{u}_2\, d\mathbf{u}$$

Hence verify that for elastic collisions, the collision term occurring on the right-hand side of the first velocity moment of the Boltzmann equation vanishes.

Make use of the results of Problem 4.10 and show that the collision term $\mathbf{P}_{\text{coll}}$ appearing on the right-hand side of the second velocity moment of the Boltzmann equation has only a component in the direction of the initial velocity of the particle with respect to particle 2 with which it collides.

6.31. In the expansion of the velocity distribution function in terms of the Legendre polynomials, the order of the terms depends on the value of n. Thus the terms corresponding to $n = 0$, $n = 1$, and $n = 2$ yield the zeroth-, the first-, and the second-order terms, respectively. Note that the zeroth-order term gives the isotropic part of the distribution function.

Show that only the isotropic part of the distribution function contributes to the number density N and the average speed $< u >$. Verify that only the first-order anisotropies contribute to the average velocity $\mathbf{v}$. Prove that the isotropic term and the anisotropies of the first two orders are involved in the determination of the kinetic pressure dyad.

For a plasma under the action of a strong magnetostatic field in the z-direction, the distribution function, although strongly anisotropic, has a rotational symmetry about the z-axis. Therefore all the terms in the expansion of $f(\mathbf{r}, \mathbf{u}, t)$ other than those corresponding to $m = 0$ are zero. Establish that for this case, the kinetic pressure dyad is diagonal with the two terms corresponding to the two directions perpendicular to the axis of symmetry being identical. Find the conditions under which the diagonal pressure dyad reduces to a scalar pressure.

6.32. Consider a mixture of two types of particles. Let m_1 be the particle mass, T_1 the kinetic temperature, and N_1 the number density of type 1 particles. Let m_2, T_2, and N_2 be the corresponding quantities for type 2 particles. Both the gases are characterized by a Maxwell-Boltzmann distribution of velocities as given by

$$f_\alpha(\mathbf{u}_\alpha) = N_\alpha\left(\frac{m_\alpha}{2\pi KT_\alpha}\right)^{3/2} \exp(-m_\alpha u_\alpha^2/2KT_\alpha) \qquad \alpha = 1, 2$$

The relative speed g between the two species of particles when averaged over both their velocity distributions is obtained as

$$<g> = \frac{1}{N_1 N_2}\int\int g f_1(\mathbf{u}_1) f_2(\mathbf{u}_2)\, d\mathbf{u}_1\, d\mathbf{u}_2 \qquad \text{(A)}$$

Let $\mathbf{u}_1 = \bar{\mathbf{u}}_C + \bar{M}_2\mathbf{g}$, $\mathbf{u}_2 = \bar{\mathbf{u}}_C - \bar{M}_1\mathbf{g}$, $\bar{M}_1 = \bar{m}_1/(\bar{m}_1 + \bar{m}_2)$, $\bar{M}_2 = \bar{m}_2/(\bar{m}_1 + \bar{m}_2)$, where $\bar{\mathbf{u}}_C$ is similar to the velocity of the center of mass and $\mathbf{g}$ is the relative velocity. Also $\bar{m}_1 = m_1/T_1$ and $\bar{m}_2 = m_2/T_2$. Show that the Jacobian

$$|J| = \frac{\partial(\bar{\mathbf{u}}_C, \mathbf{g})}{\partial(\mathbf{u}_1, \mathbf{u}_2)} = 1$$

Transform the variables $\mathbf{u}_1$ and $\mathbf{u}_2$ of integration in (A) to $\bar{\mathbf{u}}_C$ and $\mathbf{g}$. The integrals with respect to $\bar{\mathbf{u}}_C$ and $\mathbf{g}$ can be easily evaluated. Hence, prove that

$$<g> = \left(\frac{8K}{\pi}\right)^{\frac{1}{2}}\left(\frac{T_1}{m_1} + \frac{T_2}{m_2}\right)^{\frac{1}{2}}$$

If $m_1 = m_2$, $T_1 = T_2$ and $N_1 = N_2$ so that only one kind of particles is present with number density equal to N, average speed $<u>$ and mutual scattering cross section σ, show by using Eq. (4.10.8) that the collision frequency in a homogeneous Maxwellian gas is given by $\nu = \sqrt{2}\, N\sigma <u>$. Note that this result has been deduced in a slightly different manner in Problem 4.14.

6.33. Consider a plasma whose equilibrium state is characterized by the following distribution of velocities in the direction of the wave normal of the longitudinal plasma wave set up initially in the plasma:

$$f_0(\mathbf{u}) = N_0\,\delta(u_y)\delta(u_z)\frac{1}{\sqrt{2\pi}}\left(\frac{m_e}{KT}\right)^{3/2} u_x^2 \exp\left(-\frac{m_e}{2KT}u_x^2\right)$$

The propagation coefficient k is real.

Deduce the dispersion equation for the longitudinal plasma wave. Find the range of frequencies in which the longitudinal plasma wave is unstable. In the region of the instability of the longitudinal plasma wave, find the value of the growth constant.

Appendix A: Numerical Values

Physical Constants

Electron mass	9.109×10^{-31} kilogram (kg)
Proton mass	1.673×10^{-27} kilogram (kg)
Electronic charge	-1.6021×10^{-19} coulomb (C)
Vacuum permittivity ε_0	$8.854 \times 10^{-12} \approx 10^{-9}/36\pi$ farad per meter (F/m)
Vacuum permeability μ_0	$1.257 \times 10^{-6} = 4\pi \times 10^{-7}$ henry per meter (H/m)
Electromagnetic wave velocity in vacuum	2.99793×10^{8} meters per second (m/sec)
Boltzmann constant	1.3805×10^{-23} joule per degree Kelvin (J/°K)

Conversion Factors

1 gauss = 10^{-4} weber per square meter (Wb/m^2)
1 electron volt = 1.602×10^{-19} joule (J)

Numerical Formulas

(a) Electron plasma frequency $f_{pe} = 8.978\sqrt{N}$ Hertz (Hz)
(b) Electron gyromagnetic frequency $f_{ce} = 2.799 \times 10^{6} B$ Hertz (Hz)
(c) Debye length for the electrons $\lambda_{De} = 69\sqrt{T/N}$ meter (m)
N is expressed as the number of particles per cubic meter,
B in gauss, and T in degrees Kelvin.

Physical Parameters of Some Typical Plasmas

Plasma	N (m^{-3})	T (°K)	f_{pe} (Hz)	λ_{De} (m)
Interstellar space	10^6	10^3	8.98×10^3	2.19
Ionosphere	10^{12}	10^3	8.98×10^6	2.19×10^{-3}
Solar Corona	10^{13}	10^6	2.84×10^7	2.19×10^{-2}
Quiescent plasma	10^{14}	10^5	8.98×10^7	2.19×10^{-3}
Quiescent plasma, dense	10^{17}	10^4	2.84×10^9	2.19×10^{-5}
Solar atmosphere	10^{18}	10^4	8.98×10^9	6.90×10^{-6}
Arc discharge	10^{20}	10^4	8.98×10^{10}	6.90×10^{-7}
Hot plasma, dense	10^{22}	10^6	8.98×10^{11}	6.90×10^{-7}
Thermonuclear plasma	10^{22}	10^8	8.98×10^{11}	6.90×10^{-6}

Appendix B: Vector Analysis

Scalar and vector products

A scalar quantity is specified completely by its magnitude alone whereas a vector quantity requires a magnitude and a direction for its specification. There are two types of multiplication of vectors. The scalar or the dot product of two vectors **A** and **B** is a scalar whose magnitude is equal to the product of the magnitudes of the two vectors and the cosine of the angle θ between them:

$$\mathbf{A} \cdot \mathbf{B} = |A||B|\cos\theta \tag{B.1}$$

The vector or the cross product of two vectors **A** and **B** is a vector whose magnitude is equal to the product of the magnitudes of the two vectors and the sine of the smaller angle θ between them:

$$|\mathbf{A} \times \mathbf{B}| = |A||B|\sin\theta \tag{B.2}$$

The direction of $\mathbf{A} \times \mathbf{B}$ is normal to the plane containing **A** and **B**, and points towards the thumb of the right hand when the fingers curl from the first vector to the second through the smaller angle between them. If $\hat{\mathbf{x}}$, $\hat{\mathbf{y}}$, and $\hat{\mathbf{z}}$ are the unit vectors in the direction of the axes of a right-handed rectangular coordinates x, y, and z, respectively, it follows that

$$\begin{gathered}\hat{\mathbf{x}} \cdot \hat{\mathbf{x}} = \hat{\mathbf{y}} \cdot \hat{\mathbf{y}} = \hat{\mathbf{z}} \cdot \hat{\mathbf{z}} = 1 \\ \hat{\mathbf{x}} \cdot \hat{\mathbf{y}} = \hat{\mathbf{y}} \cdot \hat{\mathbf{x}} = \hat{\mathbf{y}} \cdot \hat{\mathbf{z}} = \hat{\mathbf{z}} \cdot \hat{\mathbf{y}} = \hat{\mathbf{z}} \cdot \hat{\mathbf{x}} = \hat{\mathbf{x}} \cdot \hat{\mathbf{z}} = 0\end{gathered} \tag{B.3}$$

$$\begin{gathered}\hat{\mathbf{x}} \times \hat{\mathbf{x}} = \hat{\mathbf{y}} \times \hat{\mathbf{y}} = \hat{\mathbf{z}} \times \hat{\mathbf{z}} = 0 \\ \hat{\mathbf{x}} \times \hat{\mathbf{y}} = -\hat{\mathbf{y}} \times \hat{\mathbf{x}} = \hat{\mathbf{z}} \qquad \hat{\mathbf{y}} \times \hat{\mathbf{z}} = -\hat{\mathbf{z}} \times \hat{\mathbf{y}} = \hat{\mathbf{x}} \qquad \hat{\mathbf{z}} \times \hat{\mathbf{x}} = -\hat{\mathbf{x}} \times \hat{\mathbf{z}} = \hat{\mathbf{y}}\end{gathered} \tag{B.4}$$

A vector can be written in terms of its components along any three mutually perpendicular coordinate axes. Hence

$$\mathbf{A} = \hat{\mathbf{x}}A_x + \hat{\mathbf{y}}A_y + \hat{\mathbf{z}}A_z \qquad \mathbf{B} = \hat{\mathbf{x}}B_x + \hat{\mathbf{y}}B_y + \hat{\mathbf{z}}B_z \tag{B.5}$$

where $A_x(B_x)$, $A_y(B_y)$, and $A_z(B_z)$ are, respectively, the x-, y-, and z-components of **A**(**B**). From Eqs. (B.3)–(B.5), it can be shown that

$$\mathbf{A} \cdot \mathbf{B} = A_x B_x + A_y B_y + A_z B_z \tag{B.6}$$

$$\mathbf{A} \times \mathbf{B} = \hat{\mathbf{x}}(A_y B_z - A_z B_y) + \hat{\mathbf{y}}(A_z B_x - A_x B_z) + \hat{\mathbf{z}}(A_x B_y - A_y B_x) \tag{B.7}$$

In a similar manner, unit vectors along the axes of cylindrical and spherical coordinate systems can be defined and the results of $\mathbf{A} \cdot \mathbf{B}$ and $\mathbf{A} \times \mathbf{B}$ can be expressed in terms of the cylindrical and the spherical components of the two vectors.

The addition, the subtraction, and the scalar multiplication of two vectors obey the commutative law. But the cross product of two vectors does not satisfy the commutative law since $\mathbf{A} \times \mathbf{B} = -\mathbf{B} \times \mathbf{A}$. The vectors satisfy the associative law of addition as well as the distributive law for the scalar and the vector multiplication.

Gradient, divergence, and curl

There are three important vector differential operations involving functions of position. These are the gradient, the divergence and the curl operations and are denoted by ∇, $\nabla \cdot$, and $\nabla \times$, respectively. In rectangular coordinates the vector differential operators ∇ (*del*), $\nabla \cdot$ (*del dot*), and $\nabla \times$ (*del cross*) are defined as

$$\nabla = \frac{\partial}{\partial x}\hat{\mathbf{x}} + \frac{\partial}{\partial y}\hat{\mathbf{y}} + \frac{\partial}{\partial z}\hat{\mathbf{z}} \tag{B.8}$$

$$\nabla \cdot = \frac{\partial}{\partial x}\hat{\mathbf{x}} \cdot + \frac{\partial}{\partial y}\hat{\mathbf{y}} \cdot + \frac{\partial}{\partial z}\hat{\mathbf{z}} \cdot \tag{B.9}$$

and

$$\nabla \times = \frac{\partial}{\partial x}\hat{\mathbf{x}} \times + \frac{\partial}{\partial y}\hat{\mathbf{y}} \times + \frac{\partial}{\partial z}\hat{\mathbf{z}} \times \tag{B.10}$$

The gradient of a scalar function ϕ of position is obtained by operating ∇ on ϕ; the result on noting that $\hat{\mathbf{x}}$, $\hat{\mathbf{y}}$, and $\hat{\mathbf{z}}$ do not vary with x, y, and z is

$$\nabla\phi = \text{grad } \phi = \hat{\mathbf{x}}\frac{\partial\phi}{\partial x} + \hat{\mathbf{y}}\frac{\partial\phi}{\partial y} + \hat{\mathbf{z}}\frac{\partial\phi}{\partial z} \tag{B.11}$$

The divergence and the curl of a vector function $\mathbf{A}$ of position are obtained by operating respectively $\nabla \cdot$ and $\nabla \times$ on $\mathbf{A}$; the results are

$$\nabla \cdot \mathbf{A} = \text{div } \mathbf{A} = \frac{\partial}{\partial x}A_x + \frac{\partial}{\partial y}A_y + \frac{\partial}{\partial z}A_z \tag{B.12}$$

and

$$\begin{aligned}\nabla \times \mathbf{A} &= \text{curl } \mathbf{A} \\ &= \hat{\mathbf{x}}\left(\frac{\partial}{\partial y}A_z - \frac{\partial}{\partial z}A_y\right) + \hat{\mathbf{y}}\left(\frac{\partial}{\partial z}A_x - \frac{\partial}{\partial x}A_z\right) + \hat{\mathbf{z}}\left(\frac{\partial}{\partial x}A_y - \frac{\partial}{\partial y}A_x\right)\end{aligned} \tag{B.13}$$

It is to be noted that the gradient of a scalar function is a vector, the divergence of a vector function is a scalar, and the curl of a vector function is a vector.

It is possible to give alternative definitions for the gradient, the divergence, and the curl operations, and these alternative definitions enable the expressions for $\nabla\phi$, $\nabla \cdot \mathbf{A}$, and $\nabla \times \mathbf{A}$ to be obtained in other coordinate systems. Let an elemental volume ΔV centered on the point where $\nabla\phi$, $\nabla \cdot \mathbf{A}$, and $\nabla \times \mathbf{A}$ are desired, be considered and let it be enclosed by the surface ΔS. Then

$$\nabla\phi = \lim_{\Delta V \to 0} \frac{1}{\Delta V} \int_{\Delta S} \hat{\mathbf{n}}\phi \, dS \tag{B.14}$$

$$\nabla \cdot \mathbf{A} = \lim_{\Delta V \to 0} \frac{1}{\Delta V} \int_{\Delta S} \hat{\mathbf{n}} \cdot \mathbf{A} \, dS \tag{B.15}$$

and

$$\nabla \times \mathbf{A} = \lim_{\Delta V \to 0} \frac{1}{\Delta V} \int_{\Delta S} \hat{\mathbf{n}} \times \mathbf{A} \, dS \tag{B.16}$$

where $\hat{\mathbf{n}}$ is an outwardly drawn unit normal to the surface ΔS. By taking ΔV to be the elemental volume bounded by the coordinate planes, x, $x + \Delta x$, y, $y + \Delta y$, z, and $z + \Delta z$, the right-hand sides of Eqs. (B.14)–(B.16) can be evaluated in a straightforward manner. The results so derived for $\nabla\phi$, $\nabla \cdot \mathbf{A}$, and $\nabla \times \mathbf{A}$ are respectively the same as those given by Eqs. (B.11), (B.12), and (B.13). It is thus verified that the alternative definitions for the gradient, the divergence and the curl as stated in Eqs. (B.14), (B.15), and (B.16) are equivalent to the previous definitions of these operations as contained in Eqs. (B.8), (B.9), and (B.10), respectively.

The following theorems can be proved with the help of Eqs. (B.14)–(B.16):

$$\int_V \nabla\phi \, dV = \int_S \hat{\mathbf{n}}\phi \, dS \tag{B.17}$$

$$\int_V \nabla \cdot \mathbf{A} \, dV = \int_S \hat{\mathbf{n}} \cdot \mathbf{A} \, dS \tag{B.18}$$

and

$$\int_V \nabla \times \mathbf{A} \, dV = \int_S \hat{\mathbf{n}} \times \mathbf{A} \, dS \tag{B.19}$$

where V is a finite volume and S is the surface enclosing it. As before, $\hat{\mathbf{n}}$ is the unit outwardly drawn normal to the surface S. Note that Eq. (B.18) is known as the *divergence theorem.*

Another useful interpretation can be given to the curl of a vector function. The component of a vector parallel to an elemental length $d\mathbf{l}$ integrated throughout the length of a curve defines the line integral of that vector function. The line integral of a vector taken around a closed curve C bounding an open surface S is called the circulation of the vector around the surface area S. The curl of a vector function is a vector whose component in a given direction at a point is the circulation of the vector around an elemental surface of area ΔS centered on the point under consideration divided by the area ΔS of the surface. The elemental surface is oriented perpendicular to the given direction. If the thumb of the right hand points in the given direction, the fingers point in the direction in which the line integral around the curve ΔC bounding the elemental area ΔS should be taken. Therefore

$$(\nabla \times \mathbf{A}) \cdot \hat{\mathbf{n}} = \lim_{\Delta S \to 0} \frac{1}{\Delta S} \int_{\Delta C} \mathbf{A} \cdot d\mathbf{l} \tag{B.20}$$

where $\hat{\mathbf{n}}$, the normal to the elemental surface ΔS, is related to the direction of integration around the periphery ΔC by the right-hand rule. Consider an elemental surface area ΔS in the yz-plane lying between the lines corresponding to y, $y + \Delta y$, z, and $z + \Delta z$. When the x-component of the circulation of the vector function $\mathbf{A}$ around the elemental surface area ΔS is deduced with the help of Eq. (B.20), the result is

$$(\nabla \times \mathbf{A}) \cdot \hat{\mathbf{x}} = \frac{\partial}{\partial y} A_z - \frac{\partial}{\partial z} A_y \tag{B.21}$$

In a similar manner, the y- and the z-components of $\nabla \times \mathbf{A}$ can be evaluated and the resulting expression for $\nabla \times \mathbf{A}$ is the same as that given by Eq. (B.13) thus verifying that the alternative definition of the curl as expressed by Eq. (B.20) is equivalent to the previous definitions of the curl as given by Eqs. (B.10) and (B.16).

Let S be a finite unclosed area bounded by the closed curve C. With the help of Eq. (B.20) it can be proved that

$$\int_S \nabla \times \mathbf{A} \cdot \hat{\mathbf{n}}\, dS = \int_C \mathbf{A} \cdot d\mathbf{l} \tag{B.22}$$

which is the statement of *Stokes' theorem.*

Laplacian

Another important differential operation is the Laplacian which is defined by

$$\nabla^2 \phi = \nabla \cdot (\nabla \phi) \tag{B.23}$$

With the help of Eqs. (B.8) and (B.9), it can be verified that

$$\nabla^2 \phi = \frac{\partial^2 \phi}{\partial x^2} + \frac{\partial^2 \phi}{\partial y^2} + \frac{\partial^2 \phi}{\partial z^2} \tag{B.24}$$

Appendix C: Vector Relations

$$\mathbf{A}\cdot(\mathbf{B}\times\mathbf{C}) = \mathbf{B}\cdot(\mathbf{C}\times\mathbf{A}) = \mathbf{C}\cdot(\mathbf{A}\times\mathbf{B})$$

$$\mathbf{A}\times(\mathbf{B}\times\mathbf{C}) = (\mathbf{A}\cdot\mathbf{C})\mathbf{B} - (\mathbf{A}\cdot\mathbf{B})\mathbf{C}$$

$$\nabla(\phi\psi) = \phi\nabla\psi + \psi\nabla\phi$$

$$\nabla(\mathbf{A}\cdot\mathbf{B}) = (\mathbf{A}\cdot\nabla)\mathbf{B} + (\mathbf{B}\cdot\nabla)\mathbf{A} + \mathbf{A}\times(\nabla\times\mathbf{B}) + \mathbf{B}\times(\nabla\times\mathbf{A})$$

$$\nabla\cdot(\phi\mathbf{A}) = \mathbf{A}\cdot(\nabla\phi) + \phi(\nabla\cdot\mathbf{A})$$

$$\nabla\cdot(\mathbf{A}\times\mathbf{B}) = \mathbf{B}\cdot(\nabla\times\mathbf{A}) - \mathbf{A}\cdot(\nabla\times\mathbf{B})$$

$$\nabla\cdot(\nabla\times\mathbf{A}) = 0$$

$$\nabla\times(\phi\mathbf{A}) = (\nabla\phi)\times\mathbf{A} + \phi(\nabla\times\mathbf{A})$$

$$\nabla\times(\mathbf{A}\times\mathbf{B}) = \mathbf{A}(\nabla\cdot\mathbf{B}) - \mathbf{B}(\nabla\cdot\mathbf{A}) + (\mathbf{B}\cdot\nabla)\mathbf{A} - (\mathbf{A}\cdot\nabla)\mathbf{B}$$

$$\nabla\times(\nabla\times\mathbf{A}) = \nabla(\nabla\cdot\mathbf{A}) - \nabla^2\mathbf{A}$$

$$\nabla\times(\nabla\phi) = 0$$

$$\nabla\cdot(\nabla\phi) = \nabla^2\phi$$

In the following formulas V is a volume bounded by the closed surface S and $\hat{\mathbf{n}}$ is a unit normal drawn outwardly to the surface S.

$$\int_V \nabla\phi\, dV = \int_S \phi\hat{\mathbf{n}}\, dS$$

$$\int_V \nabla\cdot\mathbf{A}\, dV = \int_S \mathbf{A}\cdot\hat{\mathbf{n}}\, dS \quad \text{(Divergence theorem)}$$

$$\int_V \nabla\times\mathbf{A}\, dV = \int_S \hat{\mathbf{n}}\times\mathbf{A}\, dS$$

In the following formulas S is an unclosed area bounded by the curve C.

$$\int_S \hat{\mathbf{n}}\times\nabla\phi\, dS = \int_C \phi\, d\mathbf{l}$$

$$\int_S \nabla\times\mathbf{A}\cdot\hat{\mathbf{n}}\, dS = \int_C \mathbf{A}\cdot d\mathbf{l} \quad \text{(Stokes' theorem)}$$

Gradient, Divergence, Curl, and Laplacian Operations in Rectangular (x, y, z), Cylindrical (ρ, φ, z) and Spherical (r, θ, φ) Coordinate Systems

Rectangular coordinates

$$\nabla a = \hat{\mathbf{x}}\frac{\partial a}{\partial x} + \hat{\mathbf{y}}\frac{\partial a}{\partial y} + \hat{\mathbf{z}}\frac{\partial a}{\partial z}$$

$$\nabla \cdot \mathbf{A} = \frac{\partial}{\partial x}A_x + \frac{\partial}{\partial y}A_y + \frac{\partial}{\partial z}A_z$$

$$\nabla \times \mathbf{A} = \hat{\mathbf{x}}\left(\frac{\partial}{\partial y}A_z - \frac{\partial}{\partial z}A_y\right) + \hat{\mathbf{y}}\left(\frac{\partial}{\partial z}A_x - \frac{\partial}{\partial x}A_z\right) + \hat{\mathbf{z}}\left(\frac{\partial}{\partial x}A_y - \frac{\partial}{\partial y}A_x\right)$$

$$\nabla^2 a = \frac{\partial^2 a}{\partial x^2} + \frac{\partial^2 a}{\partial y^2} + \frac{\partial^2 a}{\partial z^2}$$

$$\nabla^2 \mathbf{A} = \hat{\mathbf{x}}\nabla^2 A_x + \hat{\mathbf{y}}\nabla^2 A_y + \hat{\mathbf{z}}\nabla^2 A_z$$

Cylindrical coordinates

$$\nabla a = \hat{\boldsymbol{\rho}}\frac{\partial a}{\partial \rho} + \hat{\boldsymbol{\varphi}}\frac{1}{\rho}\frac{\partial a}{\partial \varphi} + \hat{\mathbf{z}}\frac{\partial a}{\partial z}$$

$$\nabla \cdot \mathbf{A} = \frac{1}{\rho}\frac{\partial}{\partial \rho}(\rho A_\rho) + \frac{1}{\rho}\frac{\partial}{\partial \varphi}A_\varphi + \frac{\partial}{\partial z}A_z$$

$$\nabla \times \mathbf{A} = \hat{\boldsymbol{\rho}}\left[\frac{1}{\rho}\frac{\partial}{\partial \varphi}A_z - \frac{\partial}{\partial z}A_\varphi\right] + \hat{\boldsymbol{\varphi}}\left[\frac{\partial}{\partial z}A_\rho - \frac{\partial}{\partial \rho}A_z\right] + \hat{\mathbf{z}}\left[\frac{1}{\rho}\frac{\partial}{\partial \rho}(\rho A_\varphi) - \frac{1}{\rho}\frac{\partial}{\partial \varphi}A_\rho\right]$$

$$\nabla^2 a = \frac{1}{\rho}\frac{\partial}{\partial \rho}\left(\rho\frac{\partial a}{\partial \rho}\right) + \frac{1}{\rho^2}\frac{\partial^2 a}{\partial \varphi^2} + \frac{\partial^2 a}{\partial z^2}$$

Spherical coordinates

$$\nabla a = \hat{\mathbf{r}}\frac{\partial a}{\partial r} + \hat{\boldsymbol{\theta}}\frac{1}{r}\frac{\partial a}{\partial \theta} + \hat{\boldsymbol{\varphi}}\frac{1}{r\sin\theta}\frac{\partial a}{\partial \varphi}$$

$$\nabla \cdot \mathbf{A} = \frac{1}{r^2}\frac{\partial}{\partial r}(r^2 A_r) + \frac{1}{r\sin\theta}\frac{\partial}{\partial \theta}(\sin\theta A_\theta) + \frac{1}{r\sin\theta}\frac{\partial}{\partial \varphi}A_\varphi$$

$$\nabla \times \mathbf{A} = \frac{\hat{\mathbf{r}}}{r\sin\theta}\left[\frac{\partial}{\partial \theta}(\sin\theta A_\varphi) - \frac{\partial}{\partial \varphi}A_\theta\right] + \frac{\hat{\boldsymbol{\theta}}}{r}\left[\frac{1}{\sin\theta}\frac{\partial}{\partial \varphi}A_r - \frac{\partial}{\partial r}(rA_\varphi)\right] + \frac{\hat{\boldsymbol{\varphi}}}{r}\left[\frac{\partial}{\partial r}(rA_\theta) - \frac{\partial}{\partial \theta}A_r\right]$$

$$\nabla^2 a = \frac{1}{r^2}\frac{\partial}{\partial r}\left(r^2\frac{\partial a}{\partial r}\right) + \frac{1}{r^2\sin\theta}\frac{\partial}{\partial \theta}\left(\sin\theta\frac{\partial a}{\partial \theta}\right) + \frac{1}{r^2\sin^2\theta}\frac{\partial^2 a}{\partial \varphi^2}$$

Appendix D: Dyads

Let **C** and **D** be two vectors. The quantity

$$\mathbf{T} = \mathbf{C}\mathbf{D} \tag{D.1}$$

without any operational sign such as +, −, •, and × is called a *dyad.* If **C** and **D** are expressed in component form in Cartesian coordinates as follows

$$\mathbf{C} = \hat{\mathbf{x}}C_x + \hat{\mathbf{y}}C_y + \hat{\mathbf{z}}C_z \qquad \mathbf{D} = \hat{\mathbf{x}}D_x + \hat{\mathbf{y}}D_y + \hat{\mathbf{z}}D_z \tag{D.2}$$

C D in Eq. (D.1) can be expanded to yield:

$$\begin{aligned} \mathbf{T} = {} & \hat{\mathbf{x}}\hat{\mathbf{x}}T_{xx} + \hat{\mathbf{x}}\hat{\mathbf{y}}T_{xy} + \hat{\mathbf{x}}\hat{\mathbf{z}}T_{xz} \\ & + \hat{\mathbf{y}}\hat{\mathbf{x}}T_{yx} + \hat{\mathbf{y}}\hat{\mathbf{y}}T_{yy} + \hat{\mathbf{y}}\hat{\mathbf{z}}T_{yz} \\ & + \hat{\mathbf{z}}\hat{\mathbf{x}}T_{zx} + \hat{\mathbf{z}}\hat{\mathbf{y}}T_{zy} + \hat{\mathbf{z}}\hat{\mathbf{z}}T_{zz} \end{aligned} \tag{D.3}$$

where

$$T_{ij} = C_i D_j \qquad i, j = x, y, \text{ and } z \tag{D.4}$$

Scalar multiplication by a vector

The scalar multiplication from the left of dyad **T** by the vector

$$\mathbf{A} = \hat{\mathbf{x}}A_x + \hat{\mathbf{y}}A_y + \hat{\mathbf{z}}A_z \tag{D.5}$$

gives

$$\begin{aligned} \mathbf{A} \cdot \mathbf{T} = {} & (\mathbf{A} \cdot \hat{\mathbf{x}})\hat{\mathbf{x}}T_{xx} + (\mathbf{A} \cdot \hat{\mathbf{x}})\hat{\mathbf{y}}T_{xy} + (\mathbf{A} \cdot \hat{\mathbf{x}})\hat{\mathbf{z}}T_{xz} \\ & + (\mathbf{A} \cdot \hat{\mathbf{y}})\hat{\mathbf{x}}T_{yx} + (\mathbf{A} \cdot \hat{\mathbf{y}})\hat{\mathbf{y}}T_{yy} + (\mathbf{A} \cdot \hat{\mathbf{y}})\hat{\mathbf{z}}T_{yz} \\ & + (\mathbf{A} \cdot \hat{\mathbf{z}})\hat{\mathbf{x}}T_{zx} + (\mathbf{A} \cdot \hat{\mathbf{z}})\hat{\mathbf{y}}T_{zy} + (\mathbf{A} \cdot \hat{\mathbf{z}})\hat{\mathbf{z}}T_{zz} \\ = {} & \hat{\mathbf{x}}(A_x T_{xx} + A_y T_{yx} + A_z T_{zx}) \\ & + \hat{\mathbf{y}}(A_x T_{xy} + A_y T_{yy} + A_z T_{zy}) \\ & + \hat{\mathbf{z}}(A_x T_{xz} + A_y T_{yz} + A_z T_{zz}) \end{aligned} \tag{D.6}$$

In a similar manner, the scalar multiplication from the right of **T** by **A** may be shown to yield:

$$\mathbf{T}\cdot\mathbf{A} = \hat{\mathbf{x}}(A_x T_{xx} + A_y T_{xy} + A_z T_{xz}) + \hat{\mathbf{y}}(A_x T_{yx} + A_y T_{yy} + A_z T_{yz}) + \hat{\mathbf{z}}(A_x T_{zx} + A_y T_{zy} + A_z T_{zz}) \qquad \text{(D.7)}$$

Thus the scalar multiplication of a dyad by a vector gives a vector which is different depending on whether the multiplication is carried out from the left or the right.

The kinetic pressure $\mathbf{\Psi}$ as defined by Eq. (1.9.5) is a dyad. As in Eq. (D.3), the dyad $\mathbf{\Psi}$ is expressed in component form in Eq. (1.9.15). From Eqs. (1.9.15) and (D.6), it can be shown that

$$\frac{\partial}{\partial x}\hat{\mathbf{x}}\cdot\mathbf{\Psi} = \frac{\partial}{\partial x}[\hat{\mathbf{x}}P_{xx} + \hat{\mathbf{y}}P_{xy} + \hat{\mathbf{z}}P_{xz}] = \hat{\mathbf{x}}\frac{\partial}{\partial x}P_{xx} + \hat{\mathbf{y}}\frac{\partial}{\partial x}P_{xy} + \hat{\mathbf{z}}\frac{\partial}{\partial x}P_{xz} \qquad \text{(D.8)}$$

Similarly the following results are obtained:

$$\frac{\partial}{\partial y}\hat{\mathbf{y}}\cdot\mathbf{\Psi} = \hat{\mathbf{x}}\frac{\partial}{\partial y}P_{yx} + \hat{\mathbf{y}}\frac{\partial}{\partial y}P_{yy} + \hat{\mathbf{z}}\frac{\partial}{\partial y}P_{yz} \qquad \text{(D.9)}$$

and

$$\frac{\partial}{\partial z}\hat{\mathbf{z}}\cdot\mathbf{\Psi} = \hat{\mathbf{x}}\frac{\partial}{\partial z}P_{zx} + \hat{\mathbf{y}}\frac{\partial}{\partial z}P_{zy} + \hat{\mathbf{z}}\frac{\partial}{\partial z}P_{zz} \qquad \text{(D.10)}$$

Addition of Eqs. (D.8)–(D.10) yields

$$\begin{aligned}\left(\frac{\partial}{\partial x}\hat{\mathbf{x}}\cdot + \frac{\partial}{\partial y}\hat{\mathbf{y}}\cdot + \frac{\partial}{\partial z}\hat{\mathbf{z}}\cdot\right)\mathbf{\Psi} &= \nabla\cdot\mathbf{\Psi} \\ &= \hat{\mathbf{x}}\left(\frac{\partial}{\partial x}P_{xx} + \frac{\partial}{\partial y}P_{yx} + \frac{\partial}{\partial z}P_{zx}\right) \\ &\quad + \hat{\mathbf{y}}\left(\frac{\partial}{\partial x}P_{xy} + \frac{\partial}{\partial y}P_{yy} + \frac{\partial}{\partial z}P_{zy}\right) \\ &\quad + \hat{\mathbf{z}}\left(\frac{\partial}{\partial x}P_{xz} + \frac{\partial}{\partial y}P_{yz} + \frac{\partial}{\partial z}P_{zz}\right)\end{aligned} \qquad \text{(D.11)}$$

The divergence of a dyad is a vector. The components of $\nabla\cdot\mathbf{\Psi}$ which appears in the momentum transport equation (1.9.14) are given explicitly in Eq. (D.11). In Eq. (1.9.7) the divergence of the dyad $Nm\mathbf{vv}$ is evaluated as detailed in Eqs. (D.8)–(D.11).

Matrix representation

Let the matrix $[c]$ be the product of two 3×3 matrices $[a]$ and $[b]$. Note that $[c]$ is also a 3×3 matrix. This relation can be expressed as

$$\begin{bmatrix} c_{11} & c_{12} & c_{13} \\ c_{21} & c_{22} & c_{23} \\ c_{31} & c_{32} & c_{33}\end{bmatrix} = \begin{bmatrix} a_{11} & a_{12} & a_{13} \\ a_{21} & a_{22} & a_{23} \\ a_{31} & a_{32} & a_{33}\end{bmatrix}\begin{bmatrix} b_{11} & b_{12} & b_{13} \\ b_{21} & b_{22} & b_{23} \\ b_{31} & b_{32} & b_{33}\end{bmatrix} \qquad \text{(D.12)}$$

Any element in $[c]$ can be determined from the elements of $[a]$ and $[b]$ in accordance with the following rule:

$$c_{ij} = a_{i1} b_{1j} + a_{i2} b_{2j} + a_{i3} b_{3j} \tag{D.13}$$

Using this rule for matrix multiplication, the dyad $\mathbf{T}$ as given by Eq. (D.3) can be expressed as

$$\mathbf{T} = \begin{bmatrix} \hat{\mathbf{x}} & \hat{\mathbf{y}} & \hat{\mathbf{z}} \end{bmatrix} \begin{bmatrix} T_{xx} & T_{xy} & T_{xz} \\ T_{yx} & T_{yy} & T_{yz} \\ T_{zx} & T_{zy} & T_{zz} \end{bmatrix} \begin{bmatrix} \hat{\mathbf{x}} \\ \hat{\mathbf{y}} \\ \hat{\mathbf{z}} \end{bmatrix} \tag{D.14}$$

It is usual to omit the pre- and the post-multiplicative vectors and denote a dyad only by the 3×3 matrix containing the coefficients of the elements. Thus T_{ij} corresponds to the ith row and jth column. In this book the dyads are commonly represented in the above-mentioned matrix notation.

The following set of three relations reproduced from Eqs. (5.2.30 a–c)

$$D_x = \varepsilon_1 E_x + i\varepsilon_2 E_y$$

$$D_y = -i\varepsilon_2 E_x + \varepsilon_1 E_y$$

$$D_z = \varepsilon_3 E_z$$

can be expressed in the matrix form as in

$$\begin{bmatrix} D_x \\ D_y \\ D_z \end{bmatrix} = \begin{bmatrix} \varepsilon_1 & i\varepsilon_2 & 0 \\ -i\varepsilon_2 & \varepsilon_1 & 0 \\ 0 & 0 & \varepsilon_3 \end{bmatrix} \begin{bmatrix} E_x \\ E_y \\ E_z \end{bmatrix} \tag{D.15}$$

or using the dyad $\boldsymbol{\varepsilon}$ as in

$$\mathbf{D} = \boldsymbol{\varepsilon} \cdot \mathbf{E} \quad \text{where } \boldsymbol{\varepsilon} = \varepsilon_1 (\hat{\mathbf{x}}\hat{\mathbf{x}} + \hat{\mathbf{y}}\hat{\mathbf{y}}) + i\varepsilon_2 (\hat{\mathbf{x}}\hat{\mathbf{y}} - \hat{\mathbf{y}}\hat{\mathbf{x}}) + \varepsilon_3 \hat{\mathbf{z}}\hat{\mathbf{z}} \tag{D.16}$$

A comparison of Eqs. (D.15) and (D.16) shows the matrix representation of the dyad $\boldsymbol{\varepsilon}$.

Symmetry and transpose

If in the matrix representation of a dyad $T_{ij} = T_{ji}$ for $i, j = x, y$, and z, the dyad is said to be *symmetric*. Then, it follows from Eqs. (D.6) and (D.7) that

$$\mathbf{A} \cdot \mathbf{T} = \mathbf{T} \cdot \mathbf{A} \tag{D.17}$$

Thus for a symmetric dyad the scalar multiplication by a vector yields the same result irrespective of whether the multiplication is carried out from the left or the right.

If in the matrix representation of a dyad the rows and the columns are interchanged, the resulting dyad is called the *transpose* of the original dyad. The transpose of $\mathbf{T}$ is denoted by $\mathbf{T}^t$.

Unit dyad

If in the matrix representation of a dyad, the diagonal terms are all equal to unity and the nondiagonal terms vanish, the resulting dyad becomes

$$\mathbf{1} = \hat{\mathbf{x}}\hat{\mathbf{x}} + \hat{\mathbf{y}}\hat{\mathbf{y}} + \hat{\mathbf{z}}\hat{\mathbf{z}} \tag{D.18}$$

From Eqs. (D.6), (D.7), and (D.18), it can be verified that

$$\mathbf{A} \cdot \mathbf{1} = \mathbf{1} \cdot \mathbf{A} = \mathbf{A} \tag{D.19}$$

Thus $\mathbf{1}$ is a special case of a symmetric dyad which on a scalar multiplication by a vector reproduces the same vector and hence is called a *unit dyad.* Note that multiplication of a dyad by a scalar is equivalent to multiplying all the components of the dyad by the same scalar. Hence

$$\mathbf{1}P = (\hat{\mathbf{x}}\hat{\mathbf{x}} + \hat{\mathbf{y}}\hat{\mathbf{y}} + \hat{\mathbf{z}}\hat{\mathbf{z}})P \tag{D.20}$$

From Eq. (D.11) it is found that

$$\nabla \cdot \mathbf{1}P = \hat{\mathbf{x}}\frac{\partial P}{\partial x} + \hat{\mathbf{y}}\frac{\partial P}{\partial y} + \hat{\mathbf{z}}\frac{\partial P}{\partial z} = \nabla P \tag{D.21}$$

which is the same as the result in Eqs. (1.9.20) and (2.9.23).

Cross multiplication by a vector

So far multiplication by a scalar and scalar multiplication by a vector have been considered. But a dyad can also be cross-multiplied by a vector in a straightforward manner. Thus

$$\begin{aligned}\mathbf{B} \times \Psi = {} & (\mathbf{B} \times \hat{\mathbf{x}})\hat{\mathbf{x}}P_{xx} + (\mathbf{B} \times \hat{\mathbf{x}})\hat{\mathbf{y}}P_{xy} + (\mathbf{B} \times \hat{\mathbf{x}})\hat{\mathbf{z}}P_{xz} \\ & + (\mathbf{B} \times \hat{\mathbf{y}})\hat{\mathbf{x}}P_{yx} + (\mathbf{B} \times \hat{\mathbf{y}})\hat{\mathbf{y}}P_{yy} + (\mathbf{B} \times \hat{\mathbf{y}})\hat{\mathbf{z}}P_{yz} \\ & + (\mathbf{B} \times \hat{\mathbf{z}})\hat{\mathbf{x}}P_{zx} + (\mathbf{B} \times \hat{\mathbf{z}})\hat{\mathbf{y}}P_{zy} + (\mathbf{B} \times \hat{\mathbf{z}})\hat{\mathbf{z}}P_{zz}\end{aligned} \tag{D.22}$$

Since

$$\mathbf{B} \times \hat{\mathbf{x}} = (\hat{\mathbf{x}}B_x + \hat{\mathbf{y}}B_y + \hat{\mathbf{z}}B_z) \times \hat{\mathbf{x}} = (-\hat{\mathbf{z}}B_y + \hat{\mathbf{y}}B_z)$$
$$\mathbf{B} \times \hat{\mathbf{y}} = (\hat{\mathbf{x}}B_x + \hat{\mathbf{y}}B_y + \hat{\mathbf{z}}B_z) \times \hat{\mathbf{y}} = (\hat{\mathbf{z}}B_x - \hat{\mathbf{x}}B_z)$$

and

$$\mathbf{B} \times \hat{\mathbf{z}} = (\hat{\mathbf{x}}B_x + \hat{\mathbf{y}}B_y + \hat{\mathbf{z}}B_z) \times \hat{\mathbf{z}} = (-\hat{\mathbf{y}}B_x + \hat{\mathbf{x}}B_y)$$

Eq. (D.22) becomes

$$\begin{aligned}
\mathbf{B} \times \Psi &= (\hat{\mathbf{y}}B_z - \hat{\mathbf{z}}B_y)\hat{\mathbf{x}}P_{xx} + (\hat{\mathbf{y}}B_z - \hat{\mathbf{z}}B_y)\hat{\mathbf{y}}P_{xy} + (\hat{\mathbf{y}}B_z - \hat{\mathbf{z}}B_y)\hat{\mathbf{z}}P_{xz} \\
&\quad + (-\hat{\mathbf{x}}B_z + \hat{\mathbf{z}}B_x)\hat{\mathbf{x}}P_{yx} + (-\hat{\mathbf{x}}B_z + \hat{\mathbf{z}}B_x)\hat{\mathbf{y}}P_{yy} + (-\hat{\mathbf{x}}B_z + \hat{\mathbf{z}}B_x)\hat{\mathbf{z}}P_{yz} \\
&\quad + (\hat{\mathbf{x}}B_y - \hat{\mathbf{y}}B_x)\hat{\mathbf{x}}P_{zx} + (\hat{\mathbf{x}}B_y - \hat{\mathbf{y}}B_x)\hat{\mathbf{y}}P_{zy} + (\hat{\mathbf{x}}B_y - \hat{\mathbf{y}}B_x)\hat{\mathbf{z}}P_{zz} \\
&= (B_y P_{zx} - B_z P_{yx})\hat{\mathbf{x}}\hat{\mathbf{x}} + (B_y P_{zy} - B_z P_{yy})\hat{\mathbf{x}}\hat{\mathbf{y}} + (B_y P_{zz} - B_z P_{yz})\hat{\mathbf{x}}\hat{\mathbf{z}} \\
&\quad + (B_z P_{xx} - B_x P_{zx})\hat{\mathbf{y}}\hat{\mathbf{x}} + (B_z P_{xy} - B_x P_{zy})\hat{\mathbf{y}}\hat{\mathbf{y}} + (B_z P_{xz} - B_x P_{zz})\hat{\mathbf{y}}\hat{\mathbf{z}} \\
&\quad + (B_x P_{yx} - B_y P_{xx})\hat{\mathbf{z}}\hat{\mathbf{x}} + (B_x P_{yy} - B_y P_{xy})\hat{\mathbf{z}}\hat{\mathbf{y}} + (B_x P_{yz} - B_y P_{xz})\hat{\mathbf{z}}\hat{\mathbf{z}}
\end{aligned} \tag{D.23}$$

Thus the cross product of a vector and a dyad is a dyad. The result of $\mathbf{B} \times \Psi$ is deduced in component form in Eq. (D.23). From Eq. (D.23) it follows that

$$\hat{\mathbf{x}} \cdot \{\hat{\mathbf{x}} \cdot (\mathbf{B} \times \Psi)\} = B_y P_{zx} - B_z P_{yx} \tag{D.24}$$

a result which is used in Eq. (1.10.26).

Scalar multiplication by a dyad

Another operation that is useful is the scalar multiplication of a dyad by another dyad. Thus

$$\begin{aligned}
\Psi \cdot (\hat{\mathbf{x}}\hat{\mathbf{x}}T_{xx}) &= \hat{\mathbf{x}}(\hat{\mathbf{x}} \cdot \hat{\mathbf{x}})\hat{\mathbf{x}}P_{xx}T_{xx} + \hat{\mathbf{x}}(\hat{\mathbf{y}} \cdot \hat{\mathbf{x}})\hat{\mathbf{x}}P_{xy}T_{xx} + \hat{\mathbf{x}}(\hat{\mathbf{z}} \cdot \hat{\mathbf{x}})\hat{\mathbf{x}}P_{xz}T_{xx} \\
&\quad + \hat{\mathbf{y}}(\hat{\mathbf{x}} \cdot \hat{\mathbf{x}})\hat{\mathbf{x}}P_{yx}T_{xx} + \hat{\mathbf{y}}(\hat{\mathbf{y}} \cdot \hat{\mathbf{x}})\hat{\mathbf{x}}P_{yy}T_{xx} + \hat{\mathbf{y}}(\hat{\mathbf{z}} \cdot \hat{\mathbf{x}})\hat{\mathbf{x}}P_{yz}T_{xx} \\
&\quad + \hat{\mathbf{z}}(\hat{\mathbf{x}} \cdot \hat{\mathbf{x}})\hat{\mathbf{x}}P_{zx}T_{xx} + \hat{\mathbf{z}}(\hat{\mathbf{y}} \cdot \hat{\mathbf{x}})\hat{\mathbf{x}}P_{zy}T_{xx} + \hat{\mathbf{z}}(\hat{\mathbf{z}} \cdot \hat{\mathbf{x}})\hat{\mathbf{x}}P_{zz}T_{xx} \\
&= \hat{\mathbf{x}}\hat{\mathbf{x}}P_{xx}T_{xx} + \hat{\mathbf{y}}\hat{\mathbf{x}}P_{yx}T_{xx} + \hat{\mathbf{z}}\hat{\mathbf{x}}P_{zx}T_{xx}
\end{aligned} \tag{D.25}$$

which shows that the scalar multiplication of two dyads yields another dyad. In a similar manner the results of the scalar multiplication of Ψ by each of the components of $\mathbf{T}$ as given by Eq. (D.3) can be obtained and by summing up these results it can be shown that

$$\begin{aligned}
\Psi \cdot \mathbf{T} &= \hat{\mathbf{x}}\hat{\mathbf{x}}S_{xx} + \hat{\mathbf{x}}\hat{\mathbf{y}}S_{xy} + \hat{\mathbf{x}}\hat{\mathbf{z}}S_{xz} \\
&\quad + \hat{\mathbf{y}}\hat{\mathbf{x}}S_{yx} + \hat{\mathbf{y}}\hat{\mathbf{y}}S_{yy} + \hat{\mathbf{y}}\hat{\mathbf{z}}S_{yz} \\
&\quad + \hat{\mathbf{z}}\hat{\mathbf{x}}S_{zx} + \hat{\mathbf{z}}\hat{\mathbf{y}}S_{zy} + \hat{\mathbf{z}}\hat{\mathbf{z}}S_{zz}
\end{aligned} \tag{D.26}$$

where

$$S_{ij} = P_{ix} T_{xj} + P_{iy} T_{yj} + P_{iz} T_{zj} \qquad \text{for } i, j = x, y, \text{ and } z \tag{D.27}$$

In particular it is seen from Eq. (D.27) that

$$S_{xx} = P_{xx} T_{xx} + P_{xy} T_{yx} + P_{xz} T_{zx} \tag{D.28}$$

Consider the gradient of a vector $\mathbf{v} = \hat{\mathbf{x}}v_x + \hat{\mathbf{y}}v_y + \hat{\mathbf{z}}v_z$:

$$\begin{aligned} \mathbf{T} = \nabla\mathbf{v} &= \left(\frac{\partial}{\partial x}\hat{\mathbf{x}} + \frac{\partial}{\partial y}\hat{\mathbf{y}} + \frac{\partial}{\partial z}\hat{\mathbf{z}}\right)(\hat{\mathbf{x}}v_x + \hat{\mathbf{y}}v_y + \hat{\mathbf{z}}v_z) \\ &= \hat{\mathbf{x}}\hat{\mathbf{x}}\frac{\partial v_x}{\partial x} + \hat{\mathbf{x}}\hat{\mathbf{y}}\frac{\partial v_y}{\partial x} + \hat{\mathbf{x}}\hat{\mathbf{z}}\frac{\partial v_z}{\partial x} \\ &+ \hat{\mathbf{y}}\hat{\mathbf{x}}\frac{\partial v_x}{\partial y} + \hat{\mathbf{y}}\hat{\mathbf{y}}\frac{\partial v_y}{\partial y} + \hat{\mathbf{y}}\hat{\mathbf{z}}\frac{\partial v_z}{\partial y} \\ &+ \hat{\mathbf{z}}\hat{\mathbf{x}}\frac{\partial v_x}{\partial z} + \hat{\mathbf{z}}\hat{\mathbf{y}}\frac{\partial v_y}{\partial z} + \hat{\mathbf{z}}\hat{\mathbf{z}}\frac{\partial v_z}{\partial z} \end{aligned} \tag{D.29}$$

Thus the gradient of a vector is seen to yield a dyad. From a comparison of Eqs. (D.3) and (D.29) each of the components of $\mathbf{T}$ can be identified with the corresponding component of the dyad $\nabla\mathbf{v}$. Then with the help of Eqs. (D.26) and (D.28), it can be established that

$$\hat{\mathbf{x}} \cdot \{\hat{\mathbf{x}} \cdot (\mathbf{\Psi} \cdot \nabla\mathbf{v})\} = P_{xx}\frac{\partial v_x}{\partial x} + P_{xy}\frac{\partial v_x}{\partial y} + P_{xz}\frac{\partial v_x}{\partial z} \tag{D.30}$$

which is the result stated in Eq. (1.10.21).

Triad

The definition given in Eq. (D.1) can be extended as follows

$$\mathbf{R} = \mathbf{BCD} \tag{D.31}$$

where, as in Eq. (D.1), there are no operational signs such as +, −, •, and × separating the vectors **B**, **C**, and **D**. The quantity defined by Eq. (D.31) is called a *triad.* The thermal energy flux density **Q** as defined by Eqs. (1.10.5) and (1.10.6) is therefore a triad. In component form **Q** can be written as follows:

$$\mathbf{Q} = \hat{\mathbf{x}}\mathbf{Q}_x + \hat{\mathbf{y}}\mathbf{Q}_y + \hat{\mathbf{z}}\mathbf{Q}_z \tag{D.32}$$

where

$$\begin{aligned} \mathbf{Q}_i = \hat{\mathbf{x}}\hat{\mathbf{x}}Q_{ixx} + \hat{\mathbf{x}}\hat{\mathbf{y}}Q_{ixy} + \hat{\mathbf{x}}\hat{\mathbf{z}}Q_{ixz} + \hat{\mathbf{y}}\hat{\mathbf{x}}Q_{iyx} + \hat{\mathbf{y}}\hat{\mathbf{y}}Q_{iyy} + \hat{\mathbf{y}}\hat{\mathbf{z}}Q_{iyz} \\ + \hat{\mathbf{z}}\hat{\mathbf{x}}Q_{izx} + \hat{\mathbf{z}}\hat{\mathbf{y}}Q_{izy} + \hat{\mathbf{z}}\hat{\mathbf{z}}Q_{izz} \qquad \text{for } i = x, y, z \end{aligned} \tag{D.33}$$

Divergence of a triad

Mathematical operations with a triad proceed in the same way as for a dyad. For example, it is obtained that

$$\begin{aligned}\nabla \cdot \mathbf{Q} &= \left(\frac{\partial}{\partial x}\hat{\mathbf{x}} \cdot + \frac{\partial}{\partial y}\hat{\mathbf{y}} \cdot + \frac{\partial}{\partial z}\hat{\mathbf{z}} \cdot\right)(\hat{\mathbf{x}}\mathbf{Q}_x + \hat{\mathbf{y}}\mathbf{Q}_y + \hat{\mathbf{z}}\mathbf{Q}_z) \\ &= \frac{\partial}{\partial x}\mathbf{Q}_x + \frac{\partial}{\partial y}\mathbf{Q}_y + \frac{\partial}{\partial z}\mathbf{Q}_z\end{aligned} \tag{D.34}$$

With the help of Eqs. (D.33) and (D.34) it is found that

$$\begin{aligned}\nabla \cdot \mathbf{Q} = \hat{\mathbf{x}}\hat{\mathbf{x}}S_{xx} + \hat{\mathbf{x}}\hat{\mathbf{y}}S_{xy} + \hat{\mathbf{x}}\hat{\mathbf{z}}S_{xz} \\ + \hat{\mathbf{y}}\hat{\mathbf{x}}S_{yx} + \hat{\mathbf{y}}\hat{\mathbf{y}}S_{yy} + \hat{\mathbf{y}}\hat{\mathbf{z}}S_{yz} \\ + \hat{\mathbf{z}}\hat{\mathbf{x}}S_{zx} + \hat{\mathbf{z}}\hat{\mathbf{y}}S_{zy} + \hat{\mathbf{z}}\hat{\mathbf{z}}S_{zz}\end{aligned} \tag{D.35}$$

where, for example,

$$S_{xx} = \frac{\partial}{\partial x}Q_{xxx} + \frac{\partial}{\partial y}Q_{yxx} + \frac{\partial}{\partial z}Q_{zxx} \tag{D.36}$$

The divergence of a triad is a dyad. Note that $\nabla \cdot \mathbf{Q}$ appears in the transport equation for the kinetic pressure as given by Eq. (1.10.30). It is found from Eqs. (D.35) and (D.36) that

$$\hat{\mathbf{x}} \cdot \{\hat{\mathbf{x}} \cdot (\nabla \cdot \mathbf{Q})\} = \frac{\partial}{\partial x}Q_{xxx} + \frac{\partial}{\partial y}Q_{yxx} + \frac{\partial}{\partial z}Q_{zxx}$$

which is the same as the result stated in Eq. (1.10.12).

Mathematical operations with a dyad and a triad are carried out in essentially the same way as for a vector. It is important to bear in mind that the order in which the component vectors appear in a dyad or a triad is significant and should not be interchanged. In general, therefore, $\hat{\mathbf{x}}\hat{\mathbf{y}}$ and $\hat{\mathbf{y}}\hat{\mathbf{x}}$ indicate two different dyads.

Appendix E: Bessel Functions

The important properties of the Bessel functions are collected together in this appendix. Consider the Bessel equation

$$\frac{d^2f}{dx^2} + \frac{1}{x}\frac{df}{dx} + \left(1 - \frac{n^2}{x^2}\right)f = 0 \tag{E.1}$$

Only the solutions of Eq. (E.1) which are finite at $x = 0$ and which are of integer order n occur in this book and therefore attention is restricted to only the properties of such solutions. The solution of the Bessel equation for $n \geqslant 0$ which is finite at $x = 0$ is the Bessel function of the first kind $J_n(x)$ whose series expression can be obtained from Eq. (E.1) as

$$J_n(x) = \left(\frac{x}{2}\right)^n \sum_{m=0}^{\infty} (-1)^m \frac{1}{m!\,(m+n)!}\left(\frac{x}{2}\right)^{2m} \tag{E.2}$$

For integer values of n the following simple relations hold:

$$J_{-n}(x) = (-1)^n J_n(x) \tag{E.3}$$

and

$$J_n(-x) = (-1)^n J_n(x) \tag{E.4}$$

The Bessel function of the first kind corresponds to the case of real arguments and hence is real.

For small arguments, $x \ll 1$, the following approximations are valid:

$$J_0(x) = 1 \qquad J_n(x) = \frac{1}{n!}\left(\frac{x}{2}\right)^n \qquad \text{for } n > 0 \tag{E.5}$$

For large arguments, $x \gg 1$, asymptotic series solution of Eq. (E.1) can be obtained and the result of retaining only the leading term is

$$J_n(x) = \sqrt{\frac{2}{\pi x}}\cos\left(x - \frac{\pi}{4} - \frac{n\pi}{4}\right) \tag{E.6}$$

The following important recursive relations are satisfied by the Bessel function of the first kind:

$$J_{n-1}(x) + J_{n+1}(x) = \frac{2n}{x} J_n(x) \tag{E.7}$$

and

$$J_{n-1}(x) - J_{n+1}(x) = 2\frac{d}{dx}J_n(x) \tag{E.8}$$

In particular Eqs. (E.7) and (E.8) yield

$$J_1(x) = -\frac{d}{dx}J_0(x) \tag{E.9}$$

and

$$\frac{d}{dx}[x^n J_n(x)] = x^n J_{n-1}(x) \qquad \frac{d}{dx}[x^{-n} J_n(x)] = -x^{-n} J_{n+1}(x) \tag{E.10}$$

When the argument of the Bessel function of the first kind is imaginary, it is convenient to introduce the Bessel function of the second kind $I_n(x)$ in accordance with the relation:

$$I_n(x) = i^{-n} J_n(ix) \tag{E.11}$$

The following series representation for $I_n(x)$ can be deduced from Eq. (E.2):

$$I_n(x) = \left(\frac{x}{2}\right)^n \sum_{m=0}^{\infty} \frac{1}{m!\,(m+n)!}\left(\frac{x}{2}\right)^{2m} \tag{E.12}$$

For integer values of n, it is found that

$$I_{-n}(x) = I_n(x) \tag{E.13}$$

When the argument x is real, Eq. (E.12) shows that the Bessel function of the second kind $I_n(x)$ is real.

For small arguments, $x \ll 1$, the following approximations are valid:

$$I_0(x) = 1 \qquad I_n(x) = \frac{1}{n!}\left(\frac{x}{2}\right)^n \qquad \text{for } n > 0 \tag{E.14}$$

For large arguments, $x \gg 1$, asymptotic series solution of Eq. (E.1) can be obtained for imaginary arguments and the result of retaining only the leading term is

$$I_n(x) = e^x/\sqrt{2\pi x} \tag{E.15}$$

The Bessel function of the second kind satisfies the following recursive relations:

$$I_{n-1}(x) + I_{n+1}(x) = 2\frac{d}{dx}I_n(x) \tag{E.16}$$

and

$$I_{n-1}(x) - I_{n+1}(x) = \frac{2n}{x}I_n(x) \tag{E.17}$$

From Eqs. (E.16) and (E.17), it can be deduced that

$$I_1(x) = \frac{d}{dx}I_0(x) \tag{E.18}$$

and

$$\frac{d}{dx}[x^n I_n(x)] = x^n I_{n-1}(x) \qquad \frac{d}{dx}[x^{-n} I_n(x)] = x^{-n} I_{n+1}(x) \tag{E.19}$$

Appendix F: Legendre Polynomials

Consider the associated Legendre equation

$$\frac{d^2f}{d\theta^2} + \frac{1}{\tan\theta}\frac{df}{d\theta} + \left[n(n+1) - \frac{m^2}{\sin^2\theta}\right]f = 0 \tag{F.1}$$

where m and n are integers. By setting

$$u = \cos\theta \tag{F.2}$$

Eq. (F.1) can be recast into the form

$$(1-u^2)\frac{d^2f}{du^2} - 2u\frac{df}{du} + \left[n(n+1) - \frac{m^2}{1-u^2}\right]f = 0 \tag{F.3}$$

When $m = 0$, Eq. (F.3) becomes the ordinary Legendre equation:

$$(1-u^2)\frac{d^2f}{du^2} - 2u\frac{df}{du} + n(n+1)f = 0 \tag{F.4}$$

In Eq. (F.1) θ is the spherical coordinate and hence $0 \leqslant \theta \leqslant \pi$. Therefore the solutions of Eqs. (F.3) and (F.4) are desired over the range $-1 \leqslant u \leqslant 1$.

The solution of the ordinary Legendre equation (F.4) which is finite over the range $-1 \leqslant u \leqslant 1$ is the Legendre polynomial of order n, $P_n(u)$, whose series solution can be obtained as

$$P_n(u) = \sum_{l=0}^{L} \frac{(-1)^l (2n-2l)!}{2^n\, l!\,(n-l)!\,(n-2l)!} u^{n-2l} \tag{F.5}$$

where $L = n/2$ or $(n-1)/2$, whichever is an integer. $P_n(u)$ is real and is a finite polynomial. The Legendre polynomials up to fourth order are given explicitly by

$$\begin{aligned} &P_0(u) = 1 \qquad P_1(u) = u \qquad P_2(u) = \tfrac{1}{2}(3u^2 - 1) \\ &P_3(u) = \tfrac{1}{2}(5u^3 - 3u) \qquad P_4(u) = \tfrac{1}{8}(35u^4 - 30u^2 + 3) \end{aligned} \tag{F.6}$$

In terms of θ, Eqs. (F.6) become

$$P_0(\cos\theta) = 1 \qquad P_1(\cos\theta) = \cos\theta$$

$$P_2(\cos\theta) = \tfrac{1}{4}(3\cos 2\theta + 1) \qquad P_3(\cos\theta) = \tfrac{1}{8}(5\cos 3\theta + 3\cos\theta)$$

$$P_4(\cos\theta) = \tfrac{1}{64}(35\cos 4\theta + 20\cos 2\theta + 9) \tag{F.7}$$

The following relation is valid in general:

$$P_n(-u) = (-1)^n P_n(u) \tag{F.8}$$

which is seen to be satisfied by the five lowest-order Legendre polynomials as given by Eqs. (F.6).

The solution of the associated Legendre equation (F.3) can be determined directly in terms of the Legendre polynomials as follows. The result of differentiating Eq. (F.4) m times with respect to u is

$$\left[(1-u^2)\frac{d^2}{du^2} - 2u(m+1)\frac{d}{du} + (n-m)(n+m+1)\right]\frac{d^m f}{du^m} = 0 \tag{F.9}$$

By setting

$$g = (1-u^2)^{m/2}(d^m f/du^m) \tag{F.10}$$

in Eq. (F.9), it can be deduced that

$$(1-u^2)\frac{d^2 g}{du^2} - 2u\frac{dg}{du} + \left[n(n+1) - \frac{m^2}{1-u^2}\right]g = 0 \tag{F.11}$$

which is the associated Legendre equation. Hence, the solutions of the associated Legendre equation (F.3) which remain finite over the range $-1 \leqslant u \leqslant 1$ can be taken as

$$P_n^m(u) = (-1)^m(1-u^2)^{m/2}\frac{d^m}{du^m}P_n(u) \tag{F.12}$$

which is the associated Legendre polynomial of order n and degree m. Note that $P_n^m(u)$ is a real, finite polynomial and that

$$P_n^m(u) = P_n(u) \quad \text{for } m = 0 \quad \text{and} \quad P_n^m(u) = 0 \quad \text{for } m > n \tag{F.13}$$

Some of the lower order associated Legendre polynomials are deduced from Eq. (F.12) to be:

$$P_1^1(u) = -(1-u^2)^{1/2} \qquad P_2^1(u) = -3(1-u^2)^{1/2}u$$

$$P_2^2(u) = 3(1-u^2) \qquad P_3^1(u) = \tfrac{3}{2}(1-u^2)^{1/2}(1-5u^2) \tag{F.14}$$

$$P_3^2(u) = 15(1-u^2)u \qquad P_3^3(u) = -15(1-u^2)^{3/2}$$

In terms of θ, Eqs. (F.14) become

$$\begin{aligned} &P_1^1(\cos\theta) = -\sin\theta && P_2^1(\cos\theta) = -3\sin\theta\cos\theta \\ &P_2^2(\cos\theta) = 3\sin^2\theta && P_3^1(\cos\theta) = \tfrac{3}{2}\sin\theta(1-5\cos^2\theta) \\ &P_3^2(\cos\theta) = 15\sin^2\theta\cos\theta && P_3^3(\cos\theta) = -15\sin^3\theta \end{aligned} \tag{F.15}$$

For the special case of $\theta = 0$, i.e., $u = 1$, it can be shown that

$$P_n^m(1) = \begin{cases} 1 & \text{for } m = 0 \\ 0 & \text{for } m \neq 0 \end{cases} \tag{F.16}$$

A function defined in the range $-1 \leqslant u \leqslant 1$ can be expanded in terms of Legendre polynomials. For this purpose, the following orthogonality relation is useful:

$$\int_{-1}^{1} P_n^m(u) P_s^m(u)\, du = \begin{cases} 0 & \text{for } n \neq s \\ \dfrac{2}{2n+1}\dfrac{(n+m)!}{(n-m)!} & \text{for } n = s \end{cases} \tag{F.17}$$

Appendix G: Integral Relations

Several integral relations appearing in the text are collected together here for the sake of convenience.

1.

$$h(n) = \int_0^\infty x^n e^{-ax^2}\,dx$$

$$h(0) = \tfrac{1}{2}\sqrt{\frac{\pi}{a}} \qquad h(1) = \frac{1}{2a} \qquad h(2) = \tfrac{1}{4}\sqrt{\frac{\pi}{a^3}}$$

$$h(3) = \frac{1}{2a^2} \qquad h(4) = \tfrac{3}{8}\sqrt{\frac{\pi}{a^5}} \qquad h(5) = \frac{1}{a^3}$$

2

$$\int_0^\infty \frac{1}{x^2 + a^2}\,dx = \frac{\pi}{2a}$$

3.

$$\int_0^{x_0} \frac{x\,dx}{(1 + ax^2)^2} = \frac{x_0^2}{2(1 + ax_0^2)}$$

4. The error function is defined by

$$\operatorname{erf}(y) = \frac{2}{\sqrt{\pi}} \int_0^y e^{-s^2}\,ds$$

and has the following properties:

$$\operatorname{erf}(0) = 0 \qquad \operatorname{erf}(\infty) = 1 \qquad \frac{d}{dy}\operatorname{erf}(y) = \frac{2}{\sqrt{\pi}} e^{-y^2}$$

5. Dirac's delta function is defined by

$$\delta(x) = 0 \quad \text{for } x \neq 0 \quad \text{and} \quad \int_{-\infty}^{\infty} \delta(x)\,dx = 1$$

and it has the following property:

$$\int_{-\infty}^{\infty} g(x')\delta(x' - x)\,dx' = g(x)$$

6. Bessel function of the first kind and order n has the following integral representation:

$$J_n(\zeta) = \frac{1}{2\pi} \int_0^{2\pi} e^{i(n\varphi - \zeta \sin\varphi)} \, d\varphi$$

From this expression the following Fourier series representation can be derived:

$$e^{i\zeta \sin\varphi} = \sum_{n=-\infty}^{n=\infty} J_n(\zeta) e^{in\varphi}$$

7. Dispersion function:

$$I(x) = 2x \int_0^x e^{w^2 - x^2} \, dw \qquad \text{[as defined in Eq. (6.2.78)]} .$$

$$I(x) = 1 + \frac{1}{2x^2} + \frac{3}{4x^4} \cdots \qquad \text{for } x \to \infty$$

$$I(x) = 2x^2 (1 - \tfrac{2}{3}x^2 + \tfrac{4}{15}x^4 \ldots) \qquad \text{for } x \to 0$$

8. Weber's second exponential integral:

$$\int_0^\infty \exp(-p^2 t^2) J_n(at) J_n(bt) t \, dt = \frac{1}{2p^2} \exp\left(-\frac{a^2 + b^2}{4p^2}\right) I_n\left(\frac{ab}{2p^2}\right)$$

where $I_n(x)$ is the Bessel function of the second kind and order n.

Bibliography

A. I. Akhiezer, I. A. Akhiezer, R. V. Polovin, A. G. Sitenko, and K. N. Stepanov, *Collective Oscillations in a Plasma*, The Massachusetts Institute of Technology Press, Cambridge, Mass., 1967.

H. Alfvén and C. G. Fälthammar, *Cosmical Electrodynamics*, 2nd ed., Oxford University Press, London, 1963.

S. Chandrasekhar, *Plasma Physics*, The University of Chicago Press, Chicago, 1960.

S. Chapman and T. G. Cowling, *The Mathematical Theory of Nonuniform Gases*, 2nd ed., Cambridge University Press, London, 1960.

J. L. Delcroix, *Plasma Physics*, John Wiley, New York, 1965.

V. C. A. Ferraro and C. Plumpton, *An Introduction to Magneto-Fluid Mechanics*, Oxford University Press, London, 1961.

R. Jancel and Th. Kahan, *Electrodynamics of Plasmas*, John Wiley, New York, 1966.

J. Jeans, *The Dynamical Theory of Gases*, Cambridge University Press, London, 1954.

D. J. Rose and M. Clark, Jr., *Plasmas and Controlled Fusion*, John Wiley, New York, 1961.

L. Spitzer, Jr., *Physics of Fully Ionized Gases*, 2nd ed., Interscience Publishers, New York, 1962.

Solutions to Problems

1.4. $u_{mp} = \sqrt{KT/m}$

1.5. 0.57 [use tables of error function and its derivative].

1.6. $N(z) = N_0 \exp(-mgz/KT)$; N_0 is the number density at the earth's surface.

1.8. $P = NKT$

1.13. $\mathbf{\Psi} = (\hat{\mathbf{x}}\hat{\mathbf{x}} + \hat{\mathbf{y}}\hat{\mathbf{y}})NKT_\perp + \hat{\mathbf{z}}\hat{\mathbf{z}}NKT_\parallel$

1.18. $\mathbf{Q} = 0$

1.19. $\mathbf{Q}$ has ten independent elements.

1.22. $f(\mathbf{r}, \mathbf{u}, t) = N/8a^3 \qquad T = ma^2/3K \qquad \mathbf{Q} = 0$

1.23. $f(\mathbf{r}, \mathbf{u}, t) = Na_r/\pi(u_x^2 + a_r^2)$; $\frac{1}{2}m\langle u_x^2\rangle = \infty$; the velocity distribution function has limited physical significance since the average kinetic energy density is infinite.

1.26. The adiabatic gas law $PN^{-5/3}$ = constant is obtained if viscosity, heat conduction, and energy interchange due to collisional interactions are neglected.

2.2.

$$\phi = \frac{q}{4\pi\varepsilon_0 r}\exp[(1 + ZT_e/T_i)^{1/2} r/\lambda_D] \qquad \lambda_D = \sqrt{\varepsilon_0 KT_e/N_0 e^2}$$

For $T_e = T_i = T$ and $Z = 1$, $\phi = (q/4\pi\varepsilon_0 r)\exp[\sqrt{2}\, r/\lambda_D]$, which is the same as that given by Eq. (2.3.18).

2.3.

$$\phi = -3.015 \times 10^5\,\text{V} \qquad T = 3.53 \times 10^{11}\,{}^\circ\text{K}$$

2.4.

$$\phi_w = -\frac{KT_e T_i}{2e(ZT_e + T_i)}\ln\left(\frac{m_i T_e}{m_e T_i}\right) \quad \text{V}$$

2.5.

$$\frac{d^2\phi(x)}{dx^2} = \frac{2}{\lambda_D^2}\phi(x) \qquad \lambda_D = \sqrt{\frac{\varepsilon_0 KT}{N_0 e^2}} \qquad \phi(x) = \phi_w \exp(-\sqrt{2}\, x/\lambda_D)$$

2.6.

$$\frac{d\Gamma_i}{dy} = N_0\left(\frac{KT}{2m_i}\right)^{1/2}[1 + \text{erf}(y)] > 0$$

2.7.

$$N_{e,w} = 0.135N_0 \qquad N_{i,w} = 0.827N_0$$

$$\tilde{v}_{x,w}\sqrt{\frac{m_i}{KT}} = 21.84 \qquad v_{x,w}\sqrt{\frac{m_i}{KT}} = 3.56$$

2.9.

$$J_{e0} + J_i = 52.7\text{A/ m}^2$$

$$\frac{d}{d\phi}\ln(J_p + J_i) = 0.103 \qquad T_e = 1.13 \times 10^5\,{}^\circ\text{K}$$

$$J_{e0} = 50.0\text{A/ m}^2 \qquad J_i = 2.7\text{A/ m}^2$$

$$\phi_w = -28.4\,\text{V} \qquad N_0 = 5.98 \times 10^{14}/\text{ m}^3$$

2.13.

$$v_g = a_e[1 - \omega_{pe}^2/\omega^2]^{1/2} \qquad v_p v_g = a_e^2$$

2.14.

$$\mathbf{v}_{e,p} = \mathbf{v}_{i,p} = -\frac{i}{\omega N_0(m_e + m_i)}\nabla(P_e + P_i)$$

$$-i\omega(P_e + P_i) + (\gamma_e KT_e + \gamma_i KT_i)N_0\nabla\cdot\mathbf{v}_{e,p} = 0$$

$$\nabla^2(P_e + P_i) + \frac{\omega^2(m_e + m_i)}{(\gamma_e KT_e + \gamma_i KT_i)}(P_e + P_i) = 0$$

$$v_{pi} = [\{\gamma_e KT_e + \gamma_i KT_i\}/(m_e + m_i)]^{1/2} \approx [\{\gamma_e KT_e + \gamma_i KT_i\}/m_i]^{1/2}$$

2.15.

$$\omega^2 = \omega_{pi}^2/(1 + \lambda^2/4\pi^2\lambda_{De}^2) \quad \lambda_{De} = \sqrt{\varepsilon_0 KT_e/N_0 e^2}$$

$$v_p^2 = \omega^2/k_p^2 = (KT_e/m_i)/(1 + 4\pi^2\lambda_{De}^2/\lambda^2)$$

$$\gamma_e = 1(\text{isothermal condition})$$

2.18.

$$P_m = 9.947 \times 10^6\text{N/m}^2 \qquad N = 2.484 \times 10^{21}/\text{m}^3 \qquad B = 10\text{W/m}^2$$

2.19.

$$\mathbf{J}_\perp = N_{e0}q_e\left[\frac{\mathbf{E}\times\mathbf{B}_0}{B_0^2} - \mathbf{V}_\perp\right] \quad \text{for } \sigma = \infty$$

2.20. For the theta pinch device: $\mathbf{V} = -\hat{\boldsymbol{\theta}}V$; $\mathbf{B}_0 = \hat{\mathbf{z}}B_0$; $\mathbf{E} = \hat{\mathbf{r}}E$; $\mathbf{J} = -\hat{\boldsymbol{\theta}}J$; $\nabla P_i = -\hat{\mathbf{r}}|\nabla P_i|$.

2.21. $B_\theta(r) = \mu_0 I_0 r/2\pi R^2$ for $0 < r < R$; $B_\theta(r) = \mu_0 I_0/2\pi r$ for $R < r < \infty$; $P(r) = (\mu_0 I_0^2/4\pi^2 R^2)(1 - r^2/R^2)$; R = radius of plasma column. Note that $\langle P(r)\rangle = B_\theta^2(R)/2\mu_0$.

2.22.

$$B_\theta(r) = \frac{\mu_0 e v_{ez} N_0 r}{2(1 + N_0 b r^2)}$$

2.23. For the equilibrium pinch of Problem 2.21:

$$E_r = \frac{\mu_0 I_0^2 r}{2\pi^2 N_{e0} q_e R^4} \text{ for } 0 < r < R \qquad E_r = \frac{\mu_0 I_0^2}{2\pi^2 N_{e0} q_e R^2 r} \text{ for } R < r < \infty$$

For the Bennet pinch of Problem 2.22:

$$E_r = -\frac{\mu_0 e v_{ez}^2 N_0 r}{2[1 + N_0 b r^2]}$$

2.24. The evaluation of the magnetic pressure tensor $-T_{ij}^{(m)}$ with the help of Eq. (2.9.25) for $\mathbf{B} = \hat{\boldsymbol{\theta}}B_\theta$ using the cylindrical coordinates r, θ, and z yields an isotropic magnetic pressure $B_\theta^2/2\mu_0$ plus a tension B_θ^2/μ_0 in the θ-direction. The magnetic pressure $P_m(r)$ given in Eq. (2.10.40) is identical to that determined from Eq. (2.9.25).

2.25. The plasma is assumed to be perfectly conducting and the kinetic pressure of the plasma is ignored. The hollow cylindrical metal tube of Fig. 2.9 is of radius r_0. Let the radius of the plasma column be r and the initial value of r is r_0. The current exists only on the surface of the plasma column. A high voltage is discharged such that the current in the metal tube is in the θ-direction. The induced current density on the surface of the plasma column is in the $-\theta$-direction and let it be given by $\mathbf{J} = -\hat{\boldsymbol{\theta}}J_\theta$. The magnetic flux density at the surface of the plasma column is specified by $\mu_0\mathbf{J} = \hat{\mathbf{r}} \times \mathbf{B}$. Therefore, the magnetic flux density B_z at the surface of the plasma column is given by $\mu_0 J_\theta$ and is in the z-direction. Hence, the magnetic pressure at the surface of the plasma column is $P_{mz} = \mu_0 J_\theta^2/2$. Then, following the procedure given in Sec. 2.10 for the dynamic pinch, the differential equation specifying the time-dependent radius of the plasma column is obtained as

$$(d/dt)[\rho_m(r_0^2 - r^2)(dr/dt)] = -\mu_0 J_\theta^2(r)r$$

where ρ_m is the initial mass density of the plasma.

2.26. Azimuthal magnetic flux density is inversely proportional to the radius of the plasma column: $(B_\theta + \Delta B_\theta)/B_\theta = R/(R - x)$. Therefore $\Delta B_\theta = B_\theta x/R$ for $x/R \ll 1$. Since $P_{m\theta} = B_\theta^2/2\mu_0$, $\Delta P_{m\theta} = B_\theta \Delta B_\theta/\mu_0 = P_{m\theta}2x/R$.

Since axial magnetic flux is frozen into the plasma column, the axial magnetic flux density is inversely proportional to the square of the radius of the plasma column: $(B_z + \Delta B_z)/B_z = R^2/(R-x)^2$. Therefore $\Delta B_z = B_z 2x/R$ for $x/R \ll 1$. Since $P_{mz} = B_z^2/2\mu_0$, $\Delta P_{mz} = B_z \Delta B_z/\mu_0 = P_{mz} 4x/R$.

For stability ΔP_{mz} should exceed $\Delta P_{m\theta}$; therefore the necessary requirement is $B_z^2 > 0.5 B_\theta^2$.

2.27. As the plasma column moves away from the axis of the hollow conducting tube and towards the conducting wall, the azimuthal magnetic flux lines are compressed between the conducting wall and the plasma column, and the resulting increase in the azimuthal magnetic flux density causes an additional force which tends to push the plasma column towards the axis of the hollow conducting tube.

There is an alternative interpretation for the stabilizing role played by the conducting tube. The current on the conducting wall is in a direction opposite to that in the plasma column and hence the conducting wall exerts a repelling force on the plasma column. As the plasma column moves towards the conducting wall, the repelling force is increased since the separation distance is decreased, and, this increase in the repelling force tends to push the plasma column away from the conducting wall.

2.28. From the knowledge of the magnetic flux lines in the neighborhood of a circular current loop, the magnetic flux lines in the neighborhood of a pair of parallel and coaxial circular current loops can be determined by superposition. The results of the magnetic flux lines in a plane passing through the centers of the two loops are qualitatively shown in Fig. Sa for the loop currents in opposite directions and in Fig. Sb for the loop currents in the same direction. The magnetic flux lines are rotationally symmetrical about the z-axis. For the case of oppositely directed loop currents, the magnetic flux lines in the region between the two loops are similar to those in Fig. 2.20 and have cusps. There are two point cusps in the axial direction and one line cusp in the midplane. For the case of loop currents in the same direction, the magnetic flux lines in the region between the loops are similar to those in Fig. 3.5, and therefore constitute a magnetic mirror field.

2.29. There are four line cusps which lie on the two mutually perpendicular midplanes. The particles moving in the general direction of the cusps with the ratio of the energy in the cusp direction to the total energy exceeding a certain maximum value escape in the direction of the cusps. The physical basis of reflection and transmission of particles in the direction of the cusp of the magnetic flux lines is discussed in Sec. 3.8 on the basis of the adiabaticity of the orbital magnetic moment of the particles.

At the center of the four line currents, the magnetic field is zero. The particles passing through the region of zero magnetic field momentarily travel in straight lines and their subsequent motion bears little relationship to those prior to their passage through the region of zero magnetic field. The adiabaticity of the magnetic moment does not hold good in the central region. As a consequence of the nonadiabatic behavior of the particles in the central region, the "loss cones" in the velocity space are continuously refilled and the particles escape continuously in the direction of the four line cusps.

2.30. $\omega_i^2 \ll \omega_r^2$ leads to $E_{x0} \ll E_{max}$ where

$$E_{max} = \frac{32}{9}\frac{m_e}{e}\frac{\omega_{pe}^4\omega_r^2}{k_x^5}\left\{\frac{1}{N_0}f'\left(\frac{\omega_r}{k_x}\right)\right\}^2$$

The most severe limitation occurs at the shortest wavelengths where E_{max} becomes

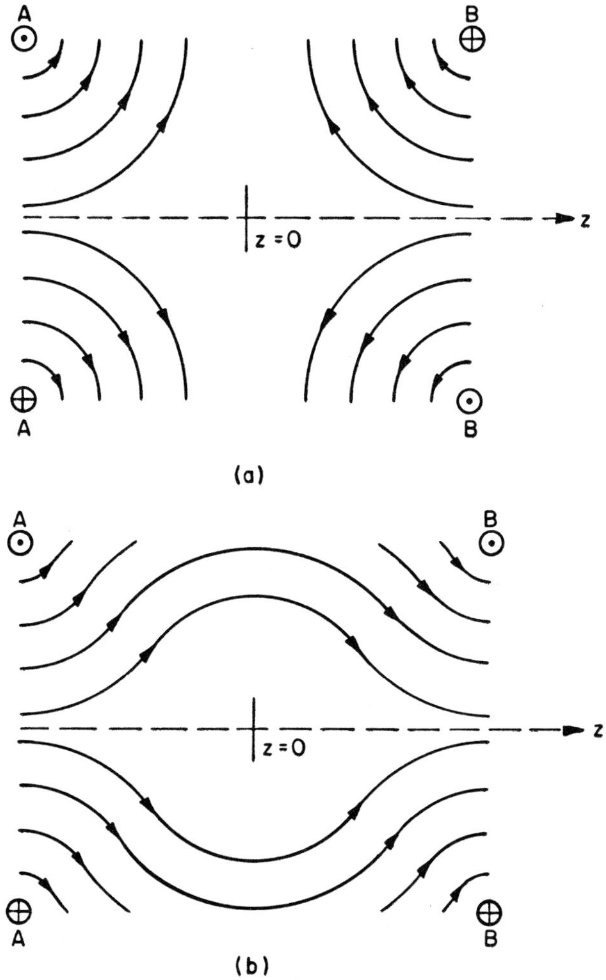

Fig. S. Magnetic flux lines in a plane passing through the centers of two current loops *AA* and *BB* and in the region between them; ⊙ current out of the plane of the paper, ⊕ current into the plane of the paper.

extremely small. The requirement on the depth D of the potential energy wells is $D \ll (2e/k_x)E_{max}$.

2.31. For $v_x = 1.5v_{ph}$,

$$\omega_i = \frac{2}{3\sqrt{\pi}}\frac{\omega_{pe}^2\omega_r^2}{k_x^3}\left(\frac{m_e}{KT}\right)^{3/2}\exp\left[-\frac{m_e v_{ph}^2}{8KT}\right]$$

For $v_x = v_{ph}$, $f'(v_{ph}) = 0$ and the wave amplitude remains constant in time since on the average there is no interchange of energy between the wave and the particles.

2.32.

$$\int_{-\infty}^{\infty}\frac{N_0}{\sqrt{2\pi}}\left(\frac{m_e}{KT}\right)^{3/2}u_x^2\exp\left(-\frac{m_e}{2KT}u_x^2\right)du_x = N_0$$

$$\tfrac{1}{2}m_e\langle u_x^2\rangle = \tfrac{1}{2}KT_k = \int_{-\infty}^{\infty}\frac{m_e}{2}\frac{1}{\sqrt{2\pi}}\left(\frac{m_e}{KT}\right)^{3/2}u_x^4\exp\left(-\frac{m_e}{2KT}u_x^2\right)du_x = \frac{3}{2}KT$$

Therefore $T_k = 3T$ where T_k is the kinetic temperature. The growth constant

$$\omega_i = \frac{4\sqrt{2}}{3}\frac{\omega_{pe}^2\omega_r}{N_0k_x^2}f'(\omega_r/k_x)$$

$$= \frac{8}{3\sqrt{\pi}}\frac{\omega_{pe}^2\omega_r^2}{k_x^3}\left(\frac{m_e}{KT}\right)^{3/2}\left[1-\frac{m_e}{2KT}\frac{\omega_r^2}{k_x^2}\right]\exp\left[-\frac{m_e}{2KT}\frac{\omega_r^2}{k_x^2}\right]$$

For $\omega_r < k_x\sqrt{2KT/m_e}$, $\omega_i > 0$ and there is instability.

3.1. Center: $\tilde{x} = 0$, $\tilde{y} = -A^2/(1-A)$; radius: $A^2/(1-A)$. For $A = u_{x0}B/E_y < 1$, the electron moves in the clockwise direction.

3.2. Let $\tilde{x}_1$, $\tilde{y}_1$ denote the values of $\tilde{x}$, $\tilde{y}$ for $A = (1-a)$ and $\tilde{x}_2$, $\tilde{y}_2$ the corresponding values for $A = (1+a)$. Then, $\tilde{x}_2 - \omega_c t = -(\tilde{x}_1 - \omega_c t)$ and $\tilde{y}_2 = -\tilde{y}_1$.

3.3.

$$\sigma_H = \frac{N_0e^2}{m_e}\frac{\omega_{ce}}{\omega^2-\omega_{ce}^2} - \frac{N_0e^2}{m_i}\frac{\omega_{ci}}{\omega^2-\omega_{ci}^2}$$

For $\omega/\omega_{ce} \ll 1$ and $\omega/\omega_{ci} \ll 1$,

$$\sigma_H = \frac{N_0e}{B_0}\left[1-\frac{\omega_{ci}^2}{\omega_{ce}^2}\right]\left(\frac{\omega}{\omega_{ci}}\right)^2$$

and hence vanishes in the limit of zero frequency.

$$\sigma_{MR} = i\frac{N_0e^2}{m_e}\frac{\omega}{\omega^2-\omega_{ce}^2} + i\frac{N_0e^2}{m_i}\frac{\omega}{\omega^2-\omega_{ci}^2}$$

For $\omega/\omega_{ce} \ll 1$ and $\omega/\omega_{ci} \ll 1$,

$$\sigma_{MR} = -i\frac{N_0e}{B_0}\left(1+\frac{\omega_{ci}}{\omega_{ce}}\right)\frac{\omega}{\omega_{ci}}$$

Therefore σ_{MR} vanishes at a slower rate than σ_H in the limit of low frequencies.

3.4. From the answer to Problem 3.3, it is seen that only the magnetoresistance terms dominate in the limit of very low frequencies. The electron and the ion contributions to the magnetoresistance terms of the conductivity are given by $\sigma_{e,MR} = -i\omega(N_0/B_0^2)m_e$ and $\sigma_{i,MR} = -i\omega(N_0/B_0^2)m_i$. Since $m_i \gg m_e$, the net current is contributed mainly by the ions.

The equivalent permittivity $\varepsilon = \varepsilon_0(1 + i\sigma_{MR}/\omega\varepsilon_0) = \varepsilon_0(1 + \rho_m/\varepsilon_0 B_0^2)$ where the mass density $\rho_m = N_0(m_e + m_i)$. The equivalent permittivity is the same as that given by Eq. (3.5.8) since the electric current density in the limit of very low frequencies is the same as the polarization current density.

3.5. If the electron is at the origin $x = y = 0$ at time $t = 0$, for $\omega/\omega_c \ll 1$, the orbit is specified by $x = (eE_{y0}/m\omega_c\omega)(1 - \cos\omega t)$; $y = -(eE_{y0}/m\omega_c^2)\sin\omega t$; therefore

$$\frac{(x - eE_{y0}/m\omega_c\omega)^2}{(eE_{y0}/m\omega_c\omega)^2} + \frac{y^2}{(eE_{y0}/m\omega_c^2)^2} = 1$$

which is an ellipse with the principal axes in the x- and y-directions. The ratio of the principal axis in the x-direction to that in the y-direction $= \omega_c/\omega$. The major axis is in the x-direction which is perpendicular to the electric field. The ratio of the minor to the major axes $= \omega/\omega_c$.

For $\omega/\omega_c \gg 1$, the orbit is specified by $x = (eE_{y0}/m\omega\omega_c)(1 - \cos\omega_c t)$; $y = -(eE_{y0}/m\omega\omega_c)\sin\omega_c t$; therefore $(x - eE_{y0}/m\omega\omega_c)^2 + y^2 = (eE_{y0}/m\omega\omega_c)^2$ which is a circle. The electron orbits in the counterclockwise direction at the gyromagnetic angular frequency.

Let $a = eE_{y0}\omega_c/m(\omega^2 - \omega_c^2)\omega$ and $b = eE_{y0}/m(\omega^2 - \omega_c^2)$. The contribution to the electron orbit which has the same angular frequency as the forcing electric field is specified by $x = a(\cos\omega t - 1)$; $y = b\sin\omega t$; therefore $(x + a)^2/a^2 + y^2/b^2 = 1$ which is an ellipse. Since $b/a = \omega/\omega_c$, for $\omega > \omega_c$, the major axis is b in the y-direction which is the direction of the electric field. Since $b/E_{y0} > 0$, the electron is in phase with the electric field for $\omega > \omega_c$.

For $\omega < \omega_c$, $|b|/|a| = \omega/\omega_c < 1$; hence the minor axis is b in the y-direction which is the direction of the electric field. Since $b/E_{y0} < 0$, the electron is out of phase with the electric field.

3.6. The finite dimension of the container is the dominant factor.

3.8. The equation of the magnetic flux line passing through the initial average position of the electron is $x = x_0(1 + \alpha z)^{-1/2} \approx x_0(1 - \frac{1}{2}\alpha z)$.

For $t \leqslant 2u_{z0}/\alpha u_{\perp 0}^2$, $\dot{z} \geqslant 0$, and for $t > 2u_{z0}/\alpha u_{\perp 0}^2$, $\dot{z} < 0$; therefore the axial velocity is reversed resulting in the magnetic mirror effect.

The leading term in the nonperiodic parts of u_x, u_y, and u_z are $u_x = -(\alpha/2)u_{z0}x_0$; $u_y = 0$; $u_z = u_{z0}$. Therefore, the nonperiodic part of the electron orbit is specified by $x = x_0(1 - (\alpha/2)u_{z0}t)$; $y = 0$; $z = u_{z0}t$. Hence $x = x_0(1 - \alpha z/2)$ which is the same as the equation of the magnetic flux line. Thus the average position of the electron is seen to follow the magnetic flux line.

3.9. $\nabla \cdot \mathbf{B} = 0$ and $\nabla \times \mathbf{B} = 0$ showing that $\mathbf{B}$ given by Eq. (a) is consistent with

Maxwell's equations. The Lorentz force equation in component form is $\dot{u}_x = -\omega_c u_y(1+\alpha x)$; $\dot{u}_y = \omega_c[u_x(1+\alpha x) - u_z \alpha z]$; $\dot{u}_z = \omega_c u_y \alpha z$. The leading terms in the nonperiodic parts of u_x, u_y, and u_z are $u_x = \alpha u_{z0}^2 t$; $u_y = -(\alpha/\omega_c)(u_{\perp 0}^2/2 + u_{z0}^2)$; $u_z = u_{z0}$. The average position of the electron is specified by $x = \alpha u_{z0}^2 t^2/2 + x_0$; $y = -(\alpha/\omega_c)(u_{\perp 0}^2/2 + u_{z0}^2)t$; $z = u_{z0} t$. Hence $x = x_0 + \alpha z^2/2$. The equation of the magnetic flux line is given by $dx/dz = B_x/B_z = \alpha z/(1+\alpha x)$ whose solution is $x + \alpha x^2/2 = \alpha z^2/2$. Since $\alpha x \ll 1$ and $\alpha z \ll 1$, the flux line passing through $\mathbf{r} = \hat{\mathbf{x}} x_0$ is specified by $x = x_0 + \alpha z^2/2$. Consequently the average position of the electron follows the magnetic flux line.

The gradient drift velocity obtained from Eq. (3.10.9) is $\mathbf{u}_G = -\hat{\mathbf{y}}(\alpha/\omega_c)u_{\perp 0}^2/2$. The curvature drift velocity determined from Eqs. (3.10.14) and (3.10.40) is $\mathbf{u}_C = -\hat{\mathbf{y}}(\alpha/\omega_c)u_{z0}^2$. Therefore $(\mathbf{u}_G + \mathbf{u}_C)\cdot\hat{\mathbf{y}} = -(\alpha/\omega_c)(u_{\perp 0}^2/2 + u_{z0}^2)$ which is precisely the nonperiodic part of the y-component of the electron velocity.

3.10. In the cylindrical coordinate system r, θ, and z, let $z = 0$ denote the midplane and $r = a$ the axis of the torus. Since from Ampère's law $2\pi r B_\theta/\mu_0$ is a constant inside the torus, the axial magnetic flux density is given by $\mathbf{B} = \hat{\boldsymbol{\theta}} Ba/r$ where B is the axial magnetic flux density at $r = a$. Assume that a is much larger than the radius of the cylindrical section of the torus with the result that the effect of curvature of B_θ can be neglected. B_θ field has a gradient in the $-r$-direction and the gradient drift causes the ions to move in the $+z$-direction and the electrons in the $-z$-direction. The electric field induced by the charge separation is in the $-z$-direction and the $\mathbf{E}\times\mathbf{B}$ drift forces the charged particles to flow out in the r-direction preventing the confinement in the magnetic field of the torus.

3.12.

$$\dot{u}_x = -\omega_c[u_y(1-\alpha t) - \alpha y/2];\ \dot{u}_y = \omega_c[u_x(1-\alpha t) - \alpha x/2];\ \dot{u}_z = 0$$

$$u_x + iu_y = u_{\perp 0}[\omega_c \alpha t^2/2 + i(1-\alpha t/2)]e^{i\omega_c t}$$

$$u_x^2 + u_y^2 = u_{\perp 0}^2[(1-\alpha t) + (\alpha^2 t^2/4)(1+\omega_c^2 t^2)]$$

$$\mathfrak{m} = \frac{1}{2}\frac{m(u_x^2+u_y^2)}{B(1-\alpha t)} = \frac{1}{2}\frac{m u_{\perp 0}^2}{B}\left[1 + \frac{\alpha^2 t^2}{4}(1+\omega_c^2 t^2)\right] \qquad \text{for } \alpha t \ll 1$$

Since $u_x + iu_y$ is accurate only up to linear term in αt, it follows that $\mathfrak{m}$ is independent of αt for the same order of accuracy. The orbit is obtained as

$$x = r_L\left[\left(1+\frac{\alpha t}{2}\right)\cos\omega_c t + \frac{\alpha}{2\omega_c}(\omega_c^2 t^2 - 1)\sin\omega_c t\right]$$

$$y = r_L\left[\left(1+\frac{\alpha t}{2}\right)\sin\omega_c t - \frac{\alpha}{2\omega_c}(\omega_c^2 t^2 - 1)\cos\omega_c t - \frac{\alpha}{2\omega_c}\right]$$

$$z = u_{z0} t$$

The orbit is also accurate up to linear term in αt.

3.14.

$$V_{\min} = \frac{1}{2}\frac{mv^2}{e} \qquad B_{\min} = \frac{1}{d}\sqrt{\frac{2mV}{e}}$$

3.15.

$$u_{\parallel} = \sqrt{\frac{2\mathrm{m}}{m}}\,[B(z_m) - B(z)]^{1/2} \qquad T = 2\int_{-z_m}^{z_m}\frac{dz}{u_{\parallel}} = 2\pi a(m/2\mathrm{m}B_0)^{1/2}$$

Equation of motion: $md^2z/dt^2 = F_z$, i.e., $d^2z/dt^2 + (2\mathrm{m}B_0/a^2m)z = 0$; therefore the oscillations are of the simple harmonic type with the angular frequency $\omega = \sqrt{2\mathrm{m}B_0/a^2m}$. Hence $T = 2\pi/\omega = 2\pi a(m/2\mathrm{m}B_0)^{1/2}$. The adiabaticity condition leads to the following approximate requirements:

$$\frac{1}{2}mu^2 \ll \frac{ma^4\omega_{c0}^2}{8z_m^2}\left(1 + \frac{z_m^2}{a^2}\right) \qquad \mathrm{m} \ll \frac{a^4e^2B_0}{8z_m^2m} \qquad \text{where } \omega_{c0} = eB_0/m$$

3.16. $\nabla \cdot \mathbf{B} = (1/r)(\partial/\partial r)(rB_r) + \partial B_z/\partial z = 0$; therefore axial variation of B_z gives rise to B_r. If Eqs. (3.8.5) and (3.11.3) are evaluated at $z = z_m$, it is obtained that $E_\theta = -(r/2)(\partial B/\partial z)u_m = B_r u_m$. Since $(d/dt)(mu_z) = -e\hat{\mathbf{z}} \cdot \mathbf{u} \times \mathbf{B} = eu_\theta B_r$, on integration it is found that: $mu_z|_0^{t_r} = -2mu_z = e\int_0^{t_r} u_\theta B_r\,dt = eu_\theta B_r t_r$ or $eu_\theta t_r = -2mu_z/B_r$. The change in energy at each reflection is $\Delta w = -eE_\theta u_\theta t_r = E_\theta 2mu_z/B_r = 2u_m mu_z$. Since $u_\perp$ is unchanged before and after reflection, it follows that $\Delta w = \Delta(\frac{1}{2}mu_z^2 + \frac{1}{2}mu_\perp^2) = mu_z\Delta u_z$. Therefore $\Delta u_z = 2u_m$. As an electron moves from the reflection region $|z| = z_m$ to $|z| = z_0$, B_z decreases and the adiabaticity of the magnetic moment shows that the transverse kinetic energy decreases. Since the total energy remains constant after reflection, the axial kinetic energy increases resulting in the increase of u_z.

3.17. $\int_{u_\perp=0}^{\infty}\int_{u_z=-\infty}^{\infty} f(\mathbf{u})2\pi u_\perp\,du_z\,du_\perp = N$. For a given mirror ratio and for a given total kinetic energy, particles with $u_\perp$ less than a minimum value $(u_\perp)_{\min}$ leak through the ends of the mirror system. Therefore in the phase space the density of the representative points decreases as $u_\perp$ becomes less than $(u_\perp)_{\min}$ and goes to zero for $u_\perp = 0$. The "loss-cone" distribution function satisfies these requirements.

4.1. Deflection angle in the laboratory system $\chi_0 = \chi/2$. Recoil angle in the laboratory system $\theta_r = \pi/2 - \chi/2$. Angle between the directions of travel of the two particles after collision is $\theta_r + \chi_0 = \pi/2$. Differential scattering cross section in the laboratory system $\sigma_0(\chi_0) = 4\cos(\chi/2)\sigma(\chi)$.

4.2. $(W_1 - \tilde{W}_1)/W_1 = [u_2^2 + 2\mathbf{u}_2 \cdot \mathbf{g} + g^2]^{-1}[2m_2\mathbf{u}_2 \cdot (\mathbf{g} - \tilde{\mathbf{g}})/(m_1 + m_2) + 2m_1m_2g^2(1 - \cos\chi)/(m_1 + m_2)^2]$. For $u_2 = 0$, $(W_1 - \tilde{W}_1)/W_1 = 2m_1m_2(1 - \cos\chi)/(m_1 + m_2)^2$ which is the same as that given in Eq. (4.6.9).

4.3. For $\chi = \pi/2$, Eq. (4.8.10) gives $p_{\pi/2} = p_0$. For $p = 0$, Eq. (4.8.3) yields $r_m = 2p_0 = 2p_{\pi/2}$. Since $\tan(\pi/8) = p_0/p_{\pi/4}$, $r_m = 2\tan(\pi/8)p_{\pi/4}$.

4.4. From Eq. (4.3.22) it is found that for $p < a$, $r_m = p/n$ and for $p > a$, $r_m = p$. For $r_m = p > a$, Eq. (4.3.21) yields $\theta_m = \int_p^\infty (p/r^2)(1 - p^2/r^2)^{-1/2}\, dr = \pi/2$ and $\chi = 0$. For $p < a$ and $r_m = p/n$, it is obtained that

$$\theta_m = \int_{p/n}^{a} \frac{p}{r^2}\left(n^2 - \frac{p^2}{r^2}\right)^{-1/2} dr + \int_a^\infty \frac{p}{r^2}\left(1 - \frac{p^2}{r^2}\right)^{-1/2} dr$$

On setting $y = p/nr$ in the first integral and $z = p/r$ in the second, θ_m can be evaluated as

$$\theta_m = \int_{p/na}^{1} (1 - y^2)^{-1/2}\, dy + \int_0^{p/a} (1 - z^2)^{-1/2}\, dz$$
$$= \pi/2 - \sin^{-1}(p/na) + \sin^{-1}(p/a)$$

Therefore $\chi = \pi - 2\theta_m = 2[\sin^{-1}(p/na) - \sin^{-1}(p/a)]$ from which it can be shown that $\cos(\chi/2) = (1/a^2 n)[p^2 + \sqrt{(a^2n^2 - p^2)(a^2 - p^2)}]$. On solving for p^2, it is found that

$$p^2 = \frac{a^2 n^2 \sin^2(\chi/2)}{[1 - 2n\cos(\chi/2) + n^2]}$$

and

$$p\frac{dp}{d\chi} = \frac{a^2 n^2 \sin(\chi/2)}{2} \frac{[n\cos(\chi/2) - 1][n - \cos(\chi/2)]}{[1 - 2n\cos(\chi/2) + n^2]^2}$$

Therefore the differential scattering cross section is obtained as

$$\sigma(\chi) = \frac{p}{\sin\chi}\frac{dp}{d\chi} = \frac{a^2 n^2 [n\cos(\chi/2) - 1][n - \cos(\chi/2)]}{4\cos(\chi/2)[1 - 2n\cos(\chi/2) + n^2]^2} \qquad \text{for } p < a$$

There is no scattering for $p > a$. Since scattering occurs only for the impact parameter in the range $0 < p < a$, the total scattering cross section is obtained as $\sigma_t = 2\pi \int_0^a p\, dp = \pi a^2$.

4.5. Since p and r both have the dimensions of length, $v = p/r$ is dimensionless. Since the force F_r has the dimension MLT^{-2}, the dimensions of K_{12} are $ML^{s+1}T^{-2}$ and those of $m_r g^2/K_{12}$ are $ML^2T^{-2}M^{-1}L^{-s-1}T^2 = L^{-(s-1)}$. Therefore the dimensions of v_0 are LL^{-1} and hence v_0 is dimensionless. In view of Eq. (4.3.12), it is obtained that $\phi(r) = -K_{12}r^{-(s-1)}/(s-1) = -m_r g^2/(s-1)(rv_0/p)^{s-1}$. Let $p/r_m = v_m$. Then Eq. (4.3.22) yields $1 - v_m^2 + [2/(s-1)](v_m/v_0)^{s-1} = 0$. For $s = 2$ the equation specifying v_m becomes $1 - v_m^2 + 2v_m/v_0 = 0$ which has the following positive real solution: $v_m = 1/v_0 + \sqrt{1/v_0^2 + 1}$. Hence there is a vertex for $s = 2$. For $s = 3$ the equation specifying v_m becomes $1 - v_m^2(1 - 1/v_0^2) = 0$. Hence $v_m^2 = 1/(1 - 1/v_0^2)$. For $v_0 > 1$ or $p > (K_{12}/m_r g^2)^{1/2}$, v_m is real and positive; consequently for $s = 3$ there is a vertex only if the impact parameter exceeds a critical value. The equation specifying v_m can be written as $f(v_m) = 1$ where $f(v_m) = v_m^2 - [2/(s-1)](v_m/v_0)^{s-1}$.

Since $f'(v_m) = 0$ for $v_m = (v_m)_{max} = [v_0]^{(s-1)/(s-3)}$ and since $f''(v_m) = 2(3 - s) < 0$ for $s > 3$, $f(v_m)$ has a maximum at $v_m = (v_m)_{max}$. It is seen that $f(v_m)$ is zero for $v_m = 0$, has a maximum at $v_m = (v_m)_{max}$ and becomes $-\infty$ for $v_m = \infty$. Therefore if $f[(v_m)_{max}] > 1$, $f(v_m) = 1$ has a real, positive solution for v_m. For $s > 3$ the trajectory has a vertex only if $f[(v_m)_{max}] > 1$ or equivalently $[(s-3)/(s-1)]$ $[v_0]^{2(s-1)/(s-3)} > 1$, i.e., if

$$p > \left(\frac{K_{12}}{m_r g^2}\right)^{\frac{1}{s-1}} \left(\frac{s-1}{s-3}\right)^{\frac{(s-3)}{2(s-1)}}$$

Initially $(dr/d\theta) < 0$. Since there is no vertex, $(dr/d\theta)$ never becomes zero. Therefore $(dr/d\theta) < 0$ always. Consequently the particle spirals in towards the center of force if the impact parameter is less than the critical value. For $s = 3$, Eq. (4.3.18a) yields

$$\theta = -\int_\infty^r \frac{p}{r^2}\left(1 + \frac{\alpha^2 p^2}{r^2}\right)^{-1/2} dr \qquad \text{where} \quad \alpha^2 = 1/v_0^2 - 1 > 0$$

which on integration gives $r = \alpha p / \sinh(\alpha\theta)$. For very large θ, it is seen that $r = 2\alpha p e^{-\alpha\theta}$ which is an exponential spiral. Thus for $v_0 < 1$ and $s = 3$, a simple expression is obtained for the trajectory of the particle.

4.6. The deflection angle is obtained as

$$\chi = \pi - 2\int_0^{p/r_m} \left[1 - y^2 - \frac{2}{(s-1)}\left(\frac{y}{v_0}\right)^{s-1}\right]^{-1/2} dy$$

where $v_0 = p(m_r g^2/K_{12})^{1/(s-1)}$ is the dimensionless parameter. Since from Eq. (4.5.4)

$$\sigma(\chi) = \left(\frac{K_{12}}{m_r g^2}\right)^{2/(s-1)} \frac{v_0}{\sin\chi}\frac{dv_0}{d\chi}$$

the total scattering cross section is obtained as:

$$\sigma_t = \left(\frac{K_{12}}{m_r g^2}\right)^{2/(s-1)} \sigma_{ND}$$

where σ_{ND} is dimensionless. Hence the dimensions of σ_t are $(K_{12}/m_r g^2)^{2/(s-1)}$. Also, since $\nu = \sigma_t N g = N\sigma_{ND}(K_{12}/m_r)^{2/(s-1)} g^{(s-5)/(s-1)}$, for $s = 5$ the collision frequency is independent of the relative speed of the particles.

4.7. $(1 - 1/e) = 0.63$ times the total number of particles have free lengths less than the mean free path.

4.8. The probability that a particle suffers no collision in the time interval between t and $t + dt$ is $(1 - \nu_1\, dt)$. Therefore $P(t + dt) = P(t)(1 - \nu_1\, dt)$ which yields $P(t) = e^{-\nu_1 t} = e^{-t/\tau_1}$. Using the momentum transfer cross section, it is found from Eqs. (4.10.8), (4.9.9), and (4.8.2) that

$$\tau_1 = \frac{1}{\nu_1} = \frac{1}{\sigma_m N g} = \frac{4\pi\varepsilon_0^2 m_r^2 g^3}{Ne^4 \ln \Lambda}$$

4.9. Let the collision frequencies ν_a and ν_b denote the frequencies at which a particle takes part in the two different processes. Since the resultant collision frequency $\nu = \nu_a + \nu_b$, it follows from Eq. (4.10.8) that $1/\lambda = 1/\lambda_a + 1/\lambda_b$.

4.10. Since $\mathbf{u}_2 = 0$, $\mathbf{g} = \mathbf{u}_1$ and is directed along the z-axis and the components of $\mathbf{g}$ are given by $g_x = 0$, $g_y = 0$, and $g_z = g$. Let the direction of $\tilde{\mathbf{g}}$ be given by the spherical coordinates θ and φ. Since the deflection angle χ is the angle between $\tilde{\mathbf{g}}$ and $\mathbf{g}$, and therefore between $\tilde{\mathbf{g}}$ and $\mathbf{u}_1$, it follows that $\theta = \chi$. The components of $\tilde{\mathbf{g}}$ are therefore given by $\tilde{g}_x = g \sin\chi \cos\varphi$, $\tilde{g}_y = g \sin\chi \sin\varphi$, and $\tilde{g}_z = g \cos\chi$ since $|\mathbf{g}| = |\tilde{\mathbf{g}}| = g$. From Eqs. (4.2.14) and (4.2.15), it is found that $\Delta\mathbf{u}_1 = \tilde{\mathbf{u}}_1 - \mathbf{u}_1 = M_2(\tilde{\mathbf{g}} - \mathbf{g})$. Hence the components of $\Delta\mathbf{u}_1$ are obtained as $\Delta u_{1x} = M_2(\tilde{g}_x - g_x) = M_2 g \sin\chi \cos\varphi$, $\Delta u_{1y} = M_2(\tilde{g}_y - g_y) = M_2 g \sin\chi \sin\varphi$, and $\Delta u_{1z} = M_2(\tilde{g}_z - g_z) = -M_2 g(1 - \cos\chi)$.

The integration over φ vanishes yielding $\langle\Delta u_{1x}\rangle = \langle\Delta u_{1y}\rangle = 0$ and $\langle\Delta u_{1x}\Delta u_{1y}\rangle = \langle\Delta u_{1x}\Delta u_{1z}\rangle = \langle\Delta u_{1y}\Delta u_{1z}\rangle = 0$. Also

$$\langle\Delta u_{1z}\rangle = \int_{\chi=0}^{\pi}\int_{\varphi=0}^{2\pi} -M_2 g^2(1 - \cos\chi)\frac{p_0^2}{4\sin^4(\chi/2)}\sin\chi\, d\chi\, d\varphi$$

$$= -M_2 g^2 4\pi p_0^2 \int_{\chi_{\min}}^{\pi} \frac{d[\sin(\chi/2)]}{\sin(\chi/2)} = M_2 g^2 4\pi p_0^2 \ln[\sin(\chi_{\min}/2)]$$

where a cut-off has been introduced at $\chi_c = \chi_{\min}$ corresponding to $p_c = \lambda_D$. From Eqs. (4.9.7) and (4.9.10), it is found that $\sin(\chi_{\min}/2) = (1 + \Lambda^2)^{-1/2}$. Hence $\langle\Delta u_{1z}\rangle$ simplifies to

$$\langle\Delta u_{1z}\rangle = -2\pi M_2 g^2 p_0^2 \ln[1 + \Lambda^2]$$

Similarly

$$\left.\begin{matrix}\langle(\Delta u_{1x})^2\rangle \\ \langle(\Delta u_{1y})^2\rangle\end{matrix}\right\} = \int_{\chi=0}^{\pi}\int_{\varphi=0}^{2\pi} M_2^2 g^3 \sin^2\chi \begin{Bmatrix}\cos^2\varphi \\ \sin^2\varphi\end{Bmatrix}\frac{p_0^2}{4\sin^4(\chi/2)}\sin\chi\, d\chi\, d\varphi$$

which on simplification yields

$$\langle(\Delta u_{1x})^2\rangle = \langle(\Delta u_{1y})^2\rangle$$

$$= 4\pi M_2^2 g^3 p_0^2 \int_{\chi_{\min}}^{\pi}\left[\frac{1}{\sin(\chi/2)} - \sin(\chi/2)\right] d[\sin(\chi/2)]$$

$$= 2\pi M_2^2 g^3 p_0^2 \left[\ln(1 + \Lambda^2) - \frac{\Lambda^2}{1 + \Lambda^2}\right]$$

In the same manner

$$\langle(\Delta u_{1z})^2\rangle = \int_{\chi=0}^{\pi}\int_{\varphi=0}^{2\pi} M_2^2 g^3(1-\cos\chi)^2 \frac{p_0^2}{4\sin^4(\chi/2)} \sin\chi\, d\chi\, d\varphi$$

$$= 8\pi M_2^2 g^3 p_0^2 \int_{\chi_{\min}}^{\pi} \sin(\chi/2)\, d[\sin(\chi/2)]$$

$$= 4\pi M_2^2 g^3 p_0^2 \frac{\Lambda^2}{(1+\Lambda^2)}$$

4.11. *Case 1*: $p > a_0$. For $r > a_0$, it is seen from Eq. (4.3.22) that $r_m = p$ and Eq. (4.3.21) then yields $\theta_m = \pi/2$. Since the minimum value of r is r_m, it follows that $p > a_0$ for this case. Also $d\theta_m/dp = 0$.

Case 2: $a_0 > p > p_c$. For $a_0 > r > a$, Eq. (4.3.22) gives $r_m = pa/p_c$ where $p_c = a(1 + 2\phi_0/m_r g^2)^{1/2}$. Since the minimum value of r is a, the minimum value of p is p_c. This case, therefore, corresponds to the range $a_0 > p > p_c$. From Eq. (4.3.21) it is found that

$$\theta_m = \int_{pa/p_c}^{a_0} \frac{p}{r^2}\left(1 - \frac{p^2}{r^2} + \frac{2\phi_0}{m_r g^2}\right)^{-1/2} dr + \int_{a_0}^{\infty} \frac{p}{r^2}\left(1 - \frac{p^2}{r^2}\right)^{-1/2} dr$$

which when evaluated yields

$$\theta_m = \cos^{-1}(ap/a_0 p_c) - \cos^{-1}(p/a_0) + \pi/2$$

Therefore

$$\frac{d\theta_m}{dp} = -\left[\left(\frac{a_0 p_c}{a}\right)^2 - p^2\right]^{-1/2} + (a_0^2 - p^2)^{-1/2}$$

Case 3: $p_c > p > 0$. For $a > r$, $r_m = a$ due to the presence of an impenetrable core of radius a. This case corresponds to the range of impact parameters given by $p_c > p > 0$. It is obtained from Eq. (4.3.21) that

$$\theta_m = \int_{a}^{a_0} \frac{p}{r^2}\left(1 - \frac{p^2}{r^2} + \frac{2\phi_0}{m_r g^2}\right)^{-1/2} dr + \int_{a_0}^{\infty} \frac{p}{r^2}\left(1 - \frac{p^2}{r^2}\right)^{-1/2} dr$$

$$= \cos^{-1}\left(\frac{ap}{a_0 p_c}\right) - \cos^{-1}\left(\frac{p}{p_c}\right) + \frac{\pi}{2} - \cos^{-1}\left(\frac{p}{a_0}\right)$$

Therefore

$$\frac{d\theta_m}{dp} = -\left[\left(\frac{a_0 p_c}{a}\right)^2 - p^2\right]^{-1/2} + (p_c^2 - p^2)^{-1/2} + (a_0^2 - p^2)^{-1/2}$$

Since $\chi = \pi - 2\theta_m$, $dp/d\chi = -\frac{1}{2}(dp/d\theta_m)$ and the differential scattering cross section is determined as

$$\sigma(\chi) = \frac{p}{\sin\chi}\left|\frac{dp}{d\chi}\right| = [p/2\sin(2\theta_m)]/|d\theta_m/dp|$$

where θ_m and $d\theta_m/dp$ have been determined separately for the three ranges of the impact parameters. Thus $\sigma(\chi)$ is evaluated for all values of the impact parameter.

4.12. Let $\mathbf{u}_1$ be in the z-direction and let the direction of $\mathbf{u}_2$ be specified by the spherical coordinates θ and φ where θ is the angle between $\mathbf{u}_2$ and the z-axis. The relative speed $g(\theta) = (u_1^2 + u_2^2 - 2u_1 u_2 \cos\theta)^{1/2}$ averaged over all directions is evaluated as

$$g = \frac{1}{4\pi}\int_{\theta=0}^{\pi}\int_{\varphi=0}^{2\pi} g(\theta)\sin\theta\, d\theta\, d\varphi$$

$$= \frac{1}{2}\int_0^{\pi} (u_1^2 + u_2^2 - 2u_1 u_2\cos\theta)^{1/2}\sin\theta\, d\theta$$

On setting $y = u_1^2 + u_2^2 - 2u_1 u_2 \cos\theta$, it is found that

$$g = \frac{1}{4u_1 u_2}\int_{|u_1-u_2|^2}^{(u_1+u_2)^2} \sqrt{y}\, dy = \frac{1}{6u_1 u_2}[(u_1 + u_2)^3 - \{|u_1 - u_2|\}^3]$$

Hence $g = u_1 + u_2^2/3u_1$ for $u_1 > u_2$, and $g = u_2 + u_1^2/3u_2$ for $u_1 < u_2$.

For $u_1 = u_2 = u$, $g = \frac{4}{3}u$. From Eq. (4.10.8), it is found that $\nu = gN\sigma = \frac{4}{3}uN\sigma$.

4.13. Since the relative speed $g(u_2)$ is a function of u_2 and $f(\mathbf{u}_2)$ is isotropic, $g(u_2)$ averaged over the velocity distribution of type 2 particles is obtained as

$$g = \frac{1}{N_2}\int_0^{\infty} g(u_2) 4\pi u_2^2 f(\mathbf{u}_2)\, du_2$$

Substitution of $g(u_2)$ from Problem 4.12 yields

$$\nu = N_2 g\sigma = 4\pi\sigma\left[\int_0^{u_1}\left(u_1 + \frac{u_2^2}{3u_1}\right)u_2^2 f(\mathbf{u}_2)\, du_2 + \int_{u_1}^{\infty}\left(u_2 + \frac{u_1^2}{3u_2}\right)u_2^2 f(\mathbf{u}_2)\, du_2\right]$$

On using the expression for $f(\mathbf{u}_2)$, and setting $y = (m_2/2KT_2)^{1/2}u_2$ and $x = (m_2/2KT_2)^{1/2}u_1$, the expression for ν becomes

$$\nu = 4N_2\sigma\left(\frac{2KT_2}{\pi m_2}\right)^{1/2}\left[x\int_0^x y^2 e^{-y^2}\, dy + \frac{1}{3x}\int_0^x y^4 e^{-y^2}\, dy + \int_x^{\infty} y^3 e^{-y^2}\, dy + \frac{x^2}{3}\int_x^{\infty} y e^{-y^2}\, dy\right]$$

Since

$$\int_0^x y^2 e^{-y^2}\, dy = -\frac{x}{2}e^{-x^2} + \frac{1}{2}I, \quad \int_0^x y^4 e^{-y^2}\, dy = -\frac{x^3}{2}e^{-x^2} - \frac{3x}{4}e^{-x^2} + \frac{3}{4}I$$

where $I = \int_0^x e^{-y^2}\, dy$ and $\int_x^{\infty} y^3 e^{-y^2}\, dy = \frac{1}{2}(x^2 + 1)e^{-x^2}$ and $\int_x^{\infty} y e^{-y^2}\, dy = \frac{1}{2}e^{-x^2}$, the expression for ν simplifies to

$$\nu = N_2\sigma\left(\frac{2KT_2}{\pi m_2}\right)^{1/2}\left[e^{-x^2} + \left(2x + \frac{1}{x}\right)\int_0^x e^{-y^2}\,dy\right]$$

4.14. Since the collision frequency $\nu(u_1)$ obtained in Problem 4.13 is a function of u_1 and $f(\mathbf{u}_1)$ is isotropic, $\nu(u_1)$ averaged over the velocity distribution of type 1 particles is obtained as $\nu = (1/N_1)\int_0^\infty \nu(u_1)4\pi u_1^2 f(\mathbf{u}_1)\,du_1$

If $\nu(u_1)$ from Problem 4.13 and the expression for $f(\mathbf{u}_1)$ are substituted, and if x and y are replaced respectively by $(m_2/2KT_2)^{1/2}u_1$ and $(m_2/2KT_2)^{1/2}u_2$, the resulting expression for ν is that given by Eq. (A) which can be rewritten for convenience as follows:

$$\nu = \frac{2}{\pi}N_2\sigma\frac{m_1}{KT_1}\left(\frac{m_1 T_2}{m_2 T_1}\right)^{1/2}[A_1 + A_2]$$

where

$$A_1 = \int_0^\infty u_1^2 e^{-(m_1/T_1+m_2/T_2)(u_1^2/2K)}\,du_1 = \sqrt{\frac{\pi}{2}}\,K^{3/2}\left(\frac{m_1}{T_1} + \frac{m_2}{T_2}\right)^{-3/2} \qquad \text{(B1)}$$

and

$$A_2 = \int_0^\infty du_1 \int_0^{u_1} du_2\left[\frac{m_2}{KT_2}u_1^3 + u_1\right]e^{-m_1u_1^2/2KT_1}e^{-m_2u_2^2/2KT_2} \qquad \text{(B2)}$$

Inverting the order of integration in Eq. (B2) and evaluating the inner integral by integration by parts, it is found that

$$\begin{aligned} A_2 &= \int_0^\infty du_2\, e^{-m_2u_2^2/2KT_2}\int_{u_2}^\infty du_1\left[\frac{m_2u_1^3}{KT_2} + u_1\right]e^{-m_1u_1^2/2KT_1} \\ &= \int_0^\infty du_2\left[\frac{T_1 m_2}{T_2 m_1}u_2^2 + \left(\frac{2T_1 m_2}{T_2 m_1} + 1\right)\frac{KT_1}{m_1}\right]e^{-(m_1/T_1+m_2/T_2)(u_2^2/2K)} \\ &= \frac{T_1 m_2}{T_2 m_1}\sqrt{\frac{\pi}{2}}\left(\frac{m_1}{T_1} + \frac{m_2}{T_2}\right)^{-3/2}K^{3/2} \\ &\quad + \frac{T_1}{m_1}\sqrt{\frac{\pi}{2}}\left(\frac{m_1}{T_1} + \frac{m_2}{T_2}\right)^{-1/2}K^{3/2}\left(1 + \frac{2T_1 m_2}{T_2 m_1}\right) \end{aligned} \qquad \text{(B3)}$$

The addition of Eqs. (B1) and (B3) gives

$$A_1 + A_2 = \sqrt{2\pi}\,\frac{T_1^2}{m_1^2}\left(\frac{m_1}{T_1} + \frac{m_2}{T_2}\right)^{1/2}K^{3/2}$$

Hence ν simplifies to $\nu = N_2\sigma\sqrt{8K/\pi}\,(T_1/m_1 + T_2/m_2)^{1/2}$.

If $T_1 = T_2 = T$, $m_1 = m_2 = m$, and $N_2 = N$, the expression for the collision frequency reduces to $\nu = \sqrt{2}\,N\sigma\langle u\rangle$.

5.2. The dispersion relation (5.3.7) is unchanged.

5.5. The time-varying field quantities are assumed to have the phase factor $\exp\{i(k_p k_z - \omega t)\}$. The linearized Langevin equation for the collisionless case is obtained as

$$\frac{\partial}{\partial t}\mathbf{v} + v_d \frac{\partial}{\partial z}\mathbf{v} = -\frac{e}{m}[\mathbf{E} + v_d \hat{\mathbf{z}} \times \mu_0 \mathbf{H} + \mathbf{v} \times \hat{\mathbf{z}} B_0]$$

or equivalently

$$\alpha \mathbf{v} = -\frac{ie}{m\omega_p}[\mathbf{E} + v_d \hat{\mathbf{z}} \times \mu_0 \mathbf{H} + \mathbf{v} \times \hat{\mathbf{z}} B_0] \qquad \text{where } \alpha = \Omega - k_z \tilde{v}_d$$

If $\nabla \times \mathbf{E} = i\omega\mu_0 \mathbf{H}$ or $\mu_0 \mathbf{H} = k_p k_z \hat{\mathbf{z}} \times \mathbf{E}/\omega$ is used to eliminate the time-varying magnetic field, the Langevin equation becomes

$$\alpha \mathbf{v} = -\frac{ie}{\omega m}[\alpha(\hat{\mathbf{x}}E_x + \hat{\mathbf{y}}E_y) + \hat{\mathbf{z}}\Omega E_z] - iR\mathbf{v} \times \hat{\mathbf{z}}$$

whose solution yields

$$v_x = -ie\alpha(\alpha E_x - iRE_y)/\omega m(\alpha^2 - R^2)$$

$$v_y = -ie\alpha(iRE_x + \alpha E_y)/\omega m(\alpha^2 - R^2) \qquad \text{(A)}$$

and

$$v_z = -ieE_z/\omega_p m\alpha$$

The linearized equation of continuity leads to the following expression for the time-varying number density: $N = k_z N_0 \sqrt{\mu_0 \varepsilon_0}\, v_z/\alpha$. The time-varying part of the linearized electric current density is then deduced as $\mathbf{J} = -e(N_0 \mathbf{v} + \hat{\mathbf{z}} v_d N) = -eN_0(\mathbf{v} + \hat{\mathbf{z}} k_z v_z \tilde{v}_d/\alpha)$. With the help of Eqs. (A), $\mathbf{J}$ can be expressed in terms of $\mathbf{E}$ and therefore the expression for the conductivity dyad can be deduced. From Eq. (5.2.14) the relative dyad $\boldsymbol{\varepsilon}_r$ can be evaluated and is found to be of the same form as in Eq. (5.2.15) with

$$\varepsilon_1 = 1 - \alpha^2/\Omega^2(\alpha^2 - R^2)$$

$$\varepsilon_2 = R\alpha/\Omega^2(\alpha^2 - R^2) \qquad \text{(B)}$$

and

$$\varepsilon_3 = 1 - 1/\alpha^2$$

When $v_d = 0$, $\alpha = \Omega$ and Eqs. (B) become identical to those in Eqs. (5.2.21) as they should.

It can be shown from Eqs. (B) that

$$\varepsilon_1 \pm \varepsilon_2 = \frac{1}{\Omega^2}\left[\Omega^2 - \frac{(\Omega - k_z \tilde{v}_d)}{(\Omega - k_z \tilde{v}_d \pm R)}\right]$$

and the use of Eqs. (5.4.15) and (5.4.16) then yields the required result:

$$k_z = \left[\Omega^2 - \frac{(\Omega - k_z \tilde{v}_d)}{(\Omega - k_z \tilde{v}_d \pm R)}\right]^{1/2}$$

where the upper and the lower signs correspond, respectively, to the left and the right circular polarizations. If $\tilde{v}_d = 0$, k_z can be verified to become equal to Eq. (5.4.19) for the upper sign and Eq. (5.4.22) for the lower sign.

5.7. The elimination of $\mathbf{H}(\mathbf{r})$ from Eqs. (5.2.18) and (5.2.19) gives the differential equation:

$$\nabla \times \nabla \times \mathbf{E}(\mathbf{r}) - k_p^2 \Omega^2 \boldsymbol{\varepsilon}_r \cdot \mathbf{E}(\mathbf{r}) = 0$$

$\nabla \times \mathbf{E}(\mathbf{r}) = ik_p k(m\hat{\mathbf{x}} + l\hat{\mathbf{z}}) \times \mathbf{E}(\mathbf{r}) = ik_p k[-\hat{\mathbf{x}} lE_y(\mathbf{r}) + \hat{\mathbf{y}}\{lE_x(\mathbf{r}) - mE_z(\mathbf{r})\} + \hat{\mathbf{z}} mE_y(\mathbf{r})]$. Hence $\nabla \times \mathbf{E}(\mathbf{r})$ can be written as

$$\nabla \times \mathbf{E}(\mathbf{r}) = ik_p k \begin{bmatrix} 0 & -l & 0 \\ l & 0 & -m \\ 0 & m & 0 \end{bmatrix} \begin{bmatrix} E_x(\mathbf{r}) \\ E_y(\mathbf{r}) \\ E_z(\mathbf{r}) \end{bmatrix}$$

When Eq. (C) is used in Eq. (D) and the indicated matrix multiplications are carried out, the result on noting that $l^2 + m^2 = 1$ is

$$\begin{bmatrix} \Omega^2\varepsilon_1 - k^2 l^2 & i\Omega^2\varepsilon_2 & k^2 lm \\ -i\Omega^2\varepsilon_2 & \Omega^2\varepsilon_1 - k^2 & 0 \\ k^2 lm & 0 & \Omega^2\varepsilon_3 - k^2 m^2 \end{bmatrix} [E_0] = 0$$

which is the required set of homogeneous equations satisfied by E_{x0}, E_{y0}, and E_{z0}. For a nontrivial solution, the determinant of the matrix operating on $[E_0]$ should vanish. When this determinant is evaluated and rearranged, the dispersion equation given by Eqs. (5.6.18) and (5.6.19) is reproduced. Note that the coefficient of k^6 in the determinant vanishes identically.

5.11. For propagation in the direction of the magnetostatic field, $l = 1$ and B_2, B_1, and B_0 as given by Eqs. (5.6.42) have a common factor $(\Omega^2 - 1)$ which can be removed with the result that the dispersion equation is the same as that given by Eq. (5.6.43) but with $B_2 = (\Omega^2 - R^2)$, $B_1 = -2\Omega^2(\Omega^2 - 1 - R^2)$ and $B_0 = \Omega^2\{\Omega^4 - \Omega^2(R^2 + 2) + 1\}$. The parametric curves are obtained as (*a*) $R^2 = \Omega^2$ for $B_2 = 0$, (*b*) $R^2 = \Omega^2 - 1$ for $B_1 = 0$, and (*c*) $R^2 = \Omega^{-2}(\Omega^2 - 1)^2$ for $B_0 = 0$, and these are shown in Fig. 5.11. The regions of propagation of the two modes can be determined by the method described in Sec. 5.6 and these results are also included in Fig. 5.11.

5.12. For propagation across the magnetostatic field, $l = 0$. The relative position of the curve *b* in Fig. 5.9 is unchanged and the curve *a* becomes a straight line specified by $R^2 = \Omega^2 - 1$ for $\Omega > 1$. The portion of the curve *a* which exists for

$\Omega < 1$ in Fig. 5.9 disappears for this limiting case and therefore the low frequency branch of mode 1 is eliminated. Hence for propagation across the magnetostatic field, mode 1 propagates for $1 < \Omega < \infty$ and mode 2 for $\Omega_l < \Omega < \Omega_u$ and $\Omega_2 < \Omega < \infty$. These results agree with those deduced in Sec. 5.5 for this special case.

5.13. For $R < 1$, as the propagation direction approaches that of the magnetostatic field, the cut-off frequency $\Omega = 1$ of mode 1 discontinuously jumps to $\Omega = \Omega_l$ which is less than 1 and the resonance of mode 2 occurring at $\Omega = \Omega_p$ which is greater than 1 disappears with the result that the dispersion in the range $1 \leqslant \Omega \leqslant \Omega_p$ undergoes very rapid changes. For $R > 1$, as the propagation direction approaches that of the magnetostatic field, the resonance of mode 1 occurring at $\Omega = \Omega_g$ which is less than 1 changes discontinuously to $\Omega = R$, the cut-off frequency $\Omega = 1$ of mode 1 jumps discontinuously to $\Omega = \Omega_l$, which is less than 1, and the resonance of mode 2 occurring at $\Omega = \Omega_p$, which is greater than R, disappears with the result that the dispersion in the range $\Omega_g \leqslant \Omega \leqslant \Omega_p$ undergoes very rapid changes. Consequently in the frequency range $\Omega_g \leqslant \Omega \leqslant \Omega_p$ where the dispersion changes discontinuously as the propagation direction approaches that of the magnetostatic field, the representation in terms of the quasicircularly polarized modes is not valid.

5.15. When the strength of the magnetostatic field is infinite, $\varepsilon_2 = 0$ and $\varepsilon_1 = 1$; therefore $\boldsymbol{\varepsilon}_r = \hat{\mathbf{x}}\hat{\mathbf{x}} + \hat{\mathbf{y}}\hat{\mathbf{y}} + \varepsilon_3 \hat{\mathbf{z}}\hat{\mathbf{z}}$. The magnetostatic field is assumed to be in the z-direction. For propagation in the direction of the magnetostatic field, Eqs. (5.4.2) and (5.4.3) show that there are two TEM modes, one consisting of E_x and H_y, and the other E_y and H_x. These two modes are not influenced by the presence of the plasma and are seen from Eqs. (5.4.15) and (5.4.16) to have the same propagation coefficient $k_z = \Omega$. For propagation across the magnetostatic field, that is in the x-direction, there is again a TEM mode consisting of E_z and H_y governed by the dispersion relation $k_x = \sqrt{\Omega^2 - 1}$ as may be seen from Eq. (5.5.7). This mode propagates for $\Omega > 1$ only. The other mode is seen from Eqs. (5.5.10) and (5.5.14) to be also a TEM mode consisting of E_y and H_z and governed by the dispersion relation $k_x = \Omega$. This mode is not influenced by the presence of the plasma.

For propagation in the z'-direction lying in the zx-plane, it is seen from Eqs. (5.6.2a), (5.6.2c), (5.6.3b), and (5.6.9) that one of the modes consists of E_y, H_x, and H_z and has the dispersion relation $k = \Omega$. This mode is unaffected by the presence of the plasma. Since always $H_{z'} = 0$, it follows that this mode is also a TEM mode with respect to the propagation direction. From Eqs. (5.6.2b), (5.6.3a), (5.6.3c), and (5.6.10), it is found that the other mode consists of H_y, E_x, and E_z and has the dispersion relation given by

$$\Omega^2 k_x^2 + (\Omega^2 - 1)k_z^2 = \Omega^2(\Omega^2 - 1) \qquad \text{or} \qquad k = \sqrt{\Omega^2(\Omega^2 - 1)/(\Omega^2 - l^2)}$$

This is a TM mode which, for $\Omega > 1$, propagates in all directions, but for $\Omega < 1$

propagates only for $l \geqslant \Omega$ or in the directions specified by the range $0 \leqslant \theta_{ph} \leqslant \cos^{-1}\Omega$. For $\Omega = l$ this mode has a resonance.

5.16. For the transverse electromagnetic modes in an isotropic plasma, the phase velocity is given by $v_{ph} = c/(1 - \omega_p^2/\omega^2)^{1/2}$. In the propagation range $\omega > \omega_p$, $v_{ph} > c$. Since the particle velocity u is always less than the electromagnetic wave velocity c in free space, $u < v_{ph}$ and hence there is no Cerenkov radiation.

5.17. From the results of Problem 2.12, in terms of the unnormalized quantities, the group velocity is obtained as $\mathbf{v}_g = \hat{\boldsymbol{\rho}}\partial\omega/\partial k_\rho + \hat{\mathbf{z}}\partial\omega/\partial k_z$. By changing k to $k_p k$ and dividing both sides by c, the expression for the normalized group velocity is found as $\mathbf{v}_g/c = \hat{\boldsymbol{\rho}}\partial\Omega/\partial k_\rho + \hat{\mathbf{z}}\partial\Omega/\partial k_z$. With the help of Eqs. (5.2.21) and (5.8.5), the dispersion equation (5.8.4) can be written as a polynomial in Ω as follows:

$$\begin{aligned}
&\Omega^8 - \Omega^6\{2k_\rho^2 + 2k_z^2 + R^2 + 3\}\\
&\quad + \Omega^4\{k_\rho^4 + 2(k_z^2 + R^2 + 2)k_\rho^2 + k_z^4 + 2(2 + R^2)k_z^2 + R^2 + 3\}\\
&\quad - \Omega^2\{k_\rho^4(1 + R^2) + 2k_z^2 k_\rho^2(1 + R^2)\\
&\quad + k_z^4(1 + R^2) + (R^2 + 2)k_\rho^2 + 2k_z^2(1 + R^2) + 1\}\\
&\quad + R^2 k_z^2 k_\rho^2 + k_z^4 R^2 = 0
\end{aligned}$$

from which it can be shown that $\partial\Omega/\partial k_\rho = (k_\rho/\Omega)P_\rho/Q$ and $\partial\Omega/\partial k_z = (k_z/\Omega)P_z/Q$. Since for each value of k_z there are two values of k_ρ, the additional subscript n is added to k_ρ, P_ρ, P_z, and Q to distinguish the group velocity of the two possible modes.

5.18. From Fig. 5.22 only mode 2 is seen to be excited near $\Omega = \sqrt{1 + R^2}$. It can be deduced from Eqs. (5.8.12), (5.8.14), and (5.8.17b) that near $\Omega = \sqrt{1 + R^2}$, $k_{\rho 2}^2 = -R^2(\gamma^2 + 1)/(\Omega^2 - 1 - R^2)$ and hence $k_{\rho 2}$ is infinite at the normalized upper hybrid resonant frequency $\Omega = \sqrt{1 + R^2}$. Since Eqs. (5.8.24) give $\tan\theta_{ph,2} = \infty$, the wave-normal direction is given by $\theta_{ph,2} = \pi/2$. The previous problem shows that for Ω near $\sqrt{1 + R^2}$, P_{z2} becomes infinite as $-R^2 k_{\rho 2}^2$ and $P_{\rho 2}$ is finite. For Ω approaching $\sqrt{1 + R^2}$ from below, P_{z2} is negative and Eqs. (5.8.25) show that $\tan\theta_{g2} = 0$ and hence $\theta_{g2} = \pi$ for $\Omega = \sqrt{1 + R^2}$.

5.19. If $k_p k = \omega_{pe} k/c$ is the wavenumber of the longitudinal electron plasma wave and $\Omega = \omega/\omega_{pe}$, it is found from Eq. (2.7.13) that

$$k^2 = (c^2/a_e^2)(\Omega^2 - 1)$$

where a_e is the electron sound velocity. The phase velocity is then obtained as $v_{ph}/c = \Omega/k$ or $v_{ph} = a_e\Omega/\sqrt{\Omega^2 - 1}$. Since the particle velocity u can be greater than v_{ph}, Cerenkov radiation of the longitudinal electron plasma wave is possible. If θ_{ph} is the angle between the wave normal and the direction of motion of the charge, the coherence condition becomes $\cos^2\theta_{ph} = a_e^2\Omega^2/u^2(\Omega^2 - 1)$. Since $0 < \cos^2\theta_{ph} < 1$, it follows that $\Omega > (1 - a_e^2/u^2)^{-1/2}$ which gives the frequency spectrum of emitted radiation. Note that u has to be greater than a_e for the existence of Cerenkov radiation.

Let z be the direction of motion of the charge. Then the dispersion equation in terms of the longitudinal (k_z) and the transverse (k_ρ) components of the wavenumber is given by

$$\Omega^2 = 1 + (a_e^2/c^2)(k_\rho^2 + k_z^2)$$

The normalized group velocity can then be deduced as

$$\mathbf{v}_g/c = \hat{\boldsymbol{\rho}}\frac{\partial\Omega}{\partial k_\rho} + \hat{\mathbf{z}}\frac{\partial\Omega}{\partial k_z} = \frac{a_e^2}{c^2\Omega}(\hat{\boldsymbol{\rho}}k_\rho + \hat{\mathbf{z}}k_z) = \frac{a_e^2}{c^2}\frac{\mathbf{k}}{\Omega}$$

Hence the wave normal and the group velocity or equivalently the ray are in the same direction.

The transverse electromagnetic wave is seen from Eqs. (2.7.5) and (2.7.6) to have the same dispersion relation as in a cold electron plasma. The results of Problem 5.16 show that Cerenkov radiation of the transverse electromagnetic wave is not possible in an isotropic warm electron plasma.

5.20. From the results of Problem 5.15, the phase velocity of the TEM mode in an uniaxially anisotropic plasma is obtained as $v_{ph} = c > u$ and hence this mode is not excited.

If z is taken as the anisotropic axis, the dispersion relation of the TM mode can be expressed as

$$\Omega^2 k_\rho^2 + (\Omega^2 - 1)k_z^2 = \Omega^2(\Omega^2 - 1)$$

Since $k_p k_z = \omega/u$ or $k_z = \Omega\gamma$ with $\gamma = c/u > 1$ by the coherence condition, it is found that $k_\rho = \sqrt{(\Omega^2 - 1)(1 - \gamma^2)}$. Since $\gamma > 1$, k_ρ is real for $\Omega < 1$. Hence the TM mode is excited by the Cerenkov mechanism for $\Omega < 1$.

5.21. For propagation in the direction of the magnetostatic field, $l = 1$ and therefore B_2, B_1, and B_0 as given by Eqs. (5.10.2) have the common factor $(\Omega^2 - 1 - m)$ which is removed. Further simplification is effected by omitting m and R^2m in comparison with unity with the following results: $B_2 = (\Omega^2 - R^2)(\Omega^2 - R^2m^2)$, $B_1 = -2\Omega^2(\Omega^2 - 1 - R^2)(\Omega^2 - R^2m/(1 + R^2))$, and $B_0 = \Omega^4\{\Omega^4 - \Omega^2(2 + R^2) + 1\}$. As in Fig. 5.26, the following curves are drawn in the Ω^2 - R^2 space: curve (a) $\Omega^2 = R^2$, and curve (b) $\Omega^2 = R^2m^2$ corresponding to $B_2 = 0$, curve (c) $\Omega^2 = 1 + R^2$, and curve (d) $R^2 = \Omega^2/(m - \Omega^2)$ corresponding to $B_1 = 0$, and curve (f) $R^2 = \Omega^{-2}(\Omega^2 - 1)^2$ corresponding to $B_0 = 0$. In accordance with the theory developed in Sec. 5.10, the various regions in which modes 1 and 2 propagate can be determined. It is found that mode 1 propagates for $0 < \Omega < Rm$ and $\Omega_1 < \Omega < \infty$ and mode 2 for $0 < \Omega < R$ and $\Omega_2 < \Omega < \infty$. These results are in agreement with those obtained in Sec. 5.9.

5.22. For propagation across the magnetostatic field, $l = 0$; therefore B_2, B_1, and B_0 as given by Eqs. (5.10.2) have the common factor Ω^2 which is removed. As before m and $R^2 m$ are omitted in comparison with unity resulting in the following expressions: $B_2 = \Omega^4 - \Omega^2(1 + R^2) + R^2 m$, $B_1 = -[2\Omega^4 - 2\Omega^2(2 + R^2) + 2 + R^2][\Omega^2 - R^2 m/(2 + R^2)]$, and $B_0 = \Omega^2(\Omega^2 - 1)\{\Omega^4 - \Omega^2(2 + R^2) + 1\}$. The following curves are drawn in the Ω^2 - R^2 space: curve (a) $R^2 = \Omega^2(\Omega^2 - 1)/(\Omega^2 - m)$ corresponding to $B_2 = 0$, curve (c) $R^2 = (\Omega^2 - 1)^2/(\Omega^2 - \frac{1}{2})$, and curve (d) $R^2 = 2\Omega^2/(m - \Omega^2)$ corresponding to $B_1 = 0$, and curve (e) $\Omega^2 = 1$, and curve (f) $R^2 = \Omega^{-2}(\Omega^2 - 1)^2$ corresponding to $B_0 = 0$. From the theory developed in Sec. 5.10, the various regions in which modes 1 and 2 propagate can be determined. It is found that mode 1 propagates for $1 < \Omega < \infty$ and mode 2 for $0 < \Omega < \Omega_L$, $\Omega_1 < \Omega < \Omega_u$ and $\Omega_2 < \Omega < \infty$. These results are the same as those deduced in Sec. 5.9.

5.23. The indicated portions of the dispersion curve occur near the electron plasma frequency; therefore, the effect of ion motion is negligible and the theory developed in Sec. 5.7 can be used. In the discussion preceding Eq. (5.7.35), it is pointed out that at the cut-off frequency $\Omega = 1$ the electric field is linearly polarized parallel to the magnetostatic field. Therefore the electric field corresponding to the portion AB of the dispersion curve approaches the direction of propagation as the propagation direction approaches that of the magnetostatic field. In the discussion following Eq. (5.7.34), it has been pointed out that at the resonant frequency $\Omega = \Omega_p$, $k = \infty$ and the electric vector is in the propagation direction except when $\theta_{ph} = 0$. Therefore the electric field corresponding to the portion CD of the dispersion curve is in the direction of propagation as the propagation direction approaches that of the magnetostatic field.

5.24. In the MHD approximation the displacement current density is neglected but in the treatment of waves in a cold, electron-ion plasma, the effect of the displacement current density is retained. If the displacement current density is ignored, Eqs. (5.9.3) and (5.9.4) become: $\varepsilon_1 \pm \varepsilon_2 = -(1 + m)/(\Omega \pm R)(\Omega \mp Rm)$. Then in the limit of zero frequency, Eqs. (5.9.5a), (5.9.6a), and (5.9.8) yield the following expression for the phase velocity: $v_{ph}^- = v_{ph}^+ = c(R^2 m/(1 + m))^{1/2} = v_\alpha$ which becomes identical to that obtained in the MHD approximation.

5.25. The magnetostatic field is assumed to be in the z-direction. For the transverse Alfvén wave propagating in the $+z$-direction, it is obtained from Eqs. (5.11.21b) and (5.11.24) that $B^{\mp} = -B_0 V^{\mp}/v_\alpha$. Similarly for the wave propagating in the $-z$-direction, it is found that $B^{\mp} = B_0 V^{\mp}/v_\alpha$. Therefore $|B^{\mp}|^2/2\mu_0 = B_0^2|V^{\mp}|^2/2\mu_0 v_\alpha^2 = \rho_{m0}|V^{\mp}|^2/2$. Hence the magnetic and the kinetic energy densities of the wave motion are equal.

5.26. Consider the Alfvén wave propagating along the magnetostatic field and governed by Eqs. (5.11.9a), (5.11.10b), (5.11.11a), and (5.11.12b). The magnetostatic field is in the z-direction and the time-varying magnetic field is in the x-direction. Since the magnetic flux density does not vary in the y-direction, it is sufficient to examine the flux lines in the zx-plane only. In the absence of the Alfvén wave, the magnetic flux lines are straight lines parallel to the z-axis as specified by $x = x_0$. From Eqs. (5.11.9a), (5.11.10b), (5.11.11a), and (5.11.12b), the time-varying magnetic flux density can be deduced as $B_x = B_{x0}\sin\{\omega(t - z/v_\alpha)\}$. The equation of the magnetic flux line is given by $dx/dz = B_x/B_z = B_{x0}\sin\{\omega(t - z/v_\alpha)\}/B_0$. On integration it is found that

$$x = (B_{x0}/\omega\sqrt{\mu_0 \rho_{m0}})\cos\{\omega(t - z/v_\alpha)\} + x_0$$

which shows that the magnetic flux lines are sinusoidal lines in the direction of the magnetostatic field.

The stress due to the magnetic field consists of an isotropic pressure $B^2/2\mu_0$ and a tension B^2/μ_0 along the magnetic flux lines. Since there are no density variations associated with the Alfvén wave, the isotropic pressure plays no role and the tension along the flux lines alone has effect on the characteristics of the Alfvén wave.

For the magnetosonic wave propagating across the magnetostatic field, Fig. 5.30 shows that the magnetic flux lines do not vary in the y-direction and therefore it is again sufficient to consider the magnetic flux lines in the zx-plane only. In the absence of the magnetosonic wave, the magnetic flux lines are straight lines parallel to the z-axis and their density is uniform in the x-direction. In the presence of the magnetosonic wave, the magnetic flux lines are seen from Fig. 5.30 to remain as straight lines parallel to the z-axis but their density varies sinusoidally in the propagation direction, that is, parallel to the x-axis.

The stress due to the magnetic field that has a role in the determination of the characteristics of the magnetosonic wave is the pressure that varies sinusoidally in the propagation direction. The density of the charged particles follows that of the magnetic flux lines in the x-direction.

5.27. For $a > v_\alpha$ and $\theta_{ph} = 0$, it is found from Eq. (5.11.67) that

$$v_{ph,f} = \{\tfrac{1}{2}[v_\alpha^2 + a^2 + (a^2 - v_\alpha^2)]\}^{1/2} = a$$

and

$$v_{ph,s} = \{\tfrac{1}{2}[v_\alpha^2 + a^2 - (a^2 - v_\alpha^2)]\}^{1/2} = v_\alpha$$

and for $\theta_{ph} = \pi/2$, the following results are obtained:

$$v_{ph,f} = \{\tfrac{1}{2}[v_\alpha^2 + a^2 + (v_\alpha^2 + a^2)]\}^{1/2} = \sqrt{v_\alpha^2 + a^2} = v_{ms}$$

and

$$v_{ph,s} = \{\tfrac{1}{2}[v_\alpha^2 + a^2 - (v_\alpha^2 + a^2)]\}^{1/2} = 0$$

Thus the fast wave is the acoustic wave for $\theta_{ph} = 0$ and becomes the magnetosonic wave for $\theta_{ph} = \pi/2$. The slow wave is the Alfvén wave for $\theta_{ph} = 0$ and it disappears for $\theta_{ph} = \pi/2$. The sketch of the phase velocity versus θ_{ph} is essentially the same as in Fig. 5.31 but with the labelling A and S interchanged.

5.28. Let $l = \cos\theta_{ph}$ and $m = \sin\theta_{ph}$; therefore $l^2 + m^2 = 1$. From Eq. (5.11.65) it is found that $V_z = k^2a^2lmV_x/(\omega^2 - k^2a^2l^2)$. Since $V_{\parallel} = V_xm + V_zl$ and $V_{\perp} = V_xl - V_zm$, it follows that

$$V_{\parallel} = V_xm\omega^2/(\omega^2 - k^2a^2l^2) \qquad V_{\perp} = V_xl(\omega^2 - k^2a^2)/(\omega^2 - k^2a^2l^2)$$

and therefore

$$V_{\parallel}/V_{\perp} = 2m\omega^2/lk^2\left(v_\alpha^2 - a^2 \pm \{(v_\alpha^2 - a^2)^2 + 4v_\alpha^2a^2m^2\}^{1/2}\right)$$

where in the denominator $\omega^2 = k^2v_{ph}^2$ has been substituted from Eq. (5.11.67). Note that the upper and the lower signs correspond to the fast wave and the slow wave, respectively. The denominator in the expression for $V_{\parallel}/V_{\perp}$ is positive for the fast wave and negative for the slow wave. Since $V_{\parallel}/V_{\perp}$ is positive real for the fast wave and negative real for the slow wave, it follows that $V_{\parallel}$ and $V_{\perp}$ are in phase for the fast wave and 180° out of phase for the slow wave.

The use of Eq. (5.11.63) in Eqs. (5.11.59) and (5.11.60) yields $B_x = -B_0klV_x/\omega$ and $B_z = B_0kmV_x/\omega$. The perturbation P_m of the magnetic pressure is obtained as

$$P_m = \frac{(\hat{\mathbf{z}}B_0 + \mathbf{B})\cdot(\hat{\mathbf{z}}B_0 + \mathbf{B})}{2\mu_0} - \frac{B_0^2}{2\mu_0} \approx \frac{B_0B_z}{\mu_0} = \frac{B_0^2kmV_x}{\mu_0\omega}$$

If Eq. (5.11.63) is used in Eq. (5.11.58), and B_x and B_z are expressed in terms of V_x, the perturbation of the kinetic pressure is found to be

$$P = a^2\rho_m = \rho_{m0}V_x(\omega^2 - k^2v_\alpha^2)/\omega mk$$

Hence

$$P/P_m = [a^2 - v_\alpha^2 \pm \{(a^2 - v_\alpha^2)^2 + 4v_\alpha^2a^2m^2\}^{1/2}]/2m^2v_\alpha^2$$

where Eq. (5.11.67) has been used. Since P/P_m is positive real for the fast wave and negative real for the slow wave, it follows that P and P_m are in phase for the fast wave and 180° out of phase for the slow wave.

5.29. It is seen from Eq. (5.11.65) that $V_x/V_z = (\omega^2 - k^2a^2l^2)/k^2a^2lm$. For $v_\alpha \gg a$, $v_{ph,s} = al$, $\omega \approx kal$ and hence $V_x/V_z \approx 0$. For the slow wave, **V** is essentially in the z-direction, that is, parallel to the magnetostatic field. From Eq. (5.11.64) it is found that $V_x/V_z = k^2a^2lm/[\omega^2 - k^2(a^2m^2 + v_\alpha^2)]$. For $v_\alpha \gg a$, $v_{ph,f} = v_\alpha$, $\omega \approx kv_\alpha$, and hence $V_x/V_z = -l/m$. Since $mV_x + lV_z = V_{\parallel} = 0$, for the fast wave, **V** is perpendicular to the propagation direction but is in the plane containing the propagation direction and the magnetostatic field.

5.30. $\rho_{m0} = 1.35 \times 10^4 \text{kg/m}^3$, $\mu_0 = 4\pi \times 10^{-7}$ F/m, and $B_0 = 1000 \times 10^{-4} = 0.1 \text{W/m}^2$, $v_\alpha = 0.768$ m/sec.

5.31. The oblique Alfvén wave is a plane wave propagating along the magnetostatic field with the Alfvén wave velocity and it has no component of electric current parallel to the magnetostatic field. A point charge moving with uniform velocity in the direction of the magnetostatic field gives rise to an electric current that is parallel to the magnetostatic field and therefore the oblique Alfvén wave cannot be generated.

If the phase factor $\exp\{i(k_x x + k_z z - \omega t)\}$ is assumed for the field quantities, Eqs. (5.11.61) and (5.11.62) lead to the following dispersion relation:

$$k_\rho^2[a^2 v_\alpha^2 k_z^2 - \omega^2 a^2 - \omega^2 v_\alpha^2] = -(\omega^2 - v_\alpha^2 k_z^2)(\omega^2 - a^2 k_z^2)$$

where in view of the cylindrical symmetry k_x has been replaced by k_ρ. When the coherence condition $k_z = \omega/u$ is substituted, it is found that

$$k_\rho^2 = \frac{\omega^2}{u^2}\frac{(v_\alpha^2 - u^2)(u^2 - a^2)}{[v_\alpha^2(a^2 - u^2) - u^2 a^2]} < 0$$

Therefore no wave is generated. Actually since the fast and the slow waves have no electric current in the direction of the magnetostatic field, they cannot be excited by a point charge moving with uniform velocity in the direction of the magnetostatic field.

6.1. With the help of Eq. (6.2.53) the dispersion equation can be written as

$$\frac{k_p^2}{\omega_{pe}^2}\frac{KT_e}{m_e} = -\frac{k_p}{N_0}\frac{KT_e}{m_e}\int_{-\infty}^{\infty} d\mathbf{u}\,\frac{1}{(\omega - k_p u_x)}\frac{\partial f_0(u)}{\partial u_x}$$

If $f_0(u)$ is substituted and the integrations with respect to u_y and u_z are carried out, the result is

$$\frac{k_p^2}{\omega_{pe}^2}\frac{KT_e}{m_e} = -\sqrt{\frac{m_e}{2\pi KT_e}}\int_{-\infty}^{\infty} du_x\,\frac{[\omega - k_p u_x - \omega]}{(\omega - k_p u_x)}e^{-m_e u_x^2/2KT_e}$$

or equivalently

$$1 + \frac{k_p^2}{\omega_{pe}^2}\frac{KT_e}{m_e} = \sqrt{\frac{m_e}{2\pi KT_e}}\,\omega\int_{-\infty}^{\infty}\frac{du_x}{(\omega - k_p u_x)}e^{-m_e u_x^2/2KT_e} \qquad \text{(A1)}$$

If $\mathrm{Im}(\omega) > 0$, the contour is along the real axis. The substitution of

$$\frac{1}{\omega - k_p u_x} = -i\int_0^{\infty} dt\, e^{i\omega t - ik_p u_x t}$$

enables the dispersion equation to be written as

$$1 + \frac{k_p^2}{\omega_{pe}^2}\frac{KT_e}{m_e} = -i\omega\sqrt{\frac{m_e}{2\pi KT_e}}\int_0^{\infty} dt\, e^{i\omega t}\int_{-\infty}^{\infty} du_x\, e^{-ik_p u_x t - m_e u_x^2/2KT_e}$$

Since

$$\int_{-\infty}^{\infty} du_x\, e^{-ik_p u_x t - m_e u_x^2/2KT_e} = \sqrt{\frac{2\pi KT_e}{m_e}}\, e^{-k_p^2 t^2 KT_e/2m_e}$$

the dispersion equation becomes

$$1 + \frac{k_p^2}{\omega_{pe}^2}\frac{KT_e}{m_e} = -i\omega \int_0^{\infty} dt\, e^{i\omega t - k_p^2 t^2 KT_e/2m_e} \tag{A2}$$

where the integral is seen to converge for $\mathrm{Im}(\omega) > 0$ as well as $\mathrm{Im}(\omega) < 0$.

For $\mathrm{Im}(\omega) < 0$, Eq. (A1) becomes

$$1 + \frac{k_p^2}{\omega_{pe}^2}\frac{KT_e}{m_e} = \sqrt{\frac{m_e}{2\pi KT_e}}\,\omega \int_{-\infty}^{\infty} \frac{du_x}{(\omega - k_p u_x)} e^{-m_e u_x^2/2KT_e}$$

$$- \frac{\omega}{k_p}\sqrt{\frac{m_e}{2\pi KT_e}}\, 2\pi i e^{-m_e \omega^2/2k_p^2 KT_e}$$

where the contour is again along the real axis. Since

$$\frac{1}{\omega - k_p u_x} = i \int_{-\infty}^{0} dt\, e^{i\omega t - ik_p u_x t}$$

the dispersion equation can be simplified to yield

$$1 + \frac{k_p^2}{\omega_{pe}^2}\frac{KT_e}{m_e} = i\omega \int_{-\infty}^{0} dt\, e^{i\omega t - k_p^2 t^2 KT_e/2m_e}$$

$$- \frac{\omega}{k_p}\sqrt{\frac{m_e}{2\pi KT_e}}\, 2\pi i e^{-m_e \omega^2/2k_p^2 KT_e}$$

Note that

$$\int_{-\infty}^{\infty} dt\, e^{i\omega t - k_p^2 t^2 KT_e/2m_e} = \frac{2\pi}{k_p}\sqrt{\frac{m_e}{2\pi KT_e}}\, e^{-m_e \omega^2/2k_p^2 KT_e}$$

Therefore, even for $\mathrm{Im}\,(\omega) < 0$, the dispersion equation reduces to that given in Eq. (A2).

6.2. In the previous problem, the dispersion equation is obtained in the form:

$$\frac{k_p^2}{\omega_{pe}^2} = k_p \frac{1}{\sqrt{2\pi}}\left(\frac{m_e}{KT_e}\right)^{3/2} \int_{-\infty}^{\infty} \frac{du_x\, u_x}{(\omega - k_p u_x)} e^{-m_e u_x^2/2KT_e} \tag{A1}$$

If $G(u_x)$ and z are substituted, Eq. (A1) simplifies to

$$H(k_p, z) = \frac{k_p^2}{\omega_{pe}^2} - \int_{-\infty}^{\infty} \frac{G(u_x)}{(u_x - z)}\, du_x = 0$$

The contour of integration in the ω-plane can be shifted up without changing $H(k_p, z)$ since no singularities are intercepted. Thus $H(k_p, z)$ is defined throughout the upper half of the ω-plane.

The real and the imaginary parts of $H(k_p, z)$ are

$$\operatorname{Re}\{H(k_p, z)\} = \frac{k_p^2}{\omega_{pe}^2} - \int_{-\infty}^{\infty} \frac{G(u_x)(u_x - x)}{[(u_x - x)^2 + y^2]}\, du_x$$

$$\operatorname{Im}\{H(k_p, z)\} = -y \int_{-\infty}^{\infty} \frac{G(u_x)}{[(u_x - x)^2 + y^2]}\, du_x$$

Therefore

$$\operatorname{Re}\{H(k_p, z)\} + \frac{x}{y}\operatorname{Im}\{H(k_p, z)\} = \frac{k_p^2}{\omega_{pe}^2} - \int_{-\infty}^{\infty} \frac{du_x\, u_x\, G(u_x)}{[(u_x - x)^2 + y^2]} \tag{A2}$$

Since $u_x G(u_x) < 0$ for $-\infty < u_x < \infty$, the right side of Eq. (A2) is positive. Also, since $\operatorname{Im}\{H(k_p, z)\} = 0$, $\operatorname{Re}\{H(k_p, z)\} > 0$. Therefore $H(k_p, z)$ has no complex solutions of z with a positive imaginary part. Consequently there are no unstable solutions to the dispersion equation.

The analytic continuation of $H(k_p, z)$ valid in the lower half of the complex ω-plane is

$$H(k_p, z) = \frac{k_p^2}{\omega_{pe}^2} - \int_{-\infty}^{\infty} \frac{G(u_x)}{(u_x - z)}\, du_x - 2\pi i G(z)$$

6.3. If $H(x) = 0$ for $x < 0$ and $H(x) = 1$ for $x \geqslant 0$,

$$f_0(\mathbf{u}) = \frac{N_0}{2a_s}\delta(u_y)\delta(u_z)[H(u_x + a_s) - H(u_x - a_s)]$$

and therefore

$$\frac{\partial f_0(\mathbf{u})}{\partial u_x} = \frac{N_0}{2a_s}\delta(u_y)\delta(u_z)[\delta(u_x + a_s) - \delta(u_x - a_s)]$$

The substitution of $(\partial f_0(\mathbf{u})/\partial u_x)$ into the dispersion equation

$$1 = -\frac{\omega_{pe}^2}{N_0 k_p} \int_{-\infty}^{\infty} d\mathbf{u}\, \frac{1}{(\omega - k_p u_x)} \frac{\partial f_0(\mathbf{u})}{\partial u_x}$$

and the evaluation of the resulting integrals gives

$$\omega = \pm\sqrt{\omega_{pe}^2 + k_p^2 a_s^2}$$

Since ω is real, there is neither instability nor Landau damping.

6.4. If $f_0(\mathbf{u})$ is substituted into the dispersion equation (6.2.55) and the integrals with respect to u_y and u_z are evaluated, the result is

$$1 = \frac{\omega_{pe}^2}{k_p^2}\frac{a_r}{\pi}\int_{-\infty}^{\infty}\frac{du_x}{(u_x - \omega/k_p)^2(u_x^2 + a_r^2)}$$

The integral with respect to u_x can be evaluated by closing the contour in the upper half plane. There is a double pole at $u_x = \omega/k_p$ and a simple pole at $u_x = ia_r$. When the contributions arising from the residues at these poles are evaluated, the result is

$$1 = \omega_{pe}^2 \Big/ k_p^2\left(\frac{\omega}{k_p} + ia_r\right)^2$$

or equivalently

$$\omega = \pm\omega_{pe} - ik_p a_r$$

Since Im $(\omega) < 0$, the longitudinal plasma wave is not unstable. The angular frequency of oscillation and the Landau damping constant are given by

$$\omega_r = \pm\omega_{pe} \qquad \text{and} \quad \omega_i = -k_p a_r$$

The Landau damping constant $(\omega_i)_{MB}$ for the Maxwell-Boltzmann distribution of velocities is given by Eq. (6.2.87). For $k_p\lambda_{De} \ll 1$, $(\omega_i)_{MB}$ is very small compared to ω_i resulting from the resonance distribution. But if $k_p\lambda_{De}$ is of the order of or greater than unity, ω_i obtained from the resonance distribution can be of the same order of magnitude as $(\omega_i)_{MB}$.

6.5. Let $f(\mathbf{r}, \mathbf{u}, t) = f_0(\mathbf{u}) + g(\mathbf{r}, \mathbf{u}, t)$ where $g(\mathbf{r}, \mathbf{u}, t) \ll f_0(\mathbf{u})$. With the help of the Boltzmann-Vlasov equation, it is found that

$$\frac{\partial}{\partial t}g(\mathbf{r}, \mathbf{u}, t) + (\mathbf{u}\cdot\nabla_r)g(\mathbf{r}, \mathbf{u}, t) = \frac{e}{m_e}E_x\frac{\partial f_0(\mathbf{u})}{\partial u_x}$$

If the phase factor $\exp\{i(k_p x - \omega t)\}$ is assumed for all the field quantities, the amplitude $g(\mathbf{u})$ can be deduced as

$$g(\mathbf{u}) = \frac{ie}{m_e}E_x\frac{(\partial f_0(\mathbf{u})/\partial u_x)}{(\omega - k_p u_x)}$$

Since the charge density

$$\rho = -e\int g(\mathbf{u})\,d\mathbf{u}$$

the electrostatic equations give

$$E_x = -ik_p\Phi \qquad \text{and} \quad \Phi = -\frac{e}{\varepsilon_0 k_p^2}\int g(\mathbf{u})\,d\mathbf{u}$$

Therefore

$$\Phi = -\frac{e}{\varepsilon_0 k_p^2}\frac{ie}{m_e}(-ik_p\Phi)\int d\mathbf{u}\,\frac{(\partial f_0(\mathbf{u})/\partial u_x)}{(\omega - k_p u_x)}$$

which yields the dispersion equation

$$1 = -\frac{\omega_{pe}^2}{N_0 k_p}\int d\mathbf{u}\,\frac{(\partial f_0(\mathbf{u})/\partial u_x)}{(\omega - k_p u_x)}$$

An integration by parts enables the reduction of the dispersion equation to the form

$$1 = \frac{\omega_{pe}^2}{N_0}\int d\mathbf{u}\,\frac{f_0(\mathbf{u})}{(\omega - k_p u_x)^2}$$

6.6. The velocity distribution function for the single electron stream is given by

$$f_0(\mathbf{u}) = N_0\,\delta(u_x - v_d)\delta(u_y)\delta(u_z)$$

When this value of $f_0(\mathbf{u})$ is substituted into the dispersion equation deduced in Problem 6.5 and the integrals are evaluated, it is obtained that

$$\omega = \pm\omega_{pe} + k_x v_d$$

which gives the frequency of oscillation. Since ω is real, there is no instability.

6.7. The velocity distribution function for the two oppositely drifting electron streams is obtained as

$$f_0(\mathbf{u}) = \frac{N_0}{2}[\delta(u_x - v_d) + \delta(u_x + v_d)]\delta(u_y)\delta(u_z)$$

When this value of $f_0(\mathbf{u})$ is substituted into the dispersion equation derived in Problem 6.5, it is found that

$$1 = \frac{\omega_{pe}^2}{2}\left[\frac{1}{(\omega - k_p v_d)^2} + \frac{1}{(\omega + k_p v_d)^2}\right]$$

which can be rearranged to yield

$$\tilde{\omega}^4 - B\tilde{\omega}^2 + C = 0$$

The two solutions of this polynomial equation are

$$\tilde{\omega}^2 = \tilde{\omega}_1^2 = \frac{B}{2} + \sqrt{\left(\frac{B}{2}\right)^2 - C}$$

and

$$\tilde{\omega}^2 = \tilde{\omega}_2^2 = \frac{B}{2} - \sqrt{\left(\frac{B}{2}\right)^2 - C}$$

Note that $(B/2)^2 - C > 0$ for all values of α. For $\alpha > 1$, $C > 0$, $|\sqrt{(B/2)^2 - C}| < |B/2|$ and hence $\tilde{\omega}_1^2$ and $\tilde{\omega}_2^2$ have the same sign as B which is positive real. Therefore, $\tilde{\omega}_1$ and $\tilde{\omega}_2$ are real for $\alpha > 1$ and there is no instability. For $\alpha < 1$, $C < 0$, $|\sqrt{(B/2)^2 - C}| > |B/2|$; hence $\tilde{\omega}_1^2$ is positive real and $\tilde{\omega}_2^2$ is negative real. Therefore $\tilde{\omega}_1$ is real and $\tilde{\omega}_2$ has one negative imaginary value and one positive imaginary value. The latter value of $\tilde{\omega}_2$ corresponds to an unstable mode. For this unstable mode, let $\tilde{\omega}_2 = i\tilde{\omega}_{2i}$; then $\tilde{\omega}_2^2 = -\tilde{\omega}_{2i}^2$. The growth constant is ω_{2i} and it is a maximum when $\tilde{\omega}_{2i}^2$ is a maximum or $\tilde{\omega}_2^2$ is a minimum. The minimum value of $\tilde{\omega}_2^2$ occurs at $\alpha = \sqrt{3/8}$ and the corresponding value of $\tilde{\omega}_2^2 = -1/8$. Hence the maximum growth constant is obtained as $(\omega_{2i})_{\max} = \omega_{pe}/\sqrt{8}$.

6.8. Let $f^\alpha(\mathbf{r}, \mathbf{u}, t) = f_0^\alpha(\mathbf{u}) + g^\alpha(\mathbf{r}, \mathbf{u}, t)$ for $\alpha = e$ and i where $g^\alpha(\mathbf{r}, \mathbf{u}, t) \ll f_0^\alpha(\mathbf{u})$. The superscripts e and i indicate that the corresponding distribution function pertains to the electrons and the ions, respectively. All the field quantities are assumed to have the phase factor of the form $\exp\{i(k_p x - \omega t)\}$. With the help of the Boltzmann-Vlasov equations for the electrons and the ions, the amplitude functions $g^e(\mathbf{u})$ and $g^i(\mathbf{u})$ can be deduced to be given by

$$g^e(\mathbf{u}) = \frac{ie}{m_e} E_x \frac{(\partial f_0^e(\mathbf{u})/\partial u_x)}{(\omega - k_p u_x)} \quad \text{and} \quad g^i(\mathbf{u}) = \frac{-ie}{m_i} E_x \frac{(\partial f_0^i(\mathbf{u})/\partial u_x)}{(\omega - k_p u_x)}$$

Since the charge density

$$\rho = -e \int g^e(\mathbf{u})\, d\mathbf{u} + e \int g^i(\mathbf{u})\, d\mathbf{u}$$

the electrostatic equations yield

$$E_x = -ik_p \Phi \quad \text{and} \quad \Phi = -\frac{e}{\varepsilon_0 k_p^2}\left[\int g^e(\mathbf{u})\, d\mathbf{u} - \int g^i(\mathbf{u})\, d\mathbf{u}\right]$$

Therefore

$$\Phi = -\frac{e}{\varepsilon_0 k_p^2} ie(-ik_p \Phi)\left[\frac{1}{m_e} \int d\mathbf{u} \frac{(\partial f_0^e(\mathbf{u})/\partial u_x)}{(\omega - k_p u_x)} + \frac{1}{m_i} \int d\mathbf{u} \frac{(\partial f_0^i(\mathbf{u})/\partial u_x)}{(\omega - k_p u_x)}\right]$$

which yields the dispersion equation:

$$1 = \frac{\omega_{pe}^2}{N_0 k_p^2} \int d\mathbf{u} \frac{(\partial f_0^e(\mathbf{u})/\partial u_x)}{(u_x - \omega/k_p)} + \frac{\omega_{pi}^2}{N_0 k_p^2} \int d\mathbf{u} \frac{(\partial f_0^i(\mathbf{u})/\partial u_x)}{(u_x - \omega/k_p)}$$

For the cold plasma model, the equilibrium distributions are obtained as

$$f_0^e(\mathbf{u}) = f_0^i(\mathbf{u}) = N_0 \delta(u_x)\delta(u_y)\delta(u_z)$$

After an integration by parts, the dispersion equation becomes

$$1 = \frac{\omega_{pe}^2}{N_0 k_p^2} \int \frac{d\mathbf{u}\, f_0^e(\mathbf{u})}{(u_x - \omega/k_p)^2} + \frac{\omega_{pi}^2}{N_0 k_p^2} \int \frac{d\mathbf{u}\, f_0^i(\mathbf{u})}{(u_x - \omega/k_p)^2}$$

If $f_0^e(\mathbf{u})$ and $f_0^i(\mathbf{u})$ are substituted the result is

$$\omega^2 = \omega_{pe}^2 + \omega_{pi}^2 = N_0 e^2/m_r \varepsilon_0 \qquad \text{with} \quad m_r = m_e m_i/(m_e + m_i)$$

Thus, for the longitudinal plasma wave, the effect of inclusion of the ion motion is to change m_e in the dispersion equation to m_r.

6.9. The equilibrium distribution functions for the electrons and the ions are obtained as

$$f_0^e(\mathbf{u}) = N_0\,\delta(u_x - v_d)\delta(u_y)\delta(u_z)$$

and

$$f_0^i(\mathbf{u}) = N_0\,\delta(u_x)\delta(u_y)\delta(u_z)$$

The result of substituting $f_0^e(\mathbf{u})$ and $f_0^i(\mathbf{u})$ in the dispersion equation derived in Problem 6.8 is

$$1 = \frac{\omega_{pe}^2}{(\omega - k_p v_d)^2} + \frac{\omega_{pi}^2}{\omega^2}$$

which can be recast as

$$F(\tilde{\omega}, \alpha) = \frac{1}{(\tilde{\omega} - \alpha)^2} + \frac{m}{\tilde{\omega}^2} = 1$$

This dispersion equation can be expressed as a polynomial equation in $\tilde{\omega}$ as follows:

$$\tilde{\omega}^4 - 2\alpha\tilde{\omega}^3 + \tilde{\omega}^2(\alpha^2 - 1 - m) + 2m\alpha\tilde{\omega} - m\alpha^2 = 0$$

If $F(\tilde{\omega}, \alpha)$ is plotted as a function of $\tilde{\omega}$, it is seen that $F(\tilde{\omega}, \alpha)$ is infinite at $\tilde{\omega} = 0$ and $\tilde{\omega} = \alpha$ and is zero at $\tilde{\omega} = \pm\infty$. There is a minimum in the range $0 < \tilde{\omega} < \alpha$. Let the minimum value be denoted by $F_{\min}(\tilde{\omega}, \alpha)$. The solutions of the dispersion equation are the values of $\tilde{\omega}$ for which the ordinate is equal to unity. There are two real solutions for $\tilde{\omega}$ one occurring in the range $-\infty < \tilde{\omega} < 0$ and the other in the range $\alpha < \tilde{\omega} < \infty$. If $F_{\min}(\tilde{\omega}, \alpha)$ is greater than unity, there are two complex solutions. Since the coefficients of the polynomial equation are real, these complex solutions are complex conjugates of each other. Therefore, if $F_{\min}(\tilde{\omega}, \alpha) > 1$, one solution of the dispersion equation yields a value of $\tilde{\omega}$ with a positive imaginary part and hence corresponds to an unstable mode. The details have been worked out by Bernstein and Trehan.

6.10. If $f_0^\alpha(\mathbf{u})$ is substituted, the integrations with respect to u_y and u_z are carried out and the integration variable u_x is changed to $a_\alpha w$, it is found that

$$\frac{\omega_{p\alpha}^2}{N_0 k_p^2}\int_{-\infty}^{\infty} d\mathbf{u}\,\frac{1}{(u_x-\omega/k_p)}\frac{\partial f_0^\alpha(\mathbf{u})}{\partial u_x}$$
$$= -\frac{\lambda_{\mathrm{D}\alpha}^{-2}k_p^{-2}}{\sqrt{\pi}}\int_{-\infty}^{\infty} dw\,\frac{w}{(w-z_\alpha)}e^{-w^2}$$
$$= -\lambda_{\mathrm{D}\alpha}^{-2}k_p^{-2}\left[1 + i\sqrt{\pi}\,z_\alpha e^{-z_\alpha^2} + \frac{z_\alpha}{\sqrt{\pi}}P\int_{-\infty}^{\infty}\frac{dw\,e^{-w^2}}{(w-z_\alpha)}\right]$$
$$= -\lambda_{\mathrm{D}\alpha}^{-2}k_p^{-2}[1 + i\sqrt{\pi}\,z_\alpha e^{-z_\alpha^2} - 2z_\alpha\int_0^{z_\alpha}\exp(w^2 - z_\alpha^2)\,dw]$$

where the principal value of the integral was obtained from Eqs. (6.2.69) and (6.2.76). Therefore, the dispersion equation for the longitudinal plasma wave in an electron-ion plasma simplifies to

$$\begin{aligned}1 &+ k_p^{-2}\lambda_{\mathrm{D}e}^{-2}[1 + i\sqrt{\pi}\,z_e e^{-z_e^2} - 2z_e\int_0^{z_e} e^{w^2-z_e^2}\,dw]\\ &+ k_p^{-2}\lambda_{\mathrm{D}i}^{-2}[1 + i\sqrt{\pi}\,z_i e^{-z_i^2} - 2z_i\int_0^{z_i} e^{w^2-z_i^2}\,dw] = 0\end{aligned} \tag{A1}$$

As in Appendix G.7

$$2z_e\int_0^{z_e} e^{w^2-z_e^2}\,dw = 2z_e^2 \qquad \text{for Re } z_e \ll 1$$

and

$$2z_i\int_0^{z_i} e^{w^2-z_i^2}\,dw = 1 + \frac{1}{2z_i^2} \qquad \text{for Re } z_i \gg 1$$

Therefore, the dispersion equation (A1) can be approximated as

$$\begin{aligned}1 &+ k_p^{-2}\lambda_{\mathrm{D}e}^{-2}(1 + i\sqrt{\pi}\,z_e) + k_p^{-2}\lambda_{\mathrm{D}i}^{-2}\left(1 - 1 - \frac{1}{2z_i^2}\right)\\ &= 1 + k_p^{-2}\lambda_{\mathrm{D}e}^{-2}\left(1 - \frac{m_e}{2m_i}z_e^{-2} + i\sqrt{\pi}\,z_e\right) = 0\end{aligned} \tag{A2}$$

If $\omega_{\text{imag}} \ll \omega_{\text{real}}$, the imaginary part of z_e can be neglected in the first approximation. The real part of Eq. (A2) then leads to

$$\omega_{\text{real}} = a_s k_p(1 + k_p^2\lambda_{\mathrm{D}e}^2)^{-1/2} \tag{A3}$$

The imaginary part of Eq. (A2) gives

$$\sqrt{\pi}\,\text{Re}\,(z_e) = \frac{m_e}{2m_i}k_p^2 a_e^2\,\text{Im}\left(\frac{1}{(\omega_{\text{real}} + i\omega_{\text{imag}})^2}\right)$$

which leads to

$$\sqrt{\pi}\,\omega_{\text{real}}/k_p a_e = -\frac{m_e}{2m_i} k_p^2 a_e^2 \frac{2\omega_{\text{imag}}}{\omega_{\text{real}}^3} \tag{A4}$$

On simplification, Eq. (A4) yields

$$\omega_{\text{imag}} = -\sqrt{\pi}\frac{m_i}{m_e}\frac{\omega_{\text{real}}^4}{k_p^3 a_e^3} = -\sqrt{\frac{\pi m_e}{8m_i}}\, a_s k_p(1 + k_p^2\lambda_{De}^2)^{-2} \tag{A5}$$

Since $m_e \ll m_i$, Eq. (A5) shows that the requirement $\omega_{\text{imag}} \ll \omega_{\text{real}}$ is fulfilled. Also, for the same reason, the condition $\omega_{\text{real}}/k_p a_e \ll 1$ is satisfied. The condition $a_i k_p/\omega_{\text{real}} \ll 1$ is fulfilled only if

$$\frac{a_s^2}{a_i^2} = \frac{T_e}{2T_i} \gg (1 + k_p^2\lambda_{De}^2)$$

therefore, the plasma has to be strongly non-isothermal with hot electrons and cold ions.

In the long-wave range, it is found approximately that $\omega_{\text{real}} = k_p a_s$. Since these long-wave, low-frequency oscillations propagate at a sound speed determined by the *electron* temperature and the *ion* mass, they are essentially the same as the ion acoustic waves discussed in Sec. 2.8.

6.11. Let

$$\begin{aligned} h(u_x) &= [(u_x - v_d)^2 + a_r^2]^{-1} + [(u_x + v_d)^2 + a_r^2]^{-1} \\ &= 2(u_x^2 + v_d^2 + a_r^2)[u_x^4 + 2u_x^2(a_r^2 - v_d^2) + (v_d^2 + a_r^2)^2]^{-1} \end{aligned}$$

The condition $(\partial/\partial u_x)h(u_x) = 0$ leads to $u_x = 0$ and

$$u_x^4 + 2Bu_x^2 + C = 0 \tag{A1}$$

where $B = v_d^2 + a_r^2$ and $C = (a_r^2 + v_d^2)(a^2 - 3v_d^2)$. The velocity distribution function has two peaks only if $h(u_x)$ has three extrema or if $(\partial/\partial u_x)h(u_x) = 0$ yields three real solutions for u_x. Since $u_x = 0$ is one solution, Eq. (A1) should give two real solutions for u_x or one positive, real solution for u_x^2. Since $B^2 - C = 4v_d^2(v_d^2 + a_r^2) > 0$, the two solutions u_{x1}^2 and u_{x2}^2 of Eq. (A1) are real. Also, since $B > 0$, one of the solutions has to be negative. Therefore, the other solution is positive only if $C < 0$, that is, if $v_d/a_r > 1/\sqrt{3}$. Consequently the velocity distribution function has two peaks only if $v_d/a_r > 1/\sqrt{3}$.

If $f_0(\mathbf{u})$ is substituted into the dispersion equation deduced in Problem 6.5 and the integrals with respect to u_y and u_z are evaluated, the result is

$$1 = \frac{\omega_{pe}^2}{k_p^2}\frac{a_r}{2\pi}\int_{-\infty}^{\infty} du_x \frac{1}{(u_x - \omega/k_p)^2}\left[\frac{1}{(u_x - v_d)^2 + a_r^2} + \frac{1}{(u_x + v_d)^2 + a_r^2}\right] \tag{A2}$$

The integral in Eq. (A2) can be evaluated by closing the contour in the lower half plane. There are two simple poles at $(v_d - ia_r)$ and $(-v_d - ia_r)$ occurring within the contour of integration. The contributions arising from the residues at these poles are evaluated with the result that Eq. (A2) becomes

$$2 = (x - y)^{-2} + (x + y)^{-2} \tag{A3}$$

where $x = (\omega + ik_p a_r)/\omega_{pe}$ and $y = k_p v_d/\omega_{pe}$.

If Eq. (A3) is arranged into a polynomial equation in x^2 and solved, it is seen that

$$x^2 = y^2 + \frac{1}{2} \pm \sqrt{2y^2 + \frac{1}{4}} \tag{A4}$$

with the result that x is real or imaginary. When x is real, ω cannot have a positive imaginary part leading to an instability. Thus the solution corresponding to the positive sign in front of the radical in Eq. (A4) cannot give rise to unstable modes. For the negative sign in Eq. (A4), x can be imaginary; hence Re $(\omega) = 0$. Let $\omega = i\omega_i$; therefore $x = i(\omega_i + k_p a_r)/\omega_{pe}$. Instability corresponds to $\omega_i > 0$ and hence $x^2 < -k_p^2 a_r^2/\omega_{pe}^2 = -y^2 a_r^2/v_d^2$. Consequently, the condition for the existence of instability is determined from Eq. (A4) as

$$y^2 + \frac{1}{2} - \sqrt{2y^2 + \frac{1}{4}} \leqslant -y^2 a_r^2/v_d^2 \tag{A5}$$

where the equality sign corresponds to the onset of instability. It is possible to rearrange the condition (A5) to yield

$$y^2 \leqslant (1 - a_r^2/v_d^2)/(1 + a_r^2/v_d^2)^2 = y_{\max}^2$$

Since y^2 is positive, it follows that $v_d > a_r$ has to be satisfied for the onset of instability. Note that if $v_d > a_r$, the condition for the existence of two peaks in the velocity distribution function is also satisfied.

Instability is absent if $v_d/a_r < 1$. The longitudinal plasma waves are unstable if $v_d/a_r > 1$ and if $k_p v_d/\omega_{pe} \leqslant y_{\max}$.

6.12. Let $\omega = \omega_r + i\omega_i$ with $\omega_i > 0$. The real and the imaginary parts of $H(\omega)$ are

$$\text{Re}\,\{H(\omega)\} = k_{em}^2 c^2 - \omega_r^2 + \omega_i^2 + \omega_{pe}^2 \int_{-\infty}^{\infty} \frac{du_x\, g(u_x)\{\omega_r^2 + \omega_i^2 - \omega_r k_{em} u_x\}}{[(\omega_r - k_{em} u_x)^2 + \omega_i^2]}$$

and

$$\text{Im}\,\{H(\omega)\} = \omega_i \left[-2\omega_r - \omega_{pe}^2 \int_{-\infty}^{\infty} \frac{du_x\, g(u_x) k_{em} u_x}{[(\omega_r - k_{em} u_x)^2 + \omega_i^2]}\right]$$

Hence,

$$\text{Re}\,\{H(\omega)\} - \frac{\omega_r}{\omega_i}\,\text{Im}\{H(\omega)\} = k_{em}^2 c^2 + \omega_r^2 + \omega_i^2 + \omega_{pe}^2 \int_{-\infty}^{\infty} \frac{du_x\, g(u_x)(\omega_r^2 + \omega_i^2)}{[(\omega_r - k_{em}u_x)^2 + \omega_i^2]} \tag{A}$$

which is always greater than zero. Let ω_r and ω_i be chosen such that $\text{Im}\{H(\omega)\} = 0$. Then Eq. (A) shows that $\text{Re}\,\{H(\omega)\} > 0$. Consequently $H(\omega) = 0$ can never be satisfied by any value of ω with a positive imaginary part. Therefore, the transverse electromagnetic wave is not unstable.

6.13. Since $f_0(\mathbf{u})$ is isotropic, the dispersion equation can be determined from Eq. (6.4.45) as

$$(k_z^+ c)^2 - \omega^2 - \frac{\omega_{pe}^2\,\omega\pi}{N_0} \int_0^{\infty} du_\rho \int_{-\infty}^{\infty} du_z \frac{u_\rho^2}{(\omega - \omega_{ce} - k_z^+ u_z)} \frac{\partial f_0(\mathbf{u})}{\partial u_\rho} = 0$$

An integration by parts with respect to u_ρ enables the dispersion equation to be written in the form

$$(k_z^+ c)^2 - \omega^2 + \frac{2\omega_{pe}^2\,\omega\pi}{N_0} \int_0^{\infty} du_\rho \int_{-\infty}^{\infty} du_z \frac{u_\rho}{(\omega - \omega_{ce} - k_z^+ u_z)} f_0(\mathbf{u}) = 0 \tag{A1}$$

Since

$$\int_0^{\infty} du_\rho\, 2\pi u_\rho \frac{a_r}{\pi^2} \frac{1}{(u^2 + a_r^2)^2} = \frac{a_r}{\pi} \frac{1}{(u_z^2 + a_r^2)}$$

the substitution of $f_0(\mathbf{u})$ in Eq. (A1) and the evaluation of the integral with respect to u_ρ yields

$$(k_z^+ c)^2 - \omega^2 + \frac{\omega_{pe}^2\,\omega a_r}{\pi} \int_{-\infty}^{\infty} du_z \frac{1}{(\omega - \omega_{ce} - k_z^+ u_z)(u_z^2 + a_r^2)} = 0 \tag{A2}$$

The integral in Eq. (A2) is evaluated by closing the contour in the lower half of the u_z-plane with the following result:

$$(k_z^+ c)^2 - \omega^2 + \frac{\omega_{pe}^2\,\omega}{(\omega - \omega_{ce} + ik_z^+ a_r)} = 0 \tag{A3}$$

If ω is not close to ω_{ce}, $k_z^+ a_r$ can be omitted in comparison with $(\omega - \omega_{ce})$. Then the following result based on the cold plasma model is recovered:

$$(k_z^+ c)^2 = \omega \frac{(\omega^2 - \omega\omega_{ce} - \omega_{pe}^2)}{(\omega - \omega_{ce})}$$

which shows that for ω not close to ω_{ce}, $k_z^+ c/\omega$ is of order unity. Therefore $k_z^+ a_r$ is of order a_r/c smaller than $(\omega - \omega_{ce})$ thus justifying the omission of $k_z^+ a_r$ in comparison to $(\omega - \omega_{ce})$. It is to be noted that ω is real in the first approximation.

For ω close to ω_{ce}, k_z^+ becomes very large and the small imaginary part of $(\omega - \omega_{ce})$ can be neglected in comparison to $k_z^+ a_r$. Hence Eq. (A3) simplifies to

$$\left(\frac{k_z^+ c}{\omega}\right)^3 - \left(\frac{k_z^+ c}{\omega}\right) - i\frac{\omega_{pe}^2}{\omega^2}\frac{c}{a_r} = 0 \qquad \text{(A4)}$$

Since $k_z^+ c/\omega \gg 1$, the second term on the left side can be neglected in comparison to the first term. Let $\omega = \omega_r + i\omega_i$. In the small third term, ω can be replaced by ω_{ce}, namely the value obtained in the first approximation. Therefore, it is found that

$$\left(\frac{\omega}{k_z^+ c}\right)^3 = -i\frac{a_r}{c}\left(\frac{\omega_{ce}}{\omega_{pe}}\right)^2$$

which can be manipulated to yield

$$\omega_r = \frac{\sqrt{3}}{2} k_z^+ (a_r c^2 \omega_{ce}^2/\omega_{pe}^2)^{1/3}$$

and

$$\omega_i = -\frac{k_z^+}{2}(a_r c^2 \omega_{ce}^2/\omega_{pe}^2)^{1/3}$$

A crude approximation to the cyclotron damping constant is given by ω_i.

6.14. If the equilibrium distribution function, instead of being isotropic, is cylindrically symmetric about the direction of the magnetostatic field, the right side of Eq. (6.4.9) becomes

$$G(\mathbf{u}) = \frac{e}{m_e \omega_{ce}}[\mathbf{E} + \mathbf{u} \times \mathbf{B}] \cdot \nabla_u f_0(u_\rho, u_z)$$

where $\mathbf{B}$ is the amplitude of the time-varying magnetic field. Maxwell's equation $\nabla \times \mathbf{E}(\mathbf{r}) = i\omega\mathbf{B}(\mathbf{r})$ leads to $\mathbf{B} = (k_z/\omega)\hat{\mathbf{z}} \times \mathbf{E}$. Let $u_x = u_\rho \cos\varphi$, $u_y = u_\rho \sin\varphi$ and $u_z = u_z$ where u_x, u_y, and u_z are the rectangular components, and u_ρ, φ, and u_z are the cylindrical components of $\mathbf{u}$. Since $\partial f_0(\mathbf{u})/\partial u_x = \cos\varphi(\partial f_0(\mathbf{u})/\partial u_\rho)$ and $\partial f_0(\mathbf{u})/\partial u_y = \sin\varphi(\partial f_0(\mathbf{u})/\partial u_\rho)$, it is found that

$$\nabla_u f_0(\mathbf{u}) = \frac{1}{u_\rho}\frac{\partial f_0(\mathbf{u})}{\partial u_\rho}(\hat{\mathbf{x}}u_x + \hat{\mathbf{y}}u_y) + \hat{\mathbf{z}}\frac{\partial f_0(\mathbf{u})}{\partial u_z}$$

Consequently, it can be shown that

$$\mathbf{E} \cdot \nabla_u f_0(\mathbf{u}) = (E_x u_x + E_y u_y)\frac{1}{u_\rho}\frac{\partial f_0(\mathbf{u})}{\partial u_\rho} + E_z\frac{\partial f_0(\mathbf{u})}{\partial u_z}$$

and

$$(\mathbf{u} \times \mathbf{B}) \cdot \nabla_u f_0(\mathbf{u}) = (E_x u_x + E_y u_y)\frac{k_z}{u_\rho \omega}\left\{-u_z\frac{\partial f_0(\mathbf{u})}{\partial u_\rho} + u_\rho\frac{\partial f_0(\mathbf{u})}{\partial u_z}\right\}$$

The inclusion of the $(\mathbf{u} \times \mathbf{B}) \cdot \nabla_u f_0(\mathbf{u})$ is equivalent to the addition of

$$\frac{k_z}{\omega}\left\{-u_z \frac{\partial f_0(\mathbf{u})}{\partial u_\rho} + u_\rho \frac{\partial f_0(\mathbf{u})}{\partial u_z}\right\} \tag{A1}$$

to $\partial f_0(\mathbf{u})/\partial u_\rho$. The remainder of the analysis proceeds as in Sec. 6.4. The dispersion equation for the right circularly polarized transverse electromagnetic wave propagating along the magnetostatic field is obtained from Eq. (6.4.45) by adding the term given by (A1) to $\partial f_0(\mathbf{u})/\partial u_\rho$. The result is

$$k_z^2 c^2 = \omega^2 + \frac{\omega_{pe}^2 \pi}{N_0} \int_0^\infty du_\rho \int_{-\infty}^\infty du_z \frac{u_\rho^2}{(\omega - \omega_{ce} - k_z u_z)}\left[(\omega - k_z u_z)\frac{\partial f_0(\mathbf{u})}{\partial u_\rho} + k_z u_\rho \frac{\partial f_0(\mathbf{u})}{\partial u_z}\right] \tag{A2}$$

For $f_0(\mathbf{u}) = N_0(m_e/2\pi K T_e)^{3/2}\exp[-(m_e/2KT_e)\{u_\rho^2 + (u_z - v_d)^2\}]$, $\partial f_0(\mathbf{u})/\partial u_z = [(u_z - v_d)/u_\rho](\partial f_0(\mathbf{u})/\partial u_\rho)$. This expression is used to eliminate $\partial f_0(\mathbf{u})/\partial u_z$ from Eq. (A2) and an integration by parts is carried out with respect to u_ρ to obtain

$$k_z^2 c^2 = \omega^2 - \frac{\omega_{pe}^2}{N_0}\int_{-\infty}^\infty d\mathbf{u} \frac{(\omega - k_z v_d)}{(\omega - \omega_{ce} - k_z u_z)} f_0(\mathbf{u})$$

For $T_e = 0$, it follows that

$$f_0(\mathbf{u}) = N_0 \delta(u_x)\delta(u_y)\delta(u_z - v_d)$$

and the corresponding dispersion equation is

$$k_z^2 c^2 = \omega^2 - \omega_{pe}^2(\omega - k_z v_d)/(\omega - \omega_{ce} - k_z v_d)$$

6.15. As in Sec. 6.5, the magnetostatic field is assumed to be in the z-direction. The field quantities including the scalar potential $\Phi(\mathbf{r}, t)$ have the form given in Eqs. (6.5.1). Since $\mathbf{E}(\mathbf{r}, t) = -\nabla\Phi(\mathbf{r}, t)$, $E_x = -ik_x \Phi$ and $E_y = E_z = 0$. The phasor amplitude $g(\mathbf{u})$ of the perturbation $g(\mathbf{r}, \mathbf{u}, t)$ in the equilibrium distribution function $f_0(u)$ is therefore obtained from Eq. (6.5.6) as

$$g(\mathbf{u}) = -\frac{iek_x \Phi}{m_e \omega_{ce}} \exp\left(-\frac{ik_x u_\rho}{\omega_{ce}} \sin \varphi\right)\frac{\partial f_0(u)}{\partial u_\rho} I_1$$

where I_1 is defined in Eq. (6.5.10) and its value is deduced in Eq. (6.5.16) as

$$I_1 = -\frac{i}{\zeta}\exp(i\zeta \sin \varphi)\left[1 - \sum_{n=-\infty}^{\infty} \frac{(\omega/\omega_{ce} - n + n)}{(\omega/\omega_{ce} - n)} J_n(\zeta) e^{i(n\varphi - \zeta \sin \varphi)}\right]$$

where ζ is defined in Eq. (6.5.14). Since

$$\sum_{n=-\infty}^{\infty} J_n(\zeta) e^{i(n\varphi - \zeta \sin\varphi)} = 1$$

it follows that

$$I_1 = \frac{i}{\zeta}\exp(i\zeta \sin\varphi) \sum_{n=-\infty}^{\infty} \frac{nJ_n(\zeta)}{(\omega/\omega_{ce} - n)} e^{i(n\varphi - \zeta\sin\varphi)}$$

The charge density is obtained as

$$\rho = -e\int_{-\infty}^{\infty} d\mathbf{u}\, g(\mathbf{u}) = -\frac{e^2\Phi}{m_e}\int_0^{\infty} du_\rho \int_{-\infty}^{\infty} du_z \frac{\partial f_0(u)}{\partial u_\rho} I_2 \qquad \text{(A1)}$$

where

$$I_2 = \int_0^{2\pi} d\varphi \sum_{n=-\infty}^{\infty} \frac{nJ_n(\zeta)}{(\omega/\omega_{ce} - n)} e^{i(n\varphi - \zeta\sin\varphi)}$$

$$= 2\pi\frac{\omega_{ce}}{\omega}\sum_{n=-\infty}^{\infty} \frac{n(\omega/\omega_{ce} - n + n)}{(\omega/\omega_{ce} - n)} J_n^2(\zeta)$$

Since $\sum_{n=-\infty}^{\infty} n\, J_n^2(\zeta) = 0$, I_2 simplifies to

$$I_2 = 2\pi\frac{\omega_{ce}}{\omega}\sum_{n=-\infty}^{\infty} \frac{n^2 J_n^2(\zeta)}{(\omega/\omega_{ce} - n)} \qquad \text{(A2)}$$

If $f_0(u)$ as given by Eq. (6.5.26) and Eq. (A2) are substituted in Eq. (A1) and the integration with respect to u_z is carried out, the result is

$$\rho = \frac{k_x^2 N_0 e^2 \Phi}{m_e \omega_{ce} \omega \tilde{\nu}^2}\int_0^{\infty} d\zeta\, \zeta e^{-\zeta^2/2\tilde{\nu}} \sum_{n=-\infty}^{\infty} \frac{n^2 J_n^2(\zeta)}{(\omega/\omega_{ce} - n)} \qquad \text{(A3)}$$

where $\tilde{\nu}$ is defined in Eq. (6.5.27). The use of Eq. (6.5.29) enables Eq. (A3) to be written as

$$\rho = \frac{k_x^2 N_0 e^2 \Phi}{m_e \omega_{ce} \omega \tilde{\nu}} e^{-\tilde{\nu}} \sum_{n=-\infty}^{\infty} \frac{n^2 I_n(\tilde{\nu})}{(\omega/\omega_{ce} - n)} \qquad \text{(A4)}$$

Since $\nabla^2\Phi = -\rho/\varepsilon_0$ gives $k_x^2\Phi = \rho/\varepsilon_0$, Eq. (A4) can be used to obtain the following dispersion equation:

$$1 = \frac{\omega_{pe}^2}{\omega_{ce}\omega}\frac{e^{-\tilde{\nu}}}{\tilde{\nu}} \sum_{n=-\infty}^{\infty} \frac{n^2 I_n(\tilde{\nu})}{(\omega/\omega_{ce} - n)}$$

Note that this dispersion equation is identical to that given in Eq. (6.5.48).

6.16. In the equilibrium state, the Boltzmann-Vlasov equation satisfied by a homogeneous velocity distribution function reduces to

$$(\mathbf{u} \times \hat{\mathbf{z}}) \cdot \nabla_u f_0(\mathbf{u}) = 0 \qquad \text{(A1)}$$

wherein the uniform magnetostatic field has been assumed to be in the z-direction. Note that in the equilibrium state, the charged particle motions do not create any electromagnetic field. If the rectangular components of $\mathbf{u}$ and $\nabla_u f_0(\mathbf{u})$ are substituted in Eq. (A1), it is found that

$$(1/u_x)\partial f_0(\mathbf{u})/\partial u_x = (1/u_y)\partial f_0(\mathbf{u})/\partial u_y$$

which shows that the dependence of $f_0(\mathbf{u})$ on u_x and u_y occurs only in the form $(u_x^2 + u_y^2) = u_\rho^2$. Hence $f_0(\mathbf{u}) = f_0(u_\rho, u_z)$ which is cylindrically symmetric about the direction of the magnetostatic field.

6.17. The phase factor $\exp\{i(kmx + klz - \omega t)\}$ is assumed for all the field quantities where $m = \sin\theta$ and $l = \cos\theta$. The magnetostatic field is in the z-direction. The Boltzmann-Vlasov equation yields the following differential equation for the phasor amplitude $g(\mathbf{u})$ of the perturbation $g(\mathbf{r}, \mathbf{u}, t)$ of the equilibrium distribution function $f_0(u)$:

$$\frac{dg(\mathbf{u})}{d\varphi} - \frac{i(\omega - kmu_x - klu_z)}{\omega_{ce}} g(\mathbf{u}) = \frac{e}{m_e\omega_{ce}} \mathbf{E} \cdot \nabla_u f_0(u) \tag{A1}$$

Since $E_x = -ikm\Phi$, $E_y = 0$, and $E_z = -ikl\Phi$ from the electrostatic equation $\mathbf{E}(\mathbf{r}, t) = -\nabla\Phi(\mathbf{r}, t)$, it follows that

$$\mathbf{E} \cdot \nabla_u f_0(u) = -ik\Phi\left[m\cos\varphi\frac{\partial f_0(u)}{\partial u_\rho} + l\frac{\partial f_0(u)}{\partial u_z}\right] \tag{A2}$$

Following the procedure used in Sec. 6.5, the solution of Eq. (A1) can be deduced to be given by

$$g(\mathbf{u}) = -\frac{ike\Phi}{m_e\omega_{ce}} \exp\left\{-\frac{ikmu_\rho \sin\varphi}{\omega_{ce}}\right\}\left[I_1 m\frac{\partial f_0(u)}{\partial u_\rho} + I_1' l\frac{\partial f_0(u)}{\partial u_z}\right]$$

where

$$\begin{aligned} I_1 &= \int_0^\infty d\varphi' \cos(\varphi - \varphi')\exp\left[\frac{i(\omega - klu_z)\varphi'}{\omega_{ce}} + \frac{ikmu_\rho\sin(\varphi - \varphi')}{\omega_{ce}}\right] \\ &= -\frac{i}{\zeta}\exp(i\zeta\sin\varphi) + \frac{i(\omega - klu_z)}{\zeta\omega_{ce}} \sum_{n=-\infty}^{\infty} \frac{J_n(\zeta)e^{in\varphi}}{[(\omega - klu_z)/\omega_{ce} - n]} \\ &= \frac{i}{\zeta}\sum_{n=-\infty}^{\infty} \frac{nJ_n(\zeta)e^{in\varphi}}{[(\omega - klu_z)/\omega_{ce} - n]} \\ I_1' &= \int_0^\infty d\varphi' \exp\left[\frac{i(\omega - klu_z)\varphi'}{\omega_{ce}} + \frac{ikmu_\rho\sin(\varphi - \varphi')}{\omega_{ce}}\right] \\ &= i\sum_{n=-\infty}^{\infty} \frac{J_n(\zeta)e^{in\varphi}}{[(\omega - klu_z)/\omega_{ce} - n]} \end{aligned}$$

and

$$\zeta = kmu_\rho/\omega_{ce}$$

The charge density is obtained as

$$\rho = -e\int_{-\infty}^{\infty} d\mathbf{u}\, g(\mathbf{u}) = -\frac{e^2\Phi}{m_e}\int_0^{\infty} du_\rho \int_{-\infty}^{\infty} du_z\left[I_2\frac{\partial f_0(u)}{\partial u_\rho} + I_2'\frac{\partial f_0(u)}{\partial u_z}\right] \quad \text{(A3)}$$

where

$$I_2 = \int_0^{2\pi} d\varphi \sum_{n=-\infty}^{\infty} \frac{nJ_n(\zeta)e^{i(n\varphi-\zeta\sin\varphi)}}{[(\omega - klu_z)/\omega_{ce} - n]} = 2\pi \sum_{n=-\infty}^{\infty} \frac{nJ_n^2(\zeta)}{[(\omega - klu_z)/\omega_{ce} - n]} \quad \text{(A4)}$$

and

$$\begin{aligned} I_2' &= \frac{klu_\rho}{\omega_{ce}}\int_0^{2\pi} d\varphi \sum_{n=-\infty}^{\infty} \frac{J_n(\zeta)e^{i(n\varphi-\zeta\sin\varphi)}}{[(\omega - klu_z)/\omega_{ce} - n]} \\ &= \frac{2\pi klu_\rho}{\omega_{ce}} \sum_{n=-\infty}^{\infty} \frac{J_n^2(\zeta)}{[(\omega - klu_z)/\omega_{ce} - n]} \end{aligned} \quad \text{(A5)}$$

The distribution function $f_0(u)$ as given by Eq. (6.5.26) and Eqs. (A4) and (A5) are substituted in Eq. (A3) and the variable of integration is changed from u_ρ to ζ. With the help of Eq. (6.5.29), the integration with respect to ζ is carried out. Then, the following substitutions are made:

$$\tilde{\nu} = \frac{KT_e}{m_e}\frac{k^2m^2}{\omega_{ce}^2} \qquad a_e = \sqrt{\frac{2KT_e}{m_e}}$$

$$z = u_z/a_e \qquad \tilde{\omega} = \omega/kla_e \qquad \tilde{\omega}_{ce} = \omega_{ce}/kla_e$$

The result is

$$\rho/\varepsilon_0 k^2\Phi = -\frac{\omega_{pe}^2 m^2}{\omega_{ce}^2}\frac{e^{-\tilde{\nu}}}{\tilde{\nu}}\sum_{n=-\infty}^{\infty} I_n(\tilde{\nu})\frac{1}{\sqrt{\pi}}\int_{-\infty}^{\infty} dz\, e^{-z^2}\frac{[z + n\tilde{\omega}_{ce} - \tilde{\omega} + \tilde{\omega}]}{[z + n\tilde{\omega}_{ce} - \tilde{\omega}]} \quad \text{(A6)}$$

The electrostatic equation $\nabla^2\Phi = -\rho/\varepsilon_0$ gives $k^2\Phi = \rho/\varepsilon_0$ which together with Eq. (A6) results in the following dispersion equation:

$$\frac{\omega_{ce}^2\tilde{\nu}}{\omega_{pe}^2 m^2} = \frac{KT_e}{m_e}\frac{k^2}{\omega_{pe}^2} = -e^{-\tilde{\nu}}\sum_{n=-\infty}^{\infty} I_n(\tilde{\nu})[1 + \tilde{\omega}I_3] \quad \text{(A7)}$$

where

$$I_3 = \frac{1}{\sqrt{\pi}}\int_{-\infty}^{\infty} dz\, e^{-z^2}\frac{1}{[z + n\tilde{\omega}_{ce} - \tilde{\omega}]} \quad \text{(A8)}$$

Since $e^{-\tilde{\nu}} \sum_{n=-\infty}^{\infty} I_n(\tilde{\nu}) = 1$, Eq. (A7) becomes

$$1 + \frac{KT_e}{m_e} \frac{k^2}{\omega_{pe}^2} = -\tilde{\omega} e^{-\tilde{\nu}} \sum_{n=-\infty}^{\infty} I_n(\tilde{\nu}) I_3 \tag{A9}$$

Following the procedure used in Problem 6.1, it can be shown that

$$I_3 = i \int_0^{\infty} dt\, e^{i(\tilde{\omega}-n\tilde{\omega}_{ce})t - t^2/4}$$

Also, since

$$e^{\tilde{\nu}\cos(\tilde{\omega}_{ce}t)} = \sum_{n=-\infty}^{\infty} I_n(\tilde{\nu}) e^{-in\tilde{\omega}_{ce}t}$$

Eq. (A9) becomes

$$1 + \frac{KT_e}{m_e} \frac{k^2}{\omega_{pe}^2} = -i\tilde{\omega} \int_0^{\infty} dt_1\, e^{i\tilde{\omega}t_1 - \tilde{\nu}\{1-\cos(\tilde{\omega}_{ce}t_1)\} - t_1^2/4} \tag{A10}$$

The variable of integration is changed to $t = t_1/kla_e$ with the following result:

$$1 + \frac{KT_e}{m_e} \frac{k^2}{\omega_{pe}^2} = -i\omega \int_0^{\infty} dt\, e^{i\omega t - \{1-\cos\omega_{ce}t\}k^2a_e^2m^2/2\omega_{ce}^2 - k^2a_e^2l^2t^2/4} \tag{A11}$$

which is the desired dispersion equation.

For the case of a very weak magnetostatic field, the following approximations are valid:

$$\{1 - \cos \omega_{ce} t\} k^2 a_e^2 m^2/2\omega_{ce}^2 + k^2 a_e^2 l^2 t^2/4 = k^2 a_e^2 t^2/4 - k^2 a_e^2 \omega_{ce}^2 m^2 t^4/48$$

and

$$\exp[-\{1 - \cos \omega_{ce} t\} k^2 a_e^2 m^2/2\omega_{ce}^2 - k^2 a_e^2 l^2 t^2/4]$$
$$= 1 - k^2 a_e^2 t^2/4 + k^2 a_e^2 \omega_{ce}^2 m^2 t^4/48 + k^4 a_e^4 t^4/32$$

If these approximate values are substituted in Eq. (A11) and the integration is carried out term by term using the relations

$$\int_0^{\infty} dt\, e^{i\omega t} = i/\omega \qquad \int_0^{\infty} dt\, t^2 e^{i\omega t} = -2i/\omega^3 \qquad \int_0^{\infty} dt\, t^4 e^{i\omega t} = 24i/\omega^5$$

it is obtained that

$$\frac{KT_e}{m_e} \frac{k^2}{\omega_{pe}^2} = \frac{k^2 a_e^2}{2\omega^2} + \frac{1}{\omega^4} \left\{ \frac{3k^4 a_e^4}{4} + \frac{k^2 a_e^2 \omega_{ce}^2 m^2}{2} \right\} \tag{A12}$$

In the first approximation, the small second term on the right side of Eq. (A12) is omitted leading to $\omega^2 = \omega_{pe}^2$ and therefore ω^2 is replaced by ω_{pe}^2 in the small second term. Then Eq. (A12) after some rearrangement becomes

$$\omega^2 = \omega_{pe}^2 + \frac{3KT_e}{m_e}k^2 + \omega_{ce}^2 \sin^2\theta \tag{A13}$$

Note that ω^2 given by Eq. (A13) attains the correct limiting form both for $\omega_{ce} = 0$ and for $\theta = 0$.

6.18. If $f_0(u)$ and $g(\mathbf{r}, \mathbf{u}, t)$ are, respectively, the equilibrium value and the perturbation in the distribution function, then

$$\frac{\partial}{\partial t}g(\mathbf{r}, \mathbf{u}, t) + (\mathbf{u} \cdot \nabla_r)g(\mathbf{r}, \mathbf{u}, t) - \frac{e}{m_e}\mathbf{E}(\mathbf{r}, t) \cdot \nabla_u f_0(u) = -\nu_r(u)g(\mathbf{r}, \mathbf{u}, t)$$

Consider a plane wave with the phase factor $\exp\{i(k_p x - \omega t)\}$. The amplitude $g(\mathbf{u})$ of $g(\mathbf{r}, \mathbf{u}, t)$ is then deduced as

$$g(\mathbf{u}) = -\frac{ie}{k_p m_e}\frac{E_x(\partial f_0(u)/\partial u_x)}{\{u_x - [\omega + i\nu_r(u)]/k_p\}}$$

Since from $\nabla \cdot \mathbf{E} = -Ne/\varepsilon_0$, $E_x = iNe/\varepsilon_0 k_p$ and since the perturbation in the number density is given by $N = \int_{-\infty}^{\infty} g(\mathbf{u})\, d\mathbf{u}$, the following dispersion relation is obtained:

$$k_p^2 = \frac{\omega_{pe}^2}{N_0}\int_{-\infty}^{\infty} d\mathbf{u} \frac{[\partial f_0(u)/\partial u_x]}{\{u_x - [\omega + i\nu_r(u)]/k_p\}}$$

In the complex u_x-plane, the pole is in the upper half plane and the contour is along the real axis. In the limit of $\nu_r(u)$ tending to zero, the pole is on the real axis and the contour is again on the real axis except for the identation from below at the pole This contour is identical to that deduced from the Boltzmann-Vlasov equation using the causality considerations. The dispersion equation obtained here is seen from Eqs. (6.2.43) and (6.2.53) to be the same as that given by Eq. (6.2.50).

6.19. If ω is the characteristic angular frequency of variation in N_e, it follows that $\omega/\nu_{ce} \ll 1$ since $1/\nu_{ce}$ is the average period between two successive collisions of the electrons with the neutral particles. In the linearized momentum transport equation

$$m_e N_{e0}\, \partial \mathbf{v}_e/\partial t = -KT_e \nabla N_e - m_e N_{e0} \nu_{ce} \mathbf{v}_e$$

since $|m_e N_{e0}\, \partial \mathbf{v}_e/\partial t|/|m_e N_{e0} \nu_{ce} \mathbf{v}_e| = \omega/\nu_{ce} \ll 1$, the left side can be neglected with the result that $\mathbf{v}_e = -(KT_e/m_e N_{e0} \nu_{ce})\nabla N_e$. When $\mathbf{v}_e$ is substituted into the linearized equation of continuity: $\partial N_e/\partial t = -N_{e0} \nabla \cdot \mathbf{v}_e$, the diffusion equation: $\partial N_e/\partial t = D_e \nabla^2 N_e$ is obtained.

6.20. Any spatial derivative is of order $1/\Lambda$. Therefore, $\nabla \cdot \mathbf{E}_s = -Ne/\varepsilon_0$ leads to $|\mathbf{E}_s| \approx Ne\Lambda/\varepsilon_0$. Hence, $|\mu_e N_e \mathbf{E}_s|/|D_e \nabla N_e| = |\mu_e| N_0 |\mathbf{E}_s| \Lambda / D_e N = N_0 e^2 \Lambda^2 / KT_e \varepsilon_0 = \Lambda^2/\lambda_{De}^2 \ll 1$ where N_0 is the ambient number density and N is the perturbation part of N_e. If $\Lambda^2 \ll \lambda_{De}^2$, the effect of the electric field due to the space charge can be neglected and the equation of free diffusion results.

6.21. In the absence of any time dependence, the momentum transport equation for the electrons can be written as

$$\boldsymbol{\Gamma}_e = -\nabla(D_{e\|} N_e) + \mu_{e\|} N_e \mathbf{E}_s - (\omega_{ce}/\nu_{ce})\boldsymbol{\Gamma}_e \times \hat{\mathbf{z}} \tag{A1}$$

This equation is written in component form and the resulting equations are solved for Γ_{ex}, Γ_{ey}, and Γ_{ez}. The solution can be expressed compactly in the following dyadic notation:

$$\boldsymbol{\Gamma}_e = -\nabla \cdot (\mathbf{D}_{fe} N_e) + N_e \boldsymbol{\mu}_e \cdot \mathbf{E}_s \tag{A2}$$

In a similar manner, in the absence of any time dependence, the momentum transport equation for the ions can be written as

$$\boldsymbol{\Gamma}_i = -\nabla(D_{i\|} N_i) + \mu_{i\|} N_i \mathbf{E}_s + (\omega_{ci}/\nu_{ci})\boldsymbol{\Gamma}_i \times \hat{\mathbf{z}} \tag{A3}$$

where $D_{i\|} = KT_i/m_i\nu_{ci}$, $\mu_{i\|} = e/m_i\nu_{ci}$, and $\omega_{ci} = eB_0/m_i$. The momentum transport equation is written in component form and the resulting equations solved for Γ_{ix}, Γ_{iy}, and Γ_{iz}. As before, the solution can be expressed compactly as

$$\boldsymbol{\Gamma}_i = -\nabla \cdot (\mathbf{D}_{fi} N_i) + N_i \boldsymbol{\mu}_i \cdot \mathbf{E}_s \tag{A4}$$

where

$$\mathbf{D}_{fi} = (\hat{\mathbf{x}}\hat{\mathbf{x}} + \hat{\mathbf{y}}\hat{\mathbf{y}})D_{i\perp} - (\hat{\mathbf{x}}\hat{\mathbf{y}} - \hat{\mathbf{y}}\hat{\mathbf{x}})D_{iH} + \hat{\mathbf{z}}\hat{\mathbf{z}}D_{i\|}$$

$$\boldsymbol{\mu}_i = (\hat{\mathbf{x}}\hat{\mathbf{x}} + \hat{\mathbf{y}}\hat{\mathbf{y}})\mu_{i\perp} + (\hat{\mathbf{x}}\hat{\mathbf{y}} - \hat{\mathbf{y}}\hat{\mathbf{x}})\mu_{iH} + \hat{\mathbf{z}}\hat{\mathbf{z}}\mu_{i\|}$$

$$D_{i\perp} = KT_i\nu_{ci}/m_i(\nu_{ci}^2 + \omega_{ci}^2) \qquad D_{iH} = KT_i\omega_{ci}/m_i(\nu_{ci}^2 + \omega_{ci}^2)$$

$$\mu_{i\perp} = e\nu_{ci}/m_i(\nu_{ci}^2 + \omega_{ci}^2) \qquad \mu_{iH} = e\omega_{ci}/m_i(\nu_{ci}^2 + \omega_{ci}^2)$$

Let $\boldsymbol{\Gamma}_e = \boldsymbol{\Gamma}_i = \boldsymbol{\Gamma}$ and $N_e = N_i = N$. If the z-components of Eqs. (A2) and (A4) are written down and if E_{sz} is eliminated from these two equations, the result is

$$\Gamma_z = -\frac{\partial}{\partial z}(D_{a\|} N) \quad \text{where } D_{a\|} = (\mu_{i\|} D_{e\|} - \mu_{e\|} D_{i\|})/(\mu_{i\|} - \mu_{e\|}) \tag{A5}$$

From Eqs. (6.8.1b), (6.8.2b), (6.8.4), (6.8.6), and (6.8.18), it is seen that the ambipolar diffusion coefficient in the direction of the B_0-field is the same as that when no B_0-field is present.

If the y-component of Eq. (A2) is multiplied by $\pm i$ and added on to the x-component, it is found that

$$\Gamma_{\mp} = -L_{\mp}(D_{e\mp} N) + N\mu_{e\mp} E_{\mp} \tag{A6}$$

where $\Gamma_{\mp} = \Gamma_x \pm i\Gamma_y$, $L_{\mp} = \partial/\partial x \pm i\partial/\partial y$, $D_{e\mp} = D_{e\perp} \pm iD_{eH}$, $\mu_{e\mp} = \mu_{e\perp} \pm i\mu_{eH}$, and $E_{\mp} = E_{sx} \pm iE_{sy}$. In a similar manner, it can be derived from Eq. (A4) that

$$\Gamma_{\mp} = -L_{\mp}(D_{i\pm} N) + N\mu_{i\pm} E_{\mp} \tag{A7}$$

where $D_{i\mp} = D_{i\perp} \pm iD_{iH}$ and $\mu_{i\mp} = \mu_{i\perp} \pm i\mu_{iH}$. The elimination of $E_{\mp}$ from Eqs. (A6) and (A7) yields

$$\Gamma_{\mp} = -L_{\mp}(D_{a\mp} N) \tag{A8}$$

where

$$D_{a\mp} = (\mu_{i\pm} D_{e\mp} - \mu_{e\mp} D_{i\pm})/(\mu_{i\pm} - \mu_{e\mp}) \tag{A9}$$

From Eq. (A8) it can be deduced that

$$\Gamma_x = -\frac{\partial}{\partial x}(D_{a\perp} N) + \frac{\partial}{\partial y}(D_{aH} N) \tag{A10}$$

$$\Gamma_y = -\frac{\partial}{\partial x}(D_{aH} N) - \frac{\partial}{\partial y}(D_{a\perp} N) \tag{A11}$$

$$D_{a\perp} = (D_{a-} + D_{a+})/2 \quad \text{and} \quad D_{aH} = (D_{a-} - D_{a+})/2i$$

Since $\mu_{e\mp} = -e/m_e(\nu_{ce} \mp i\omega_{ce})$, $\mu_{i\mp} = e/m_i(\nu_{ci} \mp i\omega_{ci})$, $D_{e\mp} = KT_e/m_e(\nu_{ce} \mp i\omega_{ce})$, and $D_{i\mp} = KT_i/m_i(\nu_{ci} \mp i\omega_{ci})$, it follows that $D_{a-} = D_{a+} = D_{a\|} = D_{a\perp} = D_a$ and $D_{aH} = 0$ with the result that $\boldsymbol{\Gamma} = -\boldsymbol{\nabla}(D_a N)$.

This result can also be deduced in a simpler manner. If Eq. (A3) is multiplied by $\mu_{e\|}$ and is then subtracted from Eq. (A1) after it is multiplied by $\mu_{i\|}$, it is obtained that $\boldsymbol{\Gamma} = -\boldsymbol{\nabla}(D_a N)$ since $\omega_{ce}\mu_{i\|}/\nu_{ce} + \omega_{ci}\mu_{e\|}/\nu_{ci} = 0$. This result should be expected from physical considerations. The electric current is zero for the case of perfect ambipolar diffusion and hence the magnetostatic field has no effect.

6.22. Let $f(\mathbf{r}, \mathbf{u}, t) = f_0(\mathbf{r}, u) + f_1(\mathbf{r}, \mathbf{u})$, where $f_1(\mathbf{r}, \mathbf{u}) \ll f_0(\mathbf{r}, u)$. In the absence of the time variation and the electromagnetic field, the Boltzmann equation with a relaxation model for the collision term yields $f_1(\mathbf{r}, \mathbf{u}) = -\mathbf{u} \cdot \boldsymbol{\nabla}_r f_0(\mathbf{r}, u)/\nu_c$. Since $P_e = N_e(\mathbf{r})KT_e(\mathbf{r})$ is a constant, $-T_e(\mathbf{r})\boldsymbol{\nabla} N_e(\mathbf{r}) = N_e(\mathbf{r})\boldsymbol{\nabla} T_e(\mathbf{r})$. Hence $\boldsymbol{\nabla}_r f_0(\mathbf{r}, u) = f_0(\mathbf{r}, u)[-5/2 + (m_e/2KT_e(\mathbf{r}))u^2]\boldsymbol{\nabla} T_e(\mathbf{r})/T_e(\mathbf{r})$. The substitution of $f_1(\mathbf{r}, \mathbf{u})$ and $\boldsymbol{\nabla}_r f_0(\mathbf{r}, u)$ in the expression for $\mathbf{q}$ yields

$$\mathbf{q} = -\frac{m_e}{2\nu_c}\int_{-\infty}^{\infty} d\mathbf{u}\, u^2 \mathbf{u}\mathbf{u} \cdot f_0(\mathbf{r}, u)\left[-\frac{5}{2} + \left(\frac{m_e}{2KT_e(\mathbf{r})}\right)u^2\right]\frac{\boldsymbol{\nabla} T_e(\mathbf{r})}{T_e(\mathbf{r})}$$

Since $f_0(\mathbf{r}, u)$ is even in u_x, u_y, and u_z, all the offdiagonal terms vanish on integration and moreover since $f_0(\mathbf{r}, u)$ is isotropic, it follows that

$$\mathbf{q} = -\frac{m_e}{6\nu_c}\int_{-\infty}^{\infty} d\mathbf{u}\, u^4 f_0(\mathbf{r}, u)\left[-\frac{5}{2} + \left(\frac{m_e}{2KT_e(\mathbf{r})}\right)u^2\right]\frac{\boldsymbol{\nabla} T_e(\mathbf{r})}{T_e(\mathbf{r})}$$

It can be verified that

$$\int_{-\infty}^{\infty} d\mathbf{u}\, u^4 f_0(\mathbf{r}, u) = 4\pi \int_0^{\infty} du\, u^6 f_0(\mathbf{r}, u) = 15KT_e(\mathbf{r})P_e/m_e^2$$

and

$$\int_{-\infty}^{\infty} d\mathbf{u}\, u^6 f_0(\mathbf{r}, u) = 4\pi \int_0^{\infty} du\, u^8 f_0(\mathbf{r}, u) = 105 K^2 T_e^2(\mathbf{r}) P_e / m_e^3$$

When the values of these integrals are substituted into the expression for $\mathbf{q}$, it is found that $\mathbf{q} = -\mathcal{K} \nabla T_e(\mathbf{r})$ where $\mathcal{K} = 5KP_e/2m_e\nu_c$.

6.23. Let $f(\mathbf{r}, \mathbf{u}, t) = f_0(\mathbf{r}, u) + f_1(\mathbf{r}, \mathbf{u})$ where $f_1(\mathbf{r}, \mathbf{u}) \ll f_0(\mathbf{r}, u)$. Since $-(\mathbf{u} \times \hat{\mathbf{z}}) \cdot \nabla_u f_1(\mathbf{r}, \mathbf{u}) = (d/d\varphi) f_1(\mathbf{r}, \mathbf{u})$, in the absence of the time variation and in the presence of the magnetostatic field in the z-direction, the Boltzmann equation with a relaxation model for the collision term gives

$$\frac{d}{d\varphi} f_1(\mathbf{r}, \mathbf{u}) + \frac{\nu_c}{\omega_{ce}} f_1(\mathbf{r}, \mathbf{u}) = -\frac{1}{\omega_{ce}} \mathbf{u} \cdot \nabla_r f_0(\mathbf{r}, u)$$

whose solution is

$$f_1(\mathbf{r}, \mathbf{u}) = -e^{-\nu_c \varphi/\omega_{ce}} \int_{-\infty}^{\varphi} d\varphi' e^{\nu_c \varphi'/\omega_{ce}} (\mathbf{u} \cdot \nabla_r) f_0(\mathbf{r}, u)/\omega_{ce} \tag{A1}$$

It can be shown that

$$(\mathbf{u} \cdot \nabla_r) f_0(\mathbf{r}, u) = f_0(\mathbf{r}, u) \left[-\frac{5}{2} + \left(\frac{m_e}{2KT_e(\mathbf{r})} \right) u^2 \right] \left\{ u_\rho \cos\varphi \frac{1}{T_e(\mathbf{r})} \frac{\partial}{\partial x} T_e(\mathbf{r}) + u_\rho \sin\varphi \frac{1}{T_e(\mathbf{r})} \frac{\partial}{\partial y} T_e(\mathbf{r}) + u_z \frac{1}{T_e(\mathbf{r})} \frac{\partial}{\partial z} T_e(\mathbf{r}) \right\} \tag{A2}$$

If Eq. (A2) is substituted into Eq. (A1) and the integration with respect to φ' is carried out, the value of $f_1(\mathbf{r}, \mathbf{u})$ is obtained. When this value of $f_1(\mathbf{r}, \mathbf{u})$ is used in the expression for q_x, it is found that

$$\begin{aligned} q_x = {} & \frac{1}{2} m_e N_e(\mathbf{r}) \left(\frac{m_e}{2\pi K T_e(\mathbf{r})} \right)^{3/2} \int_0^{\infty} du_\rho\, u_\rho \int_{-\infty}^{\infty} du_z \int_0^{2\pi} d\varphi\, (u_\rho^2 + u_z^2) \\ & \times u_\rho \cos\varphi \exp\left\{ -\frac{m_e(u_\rho^2 + u_z^2)}{2KT_e(\mathbf{r})} \right\} \left[-\frac{5}{2} + \left(\frac{m_e}{2KT_e(\mathbf{r})} \right) (u_\rho^2 + u_z^2) \right] \\ & \left[\frac{u_\rho}{\nu_c^2 + \omega_{ce}^2} \left\{ -(\omega_{ce} \sin\varphi + \nu_c \cos\varphi) \frac{1}{T_e(\mathbf{r})} \frac{\partial}{\partial x} T_e(\mathbf{r}) \right. \right. \\ & \left. \left. + (\omega_{ce} \cos\varphi - \nu_c \sin\varphi) \frac{1}{T_e(\mathbf{r})} \frac{\partial}{\partial y} T_e(\mathbf{r}) \right\} - \frac{u_z}{\nu_c} \frac{1}{T_e(\mathbf{r})} \frac{\partial}{\partial z} T_e(\mathbf{r}) \right] \end{aligned}$$

The integration with respect to φ is carried out first and the remaining integrals with respect to u_ρ and u_z are evaluated in a straightforward manner with the following result:

$$q_x = -\mathcal{K}_\perp (\partial/\partial x) T_e(\mathbf{r}) + \mathcal{K}_H (\partial/\partial y) T_e(\mathbf{r}) \tag{A3}$$

where

$$\mathcal{K}_\perp = \frac{5}{2}\frac{\nu_c}{\nu_c^2+\omega_{ce}^2}\frac{KP_e}{m_e} \quad \text{and} \quad \mathcal{K}_H = \frac{5}{2}\frac{\omega_{ce}}{\nu_c^2+\omega_{ce}^2}\frac{KP_e}{m_e} \tag{A4}$$

In a similar manner, it can be shown that

$$q_y = -\mathcal{K}_H\frac{\partial}{\partial x}T_e(\mathbf{r}) - \mathcal{K}_\perp\frac{\partial}{\partial y}T_e(\mathbf{r}) \quad \text{and} \quad q_z = -\mathcal{K}_\parallel\frac{\partial}{\partial z}T_e(\mathbf{r}) \tag{A5}$$

where

$$\mathcal{K}_\parallel = \tfrac{5}{2}KP_e/\nu_c m_e \tag{A6}$$

It is possible to write Eqs. (A3) and (A5) compactly using the dyadic notation as: $\mathbf{q} = -\boldsymbol{\mathcal{K}}\cdot\nabla T_e(\mathbf{r})$. Note that the thermal conductivity in the parallel direction is unaffected by the magnetostatic field.

6.24. Let $f(\mathbf{r},\mathbf{u}) = f_0(\mathbf{r},\mathbf{u}) + f_1(\mathbf{r},\mathbf{u})$, where $f_1(\mathbf{r},\mathbf{u}) \ll f_0(\mathbf{r},\mathbf{u})$. In the absence of the time variation and the electromagnetic field, the Boltzmann equation with a relaxation model for the collision term leads to

$$f_1(\mathbf{r},\mathbf{u}) = \frac{-(\mathbf{u}\cdot\nabla_r)f_0(\mathbf{r},\mathbf{u})}{\nu_c} = -\frac{2f_0(\mathbf{r},\mathbf{u})}{\nu_c}\left(\frac{m_e}{2KT_e}\right)u_z\{u_x - v_x(z)\}\frac{\partial}{\partial z}v_x(z)$$

Since $f_0(\mathbf{r},\mathbf{u})$ is even in u_z, it can be shown that

$$\begin{aligned}
P_{xz} &= \int_{-\infty}^{\infty} du\, m_e u_z\{u_x - v_x(z)\}f(\mathbf{r},\mathbf{u}) = \int_{-\infty}^{\infty} du\, m_e u_z\{u_x - v_x(z)\}f_1(\mathbf{r},\mathbf{u}) \\
&= -\frac{\pi m_e N_e}{\nu_c}\left(\frac{m_e}{2\pi KT_e}\right)^{5/2} 2\frac{\partial}{\partial x}v_x(z)\int_{-\infty}^{\infty} dw_x \int_{-\infty}^{\infty} du_y \int_{-\infty}^{\infty} du_z\, w_x^2 u_z^2 \\
&\quad \times \exp\left\{-\frac{m_e(w_x^2+u_y^2+u_z^2)}{2KT_e}\right\} \\
&= -\frac{N_e KT_e}{\nu_c}\frac{\partial}{\partial z}v_x(z)
\end{aligned}$$

Hence $P_{xz} = -\eta(\partial/\partial z)v_x(z)$ where the coefficient of viscosity $\eta = N_e KT_e/\nu_c$.

6.25. From Eq. (6.9.15) it is found that

$$\{\omega + i\nu_{eff}(\omega)\}^{-1} = -\frac{4\pi}{3N_0}\int_0^{\infty} du\{\omega + i\nu_r(u)\}^{-1}u^3\frac{\partial f_0(u)}{\partial u}$$

In the low-frequency limit $\omega + i\nu_{eff}(\omega) \approx i\nu_{eff}(\omega)$ and $\omega + i\nu_r(u) \approx i\nu_r(u)$; hence it is obtained that

$$\frac{1}{\nu_{eff}(\omega)} = -\frac{4\pi}{3N_0}\int_0^{\infty}\frac{du}{\nu_r(u)}u^3\frac{\partial f_0(u)}{\partial u} \tag{A1}$$

For the high-frequency limit, $\{\omega + i\nu_{eff}(\omega)\}^{-1} \approx 1/\omega - i\nu_{eff}(\omega)/\omega^2$ and $\{\omega + i\nu_r(u)\}^{-1} = 1/\omega - i\nu_r(u)/\omega^2$. Also, since $-(4\pi/3N_0)\int_0^\infty du\, u^3(\partial f_0(u)/\partial u) = 1$, it is found that

$$\nu_{eff}(\omega) = -\frac{4\pi}{3N_0}\int_0^\infty du\, \nu_r(u)u^3 \frac{\partial f_0(u)}{\partial u} \tag{A2}$$

It follows from Eqs. (A1) and (A2) that $\nu_{eff}(\omega)$ is independent of ω in the low- and the high-frequency limits.

6.26. From Eq. (A2) of Problem 6.25, it is seen that

$$\nu_{eff} = -\frac{4\pi\nu_c}{3N_0}\int_0^\infty du\, u^{n+3}\frac{\partial f_0(u)}{\partial u}$$

which by an integration by parts can be reduced to

$$\nu_{eff} = \frac{4\pi(n+3)\nu_c}{3N_0}\int_0^\infty du\, u^{n+2} f_0(u)$$

If $f_0(u)$ is substituted and the integration variable is changed to $mu^2/2KT = t$, the result is

$$\nu_{eff} = \frac{2\nu_c}{3\sqrt{\pi}}(n+3)\left(\frac{2KT}{m}\right)^{n/2}\int_0^\infty t^{(n+3)/2-1} e^{-t}\, dt$$

Since $\Gamma(z) = \int_0^\infty t^{z-1}e^{-t}\, dt$ where $\Gamma(z)$ is the gamma function and since $z\Gamma(z) = \Gamma(z+1)$, it follows that

$$\nu_{eff} = \frac{4\nu_c}{3\sqrt{\pi}}\left(\frac{2KT}{m}\right)^{n/2}\Gamma\left(\frac{n+5}{2}\right)$$

Note that $\Gamma(1/2) = \sqrt{\pi}$ and if n is an integer $\Gamma(n+1) = n!$. With these relations ν_{eff} can be evaluated for any n.

In a similar manner, it can be shown that

$$\langle \nu_r(u)\rangle = \frac{4\pi}{N_0}\int_0^\infty du\, \nu_r(u)u^2 f_0(u) = \frac{2\nu_c}{\sqrt{\pi}}\left(\frac{2KT}{m}\right)^{n/2}\Gamma\left(\frac{n+3}{2}\right)$$

Therefore

$$\nu_{eff}/\langle \nu_r(u)\rangle = 1 + n/3$$

which shows that for n not too large ν_{eff} and $\langle \nu_r(u)\rangle$ are of the same order of magnitude.

6.27. From the Fokker-Planck equation, it is found approximately that $\nu_r(u)$ is proportional to u corresponding to $n = 1$ in Problem 6.26. Therefore, the effective collision frequency behaves as $\nu_{eff} \propto 1/\sqrt{m}$ where m is the particle mass. Since the coefficient of viscosity behaves as $\eta \propto 1/\nu \propto \sqrt{m}$, the viscosity is contributed principally by the ions. Since the diffusion coefficient and the thermal conductivity behave as $1/m\nu$, that is $1/\sqrt{m}$, it follows that diffusion and thermal conduction are contributed mainly by the electrons. The high-freqency electrical conductivity behaves as $1/m$ and the direct current conductivity as $1/\nu m \propto 1/\sqrt{m}$. Hence the contribution to the electrical conduction is dominated by the electrons.

6.28. The equation of detailed balance $\tilde{f}\tilde{f}^2 = ff^2$ is satisfied only if

$$A = (\tilde{\mathbf{u}} - \mathbf{v})^2 + (\tilde{\mathbf{u}}^{(2)} - \mathbf{v})^2 - (\mathbf{u} - \mathbf{v})^2 - (\mathbf{u}^{(2)} - \mathbf{v})^2$$

vanishes identically. Note that $\mathbf{u}$ and $\tilde{\mathbf{u}}$ are the initial and the final velocities of one particle, and, $\mathbf{u}^{(2)}$ and $\tilde{\mathbf{u}}^{(2)}$ are the corresponding quantities of the colliding particle. Since the particles are of equal mass, the equations of conservation of momentum and energy become

$$\mathbf{u} + \mathbf{u}^{(2)} = \tilde{\mathbf{u}} + \tilde{\mathbf{u}}^{(2)} \tag{A1}$$

and

$$u^2 + \{u^{(2)}\}^2 = \tilde{u}^2 + \{\tilde{u}^{(2)}\}^2 \tag{A2}$$

If $2v^2$ is added to both sides of Eq. (A2) and each side of the resulting equation is added on to the corresponding side of Eq. (A1) after Eq. (A1) is scalarly multiplied by $-2\mathbf{v}$, it is found on some rearrangement that $(\mathbf{u} - \mathbf{v})^2 + (\mathbf{u}^{(2)} - \mathbf{v})^2 = (\tilde{\mathbf{u}} - \mathbf{v})^2 + (\tilde{\mathbf{u}}^{(2)} - \mathbf{v})^2$. Therefore, $A \equiv 0$ and hence the equation of detailed balance is satisfied.

6.29. The equation of detailed balance $\tilde{f}_e\tilde{f}_i = f_e f_i$ is satisfied only if

$$A = \frac{m_e}{T_e}(\tilde{\mathbf{u}}_e - \mathbf{v}_e)^2 + \frac{m_i}{T_i}(\tilde{\mathbf{u}}_i - \mathbf{v}_i)^2 - \frac{m_e}{T_e}(\mathbf{u}_e - \mathbf{v}_e)^2 - \frac{m_i}{T_i}(\mathbf{u}_i - \mathbf{v}_i)^2$$

vanishes identically. It is possible to rearrange A as follows:

$$A = \frac{m_e}{T_e}[\tilde{u}_e^2 - u_e^2 - 2\mathbf{v}_e \cdot (\tilde{\mathbf{u}}_e - \mathbf{u}_e)] + \frac{m_i}{T_i}[\tilde{u}_i^2 - u_i^2 - 2\mathbf{v}_i \cdot (\tilde{\mathbf{u}}_i - \mathbf{u}_i)] \tag{A1}$$

The equations of conservation of momentum and energy show that

$$m_e(\tilde{\mathbf{u}}_e - \mathbf{u}_e) = m_i(\mathbf{u}_i - \tilde{\mathbf{u}}_i) \tag{A2}$$

and

$$m_e(\tilde{u}_e^2 - u_e^2) = m_i(u_i^2 - \tilde{u}_i^2) \tag{A3}$$

If Eqs. (A2) and (A3) are substituted into Eq. (A1), it is obtained that

$$A = m_i(u_i^2 - \tilde{u}_i^2)\left(\frac{1}{T_e} - \frac{1}{T_i}\right) - 2m_i(\mathbf{u}_i - \tilde{\mathbf{u}}_i) \cdot \left(\frac{\mathbf{v}_e}{T_e} - \frac{\mathbf{v}_i}{T_i}\right)$$

Since A should vanish for all values of $\mathbf{u}_e$, $\mathbf{u}_i$, $\tilde{\mathbf{u}}_e$, and $\tilde{\mathbf{u}}_i$ as well as for arbitrary values of $\mathbf{v}_e$ and $\mathbf{v}_i$, it follows that $T_e = T_i$ and $\mathbf{v}_e = \mathbf{v}_i$. These are the requirements for the mixture of electrons and ions to reach an equilibrium state.

6.30. It is obtained from Eq. (6.11.23) that

$$\int G(\mathbf{u})(\partial f/\partial t)_{\text{coll}}\, d\mathbf{u} = \int_{-\infty}^{\infty} d\mathbf{u} \int_{-\infty}^{\infty} d\mathbf{u}_2 \int_0^{\infty} dp \int_0^{2\pi} d\varphi\, G(\mathbf{u}) pg(\tilde{f}\tilde{f}^2 - ff^2) \quad \text{(A1)}$$

With the help of Eq. (6.11.20), it is seen that

$$\int_{-\infty}^{\infty} d\mathbf{u} \int_{-\infty}^{\infty} d\mathbf{u}_2\, G(\mathbf{u}) \tilde{f}\tilde{f}^2 = \int_{-\infty}^{\infty} d\tilde{\mathbf{u}} \int_{-\infty}^{\infty} d\tilde{\mathbf{u}}_2\, G(\mathbf{u}) \tilde{f}\tilde{f}^2$$

If the dummy variables $\tilde{\mathbf{u}}$ and $\tilde{\mathbf{u}}_2$ are changed to $\mathbf{u}$ and $\mathbf{u}_2$, respectively, $\tilde{f}$ and $\tilde{f}^2$ change to f and f^2, respectively. Since the relations for $\mathbf{u}$ and $\mathbf{u}_2$ can be deduced from the corresponding relations for $\tilde{\mathbf{u}}$ and $\tilde{\mathbf{u}}_2$ by interchanging the quantities without and with the tildes, it follows that $G(\mathbf{u})$ changes to $\tilde{G}(\tilde{\mathbf{u}})$. Therefore

$$\int_{-\infty}^{\infty} d\mathbf{u} \int_{-\infty}^{\infty} d\mathbf{u}_2\, G(\mathbf{u}) \tilde{f}\tilde{f}^2 = \int_{-\infty}^{\infty} d\mathbf{u} \int_{-\infty}^{\infty} d\mathbf{u}_2\, \tilde{G}(\tilde{\mathbf{u}}) ff^2 \quad \text{(A2)}$$

The substitution of Eq. (A2) in Eq. (A1) yields

$$\int G(\mathbf{u})(\partial f/\partial t)_{\text{coll}}\, d\mathbf{u} = \int_{-\infty}^{\infty} d\mathbf{u} \int_{-\infty}^{\infty} d\mathbf{u}_2 \int_0^{\infty} dp \int_0^{2\pi} d\varphi\, pgff^2[\tilde{G}(\tilde{\mathbf{u}}) - G(\mathbf{u})] \quad \text{(A3)}$$

If $G(\mathbf{u}) = 1$, it is obtained from Eq. (A3) that

$$\int (\partial f/\partial t)_{\text{coll}}\, d\mathbf{u} = 0$$

which is the collision term occurring on the right-hand side of the first velocity moment of the Boltzmann equation.

If $G(\mathbf{u}) = \mathbf{u}$, $\tilde{G}(\tilde{\mathbf{u}}) - G(\mathbf{u}) = \Delta\mathbf{u} = M_2 g \sin\chi(\hat{\mathbf{x}}\cos\varphi + \hat{\mathbf{y}}\sin\varphi) - M_2 g(1 - \cos\chi)\hat{\mathbf{z}}$. If $[\tilde{G}(\tilde{\mathbf{u}}) - G(\mathbf{u})]$ is substituted into Eq. (A3) and the integration with respect to φ is carried out first, the x- and the y-components of $\int \mathbf{u}(\partial f/\partial t)_{\text{coll}}\, d\mathbf{u}$ are seen to vanish. Therefore, $\mathbf{P}_{\text{coll}} = \int m\mathbf{u}(\partial f/\partial t)_{\text{coll}}\, d\mathbf{u}$ has only a z-component which is in the direction of the initial velocity of the particle with respect to particle 2 with which it collides.

6.31. The number density is obtained as

$$N = \int_0^{\infty} du \int_{-1}^{1} d(\cos\theta) \int_0^{2\pi} d\varphi\, u^2 \sum_{m=0}^{\infty} \sum_{n=0}^{\infty} P_n^m(\cos\theta) \times [f_{mn}(\mathbf{r}, u, t)\cos m\varphi + g_{mn}(\mathbf{r}, u, t)\sin m\varphi]$$

The integration with respect to φ yields nonvanishing result only for $m = 0$. The integration with respect to $\cos\theta$ together with Eq. (F.17) in Appendix F shows that only $n = 0$ term contributes to the number density. In a similar manner, it is found from

$$\langle u \rangle = \frac{1}{N} \int_0^{\infty} du \int_{-1}^{1} d(\cos\theta) \int_0^{2\pi} d\varphi\, u^3 \sum_{m=0}^{\infty} \sum_{n=0}^{\infty} P_n^m(\cos\theta) \times [f_{mn}(\mathbf{r}, u, t)\cos m\varphi + g_{mn}(\mathbf{r}, u, t)\sin m\varphi]$$

that only the $m = 0$, $n = 0$ term contributes to the average speed.

Since $\mathbf{u} = u \sin\theta \cos\varphi \hat{\mathbf{x}} + u \sin\theta \sin\varphi \hat{\mathbf{y}} + u \cos\theta \hat{\mathbf{z}} = -uP_1^1(\cos\theta)\cos\varphi \hat{\mathbf{x}} - uP_1^1(\cos\theta)\sin\varphi \hat{\mathbf{y}} + uP_1^0(\cos\theta)\hat{\mathbf{z}}$, the average velocity is obtained as

$$\langle \mathbf{u} \rangle = \frac{1}{N}\int_0^\infty du \int_{-1}^1 d(\cos\theta) \int_0^{2\pi} d\varphi\, u^2[-\hat{\mathbf{x}}uP_1^1(\cos\theta)\cos\varphi - \hat{\mathbf{y}}uP_1^1(\cos\theta)\sin\varphi + \hat{\mathbf{z}}uP_1^0(\cos\theta)] \sum_{m=0}^{\infty}\sum_{n=0}^{\infty} P_n^m(\cos\theta) \times [f_{mn}(\mathbf{r}, u, t)\cos m\varphi + g_{mn}(\mathbf{r}, u, t)\sin m\varphi] \tag{A1}$$

The integration with respect to φ shows that $m = 1$ term contributes to the x- and y-components and that $m = 0$ term contributes to the z-component of $\langle \mathbf{u} \rangle$. The integration with respect to $\cos\theta$ together with Eq. (F.17) in Appendix F shows that only $n = 1$ term gives contribution to $\langle \mathbf{u} \rangle$. Thus only the first order anisotropies contribute to the average velocity $\langle \mathbf{u} \rangle = \mathbf{v}$.

The kinetic pressure dyad is given by

$$\boldsymbol{\Psi} = m_e N\langle(\mathbf{u} - \mathbf{v})(\mathbf{u} - \mathbf{v})\rangle = m_e N\langle \mathbf{uu} \rangle - m_e N\mathbf{vv} \tag{A2}$$

where m_e is the particle mass. From the foregoing, it is seen that the contribution to the dyad $\mathbf{vv}$ arises only from the $n = 1$ term. It can be shown that

$$\mathbf{uu} = \hat{\mathbf{x}}\hat{\mathbf{x}}\frac{u^2}{2}\sin^2\theta(1 + \cos 2\varphi) + \hat{\mathbf{y}}\hat{\mathbf{y}}\frac{u^2}{2}\sin^2\theta(1 - \cos 2\varphi) + \hat{\mathbf{z}}\hat{\mathbf{z}}u^2\cos^2\theta + (\hat{\mathbf{x}}\hat{\mathbf{y}} + \hat{\mathbf{y}}\hat{\mathbf{x}})\frac{u^2}{2}\sin^2\theta \sin 2\varphi + (\hat{\mathbf{x}}\hat{\mathbf{z}} + \hat{\mathbf{z}}\hat{\mathbf{x}})u^2\sin\theta\cos\theta\cos\varphi + (\hat{\mathbf{y}}\hat{\mathbf{z}} + \hat{\mathbf{z}}\hat{\mathbf{y}})u^2\sin\theta\cos\theta\sin\varphi$$

Also

$$N\langle \mathbf{uu} \rangle = \int_0^\infty du \int_{-1}^1 d(\cos\theta) \int_0^{2\pi} d\varphi\, u^2\mathbf{uu} \sum_{m=0}^{\infty}\sum_{n=0}^{\infty} P_n^m(\cos\theta) \times [f_{mn}(\mathbf{r}, u, t)\cos m\varphi + g_{mn}(\mathbf{r}, u, t)\sin m\varphi] \tag{A3}$$

The integration with respect to φ shows that nonvanishing contributions arise only from the terms corresponding to $m = 0$, 1, and 2. The functions of θ in $\mathbf{uu}$ for a given m are written as linear combinations of the functions $P_n^m(\cos\theta)$. It is found that only $n = 0$, $n = 1$, and $n = 2$ terms are present. Consequently, the integration with respect to $\cos\theta$ together with Eq. (F.17) in Appendix F shows that only $n = 0$, $n = 1$ and $n = 2$ terms contribute to $N\langle \mathbf{uu} \rangle$. Therefore, the isotropic term and the anisotropies of the first two orders are involved in the determination of the kinetic pressure dyad.

For $m = 0$, Eq. (A1) and Eq. (F.17) in Appendix F can be used to show that

$$\langle \mathbf{u} \rangle = \mathbf{v} = \hat{\mathbf{z}}\frac{4\pi}{3N}\int_0^\infty du\, u^3 f_{01}(\mathbf{r}, u, t) = \hat{\mathbf{z}}v_z \tag{A4}$$

The part of **uu** which gives nonvanishing contributions to $\langle \mathbf{uu} \rangle$ for $m = 0$ is

$$(\mathbf{uu})_{m=0} = (\hat{\mathbf{x}}\hat{\mathbf{x}} + \hat{\mathbf{y}}\hat{\mathbf{y}})\frac{u^2}{2}\sin^2\theta + \hat{\mathbf{z}}\hat{\mathbf{z}}u^2\cos^2\theta$$

$$= (\hat{\mathbf{x}}\hat{\mathbf{x}} + \hat{\mathbf{y}}\hat{\mathbf{y}})\frac{u^2}{3}\{P_0^0(\cos\theta) - P_2^0(\cos\theta)\}$$

$$+ \hat{\mathbf{z}}\hat{\mathbf{z}}\frac{u^2}{3}\{P_0^0(\cos\theta) + 2P_2^0(\cos\theta)\}$$

which is conveniently split up into two parts as follows:

$$(\mathbf{uu})_{m=0} = (\mathbf{uu})_{m=0,n=0} + (\mathbf{uu})_{m=0,n=2} \qquad \text{(A5)}$$

where $(\mathbf{uu})_{m=0,n=0} = \mathbf{1}(u^2/3)P_0^0(\cos\theta)$ and $(\mathbf{uu})_{m=0,n=2} = [-(\hat{\mathbf{x}}\hat{\mathbf{x}} + \hat{\mathbf{y}}\hat{\mathbf{y}}) + 2\hat{\mathbf{z}}\hat{\mathbf{z}}]$ $(u^2/3)P_2^0(\cos\theta)$. Let $\boldsymbol{\Psi}_0 = m_e N\langle(\mathbf{uu})_{m=0,n=0}\rangle$ and $\boldsymbol{\Psi}_2 = m_e N\langle(\mathbf{uu})_{m=0,n=2}\rangle$. If Eq. (A5) is substituted in Eq. (A3) and the integrations with respect to φ and $\cos\theta$ are carried out using Eq. (F.17) of Appendix F, it is found that

$$\boldsymbol{\Psi}_0 = \mathbf{1}\frac{4\pi m_e}{3}\int_0^\infty du\, u^4 f_{00}(\mathbf{r}, u, t)$$

and

$$\boldsymbol{\Psi}_2 = \frac{4\pi m_e}{15}\int_0^\infty du\, u^4[-(\hat{\mathbf{x}}\hat{\mathbf{x}} + \hat{\mathbf{y}}\hat{\mathbf{y}}) + 2\hat{\mathbf{z}}\hat{\mathbf{z}}]f_{02}(\mathbf{r}, u, t)$$

Therefore, for $m = 0$, the kinetic pressure dyad is obtained as

$$\boldsymbol{\Psi} = \boldsymbol{\Psi}_0 + \boldsymbol{\Psi}_1 + \boldsymbol{\Psi}_2$$

where $\boldsymbol{\Psi}_1 = -m_e N\hat{\mathbf{z}}\hat{\mathbf{z}}v_z^2$. The kinetic pressure dyad is seen to be diagonal with the two terms corresponding to the two directions perpendicular to the z-axis being identical. If $f_{01}(\mathbf{r}, u, t) = f_{02}(\mathbf{r}, u, t) = 0$, $\boldsymbol{\Psi} = \boldsymbol{\Psi}_0$ and hence a scalar pressure is obtained only if the distribution function is isotropic.

6.32. Since $\bar{\mathbf{u}}_C = \bar{M}_1\mathbf{u}_1 + \bar{M}_2\mathbf{u}_2$ and $\mathbf{g} = \mathbf{u}_1 - \mathbf{u}_2$, it follows from Eqs. (6.11.9) and (6.11.10) that

$$\frac{\partial(\bar{\mathbf{u}}_C, \mathbf{g})}{\partial(\mathbf{u}_1, \mathbf{u}_2)} = \begin{vmatrix} \bar{M}_1 & 0 & 0 & \bar{M}_2 & 0 & 0 \\ 0 & \bar{M}_1 & 0 & 0 & \bar{M}_2 & 0 \\ 0 & 0 & \bar{M}_1 & 0 & 0 & \bar{M}_2 \\ 1 & 0 & 0 & -1 & 0 & 0 \\ 0 & 1 & 0 & 0 & -1 & 0 \\ 0 & 0 & 1 & 0 & 0 & -1 \end{vmatrix} = -1$$

Therefore $|J| = 1$. If the variables of integration $\mathbf{u}_1$ and $\mathbf{u}_2$ are changed to $\bar{\mathbf{u}}_C$ and $\mathbf{g}$, it is obtained that

$$\langle g\rangle = (\bar{m}_1\bar{m}_2)^{3/2}(2\pi K)^{-3}$$

$$\iint d\bar{\mathbf{u}}_C\, d\mathbf{g}\, g\, \exp\left[-\frac{1}{2K}\left\{(\bar{m}_1+\bar{m}_2)\bar{u}_C^2 + \frac{\bar{m}_1\bar{m}_2}{\bar{m}_1+\bar{m}_2}g^2\right\}\right]$$

Since

$$\int d\bar{\mathbf{u}}_C \exp\left[-\frac{1}{2K}(\bar{m}_1+\bar{m}_2)\bar{u}_C^2\right] = 4\pi\int_0^\infty \bar{u}_C^2\exp\left[-\frac{1}{2K}(\bar{m}_1+\bar{m}_2)\bar{u}_C^2\right]$$
$$= \{2\pi K/(\bar{m}_1+\bar{m}_2)\}^{3/2}$$

and

$$\int d\mathbf{g}\, g\, \exp\left[-\frac{1}{2K}\frac{\bar{m}_1\bar{m}_2}{(\bar{m}_1+\bar{m}_2)}g^2\right] = 4\pi\int_0^\infty dg\, g^3\exp\left[-\frac{1}{2K}\frac{\bar{m}_1\bar{m}_2}{(\bar{m}_1+\bar{m}_2)}g^2\right]$$
$$= 8\pi K^2\{(\bar{m}_1+\bar{m}_2)/\bar{m}_1\bar{m}_2\}^2$$

it follows that

$$\langle g\rangle = \sqrt{\frac{8K}{\pi}}\left(\frac{T_1}{m_1}+\frac{T_2}{m_2}\right)^{1/2}$$

If $m_1 = m_2$, $T_1 = T_2$, and $N_1 = N_2 = N$, $\langle g\rangle = \sqrt{2}\,\langle u\rangle$ since $\langle u\rangle = (8KT/\pi m)^{1/2}$ Therefore $\nu = \sigma N\langle g\rangle = \sqrt{2}\, N\sigma\langle u\rangle$.

6.33. Note that $\int d\mathbf{u}\, f_0(\mathbf{u}) = N_0$; therefore N_0 is the number density. As in Problem 6.18, the dispersion equation for the longitudinal plasma wave can be deduced as

$$D_{lr}(k_p,\omega) + iD_{li}(k_p,\omega) = 0 \tag{A1}$$

where

$$D_{lr}(k_p,\omega) = 1 - \frac{\omega_{pe}^2}{N_0\omega^2}\int_{-\infty}^{\infty}\left(1-\frac{u_x k_p}{\omega}\right)^{-2} f_{0x}(u_x)\,du_x \tag{A2}$$

$$D_{li}(k_p,\omega) = -\frac{\pi\omega_{pe}^2}{N_0 k_p^2}f'_{0x}(\omega/k_p) \tag{A3}$$

and

$$f_{0x}(u_x) = \frac{N_0}{\sqrt{2\pi}}\left(\frac{m_e}{KT}\right)^{3/2} u_x^2\exp\left(-\frac{m_e}{2KT}u_x^2\right) \tag{A4}$$

Let $\omega = \omega_r + i\omega_i$. Assume that $\omega_i \ll \omega_r$ and $|D_{li}| \ll |D_{lr}|$. If $\omega = \omega_r + i\omega_i$ is substituted in Eq. (A1) and the left side of Eq. (A1) is expanded into a Taylor series, it is found that

$$D_{lr}(k_p, \omega_r) = 0 \tag{A5}$$

$$\omega_i = -D_{li}(k_p, \omega_r)/D'_{lr}(k_p, \omega_r) \tag{A6}$$

Assume that $|u_x k_p/\omega| \ll 1$; therefore, the following approximation is valid: $(1 - u_x k_p/\omega)^{-2} = 1 + 2u_x k_p/\omega + 3u_x^2 k_p^2/\omega^2$. Then Eqs. (A2) and (A5) can be used to show that

$$\omega_r = \omega_{pe}\left(1 + \frac{k_p^2}{\omega_{pe}^2}\frac{9KT}{2m_e}\right)$$

and

$$D'(k_p, \omega_r) = 2/\omega_r$$

Consequently, it is obtained from Eqs. (A3) and (A6) that

$$\omega_i = \frac{\pi}{2}\frac{\omega_{pe}^2 \omega_r}{N_0 k_p^2} f'_{0x}(\omega_r/k_p)$$

Instability occurs if $\omega_i > 0$, that is, if $f'_{0x}(\omega_r/k_p) > 0$. The longitudinal plasma wave is therefore unstable if $\omega_r < k_p\sqrt{2KT/m_e}$. The approximate value of the growth constant can be determined as

$$\omega_i = \sqrt{\frac{\pi}{?}}\,\frac{\omega_{pe}^4}{\kappa_p^3}\left(\frac{m_e}{KT}\right)^{3/2}\left(1 - \frac{m_e}{2KT}\frac{\omega_{pe}^2}{k_p^2}\right)\exp\left[\frac{-1}{2k_p^2\lambda_{De}^2} - \frac{9}{2}\right]$$

where λ_{De} is the Debye length.

INDEX

This book is set in Times Roman on a Harris Intertype Fototronic CRT, a third generation photocomposition machine. The composition was done by Composition Technology, Incorporated, Cambridge, Massachusetts. All typesetting and page composition was done within a Digital Equipment PDP-10 computer, which produced a drive tape for the Fototronic CRT. Only figures were pasted up manually. Copy was prepared using a special mathematics language developed by CTI. Typescript was prepared using a standard IBM Selectric typewriter (Courier 12 typing element) and then optically scanned using an ECRM, Inc., AUTOREADER. The OCR device produced a magnetic tape, which was input to the PDP-10 for typesetting. Corrections were made on-line to the computer.

This book was printed and bound by R. R. Donnelley and Sons Co. on the Cameron book production system. The heart of the system is the Cameron belt press which prints an entire one-color book at one time from flexible photopolymer relief plates. Completely folded, cut, and collated four-page sections are delivered from the press to an in-line adhesive binder which glues the signatures, applies covers, trims, and stacks finished books.